HIGH-TEMPERATURE PROPERTY DATA: Ferrous Alloys

M.F. ROTHMAN
Consulting Editor

Group Leader, High Temperature Alloys
Haynes International

Chairman, Specialty Materials Division
Heat Resistant Materials Group, Properties Committee

ASM INTERNATIONAL
Metals Park, OH 44073

First printing, December 1987
Second printing, November 1989

Comments, criticisms, and suggestions are invited, and should be forwarded to ASM INTERNATIONAL, Metals Park, Ohio 44073.

ISBN: 0-87170-243-6

PRINTED IN THE UNITED STATES OF AMERICA

Foreword

The publication of *High-Temperature Property Data: Ferrous Alloys* represents a new direction in publishing for ASM INTERNATIONAL. This volume was produced entirely on personal computers using available software. All information was captured electronically on disk so that it can be imported into searchable databases in the future.

Ten years ago, the production of a book of this magnitude and scope would have required a small army of people working full time to complete the project within an reasonable time frame. The past decade has seen the widespread use of the copy machine and the introduction of the personal computer. These advances in the work place have made this book possible.

Had this book been published ten years ago, all the artwork and tables would have been taken directly from the source material. The art would have been redrawn by a draftsman or artist, and the tables might have been retyped and marked for typesetting--two very expensive and time-consuming steps. In addition, corrections to the artwork or tables would have added to the cost and time.

The text might have been typed on an electric typewriter and subsequently edited before it was retyped. A clean copy would have been sent to the typesetter, who would have to type it into his machine one more time. And then the galleys would have been proofed against the original copy, and still more corrections would have been made.

Given this scenario, it's surprising anything got published in 1977. In fact, after looking at cost and economic viability, a publisher could well have scrapped a project of this magnitude before production started.

In contrast to the methods of the last decade, this book was produced using personal computers and a total of five available software programs--in particular, ASM's MetSel/2 and EnPlot programs. MetSel/2 is a materials selection package that records alloy information in a searchable format. EnPlot is an engineering graphics package that generates *x-y* plots. EnPlot graphs can be stored in MetSel/2 data files.

Almost no commercial typesetting was used in this volume; text, tables, and captions were printed on a laser jet printer driven by a word processing package. The figures were digitized and produced on a plotter using EnPlot.

The text was word processed in ASM's materials database program MetSel/2. Each alloy has a database file. Once the data were input, the file was saved as an ASCII text file, imported into a word processing package, and edited. This step allowed the database file for each alloy to be initiated.

Each table was input into a spreadsheet program. Using the spreadsheet, conversions were made to provide both metric and English units. Each figure was enlarged, placed on a digitizing pad, and digitized using EnPlot.

Enlarging the figure facilitated the digitizing process, thus ensuring a greater degree of accuracy.

Text and tables were imported into a technical word processing package where type sizes and faces were selected. The table titles, heads, and footnotes were input. Property abbreviations from the MetSel/2 databases were used so that the tabular information could be imported easily into the appropriate MetSel/2 database file. Figures generated from EnPlot were plotted and reduced photographically. Once the material was technically reviewed and corrected, it was keylined into pages.

In addition to speeding up production and saving on costs, it was possible to make corrections easily and inexpensively up to the last possible minute. However, the most important advantage of this new production method is that all of the information in this book has been captured on disks. This information will be integrated into MetSel/2 datadisks so that specific properties can be found easily. Building these materials properties information databases will be a priority for ASM over the next several years. Our ultimate goal is to provide all available property information for each alloy in a searchable electronic format.

Even in 1987, a project of this size needed the input of many individuals to make it successful. ASM INTERNATIONAL would like to acknowledge the contributions of the following people: Don Baxter, for his advice and guidance in the organization and first-pass editing of an endless sea of material; Terri Weintraub, Julia Talsma, and Jane McCullam of Editorial Associated Services, Inc. for undertaking the task of inputting the tables and text and digitizing the graphs; and Mike Rothman, for his able and thorough editing of each data sheet. Many thanks to all the others who assisted.

Sunniva Refsnes
ASM INTERNATIONAL

Preface

In selecting materials for high-temperature applications, it is wise to remember the old adage--"Steel will always work; it's only a question of for how long!" To determine "how long" not only steels, but stainless steels, iron-nickel-chromium alloys, or other high-temperature materials might last, some knowledge of the relative performance characteristics of these materials is required. Of obvious importance are the various high-temperature mechanical and physical properties that determine material performance in service. These are the focus of this book and of the second volume covering nonferrous alloys. Together, these volumes represent the first comprehensive comparative database of its kind.

It is the intent of this first volume to provide information that is useful in the selection of ferrous materials to be considered for high-temperature applications. It is not a detailed collection of design data, and the properties described are not suitable for direct use in final design calculations. Having said this, it should be pointed out that one of the first steps in designing components for use in elevated-temperature service is to reduce the choices of potential materials of construction to a reasonable number.

It is in this step that this book on high-temperature properties of ferrous alloys, and the companion volume on nonferrous alloys, is likely to make its most important contribution to component engineering. First, it will help users to more clearly identify which of the more well-known, traditional materials of construction are the most likely to be capable of providing the required properties for an application, as well as which should be eliminated from consideration because they are obviously inadequate. Secondly, and of key importance, this book will alert users to the capabilities of newer materials with which they may not be completely familiar. Thus, materials and design engineers can focus upon the most qualified alloys for the application in question, without searching through various diverse sources for the necessary information.

Some mention of the manner in which this valuable collection of data was assembled seems appropriate at this point. The property information presented in this volume was gathered from a variety of published sources, including previous tabulations and handbooks, as well as company brochures and data sheets. The information was carefully reviewed to ensure accurate reproduction. On the other hand, no attempt was made to verify or to critically analyze the data, or to trace the pedigree of the material from which the data were generated. Still, the information is perceived to be reasonably typical in terms of describing the relative performance capabilities of the various materials.

Although procurement of the data was a significant undertaking on the part of the ASM International staff, once in hand it was efficiently organized, edited and printed entirely through the use of personal computers; text, tables and captions were all printed on a laser jet printer driven by a word processor, whereas graphs were produced on a computer plotter. The use of typesetting in the production of this book was virtually eliminated, thus reducing the cost of its publication by half. Without this savings, this book would have been too expensive to publish at all, and you wouldn't be reading this now.

The book itself is organized into sections representing basic ferrous alloy families. These include Irons, Carbon Steels, Alloy Steels, ASTM Steels, Low-Alloy Steels, Ultrahigh Strength Steels, Tool Steels, Maraging Steels, Wrought Stainless Steels, ACI Casting Alloys, and Wrought Iron-Nickel Alloys and Iron-Nickel Superalloys. In most cases, each section has an introduction that briefly describes the nature of the compositions described, the roles of various elements in the alloys, the metallurgical characteristics of the alloy type, the typical temperature limits or range of operation, and examples of suitable applications. Following the introduction and general information for some groups, the most common alloys within an alloy group are summarized individually. Summaries include alloy name, alternative designations, relevant specifications, and typical alloy composition or composition ranges. Many descriptions also include some discussion of individual alloy characteristics and pertinent applications.

Following each individual alloy summary, physical and mechanical property data that has been located for that alloy are presented in tabular and/or graphical format. The extent of the data presented varies considerably from alloy to alloy, reflecting in the main the extent to which each alloy is used. It is expected that future editions of this book may contain expanded information on at least some of the alloys as additional data are uncovered through on-going efforts.

One area in which the data are very complete is the realm of physical properties. In most all cases, data are presented for modulus of elasticity, mean coefficient of expansion, thermal conductivity and electrical resistivity, all as a function of temperature. Tensile properties as a function of temperature are also reasonably complete for many product forms.

For many of the most common alloys, and almost all of the more sophisticated materials, data are presented detailing fatigue and creep strength properties. Considerable effort has been expended to present this data in a manner that allows for ready comparison between alloys. Care has been exercised where possible, for example, to avoid inappropriate extrapolation of minimum creep rate data to calculate total creep lives (i.e., time to 1% creep under set temperature and stress conditions), as doing so ignores the considerable impact of primary stage creep.

In examining the data embodied in this book, it will be noticed immediately that there is a singular absence of oxidation-resistance data, or environment resistance data of any kind. This is intentional. Whereas mechanical and physical property measurements more or less conform to relatively well established testing procedures, the same cannot be said for environment resistance tests at elevated temperature. Despite many attempts by the technical community to standardize such seemingly "elementary" tests as oxidation resistance, for example, the fact remains that comparison of the results from different test sources is often dangerously misleading. Rather than present an assortment of apples and oranges, it was felt that environment resistance data be best left to another time.

One final word on the future work of ASM International and its technical committee activity in the area of high-temperature materials properties. This book, and the second volume on nonferrous alloys, although of great utility as they are, in fact provide a framework upon which to build. Efforts are currently underway by the Heat Resistant Materials Committee of ASM to develop sources of input to a more carefully pedigreed database of properties relevant to high-temperature applications. As such information becomes available, these books will be updated, and made even more valuable to designers and materials engineers as definitive sources of the data required for optimum material selection and component design.

Michael F. Rothman
Consulting Editor
October, 1987

Contents

1. Irons

3. Alloy Steels

4. ASTM Steels

8. Maraging Steels

9. Wrought Stainless Steels

10. ACI Casting Alloys

11. Wrought Iron-Nickel Alloys and Iron-Nickel Superalloys

Abbreviations

.%CW	Percent Cold Work
BLKMD	Bulk Modulus
CORR	Corrosion Rate
CRIS	Initial Stress before Creep
CRSRD	Residual Stress after Creep
CRSTN	Creep Strain
CRSTR	Creep Strength
CRT	Creep Rate
CRTA	Average Creep Rate
CRTMN	Minimum Creep Rate
CSHX	Cyclic Strain Hardening Exponent
CSTRC	Cyclic Strength Coefficient
DENS	Density
DMPCP	Specific Damping Capacity
da/dN	Fatigue Crack Propagation Rate
E	Modulus of Elasticity
EL	Tensile Elongation
ELCON	Electrical Conductivity
ELRES	Electrical Resistivity
EMITT	Emittance
FAT.n	Fatigue Cycles
FATDC	Fatigue Ductility Coefficient

FATDX	Fatigue Ductility Exponent
FATL	Fatigue Resistance
FATSC	Fatigue Strength Coefficient
FATSR	Fatigue Stress Ratio
FATST	Fatigue Strength at N Cycles
FATSX	Fatigue Strength Exponent
FRSTN	Elastic Strain at Fracture
FRSTR	Fracture Strength
G	Modulus of Rigidity
GRT	Growth Rate
HARD	Hardness
IS	Impact Strength
KIC	Fracture Toughness
NTS	Notch Tensile Strength
PR	Poisson's Ratio
RA	Tensile Reduction in Area
RMIBR	Residual Magnetic Inductance
RPSTR	Stress-Rupture Strength
RPT	Rupture Life
SHTCP	Specific Heat
SHX	Strain Hardening Exponent
SRX	Strain Rate Exponent
STREX	Exposure Stress
STRN	Natural Stress at Maximum Load
STRTR	True Stress at Maximum Load
THCON	Thermal Conductivity
THEXP	Thermal Coefficient of Expansion
TSR	True Total Strain Rate
TSRA	Axial Strain Rate
TSRGX	Maximum Total Strain Range
USB	Ultimate Bearing Strength
USS	Ultimate Strength in Shear
UTS	Tensile Ultimate Strength

WTGN	Weight Gain
YS	Tensile Yield Strength
YSB	Bearing Yield Strength
YSC	Compressive Yield Strength
YSS	Yield Strength in Shear

References

1. *Metals Handbook,* Vol 1, *Properties and Selection: Irons and Steels,* 9th ed., American Society for Metals, 1978.

2. *Metals Handbook,* Vol 2, *Properties and Selection: Nonferrous Alloys and Pure Metals,* 9th ed., American Society for Metals, 1979.

3. *Metals Handbook,* Vol 3, *Properties and Selection: Stainless Steels, Tool Materials and Special-Purpose Metals,* 9th ed., American Society for Metals, 1980.

4. *Metals Handbook,* Vol 4, *Heat Treating,* 9th ed., American Society for Metals, 1981.

5. *Metals Handbook,* Vol 5, *Surface Cleaning, Finishing, and Coating,* 9th ed., American Society for Metals, 1982.

6. *Metals Handbook,* Vol 6, *Welding, Brazing, and Soldering,* 9th ed., American Society for Metals, 1983.

7. *Metals Handbook,* Vol 7, *Powder Metallurgy,* 9th ed., American Society for Metals, 1984.

8. *Metals Handbook,* Vol 8, *Mechanical Testing,* 9th ed., American Society for Metals, 1985.

9. *Source Book on Materials for Elevated Temperature Applications,* E.F. Bradley, Ed., American Society for Metals, 1979.

10. *Source Book on Stainless Steels,* American Society for Metals, 1976.

11. *Source Book on the Ferritic Stainless Steels,* R.A. Lula, Ed., American Society for Metals, 1982.

12. *Superalloys: Source Book,* M.J. Donachie, Ed., American Society for Metals, 1983.

13. *Titanium and Titanium Alloys: Source Book,* M.J. Donachie, Ed., American Society for Metals, 1982.

14. *Source Book on Industrial Alloy and Engineering Data,* American Society for Metals, 1978.

15. *Engineering Properties of Steel,* P.D. Harvey, Ed., American Society for Metals, 1982.

16. *Flow and Fracture at Elevated Temperatures,* R. Raj, Ed., American Society for Metals, 1985.

17. *Principles of Physical and Chemical Metallurgy,* G.F. Carter, American Society for Metals, 1979.

18. *Heat Treatment, Structure and Properties of Nonferrous Alloys,* C.R. Brooks, American Society for Metals, June 1982.

19. *Metal Progress 1985 Materials and Processing Data Book,* Vol 128, No. 1, American Society for Metals, June 1985.

20. *Forging Handbook,* T.G. Byrer, Ed., Forging Industry Association and American Society for Metals, 1985.

21. *Structural Alloys Handbook,* Vol 1 and 2, D.J. Maykuth, Ed., Mechanical Properties Data Center, Battelle Columbus Laboratories, 1980.

22. *Aerospace Structural Materials Handbook,* Vol 1 to 5, W.F. Brown, Ed., Mechanical Properties Data Center, Battelle Columbus Laboratories, 1981.

23. *Atlas of Fatigue Curves,* H. Boyer, Ed., American Society for Metals, 1986.

24. *Duplex Stainless Steels,* R.A. Lula, Ed., American Society for Metals, 1983.

25. *Standards Handbook,* Part 2 -- Alloy Data, "Wrought Copper and Copper Alloy Mill Products," 8th ed., Copper Development Association, Inc., Greenwich, CT, 1985.

26. *Steel Products Manual,* "Tool Steels," American Iron and Steel Institute, New York, January 1970.

27. *Steel Products Manual,* "Stainless and Heat Resisting Steels," American Iron and Steel Institute, Washington DC, December 1974.

28. *Steel Products Manual,* "High Temperature High Strength Alloys," American Iron and Steel Institute, New York, February 1963.

29. *Alloy Digest,* Engineering Alloys Digest, Inc., Upper Montclair, NJ.

30. *Metals Handbook Desk Edition,* H.E. Boyer and T.L. Gall, Ed., American Society for Metals, 1985.

1. Irons

Irons

Gray Irons

In gray cast iron, carbon that exceeds the solubility in austenite precipitates as flake graphite. Gray irons usually contain 2 to 4% C and 1 to 3% Si. If the service temperature of a gray iron casting is to exceed 480 C (900 F), alloying with a carbide-stabilizing element is required to prevent the breakdown of iron carbide and prevent dimensional growth of the part. For example, gray irons with a chromium content of about 1% commonly are used for castings that operate at 480 to 760 C (900 to 1400 F). For temperatures above 760 C (1400 F), chromium content should exceed 15% for long-term protection against oxidation. This high percentage of chromium actually suppresses the formation of graphite, forcing the alloy to solidify as a white cast iron.

Elevated-temperature creep resistance of gray irons can be enhanced by alloying with molybdenum. Heat-resistant, high-alloy gray irons include the Ni-Resist grades covered by ASTM A436. These austenitic, high-nickel (18 to 36% Ni) irons are not as strong or as tough as Ni-Resist ductile irons, but they can be used in most oxidizing atmospheres at temperatures up to 815 C (1500 F), in steam service up to 530 C (990 F), and in sour gases and liquids up to 400 C (750 F). Data on austenitic, high-nickel gray irons are given in the section on Alloy Irons.

Gray irons: Typical compositions

Iron No.	Type	Condition	C %	CE %	Si %	Mn %	P %	S %	Cr %	Ni %	Mo %	Cu %	V %
1	Mo	Stress relieved	3.31	3.83	1.56	0.68	0.19	0.11	0.08	0.08	0.73	0.15	0.02
2	NiMo	Stress relieved	2.98	3.46	1.51	0.57	0.11	0.1	0.05	1.56	0.58	0.22	<0.01
3	CrCu	Annealed	3.22	3.85	1.97	0.74	0.13	0.1	0.57	0.03	0.01	0.37	0.02
4	NiMoCrV	Stress relieved	3.2	3.82	1.96	0.64	0.11	0.09	0.41	1.12	0.44	0.1	0.12
5	Unalloyed	Stress relieved	3.27	3.87	1.74	0.72	0.26	0.16	0.08	0.15	0.07	0.12	0.02
6	MoCr	Annealed	3.06	3.61	1.79	0.7	0.04	0.09	0.61	0.04	0.84	0.05	<0.01

See Mechanical Properties section for the tensile properties of these irons
Source: Ref 14, p 14

Physical Properties

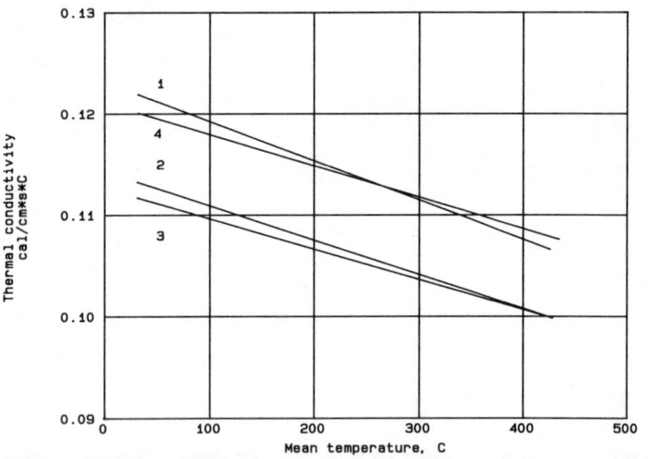

Iron No.	Composition, %
1	3.19% TC, 1.57% Si, 0.70% Mn
2	3.19% TC, 1.57 % Si, 1.58 % Cu
3	3.12% TC, 2.28% Si, 0.38% Mn
4	3.12% TC, 2.28% Si, 0.38% Mn, 0.54% Cr, 0.77% Mo

Gray irons: Thermal conductivity of high-strength gray irons. Source: Ref 21, p 11

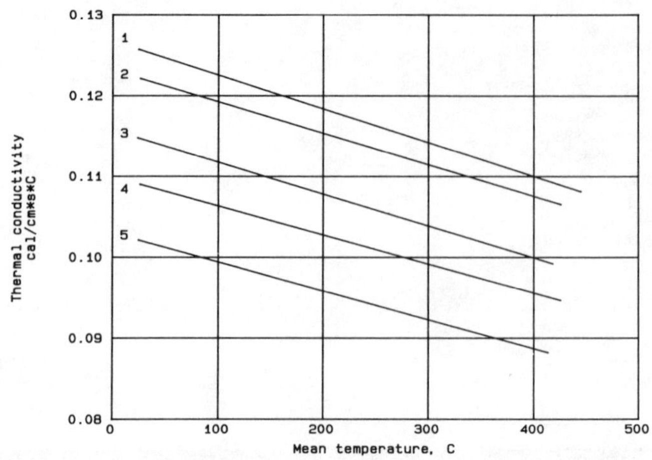

Iron No.	Composition, %	Tensile strength MPa
1	2.56% TC, 2.20% Si, 0.63% Mn, 0.58% Mo	392
2	3.41% TC, 1.03% Si, 0.65% Mn, 1.49% Ni, 0.54% Cr	332
3	2.61% TC, 2,46% Si, 0.45% Mn	350
4	3.10% TC, 2.51% Si, 3.11% Mn, 1.00% Ni	381
5	2.80% TC, 2.51% Si, 0.68% Mn, 1.71% Ni, 0.54% Cr	386

Gray irons: Thermal conductivity. Source: Ref 21, p 11

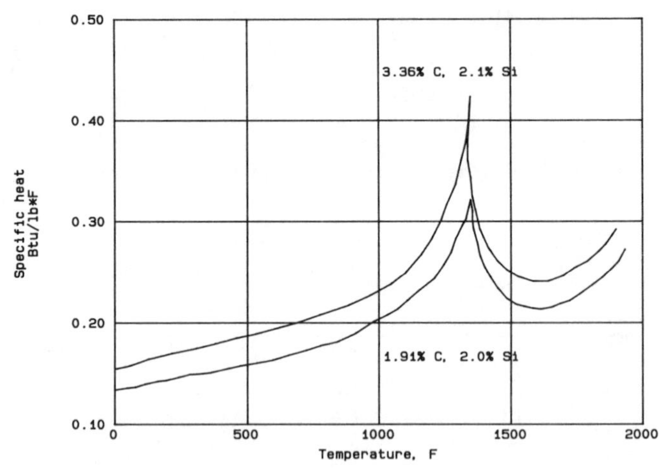

Gray irons: Specific heat. Source: Ref 21, p 13

Gray irons: Thermal conductivity of select gray irons

Iron	TC %	Si %	Mn %	Ni %	Cr %	Mo %	Cu %	W %	THCON (a) W/m*K	THCON (b) W/m*K
Class 40+low Si	3.35	0.65	0.85	...	...	...	...	...	56.48	47.70
Class 40+Cr	3.17	1.40	0.97	...	0.392	...	...	...	54.81	48.12
Class 40+Ni	3.16	1.56	0.94	0.746	...	...	...	...	45.19	42.26
Class 40+W	3.02	1.89	0.76	...	...	...	...	0.475	49.37	43.93
Med strength	3.20	1.56	0.72	...	...	...	...	...	50.63	45.19
Med strength +Cu	3.18	1.58	0.69	...	...	...	1.58	...	46.86	42.26
Med strength	3.11	2.26	0.39	...	...	...	...	...	46.44	42.26
Med strength +Cr-Mo	3.12	2.31	0.39	...	0.54	0.77	...	...	49.79	45.61
Low carbon	2.61	2.46	0.45	...	...	...	...	...	46.02	41.84
Class 50+NiCr	2.80	2.51	0.68	1.71	0.54	...	...	...	42.26	38.49
High Mn+Ni	3.10	2.51	3.11	1.00	...	...	...	...	44.35	40.58
Low carbon+Mo	2.56	2.20	0.63	...	...	0.58	...	...	49.37	45.19
Class 45+NiCr	3.41	1.03	0.65	1.49	0.54	...	...	...	48.53	44.35

Note: TC = total carbon
(a) At 100 C (212 F). (b) At 400 C (752 F)
Source: Ref 21, p 11

Gray irons: Mean coefficients of linear thermal expansion

TEMP C	TEMP F	THEXP(a) 10^{-6}/C	THEXP(a) 10^{-6}/F	THEXP(b) 10^{-6}/C	THEXP(b) 10^{-6}/F
21-100	70-212	10.1	5.6	10.1	5.6
21-200	70-392	10.6	5.9	11.0	6.1
21-300	70-572	11.9	6.6	11.9	6.6
21-400	70-752	12.2	6.8	12.6	7.0
21-500	70-932	13.0	7.2	13.14	7.3
21-600	70-1112	13.3	7.4	13.5	7.5
21-700	70-1292	13.7	7.6	13.7	7.6

Note: TC = total carbon
(a) Composition: 3.13% TC, 0.80% combined C, 2.31% Si, 0.69% Mn, 0.12% Ni, 0.12% Cr, 0.12% Mo, 0.67% P, 0.14% S. (b) Composition: 3.02% TC, 0.77% combined C, 1.57% Si, 0.87% Mn, 0.16% Ni, 0.10% Cr, 0.57% Mo, 0.11% Cu, 0.17% P, 0.10% S
Source: Ref 21, p 10

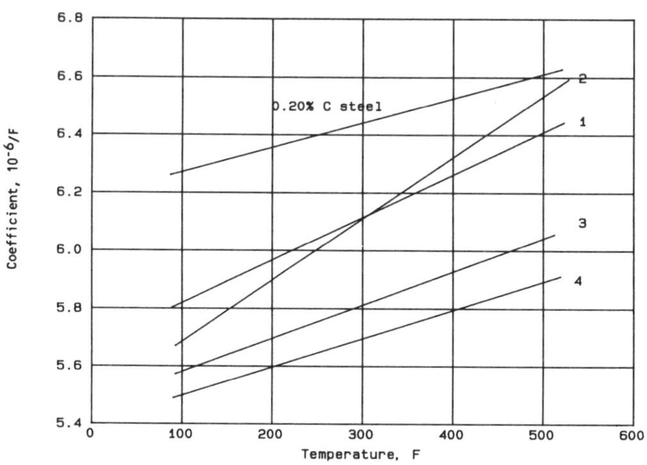

Gray irons: Mean coefficient of linear thermal expansion. Source: Ref 21, p 10

Iron No.	Composition, %
1	3.25% TC, 2.22% Si, 0.59% Mn, 0.093% P, 0.091% S
2	3.32% TC, 1.93% Si, 0.44% Mn, 0.127% P, 0.065% S, 0.37% Mo
3	3.26% TC, 1.89% Si, 0.43% Mn, 0.116% P, 0.064% S, 0.73 % Mo
4	3.14% TC, 1.99% Si, 0.40% Mn, 0.126% P, 0.080% S, 1.41 Ni

Mechanical Properties

Gray irons: Tensile property data for gray irons

TEMP C	TEMP F	HARD HB	UTS MPa	UTS ksi	EL %
Iron 1(a)					
27	80	225	329	47.7	1.0
427	800	...	310	45.0	1.5
538	1000	...	228	33.1	0.5
Iron 2(a)					
27	80	258	405	58.8	1.0
427	800	...	365	52.9	1.7
538	1000	...	217	31.4	0.5
Iron 3(b)					
27	80	192	258	37.4	1.5
427	800	...	229	33.2	1.0
538	1000	...	217	31.4	1.0
Iron 4(a)					
27	80	255	378	54.8	1.5
427	800	...	327	47.4	1.5
538	1000	...	219	31.8	0.5
Iron 5(a)					
27	80	217	294	42.7	1.0
427	800	...	252	36.5	1.2
538	1000	...	170	24.7	0.5
Iron 6(b)					
27	80	217	394	57.2	1.0
427	800	...	374	54.3	1.5
538	1000	...	314	45.6	2.0

Note: See accompanying table in this section for composition of irons
(a) Stress relieved. (b) Heat treated for 4 h at 899 C (1650 F) and furnace cooled
Source: Ref 14, p 14

Gray irons: Tensile strength. Source: Ref 21, p 16

Iron No.	Composition, %
1	3.41% CE, 2.67% TC, 0.79% CC, 2.39% Si, 0.67% Mn, 0.76% Ni, 0.45% Cr, 0.07% P, 0.09% S
2	3.74% CE, 2.96% TC, 0.73% CC, 2.53% Si, 0.78 Mn, 0.96% Ni, 0.02% Cr, 0.06% P, 0.11% S
3	3.99% CE, 3.46% TC, 0.87% CC, 1.36% Si, 0.68% Mn, 0.01% Ni, 0.54% Cr, 0.39% P, 0.08% S
4	4.03% TC, 3.63% TC, 0.73% CC, 1.15 Si, 0.54 Mn, 0.03% Ni, 0.03% Cr, 0.17% P, 0.09% S

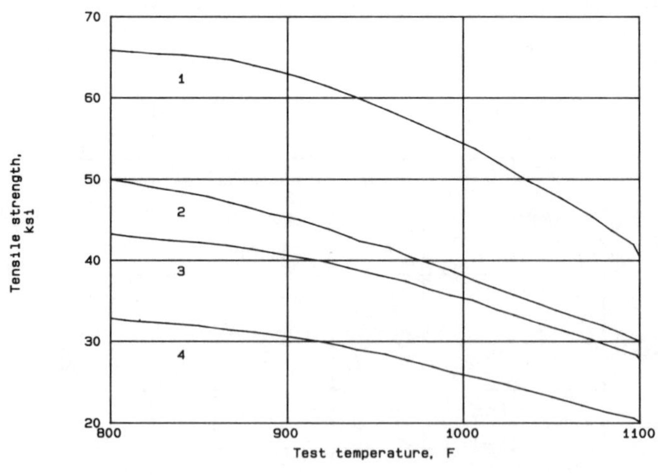

Gray irons: Tensile properties at 400 C (750 F)

HARD HB	UTS MPa	UTS ksi	YS MPa	YS ksi	E GPa	E 10^3 ksi	FRSTR MPa	FRSTR ksi	EL(a) %
235	324	47.0	87	12.6	125	18.1	346.1	50.2	0.6
255	383	55.5	178	25.8	137	19.9	355.1	51.5	0.6
269	441	64.0	133	19.3	103	15.0	346.1	50.2	0.9
285	386	56.0	154	22.4	114	16.5	359.9	52.2	0.6
207	299	43.3	99	14.3	78	11.3	257.9	37.4	1.8

(a) Elongation in 125 mm (5 in.)
Source: Ref 21, p 17

Gray irons: Creep properties at 370 C (700 F) with an exposure stress of 72 MPa (10.5 ksi)

	HARD HB	UTS MPa	UTS ksi	FATST MPa	FATST ksi	CRT 0-150 h 0.1%/h	CRT 150-450 h 0.1%/h	CRT 450-900 h 0.1%/h	CRT 900-1200 h 0.1%/h	CRT 1200-2000 h 0.1%/h
Iron 1	197	217	31.5	100	14.5	12.1	12.1	5.9	3.4	3.0
Iron 2	237	310	45.0	155	22.5	5.3	5.3	5.3	3.4	0.0
Iron 3	230	307	44.5	121	17.5	...	8.2(a)	5.3	0.0	0.0
Iron 4	241	410	59.5	...	...	0.8	0.8	2.2	0.0	0.0

Note: Iron 1: 3.40% total carbon, 2.70% graphitic carbon, 1.50% Si. Iron 2: 2.95% total carbon, 2.10% graphitic carbon, 2.45% Si. Iron 3: 3.20% total carbon, 2.60% graphitic carbon, 1.15% Si, 1.50% Ni. Iron 4: 2.75% total carbon, 2.27% graphitic carbon, 2.10% Si, 0.83% Mo. All iron compositions also have about 0.2% max P, 0.1% max S, and 0.7% Mn, except Iron 3, which has 0.15% S
(a) 0 to 500 h
Source: Ref 1, p 26-27

Gray irons: Stress-rupture properties at 400 C (750 F)

STREX MPa	STREX ksi	RPT h	EL %
Class 45			
346	50.2	...	0.63
270	39.2	1075	2.79
255	37.0	5900	3.80
Class 50			
355	51.5	...	0.62
309	44.8	64	1.70
294	42.6	895	1.95
278	40.3	5000	2.20
Class 60			
346	50.2	...	0.94
247	35.8	10050	3.40
Class 50			
362	52.5	...	0.63
309	44.8	23	1.70
309	44.8	25	2.10
294	42.6	215	2.70
278	40.3	2005	2.75
Class 40			
258	37.4	...	1.80
232	33.6	1875	3.00

Source: Ref 21, p 29

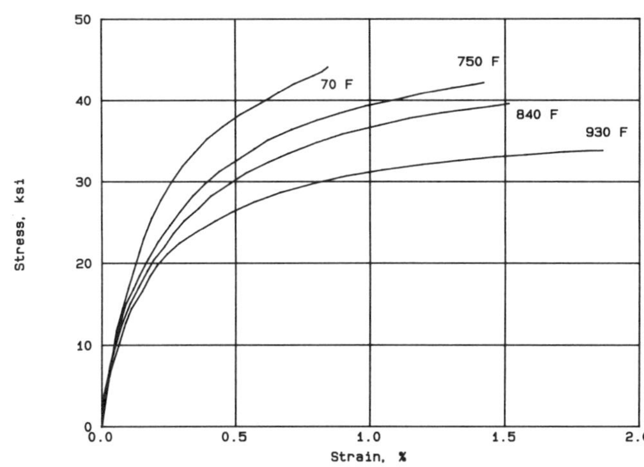

Gray irons: Effect of temperature on stress-strain relation in tension for a class 40 gray iron. Composition: 3.19% TC, 0.85% CC, 1.66% Si, 0.91% Mn, 0.077% P, 0.089% S. Source: Ref 14, p 14

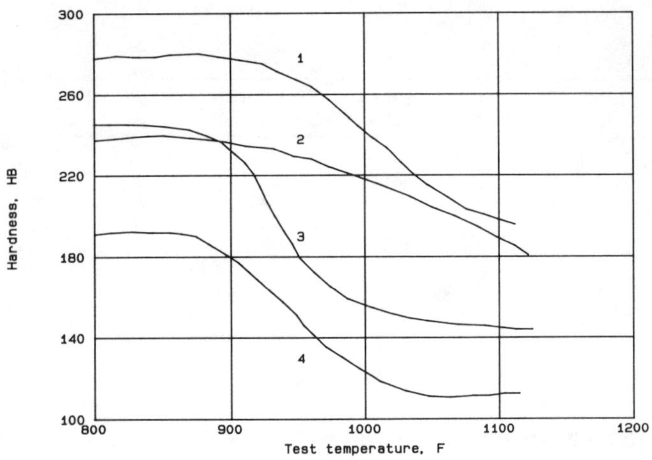

Iron No.	Composition, %
1	3.41% CE, 2.67% TC, 0.79% CC, 2.39% Si, 0.67% Mn, 0.76% Ni, 0.45% Cr, 0.07% P, 0.09% S
2	3.74% CE, 2.96% TC, 0.73% CC, 2.53% Si, 0.78 Mn, 0.96% Ni, 0.02% Cr, 0.06% P, 0.11% S
3	3.99% CE, 3.46% TC, 0.87% CC, 1.36% Si, 0.68% Mn, 0.01% Ni, 0.54% Cr, 0.39% P, 0.08% S
4	4.03% TC, 3.63% TC, 0.73% CC, 1.15 Si, 0.54 Mn, 0.03% Ni, 0.03% Cr, 0.17% P, 0.09% S

Gray irons: Effect of test temperature on hardness. Source: Ref 21, p 16

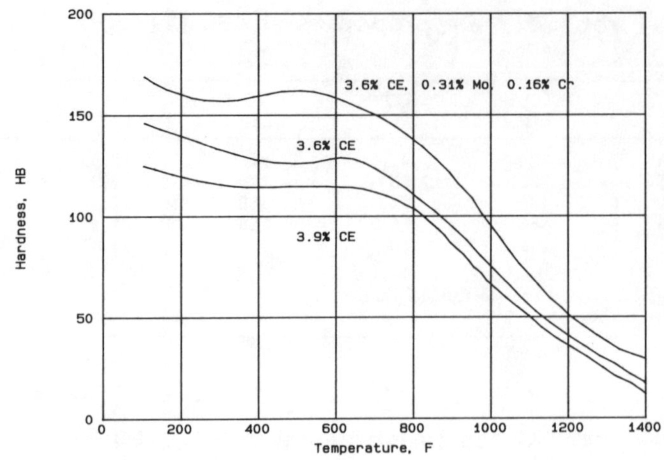

Gray irons: Brinell hardness of previously annealed gray irons. Source: Ref 14, p 159

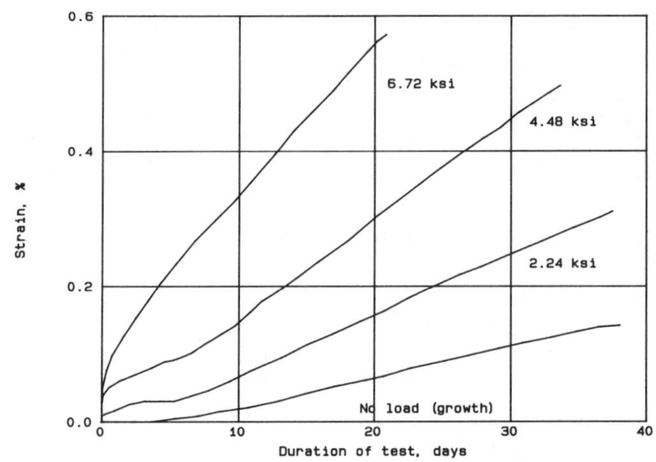

Gray irons: Growth and creep of a gray iron at 930 F. Composition: 3.19% TC, 0.85% CC, 1.66% Si, 0.91% Mn, 0.09% S, 0.08% P. Tensile strength, 42 ksi. Source: Ref 21, p 34

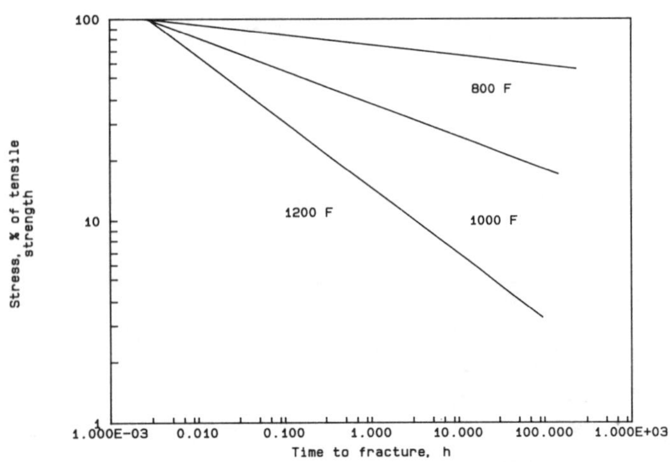

Gray irons: Stress-rupture characteristics in tension of a low-alloy iron containing 0.55% Mo and 0.75% Cr. Exposure stress was expressed as percent of room-temperature tensile strength. Source: Ref 14, p 160

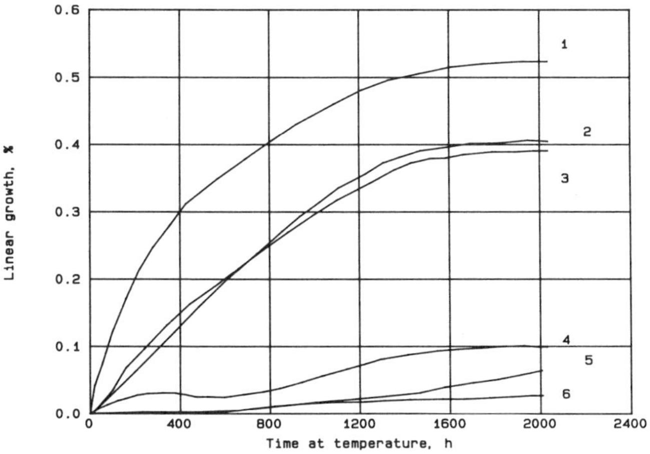

Gray irons: Growth due to graphitization at 1000 F. Curve 1: 13% unalloyed. Curve 2: 1% Mo. Curve 3: 5% Ni-Mo. Curve 4: 10% Cr-Ni-Mo-V. Curve 5: 14% Cr-Mo. Curve 6: 7% Cr annealed. Source: Ref 21, p 34

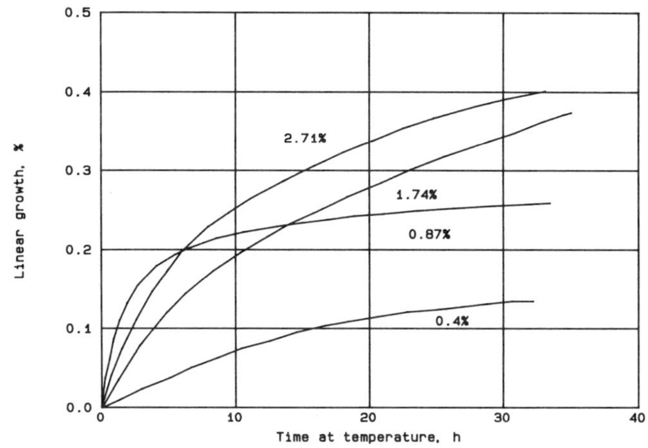

Gray irons: Graphitization growth in a vacuum at 1200 F for irons with indicated combined carbon content. Source: Ref 21, p 34

Ductile Irons

Ductile iron, also known as nodular or spheroidal graphite iron, is cast iron in which the graphite is present as tiny spheres or nodules. Iron with nodular graphite is much stronger and more ductile than gray iron of similar composition. Unalloyed ductile iron typically contains 3.50 to 3.80% total carbon, 0.30 to 1.00% Mn, and 2.00 to 2.80% Si.

Hardness and strength of all standard grades of ductile iron are relatively constant up to about 425 C (800 F), with the pearlitic ductile irons surpassing the ferritic grades in terms of elevated-temperature properties. Use of molybdenum for strength retention at high temperatures is effective in unalloyed or low-alloy ductile irons up to about 650 C (1200 F) and in Si-0.5% Al ductile irons up to 800 C (1500 F). Molybdenum also can be used to boost rupture strength, and additions of molybdenum or copper can enhance creep strength.

Oxidation resistance of ductile irons increases with increasing silicon content, and the standard grades of ductile iron are more oxidation resistant than cast steel, gray iron, or pearlitic malleable iron. Dimensional growth of ductile iron is much less than that of gray iron at elevated temperatures as well.

Heat-resistant, high-alloy ductile irons include the austenitic Ni-Resist grades covered by ASTM A439. These materials contain 18 to 36% Ni, up to 7% Cu, and 1.75 to 4% Cr. They exhibit (1) good resistance to high-temperature scaling and growth up to 815 C (1500 F) in most oxidizing atmospheres; (2) good performance in steam service up to 530 C (990 F); and (3) good resistance to sour gases and liquids up to 400 C (750 F). Maximum use temperature is 540 C (1000 F), if appreciable sulfur is present in the atmosphere.

Ductile Ni-Resist grades are considerably stronger and tougher than their gray iron counterparts. Data on both types of austenitic, high-nickel irons are given in the section on Alloy Irons.

Ductile irons: Composition of standard grades of ductile iron

	Grade or Class	UNS No.	TC(a) %	Si %	P %	S %
ASTM A395, ASME SA395	60-40-18	F32800	3 min	2.5 max(b)	0.08 max	...
ASTM A476, SAE AMS5316	80-60-03	F34100	3 min(c)	3 max	0.08 max	0.05 max
ASTM A536, MIL-I-11466B(MR)	60-40-03(d)	F32800	...	...	...	...
	65-45-12(d)	F33100	...	...	...	...
	80-55-06(d)	F33800	...	...	...	...
	100-70-03(d)	F34800	...	...	...	...
	120-90-02(d)	F36200	...	...	...	...

(a) Total carbon. (b) The silicon limit may be increased by 0.08% for each 0.01% reduction in phosphorus content, up to 2.75% Si. (c) Carbon equivalent (CE), 3.8 to 4.5; CE = TC + 0.3 (Si + P). (d) Composition subordinate to mechanical properties; composition range for any element may be specified by agreement between supplier and purchaser
Source: Ref 1, p 53

Physical Properties

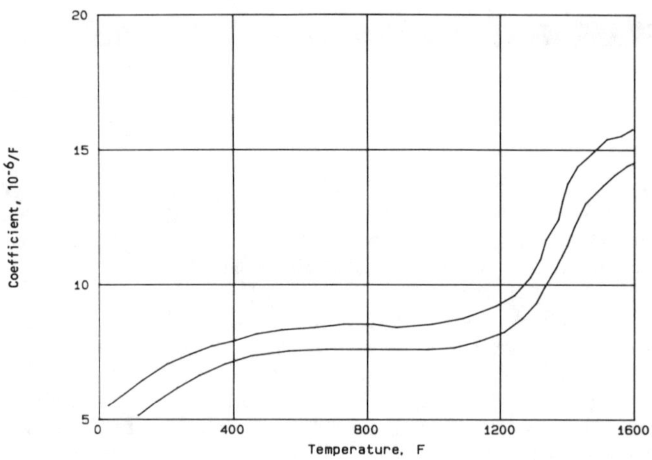

Ductile irons: Average linear thermal expansion of typical ductile irons. See figure on specific heat for compositions and types. Range of values falls within plotted curves. Source: Ref 21, p 8

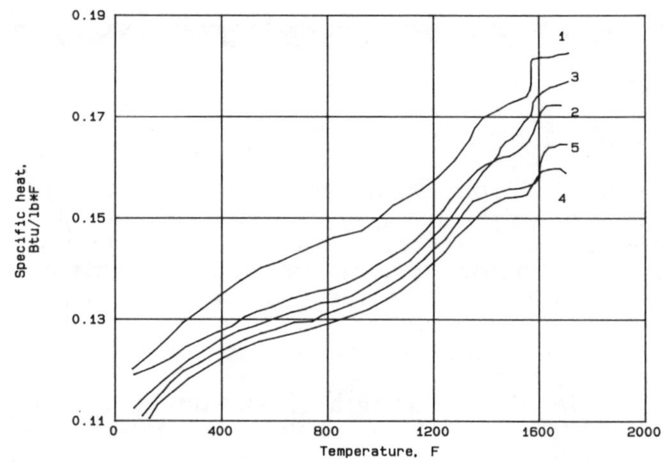

Iron No.	Pearlite %	Ferrite %	Composition %
1	50	50	3.72 TC, 2.60 Si, 0.24 Mn
2	10	90	3.05 TC, 3.47 Si, 0.59 Mn
3	25	75	2.65 TC, 2.59 Si, 0.28 Mn
4	20	80	3.45 TC, 2.65 Si, 0.29 Mn
5	20	80	3.30 TC, 2.63 Si, 0.26 Mn

Ductile irons: Specific heat of several ductile irons. Curve 1: Nodular + flake. Curve 2: Nodular + 5% flake. Curve 3: Nodular. Curve 4: Flake. Curve 5: Flake. Source: Ref 21, p 8

Ductile irons: Mean coefficients of linear thermal expansion

TEMP C	TEMP F	THEXP(a) 10^{-6}/C	THEXP(a) 10^{-6}/F	THEXP(b) 10^{-6}/C	THEXP(b) 10^{-6}/F
20-100	70-212	11.2	6.2	10.6	5.9
20-200	70-392	12.2	6.8	11.7	6.5
20-300	70-568	12.8	7.1	12.4	6.9
20-400	70-752	13.2	7.3	13.0	7.2
20-500	70-932	13.5	7.5	13.3	7.4
20-600	70-1112	13.7	7.6	13.6	7.6
20-700	70-1292	13.8	7.7	...	...
20-760	70-1472	14.8	8.2	14.8	8.2
20-870	70-1652	15.3	8.5	15.3	8.5

(a) Ferritic ductile iron. (b) Pearlitic ductile iron
Source: Ref 1, p 53

Mechanical Properties

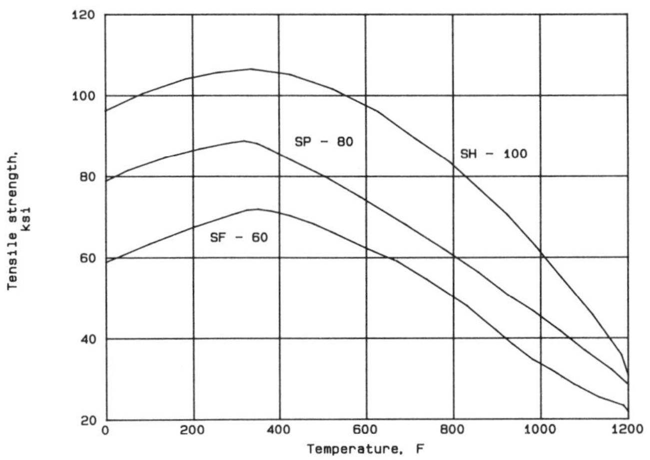

Ductile irons: Tensile strength of Meehanite ductile iron. Source: Ref 21, p 13

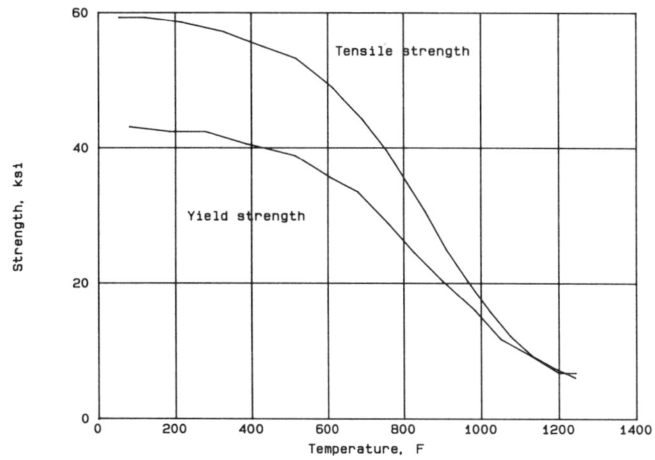

Ductile irons: Tensile and yield strengths of ferritic, annealed ductile iron. Source: Ref 21, p 13

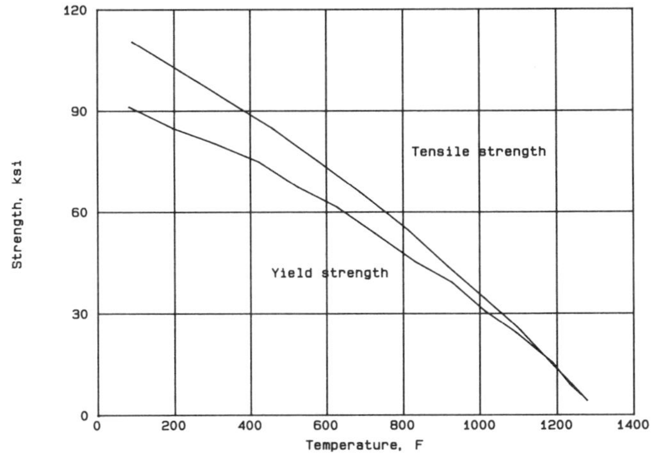

Ductile irons: Tensile and yield strengths of pearlitic, normalized iron. Source: Ref 21, p 13

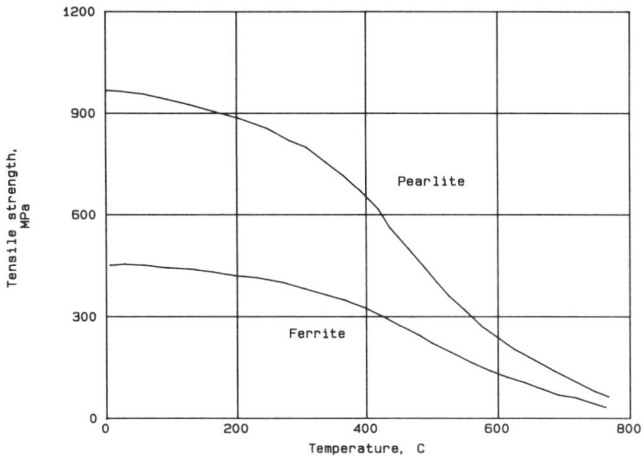

Ductile irons: Tensile strength. Source: Ref 1, p 52

Ductile irons: Creep strength of several ductile irons at 425 C (800 F)

Iron No.	Annealed	Pearlite %	CRTMN 0.0001%/h MPa	CRTMN 0.0001%/h ksi	CRTMN 0.00001%/h MPa	CRTMN 0.00001%/h ksi
1	950 C	1	48	6.96	...	...
1	870 C	10	58.5	8.48	...	...
1	(a)	30	79	11.46	...	...
2	870 C	20	141	20.45	...	...
2	950 C	10	172	24.95	103	14.94
2	(a)	100	172	24.95	114	16.53
3	950 C	3	186	26.98	152	22.04
3	(a)	50	...	...	145	21.03
4	870 C	2	124	17.98	86	12.47
5	(a)	94	145	21.03	...	...

Note: Iron 1: 3.54% TC, 0.40% Mn, 0.017% P, 2.26% Si, 0.56% Ni, 0.15% Cu, 0.050% Mg. Iron 2: 3.49% TC, 0.78% Mn, 0.086% P, 2.46% Si, 1.08% Ni, 0.56% Cu, 0.072% Mg. Iron 3: 3.61% TC, 0.47% Mn, 0.020% P, 2.43% Si, 1.19% Ni, 0.81% Mo, 0.10% Cu, 0.050% Mg. Iron 4: 3.59% TC, 0.40% Mn, 0.020% P, 2.43% Si, 1.08% Ni, 0.24% Mo, 0.10% Cu, 0.050% Mg. Iron 5: 3.49% TC, 0.37% Mn, 0.085% P, 2.50% Si, 1.22% Ni, 0.10% Cu, 0.064% Mg.
(a) As cast and stress relieved
Source: Ref 1, p 48

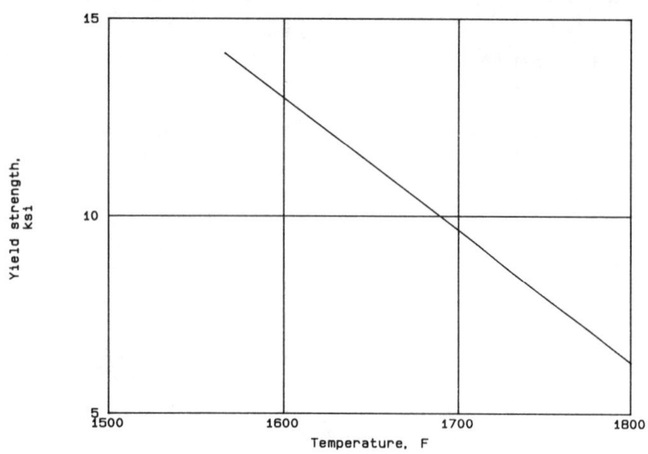

Ductile irons: Compressive yield strength of type HS Meehanite ductile iron. Source: Ref 21, p 29

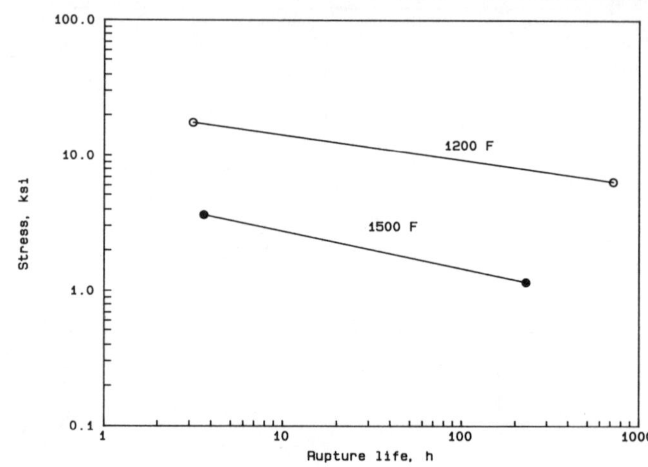

Ductile irons: Stress-rupture of 0.875-in. diam bar, normalized at 1750 F for 2 h, stabilized at test temperature for 4 h. Composition: 2.84% C, 3.94% Si, 2.05% Mo. Source: Ref 21, p 21

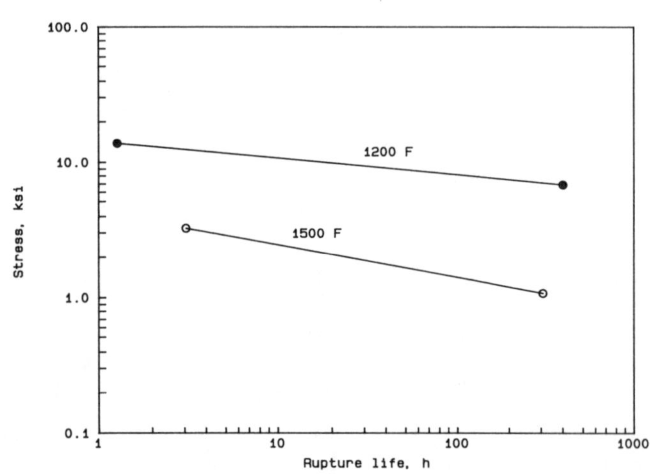

Ductile irons: Stress-rupture of 0.875-in. diam bar, normalized at 1750 F for 2 h, stabilized at test temperature for 4 h. Composition: 2.88% C, 3.94% Si, 2.08% Mo, 2.0% Co. Source: Ref 21, p 21

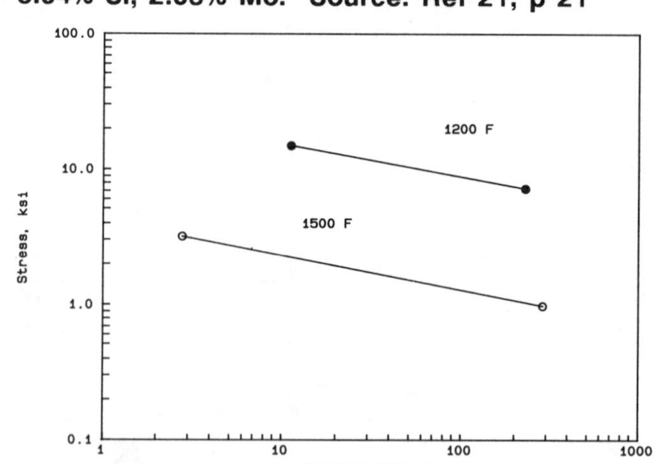

Ductile irons: Stress-rupture of 0.875-in. diam bar, normalized at 1750 F for 2 h, stabilized at test temperature for 4 h. Composition: 2.92% C, 3.83% Si, 2.09% Mo, 1.08% Cr. Source: Ref 21, p 21

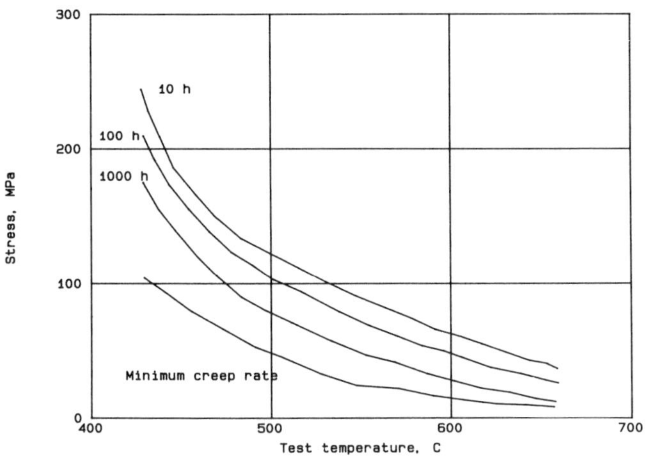

Ductile irons: Stress-rupture of annealed, ferritic iron. The minimum creep rate was 0.0001%/h. Source: Ref 1, p 49

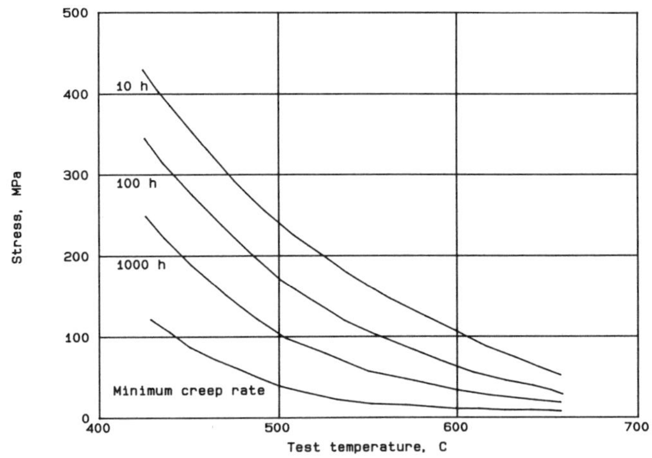

Ductile irons: Stress-rupture of normalized, pearlitic iron. The minimum creep rate was 0.0001%/h. Source: Ref 1, p 49

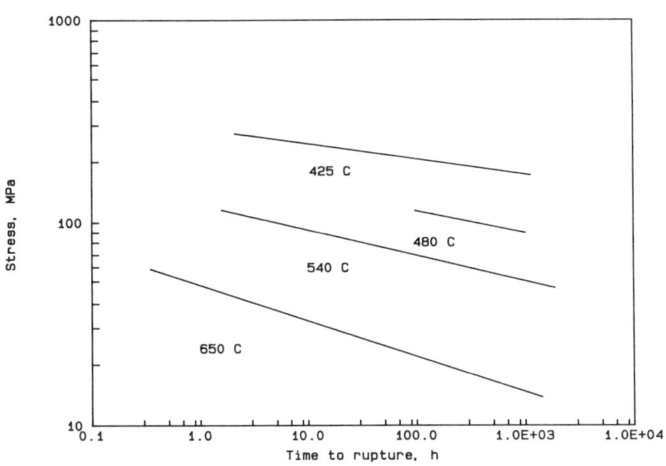

Ductile irons: Stress-rupture properties of 2.5% Si, 1.0% Ni ductile iron. Source: Ref 1, p 50

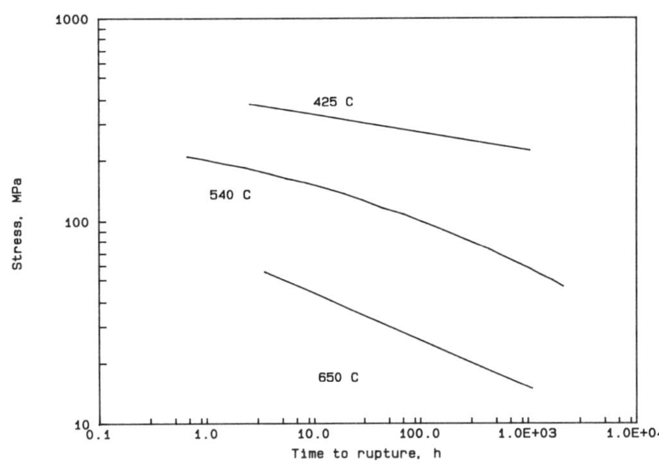

Ductile irons: Stress-rupture properties of 2.5% Si, 1.0% Ni ductile iron. Source: Ref 1, p 50

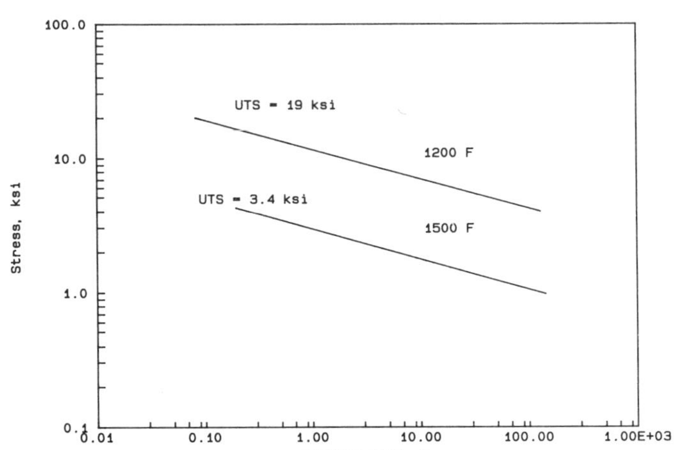

Ductile irons: Stress-rupture in Mo-free 1-in. ductile iron Y-blocks, normalized at 1750 F. Composition: 3.24% C, 3.85% Si, 0.77% Al. Source: Ref 21, p 20

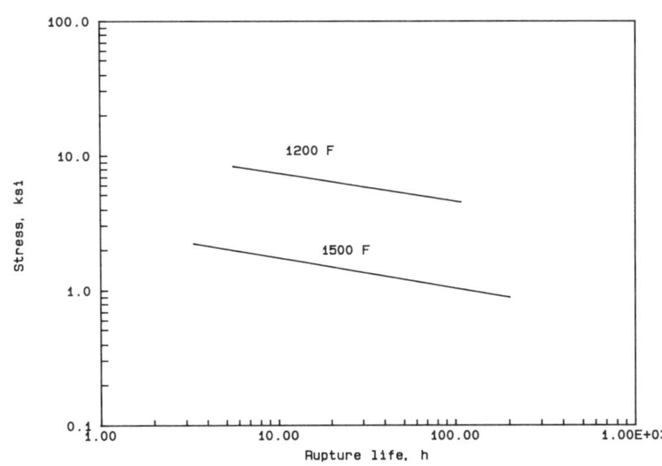

Ductile irons: Stress-rupture in Mo-free 1-in. ductile iron Y-blocks, normalized at 1750 F. Composition: 3.13% C, 4.00% Si, 0.77% Al. Source: Ref 21, p 20

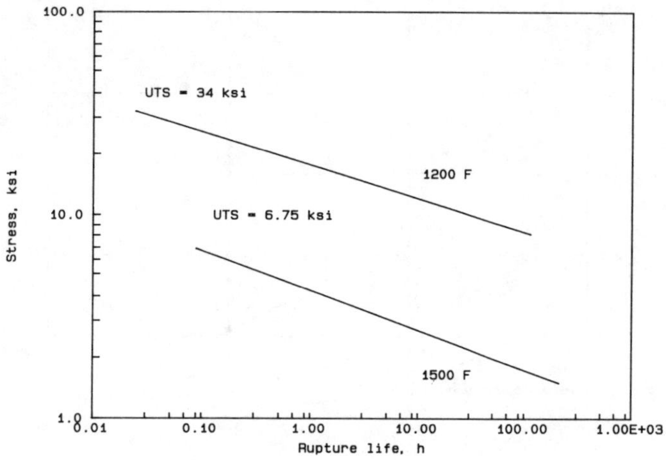

Ductile irons: Stress-rupture properties of cast specimens with indicated tensile strengths. Composition: 3.26% C, 3.94% Si, 2.01% Mo, 0.63% Al. Source: Ref 21, p 20

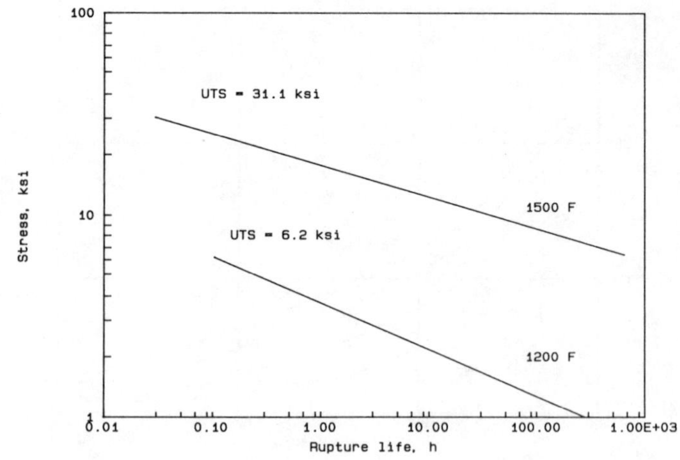

Ductile irons: Stress-rupture properties of cast specimens with indicated tensile strengths. Composition: 3.27% C, 3.72% Si, 2.05% Mo, 0.55% Al. Source: Ref 21, p 20

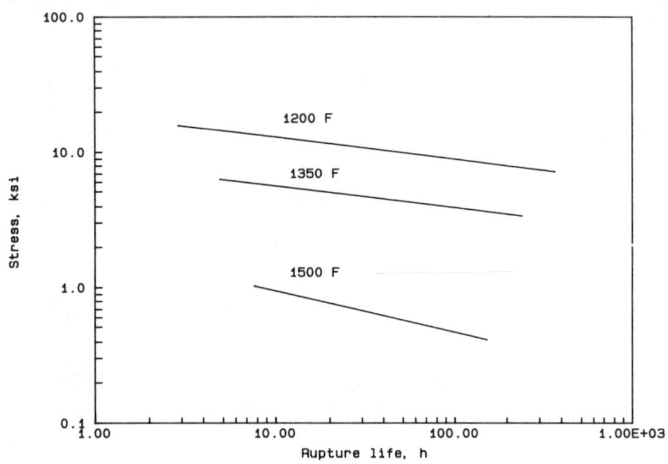

Ductile irons: Stress-rupture properties of cast specimens. Composition: 2.92% C, 4.25% Si, 1.94% Mo, 0.55% Al. Source: Ref 21, p 20

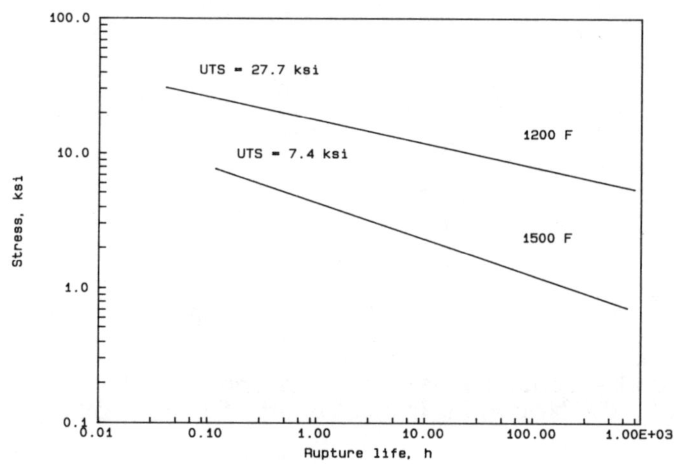

Ductile irons: Stress-rupture properties of cast specimens with indicated tensile strengths. Composition: 3.36% C, 3.84% Si, 1.97% Mo, 0.49% Al. Source: Ref 21, p 20

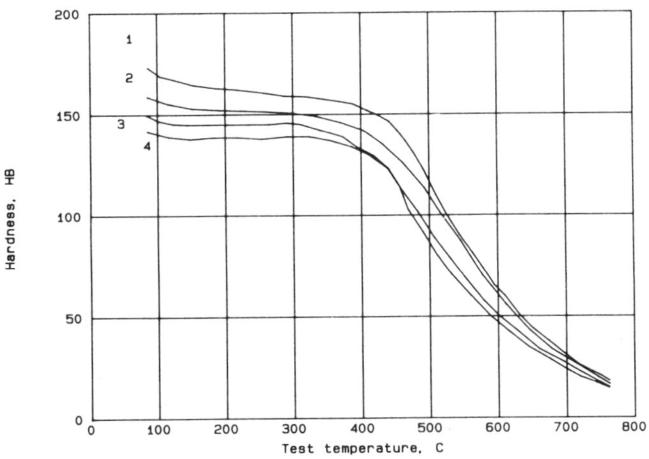

Ductile irons: Brinell hardness of annealed, ferritic irons. Curve 1: 0.59% Mn, 2.63% Si, 1.45% Ni. Curve 2: 0.42% Mn, 2.41% Si, 0.72% Ni. Curve 3: 0.26% Mn, 2.30% Si, 0.96% Ni. Curve 4: 0.57% Mn, 1.85% Si. Source: Ref 1, p 46

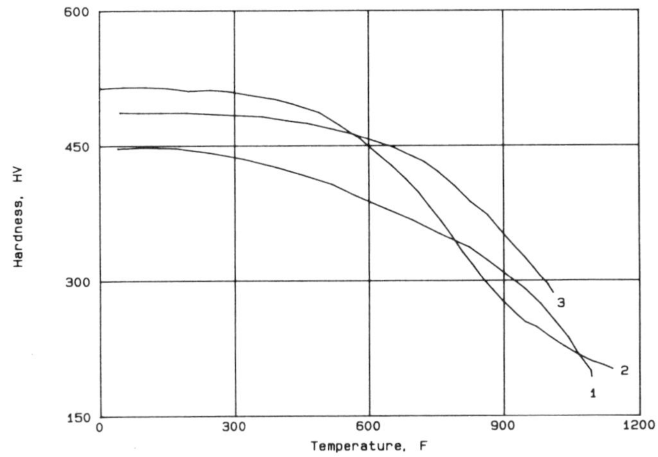

Ductile irons: Vickers hardness of three ductile irons. Curve 1: Unalloyed. Curve 2: Acicular, 0.92% Mo, 3.68% Ni. Curve 3: Acicular, 0.75% Mo, 3.68% Ni. Source: Ref 21, p 31

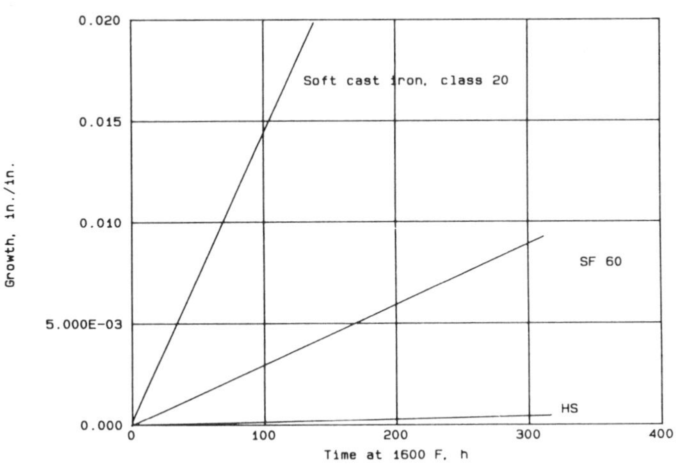

Ductile irons: Permanent growth of Meehanite iron after elevated-temperature exposure. Source: Ref 21, p 28

Malleable Irons

Malleable iron initially is produced as white cast iron, which is then heat treated to convert all or most of the carbon-containing phase from iron carbide to temper carbon, a nodular form of graphite. Ferritic malleable iron has temper carbon dispersed in a ferrite matrix, whereas pearlitic and martensitic malleable irons contain controlled amounts of combined carbon. The martensitic grades are produced by quenching and tempering pearlitic grades.

Ferritic malleable irons typically contain 2.00 to 2.70% total carbon, 0.25 to 0.55% Mn, and 1.00 to 1.75% Si. The composition of pearlitic malleable irons is 2.00 to 2.70% total carbon, 0.25 to 1.25% Mn, and 1.00 to 1.75% Si.

Malleable irons (like ductile irons) combine ductility with toughness. In many instances, the choice between malleable and ductile iron is based on economy or availability rather than on properties. However, none of the malleable irons are similar to the heat-resistant Ni-Resist ductile irons. Among the malleable irons, the pearlitic grades are stronger and harder than the ferritic irons. Martensitic irons have particularly high wear resistance.

Generally, the ferritic malleable irons covered by ASTM A47 and the pearlitic grades covered by A197 are ideal candidates for elevated temperature service. Heavy-duty application at temperatures up to 345 C (650 F) are best suited for A338 ferritic malleable irons.

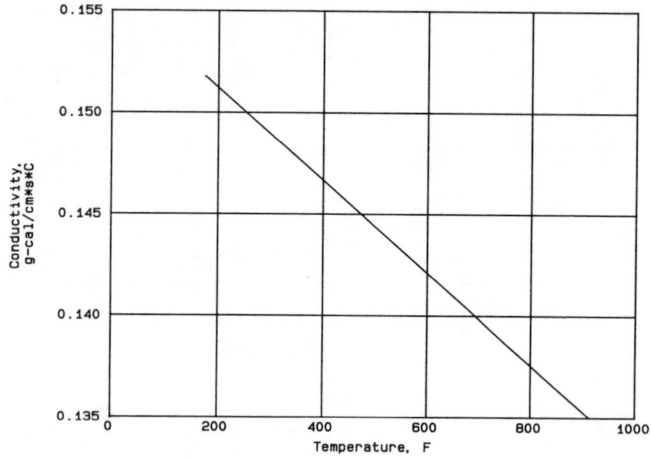

Malleable irons: Thermal conductivity. Source: Ref 21, p 6

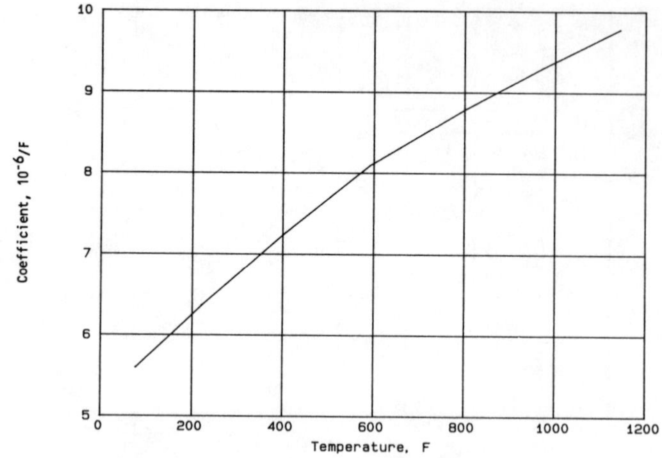

Malleable irons: Mean coefficient of thermal expansion. Source: Ref 21, p 6

Malleable irons: Comparison of electrical resistivity for high-carbon (1.22% C, 0.35% Mn), low-carbon (0.06% C, 0.38% Mn), and austenitic (1.22% C, 13.00% Mn, 0.22% Si) steels, and malleable iron. Source: Ref 21, p 6

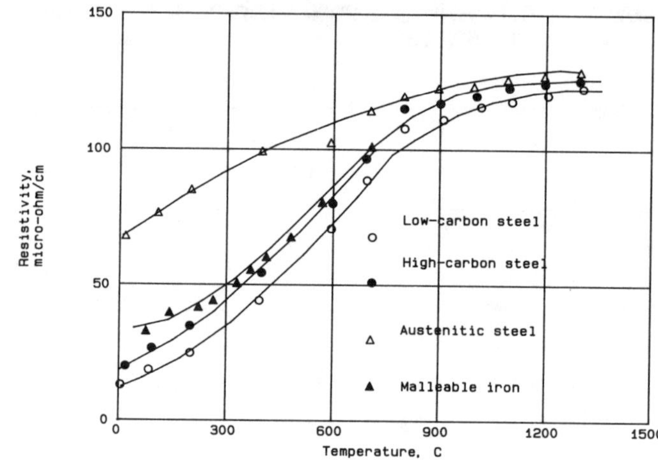

Mechanical Properties

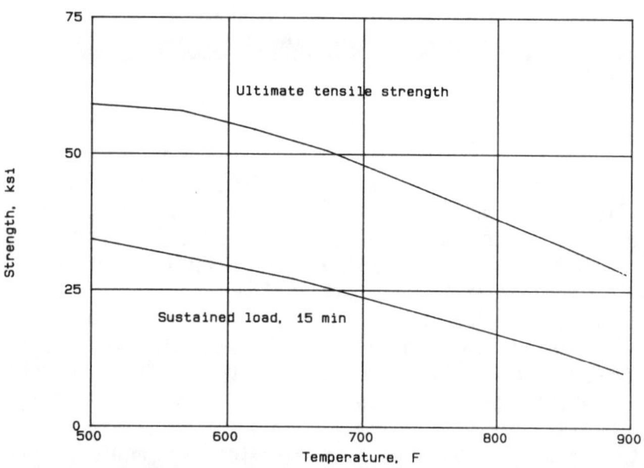

Malleable irons: Tensile strength (crosshead speed of 0.8 mm/min) and 15-min sustained load data for ferritic malleable iron. Source: Ref 1, p 66

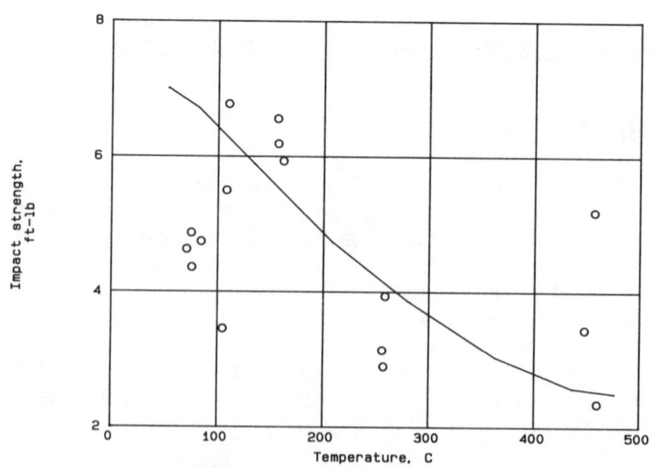

Malleable irons: Impact strength of sand cast pearlitic malleable iron specimens from 0.505-in. Charpy-V notch test bars. Source: Ref 21, p 12

Malleable irons: High-temperature property data for air quenched pearlitic malleable irons

HARD HB	YS MPa	YS ksi	UTS MPa	UTS ksi	EL %
At 21 C (70 F)					
147	...	...	...	...	...
163	...	...	...	...	...
223	...	...	...	...	...
At 204 C (400 F)					
135	214	31	338	49	12
163	234	34	400	58	8
205	427	62	545	79	5
At 260 C (500 F)					
134	214	31	352	51	10
155	303	44	414	60	8
220	434	63	614	89	7
At 316 C (600 F)					
140	248	36	365	53	10
150	303	44	427	62	8
212	427	62	648	94	10
At 371 C (700 F)					
127	...	...	...	...	...
133	303	44	414	60	10
178	400	58	558	81	10
At 427 C (800 F)					
110	228	33	303	44	16
118	290	42	359	52	11
170	365	53	421	61	15
At 482 C (900 F)					
82	179	26	207	30	16
85	248	36	269	39	8
123	331	48	372	54	16
At 538 C (1000 F)					
68	138	20	152	22	19
81	159	23	179	26	17
100	255	37	276	40	14

Source: Ref 21, p 9

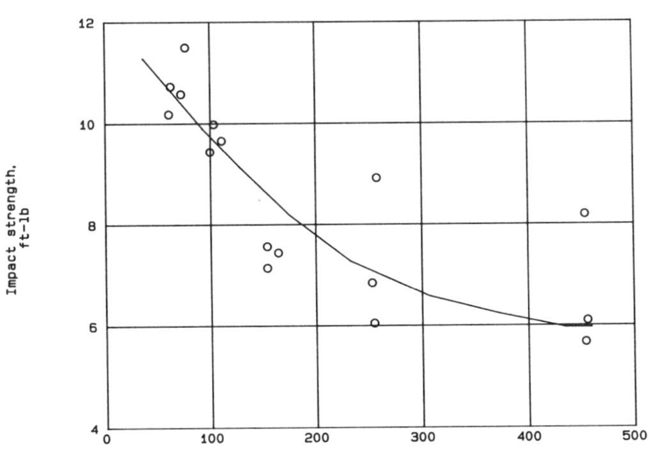

Malleable irons: Impact strength of sand cast ferritic malleable iron specimens from 0.505-in. Charpy V-notch test bars. Source: Ref 21, p 12

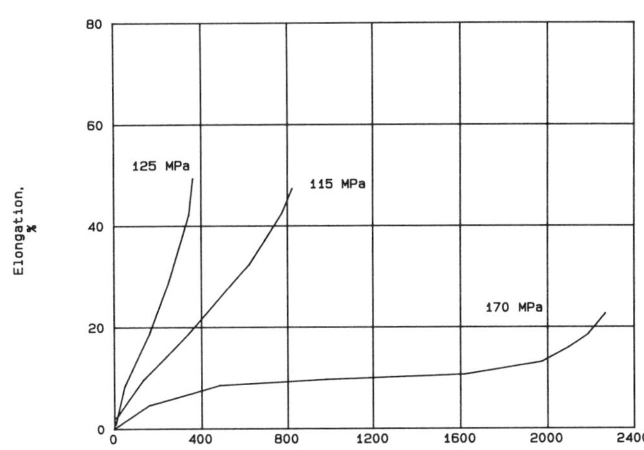

Malleable irons: Creep curves at 425 C for ferritic malleable iron. Source: Ref 1, p 66

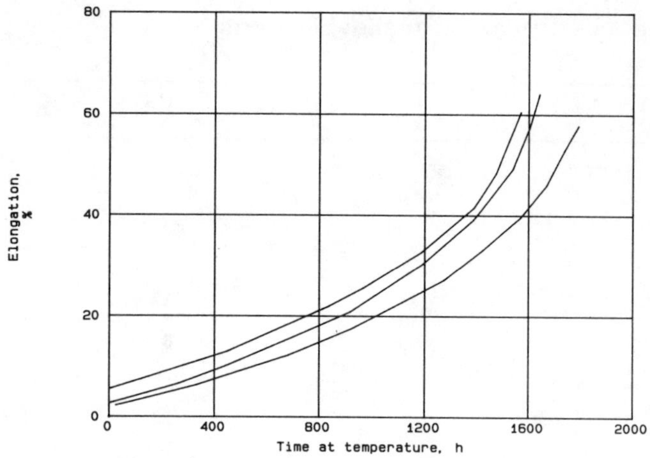

Malleable irons: Creep curves at 540 C and 40 MPa for ferritic, malleable iron. Source: Ref 1, p 66

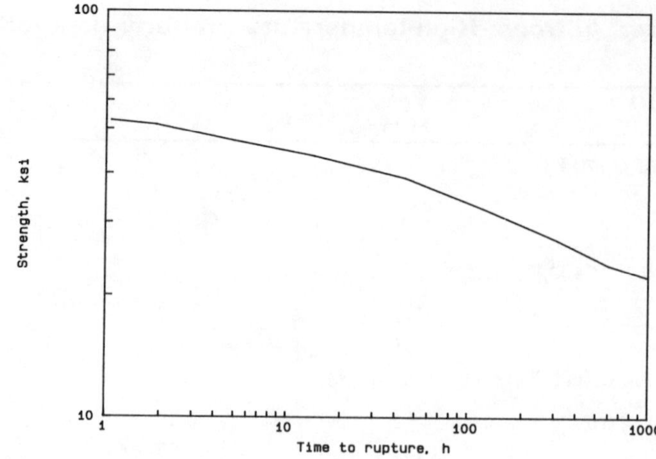

Malleable irons: Stress-rupture strength at 800 F of oil quenched pearlitic, malleable grade 80002 drawn to 269 HB. Source: Ref 21, p 16

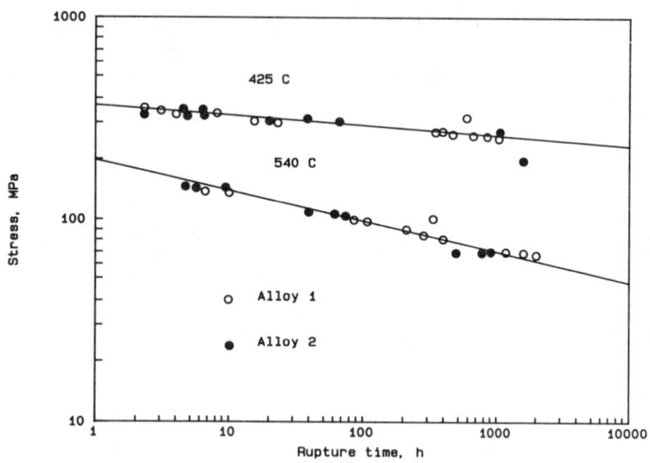

Malleable irons: Stress-rupture strength of pearlitic, malleable iron. Alloy 1: 2.27% C, 1.15% Si, 0.89% Mn, 0.098% S, 0.135% P, 0.019% Cr. Alloy 2: 2.29% C, 1.01% Si, 0.75% Mn, 0.086% S, 0.11% P. Source: Ref 1, p 72

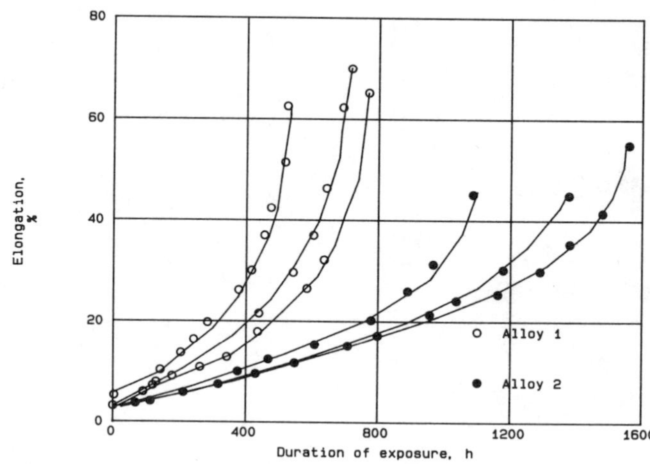

Malleable irons: Creep curves at 540 C and 70 MPa for pearlitic, malleable iron. Alloy 1: 2.27% C, 1.15% Si, 0.89% Mn, 0.098% S, 0.135% P, 0.019% Cr. Alloy 2: 2.29% C, 1.01% Si, 0.75% Mn, 0.086% S, 0.11% P. Source: Ref 1, p 72

Malleable irons: Stress-rupture data for ferritic, malleable irons. Source: Ref 1, p 66

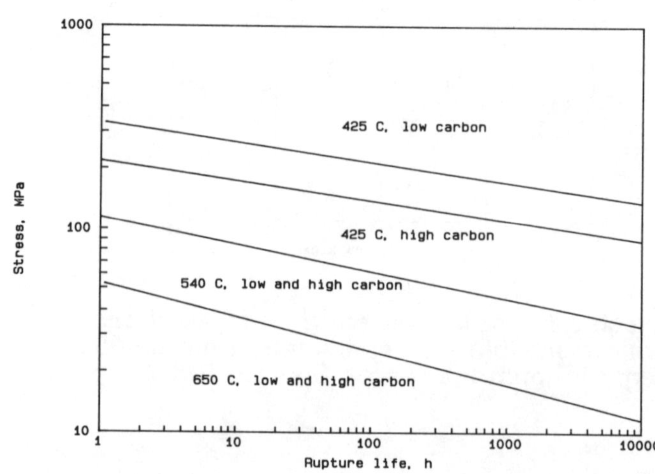

Malleable irons: Hardness data for iron castings

	Class or grade	HARD HB	Treatment	Microstructure
Pearlitic				
ASTM A 338-84	M3210	156 max	(a)	Ferritic
	M4504	163-217	(b)	(d)
Automotive				
ASTM A602-70 (1982)	M5003	187-241	(b)	(d)
SAE J158a	M5503	187-241	(c)	(e)
	M7002	229-269	(c)	(e)
	M8501	269-302	(c)	(e)

(a) Annealed. (b) Air quenched and tempered. (c) Liquid quenched and tempered. (d) Ferritic and tempered pearlite. (e) Tempered martensitic
Source: Ref 19, p 41

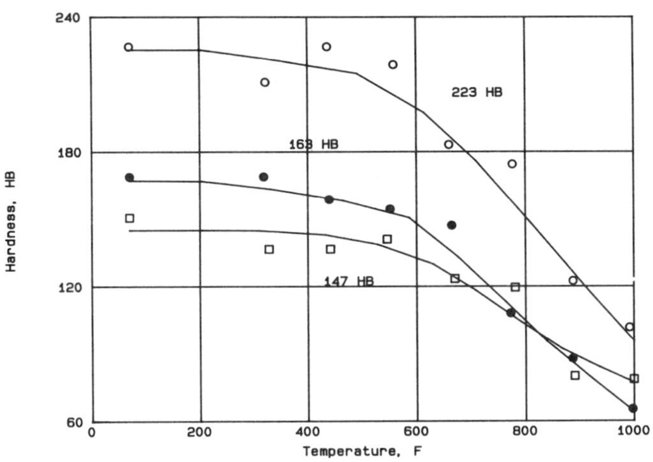

Malleable irons: Brinell hardness of air quenched pearlitic, malleable iron. Values shown for each curve are original hardness levels at room temperature. Source: Ref 21, p 7

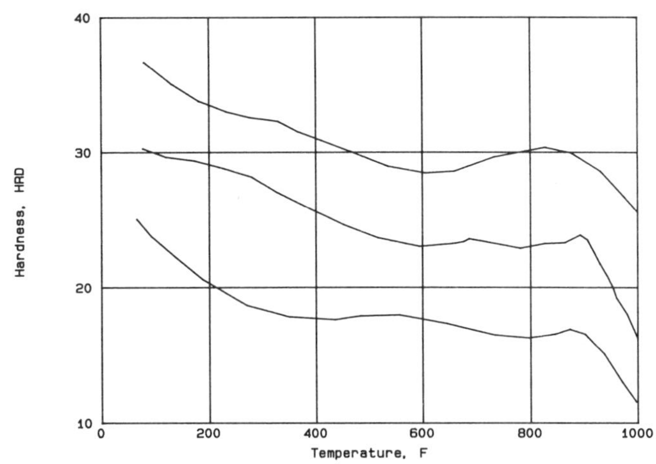

Malleable irons: Rockwell D hardness for three grades of air quenched and tempered pearlitic, malleable iron. Source: Ref 21, p 7

Alloy Irons

The general classification "alloy cast irons" includes abrasion-resistant white irons, corrosion-resistant irons, and heat-resistant irons.

Heat-resistant irons usually are ferritic or austenitic as cast. Carbon exists in these irons predominately as graphite flakes or as nodules, which subdivides them into either gray or ductile iron categories. These alloys derive their heat resistance largely from additions of chromium, nickel, molybdenum, aluminum, and silicon (in excess of 3%), either singly or in combination. Silicon and chromium increase scaling resistance, but also reduce toughness and thermal shock resistance. Nickel enhances elevated-temperature strength and toughness, whereas molybdenum increases strength at high temperatures. Aluminum helps prevent scaling and dimensional growth, but has an adverse effect on mechanical properties at room temperature.

High-Nickel Irons. Austenitic irons containing 18 to 36% Ni, 7% Cu max, and 1.75 to 4% Cr are known as "Ni-Resist" irons and are used for both heat and corrosion resistance. They are available in gray and ductile forms. These irons are tougher and more thermal shock resistant than other heat-resistant irons, but their strength is relatively low.

Austenitic irons can be used at temperatures up to 950 C (1740 F). They resist scaling and growth up to 815 C (1500 F) in most oxidizing atmospheres, offer good performance in steam service up to 530 C (990 F), and are suitable for applications involving sour gases and liquids up to 400 C (750 F). High-nickel ductile irons are considerably stronger than their gray iron counterparts.

High-Chromium Irons. Additions of 0.5 to 2% Cr in cast irons can increase resistance to oxidation, scaling, and growth in many applications at temperatures up to 760 C (1400 F). Chromium additions of 15 to 35% provide excellent oxidation and growth resistance up to 1000 C (1800 F) and up to 1100 C (2000 F) in oxidizing atmospheres, or in the presence of certain chemicals. Thermal shock resistance of these irons is good, but toughness is limited.

High-Silicon Irons. Ferritic, high-silicon gray iron containing 4 to 7% Si (Silal) is brittle and has a low resistance to thermal shock at room temperature. However, it possesses superior properties compared to conventional gray iron above about 260 C (500 F). An austenitic gray iron known as "Nicrosilal" is tougher and more thermal shock resistant. It contains 5% Si, 18% Ni, and 2 to 5% Cr.

Both alloys exhibit excellent resistance to scaling in air up to 815 C (1500 F), and the Ni-Cr-Si material can be used in sulfurous atmospheres. Typical maximum use temperatures are 900 C (1650 F) for Silal and 950 C (1750 F) for Nicrosilal. High-silicon compositions also are available in ferritic ductile irons.

High-Aluminum Irons. Scaling resistance of alloy irons containing 4 to 25% Al is superior to that of other alloy irons, including the high-silicon types. However, these materials are also very brittle and hard to cast.

Alloy irons: Typical compositions and microstructures

Iron	TC wt.%	Mn wt.%	P wt.%	S wt.%	Si wt.%	Ni wt.%	Cr wt.%	Mo wt.%	Cu wt.%	As-cast matrix microstructure
Abrasion-resistant white irons										
Low C white iron(a)	2.2-2.8	0.2-0.6	0.15	0.15	1.0-1.6	1.5	1.0	0.5	(b)	Coarse pearlite
High C, low Si white iron	2.8-3.6	0.3-2.0	0.30	0.15	0.3-1.0	2.5	3.0	1.0	(b)	Coarse pearlite
Malleable white iron	2.2-2.5	0.3-0.5	0.15	0.15	1.0-1.6	...	...	...	...	Coarse pearlite
Martensitic Ni-Cr	2.5-3.7	1.3	0.30	0.15	0.8	1.1-4.0	1.1-4.0	1.0	...	Martensite, austenite
Martensitic Ni, high Cr	2.5-3.6	1.3	0.10	0.15	1.0-2.2	7-11	7-11	1.0	...	Martensite, austenite
Martensitic Cr-Mo	2.0-3.6	0.5-1.5	0.10	0.06	1.0	11-23	11-23	0.5-3.5	1.2	Martensite, austenite
High Cr iron	2.3-3.0	0.5-1.5	0.10	0.06	1.0	23-28	23-28	1.5	1.2	Martensite
Corrosion-resistant irons										
High-Si iron(c)	0.4-1.1	1.5	0.15	0.15	14-17	5.0	5.0	1.0	0.5	Ferrite
High-Cr iron	1.2-4.0	0.3-1.5	0.15	0.15	0.5-3.0	12-35	12-35	4.0	3.0	Martensite, austenite
Ni-Cr gray iron(d)	3.0	0.5-1.5	0.08	0.12	1.0-2.8	1.5-6.0	1.5-6.0	1.0	7.0	Austenite
Ni-Cr ductile iron(e)	3.0	0.7-4.5	0.08	0.12	1.0-3.0	1.0-5.5	1.0-5.5	1.0	...	Austenite
Heat-resistant gray irons										
Medium Si iron(f)	1.6-2.5	0.4-0.8	0.30	0.10	4.0-7.0	...	...	...	...	Ferrite
High Cr iron	1.8-3.0	0.3-1.5	0.15	0.15	0.5-2.5	5.0	15-35	...	...	Ferrite, coarse pearlite
Ni-Cr iron(d)	1.8-3.0	0.4-1.5	0.15	0.15	1.0-2.75	13.5-36	1.8-6.0	1.0	7.0	Austenite
Ni-Cr-Si iron(g)	1.8-2.6	0.4-1.0	0.10	0.10	5.0-6.0	13-43	1.8-5.5	1.0	10.0	Austenite
High Al iron	1.3-2.0	0.4-1.0	0.15	0.15	1.3-6.0	...	20-25 Al	...	...	Ferrite
Heat-resistant ductile irons										
Medium Si ductile iron	2.8-3.8	0.2-0.6	0.08	0.12	2.5-6.0	1.5	...	...	...	Ferrite
Ni-Cr ductile iron(e)	3.0	0.7-2.4	0.08	0.12	1.75-5.5	18-36	1.75-3.5	1.0	...	Austenite

Note: In composition, when a single value is given rather than a range, that value is a maximum limit.
(a) May be produced from a malleable iron-base composition. (b) Cu may replace all or part of the Ni. (c) Such as Duriron, Durichlor 51, Superchlor. (d) Such as Ni-Resist austenitic iron (ASTM A436). (e) Such as Ni-Resist austenitic ductile (ASTM A439). (f) Such as Silal. (g) Such as Nicrosilal
Source: Ref 30, p 5.16

Physical Properties

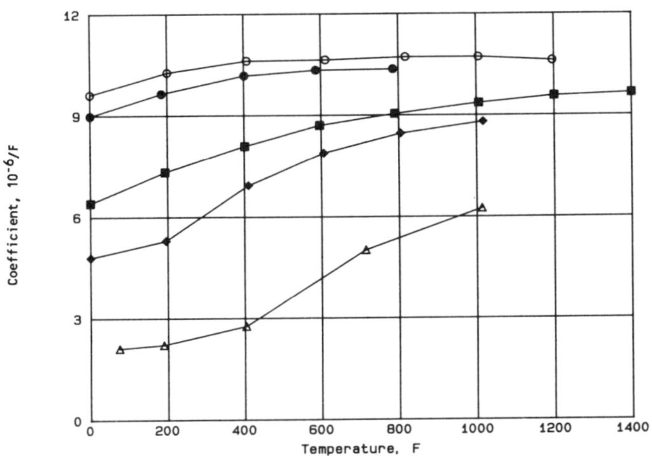

Alloy irons: Mean coefficient of linear thermal expansion for Ni-Resist cast iron from room temperature to indicated temperature. Open circle: Type 1. Solid circle: Type 2. Square: Type 3. Diamond: Type 3 (average). Triangle: Type 5. Source: Ref 21, p 3

Mechanical Properties

Alloy irons: Elevated-temperature tensile and Charpy V-notch properties of Ni-resist irons

Iron	UTS MPa	UTS ksi	YS MPa	YS ksi	IS J	IS ft-lb	RPSTR 1000 h MPa	RPSTR 1000 h ksi	CRSTR 0.0001%/h MPa	CRSTR 0.0001%/h ksi	CRSTR 0.00001%/h MPa	CRSTR 0.00001%/h ksi	EL(a) %
At 21 C (70 F)													
D-2	410	59.4	241	35.0	35	26	...	...	...	...	...	...	...
D-2w/Mo	424	61.5	255	37.0	31(b)	23(b)	...	...	...	...	...	...	...
D-2C	429	62.2	235	34.1	...	...	...	...	...	...	...	...	...
D-3	402	58.3	271	39.3	...	...	...	...	...	...	...	...	...
D-3w/Mo	421	61.0	276	40.0	...	...	...	...	...	...	...	...	...
D-4	441	64.0	305	44.3	...	...	...	...	...	...	...	...	...
D-4w/Mo	417	60.5	296	43.0	...	...	...	...	...	...	...	...	...
D-5B	419	60.8	283	41.0	...	...	...	...	...	...	...	...	...
D-5Bw/Mo	422	61.2	283	41.0	...	...	...	...	...	...	...	...	...
At 427 C (800 F)													
D-2	372	54.0	193	28.0	...	...	...	...	...	...	...	...	...
D-2w/Mo	...	...	...	...	...	...	...	...	...	...	...	...	...
D-2C	361	52.4	181	26.2	...	...	...	...	...	...	...	...	...
At 538 C (1000 F)													
D-2	329	47.7	193	28.0	47.0	35.0	193	28	159	23	62	9	6
D-2w/Mo	255	37.0	201	29.2	29.0(b)	21.0(b)	...	...	186	27	124	18	...
D-2C	290	42.0	159	23.1	...	...	145	21	90	13	...	...	14
D-3	331	48.0	196	28.4	...	...	...	...	...	...	...	...	...
D-3w/Mo	323	46.8	197	28.5	...	...	...	...	200	29	145	21	...
D-4	419	60.7	285	41.4	...	...	...	...	...	...	...	...	...
D-4w/Mo	376	54.5	265	38.4	...	...	...	...	...	...	...	...	...
D-5B	325	47.2	178	25.8	...	...	...	...	186	27	...	...	...
D-5Bw/Mo	336	48.8	197	28.6	...	...	...	...	193	28	124	18	...
At 593 C (1100 F)													
D-2	...	...	...	...	39.0	29.0	124	18	90	13.1	39	5.6	...
D-2w/Mo	...	...	...	...	31.0(b)	23.0(b)	179	26	110	16.0	76	11	5.5
D-2C	...	...	...	...	...	...	93	13.5	62	9.0	...	...	...
D-3	...	...	...	...	...	...	162	23.5	...	...	...	...	7.0
D-3w/Mo	...	...	...	...	...	...	210	30.5	124	18.0	90	13	5.0
D-4	...	...	...	...	...	...	117	17	...	...	...	...	10.5
D-4w/Mo	...	...	...	...	...	...	131	19	...	...	...	...	10.0
D-5B	...	...	...	...	...	...	172	25	110	16.0	69	10	6.5
D-5Bw/Mo	...	...	...	...	...	...	221	32	117	17.0	76	11	6.5

(continued)

Alloy irons: Elevated-temperature tensile and Charpy V-notch properties of Ni-resist irons (continued)

Iron	UTS MPa	UTS ksi	YS MPa	YS ksi	IS J	IS ft-lb	RPSTR 1000 h MPa	RPSTR 1000 h ksi	CRSTR 0.0001%/h MPa	CRSTR 0.0001%/h ksi	CRSTR 0.00001%/h MPa	CRSTR 0.00001%/h ksi	EL(a) %
At 649 C (1200 F)													
D-2	245	35.6	172	25.0	43.0	32.0	83	12.0	55	8.0	23	3.4	13.0
D-2w/Mo	265	38.5	174	25.2	33.0(b)	24.0(b)	107	15.5	66	9.6	41	6.0	...
D-2C	193	28.0	167	24.2	...	...	62	9.0	39	5.7	...	...	13.0
D-3	288	41.7	189	27.4	...	...	103	15.0	...	...	...	...	...
D-3w/Mo	303	44.0	200	29.0	...	...	124	18.0	83	12.0	55	8.0	...
D-4	331	48.0	234	34.0	...	...	66	9.5	...	...	...	...	...
D-4w/Mo	314	45.5	245	35.5	...	...	76	11.0	...	...	...	...	...
D-5b	280	40.6	167	24.2	...	...	103	15.0	66	9.6	...	...	...
D-5Bw/Mo	320	46.4	206	29.9	...	...	128	18.5	72	10.5	46	6.7	...
At 704 C (1300 F)													
D-2	...	...	...	...	...	...	59	8.5	33	4.8	14	2.1	...
D-2w/Mo	...	...	...	...	...	...	76	11.0	39	5.6	25	3.6	11.5
D-2C	...	...	...	...	...	...	41	6.0	23	3.4	...	...	...
D-3	...	...	...	...	...	...	67	9.7	...	...	...	...	12.5
D-3w/Mo	...	...	...	...	...	...	76	11.0	48	7.0	34	5.0	16.0
D-4	...	...	...	...	...	...	43	6.2	...	...	...	...	25.0
D-4w/Mo	...	...	...	...	...	...	52	7.5	...	...	...	...	21.0
D-5B	...	...	...	...	...	...	69	10.0	55	8.0	39	5.6	13.5
D-5Bw/Mo	...	...	...	...	...	...	83	12.0	41	6.0	28	4.0	11.5
At 732 C (1350 F)													
D-2	...	...	...	...	57.0	42.0	...	...	...	...	...	...	...
D-2w/Mo	...	...	...	...	37.0(b)	27.0(b)	...	...	...	...	...	...	...
At 760 C (1400 F)													
D-2	152	22.1	117	17.0	...	...	38	5.5	...	...	...	...	...
D-2w/Mo	172	24.9	118	17.1	...	...	41	6.0	...	...	...	...	...
D-2C	119	17.2	114	16.6	...	...	28	4.0	...	...	...	...	...
D-3	183	26.5	105	15.2	...	...	41	6.0	...	...	...	...	...
D-3w/Mo	199	28.8	150	21.8	...	...	48	7.0	...	...	...	...	...
D-4	150	21.8	128	18.5	...	...	21	3.0	...	...	...	...	...
D-4w/Mo	154	22.3	128	18.6	...	...	28	4.0	...	...	...	...	...
D-5B	172	24.9	128	18.6	...	...	38	5.5	...	...	...	...	...
D-5Bw/Mo	214	31.1	168	24.3	...	...	45	6.5	...	...	...	...	...
At 816 C (1500 F)													
D-2	...	...	...	...	47.0	35.0	...	...	...	...	...	...	...
D-2w/Mo	...	...	...	...	37.0(b)	27.0(b)	...	...	...	...	...	...	...

(a) Elongation at rupture. (b) Type D-2B without Mo
Source: Ref 21, p 14

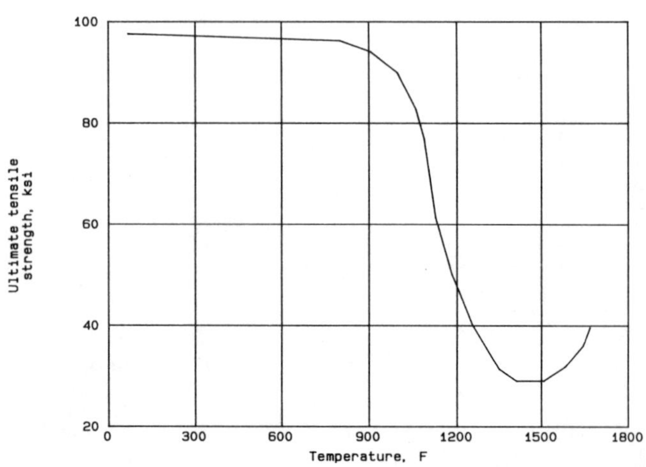

Alloy irons: Tensile strength of high-chromium iron containing 24.5% Cr, 2.0% C, 1.1% Si after annealing at 1830 F for 4 h. Source: Ref 14, p 173

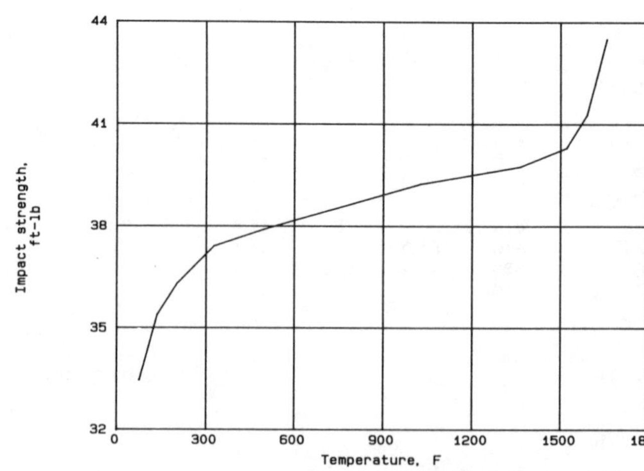

Alloy irons: Impact strength of a high-chromium iron containing 24.5% Cr, 2.0% C, 1.1% Si after annealing at 1830 F for 4 h. Source: Ref 14, p 173

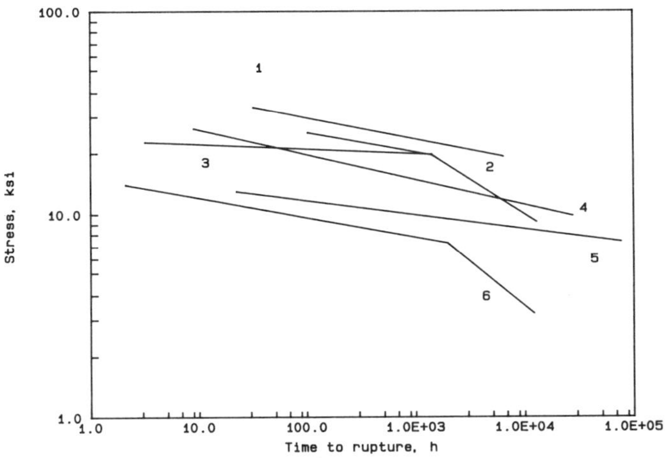

Alloy irons: Stress-rupture strength of high-chromium irons at 1300 F. Curves 1, 2, 3, and 4: austenitic irons. Curves 5 and 6: ferritic irons. Source: Ref 14, p 175

Iron No.	C %	Si %	Mn %	S %	P %	Cr %	Ni %	Al %	Ti %
1	0.99	0.64	0.29	0.021	0.022	31.2	15.2	...	...
2	0.92	2.23	0.48	0.024	0.024	32.0	14.9	...	...
3	0.98	0.66	0.29	0.014	0.025	30.2	15.2	1.0	...
4	0.95	2.55	0.47	0.018	0.020	31.4	15.2	0.9	...
5	1.88	0.39	0.39	0.025	0.024	24.8	...	...	0.38
6	0.98	0.73	0.29	0.039	0.024	33.7	...	...	0.16

Note: Ti added to ladle for grain refinements

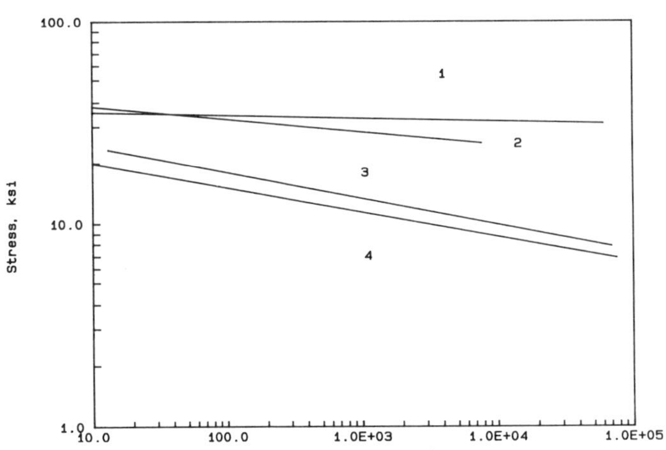

Alloy irons: Stress-rupture strength of high-chromium irons at 1100 F. Curves 1 and 2: austenitic irons. Curves 3 and 4: ferritic irons. Source: Ref 14, p 175

Iron	C %	Si %	Mn %	P %	S %	Cr %	Ni %	Mo %	Cu %	Al %
1	0.98	0.66	0.29	0.025	0.014	30.2	15.2	...	...	1.0
2	0.95	2.55	0.47	0.020	0.018	31.4	15.2	...	...	0.9
3(a)	1.88	0.39	0.39	0.024	0.025	24.8	...	...	...	...
4(b)	0.98	0.73	0.29	0.024	0.039	33.7	...	...	...	...

Note: Ti added to ladle for grain refinements.
(a) 0.38% Ti, (b) 0.16% Ti

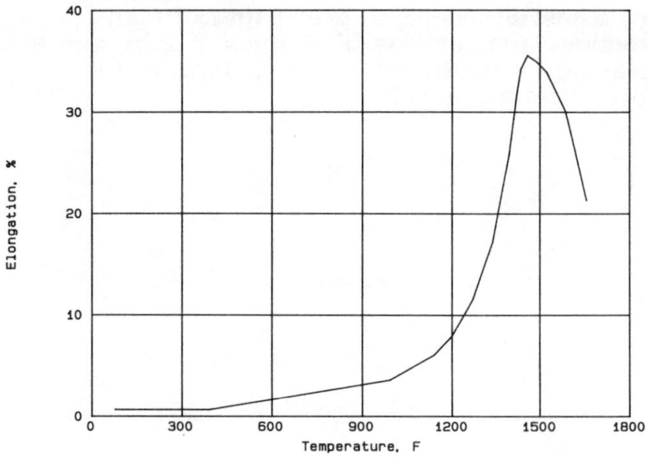

Alloy irons: Tensile elongation of a high-chromium iron containing 24.5% Cr, 2.0% C, 1.1% Si after annealing at 1830 F for 4 h. Source: Ref 14, p 173

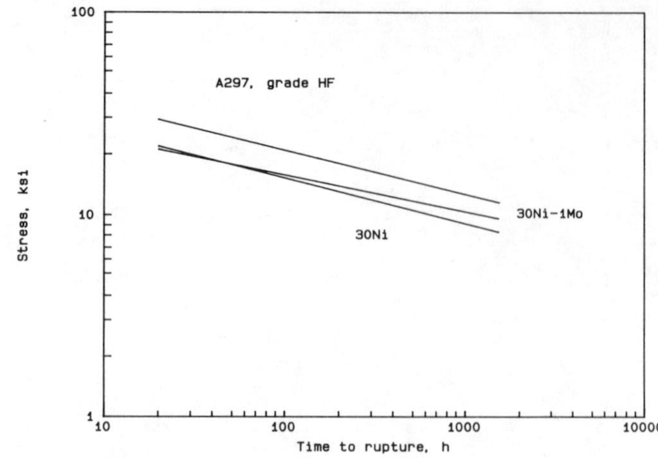

Alloy irons: Stress-rupture curves for high-nickel heat-resistant irons at 705 C. 19Cr-9Ni cast stainless steel (ASTM A297, grade HF) is shown for reference. Source: Ref 1, p 95

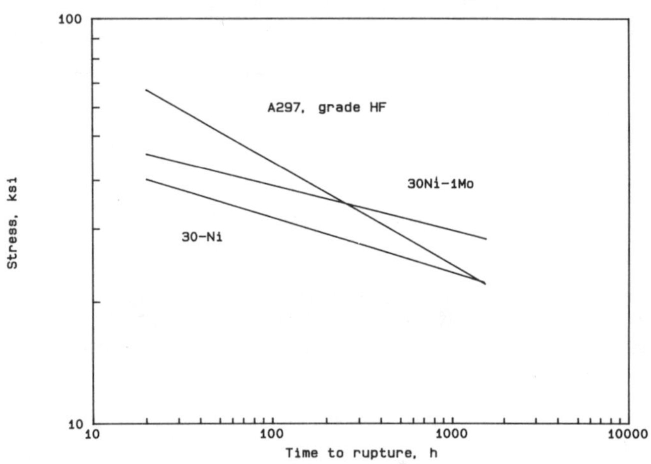

Alloy irons: Stress-rupture curves for high-nickel heat-resistant irons at 595 C. 19Cr-9Ni cast stainless steel (ASTM A297, grade HF) is shown for reference. Source: Ref 1, p 95

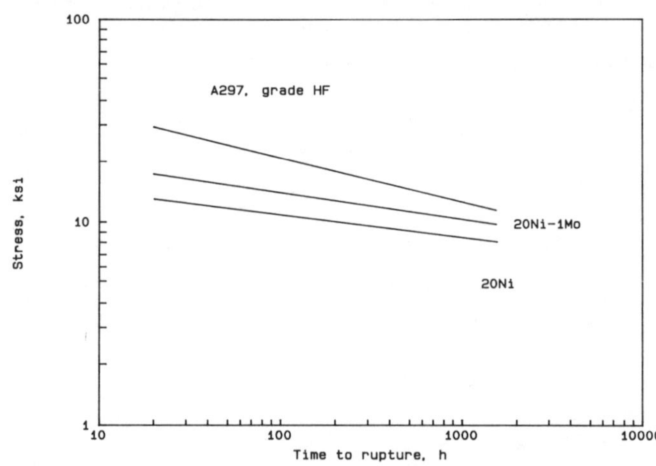

Alloy irons: Stress-rupture curves for high-nickel heat-resistant irons at 705 C. 19Cr-9Ni cast stainless steel (ASTM A297, grade HF) is shown for reference. Source: Ref 1, p 95

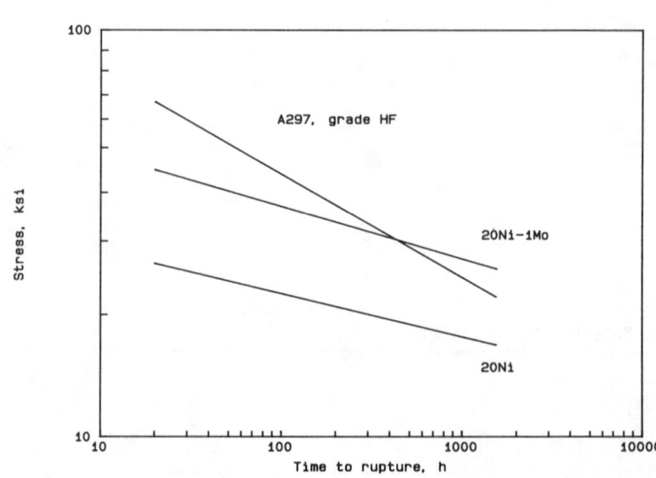

Alloy irons: Stress-rupture curves for high-nickel heat-resistant irons at 595 C. 19Cr-9Ni cast stainless steel (ASTM A297, grade HF) is shown for reference. Source: Ref 1, p 95

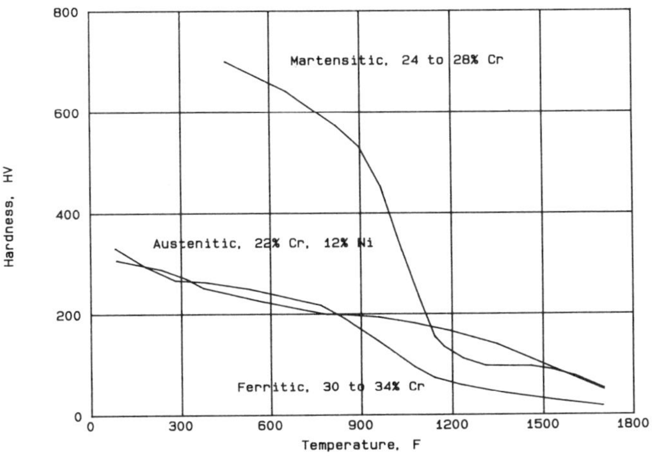

Alloy irons: Vickers hardness of three high-chromium irons. Source: Ref 14, p 172

Alloy irons: Growth of high-nickel irons in superheated steam at 480 C (900 F)

Type of iron	GRT % after 500 h	GRT % after 1000 h	GRT % after 2500 h
Gray iron (unalloyed)	0.23	0.52	1.4
High Ni gray iron (20% Ni)	0.05	0.10	0.15
High Ni gray iron (30% Ni)	0.03	0.045	0.048
High Ni ductile iron (20% Ni)	0.03	0.05	0.05
High Ni ductile iron (30% Ni)	0.03	nil	nil

Source: Ref 1, p 93

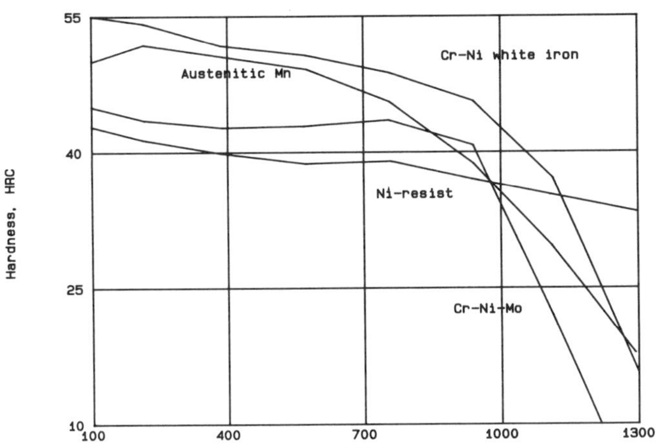

Alloy irons: Rockwell C hardness for several high-alloy irons. Source: Ref 14, p 172

Iron	C %	Si %	Mn %	P %	S %	Cr %	Ni %	Mo %	Cu %	Al %
Cr-Ni-Mo	3.10	2.40	0.80	0.10	0.08	0.6	1.1	0.6	...	...
Ni-Resist	3.10	2.20	0.98	0.31	0.12	2.6	12.8	...	6.1	...
Cr-Ni white	2.55	2.10	0.75	0.10	0.12	15.0	3.8	...	...	...
Austenitic Mn	3.52	2.35	8.30	0.65	0.04	0.28	0.15	1.6	1.7	1.5

Source: Ref 14, p 172

2. Carbon Steels

Carbon Steels

Carbon steels are not recommended for applications at temperatures above 370 C (700 F), where they begin to scale and their tensile strength begins to decrease. Long-term exposure to elevated temperatures, 650 C (1200 F), for example can cause overtempering or spheroidization and subsequent embrittlement. Generally, for heat treated medium- and high-carbon steels, maximum operating temperature should be kept below the tempering temperature.

Composition Effects. Carbon increases the strength and hardenability of steel at room temperature. In plain-carbon steels intended for elevated-temperature service, carbon content usually is limited to 0.20%, but may be as high as 0.35% in some boiler tubing. Carbon increases short-term tensile strength at elevated temperatures, but does not appreciably increase creep resistance at temperatures above 540 C (1000 F), because of the eventual spheroidization of carbides.

Manganese usually is added to carbon steels to prevent hot shortness. However, it may also enhance the effectiveness of nitrogen in increasing the strength of carbon steels at elevated temperatures.

Low-Carbon Steels (AISI 1005 to 1026). Because formability is directly related to hardness (carbon content), these softer steels are more ductile and thus better suited for applications involving cold forming. They are also commonly found in machined and welded components. Tensile strengths range from 300 to 440 MPa (43 to 64 ksi). Low-carbon steel can be hardened by surface alloying processes such as carburizing.

Medium-Carbon Steels (AISI 1029 to 1053). These grades are often selected where higher strength is required. Tensile strengths range from 470 to 620 MPa (68 to 90) ksi). These steels are heat treatable.

High-Carbon Steels (AISI 1055 to 1095). These grades are specified when the highest available strength is needed. Wear resistance is another advantage and may be very high when steels are properly heat treated. Tensile strength ranges from 650 to 830 MPa (94 to 120 ksi).

Carbon steels with enhanced hardenability are represented by members of the 15XX series (AISI 1513 to 1572), which have higher manganese contents (1.05 to 1.65% Mn) than the 10XX grades.

Standard carbon steels with improved machinability include resulfurized (11XX), resulfurized and rephosphorized (12XX), and leaded (XXLXX) grades. Also produced are boron steels (XXBXX) and steels made to prescribed hardenability limits (XXXXH).

Physical Properties

Composition of carbon steels

AISI No.	UNS No.	C, %	Mn, %	P, %	S, %
Nonresulfurized grades, Mn 1.00% max					
1005	G10050	0.06 max	0.35 max	0.040 max	0.050 max
1006	G10060	0.08 max	0.25-0.40	0.040 max	0.050 max
1008	G10080	0.10 max	0.30-0.50	0.040 max	0.050 max
1010	G10100	0.08-0.13	0.30-0.60	0.040 max	0.050 max
1012	G10120	0.10-0.15	0.30-0.60	0.040 max	0.050 max
1015	G10150	0.13-0.18	0.30-0.60	0.040 max	0.050 max
1016	G10160	0.13-0.18	0.30-0.90	0.040 max	0.050 max
1017	G10170	0.15-0.20	0.30-0.60	0.040 max	0.050 max
1018	G10180	0.15-0.20	0.30-0.90	0.040 max	0.050 max
1019	G10190	0.15-0.20	0.70-1.00	0.040 max	0.050 max
1020	G10200	0.18-0.23	0.30-0.60	0.040 max	0.050 max
1021	G10210	0.18-0.23	0.60-0.90	0.040 max	0.050 max
1022	G10220	0.18-0.23	0.70-1.00	0.040 max	0.050 max
1023	G10230	0.20-0.25	0.30-0.60	0.040 max	0.050 max
1025	G10250	0.22-0.28	0.30-0.60	0.040 max	0.050 max
1026	G10260	0.22-0.28	0.60-0.90	0.040 max	0.050 max
1029	G10290	0.25-0.31	0.60-0.90	0.040 max	0.050 max
1030	G10300	0.28-0.34	0.60-0.90	0.040 max	0.050 max
1034	G10340	0.32-0.38	0.50-0.80	0.040 max	0.050 max
1035	G10350	0.32-0.38	0.60-0.90	0.040 max	0.050 max
1037	G10370	0.32-0.38	0.70-1.00	0.040 max	0.050 max
1038	G10380	0.35-0.42	0.60-0.90	0.040 max	0.050 max
1039	G10390	0.37-0.44	0.70-1.00	0.040 max	0.050 max
1040	G10400	0.37-0.44	0.60-0.90	0.040 max	0.050 max
1042	G10420	0.40-0.47	0.60-0.90	0.040 max	0.050 max
1043	G10430	0.40-0.47	0.70-1.00	0.040 max	0.050 max
1044	G10440	0.43-0.50	0.30-0.60	0.040 max	0.050 max
1045	G10450	0.43-0.50	0.60-0.90	0.040 max	0.050 max
1046	G10460	0.43-0.50	0.70-1.00	0.040 max	0.050 max
1049	G10490	0.46-0.53	0.60-0.90	0.040 max	0.050 max
1050	G10500	0.48-0.55	0.60-0.90	0.040 max	0.050 max
1053	G10530	0.48-0.55	0.70-1.00	0.040 max	0.050 max
1055	G10550	0.50-0.60	0.60-0.90	0.040 max	0.050 max
1059	G10590	0.55-0.65	0.50-0.80	0.040 max	0.050 max
1060	G10600	0.55-0.65	0.60-0.90	0.040 max	0.050 max
1064	G10640	0.60-0.70	0.50-0.80	0.040 max	0.050 max
1065	G10650	0.60-0.70	0.60-0.90	0.040 max	0.050 max
1069	G10690	0.65-0.75	0.40-0.70	0.040 max	0.050 max
1070	G10700	0.65-0.75	0.60-0.90	0.040 max	0.050 max
1074	G10740	0.70-0.80	0.50-0.80	0.040 max	0.050 max
1078	G10780	0.72-0.85	0.30-0.60	0.040 max	0.050 max
1080	G10800	0.75-0.88	0.60-0.90	0.040 max	0.050 max
1084	G10840	0.80-0.93	0.60-0.90	0.040 max	0.050 max
1086(a)	G10860	0.80-0.93	0.30-0.50	0.040 max	0.050 max
1090	G10900	0.85-0.98	0.60-0.90	0.040 max	0.050 max
1095	G10950	0.90-1.03	0.30-0.50	0.040 max	0.050 max
Nonresulfurized grades, Mn greater than 1.00% max					
1513	G15130	0.10-0.16	1.10-1.40	0.040 max	0.050 max
1522	G15220	0.18-0.24	1.10-1.40	0.040 max	0.050 max
1524	G15240	0.19-0.25	1.35-1.65	0.040 max	0.050 max
1526	G15260	0.22-0.29	1.10-1.40	0.040 max	0.050 max
1527	G15270	0.22-0.29	1.20-1.50	0.040 max	0.050 max
1541	G15410	0.36-0.44	1.35-1.65	0.040 max	0.050 max
1547	G15470	0.43-0.51	1.35-1.65	0.040 max	0.050 max
1548	G15480	0.44-0.52	1.10-1.40	0.040 max	0.050 max
1551	G15510	0.45-0.56	0.85-1.15	0.040 max	0.050 max
1552	G15520	0.47-0.55	1.20-1.50	0.040 max	0.050 max
1561	G15610	0.55-0.65	0.75-1.05	0.040 max	0.050 max
1566	G15660	0.60-0.71	0.80-1.15	0.040 max	0.050 max

(continued)

Composition of carbon steels (continued)

AISI No.	UNS No.	C, %	Mn, %	P, %	S, %
Free-machining grades, resulfurized					
1108	G11080	0.08-0.13	0.50-0.80	0.040 max	0.08-0.13
1109	G11090	0.08-0.13	0.60-0.90	0.040 max	0.08-0.13
1110	G11100	0.08-0.13	0.30-0.60	0.040 max	0.08-0.13
1116	G11160	0.14-0.20	1.10-1.40	0.040 max	0.16-0.23
1117	G11170	0.14-0.20	1.00-1.30	0.040 max	0.08-0.13
1118	G11180	0.14-0.20	1.30-1.60	0.040 max	0.08-0.13
1119	G11190	0.14-0.20	1.00-1.30	0.040 max	0.24-0.33
1137	G11370	0.32-0.39	1.35-1.65	0.040 max	0.08-0.13
1139	G11390	0.35-0.43	1.35-1.65	0.040 max	0.13-0.20
1140	G11400	0.37-0.44	0.70-1.00	0.040 max	0.08-0.13
1141	G11410	0.37-0.45	1.35-1.65	0.040 max	0.08-0.13
1144	G11440	0.40-0.48	1.35-1.65	0.040 max	0.24-0.33
1146	G11460	0.42-0.49	0.70-1.00	0.040 max	0.08-0.13
1151	G11510	0.48-0.55	0.70-1.00	0.040 max	0.08-0.13
Free-machining grades, resulfurized and rephosphorized					
1211	G12110	0.13 max	0.60-0.90	0.07-0.12	0.10-0.15
1212	G12120	0.13 max	0.70-1.00	0.07-0.12	0.16-0.23
1213	G12130	0.13 max	0.70-1.00	0.07-0.12	0.24-0.33
1215	G12150	0.09 max	0.75-1.05	0.04-0.09	0.26-0.35
12L4(b)	G12144	0.15 max	0.85-1.15	0.04-0.09	0.26-0.35

Note: The following notes refer to boron, copper, lead, and silicon additions. Boron: standard killed carbon steels, which are generally fine grained, may be produced with a boron treatment addition to improve hardenability. Such steels are produced to a range of 0.0005 to 0.003% B. These steels are identified by inserting the letter B between the second and third numerals of the AISI or SAE number, such as 10B46. Copper: when copper is required, 0.20% min is generally specified. Lead: standard carbon steels can be produced with a lead range of 0.15 to 0.35% to improve machinability. Such steels are identified by inserting the letter L between the second and third numbers of the AISI or SAE number, such as 12L15 or 10L45. Silicon: It is not common practice to produce the 12XX series of resulfurized and rephosphorized steels to specified limits for silicon because of its adverse effect on machinability. When silicon ranges or limits are required for resulfurized or nonphosphorized steels, however, these values apply: a range of 0.08% for maximum silicon contents up to 0.15% inclusive, a range of 0.10% for maximum silicon contents over 0.15 to 0.20% inclusive, a range of 0.15% for maximum silicon contents over 0.20 to 0.30% inclusive, and a range of 0.20% for maximum silicon contents over 0.30 to 0.60% inclusive. Example: maximum silicon content is 0.25%, range is 0.10 to 0.25%.
(a) Standard grades for wire rod and wire only. (b) 0.15 to 0.35% lead
Source: Ref 15, p 3

Coefficients of linear thermal expansion of carbon steels

TEMP C	TEMP F	THEXP(a) 10^{-6}/C	THEXP(a) 10^{-6}/F	THEXP(b) 10^{-6}/C	THEXP(b) 10^{-6}/F
21	70	11.5	6.39	11.5	6.39
38	100	11.8	6.56	11.7	6.5
93	200	12.5	6.94	12.0	6.67
149	300	13.1	7.28	12.4	6.89
204	400	13.8	7.67	12.7	7.06
260	500	14.5	8.06	13.1	7.28
316	600	15.0	8.33	13.4	7.44
371	700	15.6	8.67	13.7	7.61
427	800	16.0	8.89	14.0	7.78

(a) Instantaneous coefficient of thermal expansion at indicated temperature. (b) Mean value of the coefficient of thermal expansion between 21 C (70 F) and the indicated temperature
Source: Ref 1, p 653

Effect of temperature on the modulus of elasticity of carbon steels

TEMP C	TEMP F	E(a) GPa	E(a) 10^6 psi	E(b) GPa	E(b) 10^6 psi
21	70	192	27.9	206	29.9
95	200	191	27.7	203	29.5
150	300	189	27.4	200	29.0
200	400	186	27.0	195	28.3
260	500	182	26.4	189	27.4
315	600	177	25.7	184	26.7
370	700	171	24.8	175	25.4

(a) Carbon contents of 0.3% or less. (b) Carbon contents greater that 0.3%
Source: Ref 1, p 650

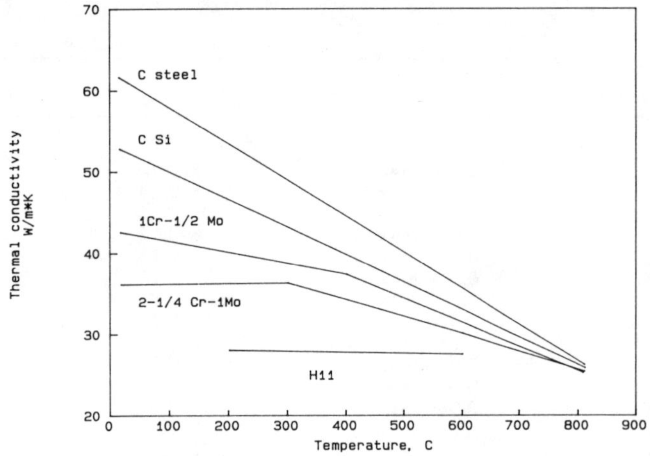

Carbon steels: Thermal conductivity of carbon and low-alloy steels. Source: Ref 1, p 652

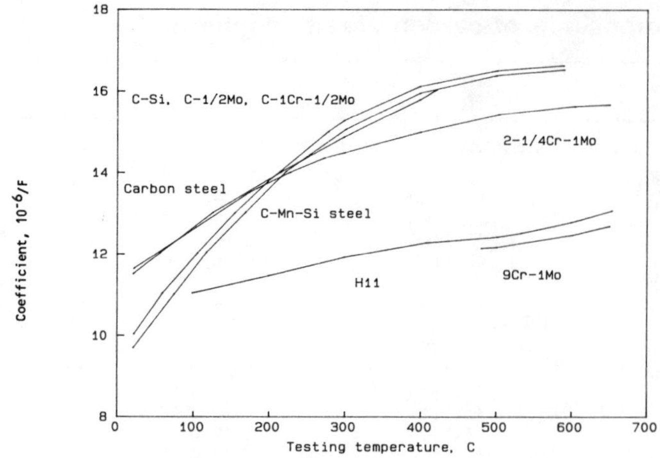

Carbon steels: Instantaneous coefficients of linear thermal expansion at indicated temperatures; these are not mean values of the coefficient over a range of temperatures. Source: Ref 1, p 653

AISI 1005

UNS number: G10050
Specifications: ASTM A29, ASTM A510, MIL S-11310, SAE J403, SAE J412
Nominal composition: 0.060% C max, 0.350% Mn max, 0.040% P max, 0.050% S max,
Product forms: Wrought, bar, rod, wire, plate, sheet, strip, tube, pipe, billets, forgings, blooms, cast, continuous, powder, welded
Description: AISI 1005 usually is produced as rimmed, capped, semikilled, and fully killed steel. Rimmed steels of this grade have exceptionally good cold formability. The soft rim enhances drawability and surface finish. Because it has relatively low tensile strengths, 1005 is used where strength is of minor importance. Aluminum-killed steels are very soft low-carbon steels of low strength and high ductility, which are easy to forge or to form and draw cold. AISI 1005 is not heat treatable in the usual manner (by quenching and tempering), but can be hardened or strengthened by cold working. These steels can be restored to their initial soft condition by annealing. Weldability (spot, projection, butt, and fusion) and brazeability are excellent.

Typical applications: Parts made from grade 1005 include extruded, cold headed, cold upset, and cold pressed parts and forms and a large variety of parts requiring severe bending and welding. For more severe drawing operations, fully killed steel should be used to eliminate or minimize the strain-aging behavior of rimmed steels.

Physical Properties

AISI 1005: Mean apparent specific heat

TEMP C	TEMP F	SHTCP(a) J/kg*K	SHTCP(a) Btu/lb*F	SHTCP(b) J/kg*K	SHTCP(b) Btu/lb*F
50-100	120-212	481	0.115	481	0.115
150-200	300-390	519	0.124	523	0.125
200-250	390-480	536	0.128	544	0.13
250 300	480-570	553	0.132	557	0.133
300-350	570-660	574	0.137	569	0.136
350-400	660-750	595	0.142	595	0.142
450-500	840-930	662	0.158	662	0.158
550-600	1020-1110	754	0.18	858	0.205
650-700	1200-1290	867	0.207	1139	0.272
700-750	1290-1380	1105	0.264	959	0.229
750-800	1380-1470	875	0.209	816	0.195
850-900	1560-1650	846	0.202	...	...

(a) Material composition: 0.06% carbon, 0.38% manganese. (b) Material composition: 0.08% carbon, 0.31% manganese
Source: Ref 15, p 6

AISI 1005: Mean coefficients of linear thermal expansion

TEMP C	TEMP F	THEXP 10^{-6}/C	THEXP 10^{-6}/F
0-100	32-212	12.6	7.0
0-200	32-390	13.1	7.3
0-300	32-570	13.5	7.5
0-400	32-750	13.7	7.6
0-500	32-930	14.2	7.9
0-600	32-1110	14.6	8.1
0-700	32-1290	14.9	8.3
0-800	32-1470	16.6	9.2
0-1000	32-1830	13.7	7.6

Note: Material composition: 0.06% carbon, 0.38% manganese, 0.55% nickel, 0.02% chromium, 0.03% molybdenum
Source: Ref 15, p 6

AISI 1006

UNS number: G10060
Specifications: ASME 5041, ASTM A29, ASTM A510, ASTM A545, FED QQ W-461, MIL S-11310, SAE J403, SAE J412, SAE J414
Nominal composition: 0.080% C max, 0.250 to 0.400% Mn, 0.040% P max, 0.050% S max
Product forms: Wrought, bar, rod, wire, plate, sheet, strip, tube, pipe, billets, forgings, blooms, cast, continuous, powder, welded
Description: AISI 1006 usually is produced as rimmed, capped, semikilled, and fully killed steels. Rimmed steels of this grade have exceptionally good cold formability. The soft rim enhances drawability and surface finish. Because it has relatively low tensile strengths, 1006 is used where strength is of minor importance. Aluminum-killed steels are very soft low-carbon steels of low strength and high ductility, which are easy to forge or to form and draw cold. AISI 1006 is not heat treatable in the usual manner (by quenching and tempering), but can be hardened or strengthened by cold working. These steels can be restored to their initial soft condition by annealing. Weldability (spot, projection, butt, and fusion) and brazeability are excellent.

Typical applications: Parts made from grade 1006 include extruded, cold headed, cold upset, and cold pressed parts and forms and a large variety of parts requiring severe bending and welding. For more severe drawing operations, fully killed steel should be used to eliminate or minimize the strain-aging behavior of rimmed steels. AISI grade 1006 steel can also be used in magnet core applications.

Physical Properties

AISI 1006: Mean apparent specific heat

TEMP C	TEMP F	SHTCP(a) J/kg*K	SHTCP(a) Btu/lb*F	SHTCP(b) J/kg*K	SHTCP(b) Btu/lb*F
50-100	120-212	481	0.115	481	0.115
150-200	300-390	519	0.124	523	0.125
200-250	390-480	536	0.128	544	0.13
250-300	480-570	553	0.132	557	0.133
300-350	570-660	574	0.137	569	0.136
350-400	660-750	595	0.142	595	0.142
450-500	840-930	662	0.158	662	0.158
550-600	1020-1110	754	0.18	858	0.205
650-700	1200-1290	867	0.207	1139	0.272
700-750	1290-1380	1105	0.264	959	0.229
750-800	1380-1470	875	0.209	816	0.195
850-900	1560-1650	846	0.202	...	...

(a) Material composition: 0.06% carbon, 0.38% manganese. (b) Material composition: 0.08% carbon, 0.31% manganese
Source: Ref 15, p 6

AISI 1006: Mean coefficients of linear thermal expansion

TEMP C	TEMP F	THEXP 10^{-6}/C	THEXP 10^{-6}/F
0-100	32-212	12.6	7.0
0-200	32-390	13.1	7.3
0-300	32-570	13.5	7.5
0-400	32-750	13.7	7.6
0-500	32-930	14.2	7.9
0-600	32-1110	14.6	8.1
0-700	32-1290	14.9	8.3
0-800	32-1470	16.6	9.2
0-1000	32-1830	13.7	7.6

Note: Material composition: 0.06% carbon, 0.38% manganese, 0.55% nickel, 0.02% chromium, 0.03% molybdenum
Source: Ref 15, p 6

AISI 1008

UNS number: G10080
Specifications: ASTM A108, ASTM A510, ASTM A519, ASTM A545, ASTM A549, ASTM A575, ASTM A576, FED QQ S-637(1008), FED QQ S-698(1008), MIL S-11310, SAE J403, SAE J412, SAE J414, DIN 1.0204, UNI CB 10 FU
Nominal composition: 0.100% C max, 0.300 to 0.500% Mn, 0.040% P max, 0.050% S max
Product forms: Wrought, bar, rod, wire, plate sheet, strip, tube, pipe, billets, forgings, blooms, cast, continuous, powder, welded
Description: AISI 1008 is usually produced as rimmed, capped, semikilled, and fully killed steels. Rimmed steels have exceptionally good cold formability. The soft rim enhances drawability and surface finish. Because they have relatively low tensile strengths, these steels are used where strength is of minor importance. Aluminum killed steels are very soft steels of low strength and high ductility, which are easy to forge or to form and draw cold. Grade 1008 is not heat treatable in the usual manner, but can be hardened or strengthened by cold working. These steels can be restored to their initial soft condition by annealing. Weldability (spot, projection, butt, and fusion) and brazeability are excellent.
Typical applications: Parts made from grade 1008 include extruded, cold headed, cold upset, and cold pressed parts and forms, and a large variety of parts requiring severe bending and welding. For more severe drawing operations, fully killed steel should be used to eliminate or minimize the strain-aging behavior of rimmed steels.

Physical Properties

AISI 1008: Mean apparent specific heat

TEMP C	TEMP F	SHTCP(a) J/kg*K	SHTCP(a) Btu/lb*F	SHTCP(b) J/kg*K	SHTCP(b) Btu/lb*F
50-100	120-212	481	0.115	481	0.115
150-200	300-390	519	0.124	523	0.125
200-250	390-480	536	0.128	544	0.13
250-300	480-570	553	0.132	557	0.133
300-350	570-660	574	0.137	569	0.136
350-400	660-750	595	0.142	595	0.142
450-500	840-930	662	0.158	662	0.158
550-600	1020-1110	754	0.18	858	0.205
650-700	1200-1290	867	0.207	1139	0.272
700-750	1290-1380	1105	0.264	959	0.229
750-800	1380-1470	875	0.209	816	0.195
850-900	1560-1650	846	0.202	...	...

(a) Material composition: 0.06% carbon, 0.38% manganese. (b) Material composition: 0.08% carbon, 0.31% manganese
Source: Ref 15, p 6

AISI 1008: Thermal conductivity

TEMP C	TEMP F	THCON(a) W/m*K	THCON(b) W/m*K	THCON(a) Btu/ft*h*F	THCON(b) Btu/ft*h*F
21	70	...	...	...	...
100	212	57.6	60.2	33.3	34.8
200	390	53.1	54.7	30.7	31.6
300	570	49.1	...	28.4	...
400	750	45.7	45.2	26.4	26.1
500	930	40.8	...	23.6	...
600	1110	36.7	36.3	21.2	21
700	1290	33.2	...	19.2	...
800	1470	28.4	28.4	16.4	16.4
900	1650	...	...	...	...
1000	1830	27.3	27.3	15.8	15.8
1100	2010	...	...	...	...
1200	2190	29.9	...	17.3	...
1300	2370	...	...	...	...

(a) Material composition: 0.08% carbon, 0.31% manganese. (b) Material composition: 0.06% carbon, 0.4% manganese
Source: Ref 15, p 6

AISI 1008: Mean coefficients of linear thermal expansion

TEMP C	TEMP F	THEXP 10^{-6}/C	THEXP 10^{-6}/F
0-100	32-212	12.6	7.0
0-200	32-390	13.1	7.3
0-300	32-570	13.5	7.5
0-400	32-750	13.7	7.6
0-500	32-930	14.2	7.9
0-600	32-1110	14.6	8.1
0-700	32-1290	14.9	8.3
0-800	32-1470	16.6	9.2
0-1000	32-1830	13.7	7.6

Material composition: 0.06% carbon, 0.38% manganese, 0.55% nickel, 0.02% chromium, 0.03% molybdenum
Source: Ref 15, p 6

AISI 1008: Electrical resistivity

TEMP C	TEMP F	ELRES(a) n{o}*m	ELRES(b) n{o}*m
21	70	0.13	142
100	212	0.178	19
200	390	0.252	263
300	570	...	...
400	750	0.448	458
500	930	...	...
600	1110	0.725	734
700	1290	0.898	905
800	1470	1.073	1081
900	1650	1.124	113
1000	1830	1.16	1165
1100	2010	1.189	1193
1200	2190	1.126	122
1300	2370	1.241	1244

(a) Material composition: 0.06% carbon, 0.38% manganese.
(b) Material composition: 0.08% carbon, 0.31% manganese
Source: Ref 15, p 6

AISI 1010

UNS number: G10100
Specifications: AMS 5040, AMS 5042, AMS 5044, AMS 5050, AMS 5053, ASTM A29, ASTM A108, ASTM A510, ASTM A519, ASTM A545, ASTM A549, ASTM A575, ASTM A576, FED QQ W-461, SAE J403, SAE J412, SAE J414, DIN 1.1121, NF XC 10, JIS S-10 C
Nominal composition: 0.080 to 0.130% C, 0.300 to 0.600% Mn, 0.040% P max, 0.050% S max
Product forms: Wrought, bar, rod, wire, plate, sheet, strip, tube, pipe, billets, forgings, blooms, cast, continuous, powder, welded, ingot, structural shapes
Description: AISI 1010 is widely used for low-strength applications because of good formability and fair response to moderate machining, particularly when following cold drawing. These steels are less expensive than AISI 1008 and can be used when the requirements for drawing and forming are less exacting. Deep-drawn parts show a marked increase in strength and hardness due to cold working and strain aging. Strength may increase from 25 ksi (softest) to 100 ksi (severely cold worked). AISI 1010 can be hardened by cyaniding, and it can be welded by all commercial welding processes with excellent results.

Typical applications: AISI 1010 is easily cold formed by heading, extruding, upsetting, bending, and other deforming processes. End uses for wires of these grades include electroplated products such as racks, storage bins, shopping carts, fan guards, and jewelry, as well as unplated products such as wires, staples, hardware items, welded wire fabric, automobile body parts, rivets, nails, and barbed wire. Wires of this carbon range are also used for shielded-metal-arc welding rods. AISI 1010 wire is also suitable for magnet core applications.

Physical Properties

AISI 1010: Mean coefficients of linear thermal expansion

TEMP C	TEMP F	THEXP $10^{-6}/C$	THEXP $10^{-6}/F$
0-100	32-212	12.2	6.8
0-200	32-390	13.0	7.2
0-300	32-570	13.5	7.5
0-400	32-750	13.8	7.7
0-500	32-930	14.2	7.9
0-600	32-1110	14.6	8.1
0-700	32-1290	14.9	8.3

Source: Ref 15, p 10

Mechanical Properties

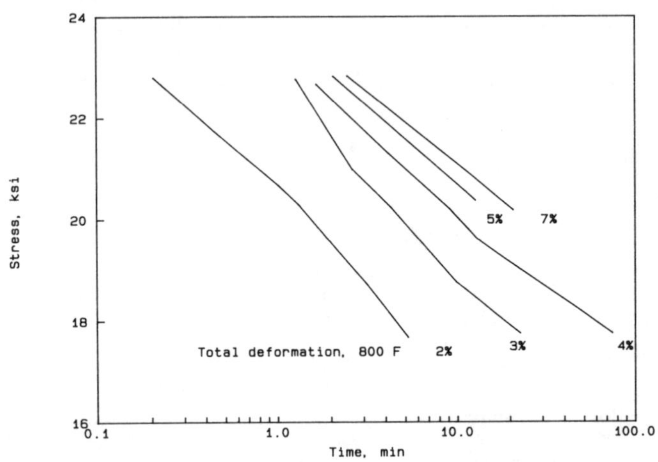

AISI 1010: Short-time creep. Deformation includes thermal expansion of 0.61%; heating rate, 115 F/s. Hot rolled commercial quality (0.058-in. thick) specimens were 17 in. by 2 in.; gage section 8 in. by 0.5 in. Tensile strength: 43 ksi. Yield strength: 28.2 ksi. Elongation: 64.4%. Source: Ref 21, p 24

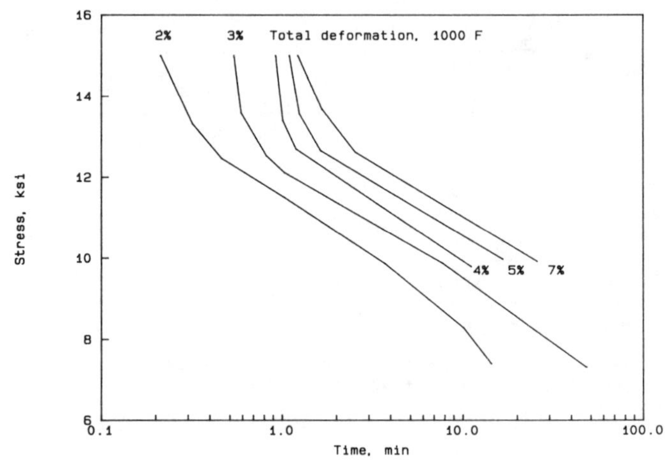

AISI 1010: Short-time creep. Deformation includes thermal expansion of 0.80%; heating rate, 125 F/s. Hot rolled commercial quality (0.058-in. thick) specimens were 17 in. by 2 in.; gage section was 8 in. by 0.5 in. Tensile strength: 43 ksi. Yield strength: 28.2 ksi. Elongation: 64.4%. Source: Ref 21, p 24

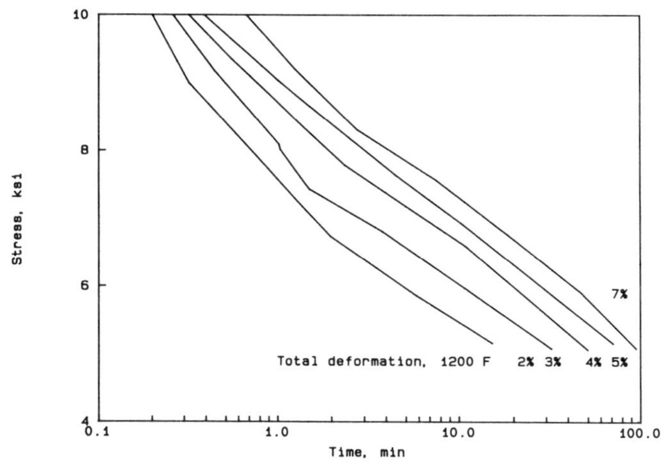

AISI 1010: Short-time creep. Deformation includes thermal expansion of 0.90%; heating rate, 115 F/s. Hot rolled commercial quality (0.058-in. thick) specimens were 17 in. by 2 in.; gage section 8 in. by 0.5 in. Tensile strength: 43 ksi. Yield strength: 28.2 ksi. Elongation: 64.4%. Source: Ref 21, p 24

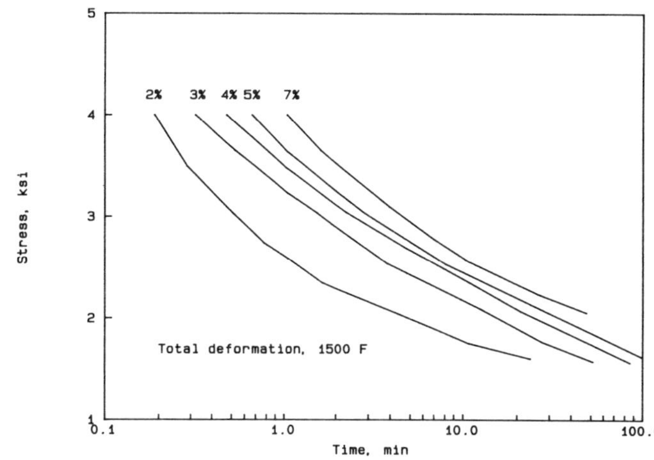

AISI 1010: Short-time creep. Deformation includes thermal expansion of 1.14%; heating rate, 125 F/s. Hot rolled commercial quality (0.058-in. thick) specimens were 17 in. by 2 in.; gage section 8 in. by 0.5 in. Tensile strength: 43 ksi. Yield strength: 28.2 ksi. Elongation: 64.4%. Source: Ref 21, p 24

AISI 1012

UNS number: G10120
Specifications: AMS 5040, AMS 5042, AMS 5044, AMS 5050, AMS 5053, ASTM A29, ASTM A108, ASTM A510, ASTM A519, ASTM A545, ASTM A549, ASTM A575, ASTM A576, FED QQ W-461, SAE J403, SAE J412, SAE J414, DIN 1.1121, NF XC 10, JIS S-10 C
Nominal composition: 0.080 to 0.130% C, 0.300 to 0.600% Mn, 0.040% P, 0.050% S
Product forms: Wrought, bar, rod, wire, plate, sheet, strip, tube, pipe, billets, forgings, blooms, cast, continuous, powder, welded, ingot, structural shapes
Description: AISI 1012 is widely used for low-strength applications because of good formability and fair response to moderate machining, particularly when following cold drawing. These steels are less expensive than AISI 1008 and can be used when the requirements for drawing and forming are less exacting. Deep-drawn parts show a marked increase in strength and hardness due to cold working and strain aging. Strength may increase from 25 ksi (softest) to 100 ksi (severely cold worked). AISI 1012 can be hardened by cyaniding, and it can be welded by all commercial welding processes with excellent results.

Typical applications: AISI 1012 is easily cold formed by heading, extruding, upsetting, bending, and other deforming processes. End uses for wires of these grades include electroplated products such as racks, storage bins, shopping carts, fan guards, and jewelry, as well as unplated products such as wires, staples, hardware items, welded wire fabric, automobile body parts, rivets, nails, and barbed wire. Wires of this carbon range are also used for shielded-metal-arc welding rods. AISI 1012 wire is also suitable for magnet core applications.

Mechanical Properties

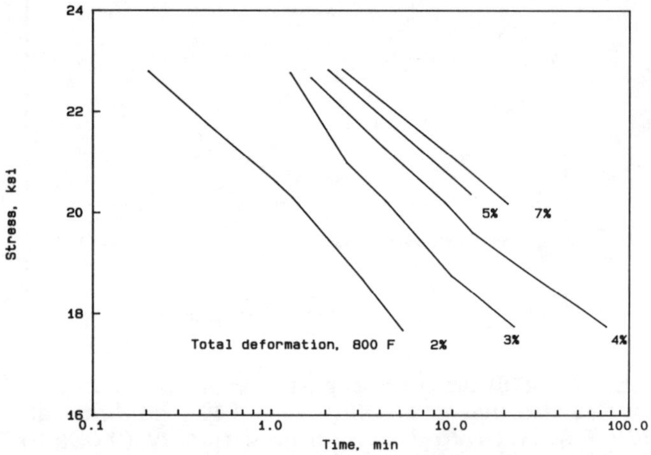

AISI 1012: Short-time creep. Deformation includes thermal expansion of 0.61%; heating rate, 115 F/s. Hot rolled commercial quality (0.058-in. thick) specimens were 17 in. by 2 in.; gage section 8 in. by 0.5 in. Tensile strength: 43 ksi. Yield strength: 28.2 ksi. Elongation: 64.4%. Source: Ref 21, p 24

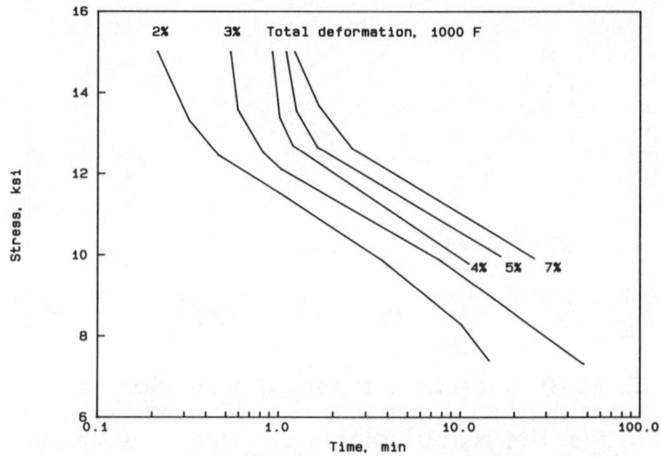

AISI 1012: Short-time creep. Deformation includes thermal expansion of 0.80%; heating rate, 125 F/s. Hot rolled commercial quality (0.058-in. thick) specimens were 17 in. by 2 in.; gage section 8 in. by 0.5 in. Tensile strength: 43 ksi. Yield strength: 28.2 ksi. Elongation: 64.4%. Source: Ref 21, p 24

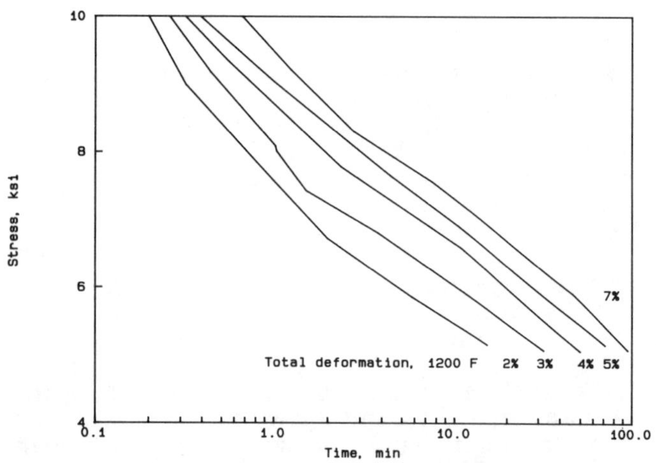

AISI 1012: Short-time creep. Deformation includes thermal expansion of 0.90%; heating rate, 115 F/s. Hot rolled commercial quality (0.058-in. thick) specimens were 17 in. by 2 in.; gage section 8 in. by 0.5 in. Tensile strength: 43 ksi. Yield strength: 28.2 ksi. Elongation: 64.4%. Source: Ref 21, p 24

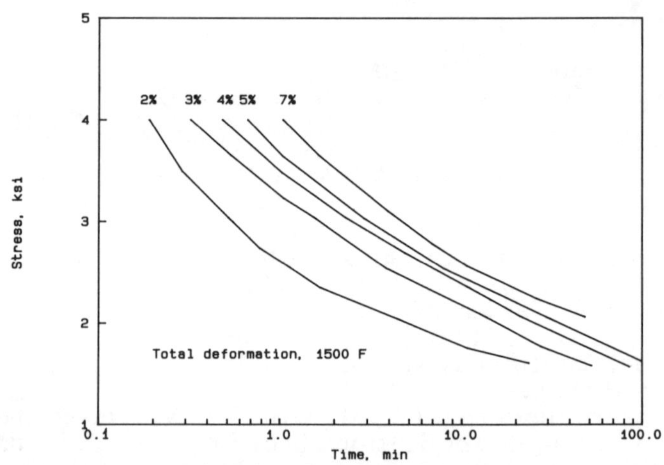

AISI 1012: Short-time creep. Deformation includes thermal expansion of 1.14%; heating rate, 125 F/s. Hot rolled commercial quality (0.058-in. thick) specimens were 17 in. by 2 in.; gage section 8 in. by 0.5 in. Tensile strength: 43 ksi. Yield strength: 28.2 ksi. Elongation: 64.4%. Source: Ref 21, p 24

AISI 1015

UNS number: G10150
Specifications: AMS 5060, ASTM A510, ASTM A519, ASTM A545, ASTM A549, ASTM A576, ASTM A659, FED QQ S-698, FED QQ S-16974, SAE J403, SAE J412, SAE J414, DIN 1.1141, NF XC 15, NF XC 18, JIS S 15 C, JIS S 17 C, JIS S 15 CK, SS 14 13 70
Nominal composition: 0.130 to 0.180% C, 0.300 to 0.600% Mn, 0.040% P max, 0.050% S max
Product forms: Wrought, bar, rod, wire, plate, sheet, strip, tube, pipe, billets, forgings, blooms, cast, continuous, powder, welded, ingot, structural shapes
Description: AISI 1015 steel has better machinability and carburizing characteristics than the more formable extra-low carbon grades. It is a carburizing grade and can be strengthened by cold working, or surface hardened by carburizing and cyaniding. AISI 1015 can be hardened by water quenching in very thin sections only. Hot rolled sheet has excellent weldability with either the arc or resistance method. Killed steel is best suited for arc welding. Rimmed and capped steel can be welded with proper welding techniques. Atmospheric corrosion resistance can be increased by small additions of copper.
Typical applications: AISI 1015 can be used in the as-rolled condition where ease of forming and joining is important. In general, applications for 1015 are similar to those for grade 1010, such as plate applications, building sections, trusses, pins, bolts, rods, and machinery components.

Physical Properties

AISI 1015: Mean coefficients of linear thermal expansion

TEMP C	TEMP F	THEXP $10^{-6}/C$	THEXP $10^{-6}/F$
0-100	32-212	11.9	6.6
0-200	32-390	12.4	6.9
0-300	32-570	13.0	7.2
0-400	32-750	13.5	7.5
0-500	32-930	14.2	7.9

Source: Ref 15, p 12

Mechanical Properties

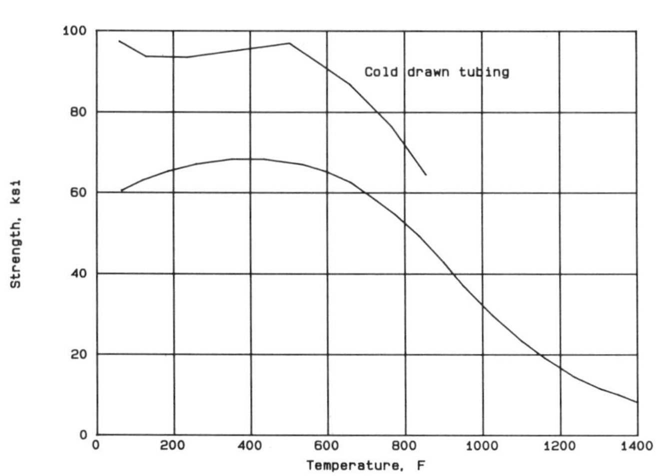

AISI 1015: Tensile strength. Data from several test programs on specimens in the annealed, normalized, hot rolled, stress relieved, oil quenched and tempered, and cold worked conditions. Cold drawn tubing data shown separately. Source: Ref 21, p 12

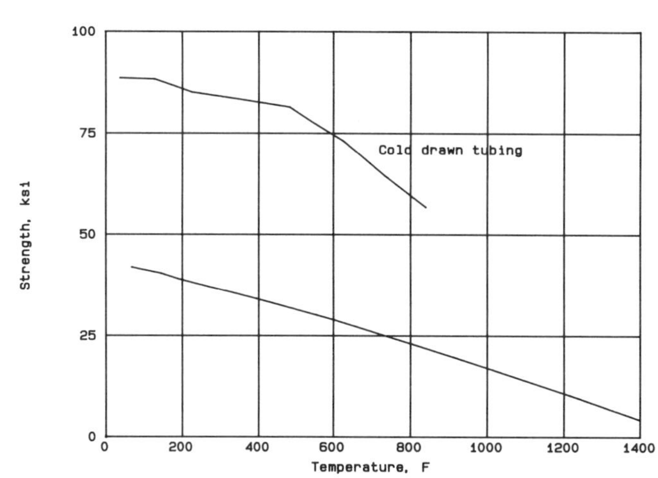

AISI 1015: Yield strength. Data from several test programs on specimens in the annealed, normalized, hot rolled, stress relieved, oil quenched and tempered, and cold worked conditions. Cold drawn tubing data shown separately. Source: Ref 21, p 12

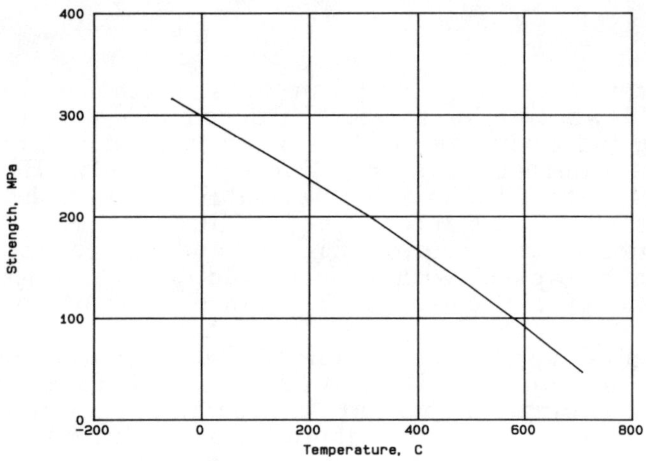

AISI 1015: Yield strength (0.2% offset). Source: Ref 8, p 36

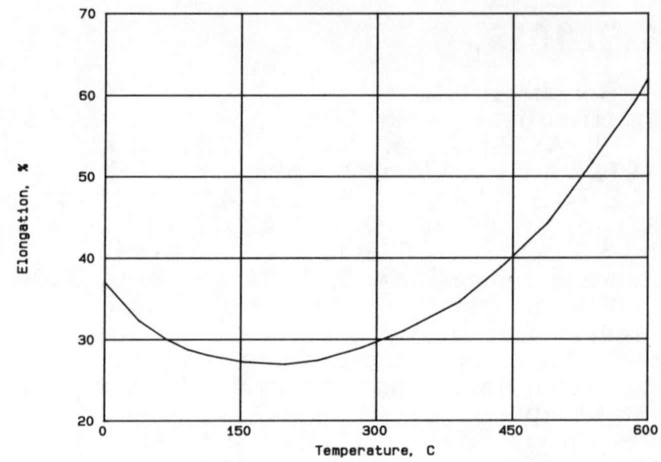

AISI 1015: Effect of temperature on tensile elongation. Source: Ref 8, p 36

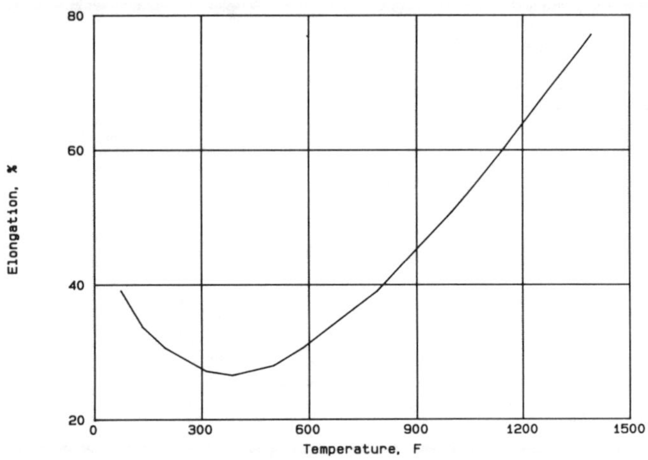

AISI 1015: Tensile elongation. Data from several test programs on specimens in the annealed, normalized, hot rolled, stress relieved, oil quenched and tempered, and cold worked conditions. Source: Ref 21, p 12

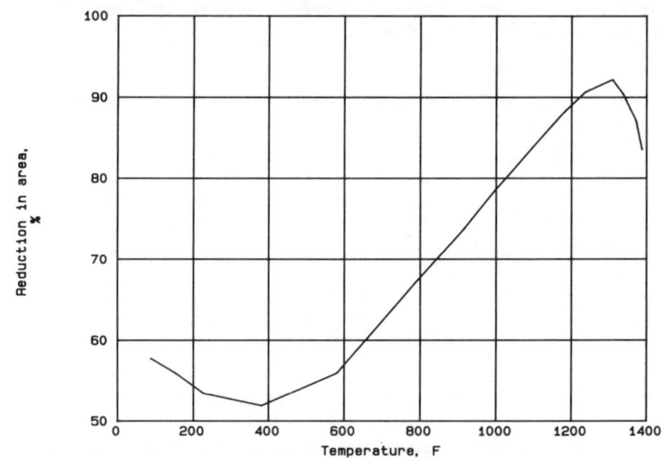

AISI 1015: Tensile reduction in area. Data from several test programs on specimens in the annealed, normalized, hot rolled, stress relieved, oil quenched and tempered, and cold worked conditions. Source: Ref 21, p 12

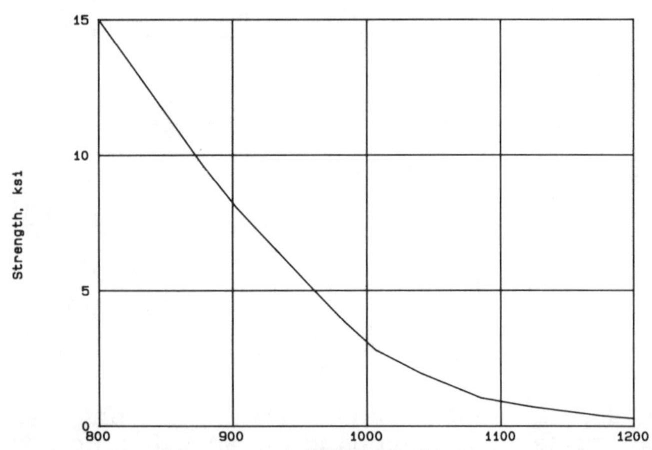

AISI 1015: Creep strength. Creep rate of 0.00001%/h (1% in 100,000 h). Specimens were annealed, normalized, hot rolled, stress relieved, oil quenched and tempered, and cold worked. Source: Ref 21, p 24

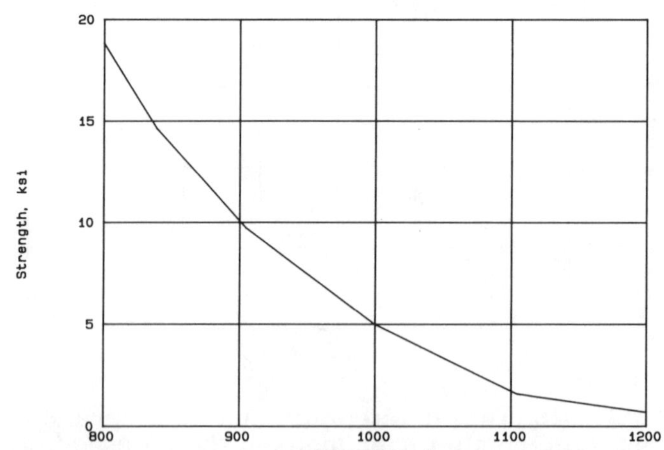

AISI 1015: Creep strength. Creep rate of 0.0001%/h (1% in 10,000 h). Specimens were annealed, normalized, hot rolled, stress relieved, oil quenched and tempered, and cold worked. Source: Ref 21, p 24

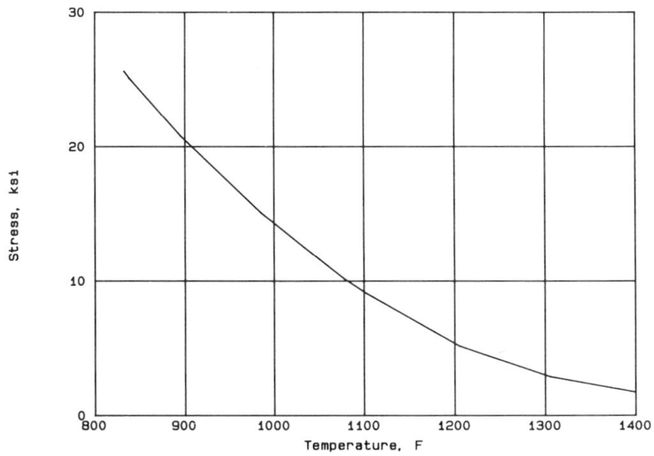

AISI 1015: Stress to cause rupture in 100 h. Specimens were annealed, normalized, hot rolled, stress relieved, oil quenched and tempered, and cold worked. Source: Ref 21, p 25

AISI 1018

UNS number: G10180
Specifications: AMS 5069, ASTM A108, ASTM A510, ASTM A513, ASTM A519, ASTM A544, ASTM A545, ASTM A548, ASTM A549, ASTM A576, ASTM A659, MIL S-11310, SAE J403, SAE J412, SAE J414
Nominal composition: 0.150 to 0.200% C, 0.600 to 0.900% Mn, 0.040% P max, 0.050% S max
Product forms: Wrought, bar, rod, wire plate, strip, tube, pipe, billets, structural shapes, blooms, cast, continuous, powder, welded, ingot
Description: AISI 1018 is supplied as killed, semikilled, rimmed and capped steel. This medium-low carbon steel has good weldability and slightly better machinability than the lower carbon grades. It is relatively soft. The higher carbon and manganese contents result in slightly increased strength. Has higher hardenability than AISI 1015 or 1020 because of a higher manganese content.
Typical applications: Is most widely used in cold forming operations such as heading, upsetting, and extrusion. It is used in the as-rolled condition as bar, shapes, sheet, and strip where ease of forming and joining, as well as stiffness, are important, such as for machinery components, carbonized parts, wire, pins, bolts, rods, sprockets, and plate applications.

Mechanical Properties

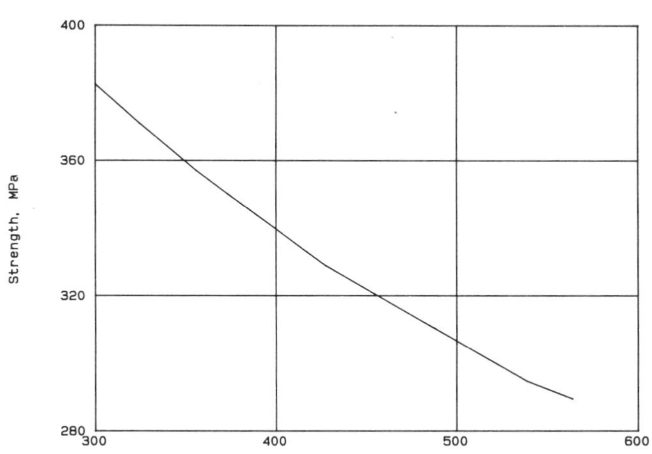

AISI 1018: Tensile strength of normalized steel after exposure (without stress) to indicated temperature for 83,000 h. Source: Ref 1, p 651

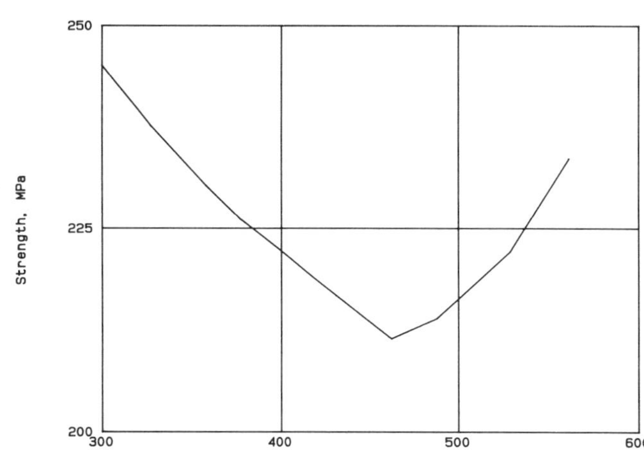

AISI 1018: Yield strength of normalized steel after exposure (without stress) to indicated temperature for 83,000 h. Source: Ref 1, p 651

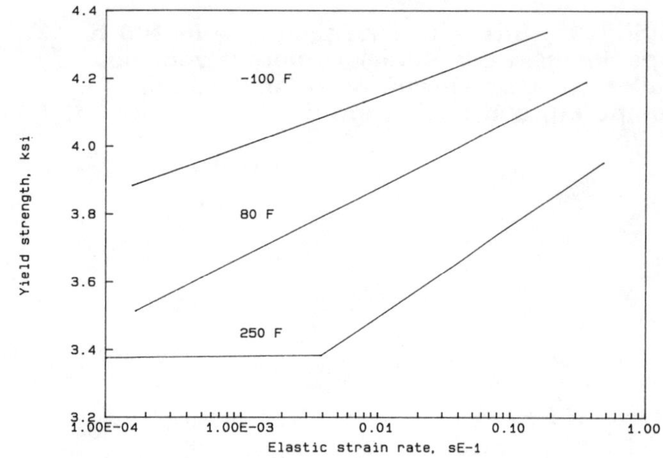

AISI 1018: Effect of strain rate and temperature on yield strength of 1-in. diam bar stock. Specimens were 0.505 in. in diameter and annealed at 1550 F for 1 h. Source: Ref 21, p 28

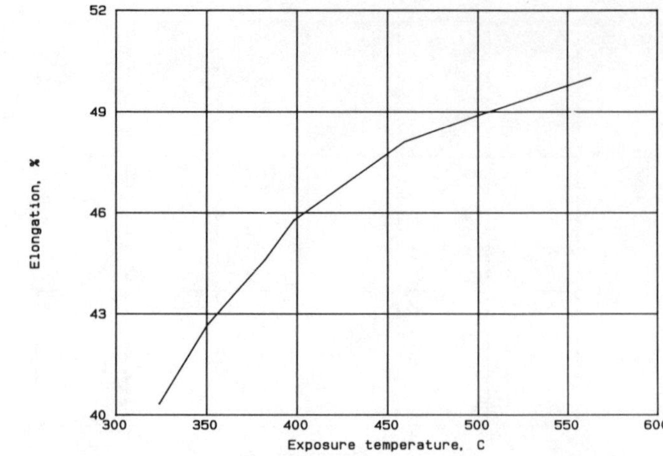

AISI 1018: Tensile elongation of normalized steel after exposure (without stress) to indicated temperature for 83,000 h. Source: Ref 1, p 651

AISI 1018: Stress-to-rupture. Circles: 1000 h. Triangles: 10,000 h. Solid symbols indicate specimens that were held at test temperature for 83,000 h prior to testing. Source: Ref 1, p 652

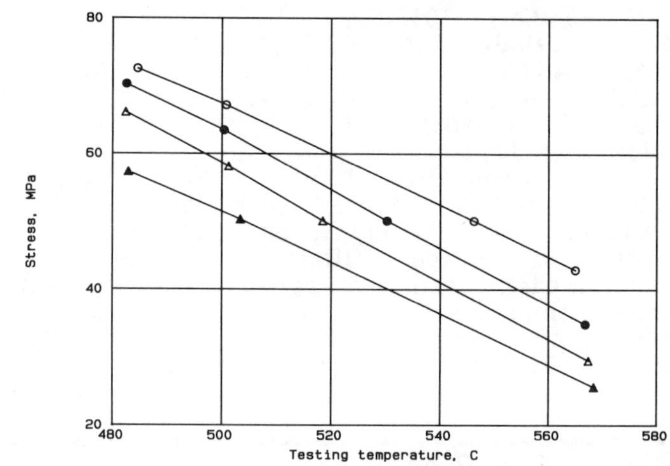

AISI 1020

UNS number: G10200
Specifications: AMS 5032, AMS 5045, ASTM A510, ASTM A519, ASTM A544, ASTM A576, ASTM 659, MIL S-11310, SAE J403, SAE J412, SAE J414, DIN 1.0402, NF CC 20, UNI C 20, SS 14 14 50, BS 040 A 20, BS 070 M 20
Nominal composition: 0.180 to 0.230% C, 0.300 to 0.600% Mn, 0.400% P max, 0.050% S max
Product forms: Wrought, bar, rod, wire, plate, strip, tube, pipe, billets, forgings, structural shapes, blooms, cast, continuous, powder, welded, ingot
Description: AISI 1020 is stronger and less easily formed than grade 1018. This steel responds well to cold work and heat treating. To minimize soft spots in the case, when case hardened, orders should specify that the steel be "normal" in the McQuaid-Ehn test. Weldability is fair. It is commonly used in the carburized condition where wear resistance rather than core strength is important. It can be fully hardened only in very thin sections. As with other plain carbon steels, cold work produces sizeable increases in hardness and strength.

Typical applications: Suitable for parts in the case-hardened condition and for shafts of larger cross section that are not highly stressed. Other uses include lightly stressed gears with hard wearing surfaces, case-hardened pins and chains, fan blades, clamps, cams, sprockets, bolts, machinery components, and forming tools.

Physical Properties

AISI 1020: Mean coefficients of linear thermal expansion

TEMP C	TEMP F	THEXP 10^{-6}/C	THEXP 10^{-6}/F
21-100	70-212	11.7	6.5
21-200	70-390	12.1	6.7
21-300	70-570	12.8	7.1
21-400	70-750	13.3	7.4
21-500	70-930	13.9	7.7
21-600	70-1110	14.4	8.0
21-700	70-1290	14.8	8.2

Source: Ref 15, p 16

Mechanical Properties

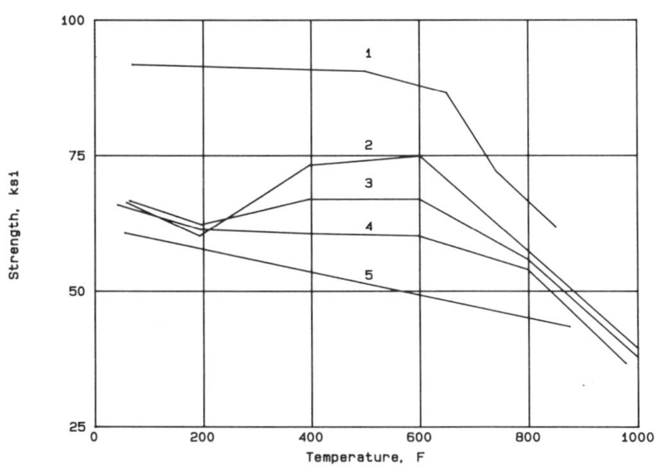

AISI 1020: Tensile strength. Curve 1: cold drawn seamless tubing. Curves 2 and 4: 1-in. plate, stress relieved for 2 h at 1150 F. Curve 3: rolled 1-in. plate. Curve 5: annealed seamless tubing. Source: Ref 21, p 6

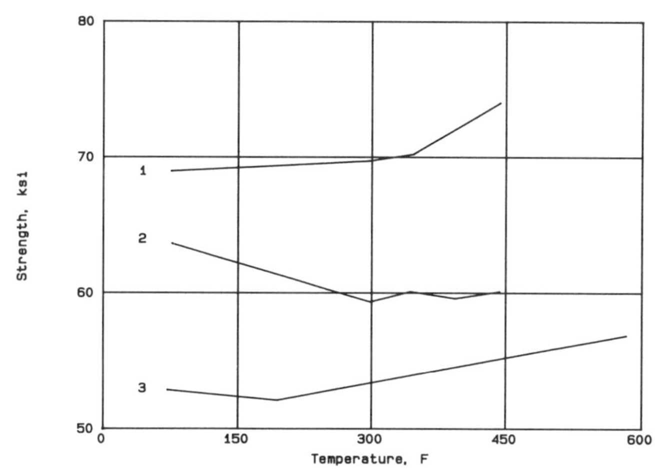

AISI 1020: Tensile strength of sheet. Curve 1: 0.065-in. hot rolled sheet. Curve 2: 0.065-in. hot rolled sheet with 0.5-in. fatigue crack. Curve 3: 0.120-in. normalized sheet. Source: Ref 21, p 5

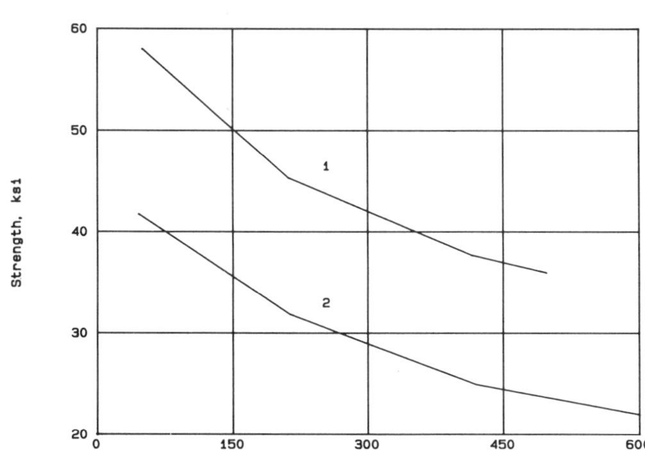

AISI 1020: Effect of temperature on 0.2% offset yield strength. Curve 1: 0.065-in. sheet tested at a strain rate of 10^{-5}/s. Curve 2: 0.120-in. normalized sheet. Source Ref 21, p 5

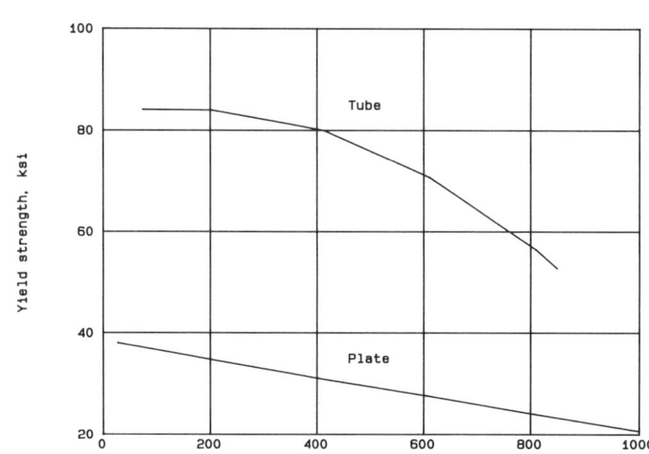

AISI 1020: Yield strength of 1-in. plate and 4.50-in. outside diameter cold drawn tubing. Source: Ref 21, p 6

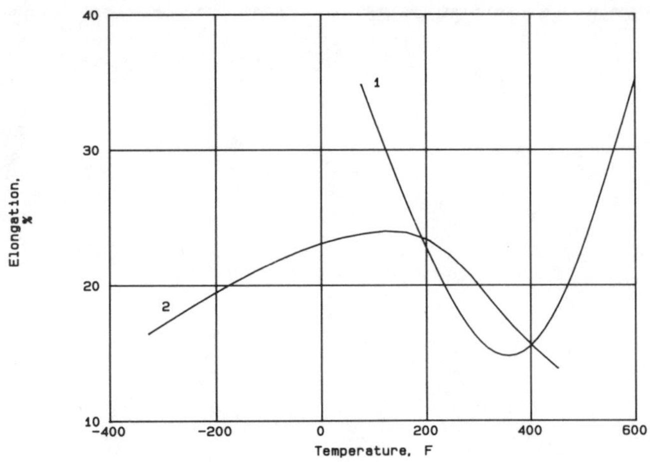

AISI 1020: Tensile elongation at failure. Curve 1: 0.120-in. normalized sheet. Curve 2: 0.065-in. hot rolled sheet. Source: Ref 21, p 5

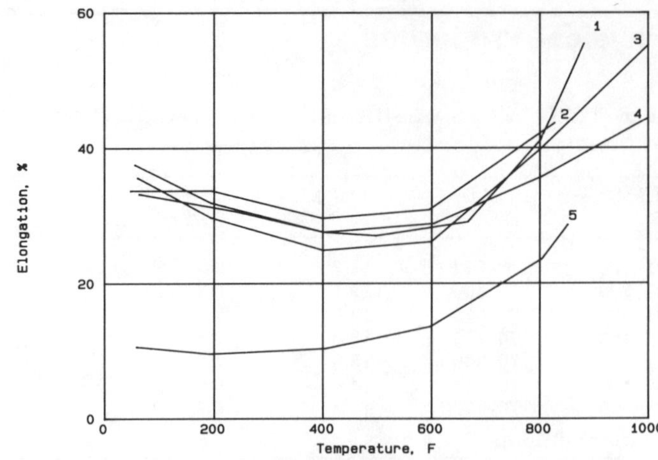

AISI 1020: Tensile elongation of plate and tube. Curve 1: annealed seamless tubing. Curves 2 and 3: 1-in. thick plate, stress relieved for 2 h at 1150 F and air cooled. Curve 4: 1-in. plate, as rolled. Curve 5: cold drawn seamless tubing. Source: Ref 21, p 6

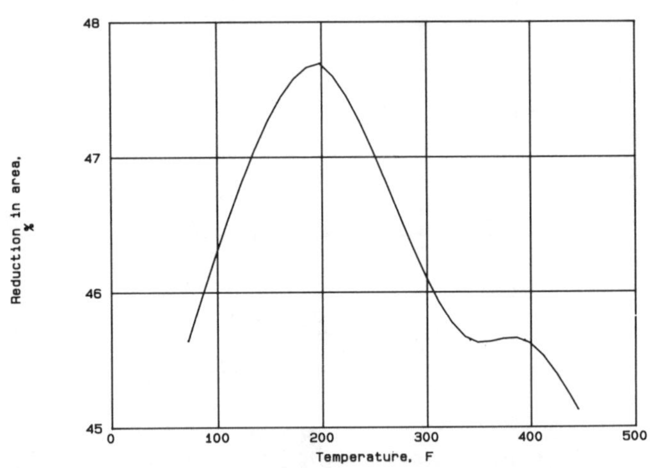

AISI 1020: Tensile reduction in area at failure of 0.065-in. hot rolled sheet. Source: Ref 21, p 5

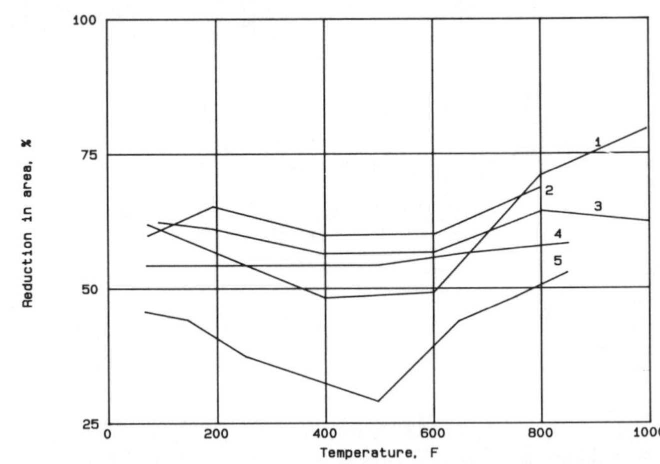

AISI 1020: Tensile reduction in area of plate and tube. Curves 1 and 2: 1-in. plate, stress relieved 2 h at 1150 h and air cooled. Curve 3: 1-in. plate, as rolled. Curve 4: annealed seamless tubing, 125 HB, 2.38-in. outside diameter. Curve 5: cold drawn seamless tubing, 179 HB, 4.5 in. outside diameter. Source: Ref 21, p 6

AISI 1020: Stress required to produce 1% creep in 10,000 h. Source: Ref 21, p 12

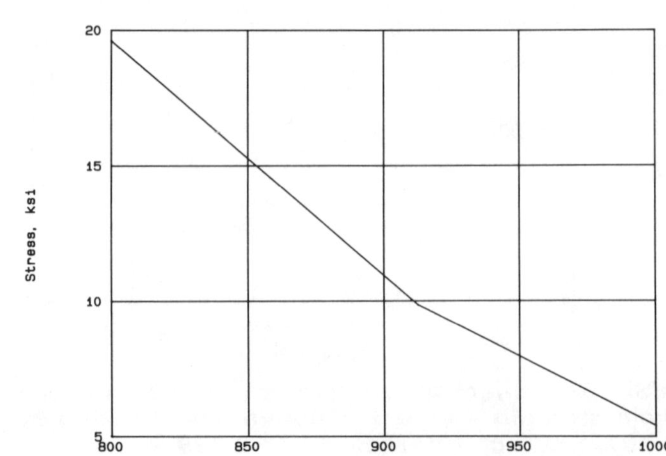

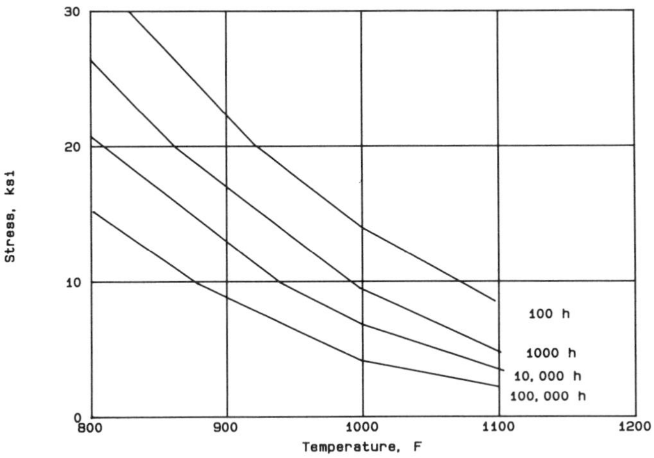

AISI 1020: Effect of temperature and exposure time on stress rupture. Source: Ref 21, p 12

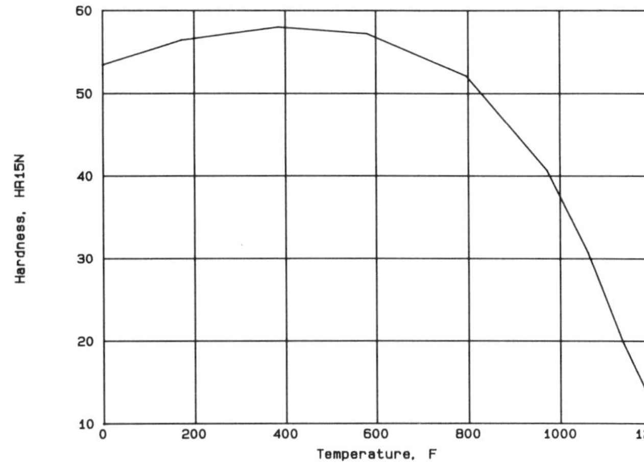

AISI 1020: Hardness. Specimens were 1 in. diameter and 0.125 to 0.250 in. thick, tested in argon atmosphere. Source: Ref 21, p 13

AISI 1023

UNS number: G10230
Specifications: ASTM A510, ASTM A575, ASTM A576, SAE J403, SAE J412, SAE J414, DIN 1.1151, NF XC 18 X, NF XC 25, JIS S-20C, JIS S-22C, JIS S-20CK
Nominal composition: 0.200 to 0.250% C, 0.300 to 0.600% Mn, 0.040% P max, 0.050% S max
Product forms: Wrought, bar, rod, wire, plate, strip, tube, pipe, billets, forgings, structural shapes, blooms, cast, continuous, powder, welded, ingot

Description: Similar to AISI 1020, AISI 1023 responds favorably to heat treating and cold work. Machinability and weldability are fair. Cold formability is good to excellent.

Typical applications: AISI 1023 has similar uses to those described for 1020 steel.

Physical Properties

AISI 1023: Mean apparent specific heat

TEMP C	TEMP F	SHTCP J/kg*K	SHTCP Btu/lb*F
50-100	120-212	486	0.116
150-200	300-390	519	0.124
200-250	390-480	532	0.127
250-300	480-570	557	0.133
300-350	570-660	574	0.137
350-400	660-750	599	0.143
450-500	840-930	662	0.158
550-600	1020-1110	749	0.179
650-700	1200-1290	846	0.202
700-750	1290-1380	1432	0.342
750-800	1380-1470	950	0.227

Note: Material composition: 0.23% carbon, 0.635% manganese
Source: Ref 15, p 20

AISI 1023: Thermal conductivity

TEMP C	TEMP F	THCON W/m*K	THCON Btu/ft*h*F
21	70	...	...
100	212	51.2	29.6
200	390	49.0	28.3
300	570	46.0	26.7
400	750	42.7	24.7
500	930	39.4	22.8
600	1110	35.6	20.6
700	1290	31.8	18.4
800	1470	26.1	15.1
900	1650	...	...
1000	1830	27.2	15.7
1100	2010	...	...
1200	2190	29.8	17.2
1300	2370	...	...

Note: Material composition: 0.23% carbon, 0.635% manganese
Source: Ref 15, p 20

AISI 1023: Mean coefficients of linear thermal expansion

TEMP C	TEMP F	THEXP 10^{-6}/C	THEXP 10^{-6}/F
21-100	70-212	12.1	6.7
21-200	70-390	12.8	7.1
21-300	70-570	13.3	7.4
21-400	70-750	13.9	7.7
21-500	70-930	14.4	8.0
21-600	70-1110	14.8	8.2

Source: Ref 15, p 20

AISI 1023: Electrical resistivity

TEMP C	TEMP F	ELRES n{o}*m
21	70	169
100	212	219
200	390	292
300	570	...
400	750	487
500	930	...
600	1110	758
700	1290	925
800	1470	1094
900	1650	1136
1000	1830	1167
1100	2010	1207
1200	2190	1219
1300	2370	1239

Note: Material composition: 0.23% carbon, 0.635% manganese
Source: Ref 15, p 20

AISI 1025

UNS number: G10250
Specifications: AMS 5075, AMS 5077, ASTM A510, ASTM A512, ASTM A519, ASTM A576, FED QQ S-700, MIL S-11310, SAE J403, SAE J412, SAE J414, DIN 1.1158, JIS S 25C, JIS S 28C
Nominal composition: 0.220 to 0.280% C, 0.300 to 0.600% Mn, 0.040% P max, 0.050% S max
Product forms: Wrought, bar, rod, wire, plate, strip, tube, pipe, billets, forgings, structural shapes, blooms, cast, continuous, powder, welded, ingot

Description: Similar to AISI 1020, AISI 1025, responds favorably to heat treating and cold work. Machinability and weldability are fair. Cold formability is good to excellent.

Typical applications: AISI 1025 has similar uses to those described for 1020 steel.

Physical Properties

AISI 1025: Mean apparent specific heat

TEMP C	TEMP F	SHTCP J/kg*K	SHTCP Btu/lb*F
50-100	120-212	486	0.116
150-200	300-390	519	0.124
200-250	390-480	532	0.127
250-300	480-570	557	0.133
300-350	570-660	574	0.137
350-400	660-750	599	0.143
450-500	840-930	662	0.158
550-600	1020-1110	749	0.179
650-700	1200-1290	846	0.202
700-750	1290-1380	1432	0.342
750-800	1380-1470	950	0.227

Note: Material composition: 0.23% carbon, 0.635% manganese
Source: Ref 15, p 20

AISI 1025: Thermal conductivity

TEMP C	TEMP F	THCON W/m*K	THCON Btu/ft*h*F
21	70	...	...
100	212	51.2	29.6
200	390	49.0	28.3
300	570	46.0	26.7
400	750	42.7	24.7
500	930	39.4	22.8
600	1110	35.6	20.6
700	1290	31.8	18.4
800	1470	26.1	15.1
900	1650	...	...
1000	1830	27.2	15.7
1100	2010	...	...
1200	2190	29.8	17.2
1300	2370	...	...

Note: Material composition: 0.23% carbon, 0.635% manganese
Source: Ref 15, p 20

AISI 1025: Mean coefficients of linear thermal expansion

TEMP C	TEMP F	THEXP $10^{-6}/C$	THEXP $10^{-6}/F$
21-100	70-212	12.1	6.7
21-200	70-390	12.8	7.1
21-300	70-570	13.3	7.4
21-400	70-750	13.9	7.7
21-500	70-930	14.4	8.0
21-600	70-1110	14.8	8.2

Source: Ref 15, p 20

AISI 1025: Electrical resistivity

TEMP C	TEMP F	ELRES n{o}*m
21	70	0.169
100	212	0.219
200	390	0.292
300	570	...
400	750	0.487
500	930	..
600	1110	0.758
700	1290	0.925
800	1470	1.094
900	1650	1.136
1000	1830	1.167
1100	2010	1.207
1200	2190	1.219
1300	2370	1.239

Note: Material composition: 0.23% carbon, 0.635% manganese
Source: Ref 15, p 20

AISI 1026

UNS number: G10260
Specifications: ASTM A273, ASTM A510, ASTM A519, ASTM A545, ASTM A576, SAE J403, SAE J412, SAE J414
Nominal composition: 0.220 to 0.280% C, 0.600 to 0.900% Mn, 0.040% P max, 0.050% S max
Product forms: Wrought, bar, rod, wire, plate, strip, tube, pipe, billets, forgings, structural shapes, blooms, cast, continuous, powder, welded, ingot
Description: Similar to AISI 1020, AISI 1026 responds favorably to heat treating and cold work. Machinability and weldability are fair. Cold formability is good to excellent.
Typical applications: AISI 1026 has similar uses to those described for 1020 steel.

Physical Properties

AISI 1026: Mean apparent specific heat

TEMP C	TEMP F	SHTCP J/kg*K	SHTCP Btu/lb*F
50-100	120-212	486	0.116
150-200	300-390	519	0.124
200-250	390-480	532	0.127
250-300	480-570	557	0.133
300-350	570-660	574	0.137
350-400	660-750	599	0.143
450-500	840-930	662	0.158
550-600	1020-1110	749	0.179
650-700	1200-1290	846	0.202
700-750	1290-1380	1432	0.342
750-800	1380-1470	950	0.227

Note: Material composition: 0.23% carbon, 0.635% manganese
Source: Ref 15, p 20

AISI 1026: Thermal conductivity

TEMP C	TEMP F	THCON W/m*K	THCON Btu/ft*h*F
21	70	...	...
100	212	51.2	29.6
200	390	49.0	28.3
300	570	46.0	26.7
400	750	42.7	24.7
500	930	39.4	22.8
600	1110	35.6	20.6
700	1290	31.8	18.4
800	1470	26.1	15.1
900	1650	...	...
1000	1830	27.2	15.7
1100	2010	...	...
1200	2190	29.8	17.2
1300	2370	...	...

Note: Material composition: 0.23% carbon, 0.635% manganese
Source: Ref 15, p 20

AISI 1026: Mean coefficients of linear thermal expansion

TEMP C	TEMP F	THEXP 10^{-6}/C	THEXP 10^{-6}/F
21-100	70-212	12.1	6.7
21-200	70-390	12.8	7.1
21-300	70-570	13.3	7.4
21-400	70-750	13.9	7.7
21-500	70-930	14.4	8.0
21-600	70-1110	14.8	8.2

Source: Ref 15, p 20

AISI 1026: Electrical resistivity

TEMP C	TEMP F	ELRES n{o}*m
21	70	169
100	212	219
200	390	292
300	570	...
400	750	487
500	930	...
600	1110	758
700	1290	925
800	1470	1094
900	1650	1136
1000	1830	1167
1100	2010	1207
1200	2190	1219
1300	2370	1239

Note: Material composition: 0.23% carbon, 0.635% manganese
Source: Ref 15, p 20

AISI 1030

UNS number: G10300
Specifications: ASTM A510, ASTM A512, ASTM A519, ASTM A544, ASTM A546, ASTM A576, ASTM A682, FED QQ S-635 (C1030), FED QQ S-700 (C1030), MIL S-11310 (CS1030), SAE J403, SAE J412, SAE J414, DIN 1.1172, UNI CB 35
Nominal composition: 0.280 to 0.340% C, 0.600 to 0.900% Mn, 0.040% P max, 0.050% S max
Product forms: Wrought, bar, rod, wire, plate, strip, pipe, billets, forgings, structural shapes, blooms, cast, continuous, powder, welded, ingot
Description: AISI 1030 is a water-hardening steel suitable for small parts of moderate strength. Grade 1035 has a slightly greater hardenability and strength than AISI 1036. It has moderate strength and hardness in the as-rolled condition and can be strengthened and hardened by cold work. Machinability is only fair, as is weldability. Precautions must be taken when welding to avoid cracking from cooling too rapidly.

Typical applications: AISI 1030 is used in manufacturing levers, bolts, studs, nuts, and similar parts that are headed, upset, or extruded. Applications for wires made of 1030 include nails requiring high shank rigidity or strength and quench-hardened nails.

Physical Properties

AISI 1030: Mean coefficient of linear thermal expansion

TEMP C	TEMP F	THEXP 10^{-6}/C	THEXP 10^{-6}/F
15-75	59-165	11.7	6.5

Source: Ref 15, p 22

AISI 1035

UNS number: G10350
Specifications: AMS 5080, AMS 5082, ASTM A510, ASTM A519, ASTM A544, ASTM A545, ASTM A546, ASTM A576, ASTM A682, FED QQ S-635, FED QQ S-700, SAE J403, SAE J412, SAE J414, DIN 1.0501, NF CC 35, UNI C 35, SS 4 15 50, BS 060 A 35, BS 080 A 35
Nominal composition: 0.320 to 0.380% C, 0.600 to 0.900% Mn, 0.040% P max, 0.050% S max
Product forms: Wrought, bar, rod, wire, plate, strip, pipe, billets, forgings, structural shapes, blooms, cast, continuous, powder, welded, ingot
Description: AISI 1035 is a water-hardening medium-carbon steel commonly used for small machinery parts of moderate strength. It may be used in annealed sheet form for deep drawing; however, permissible wall thickness reductions are not as great as for 1020 or 1010. This steel provides medium strength and toughness at low cost. AISI 1035 has a slightly greater hardenability and strength than AISI 1030. Both have moderate strength and hardenability in the as-rolled condition and can be strengthened and hardened by cold work. Machinability is only fair, as is weldability. Precautions must be taken when welding, to avoid cracking from cooling too rapidly.

Typical applications: AISI 1035 is used in manufacturing levers, bolts, studs, nuts, and similar parts that are headed, upset, or extruded. Applications for wires of these grades include nails requiring high shank rigidity or strength and quench-hardened nails.

Physical Properties

AISI 1035: Mean coefficients of thermal expansion

TEMP C	TEMP F	THEXP 10^{-6}/C	THEXP 10^{-6}/F
21-100	70-212	11.0	6.1
21-200	70-390	11.9	6.6
21-300	70-570	12.6	7.0
21-400	70-750	13.3	7.4
21-500	70-930	13.9	7.7
21-600	70-1110	14.4	8.0
21-700	70-1290	13.9	8.2

Source: Ref 15, p 22

AISI 1038

UNS number: G10380
Specifications: ASTM A510, ASTM A544, ASTM A545
Nominal composition: 0.350 to 0.420% C, 0.600 to 0.900% Mn, 0.040% P max, 0.050% S max
Product forms: Rod, springs, wrought, bar, wire, plate, strip, pipe, billets, forgings, structural shapes, blooms, cast, continuous, powder, ingot
Description: AISI 1038 is a medium-high carbon steel that can be strengthened by heat treating after forming. Machinability and weldability are fair.
Typical applications: Machine, plow, and carriage bolts, tie wire, cylinder head studs, and machined parts are made from 1038 steel. It is also used for U-bolts, concrete reinforcing rods, forgings, and noncritical springs. SAE J429 grade 5 fasteners are also made from 1038 steel.

Physical Properties

AISI 1038: Electrical resistivity and thermal conductivity

TEMP C	TEMP F	ELRES n{o}*m	THCON W/m*K	THCON Btu/ft*h*F
21	70	171	...	...
100	212	221	50.7	29.3
200	390	296	48.1	27.8
300	570	...	45.7	26.4
400	750	493	41.7	24.1
500	930	...	38.2	22.1
600	1110	763	33.9	19.6
700	1290	932	30.1	17.4
800	1470	1111	24.7	14.3
900	1650	1149	...	...
1000	1830	1179	32.9	19
1100	2010	1207	...	...
1200	2190	1230	29.8	17.2

Note: Material composition: 0.415% carbon, 0.643% manganese
Source: Ref 15, p 27

AISI 1038: Mean apparent specific heat

TEMP C	TEMP F	SHTCP J/kg*K	SHTCP Btu/lb*F
50-100	20-212	486	0.116
150-200	300-390	515	0.123
200-250	390-480	528	0.126
250-300	480-570	548	0.131
300-350	570-660	569	0.136
350-400	660-750	586	0.14
450-500	840-930	649	0.155
550-600	1020-1110	708	0.169
650-700	1200-1290	770	0.184
700-750	1290-1380	1583	0.378
750-800	1380-1470	624	0.149
850-900	1560-1650	548	0.131

Note: Material composition: 0.415% carbon, 0.643% manganese
Source: Ref 15, p 27

Mechanical Properties

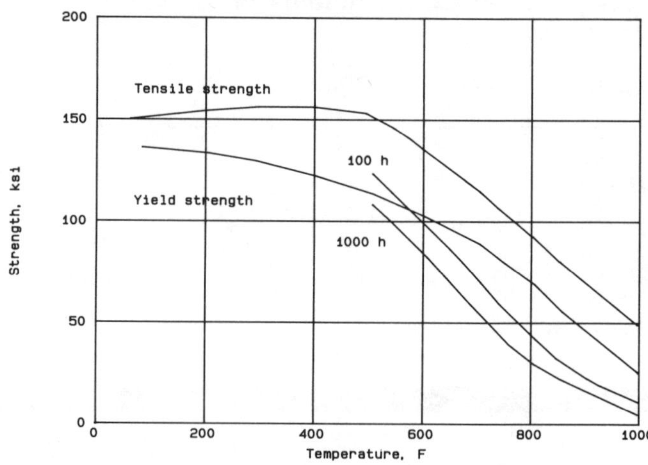

AISI 1038: Tensile, yield (0.20% offset), and rupture strength of quenched and tempered specimens. Source: Ref 21, p 43

AISI 1038H

UNS number: H10380
Specifications: ASTM A304, SAE J776, DIN 1.1176, NF XC 38 TX
Nominal composition: 0.340 to 0.430% C, 0.500 to 1.000% Mn, 0.040% P max, 0.050% S max, 0.150 to 0.350% Si max
Product forms: Wrought, bar, rod, wire, plate, strip, pipe, billets, forgings, structural shapes, blooms, cast, continuous, powder, ingot

Description: AISI 1038H is a medium-high carbon steel that can be strengthened by heat treating after forming. Machinability and weldability are fair.
Typical applications: Machine, plow, and carriage bolts, tie wire, cylinder head studs, and machined parts are made from 1038H. It is also used for U-bolts, concrete reinforcing rods, forgings, and non-critical springs.

Physical Properties

AISI 1038H: Electrical resistivity and thermal conductivity

TEMP C	TEMP F	ELRES n{o}*m	THCON W/m*K	THCON Btu/ft*h*F
21	70	0.171	...	...0
100	212	0.221	50.7	29.3
200	390	0.296	48.1	27.8
300	570	...	45.7	26.4
400	750	0.493	41.7	24.1
500	930	...	38.2	22.1
600	1110	0.763	33.9	19.6
700	1290	0.932	30.1	17.4
800	1470	1.111	24.7	14.3
900	1650	1.149	...	...
1000	1830	1.179	32.9	19
1100	2010	1.207	...	...
1200	2190	1.23	29.8	17.2

Note: Material composition: 0.415% carbon, 0.643% manganese
Source: Ref 15, p 27

AISI 1038H: Mean apparent specific heat

TEMP C	TEMP F	SHTCP J/kg*K	SHTCP Btu/lb*F
50-100	120-212	486	0.116
150-200	300-390	515	0.123
200-250	390-480	528	0.126
250-300	480-570	548	0.131
300-350	570-660	569	0.136
350-400	660-750	586	0.14
450-500	840-930	649	0.155
550-600	1020-1110	708	0.169
650-700	1200-1290	770	0.184
700-750	1290-1380	1583	0.378
750-800	1380-1470	624	0.149
850-900	1560-1650	548	0.131

Note: Material composition: 0.415% carbon, 0.643% manganese
Source: Ref 15, p 27

Mechanical Properties

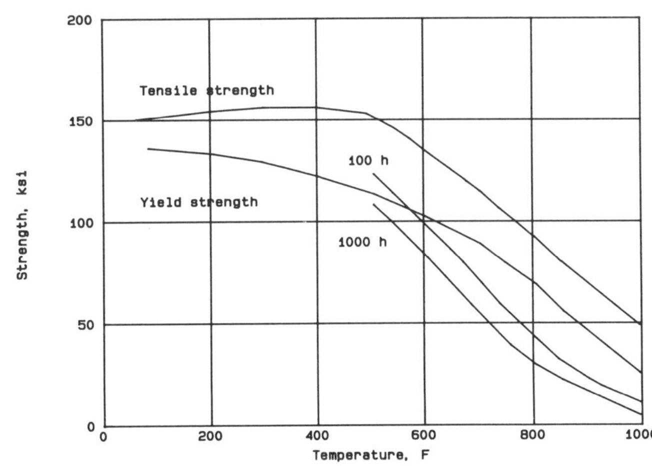

AISI 1038H: Tensile, yield (0.20% offset), and rupture strength of quenched and tempered specimens. Source: Ref 21, p 43

AISI 1039

UNS number: G10390
Specifications: ASTM A510, ASTM A546, ASTM A576, SAE J403, SAE J412, SAE J414, DIN 1.1157, NF 35 M 5, BS 120M36, BS 150M36
Nominal composition: 0.370 to 0.440% C, 0.700 to 1.000% Mn, 0.040% P max, 0.050% S max
Product forms: Wrought, bar, rod, wire, plate, strip, pipe, billets, forgings, structural shapes, blooms, cast, continuous, ingot

Description: AISI 1039 is a medium-high carbon steel that can be strengthened by heat treating after forming. Machinability and weldability are fair.
Typical applications: Machine, plow, and carriage bolts, tie wire, cylinder head studs, and machined parts are made from 1038 steel. This grade is also used for U-bolts, concrete reinforcing rods, forgings, and noncritical springs.

AISI 1039: Electrical resistivity and thermal conductivity

TEMP C	TEMP F	ELRES n{o}*m	THCON W/m*K	THCON Btu/ft*h*F
21	70	171	...	...
100	212	221	50.7	29.3
200	390	296	48.1	27.8
300	570	...	45.7	26.4
400	750	493	41.7	24.1
500	930	...	38.2	22.1
600	1110	763	33.9	19.6
700	1290	932	30.1	17.4
800	1470	1111	24.7	14.3
900	1650	1149	...	...
1000	1830	1179	32.9	19.0
1100	2010	1207	...	...
1200	2190	1230	29.8	17.2

Note: Material composition: 0.415% carbon, 0.643% manganese
Source: Ref 15, p 27

AISI 1039: Mean apparent specific heat

TEMP C	TEMP F	SHTCP J/kg*K	SHTCP Btu/lb*F
50-100	120-212	486	0.116
150-200	300-390	515	0.123
200-250	390-480	528	0.126
250-300	480-570	548	0.131
300-350	570-660	569	0.136
350-400	660-750	586	0.14
450-500	840-930	649	0.155
550-600	1020-1110	708	0.169
650-700	1200-1290	770	0.184
700-750	1290-1380	1583	0.378
750-800	1380-1470	624	0.149
850-900	1560-1650	548	0.131

Note: Material composition: 0.415% carbon, 0.643% manganese
Source: Ref 15, p 27

Mechanical Properties

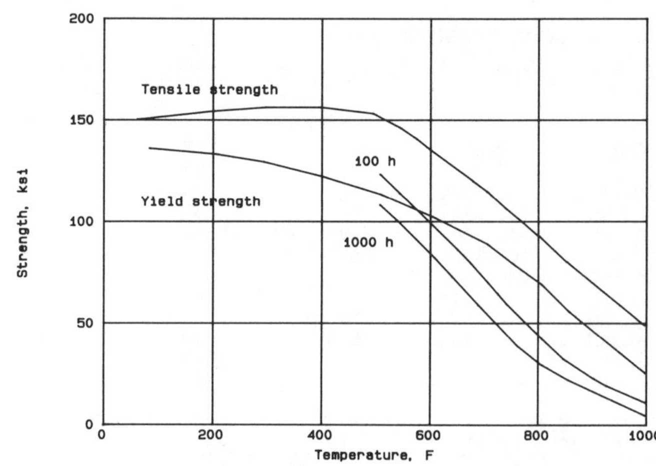

AISI 1039: Tensile, yield (0.20% offset), and rupture strength of quenched and tempered specimens. Source: Ref 21, p 43

AISI 1040

UNS number: G10400
Specifications: ASTM A29, ASTM A108, ASTM A449, ASTM A510, ASTM A519, ASTM A546, ASTM A576, ASTM A682, MIL S-11310 (CS1040), MIL S-22146, SAE J403, SAE J412, SAE J414, DIN 1.1186, JIS S 40C, BS 080 A 40, BS 2 S.93
Nominal composition: 0.370 to 0.440% C, 0.600 to 0.900% Mn, 0.040% P max, 0.050% S max
Product forms: Wrought, bar, rod, wire, plate, sheet, strip, pipe, billets, forgings, blooms, cast, continuous, ingot

Description: AISI 1040 is a medium-high carbon steel that can be strengthened by heat treating after forming. Machinability and weldability are fair. May be induction or flame hardened. However, water quenching produces full hardening only in thin sections.
Typical applications: Machine, plow, and carriage bolts, tie wire, cylinder head studs, hand tools, screw drivers, wrenches, pliers, cold drawn bolts, shafts, axles, gears, wire, bulldozer edges, and medium-strength heat treated forgings.

Physical Properties

AISI 1040: Electrical resistivity and thermal conductivity

TEMP C	TEMP F	ELRES n{o}*m	THCON W/m*K	THCON Btu/ft*h*F
21	70	171	...	...
100	212	221	50.7	29.3
200	390	296	48.1	27.8
300	570	...	45.7	26.4
400	750	493	41.7	24.1
500	930	...	38.2	22.1
600	1110	763	33.9	19.6
700	1290	932	30.1	17.4
800	1470	1111	24.7	14.3
900	1650	1149	...	...
1000	1830	1179	32.9	19.0
1100	2010	1207	...	...
1200	2190	1230	29.8	17.2

Note: Material composition: 0.415% carbon, 0.643% manganese
Source: Ref 15, p 27

AISI 1040: Mean apparent specific heat

TEMP C	TEMP F	SHTCP J/kg*K	SHTCP Btu/lb*F
50-100	120-212	486	0.116
150-200	300-390	515	0.123
200-250	390-480	528	0.126
250-300	480-570	548	0.131
300-350	570-660	569	0.136
350-400	660-750	586	0.14
450-500	840-930	649	0.155
550-600	1020-1110	708	0.169
650-700	1200-1290	770	0.184
700-750	1290-1380	1583	0.378
750-800	1380-1470	624	0.149
850-900	1560-1650	548	0.131

Note: Material composition: 0.415% carbon, 0.643% manganese
Source: Ref 15, p 27

Mechanical Properties

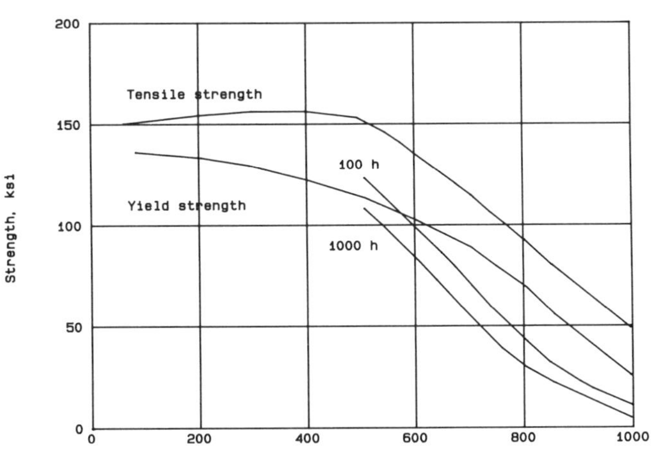

AISI 1040: Tensile, yield (0.20% offset), and rupture strength of quenched and tempered specimens. Source: Ref 21, p 43

AISI 1040: Mean coefficients of linear thermal expansion

TEMP C	TEMP F	THEXP 10^{-6}/C	THEXP 10^{-6}/F
Material A(a)			
21-100	70-212	11.3	6.3
21-200	70-390	12.1	6.7
21-300	70-570	12.2	6.8
21-400	70-750	13.3	7.4
21-500	70-930	13.9	7.7
21-600	70-1110	14.2	7.9
21-700	70-1290	14.8	8.2
Material B(b)			
0-100	32-212	11.2	6.2
0-200	32-390	12.1	6.7
0-300	32-570	13.0	7.2
0-400	32-750	13.5	7.5
0-500	32-930	14.0	7.8
0-600	32-1110	14.6	8.1
0-700	32-1290	14.8	8.2

(a) Material composition: 0.40% carbon, 0.11% manganese, 0.01% phosphorus 0.03% sulfur, 0.03% silicon, 0.03% copper.
(b) Material composition: 0.42% carbon, 0.64% manganese, 0.031% phosphorus, 0.29% sulfur, 0.11% silicon, 0.06% nickel, 0.12% copper, 0.006% aluminum, 0.033% arsenic
Source: Ref 15, p 27

AISI 1045

UNS number: G10450
Specifications: ASTM A29, ASTM A510, ASTM A519, ASTM A576, ASTM A682, FED QQ-S-635 (C1045), FED QQ-S-700 (C1045), SAE J403, SAE J412, SAE J414, DIN 1.1191, NF XC42, NF XC 42 TS, NF XC 45, NF XC 48, JIS S 45 C, JIS S 48 C, SS 14 16 72
Nominal composition: 0.430 to 0.500% C, 0.600 to 0.900% Mn, 0.040% P max, 0.050% S max
Description: AISI 1045 is used when greater strength and hardness are desired in the as-rolled condition. This medium-carbon steel can be hammer forged. It responds to heat treatment and flame and induction hardening, but is not recommended for carburizing or cyaniding. When special practices are employed, weldability is fair. Die forging and hot upsetting of this steel is good to excellent.

Typical applications: AISI 1045 is used for gears, shafts, axles, bolts, studs, and machine parts.

Physical Properties

AISI 1045: Mean coefficients of linear thermal expansion

TEMP C	TEMP F	THEXP 10^{-6}/C	THEXP 10^{-6}/F
0-100	32-212	11.5	6.4
0-200	32-390	12.2	6.8
0-300	32-570	13.0	7.2
0-400	32-750	13.7	7.6
0-500	32-930	14.0	7.8
0-600	32-1110	14.6	8.1
0-700	32-1290	15.1	8.4
25-100	75-212	11.2	6.2
25-200	75-390	11.9	6.6
25-300	75-570	12.6	7.0
25-400	75-750	13.5	7.5
25-500	75-930	14.0	7.8
25-600	75-1110	14.4	8.0
25-700	75-1290	14.8	8.2

Source: Ref 15, p 28

Mechanical Properties

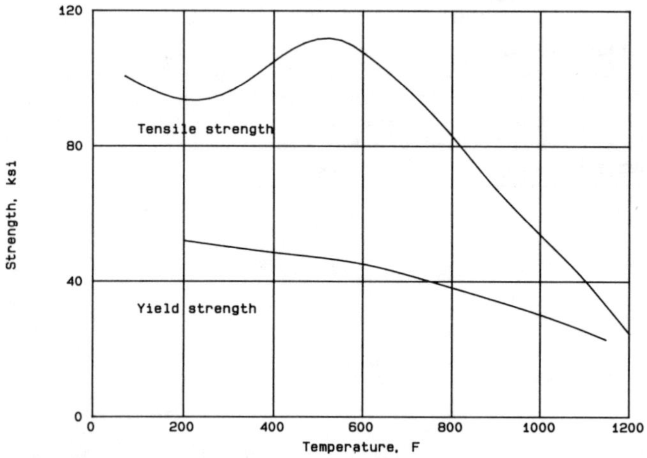

AISI 1045: Tensile and yield strength. Source: Ref 21, p 22

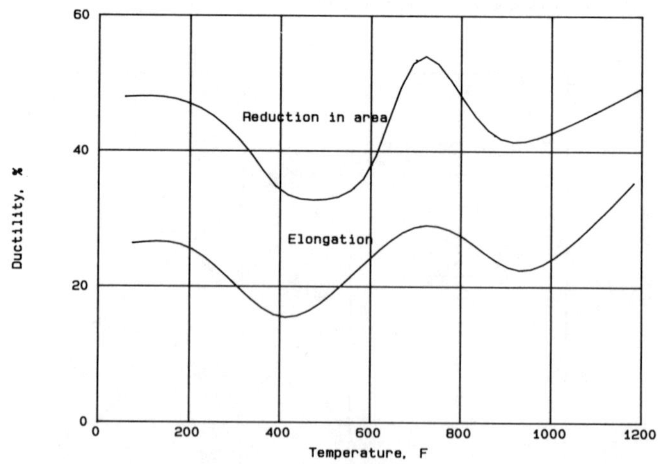

AISI 1045: Tensile elongation and reduction in area. Source: Ref 21, p 22

AISI 1046

UNS number: 10460
Specifications: ASTM A510, ASTM A576, SAE J403, SAE J412, SAE J414
Nominal composition: 0.430 to 0.500% C, 0.700 to 1.000% Mn, 0.040% P max, 0.050% S max
Product forms: Wrought, bar, rod, wire, plate, strip, pipe, billets, forgings, blooms, cast, continuous, ingot
Description: Used where greater strength and hardness in as-rolled condition are required. Responds to heat treatment and flame and induction hardening, but not recommended for carburizing or cyaniding. Can be hammer forged. When special practices are employed, weldability is fair. Die forging and hot upsetting are good to excellent.

Typical applications: AISI 1046 is used for gears, shafts, axles, bolts, studs, and machine parts.

Physical Properties

AISI 1046: Mean coefficients of linear thermal expansion

TEMP C	TEMP F	THEXP $10^{-6}/C$	THEXP $10^{-6}/F$
0-100	32-212	11.5	6.4
0-200	32-390	12.2	6.8
0-300	32-570	13.0	7.2
0-400	32-750	13.7	7.6
0-500	32-930	14.0	7.8
0-600	32-1110	14.6	8.1
0-700	32-1290	15.1	8.4
25-100	75-212	11.2	6.2
25-200	75-390	11.9	6.6
25-300	75-570	12.6	7.0
25-400	75-750	13.5	7.5
25-500	75-930	14.0	7.8
25-600	75-1110	14.4	8.0
25-700	75-1290	14.8	8.2

Source: Ref 15, p 28

Mechanical Properties

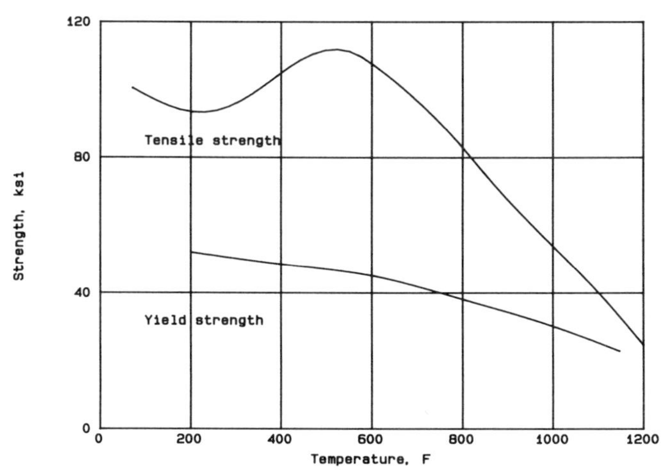

AISI 1046: Tensile and yield strength (0.20% offset). Source: Ref 21, p 22

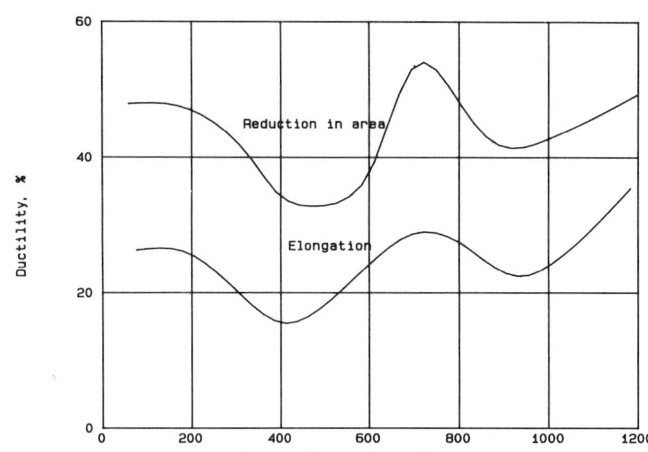

AISI 1046: Tensile elongation and reduction in area. Source: Ref 21, p 22

AISI 1055

UNS number: G10550
Specifications: ASTM A510, ASTM A576, ASTM A682, FED QQ S-700, SAE J403, SAE J412, SAE J414, DIN 1.1209
Nominal composition: 0.500 to 0.600% C, 0.600 to 0.900% Mn, 0.040% P max, 0.050% S max
Product forms: Wrought, bar, rod, wire, plate, strip, billets, forgings, blooms, cast, continuous, ingot
Description: When heat treated, yields high surface hardness and relatively good toughness. Has good forging characteristics. It is shallow hardening, however, and useful section size is limited. Parts requiring strength are oil quenched; parts requiring high hardness are water quenched.

Typical applications: AISI 1055 is used for battering tools, hot upset forging dies, ring-rolling tools, wear-resistant parts, hand tools, and parts for agricultural implements that require high strength at low cost.

Physical Properties

AISI 1055: Mean coefficients of linear thermal expansion

TEMP C	TEMP F	THEXP 10^{-6}/C	THEXP 10^{-6}/F
21-100	70-212	11.0	6.1
21-200	70-390	12.1	6.7
21-300	70-570	12.4	6.9
21-400	70-750	13.3	7.4
21-500	70-930	13.9	7.7
21-600	70-1110	14.4	8.0
21-700	70-1290	14.8	8.2

Source: Ref 15, p 34

AISI 1060

UNS number: G10600
Specifications: AMS 7240, ASTM A29, ASTM A510, ASTM A576, ASTM A682, MIL S-16974, MIL S-16974, FED QQ S-633, FED QQ S-00640, SAE J403, SAE J412, SAE J414, DIN 1.0601, NF CC 55, UNI C 60, BS 060 A 62
Nominal composition: 0.550 to 0.650% C, 0.600 to 0.900% Mn, 0.040% P max, 0.050% S max
Product forms: Wrought, bar, rod, wire, plate, strip, billets, forgings, blooms, cast, continuous, ingot, joint
Description: As with other high-carbon steels, AISI 1060 is more restricted in its applications than the lower carbon steels. Items fabricated from high carbon steels are more costly due to decreased machinability, poor formability, and poor weldability. They are also more brittle when heat treated. This steel also distorts more than low-carbon steel, due to heat treatment. Surface decarburization can also occur as a result of heat treatment.
Typical applications: Heavy machinery parts, shafts, springs, and tools requiring good strength and wear resistance such as hammers, garden hoes, wrenches, axes, agricultural implements, cold-formed flat springs, torsion bars, induction-hardened gears, shafts, and blades for agricultural and earth-moving equipment.

Physical Properties

AISI 1060: Mean coefficients of linear thermal expansion

TEMP C	TEMP F	THEXP 10^{-6}/C	THEXP 10^{-6}/F
21-100	70-212	11.0	6.1
21-200	70-390	11.5	6.4
21-300	70-570	12.2	6.8
21-400	70-750	13.1	7.3
21-500	70-930	13.7	7.6
21-600	70-1110	14.0	7.8
21-700	70-1290	14.4	8.0

Source: Ref 15, p 34

AISI 1070

UNS number: G10700
Specifications: AMS 5115, ASTM A510, ASTM A576, ASTM A682, MIL S-11713, SAE J403, SAE J412, SAE J414, DIN 1.1231, NF XC 68, SS 14 17 70, SS 14 17 78
Nominal composition: 0.650 to 0.750% C, 0.600 to 0.900% Mn, 0.040% P max, 0.050% S max
Product forms: Wrought, bar, rod, wire, plate, strip, billets, forgings, blooms, cast, continuous
Description: AISI 1070 is made into wire that is high in tensile strength and is heat treatable. Cold forming generally is limited to coiling in smaller diameters for coil-type springs, or for sinusoidal forming for springs used in upholstered furniture and automotive seats and backs. Material for mechanical and upholstery springs is ordered by tensile strength and mechanical properties, rather than by grade. AISI 1070 is produced as cold rolled strip for flat springs in the annealed or tempered condition.
Typical applications: Oil tempered wire and strip are suitable for springs subject to high and frequent stress. For springs subject to static loads or relatively infrequent stress cycles, however, hard drawn or cold rolled steels generally are used. High-carbon spring wire, which is heat treatable after forming, is used when the end product is quenched and tempered after forming. Other applications include wire that may be subject to severe forming, or for which high hardness is required. This spring wire is available as untempered, spheroidized annealed in process, and as spheroidized annealed at finished size. Composition of the wire is critically important to ensure uniform response to subsequent heat treatment.

Physical Properties

AISI 1070: Mean coefficients of linear thermal expansion

TEMP C	TEMP F	THEXP 10^{-6}/C	THEXP 10^{-6}/F
0-100	32-212	11.6	6.4
0-200	32-390	12.6	7.0
0-300	32-570	13.3	7.4
0-400	32-750	14.0	7.8

Source: Ref 15, p 41

Mechanical Properties

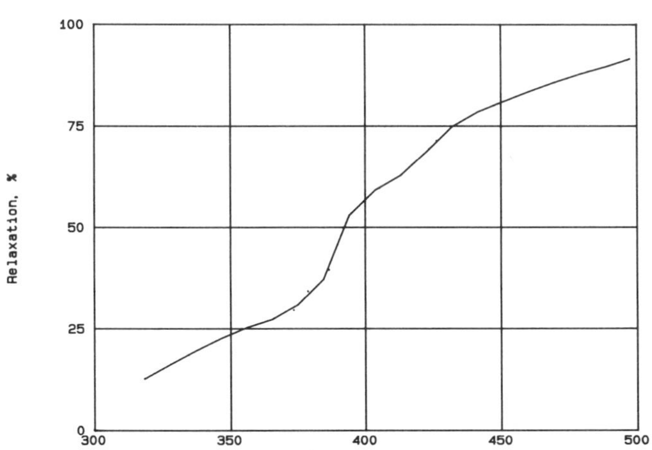

AISI 1070: Relaxation of circular flat springs (piston rings). Spring hardness, 35 HRC. Source: Ref 1, p 299

AISI 1080

UNS number: G10800

Specifications: AMS 5110, ASTM A29, ASTM A510, ASTM A576, ASTM A682, FED QQ S-700, FED QQ W-470, MIL S-16974, SAE J403, SAE J412, SAE J414

Nominal composition: 0.750 to 0.880% C, 0.600 to 0.900% Mn, 0.040% P max, 0.050% S max

Product forms: Wrought, bar, rod, wire, plate, strip, billets, forgings, blooms, cast, continuous

Description: AISI 1080 is made into wire that is high in tensile strength and is heat treatable. Cold forming is generally limited to coiling in smaller diameters for coil-type springs, or for sinusoidal forming for springs used in upholstered furniture and automotive seats and backs. Material for mechanical and upholstery springs is ordered by tensile strength and mechanical properties rather than by grade. Thus, the producer provides a steel that will meet the requirements.

Typical applications: These high-carbon steels are produced as cold rolled strip for flat springs in the annealed or tempered condition. Oil tempered wire and strip are suitable for springs subject to high and frequent stress. For springs subject to static loads or relatively infrequent stress cycles, however, hard drawn or cold rolled steels generally are used. High-carbon spring wire, which is heat treatable after forming, is used when the end product is quenched and tempered after forming. Other applications include wire that may be subject to severe forming, or for which high hardness is required. This spring wire is available as untempered, spheroidized annealed in process, and is spheroidized annealed at finished size. Composition of the wire is critically important to ensure uniform response to subsequent heat treatment. Typical uses also include springs, hand tools, thrust washers, lock washers, drill shanks, dowel pins, bushings, ball bearings, hammer drills, punches, dies, cultivator discs, coulter blades, plows, plow shares, springs, shovels, and harrows.

Physical Properties

AISI 1080: Mean apparent specific heat

TEMP C	TEMP F	SHTCP J/kg*K	SHTCP Btu/lb*F
50-100	120-212	490	0.117
150-200	300-390	531	0.127
200-250	390-480	548	0.131
250-300	480-470	565	0.135
300-350	570-660	586	0.140
350-400	660-750	607	0.145
450-500	840-930	670	0.160
550-600	1020-1110	712	0.170
650-700	1200-1290	770	0.184

Note: Material composition: 0.80% carbon, 0.32% manganese
Source: Ref 15, p 41

AISI 1080: Mean coefficients of linear thermal expansion

TEMP C	TEMP F	THEXP 10^{-6}/C	THEXP 10^{-6}/F
0-100	32-212	11.0	6.1
0-200	32-390	11.5	6.4
0-300	32-570	12.2	6.8
0-400	32-750	13.1	7.3
0-500	32-930	13.7	7.6
0-600	32-1110	14.0	7.8
0-700	32-1290	14.6	8.1

Source: Ref 15, p 41

AISI 1080: Thermal conductivity

TEMP C	TEMP F	THCON W/m*K	THCON Btu/ft*h*F
21	70	...	...
100	212	48.1	27.8
200	390	45.2	26.1
300	570	41.3	23.9
400	750	38.1	22.0
500	930	35.1	20.3
600	1110	32.7	18.9
700	1290	30.1	17.4
800	1470	24.4	14.1
900	1650	...	...
1000	1830	26.8	15.5
1100	2010	...	...
1200	2190	30.1	17.4
1250	2270	...	...

Note: Material composition: 0.08% carbon, 0.32% manganese
Source: Ref 15, p 41

AISI 1080: Electrical resistivity

TEMP C	TEMP F	ELRES n{o}*m
21	70	180
100	212	232
200	390	308
300	570	...
400	750	505
500	930	...
600	1110	772
700	1290	935
800	1470	1129
900	1650	1164
1000	1830	1191
1100	2010	1214
1200	2190	1234
1250	2270	1246

Note: Material composition: 0.08% carbon, 0.32% manganese
Source: Ref 15, p 41

AISI 1090

UNS number: G10900
Specifications: AMS 5121, AMS 5122, AMS 5132, AMS 7304, ASTM A290, ASTM A510, ASTM A576, ASTM A682, FED QQ S-700, MIL S-7947, MIL S-8559, MIL S-16788, SAE J403, SAE J412, SAE J414
Nominal composition: 0.900 to 1.030% C, 0.300 to 0.500% Mn, 0.040% P, 0.050% S
Product forms: Wrought, bar, rod, wire, plate, strip, billets, forgings, blooms, cast, continuous
Description: This high-carbon steel provides maximum surface hardness with improved wear resistance and high strength. There is, however, a loss of toughness. Because cold-forming methods generally are not practical with this steel, uses are limited to flat stampings and springs coiled from small-diameter wire. This high-carbon steel is sometimes classified as a tool steel. If good notch toughness or shock resistance are required, it must be tempered at a fairly high temperature, which in turn causes a decrease in hardness and wear resistance. This steel may be forged; however, the die life will be about one half that obtained from forged low-carbon steels.
Typical applications: AISI 1090 is used for edge tools, wear-resistant parts, high-stress flat springs, hot coiled springs, plow beams, plow shares, scraper blades, discs, mower knives, harrow teeth, leather and paper punches, surgical and dental instruments, grass cutting tools, general-purpose woodworking cutting tools, knife blades, wrenches, and hammers.

Mechanical Properties

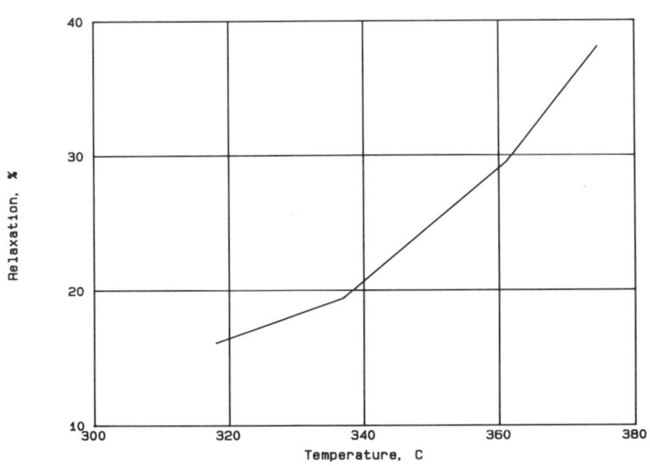

AISI 1090: Relaxation of circular flat springs (piston rings). Spring hardness, 35 HRC. Source: Ref 1, p 299

AISI 1095

UNS number: G10950
Specifications: AMS 5121, AMS 5122, AMS 5132, AMS 7304, ASTM A290, ASTM A510, ASTM A576, ASTM A682, FED QQ S-700, MIL S-7947, MIL S-8559, MIL S-16788, SAE J403, SAE J412, SAE J414
Nominal composition: 0.900 to 1.030% C, 0.300 to 0.500% Mn, 0.040% P max, 0.050% S max
Product forms: Wrought, bar, rod, wire, plate, strip, billets, forgings, blooms, cast, continuous
Description: This high-carbon steel provides maximum surface hardness with improved wear resistance and high strength. There is, however, a loss of toughness. Because cold-forming methods generally are not practical with this steel, uses are limited to flat stampings and springs coiled from small-diameter wire. This high-carbon steel is sometimes classified as a tool steel. If good notch toughness or shock resistance are required, it must be tempered at a fairly high temperature, which in turn causes a decrease in hardness and wear resistance. This steel may be forged; however, the die life will be about one half that obtained from forged low-carbon steels.
Typical applications: AISI 1095 is used for edge tools, wear-resistant parts, high-stress flat springs, hot coiled springs, plow beams, plow shares, scraper blades, discs, mower knives, harrow teeth, leather and paper punches, surgical and dental instruments, grass cutting tools, general-purpose woodworking cutting tools, knife blades, wrenches, and hammers.

Physical Properties

AISI 1095: Mean coefficients of linear thermal expansion

TEMP C	TEMP F	THEXP 10^{-6}/C	THEXP 10^{-6}/F
0-100	32-212	11.0	6.1
0-200	32-390	11.7	6.5
0-300	32-570	12.4	6.9
0-400	32-750	13.1	7.3
0-500	32-930	13.5	7.5
0-600	32-1110	14.2	7.9
0-700	32-1290	14.6	8.1

Source: Ref 15, p 41

Mechanical Properties

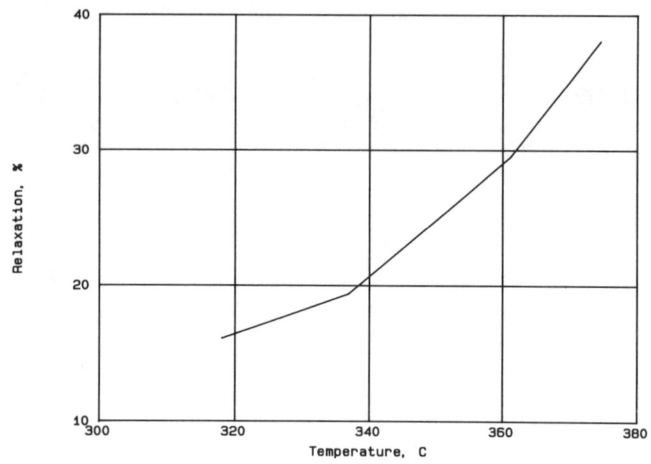

AISI 1095: Relaxation of circular flat springs (piston rings). Spring hardness, 35 HRC. Source: Ref 1, p 299

3. Alloy Steels

Alloy Steels

Steel is considered an alloy steel when the maximum content range of alloying elements exceeds one or more of the following limits: 1.65% Mn, 0.60% Si, or 0.60% Cu. Also included in the recognized field of alloy steels are steels with a specified or required range or minimum quantity of the following elements: aluminum, boron, chromium (up to 3.99%), cobalt, niobium, molybdenum, nickel, titanium, tungsten, vanadium, zirconium, or any other alloying element added to obtain a desired alloying effect.

Alloy sheet and strip are available as hot rolled and cold rolled steel in coils and cut lengths. Other product forms of alloy steel include hot rolled plate and hot rolled, cold rolled, and cold drawn bar, rod, and wire.

Alloy steels are either surface hardenable or through hardenable.

AISI-SAE Designations

The most widely used system for designating alloy steels combines the designations of the American Iron and Steel Institute (AISI) and the Society of Automotive Engineers (SAE). Although two separate systems, they are nearly identical and are carefully coordinated by the two groups.

H-steels incorporate slightly different carbon ranges, generally broader than the standard alloy steels. Also, the chemical composition limits for manganese and chromium in these steels have been widened in all except the triple alloy grades to permit better control of hardenability. The standard alloy steels and H-steels are similar in all other aspects.

Boron-modified alloy steels generally contain 0.0005 to 0.003% boron. There are five families of boron-modified alloy steels beginning with 50BOO chromium steels and concluding with the 94BOO nickel-chromium-molybdenum steels. Boron is used to increase hardenability. Boron-treated low-alloy steels typically cost significantly less than higher alloy grades of equivalent hardenability.

Hardening Characteristics

Alloy steels are either surface hardenable or through hardenable.

Surface-hardenable alloys, usually with a nominal carbon content of up to 0.30%, are specified when a tough core, wear resistance, and case hardness to a controlled depth are necessary. Frequently these grades are selected with a subsequent surface treatment, such as nitriding or carburizing, in mind.

Through hardening grades, usually quenched and tempered steels with a carbon content of 0.35% and above, are used when maximum hardness and strength must extend deep within the part, even from surface to center, to provide resistance to critical stresses.

Elevated-Temperature Service

The mechanical properties of alloy steels are determined primarily by composition and heat treatment. The effects of alloying elements on the elevated temperature service characteristics of alloy steels are as follows.

Carbon increases both strength and hardenability of steel at room temperature. In plain carbon and carbon-molybdenum steels intended for elevated temperature service, carbon content usually is limited to about 0.20%; in some classes of tubing for boilers, however, carbon content may be as high as 0.35%. For chromium-containing steels, carbon content usually is limited to 0.15%. Carbon increases short-term tensile strength, but does not add appreciably to creep resistance at temperatures above 540 C (1000 F) because carbides eventually become spheroidized at such temperatures.

Manganese, in addition to its normal function of preventing hot shortness by forming dispersed manganese sulfide inclusions, appears to enhance the effectiveness of nitrogen in increasing the strength of plain carbon steels at elevated temperatures.

Phosphorus and sulfur are considered undesirable because they reduce the elevated temperature ductility of steel. This reduction in ductility is demonstrated by reductions in stress-rupture life and thermal fatigue life.

Silicon increases the elevated temperature strength of steel; it also increases the resistance to scaling of the low-chromium steels in air at elevated temperatures.

Chromium, in small amounts (about 0.5%), is a carbide former and stabilizer; in larger amounts (up to 9% or more), it increases the resistance of steels to scaling in air. In the lower carbon grades, increasing chromium content does not increase resistance to deformation at elevated temperatures.

Molybdenum, even in small amounts (0.1 to 0.5%), increases the resistance of these steels to deformation at elevated temperatures. It is a carbide stabilizer and prevents graphitization. For certain ranges of stress and temperature, the dissolving of iron carbide and the concurrent precipitation of molybdendum carbide cause strain hardening in these steels.

Vanadium is added to some of the higher -carbon constructural steels to provide additional resistance to tempering.

Niobium sometimes is added to these steels to increase their strength through formation of carbides.

High-Temperature Properties

Although the values given for elevated temperature properties usually are based on results obtained from specimens from more than one heat, for only a very few of these steels has the number of tests and heats been sufficient to permit the determination of upper and lower confidence limits for design.

Short-term elevated temperature mechanical properties of primary interest in the design of structures for elevated temperature service are ultimate tensile strength, yield strength and elastic modulus. Reduction in area during tensile testing and compressive yield strength sometimes are useful, particularly in analysis of hot forming operations.

Long-term elevated temperature mechanical properties of primary interest in the design of structures for elevated temperature service are creep rate, rupture life and relaxation behavior.

Examples of alloy steels that boast above-average elevated temperature properties are AISI 4130, with good tensile, fatigue, and impact properties at temperatures up to 370 C (700 F), and AISI 4140, with a maximum use temperature of 480 C (900 F). AISI 4340 also has good strength retention at elevated temperatures.

AISI-SAE system of alloy steel designations

Type of steel	Numerals and digits	Type of steel	Numerals and digits
Mn	13XX....Mn 1.75	Ni-Cr-Mo	43XX....Ni 1.82; Cr 0.50 and 0.80; Mo 0.25
Ni	23XX....Ni 3.50		43BVXX..Ni 1.82; Cr 0.50; Mo 0.12 and 0.25; V 0.03 min
	25XX....Ni 5.00		47XX....Ni 1.05; Cr 0.45; Mo 0.20 and 0.35
Ni-Cr	31XX....Ni 1.25; Cr 0.65 and 0.80		81XX....Ni 0.30; Cr 0.40; Mo 0.12
	32XX....Ni 1.75; Cr 1.07		86XX....Ni 0.55; Cr 0.50; Mo 0.20
	33XX....Ni 3.50; Cr 1.50 and 1.57		87XX....Ni 0.55; Cr 0.50; Mo 0.25
	34XX....Ni 3.00; Cr 0.77		88XX....Ni 0.55; Cr 0.55; Mo 0.35
Ni-Mo	46XX....Ni 0.85 and 1.82; Mo 0.20 and 0.25		93XX....Ni 3.25; Cr 1.20; Mo 0.12
	48XX....Ni 3.50; Mo 0.25		94XX...Ni 0.45; Cr 0.40; Mo 0.12
Mo	40XX....Mo 0.20 and 0.25		97XX....Ni 0.55; Cr 0.20; Mo 0.20
	44XX....Mo 0.40 and 0.52		98XX....Ni 1.00; Cr 0.80; Mo 0.25
Cr	50XX....Cr 0.27, 0.40, 0.50 and 0.65	Cr-V	61XX....Cr 0.60, 0.80, and 0.95; V 0.10 and 0.15 min
	51XX....Cr 0.80, 0.87, 0.92, 0.95, 1.00, and 1.05	Cr-Mo-Al	71XX....Cr 1.60; Mo 0.35; Al 1.13
	50XXX...Cr 0.50; C 1.00 min	W-Cr	72XX....W 1.75; Cr 0.75
	51XXX...Cr 1.02; C 1.00 min	Si-Mn	92XX....Si 1.40 and 2.00; Mn 0.65, 0.82 and Cr 0.00 amd 0.65
	52XXX...Cr 1.45; C 1.00 min		
Cr-Mo	41XX....Cr 0.50, 0.80, and 0.95; Mo 0.12, 0.20, 0.25, and 0.30		

Note: XX in the last two digits of the designation indicates that the carbon content (in hundredths of a percent) is to be inserted; in the chromium steels with five digits in the designation, the last three digits indicate carbon content
Source: Ref 15, p 76

Compositions of AISI-SAE standard alloy steels

AISI-SAE No.	UNS No.	C %	Mn %	P % max	S % max	Si %	Ni %	Cr %	Mo %
1330	G13300	0.28-0.33	1.60-1.90	0.035	0.04	0.15-0.35	...	...	...
1335	G13350	0.33-0.38	1.60-1.90	0.035	0.04	0.15-0.35	...	...	...
1340	G13400	0.38-0.43	1.60-1.90	0.035	0.04	0.15-0.35	...	...	...
1345	G13450	0.43-0.48	1.60-1.90	0.035	0.04	0.15-0.35	...	...	...
4023	G40230	0.20-0.25	0.70-0.90	0.035	0.04	0.15-0.35	...	...	0.20-0.30
4024	G40240	0.20-0.25	0.70-0.90	0.035	0.035-0.050	0.15-0.35	...	...	0.20-0.30
4027	G40270	0.25-0.30	0.70-0.90	0.035	0.040	0.15-0.35	...	...	0.20-0.30
4028	G40280	0.25-0.30	0.70-0.90	0.035	0.035-0.050	0.15-0.35	...	...	0.20-0.30
4037	G40370	0.35-0.40	0.70-0.90	0.035	0.04	0.15-0.35	...	...	0.20-0.30
4047	G40470	0.45-0.50	0.70-0.90	0.035	0.04	0.15-0.35	...	...	0.20-0.30
4118	G41180	0.18-0.23	0.70-0.90	0.035	0.04	0.15-0.35	...	0.40-0.60	0.08-0.15
4130	G41300	0.28-0.33	0.40-0.60	0.035	0.04	0.15-0.35	...	0.80-1.10	0.15-0.25
4137	G41370	0.35-0.40	0.70-0.90	0.035	0.04	0.15-0.35	...	0.80-1.10	0.15-0.25
4140	G41400	0.38-0.43	0.75-1.00	0.035	0.04	0.15-0.35	...	0.80-1.10	0.15-0.25
4142	G41420	0.40-0.45	0.75-1.00	0.035	0.04	0.15-0.35	...	0.80-1.10	0.15-0.25
4145	G41450	0.43-0.48	0.75-1.00	0.035	0.04	0.15-0.35	...	0.80-1.10	0.15-0.25
4147	G41470	0.45-0.50	0.75-1.00	0.035	0.04	0.15-0.35	...	0.80-1.10	0.15-0.25
4150	G41500	0.48-0.53	0.75-1.00	0.035	0.04	0.15-0.35	...	0.80-1.10	0.15-0.25
4161	G41610	0.56-0.64	0.75-1.00	0.035	0.04	0.15-0.35	...	0.70-0.90	0.25-0.35
4320	G43200	0.17-0.22	0.45-0.65	0.035	0.04	0.15-0.35	1.65-2.00	0.40-0.60	0.20-0.30
4340	G43400	0.38-0.43	0.60-0.80	0.035	0.04	0.15-0.35	1.65-2.00	0.70-0.90	0.20-0.30
E4340(a)	G43406	0.38-0.43	0.65-0.85	0.025	0.025	0.15-0.35	1.65-2.00	0.70-0.90	0.20-0.30
4615	G46150	0.13-0.18	0.45-0.65	0.035	0.04	0.15-0.35	1.65-2.00	...	0.20-0.30
4620	G46200	0.17-0.22	0.45-0.65	0.035	0.04	0.15-0.35	1.65-2.00	...	0.20-0.30
4626	G46260	0.24-0.29	0.45-0.65	0.035	0.04	0.15-0.35	0.70-1.00	...	0.15-0.25
4720	G47200	0.17-0.22	0.50-0.70	0.035	0.04	0.15-0.35	0.90-1.20	0.35-0.55	0.15-0.25
4815	G48150	0.13-0.18	0.40-0.60	0.035	0.04	0.15-0.35	3.25-3.75	...	0.20-0.30
4817	G48170	0.15-0.20	0.40-0.60	0.035	0.04	0.15-0.35	3.25-3.75	...	0.20-0.30
4820	G48200	0.18-0.23	0.50-0.70	0.035	0.04	0.15-0.35	3.25-3.75	...	0.20-0.30
5117	G51170	0.15-0.20	0.70-0.90	0.035	0.04	0.15-0.35	...	0.70-0.90	...
5120	G51200	0.17-0.22	0.70-0.90	0.035	0.04	0.15-0.35	...	0.70-0.90	...
5130	G51300	0.28-0.33	0.70-0.90	0.035	0.04	0.15-0.35	...	0.80-1.10	...
5132	G51320	0.30-0.35	0.60-0.80	0.035	0.04	0.15-0.35	...	0.75-1.00	...
5135	G51350	0.33-0.38	0.60-0.80	0.035	0.04	0.15-0.35	...	0.80-1.05	...
5140	G51400	0.38-0.43	0.70-0.90	0.035	0.04	0.15-0.35	...	0.70-0.90	...
5150	G51500	0.40-0.53	0.70-0.90	0.035	0.04	0.15-0.35	...	0.70-0.90	...
5155	G51550	0.51-0.59	0.70-0.90	0.035	0.04	0.15-0.35	...	0.70-0.90	...
5160	G51600	0.56-0.64	0.75-1.00	0.035	0.04	0.15-0.35	...	0.70-0.90	...
E51100(a)	G51986	0.98-1.10	0.25-0.45	0.025	0.025	0.15-0.35	...	0.90-1.15	...
E52100(a)	G52986	0.98-1.10	0.25-0.45	0.025	0.025	0.15-0.35	...	1.30-1.60	...
6118(b)	G61180	0.16-0.21	0.50-0.70	0.035	0.04	0.15-0.35	...	0.50-0.70	...
6150(c)	G61500	0.48-0.53	0.70-0.90	0.035	0.04	0.15-0.35	...	0.80-1.10	...
8615	G86150	0.13-0.18	0.70-0.90	0.035	0.04	0.15-0.35	0.40-0.70	0.40-0.60	0.15-0.25
8617	G86170	0.15-0.20	0.70-0.90	0.035	0.04	0.15-0.35	0.40-0.70	0.40-0.60	0.15-0.25
8620	G86200	0.18-0.23	0.70-0.90	0.035	0.04	0.15-0.35	0.40-0.70	0.40-0.60	0.15-0.25
8622	G86220	0.20-0.25	0.70-0.90	0.035	0.04	0.15-0.35	0.40-0.70	0.40-0.60	0.15-0.25
8625	G86250	0.23-0.28	0.70-0.90	0.035	0.04	0.15-0.35	0.40-0.70	0.40-0.60	0.15-0.25
8627	G86270	0.25-0.30	0.70-0.90	0.035	0.04	0.15-0.35	0.40-0.70	0.40-0.60	0.15-0.25
8630	G86300	0.28-0.33	0.70-0.90	0.035	0.04	0.15-0.35	0.40-0.70	0.40-0.60	0.15-0.25
8637	G86370	0.35-0.40	0.75-1.00	0.035	0.04	0.15-0.35	0.40-0.70	0.40-0.60	0.15-0.25
8640	G86400	0.38-0.43	0.75-1.00	0.035	0.04	0.15-0.35	0.40-0.70	0.40-0.60	0.15-0.25
8642	G86420	0.40-0.45	0.75-1.00	0.035	0.04	0.15-0.35	0.40-0.70	0.40-0.60	0.15-0.25
8645	G86450	0.43-0.48	0.75-1.00	0.035	0.04	0.15-0.35	0.40-0.70	0.40-0.60	0.15-0.25
8655	G86550	0.51-0.59	0.75-1.00	0.035	0.04	0.15-0.35	0.40-0.70	0.40-0.60	0.15-0.25
8720	G87200	0.18-0.23	0.70-0.90	0.035	0.04	0.15-0.35	0.40-0.70	0.40-0.60	0.20-0.30
8740	G87400	0.38-0.43	0.75-1.00	0.035	0.04	0.15-0.35	0.40-0.70	0.40-0.60	0.20-0.30
8822	G88220	0.20-0.25	0.75-1.00	0.035	0.04	0.15-0.35	0.40-0.70	0.40-0.60	0.30-0.40
9260	G92600	0.56-0.64	0.75-1.00	0.035	0.04	1.80-2.20	...	...	...
Standard boron grades									
50B44	G50441	0.43-0.48	0.75-1.00	0.035	0.04	0.15-0.35	...	0.40-0.60	...
50B46	G50461	0.44-0.49	0.75-1.00	0.035	0.04	0.15-0.35	...	0.20-0.35	...
50B50	G50501	0.48-0.53	0.75-1.00	0.035	0.04	0.15-0.35	...	0.40-0.60	...
50B60	G50601	0.56-0.64	0.75-1.00	0.035	0.04	0.15-0.35	...	0.40-0.60	...
51B60	G51601	0.56-0.64	0.75-1.00	0.035	0.04	0.15-0.35	...	0.70-0.90	...
81B45	G81451	0.43-0.48	0.75-1.00	0.035	0.04	0.15-0.35	0.20-0.40	0.35-0.55	0.08-0.15
94B17	G94171	0.15-0.20	0.75-1.00	0.035	0.04	0.15-0.35	0.30-0.60	0.30-0.50	0.08-0.15
94B30	G94301	0.28-0.33	0.75-1.00	0.035	0.04	0.15-0.35	0.30-0.60	0.30-0.50	0.08-0.15

Note: Small quantities of certain elements not specified or required are present; maximum contents of incidental elements are: 0.35% copper; 0.25% nickel, 0.20% chromium, and 0.06% molybdenum; standard alloy steels may also be produced with 0.15 to 0.35% lead and are identified with an L inserted between the second and third numerals of the AISI-SAE number (for example, 41L40) (a) Electric furnace steel. (b) Includes 0.10 to 0.15% vanadium. (c) Includes 0.15% minimum vanadium. (d) 0.0005 to 0.003% boron
Source Ref 15, p 75

Steels recommended for bolts to be used at elevated temperatures

Bolt diameter, in.	Proof stress of 75 ksi	Proof stress of 100 ksi	Proof stress of 125 ksi
0.25-0.75	1038	4037	4037
0.75-1.25	1038	4140	4140
1.25-2.00	4140	4140	4145

Note: Temperatures used are between 200 and 370 C (400 and 700 F). All selections are based on a minimum tempering temperature of 455 C (850 F). Proof stresses are for room temperature values
Source: Ref 1, p 113

AISI 4037

UNS number: G40370
Specifications: ASTM A322, ASTM A331, ASTM A519, ASTM A547, SAE J404, SAE J412, SAE J770
Nominal composition: 0.350 to 0.400% C, 0.700 to 0.900% Mn, 0.035% P max, 0.040% S max, 0.150 to 0.300% Si, 0.200 to 0.300% Mo
Product forms: Wrought, bar, rod, wire, plate, sheet, strip, tube, pipe, billets, forgings, blooms, cast, continuous, ingot
Description: AISI 4037 contains 0.20 to 0.30% Mo. It is available as hot rolled and cold finished bar and cold heading quality wire.
Typical applications: Wire is drawn from rod rolled of AISI 4037 steel for use in manufacturing high-strength screws, bolts, cap screws, socket-head screws, and self-tapping screws. These parts can be heat treated to meet the mechanical-property requirements of SAE grade 8 alloy steel fasteners. Grade 4037 is also used for machined parts. It is recommended for 1/4 to 3/4 in. bolts with a room temperature proof stress of 100 to 125 ksi to be used at temperatures between 205 and 370 C (400 and 700 F).

Physical Properties

AISI 4037: Thermal conductivity

TEMP C	TEMP F	THCON W/m*K	THCON Btu/ft*h*F
100	212	48.1	27.8
200	390	45.3	26.2
400	750	39.4	22.8
600	1110	33.9	19.6

Source: Ref 15, p 101

AISI 4037H

UNS number: H40370
Specifications: ASTM A304, SAE J1268
Nominal composition: 0.340 to 0.410% C, 0.600 to 1.000% Mn, 0.035% P max, 0.040% S max, 0.150 to 0.300% Si, 0.200 to 0.300% Mo
Product forms: Wrought, bar, rod, wire, plate, sheet, strip, tube, pipe, billets, forgings, blooms, cast, continuous, ingot
Description: AISI 4037H steel contains 0.20 to 0.30% Mo. It is available as hot rolled and cold finished bar and cold-heading-quality wire.
Typical applications: Wire is drawn from rod rolled of AISI 4037H steel for use in manufacturing high-strength screws, bolts, cap screws, socket-head screws, and self-tapping screws. These parts can be heat treated to meet the mechanical-property requirements of SAE grade 8 alloy steel fasteners. Grades 4037H is also used for machined parts.

Physical Properties

AISI 4037H: Thermal conductivity

TEMP C	TEMP F	THCON W/m*K	THCON Btu/ft*h*F
100	212	48.1	27.8
200	390	45.3	26.2
400	750	39.4	22.8
600	1110	33.9	19.6

Source: Ref 15, p 101

AISI 4130

UNS number: G41300
Specifications: AMS 6350, AMS 6356, AMS 6360, ASTM A322, ASTM A331, ASTM A505, ASTM A513, MIL S-16974, MIL S-6758, MIL S-18729, SAE J404, SAE J412, SAE J770, DIN 1.7218, NF 25 CD 4(S), UNI 25 Cr Mo 4, UNI 25 Cr Mo 4, KB JIS SCM 2, SS 14 22 25, BS CDS 110
Nominal composition: 0.280 to 0.330% C, 0.800 to 1.100% Cr, 0.400 to 0.600% Mn, 0.150 to 0.250% Mo, 0.035% P max, 0.150 to 0.350% Si, 0.040% S max
Product forms: Wrought, bar, rod, wire, plate, sheet, strip, tube, pipe, billets, forgings, blooms, cast, continuous, welded, ingot
Description: AISI 4130 is a medium-carbon, chromium-molybdenum steel. Variations in heat treatment can produce a broad range of strength and toughness. These steels have good hardenability, strength, wear resistance, toughness, and ductility. The water-hardening alloy steel retains good tensile, fatigue, and impact properties up to 370 C (700 F).

Typical applications: Available as hot rolled and cold finished bar and seamless tube, rod, forgings, sheets and plate. These steels are intended for general-purpose applications. In the heat treated condition, they offer good strength and toughness for moderately stressed parts. 4130 is available in forging quality, aircraft-quality, and premium aircraft-quality steels.

Physical Properties

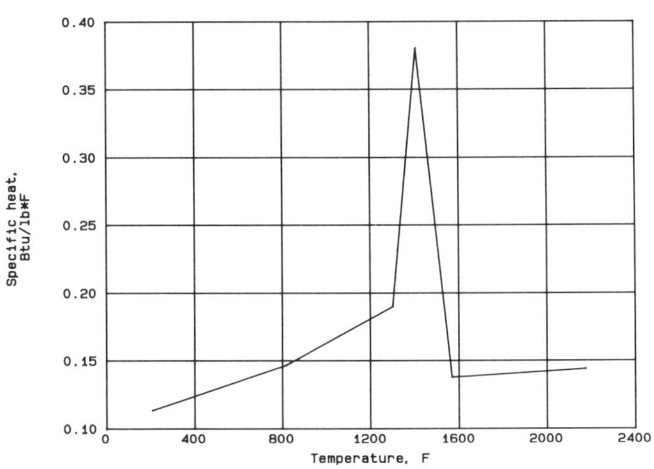

AISI 4130: Specific heat. Source: Ref 22, p 5

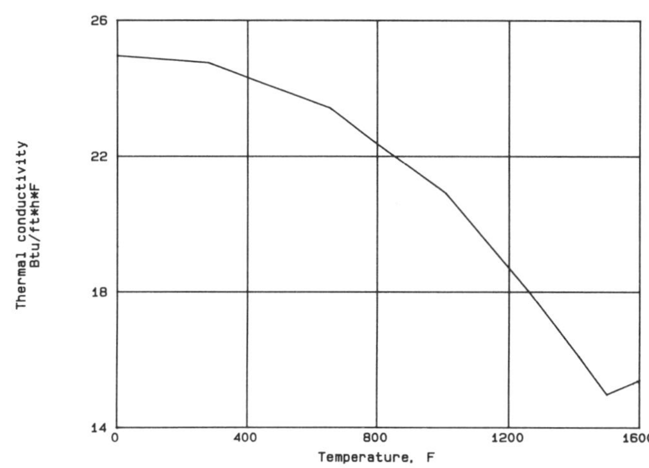

AISI 4130: Thermal conductivity. Source: Ref 22, p 5

AISI 4130: Mean apparent specific heat

TEMP C	TEMP F	SHTCP J/kg*K	SHTCP Btu/lb*F
50-100	120-212	477	0.114
150-200	300-390	523	0.125
250-300	480-570	544	0.130
350-400	660-750	607	0.145
450-500	840-930	657	0.157
550-600	1020-1110	741	0.177
650-700	1200-1290	829	0.198
750-800	1380-1470	837	0.200

Source: Ref 15, p 108

AISI 4130: Thermal conductivity

TEMP C	TEMP F	THCON W/m*K	THCON Btu/ft*h*F
100	212	42.7	24.7
300	570	40.7	23.5
500	930	37.2	21.5
700	1290	31.0	17.9
1000	1830	28.0	16.2
1200	2190	30.1	17.4

Source: Ref 15, p 108

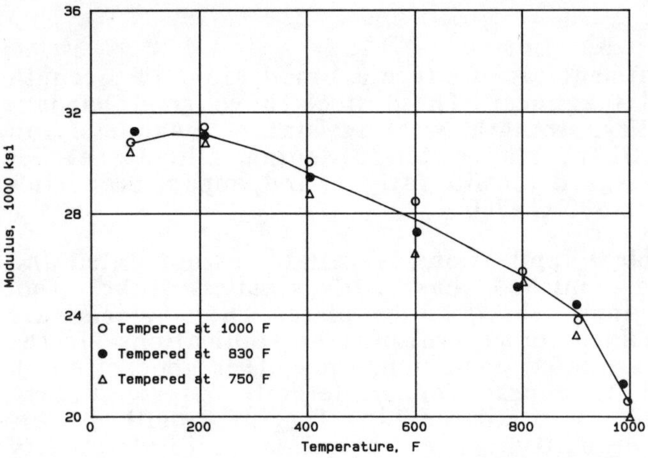

AISI 4130: Modulus of elasticity of 0.064-in. sheet austenitized at 1575 F, oil quenched, and tempered at various temperatures. Source: Ref 22, p 18

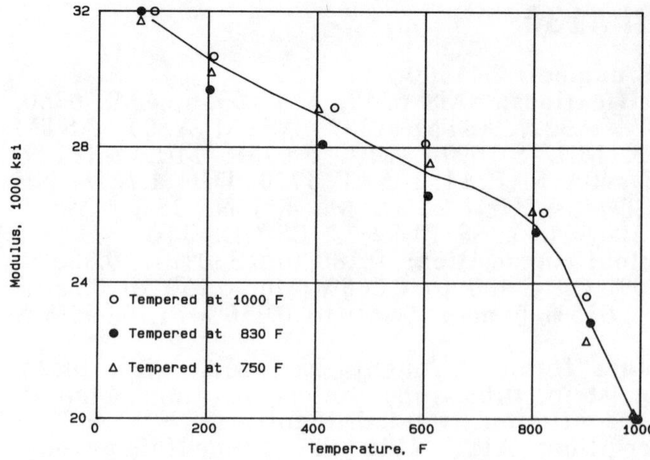

AISI 4130: Modulus of elasticity in compression of 0.064-in. sheet austenitized at 1575 F, oil quenched, and tempered to various temperatures. Source: Ref 22, p 18

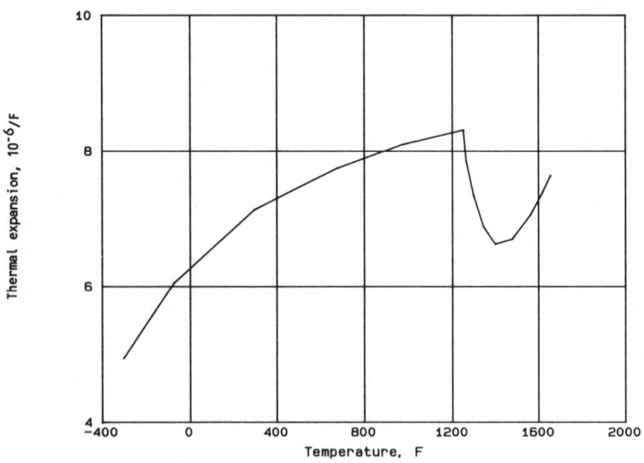

AISI 4130: Mean coefficient of linear thermal expansion from room temperature to temperature indicated. Source Ref 21, p 3

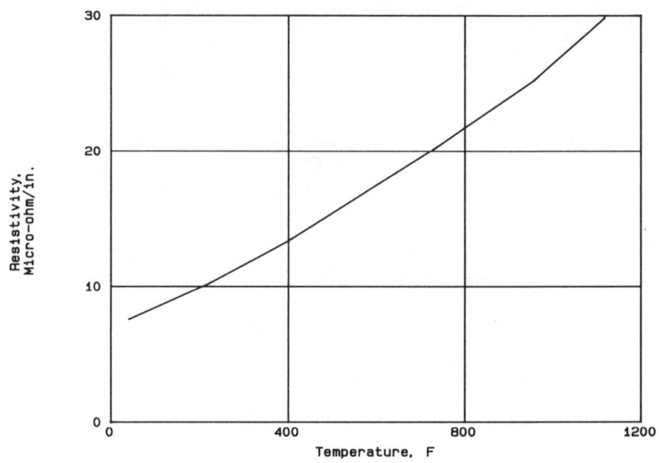

AISI 4130: Electrical resistivity. Source: Ref 21, p 3

AISI 4130: Electrical resistivity

TEMP C	TEMP F	ELRES n{o}*m
21	70	223
100	212	271
200	390	342
400	750	529
600	1110	786
800	1470	1103
1000	1830	1171
1200	2190	1222

Source: Ref 15, p 108

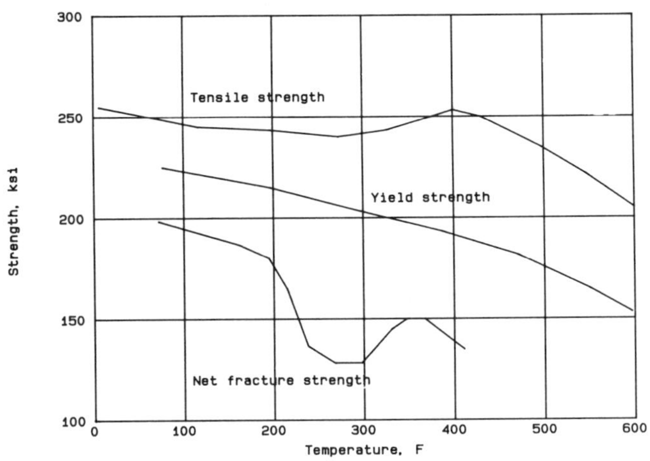

AISI 4130: Tensile properties and fracture strength of sheet. Strain rate, 5 X 10E-3 in./in.*s. Specimens were austenitized in argon at 1570 F for 0.5 h, oil quenched, tempered at 400 F in air for 2 h, and air cooled. Source: Ref 21, p 15

Mechanical Properties

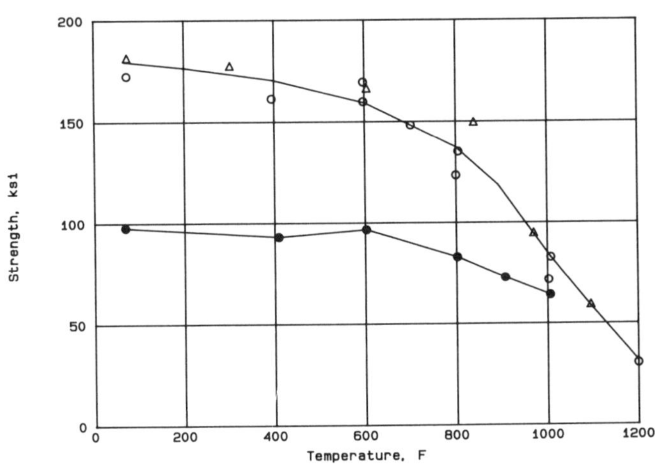

AISI 4130: Tensile strength of quenched and tempered specimens. Open circle: 0.109-in. sheet. Solid circle: 0.125-in. sheet. Open triangle: bar, normalized at 1600 F, oil quenched, and tempered at 900 F. Source: Ref 21, p 12

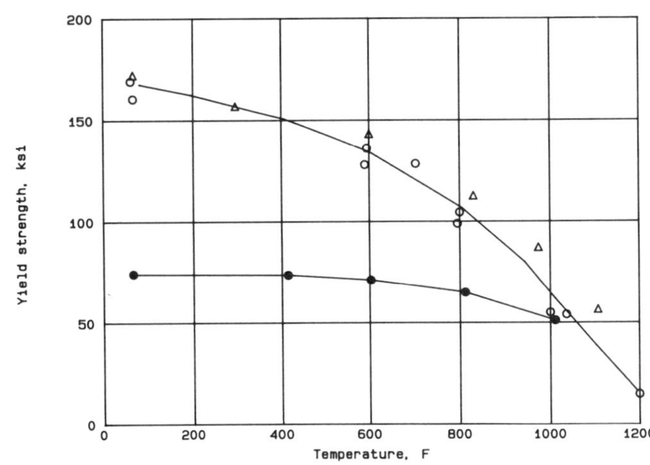

AISI 4130: Yield strength of quenched and tempered specimens. Open circle: 0.109-in. sheet. Solid circle: 0.125-in. sheet. Open triangle: bar, normalized at 1600 F, oil quenched, and tempered at 900 F. Source: Ref 21, p 12

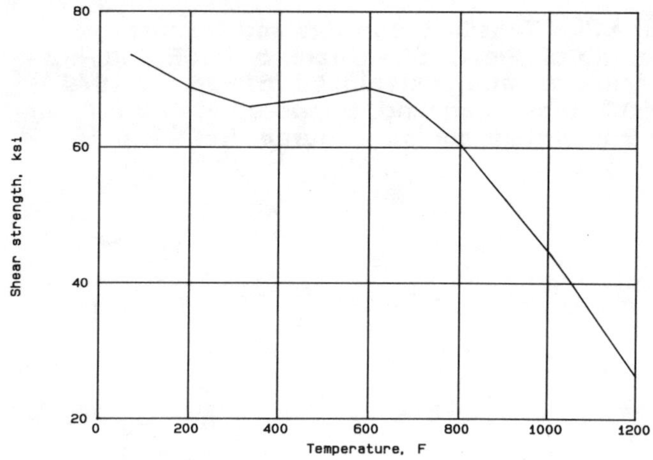

AISI 4130: Shear strength of 0.125-in. diam bar heat treated to a tensile strength of 125 ksi. Source: Ref 21, p 48

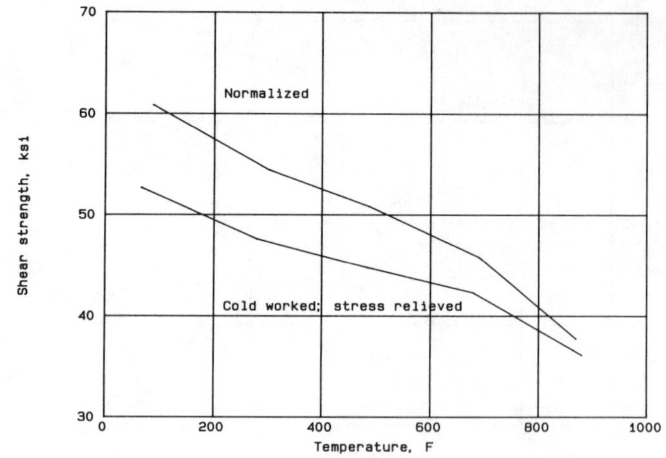

AISI 4130: Shear strength of normalized and cold worked, stress relieved specimens. Source: Ref 21, p 48

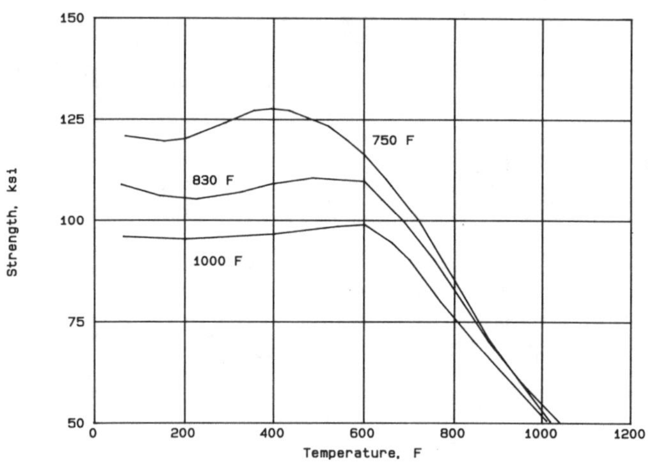

AISI 4130: Effect of tempering temperature on shear strength of 0.1875-in. sheet. Specimens were heated to 1575 F, oil quenched, and tempered at the indicated temperatures. Source: Ref 22, p 14

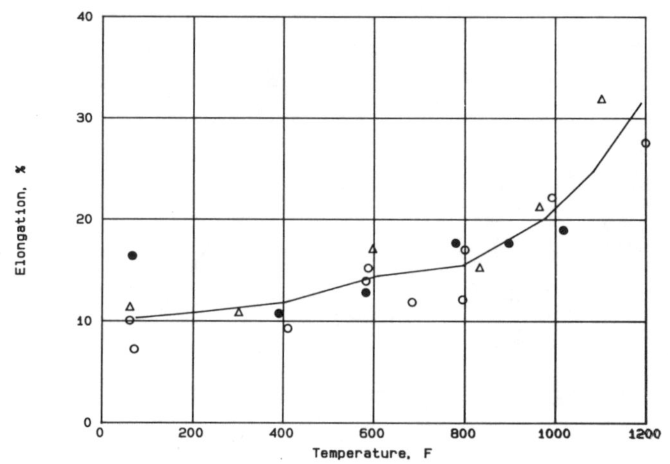

AISI 4130: Tensile elongation of quenched and tempered specimens. Open circle: 0.109-in. sheet. Solid circle: 0.125-in. sheet. Open triangle: bar, normalized at 1600 F, oil quenched, and tempered at 900 F. Source: Ref 21, p 12

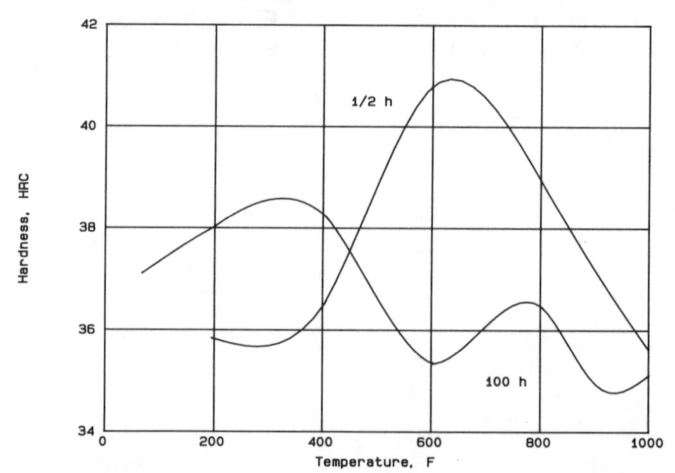

AISI 4130: Rockwell C hardness of 0.1875-in. sheet normalized at 1575 F. Specimens were oil quenched and tempered at 1000 F. Exposure time at test temperature was 0.5 and 100 h. Source Ref 22, p 4

AISI 4130H

UNS number: H41300
Specifications: ASTM A304, SAE J1268, DIN 1.7218, AFNOR 25 CD 4 (S), UNI 25 CrMo 4, UNI 25 CrMo 4 KB, JIS SCM 2, JIS SCCrM 1, SS 14 22 25
Nominal composition: 0.270 to 0.330% C, 0.300 to 0.700% Mn, 0.035% P max, 0.040% S max, 0.150 to 0.300% Si, 0.750 to 1.200% Cr, 0.150 to 0.250% Mo
Product forms: Wrought, bar, rod, wire, plate, sheet, strip, tube, pipe, billets, forgings, blooms, cast, continuous, welded, ingot
Description: AISI 4130H is a medium-carbon, chromium-molybdenum steel. Variations in heat treatment can produce a broad range of strength and toughness. These steels have good hardenability, strength, wear resistance, toughness, and ductility.
Typical applications: Available as hot rolled and cold finished bar and seamless tube, rod, forgings, sheet, and plate. These steels are intended for general-purpose applications. In the heat treated condition, they offer good strength and toughness for moderately stressed parts. 4130H is available in forging-quality, aircraft-quality, and premium aircraft-quality steels.

Physical Properties

AISI 4130H: Mean apparent specific heat

TEMP C	TEMP F	SHTCP J/kg*K	SHTCP Btu/lb*F
50-100	120-212	477	0.114
150-200	300-390	523	0.125
250-300	480-570	544	0.130
350-400	660-750	607	0.145
450-500	840-930	657	0.157
550-600	1020-1110	741	0.177
650-700	1200-1290	829	0.198
750-800	1380-1470	837	0.200

Source: Ref 15, p 108

AISI 4130H: Electrical resistivity

TEMP C	TEMP F	ELRES n{o}*m
21	70	223
100	212	271
200	390	342
400	750	529
600	1110	786
800	1470	1103
1000	1830	1171
1200	2190	1222

Source: Ref 15, p 108

AISI 4130H: Thermal conductivity

TEMP C	TEMP F	THCON W/m*K	THCON Btu/ft*h*F
100	212	42.7	24.7
300	570	40.7	23.5
500	930	37.2	21.5
700	1290	31.0	17.9
1000	1830	28.0	16.2
1200	2190	30.1	17.4

Source: Ref 15, p 108

AISI 4140

UNS number: G41400
Specifications: AMS 6381, AMS 6382, AMS 6390, AMS 6395, ASTM A322, ASTM A331, ASTM A505, ASTM A519, ASTM A547, ASTM A646, MIL S-16974, SAE J404, SAE J412, SAE J770, DIN 1.7225, NF 40 CD 4, NF 42 CD 4, UNI 40 CrMo 4, G 40 CrMo 4
Nominal composition: 0.380 to 0.430% C, 0.800 to 1.100% Cr, 0.750 to 1.000% Mn, 0.150 to 0.250% Mo, 0.035% P max, 0.040% S max, 0.150 to 0.300% Si
Product forms: Wrought, bar, rod, wire, plate, sheet, strip, tube, pipe, billets, forgings, blooms, cast, continuous, ingot
Description: AISI 4140 is a chromium-molybdenum, medium-carbon steel with high hardenability and good fatigue, abrasion, and impact resistance. These steels can be successfully nitrided for maximum wear and abrasion resistance. They are deep-hardening alloys that are suitable for severe service characterized by fatigue, abrasion, impact, high-temperature stresses, or combinations of such stresses in small and large sections. When fully hardened, 4140 demonstrates the outstanding property of relatively high impact strength at high-hardness, tensile strength. AISI 4140 is available as hot and cold rolled sheet and strip; hot rolled and cold finished bar; cold-heading-quality, aircraft-quality, and premium aircraft-quality bar; and seamless tube.

Typical applications: AISI 4140 is used for small gears, pinions, and ball studs, and for high-strength bolts, cap screws, and socket- and recessed-head screws. These products can be heat treated after machining or forming. AISI 4140 can also be used at temperatures as high as 480 C (900 F), above which its strength decreases rapidly with increasing temperature.

Physical Properties

AISI 4140: Mean apparent specific heat

TEMP C	TEMP F	SHTCP J/kg*K	SHTCP Btu/lb*F
150-200	300-390	473	0.113
350-400	660-750	519	0.124
550-600	1020-1110	561	0.134

Source: Ref 15, p 113

AISI 4140: Mean coefficients of linear thermal expansion

TEMP C	TEMP F	THEXP 10^{-6}/C	THEXP 10^{-6}/F
21-100	70-212	12.2	6.8
21-200	70-390	12.6	7.0
21-400	70-750	13.7	7.6
21-600	70-1110	14.6	8.1

Source: Ref 15, p 113

AISI 4140: Thermal conductivity

TEMP C	TEMP F	THCON W/m*K	THCON Btu/ft*h*F
100	212	42.6	24.6
200	390	42.2	24.4
400	750	37.7	21.8
600	1110	33.0	19.1

Source: Ref 15, p 113

AISI 4140: Electrical resistivity

TEMP C	TEMP F	ELRES n{o}*m
100	212	263
200	390	326
400	750	475
600	1110	646

Source: Ref 15, p 113

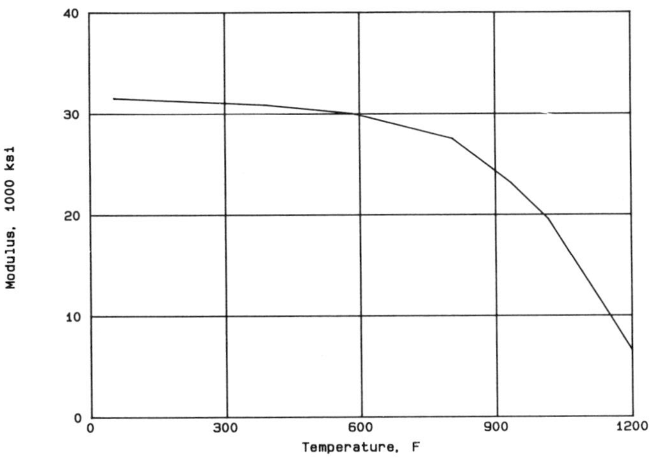

AISI 4140: Modulus of elasticity. Source: Ref 21, p 5

Mechanical Properties

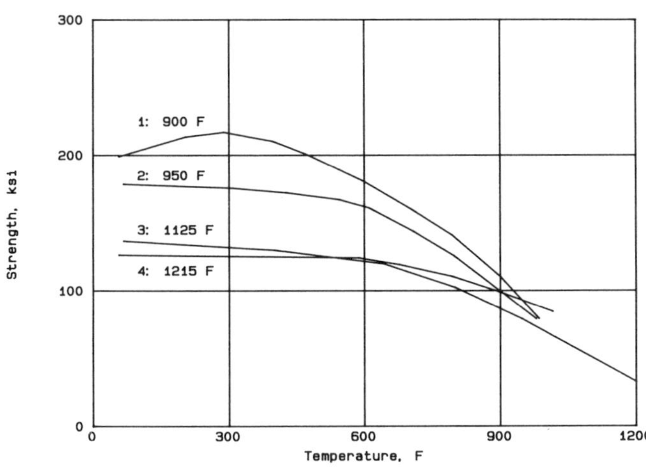

AISI 4140: Tensile strength of specimens tempered at indicated temperature. Curves 1 and 2: bar, heat treated at 1550 F and oil quenched. Curve 3: rolled 0.75-in. bar, heat treated at 1525 F and oil quenched. Curve 4: rolled specimens normalized at 1650 F, oil quenched at 1575 F and tempered for 2 h. Source: Ref 21, p 5

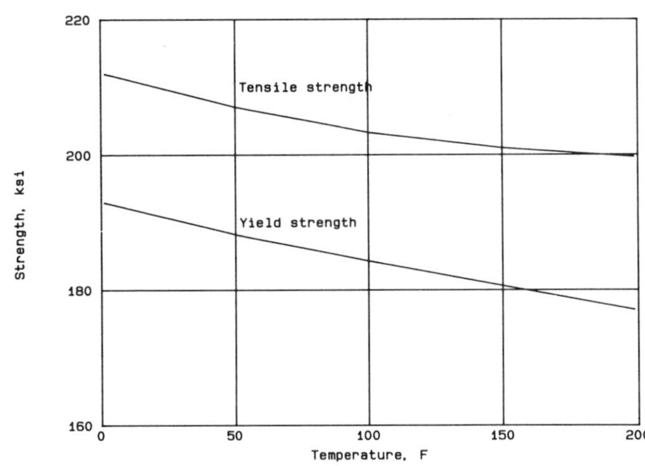

AISI 4140: Tensile and yield strength of bar austenitized at 1550 F, oil quenched, and tempered at 800 F for 2 h. Source: Ref 22, p 14

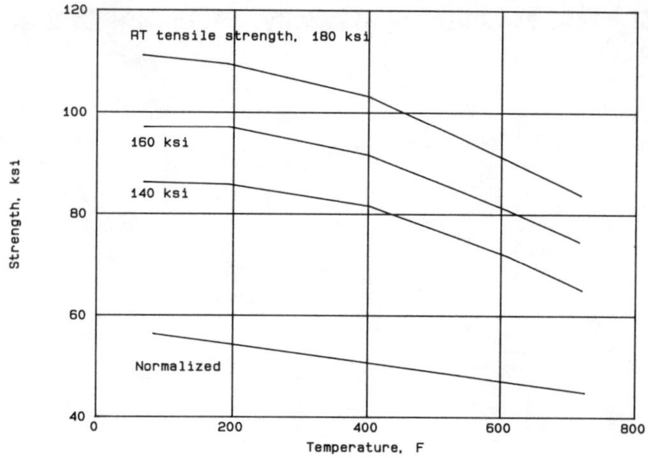

AISI 4140: Shear strength of bar and forgings heat treated to various tensile strength levels. Source: Ref 21, p 11

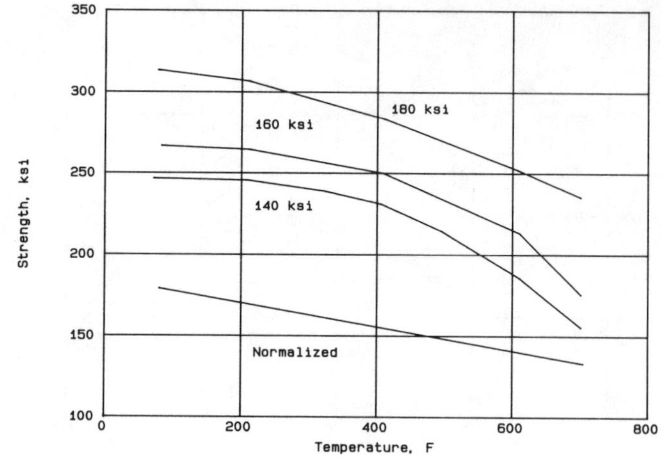

AISI 4140: Ultimate bearing strength of bar and forgings heat treated to various tensile strength levels. e/D = 1.5 to 2.0. Source: Ref 22, p 15

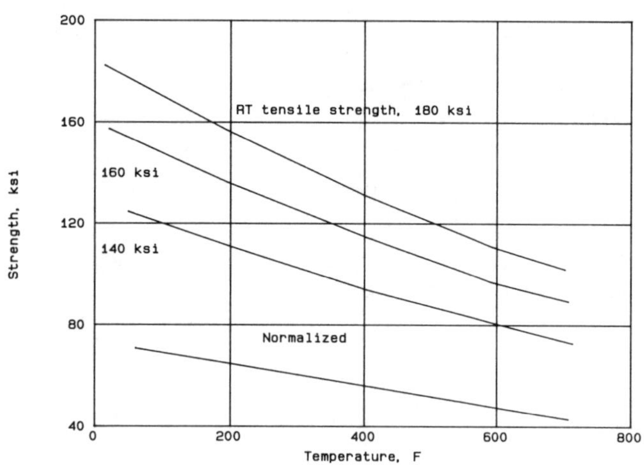

AISI 4140: Compressive yield strength of bar and forgings. Source: Ref 21, p 11

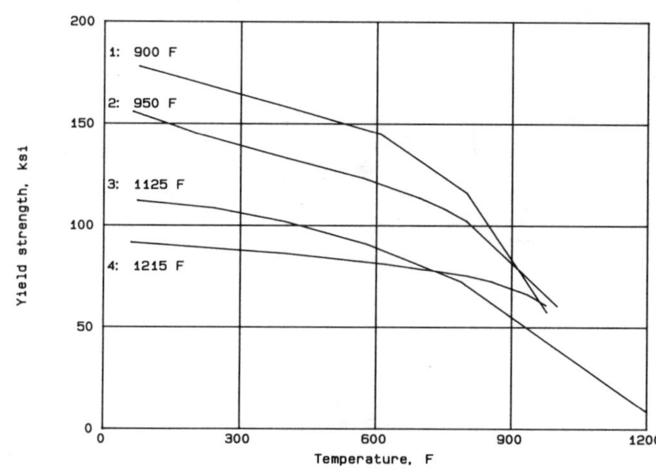

AISI 4140: Yield strength of specimens tempered at indicated temperature. Curves 1 and 2: bar, heat treated at 1550 F and oil quenched. Curve 3: rolled 0.75-in. bar, heat treated at 1525 F and oil quenched. Curve 4: rolled specimens normalized at 1650 F, oil quenched at 1575 F and tempered for 2 h. Source: Ref 21, p 5

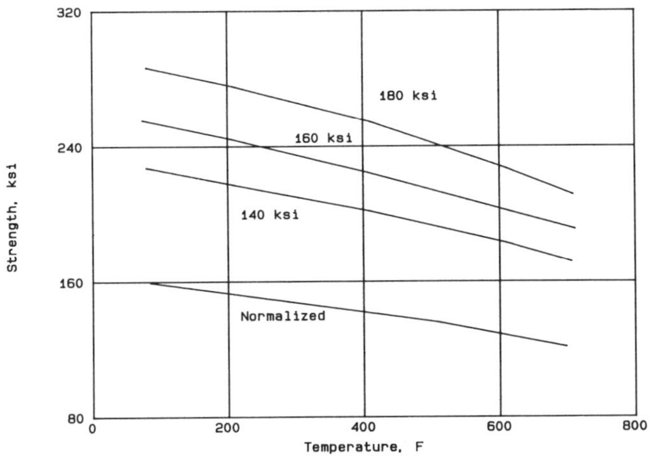

AISI 4140: Bearing yield strength of bar and forgings heat treated to various tensile strength levels. e/D = 1.5 to 2.0. Source: Ref 22, p 15

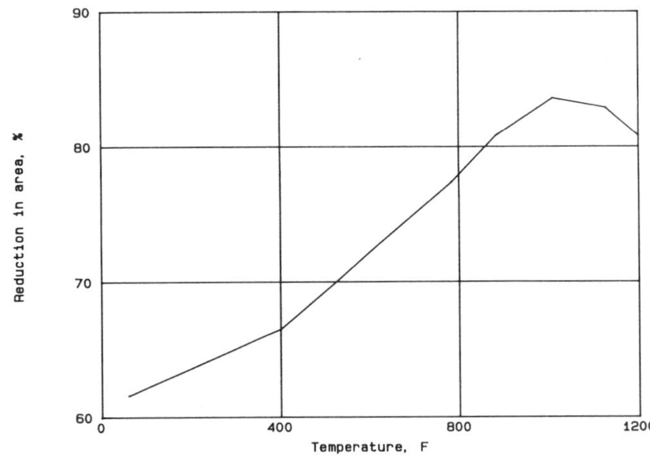

AISI 4140: Tensile reduction in area of rolled 0.75-in. bar, heated to 1575 F, oil quenched, and tempered at 1125 F. Source: Ref 21, p 5

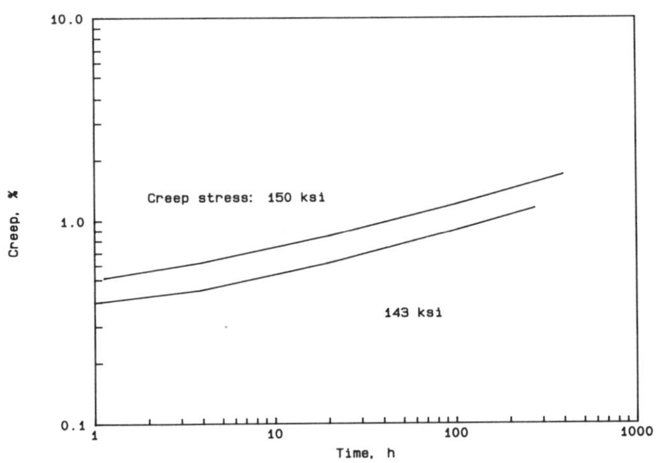

AISI 4140: Creep curves at 600 F for specimens with tensile strengths from 180 to 200 ksi. Source: Ref 21, p 13

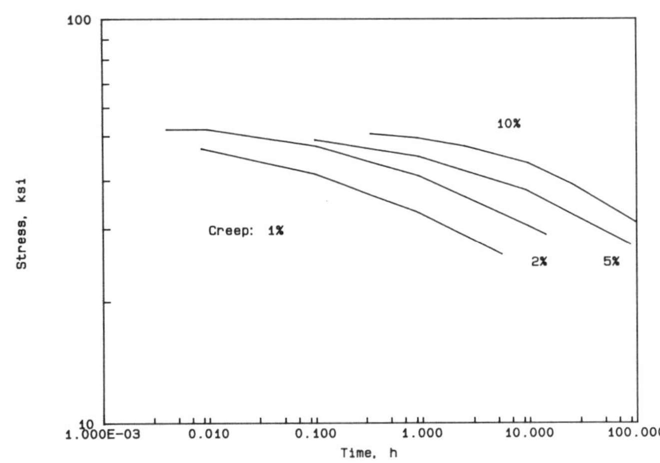

AISI 4140: Creep properties at 1000 F for 0.050-in. sheet, annealed at 1550 F. Source: Ref 21, p 16

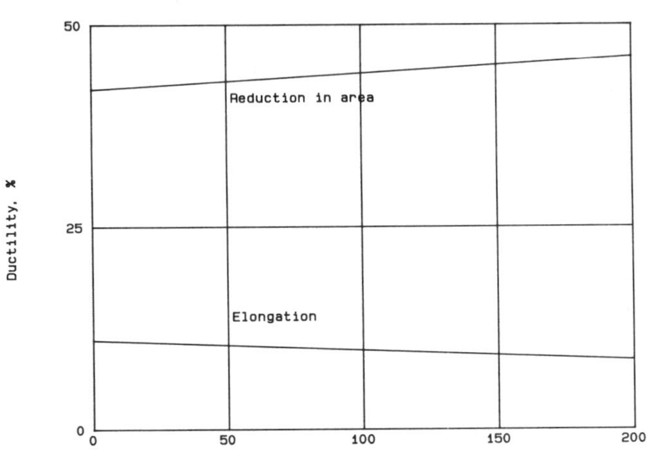

AISI 4140: Tensile elongation and reduction in area of bar austenitized at 1550 F, oil quenched, and tempered at 800 F for 2 h. Source: Ref 22, p 14

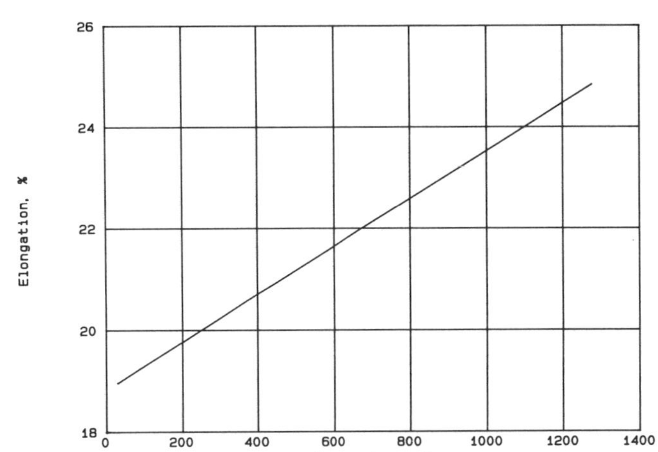

AISI 4140: Tensile elongation of bar. Source: Ref 21, p 5

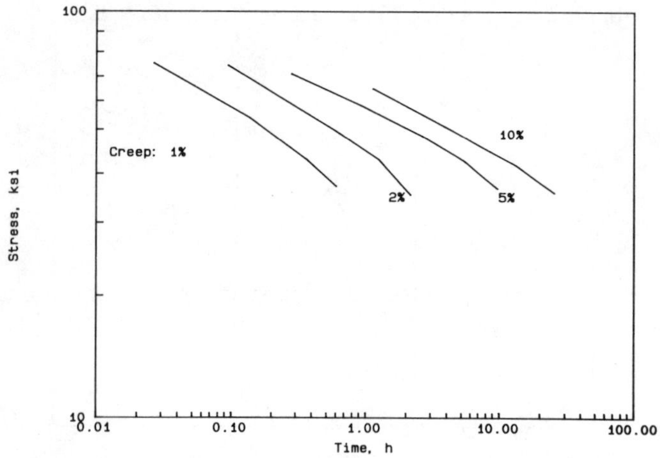

AISI 4140: Creep properties at 1000 F of 0.050-in. sheet, austenitized at 1575 F, oil quenched, and tempered at 1000 F. Source: Ref 22, p 16

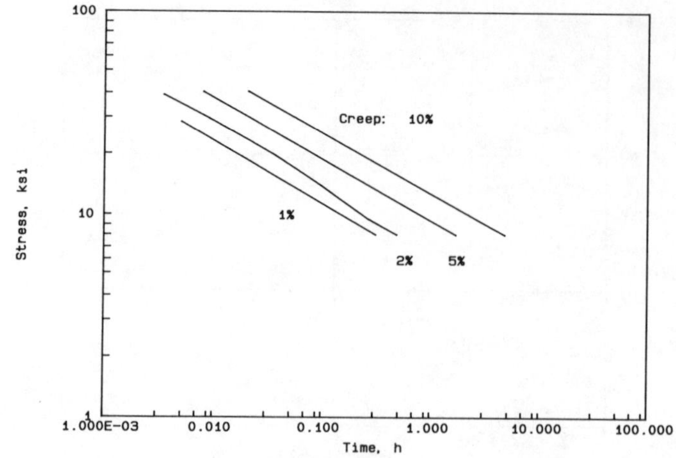

AISI 4140: Creep properties at 1200 F of 0.050-in sheet, austenitized at 1575 F, oil quenched, and tempered at 1200 F. Source: Ref 22, p 16

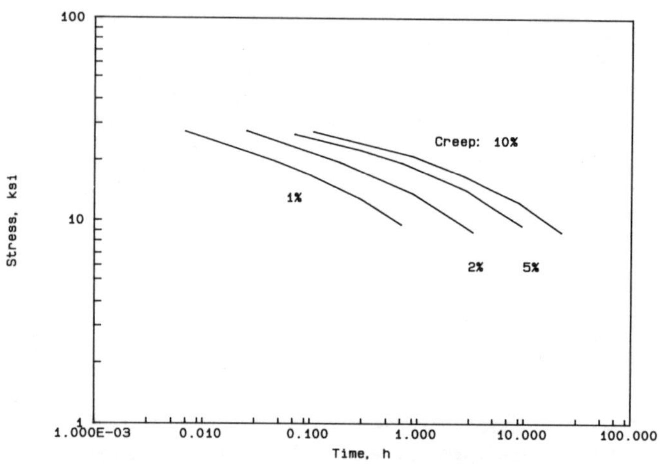

AISI 4140: Creep properties at 1200 F of 0.050-in. sheet, annealed at 1550 F. Source: Ref 21, p 16

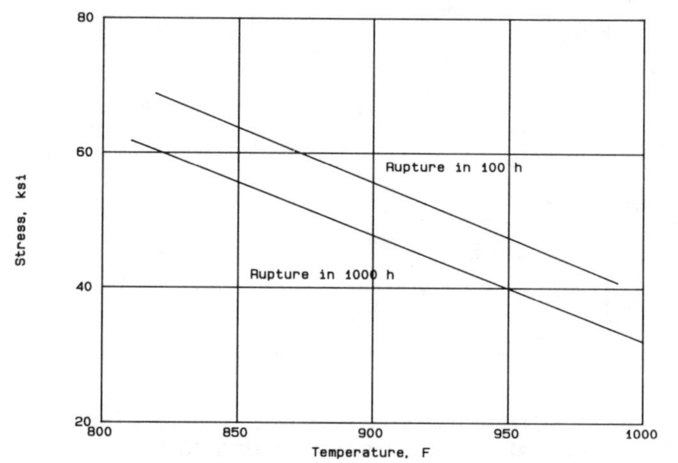

AISI 4140: Stress-rupture of 1-in. diam specimens heated to 1650 F for 1 h, air cooled, heated to 1200 F for 1 h, and furnace cooled. Hardness was 233 HB. Source: Ref 21, p 13

AISI 4140: Effect of elevated-temperature exposure of up to 30 h on room-temperature hardness. Forgings were heated to 1550 F, oil quenched, and tempered at 1100 F for 1.5 h. Source: Ref 22, p 5

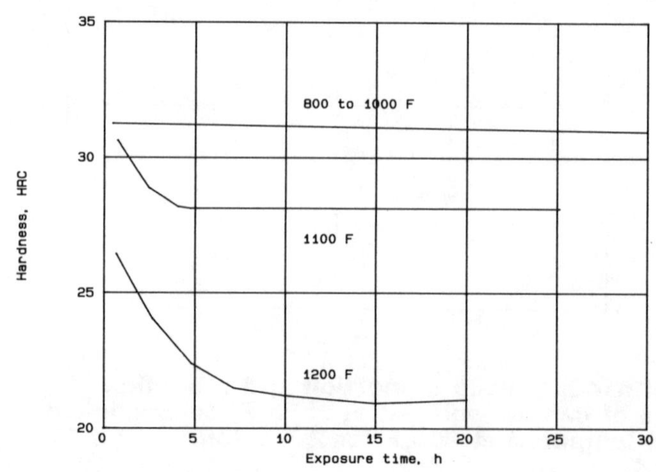

AISI 4140H

UNS number: H41400
Specifications: ASTM A304, SAE J1268, DIN 1.7225, NF 40 CD 4, NF 42 CD 4, UNI G 40 CrMo4, UNI 40 CrMo 4, UNI 38 CrMo 4, JIS SCM 4 H, JIS SCM 4, SS14 2244, BS 708 A 42, BS 708 M 40, BS 709 M 40
Nominal composition: 0.370 to 0.440% C, 0.750 to 1.200% Cr, 0.650 to 1.100% Mn, 0.150 to 0.250% Mo, 0.035% P max, 0.040% S max, 0.150 to 0.300% Si
Product forms: Wrought, bar, rod, wire, plate, sheet, strip, tube, pipe, billets, forgings, blooms, cast, continuous, ingot
Description: AISI 4140H is a chromium-molybdenum, medium-carbon steel with high hardenability and good fatigue, abrasion, and impact resistance. These steels can be successfully nitrided for maximum wear and abrasion resistance. They are deep-hardening alloys suitable for severe service characterized by fatigue, abrasion, impact, high-temperature stresses, or combinations of such stresses in small and large sections. When fully hardened, grades 4140 and 4140H demonstrate the outstanding property of relatively high impact strength at high-hardness. AISI 4140H is available as hot and cold rolled sheet and strip; hot rolled and cold finished bar; cold-heading-quality, aircraft-quality, and premium aircraft-quality bar; and seamless tube.

Typical applications: 4140H is used for small gears, pinions, and ball studs, and for high-strength bolts, cap screws, and socket- and recessed-head screws. These products can be heat treated after machining or forming. AISI 4140H can be used at temperatures as high as 480 C (900 F), above which its strength decreases rapidly with increasing temperature.

Physical Properties

AISI 4140H: Mean apparent specific heat

TEMP C	TEMP F	SHTCP J/kg*K	SHTCP Btu/lb*F
150-200	300-390	473	0.113
350-400	660-750	519	0.124
550-600	1020-1110	561	0.134

Source: Ref 15, p 113

AISI 4140H: Thermal conductivity

TEMP C	TEMP F	THCON W/m*K	THCON Btu/ft*h*F
100	212	42.6	24.6
200	390	42.2	24.4
400	750	37.7	21.8
600	1110	33.0	19.1

Source: Ref 15, p 113

AISI 4140H: Mean coefficients of linear thermal expansion

TEMP C	TEMP F	THEXP 10^{-6}/C	THEXP 10^{-6}/F
21-100	70-212	12.2	6.8
21-200	70-390	12.6	7.0
21-400	70-750	13.7	7.6
21-600	70-1110	14.6	8.1

Source: Ref 15, p 113

AISI 4140H: Electrical resistivity

TEMP C	TEMP F	ELRES n{o}*m
100	212	263
200	390	326
400	750	475
600	1110	646

Source: Ref 15, p 113

AISI 4340

UNS number: G43400
Specifications: AMS 5331, AMS 6359, AMS 6414, AMS 6415, ASTM A322, ASTM A331, ASTM A505, ASTM A519, ASTM A547, ASTM A646, MIL S-16974, SAE J404, SAE J412, SAE J770, DIN 1.6565, JIS S-NCM 8, BS 817 M 40, BS 3111 Type 6, BS 2 S 119, BS 3 S 95
Nominal composition: 0.380 to 0.430% C, 0.700 to 0.900% Cr, 0.600 to 0.800% Mn, 0.200 to 0.300% Mo, 1.650 to 2.000% Ni, 0.035% P max, 0.040% S max, 0.150 to 0.300% Si
Product forms: Wrought, bar, rod, wire, plate, sheet, strip, tube, pipe, billets, forgings, blooms, cast, continuous, ingot
Description: AISI 4340 is a nickel-chromium-molybdenum, constructional alloy steel. It is normally heat treated by quenching in oil and tempering to the desired hardness. This grade exhibits good response to heat treatment, especially in large sections, and possesses a good combination of strength, ductility, and toughness in the quenched and tempered condition. The high depth of hardness is reflected in excellent torque properties. These steels are essentially immune to temperature embrittlement and do not soften readily at elevated temperatures.

Typical applications: Because AISI 4340 is also a carburizing steel, it is used where superior case hardness or core properties are desired. Its choice is determined primarily by the hardenability necessary to obtain the desired core properties under the given conditions of section size and heat treatment. It is suitable for applications in the production of piston pins, bearings, and similar parts for relatively severe service. Rods approximately 13-mm (0.5-in.) in diameter and rod mill rounds up to 28.6-mm (1.125-in.) in diameter are used for heavy sections, or for parts subject to particularly severe service conditions that require a very mild quench to prevent distortion. AISI 4340 steel is found in aircraft and truck parts and in some ordnance materials. Additional uses include gears, fittings, forging dies, machine tool arbors, pressure vessels, and die blocks.

Physical Properties

AISI 4340: Mean coefficients of linear thermal expansion

TEMP C	TEMP F	THEXP(a) 10^{-6}/C	THEXP(a) 10^{-6}/F	THEXP(b) 10^{-6}/C	THEXP(b) 10^{-6}/F
21-95	70-200	11.5	6.4	11.5	6.4
21-150	70-300	11.9	6.6	11.9	6.6
21-205	70-400	12.1	6.7	12.2	6.8
21-260	70-500	12.4	6.9	12.6	7.0
21-315	70-600	12.6	7.0	13.0	7.2
21-370	70-700	13.0	7.2	13.1	7.3
21-425	70-800	13.3	7.4	13.5	7.5
21-480	70-900	13.5	7.5	13.7	7.6
21-540	70-1000	13.7	7.6	13.9	7.7

Note: Specimens were austenitized at 870 C (1600 F), tempered at 545 C (1010 F) for 2 h
(a) 1.88% nickel, normalized and tempered. (b) 1.90% nickel, quenched and tempered
Source: Ref 15, p 130

AISI 4340: Electrical resistivity

TEMP C	TEMP F	ELRES n{o}*m
21	70	248
100	212	298
200	390	367
400	750	552
600	1110	797

Source: Ref 15, p 130

Mechanical Properties

AISI 4340: Transverse tensile and Charpy V-notch property data

TEMP C	TEMP F	YS(a) MPa	YS(a) ksi	UTS MPa	UTS ksi	EL %	RA %	IS J	IS ft-lb
4340 VAR (1.5% Si modified)									
24	75	1524	221.0	1972	286.0	7.8	27.2	...	...
93	200	1569	227.5	2027	294.0	9.3	27.4	11.5	8.5
149	300	1491	216.3	2081	301.8	9.5	20.6	13.3	9.8
204	400	1372	199.0	2053	297.8	12.5	22.2	14.6	10.8
260	500	1227	178.0	1903	276.0	13.4	23.2	12.0	8.8
4340 VAR									
24	75	1460	211.7	1683	244.1	8.2	26.5	...	...
93	200	1331	193.0	1650	239.3	9.8	29.1	16.2	12.0
149	300	1163	168.7	1684	244.3	9.0	17.3	15.6	11.5
204	400	1048	152.0	1624	235.6	13.0	32.5	16.2	12.0
260	500	1103	160.0	1613	234.0	14.3	34.5	15.3	11.3
4340 air melted (aircraft quality)									
24	75	1379	200.0	1703	247.0	8.2	28.0	...	...
93	200	1267	183.7	1655	240.0	9.0	24.0	13.0	9.6(b)
149	300	1160	168.2	1702	246.8	11.0	21.9	14.7	10.9
204	400	1051	152.5	1623	235.4	12.5	33.7	14.9	11.0
260	500	1040	150.8	1544	223.9	14.5	41.9	15.3	11.3

Note: Test results were obtained using 175-mm (7-in.) diameter tubular extrusions. Specimens were held at test temperature for 30 min. Tensile specimens were 6.4 mm (0.252 in.) in diameter
(a) 0.10%. (b) Shattered
Source: Ref 21, p 19

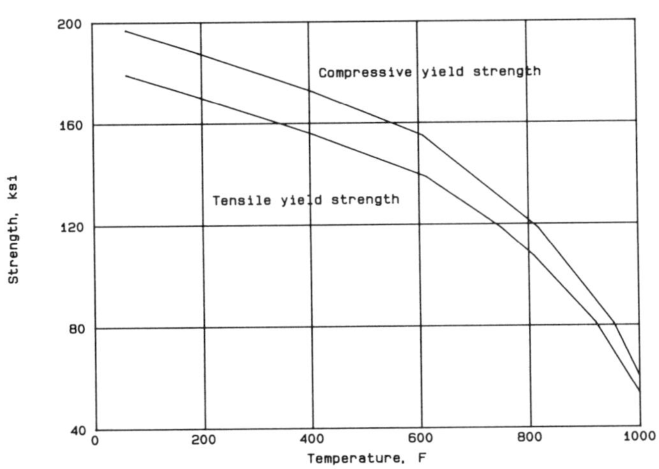

AISI 4340: Tensile yield strength and compressive yield strength of 0.064-in. sheet, heat treated to a yield strength of 200 ksi. Specimens were heated to 1525 F for 10 min, air cooled, and held at 800 F for 1 h. Source: Ref 22, p 15

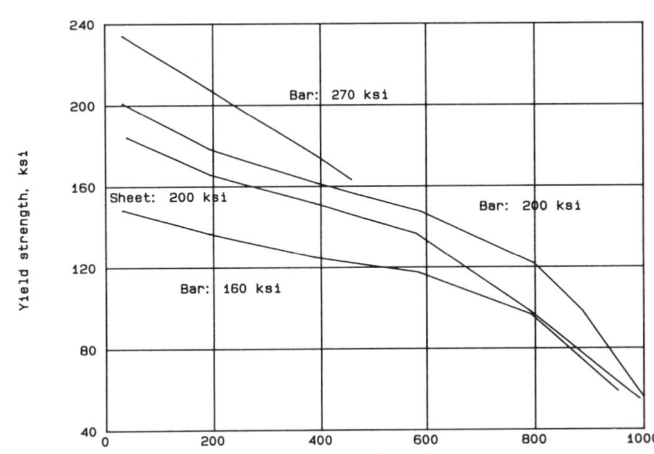

AISI 4340: Effect of prior tensile strength on yield strength of bar and 0.064-in. sheet. Source: Ref 21, p 22

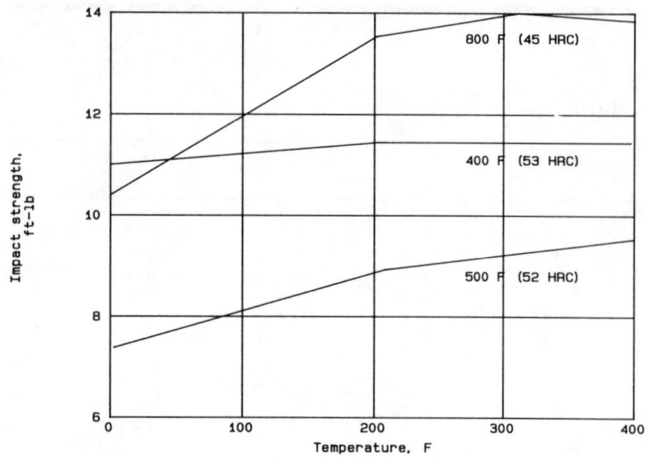

AISI 4340: Charpy V-notch impact strength. Specimens were oil quenched and tempered for 1 h at the indicated temperature. Source: Ref 22, p 15

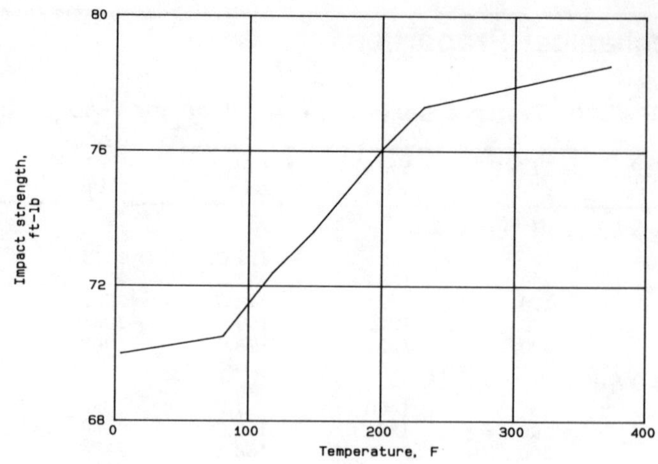

AISI 4340: Charpy V-notch impact strength of specimens with a tensile strength of 150 ksi. Source: Ref 22, p 15

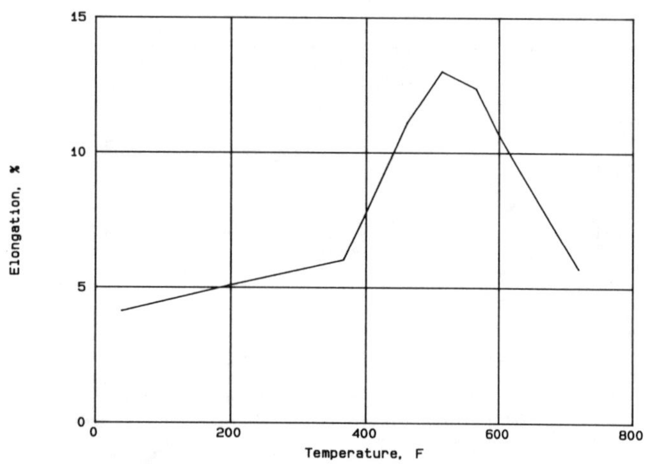

AISI 4340: Tensile elongation of air melted, tempered sheet, 0.080 in thick. Specimens were austenitized at 1550 F in a salt bath, air cooled, double tempered at 750 F for a 1-h cycle. Crosshead speed of 0.5 in./min. Source: Ref 22, p 16

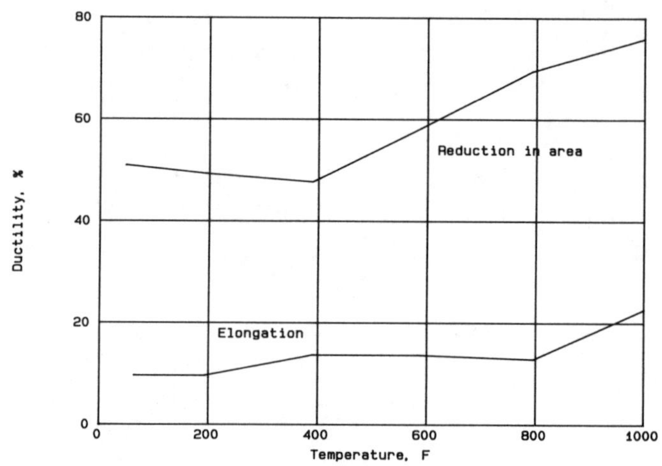

AISI 4340: Tensile elongation and reduction in area of bar. Source: Ref 21, p 22

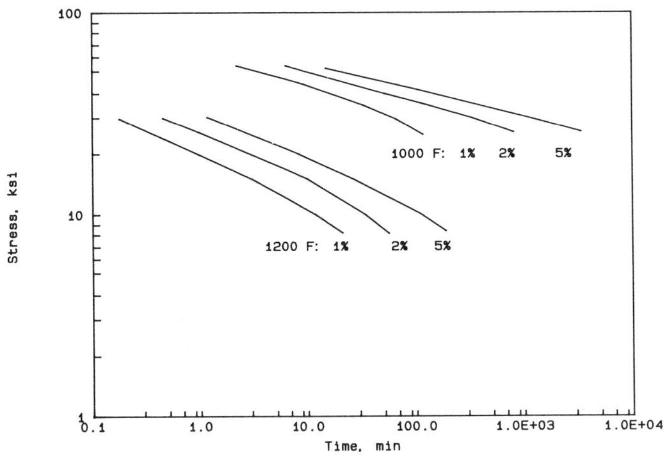

AISI 4340: Creep strength of 0.050-in. sheet. Specimens were heated to 1600 F, air cooled, and tempered. Source: Ref 21, p 106

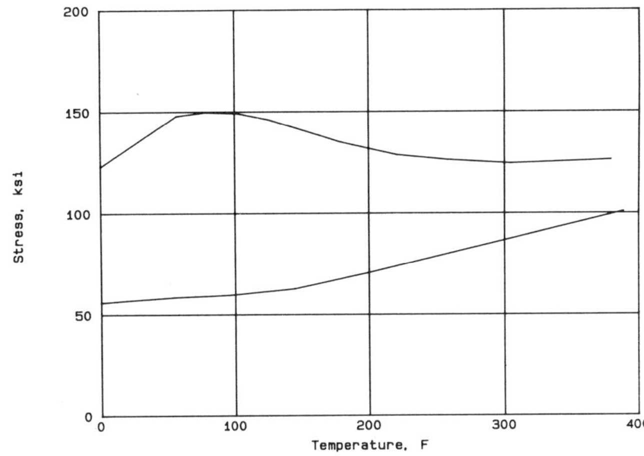

AISI 4340: Range of values for fracture toughness of air and vacuum melted sheet. Specimens were austenitized at 1550 F for 0.5 h, oil quenched to -320 F, and heated to 400 F for 2 h. Source: Ref 22, p 17

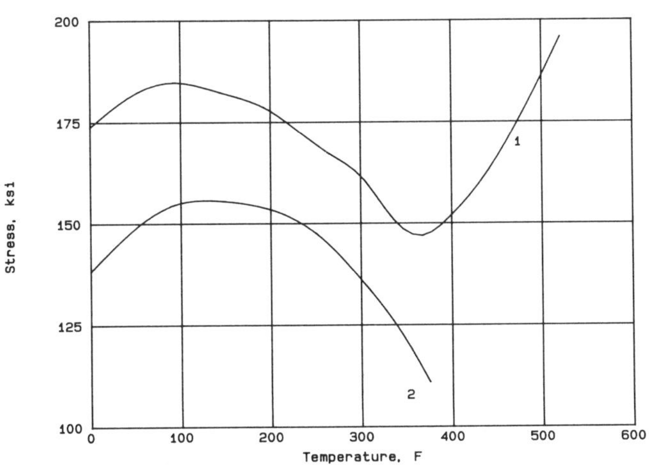

AISI 4340: Net fracture stress of 0.063-in. sheet, heat treated to a tensile strength of 280 ksi. Specimens were austenitized in argon at 1600 F for 0.5 h, oil quenched, double tempered at 400 F for 2 h. Curve 1: regular shear-crack specimen. Curve 2: flattened shear-crack specimen. Source: Ref 22, p 16

AISI 4340: Hot hardness data

TEMP C	TEMP F	HARD (a) HRC	HARD (b) HRC	HARD (c) HRC
21	70	49	40	34
38	100	49	39	33
93	200	48	38	32
149	300	48	37	31
204	400	48	36	30
260	500	47	35	29
315	600	45	34	27
371	700	41	33	26
427	800	34	31	23
482	900	26	26	18
538	1000	17	17	12
593	1100	9	9	5

Note: Vacuum melted specimens were normalized at 900 C (1650 F), 1 h, air cooled; austenitized at 815 C (1500 F), 30 min; oil quenched and tempered as indicated
(a) Tempered at 245 C (475 F), 2+2 h, air cooled. (b) Tempered at 480 C (900 F), 2+2 h, air cooled. (c) Tempered at 595 C (1100 F), 2 h, air cooled
Source: Ref 29, Jul 1979

AISI 4340H

UNS number: H43400

Specifications: ASTM A304, SAE J1268, DIN 1.6565, JIS SNCM 8, BS 817 M 40, BS 3111 Type 6, BS 2 S 119, BS 3 S 95

Nominal composition: 0.370 to 0.440% C, 0.550 to 0.900% Mn, 0.035% P max, 0.040% S max, 0.150 to 0.300% Si, 1.550 to 2.000% Ni, 0.650 to 0.950% Cr, 0.200 to 0.300% Mo

Product forms: Wrought, bar, rod, wire, plate, sheet, strip, tube, pipe, billets, forgings, blooms, cast, continuous, ingot

Description: AISI 4340H is a nickel-chromium-molybdenum constructional alloy steel. These steels normally are heat treated by quenching in oil and tempering to the desired hardness. This grade exhibits good response to heat treatment, especially in large sections, and possesses a good combination of strength, ductility, and toughness in the quenched and tempered condition. The high depth in hardness is reflected in excellent torque properties. These steels are essentially immune to temper embrittlement and do not soften readily at elevated temperatures.

Typical applications: Because AISI 4340H is also a carburizing steel, it is used where superior case hardness or core properties are desired. Their choice is determined primarily by the hardenability necessary to obtain the desired core properties under the given conditions of section size and heat treatment. This grade is suitable for applications in the production of piston pins, bearings, and similar parts for relatively severe service. Rods approximately 13-mm (0.5-in.) in diameter and rod mill rounds up to 28.6-mm (1.25-in.) in diameter are used for heavy sections, or for parts subject to particularly severe service conditions that require a very mild quench to prevent distortion. AISI 4340H steels are found in aircraft and truck parts and in some ordnance materials. Additional uses include gears, fittings, forging dies, machine tool arbors, pressure vessels, and die blocks.

Physical Properties

AISI 4340H: Mean coefficients of linear thermal expansion

TEMP C	TEMP F	THEXP(a) 10^{-6}/C	THEXP(a) 10^{-6}/F	THEXP(b) 10^{-6}/C	THEXP(b) 10^{-6}/F
21-95	70-200	11.5	6.4	11.5	6.4
21-150	70-300	11.9	6.6	11.9	6.6
21-205	70-400	12.1	6.7	12.2	6.8
21-260	70-500	12.4	6.9	12.6	7.0
21-315	70-600	12.6	7.0	13.0	7.2
21-370	70-700	13.0	7.2	13.1	7.3
21-425	70-800	13.3	7.4	13.5	7.5
21-480	70-900	13.5	7.5	13.7	7.6
21-540	70-1000	13.7	7.6	13.9	7.7

Note: Specimens were austenitized at 870 C (1600 F), tempered at 545 C (1010 F) for 2 h
(a) 1.88% nickel, normalized and tempered. (b) 1.90% nickel, quenched and tempered
Source: Ref 15, p 130

AISI 4340H: Electrical resistivity

TEMP C	TEMP F	ELRES n{o}*m
21	70	248
100	212	298
200	390	367
400	750	552
600	1110	797

Source: Ref 15, p 130

Mechanical Properties

AISI 4340H: Transverse tensile and Charpy V-notch property data

TEMP C	TEMP F	YS(a) MPa	YS(a) ksi	UTS MPa	UTS ksi	EL %	RA %	IS J	IS ft-lb
4340 VAR (1.5% Si modified)									
24	75	1524	221.0	1972	286.0	7.8	27.2	...	...
93	200	1569	227.5	2027	294.0	9.3	27.4	11.5	8.5
149	300	1491	216.3	2081	301.8	9.5	20.6	13.3	9.8
204	400	1372	199.0	2053	297.8	12.5	22.2	14.6	10.8
260	500	1227	178.0	1903	276.0	13.4	23.2	12.0	8.8
4340 VAR									
24	75	1460	211.7	1683	244.1	8.2	26.5	...	...
93	200	1331	193.0	1650	239.3	9.8	29.1	16.2	12.0
149	300	1163	168.7	1684	244.3	9.0	17.3	15.6	11.5
204	400	1048	152.0	1624	235.6	13.0	32.5	16.2	12.0
260	500	1103	160.0	1613	234.0	14.3	34.5	15.3	11.3
4340 air melted (aircraft quality)									
24	75	1379	200.0	1703	247.0	8.2	28.0	...	...
93	200	1267	183.7	1655	240.0	9.0	24.0	13.0	9.6(b)
149	300	1160	168.2	1702	246.8	11.0	21.9	14.7	10.9
204	400	1051	152.5	1623	235.4	12.5	33.7	14.9	11.0
260	500	1040	150.8	1544	223.9	14.5	41.9	15.3	11.3

Note: Test results were obtained using 175-mm (7-in.) diameter tubular extrusions. Specimens were held at test temperature for 30 min. Tensile specimens were 6.4 mm (0.252 in.) in diameter
(a) 0.10%. (b) Shattered
Source: Ref 21, p 19

AISI 4340H: Hot hardness data

TEMP C	TEMP F	HARD (a) HRC	HARD (b) HRC	HARD (c) HRC
21	70	49	40	34
38	100	49	39	33
93	200	48	38	32
149	300	48	37	31
204	400	48	36	30
260	500	47	35	29
315	600	45	34	27
371	700	41	33	26
427	800	34	31	23
482	900	26	26	18
538	1000	17	17	12
593	1100	9	9	5

Note: Vacuum melted specimens were normalized at 900 C (1650 F), 1 h, air cooled; austenitized at 815 C (1500 F), 30 min; oil quenched and tempered as indicated
(a) Tempered at 245 C (475 F), 2+2 h, air cooled. (b) Tempered at 480 C (900 F), 2+2 h, air cooled. (c) Tempered at 595 C (1100 F), 2 h, air cooled
Source: Ref 29, Jul 1979

AISI E4340

UNS number: G43406
Specifications: ASTM A304, ASTM A331, ASTM A505, ASTM A519, MIL S-5000, SAE J404, SAE J770, DIN 1.6562, UNI 40 NiCrMo 7, UNI 40 NiCrMo 7 KB, BS Type 8, BS S 139
Nominal composition: 0.380 to 0.430% C, 0.650 to 0.850% Mn, 0.025% P max, 0.020% S max, 0.150 to 0.300% Si, 1.650 to 2.000% Ni, 0.700 to 0.900% Cr, 0.200 to 0.300% Mo
Product forms: Wrought, bar, rod, wire, plate, sheet, strip, tube, pipe, billets, forgings, blooms, cast, continuous, ingot
Description: AISI E4340 is a nickel-chromium-molybdenum, constructional, alloy steel. They are normally heat treated by quenching in oil and tempering to the desired hardness. This grade exhibits good response to heat treatment, especially in large sections, and possesses a good combination of strength, ductility, and toughness in the quenched and tempered condition. The high depth of hardness is reflected in excellent torque properties. These steels are essentially immune to temper embrittlement and do not soften readily at elevated temperatures.

Typical applications: Because AISI E4340 is also a carburizing steel, it is used where superior case hardness or core properties are desired. Its choice is determined primarily by the hardenability necessary to obtain the desired core properties under the given conditions of section size and heat treatment. This grade is suitable for application in the production of piston pins, bearings, and similar parts for relatively severe service. Rods approximately 13 mm (0.5 in.) in diameter and rod mill rounds up to 28.6 mm (1.125 in.) in diameter are used for heavy sections, or for parts subject to particularly severe service conditions that require a very mild quench to prevent distortion. AISI E4340 steel is found in aircraft and truck parts and some ordnance materials. Additional uses include gears, fittings, forging dies, machine tool arbors, pressure vessels, and die blocks.

Physical Properties

AISI E4340: Mean coefficients of linear thermal expansion

TEMP C	TEMP F	THEXP(a) 10^{-6}/C	THEXP(a) 10^{-6}/F	THEXP(b) 10^{-6}/C	THEXP(b) 10^{-6}/F
21-95	70-200	11.5	6.4	11.5	6.4
21-150	70-300	11.9	6.6	11.9	6.6
21-205	70-400	12.1	6.7	12.2	6.8
21-260	70-500	12.4	6.9	12.6	7.0
21-315	70-600	12.6	7.0	13.0	7.2
21-370	70-700	13.0	7.2	13.1	7.3
21-425	70-800	13.3	7.4	13.5	7.5
21-480	70-900	13.5	7.5	13.7	7.6
21-540	70-1000	13.7	7.6	13.9	7.7

Note: Specimens were austenitized at 870 C (1600 F), tempered at 545 C (1010 F) for 2 h
(a) 1.88% nickel, normalized and tempered. (b) 1.90% nickel, quenched and tempered
Source: Ref 15, p 130

AISI E4340: Electrical resistivity

TEMP C	TEMP F	ELRES n{o}*m
21	70	248
100	212	298
200	390	367
400	750	552
600	1110	797

Source: Ref 15, p 130

Mechanical Properties

AISI E4340: Transverse tensile and Charpy V-notch property data

TEMP C	TEMP F	YS(a) MPa	YS(a) ksi	UTS MPa	UTS ksi	EL %	RA %	IS J	IS ft-lb
4340 VAR (1.5% Si modified)									
24	75	1524	221.0	1972	286.0	7.8	27.2	...	...
93	200	1569	227.5	2027	294.0	9.3	27.4	11.5	8.5
149	300	1491	216.3	2081	301.8	9.5	20.6	13.3	9.8
204	400	1372	199.0	2053	297.8	12.5	22.2	14.6	10.8
260	500	1227	178.0	1903	276.0	13.4	23.2	12.0	8.8
4340 VAR									
24	75	1460	211.7	1683	244.1	8.2	26.5	...	...
93	200	1331	193.0	1650	239.3	9.8	29.1	16.2	12.0
149	300	1163	168.7	1684	244.3	9.0	17.3	15.6	11.5
204	400	1048	152.0	1624	235.6	13.0	32.5	16.2	12.0
260	500	1103	160.0	1613	234.0	14.3	34.5	15.3	11.3
4340 air melted (aircraft quality)									
24	75	1379	200.0	1703	247.0	8.2	28.0	...	...
93	200	1267	183.7	1655	240.0	9.0	24.0	13.0	9.6(b)
149	300	1160	168.2	1702	246.8	11.0	21.9	14.7	10.9
204	400	1051	152.5	1623	235.4	12.5	33.7	14.9	11.0
260	500	1040	150.8	1544	223.9	14.5	41.9	15.3	11.3

Note: Test results were obtained using 175-mm (7-in.) diameter tubular extrusions. Specimens were held at test temperature for 30 min. Tensile specimens were 6.4 mm (0.252 in.) in diameter
(a) 0.10%. (b) Shattered
Source: Ref 21, p 19

AISI E4340: Hot hardness data

TEMP C	TEMP F	HARD (a) HRC	HARD (b) HRC	HARD (c) HRC
21	70	49	40	34
38	100	49	39	33
93	200	48	38	32
149	300	48	37	31
204	400	48	36	30
260	500	47	35	29
315	600	45	34	27
371	700	41	33	26
427	800	34	31	23
482	900	26	26	18
538	1000	17	17	12
593	1100	9	9	5

Note: Vacuum melted specimens were normalized at 900 C (1650 F), 1 h, air cooled; austenitized at 815 C (1500 F), 30 min; oil quenched and tempered as indicated
(a) Tempered at 245 C (475 F), 2+2 h, air cooled. (b) Tempered at 480 C (900 F), 2+2 h, air cooled. (c) Tempered at 595 C (1100 F), 2 h, air cooled
Source: Ref 29, Jul 1979

AISI E4340H

UNS number: H43406

Specifications: ASTM A304, SAE J1268, DIN 1.6562, UNI 40 NiCrMo 7, UNI 40 NiCrMo 7 KB, BS Type 8, BS S 139

Nominal composition: 0.370 to 0.440% C, 0.600 to 0.950% Mn, 0.025% P max, 0.020% S max, 0.150 to 0.300% Si, 1.550 to 2.000% Ni, 0.650 to 0.950% Cr, 0.200 to 0.300% Mo

Product forms: Wrought, bar, rod, wire, plate, sheet, strip, tube, pipe, billets, forgings, cast, continuous, blooms, ingot

Description: AISI E4340H is a nickel-chromium-molybdenum constructional alloy steel. It is normally heat treated by quenching in oil and tempering to the desired hardness. This grade exhibits good response to heat treatment, especially in large sections, and possesses a good combination of strength, ductility, and toughness in the quenched and tempered condition. The high depth of hardness is reflected in excellent torque properties.

Typical applications: Because AISI E4340H is also a carburizing steel, it is used where superior case hardness or core properties are desired. Its choice is determined primarily by the hardenability necessary to obtain the desired core properties under the given conditions of section size and heat treatment. These steels are suitable for applications in the production of piston pins, bearings, and similar parts for relatively severe service. Rods approximately 13 mm (0.5 in.) in diameter and rod mill rounds up to 28.6 mm (1.125 in.) in diameter are used for heavy sections, or for parts subject to particularly severe service conditions that require a very mild quench to prevent distortion. AISI E4340H steel is found in aircraft and truck parts and in some ordnance materials. Additional uses include gears, fittings, forging dies, machine tool arbors, pressure vessels, and die blocks.

Physical Properties

AISI E4340H: Mean coefficients of linear thermal expansion

TEMP C	TEMP F	THEXP(a) 10^{-6}/C	THEXP(a) 10^{-6}/F	THEXP(b) 10^{-6}/C	THEXP(b) 10^{-6}/F
21-95	70-200	11.5	6.4	11.5	6.4
21-150	70-300	11.9	6.6	11.9	6.6
21-205	70-400	12.1	6.7	12.2	6.8
21-260	70-500	12.4	6.9	12.6	7.0
21-315	70-600	12.6	7.0	13.0	7.2
21-370	70-700	13.0	7.2	13.1	7.3
21-425	70-800	13.3	7.4	13.5	7.5
21-480	70-900	13.5	7.5	13.7	7.6
21-540	70-1000	13.7	7.6	13.9	7.7

Note: Specimens were austenitized at 870 C (1600 F), tempered at 545 C (1010 F) for 2 h
(a) 1.88% nickel, normalized and tempered. (b) 1.90% nickel, quenched and tempered
Source: Ref 15, p 130

AISI E4340H: Electrical resistivity

TEMP C	TEMP F	ELRES n{o}*m
21	70	248
100	212	298
200	390	367
400	750	552
600	1110	797

Source: Ref 15, p 130

AISI E4340H: Transverse tensile and Charpy V-notch property data

TEMP C	TEMP F	YS(a) MPa	YS(a) ksi	UTS MPa	UTS ksi	EL %	RA %	IS J	IS ft-lb
4340 VAR (1.5% Si modified)									
24	75	1524	221.0	1972	286.0	7.8	27.2	...	...
93	200	1569	227.5	2027	294.0	9.3	27.4	11.5	8.5
149	300	1491	216.3	2081	301.8	9.5	20.6	13.3	9.8
204	400	1372	199.0	2053	297.8	12.5	22.2	14.6	10.8
260	500	1227	178.0	1903	276.0	13.4	23.2	12.0	8.8
4340 VAR									
24	75	1460	211.7	1683	244.1	8.2	26.5	...	...
93	200	1331	193.0	1650	239.3	9.8	29.1	16.2	12.0
149	300	1163	168.7	1684	244.3	9.0	17.3	15.6	11.5
204	400	1048	152.0	1624	235.6	13.0	32.5	16.2	12.0
260	500	1103	160.0	1613	234.0	14.3	34.5	15.3	11.3
4340 air melted (aircraft quality)									
24	75	1379	200.0	1703	247.0	8.2	28.0	...	...
93	200	1267	183.7	1655	240.0	9.0	24.0	13.0	9.6(b)
149	300	1160	168.2	1702	246.8	11.0	21.9	14.7	10.9
204	400	1051	152.5	1623	235.4	12.5	33.7	14.9	11.0
260	500	1040	150.8	1544	223.9	14.5	41.9	15.3	11.3

Note: Test results were obtained using 175-mm (7-in.) diameter tubular extrusions. Specimens were held at test temperature for 30 min. Tensile specimens were 6.4 mm (0.252 in.) in diameter
(a) 0.10%. (b) Shattered
Source: Ref 21, p 19

AISI E4340H: Hot hardness data

TEMP C	TEMP F	HARD (a) HRC	HARD (b) HRC	HARD (c) HRC
21	70	49	40	34
38	100	49	39	33
93	200	48	38	32
149	300	48	37	31
204	400	48	36	30
260	500	47	35	29
315	600	45	34	27
371	700	41	33	26
427	800	34	31	23
482	900	26	26	18
538	1000	17	17	12
593	1100	9	9	5

Note: Vacuum melted specimens were normalized at 900 C (1650 F), 1 h, air cooled; austenitized at 815 C (1500 F), 30 min; oil quenched and tempered as indicated
(a) Tempered at 245 C (475 F), 2+2 h, air cooled. (b) Tempered at 480 C (900 F), 2+2 h, air cooled. (c) Tempered at 595 C (1100 F), 2 h, air cooled
Source: Ref 29, Jul 1979

AISI 4615

UNS number: G46150
Specifications: AMS 6290, ASTM A29, ASTM A322, ASTM A331, ASTM A505, MIL S-7493, SAE J404, SAE J412, SAE J770
Nominal composition: 0.130 to 0.180% C, 0.200 to 0.300% Mo, 1.650 to 2.000% Ni, 0.035% P max, 0.040% S max, 0.150 to 0.300% Si, 0.450 to 0.650% Mn
Product forms: Wrought, bar, rod, wire, plate, sheet, strip, tube, pipe, billets, forgings, cast, continuous, welded, blooms, ingot
Description: AISI 4615 is a nickel-molybdenum, carburizing alloy steel with intermediate case hardenability and low core hardenability.
Typical applications: This grade is available as hot rolled and cold rolled sheet and strip; hot rolled and cold finished bar, rod, wire, and tube; and aircraft quality steel.

Physical Properties

AISI 4615: Electrical resistivity

TEMP C	TEMP F	ELRES n{o}*m
21	70	249
100	212	296
200	390	371

Source: Ref 15, p 140

AISI 4615: Mean coefficients of linear thermal expansion

TEMP C	TEMP F	THEXP 10^{-6}/C	THEXP 10^{-6}/F
21-100	70-212	11	6.3
21-205	70-400	12	6.7
21-300	70-570	13	7.1
21-425	70-800	13	7.4
21-500	70-930	14	7.8
21-700	70-1290	15	8.1

Source: Ref 15, p 141

AISI 5132

UNS number: G51320

Specifications: ASTM A322, ASTM A331, ASTM A505, ASTM A519, SAE J404, SAE J412, SAE J770, DIN 1.7033, NF 32 C 4, UNI 34 Cr 4 KB, JIS SCr 2 H, JIS SCR 2, BS 530 A 32, BS 530 H 32

Nominal composition: 0.300 to 0.350% C, 0.600 to 0.800% Mn, 0.035% P max, 0.040% S max, 0.150 to 0.300% Si, 0.750 to 1.000% Cr

Product forms: Wrought, bar, rod, wire, plate, sheet, strip, tube, pipe, billets, forgings, blooms, cast, continuous, welded, ingot

Description: AISI 5132 is a medium-carbon, chromium alloy steel with low hardenability for water or caustic solution quenching. Parts of small section size made from these steels can be oil quenched where stability is required.

Typical applications: Available as hot and cold rolled sheet and strip; seamless mechanical tubing; hot rolled and cold finished bar, rod, and wire; and bar subject to end-quench hardenability requirements.

Physical Properties

AISI 5132: Mean apparent specific heat of annealed specimens

TEMP C	TEMP F	SHTCP J/kg*K	SHTCP Btu/lb*F
50-100	10-38	494	0.118
150-200	66-93	502	0.120
200-250	93-120	536	0.128
250-300	120-150	553	0.132
300-350	150-175	578	0.138
350-400	175-205	599	0.143
450-500	230-260	650	0.155
550-600	290-315	733	0.175
650-700	345-370	842	0.201
700-750	370-400	1499	0.358
750-800	400-425	938	0.224
850-900	455-480	578	0.138

Source: Ref 15, p 171

AISI 5132: Thermal conductivity of annealed specimens

TEMP C	TEMP F	THCON W/m*K	THCON Btu/ft*h*F
100	212	46.5	26.9
200	390	44.3	25.6
300	570	42.2	24.4
400	750	38.4	22.2
500	930	35.6	20.6
600	1110	31.8	18.4
700	1290	28.9	16.7
800	1470	31.1	18.0
1000	1830	28.0	16.2
1200	2190	30.6	17.7

Source: Ref 15, p 171

AISI 5132: Electrical resistivity of hardened and tempered specimens

TEMP C	TEMP F	ELRES n{o}*m
21	70	210
100	212	259
200	390	330
400	750	517
600	1110	934
800	1470	1106
900	1650	1145
1000	1830	1177
1100	2010	1205
1200	2190	1230
1300	2370	1251

Source: Ref 15, p 171

AISI 5132H

UNS number: H51320
Specifications: ASTM A304, SAE J1268, DIN 1.7034, AFNOR 38 C 4, UNI 38 Cr 4 KB, JIS SCr 3 H, BS 530 A 36, BS 530 H 36, Type 3
Nominal composition: 0.290 to 0.350% C, 0.500 to 0.900% Mn, 0.035% P max, 0.040% S max, 0.150 to 0.300% Si, 0.650 to 1.100% Cr
Product forms: Wrought, bar, rod, wire, plate, sheet, strip, tube, pipe, billets, forgings, blooms, cast, continuous, welded, ingot

Description: AISI 5132H is a medium-carbon, chromium alloy steel with low hardenability for water or caustic solution quenching. Parts of small section size made from these steels can be oil quenched where stability is required.
Typical applications: Available as hot and cold rolled sheet and strip; seamless mechanical tubing; hot rolled and cold finished bar, rod, and wire; and bar subject to end-quench hardenability requirements.

Physical Properties

AISI 5132H: Mean apparent specific heat of annealed specimens

TEMP C	TEMP F	SHTCP J/kg*K	SHTCP Btu/lb*F
50-100	10-38	494	0.118
150-200	66-93	502	0.120
200-250	93-120	536	0.128
250-300	120-150	553	0.132
300-350	150-175	578	0.138
350-400	175-205	599	0.143
450-500	230-260	650	0.155
550-600	290-315	733	0.175
650-700	345-370	842	0.201
700-750	370-400	1499	0.358
750-800	400-425	938	0.224
850-900	455-480	578	0.138

Source: Ref 15, p 171

AISI 5132H: Thermal conductivity of annealed specimens

TEMP C	TEMP F	THCON W/m*K	THCON Btu/ft*h*F
100	212	46.5	26.9
200	390	44.3	25.6
300	570	42.2	24.4
400	750	38.4	22.2
500	930	35.6	20.6
600	1110	31.8	18.4
700	1290	28.9	16.7
800	1470	31.1	18.0
1000	1830	28.0	16.2
1200	2190	30.6	17.7

Source: Ref 15, p 171

AISI 5132H: Electrical resistivity of hardened and tempered specimens

TEMP C	TEMP F	ELRES n{o}*m
21	70	0.210
100	212	0.259
200	390	0.330
400	750	0.517
600	1110	0.934
800	1470	1.106
900	1650	1.145
1000	1830	1.177
1100	2010	1.205
1200	2190	1.230
1300	2370	1.251

Source: Ref 15, p 171

AISI 5140

UNS number: G51400

Specifications: ASTM A322, ASTM A331, ASTM A505, ASTM A519, SAE J404, SAE J412, SAE J770, DIN 1.7035, NF 42 C 4, UNI 40 Cr 4, UNI 41 Cr 4 KB, JIS SCr 4 H, BS 530 A 40, BS 530 H 40, BS 530 M 40, BS 2 S 117

Nominal composition: 0.380 to 0.430% C, 0.700 to 0.900% Cr, 0.700 to 0.900% Mn, 0.035% P max, 0.040% S max, 0.150 to 0.300% Si

Product forms: Wrought, bar, rod, wire, plate, sheet, strip, tube, pipe, billets, forgings, blooms, cast, continuous, ingot

Description: AISI 5140 is a medium-carbon, chromium steel with low, direct hardenability. These steels are used when superior properties are desired.

Typical applications: These steels are used for fasteners, such as high-strength screws, bolts, cap screws, and recessed-head screws, and for severely cold headed, upset, and extruded parts. 5140 is available as hot rolled and cold finished bar, rod, and wire; hot and cold rolled sheet and strip; seamless mechanical tubing; and bar subject to end-quench hardenability requirements.

Physical Properties

AISI 5140: Mean apparent specific heat of hardened and tempered specimens

TEMP C	TEMP F	SHTCP J/kg*K	SHTCP Btu/lb*F
21-100	70-212	452	0.108
21-200	70-390	473	0.113
21-400	70-750	519	0.124
21-600	70-1110	565	0.135

Source: Ref 15, p 173

AISI 5140: Mean coefficients of linear thermal expansion

TEMP C	TEMP F	THEXP 10^{-6}/C	THEXP 10^{-6}/F
21-200	70-390	12.6	7.0
21-300	70-570	13.3	7.4
21-400	70-750	13.9	7.7
21-500	70-930	14.0	7.8
21-600	70-1110	14.6	8.1
21-700	70-1290	14.9	8.3

Source: Ref 15, p 173

AISI 5140: Thermal conductivity of hardened and tempered specimens

TEMP C	TEMP F	THCON W/m*K	THCON Btu/ft*h*F
100	212	44.6	25.8
200	390	43.4	25.1
400	750	37.7	21.8
600	1110	31.3	18.1

Source: Ref 15, p 173

AISI 5140: Electrical resistivity

TEMP C	TEMP F	ELRES n{o}*m
21	70	228
100	212	281
200	390	352
400	750	530
600	1110	785

Note: Specimens were hardened and tempered
Source: Ref 15, p 173

AISI 5140H

UNS number: H51400
Specifications: ASTM A304, SAE J1268, DIN 1.7006, NF 42 C 2, FN 45 C 2
Nominal composition: 0.370 to 0.440% C, 0.600 to 1.000% Mn, 0.035% P max, 0.040% S max, 0.150 to 0.300% Si, 0.600 to 1.000% Cr
Product forms: Wrought, bar, rod, wire, plate, sheet, strip, tube, pipe, billets, forgings, blooms, cast, continuous, ingot
Description: AISI 5140H is a medium-carbon, chromium steel with low, direct hardenability. These steels are used when superior properties are desired.
Typical applications: AISI 5140H is used for fasteners, such as high-strength screws, bolts, cap screws, and recessed-head screws, as well as severely cold headed, upset, and extruded parts. 5140H is available as hot rolled and cold finished bar, rod, and wire; hot and cold rolled sheet and strip; seamless mechanical tubing; and bar subject to end-quench hardenability requirements.

Physical Properties

AISI 5140H: Mean apparent specific heat of hardened and tempered specimens

TEMP C	TEMP F	SHTCP J/kg*K	SHTCP Btu/lb*F
21-100	70-212	452	0.108
21-200	70-390	473	0.113
21-400	70-750	519	0.124
21-600	70-1110	565	0.135

Source: Ref 15, p 173

AISI 5140H: Thermal conductivity of hardened and tempered specimens

TEMP C	TEMP F	THCON W/m*K	THCON Btu/ft*h*F
100	212	44.6	25.8
200	390	43.4	25.1
400	750	37.7	21.8
600	1110	31.3	18.1

Source: Ref 15, p 173

AISI 5140H: Mean coefficients of linear thermal expansion

TEMP C	TEMP F	THEXP 10^{-6}/C	THEXP 10^{-6}/F
21-200	70-390	12.6	7.0
21-300	70-570	13.3	7.4
21-400	70-750	13.9	7.7
21-500	70-930	14.0	7.8
21-600	70-1110	14.6	8.1
21-700	70-1290	14.9	8.3

Source: Ref 15, p 173

AISI 5140H: Electrical resistivity of hardened and tempered specimens

TEMP C	TEMP F	ELRES n{o}*m
21	70	228
100	212	281
200	390	352
400	750	530
600	1110	785

Source: Ref 15, p 173

AISI E52100

UNS number: G52986
Specifications: AMS 6440, AMS 6441, AMS 6444, AMS 6447, ASTM A322, ASTM A331, ASTM A505, ASTM A519, ASTM A535, ASTM A646, ASTM A711, MIL S-980, MIL S-7420, MIL S-22141, SAE J404, SAE J412, SAE J770, DIN 1.3505, NF 100 C 6, UNI 100 Cr 6
Nominal composition: 0.980 to 1.100% C, 0.250 to 0.450% Mn, 0.025% P max, 0.025% S max, 0.150 to 0.300% Si, 1.300 to 1.600% Cr
Product forms: Wrought, bar, tube, billets, forgings, blooms, cast, continuous, ingot

Description: AISI E52100 is a high-carbon, chromium electric furnace alloy steel.

Typical applications: AISI E52100 is used primarily for the races and balls or rollers of rolling-element (antifriction) bearings. It is also suitable for other parts requiring high hardness and wear resistance.

AISI E52100: Mean coefficients of linear thermal expansion

Condition	TEMP C	TEMP F	THEXP 10^{-6}/C	THEXP 10^{-6}/F
Annealed	23-280	74-540	11.9	6.60
Hardened	23-280	74-540	12.5	6.96

Source: Ref 5, p 155

AISI 8615

UNS number: G86150

Specifications: AMS 5333, ASTM A322, ASTM A505, MIL S-866, SAE J404, SAE J770

Nominal composition: 0.130 to 0.180% C, 0.700 to 0.900% Mn, 0.035% P max, 0.040% S max, 0.150 to 0.300% Si, 0.400 to 0.700% Ni, 0.400 to 0.600% Cr, 0.150 to 0.250% Mo

Product forms: Wrought, bar, rod, wire, plate, sheet, strip, tube, billets, forgings, blooms, cast, continuous, welded, ingot

Description: AISI 8615 is a nickel-chromium-molybdenum carburizing alloy steel with medium case and core hardenability.

Typical applications: Available as hot rolled bar and aircraft-quality stock. Uses of AISI 8615 include automotive gears and shafts, knuckle pins, piston pins, pump shafts, and spline shafts.

Mechanical Properties

AISI 8615: Effect of melt practice on tensile properties

TEMP C	TEMP F	YS MPa	YS ksi	UTS MPa	UTS ksi	EL %	RA %
Open hearth process(a)							
21	70	338	49	476	69	35.0	75.4
371	700	214	31	455	66	32.0	72.2
482	900	179	26	372	54	34.0	80.6
Basic oxygen process(b)							
21	70	345	50	490	71	34.0	74.6
321	700	207	30	462	67	33.0	78.0
482	900	193	28	379	55	36.8	85.7

(a) Specimens were 30-mm (1-3/16 in.) diameter bar normalized at 925 C (1700 F), 1 h and 15 min, air cooled, tempered at 650 C (1200 F), 1 h and 30 min. Material composition: 0.14% C, 0.68% Mn, 0.15% P, 0.031% S, 0.24% Si, 0.56% Ni, 0.50% Cr, 0.20% Mo, and 0.028% Al. (b) Specimens were 30-mm (1-3/16 in.) diameter bars normalized at 925 C (1700 F), 1 h and 15 min, air cooled, tempered at 650 C (1200 F), 1 h and 30 min. Material composition: 0.17% C, 0.81% Mn, 0.012% P, 0.022% S, 0.17% Si, 0.57% Ni, 0.54% Cr, 0.19% Mo. and 0.056% Al
Source: Ref 21, p 19

AISI 8630

UNS number: G86300

Specifications: AMS 6280, AMS 6281, AMS 6355, AMS 6530, AMS 6550, ASTM A322, ASTM A331, ASTM A505, ASTM A519, MIL S-16974, SAE J404, SAE J412, SAE J770, DIN 1.6545, UNI 30 NiCrMo 2 KB

Nominal composition: 0.280 to 0.330% C, 0.400 to 0.600% Cr, 0.700 to 0.900% Mn, 0.150 to 0.250% Mo, 0.400 to 0.700% Ni, 0.035% P max, 0.040% S max, 0.150 to 0.300% Si

Product forms: Wrought, bar, rod, wire, plate, sheet, strip, tube, forgings, blooms, cast, continuous, welded, ingot

Description: AISI 8630 is a directly hardenable, triple-alloy steel with low hardenability. These steels have good mechanical properties when normalized in light sections. They are air hardening after welding.

Typical applications: AISI 8630 is available as hot rolled and cold finished bar, rod, and wire; mechanical seamless tubing; and aircraft-quality stock. Applications include connecting rods, engine bolts and studs, shapes, and tubing.

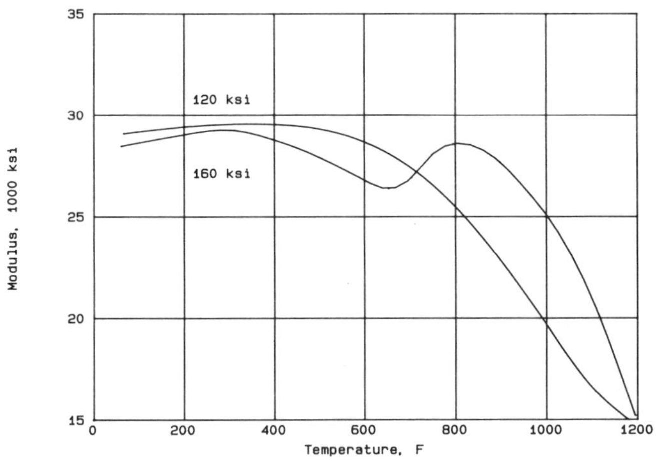

AISI 8630: Modulus of elasticity of 0.065-in. sheet, quenched and tempered to two room-temperature tensile strengths. Source: Ref 22, p 18

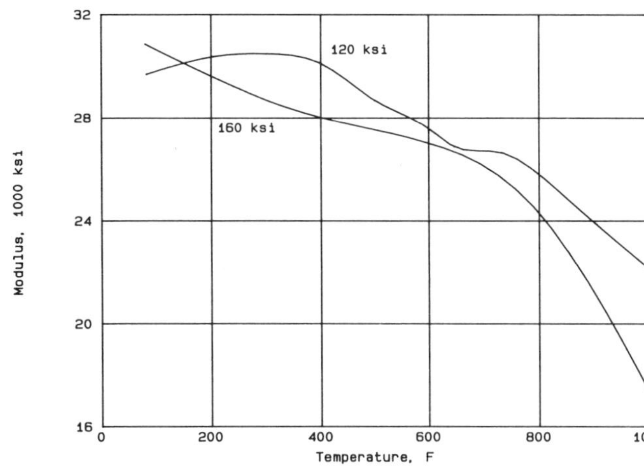

AISI 8630: Modulus of elasticity in compression of sheet quenched and tempered to two room-temperature tensile strength levels. Source: Ref 22, p 18

AISI 8630: Mean coefficients of linear thermal expansion

TEMP C	TEMP F	THEXP 10^{-6}/C	THEXP 10^{-6}/F
21-100	70-212	11.2	6.2
21-200	70-390	12.2	6.8
21-300	70-570	13.0	7.2
21-400	70-750	13.5	7.5
21-500	70-930	14.0	7.8
21-600	70-1110	14.4	8.0
21-700	70-1290	14.9	8.3

Source: Ref 15, p 197

Mechanical Properties

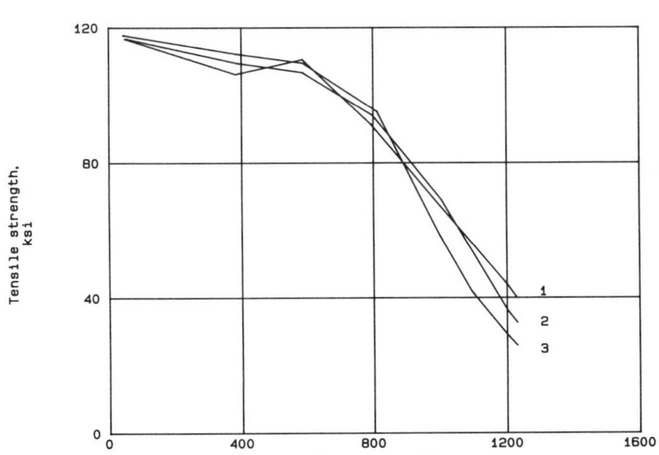

AISI 8630: Tensile strength of 0.064-in. sheet quenched and tempered to a room-temperature tensile strength of 120 ksi. Exposure times at test temperature: 0.05 h (1), 10 h (2), 100 h (3). Source: Ref 22, p 12

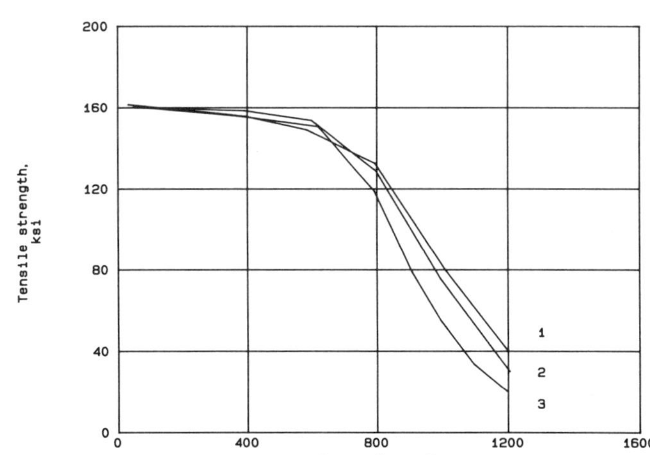

AISI 8630: Tensile strength of 0.064-in. sheet quenched and tempered to a room-temperature tensile strength of 160 ksi. Exposure times at test temperature: 0.5 h (1), 10 h (2), 100 h (3). Source: Ref 22, p 12

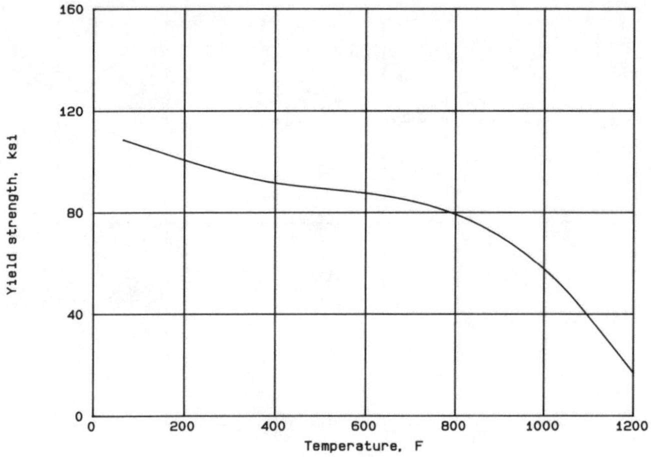

AISI 8630: Effect of 0.5-h exposure at test temperature on yield strength of 0.064-in. annealed sheet, quenched and tempered to a room-temperature tensile strength of 120 ksi. Source: Ref 22, p 12

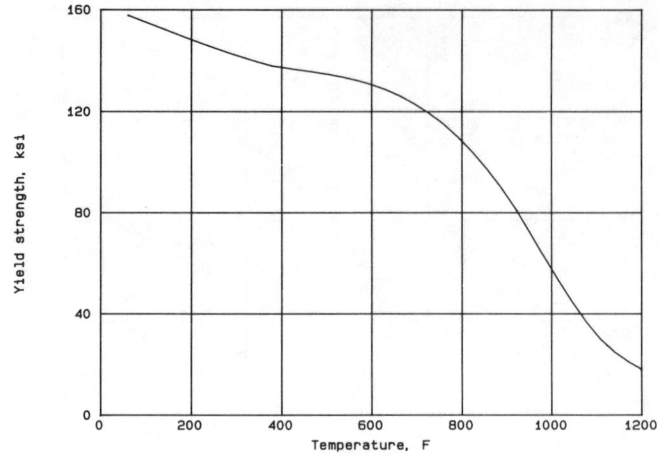

AISI 8630: Effect of 0.5-h exposure at test temperature on yield strength of 0.064-in. annealed sheet, quenched and tempered to a room-temperature tensile strength of 160 ksi. Source: Ref 22, p 12

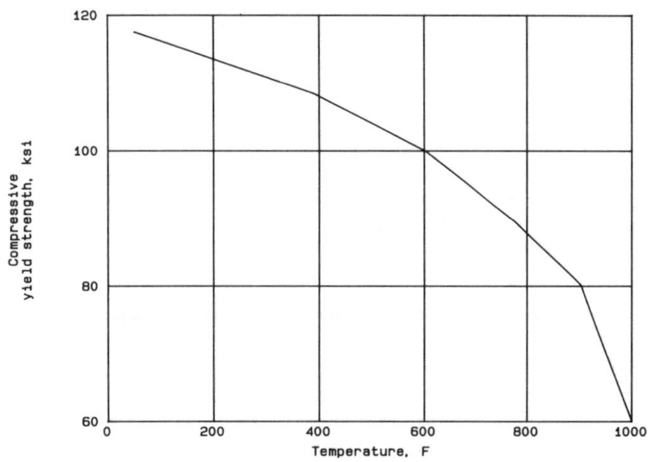

AISI 8630: Effect of 0.5-h exposure at test temperature on compressive yield strength of 0.064-in. annealed sheet, quenched and tempered to a room-temperature tensile strength of 120 ksi. Source: Ref 22, p 14

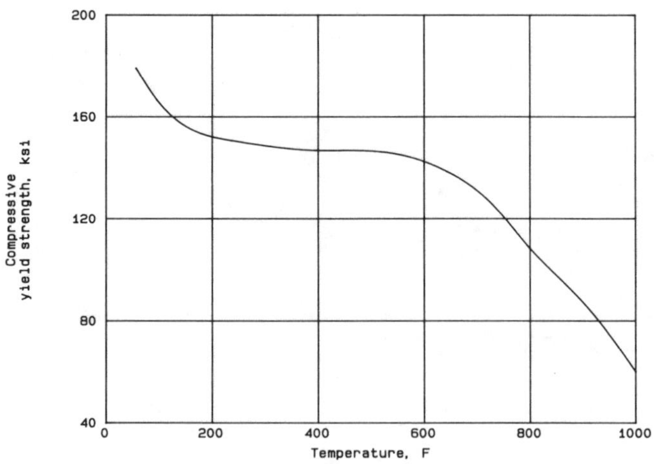

AISI 8630: Effect of 0.5-h exposure at test temperature on compressive yield strength of 0.064-in. annealed sheet, quenched and tempered to a room-temperature tensile strength of 160 ksi. Source Ref 22, p 14

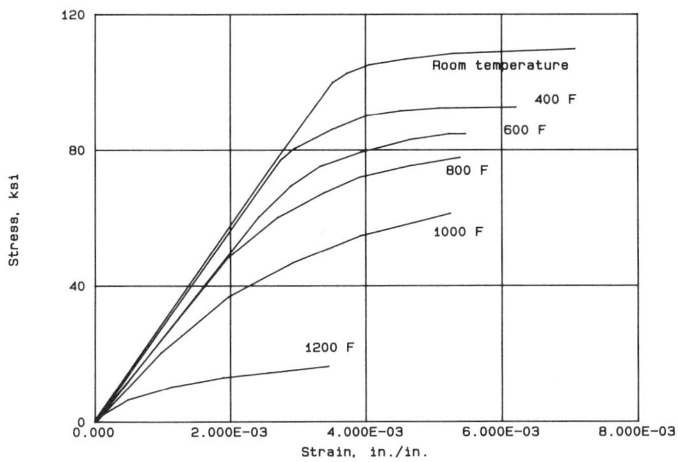

AISI 8630: Stress-strain curves for 0.064-in. sheet, quenched and tempered to a room-temperature tensile strength of 125 ksi. Source: Ref 22, p 11

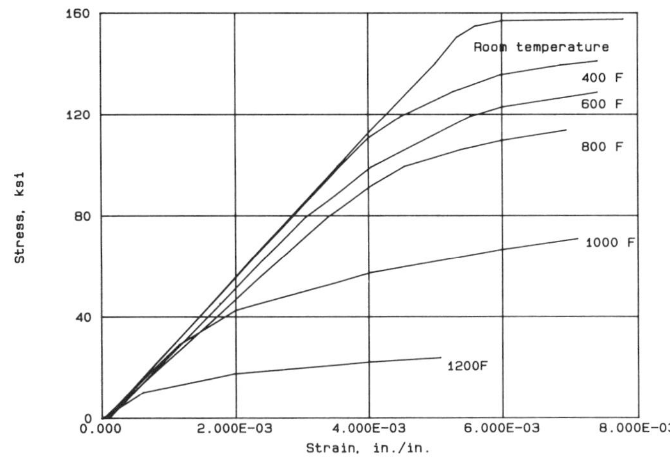

AISI 8630: Stress-strain curves for 0.064-in. sheet, quenched and tempered to a room-temperature tensile strength of 160 ksi. Source: Ref 22, p 13

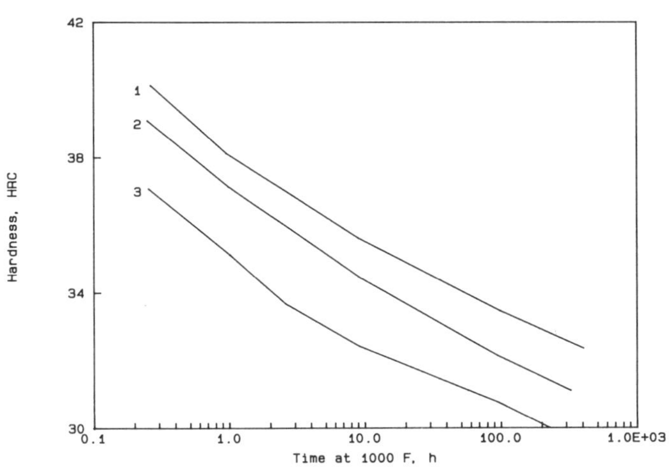

AISI 8630: Effect of exposure time at 1000 F on room-temperature Rockwell C hardness. Curve 1: 1600 F, water cooled; 55 HRC. Curve 2: 1600 F, water cooled, held at 700 F for 1 h; 44 HRC. Curve 3: 1600 F, quenched to 700 F and held for 0.25 h; 41 HRC. Source Ref 22, p 4

AISI 8630H

UNS number: H86300
Specifications: ASTM A304, SAE J1268, DIN 1.6545, UNI 30 NiCrMo 2 KB
Nominal composition: 0.270 to 0.330% C, 0.600 to 0.950% Mn, 0.035% P max, 0.040% S max, 0.150 to 0.300% Si, 0.350 to 0.750% Ni, 0.350 to 0.650% Cr, 0.150 to 0.250% Mo
Product forms: Wrought, bar, rod, wire, plate, sheet, strip, tube, billets, forgings, blooms, cast, continuous, welded

Description: AISI 8630H is a directly hardenable, triple-alloy steel with low hardenability. These steels have good mechanical properties when normalized in light sections. They are air hardening after welding.
Typical applications: AISI 8630H is available as hot rolled and cold finished bar, rod, and wire; mechanical seamless tubing; and aircraft-quality stock. Applications include connecting rods, engine bolts and studs, shapes, and tubing.

Physical Properties

AISI 8630H: Mean coefficients of linear thermal expansion

TEMP C	TEMP F	THEXP $10^{-6}/C$	THEXP $10^{-6}/F$
21-100	70-212	11.2	6.2
21-200	70-390	12.2	6.8
21-300	70-570	13.0	7.2
21-400	70-750	13.5	7.5
21-500	70-930	14.0	7.8
21-600	70-1110	14.4	8.0
21-700	70-1290	14.9	8.3

Source: Ref 15, p 197

AISI 86B30H

UNS number: H86301
Specifications: ASTM A304, SAE J1268
Nominal composition: 0.270 to 0.330% C, 0.600 to 0.950% Mn, 0.035% P max, 0.040% S max, 0.150 to 0.300% Si, 0.350 to 0.750% Ni, 0.350 to 0.650% Cr, 0.150 to 0.250% Mo
Product forms: Wrought, bar, rod, wire, plate, sheet, strip, tube, billets, forgings, blooms, cast, continuous, ingot
Description: AISI 86B30H is a directly hardenable, triple-alloy steel. The addition of boron improves its hardenability. It has good mechanical properties when normalized in light sections and is air hardened after welding.
Typical applications: AISI 86B30H is available as hot rolled and cold finished bar, rod, and wire; mechanical seamless tubing; and aircraft-quality stock. Applications include connecting rods, engine bolts and studs, shapes, and tubing.

Physical Properties

AISI 86B30H: Mean coefficients of linear thermal expansion

TEMP C	TEMP F	THEXP $10^{-6}/C$	THEXP $10^{-6}/F$
21-100	70-212	11.2	6.2
21-200	70-390	12.2	6.8
21-300	70-570	13.0	7.2
21-400	70-750	13.5	7.5
21-500	70-930	14.0	7.8
21-600	70-1110	14.4	8.0
21-700	70-1290	14.9	8.3

Source: Ref 15, p 197

4. ASTM Steels

ASTM Steels

The steels listed by the Standard Specifications of the American Society for Testing and Materials (ASTM) are those carbon and alloy steels that are commonly used, but have not been assigned AISI/SAE numbers. ASTM steels are available as structural shapes, plate, and bar intended for riveted, bolted, or welded construction of bridges and buildings, and for ship construction. ASTM grades of hot rolled commercial quality, and hot and cold rolled structural and drawing quality carbon steel sheet and strip are used for fabricating industrial parts by bending, forming, drawing, and welding.

ASTM Standard Specifications also include low-carbon sheet and coiled steel used in porcelain enameling and for coated stock. The coated steels are aluminum coated, lead-tin (terne) coated, electrodeposited tin or zinc coated, and galvanized (hot dipped, zinc coated) sheet and coil.

Carbon and alloy steel spring wire and carbon steel used in the manufacture of electric-resistance-welded (ERW), arc-welded, and seamless pipe and tubing follow ASTM specifications. ASTM specifications for carbon and alloy steels that apply to the very specialized category of pressure vessels are not included in this handbook. Pressure vessel steels are often specified by either ASTM or the American Society of Mechanical Engineers (ASME).

ASTM A36

UNS number: K02600
Specification: ASTM A36
Nominal composition: 0.290% C max, 0.850 to 1.200% Mn, 0.040% P max, 0.050% S max, 0.150 to 0.400% Si
Product forms: Wrought, bar, plate, structural shapes
Description: ASTM A36 is available in carbon steel shapes, plate, and bar of structural quality for use in riveted, bolted, or welded construction of bridges and buildings and for general structural purposes. When this steel is used in welded construction, welding procedures should be suitable for the steel and the intended service.
Typical applications: Bridges, buildings, and general structural purposes.

Mechanical Properties

ASTM A36: Tensile property data

TEMP C	TEMP F	YS MPa	YS ksi	UTS MPa	UTS ksi	EL %	RA %
27	80	289.6	42.0	496.4	72.0	31.0	60.0
149	300	262.0	38.0	555.0	80.5	24.0	44.0
260	500	234.4	34.0	548.2	79.5	22.0	41.0
371	700	206.9	30.0	483.3	70.1	32.0	54.0
482	900	172.4	25.0	372.0	54.0	38.0	71.0

Note: Specimens were normalized at 899 C (1650 F) and tempered
Source: Ref 15, p 483

ASTM A36: Tensile property data

TEMP C	TEMP F	YS MPa	YS ksi	UTS MPa	UTS ksi	EL %	RA %
27	80	248.2	36.0	441.3	64.0	37	60
149	300	208.2	30.2	441.3	64.0	25	56
260	500	191.7	27.8	439.9	63.8	28	55
371	700	175.1	25.4	393.0	57.0	35	63
482	900	148.2	21.5	303.4	44.0	42	65
593	1100	112.4	16.3	173.8	25.2	50	65
704	1300	53.1	7.7	62.1	9.0	71	65

Note: Specimens were annealed
Source: Ref 15, 483

ASTM A36: Room-temperature tensile property data after elevated-temperature exposure

EXP TEMP(a) C	EXP TEMP(a) F	YS MPa	YS ksi	UTS MPa	UTS ksi	EL(b) %	RA %
...	...	316.5	45.9	453.0	65.7	39	67
482	900	217.2	31.5	328.2	47.6	45	75
566	1050	245.5	35.6	306.1	44.4	46	77

(a) 83,000 h without stress. (b) In 25 mm (1 in.)
Source: Ref 15, p 483

ASTM A36: Creep and rupture property data for annealed steel

TEMP C	CRSTR 1%/ 10000 h MPa	CRSTR 1%/ 10000 h ksi	CRSTR 1%/ 100000 h MPa	CRSTR 1%/ 100000 h ksi	RPSTR 1000 h MPa	RPSTR 1000 h ksi	RPSTR 10000 h MPa	RPSTR 10000 h ksi	RPSTR 100000 h MPa	RPSTR 100000 h ksi
427	147.6	21.4	95.2	13.8	262,0	38.0	171.0	24.8	110.3	16.0
482	68.3	9.9	41.4	6.0	127.6	18.5	85.5	12.4	56.5	8.2
538	31.7	4.6	17.9	2.6	65.5	9.5	43.4	6.3	29.0	4.2

Source: Ref 15, p 484

ASTM A36: Effect of temperature on Charpy keyhole impact values and hardness

TEMP C	TEMP F	IS J	IS ft-lb	HARD HB
Unexposed				
...	...	50.9	37.5	128
Exposed 1000 h				
482	900	52.9	39.0	126
566	1050	55.3	40.8	115
649	1200	21.7	16.0	104
Exposed 10000 h				
482	900	56.8	41.9	109
566	1050	60.9	44.9	93
649	1200	71.3	52.6	86

Note: Specimens were normalized at 899 C (1650 F)
Source: Ref 15, p 484

ASTM A36: Stress-to-rupture property data

TEMP C	TEMP F	RPSTR 1000 h MPa	RPSTR 1000 h ksi	RPSTR 10000 h MPa	RPSTR 10000 h ksi
Unexposed					
482	900	94.5	13.7	66.2(a)	9.6(a)
566	1050	38.6	5.6	24.8(a)	3.6(a)
Exposed without stress for 83000 h					
482	900	82.1	11.9	63.4(a)	9.2(a)
566	1050	35.9	5.2	24.1(a)	3.5(a)

Note: Specimens were normalized at 899 C (1650 F)
(a) Values obtained by extrapolation
Source: Ref 15, p 484

ASTM A106(B)

UNS number: K03006
Nominal composition: 0.300% C max, 0.290 to 1.060% Mn, 0.048% P max, 0.058% S max, 0.100% Si min

Product form: Pipe
Description: ASTM A106 is used for seamless carbon steel pipe for high-temperature service.

Mechanical Properties

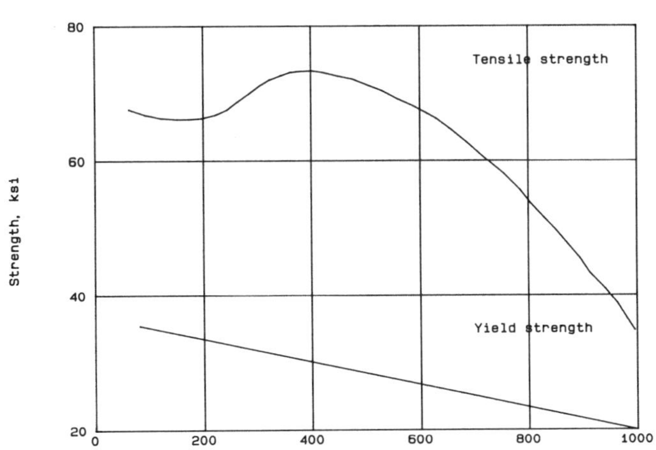

ASTM A106(B): Tensile and yield strength of grade B pipe. Source: Ref 21, p 23

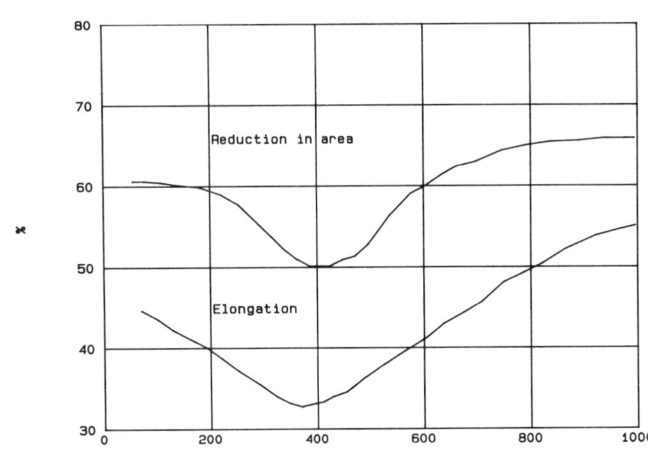

ASTM A106(B): Elongation and reduction in area of grade B pipe. Source: Ref 21, p 23

ASTM A192

UNS number: K01201
Nominal composition: 0.060 to 0.180% C, 0.270 to 0.630% Mn, 0.048% P max, 0.058% S max, 0.250% Si max

Product form: Tube
Description: ASTM A192 is used for seamless carbon steel boiler tubes for high-pressure service.

Mechanical Properties

ASTM A192: Creep relaxation of steel tubing

TEMP C	TEMP F	TIME h	CRIS MPa	CRIS ksi	CRSRD MPa	CRSRD ksi	RA %
Condition 1							
260	500	10	548.8	79.6	460.6	66.8	16
399	750	10	375.1	54.4	168.9	24.5	55
Condition 2							
260	500	10	138.6	20.1	83.4	12.1	49
538	750	10	117.9	17.1	29.6	4.3	75
Condition 3							
260	500	10	364.1	52.8	313.7	45.5	15
399	750	10	315.1	45.7	197.9	28.7	37
538	1000	10	171.0	24.8	19.3	2.8	77

Note: Specimens were 6.35-mm (0.25-in.) diam, 101.6-mm (4-in.) gage length machined from tubing. Condition 1: As received with an ultimate tensile strength of 616.4 MPa (89.4 ksi) and a tensile yield strength of 562.6 MPa (81.6 ksi). Condition 2: Annealed at 954 C (1750 F) for 30 min, and furnace cooled. Condition 3: Water quenched from 954 C (1750 F), tempered for 30 min at 538 C (1000 F)
Source: Ref 21, p 25

ASTM A192: Creep relaxation of steel tubing. Curve 1: Quenched and tempered at 1750 F 30 min, water quenched, 1100 F 90 min, air cooled. Tested at 500 F. Curve 2: Annealed 1750 F 30 min, furnace cooled. Tested at 500 F. Curve 3: Same as Curve 2, but tested at 1000 F. Source: Ref 21, p 25

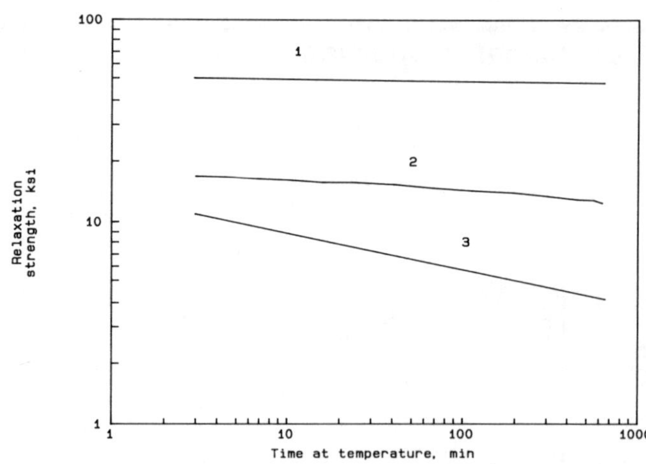

ASTM A227

UNS number: K06501
Specification: ASTM A227
Nominal composition: 0.450 to 1.300% C, 0.300 to 1.300% Mn, 0.040% P max, 0.050% S max, 0.150 to 0.350% Si
Product forms: Rod, wire
Description: ASTM A227 pertains to two classes of round, hard drawn, spring steel wire with properties and qualities suited for the manufacture of mechanical springs and wire forms. ASTM A227M is a metric companion to ASTM A227. These steels are made by the open-hearth, basic-oxygen, or electric-furnace process. A sufficient discard is made to ensure freedom from injurious piping and undue segregation. The wire is cold drawn to produce the desired mechanical properties. Test specimens taken from either end of the coils are required to meet tensile and wrap test requirements. The surface of the wire as received is to be smooth and free of rust. No serious die marks, scratches, or seams may be present.

Mechanical Properties

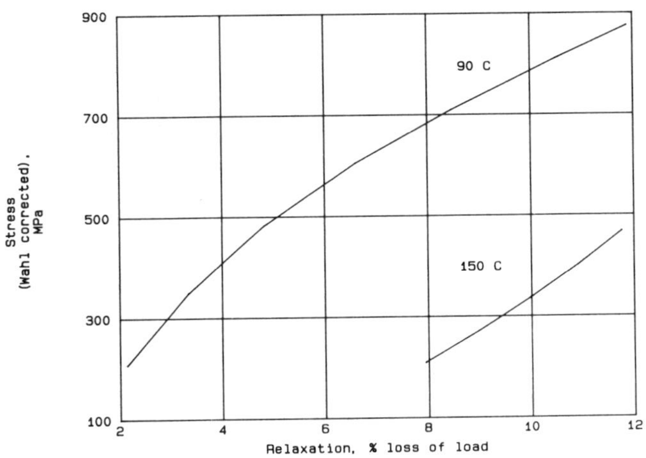

ASTM A227: Relaxation curves for steel helical compression springs after 72 h at temperature. Source: Ref 1, p 298

ASTM A228

UNS number: K08500
Specification: ASTM A228
Nominal composition: 0.700 to 1.000% C, 0.200 to 0.600% Mn, 0.025% P max, 0.030% S max, 0.100 to 0.300% Si
Product forms: Rod, wire
Description: ASTM A228 describes a high-quality, round, hard drawn, music steel, spring wire that is uniform in mechanical properties. This wire is intended for use in manufacturing springs subject to high stresses or requiring good fatigue properties. ASTM A228 steel is made by the open-hearth, basic-oxygen, or electric-furnace process. A sufficient discard is made to ensure freedom from injurious piping and undue segregation. The wire is cold drawn to produce the desired mechanical properties. Music wire is supplied with many different types of finishes, including bright, phosphate, tin, cadmium, and others. The finish desired is specified on the purchase order. The surface of the wire must be smooth and free from defects, such as seams, pits, die marks, and other defects tending to impair the use of the wire for springs. Surface requirements are negotiated when the wire is ordered.

Mechanical Properties

ASTM A228: Relaxation curves for steel helical compression springs after 72 h at temperature. Source: Ref 1, p 298

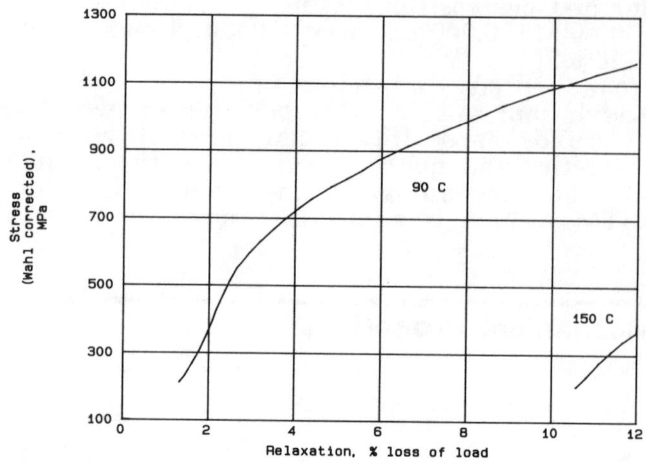

ASTM A229

UNS number: K07001

Specifications: ASTM A229

Nominal composition: 0.550 to 0.850% C, 0.300 to 1.200% Mn, 0.040 P max, 0.030% S max, 0.100 to 0.350% Si

Product forms: Rod, wire

Description: ASTM A229 consists of two classes of oil-tempered spring wire. The material is similar to AISI/SAE 1065 and is used in the pretempered condition for manufacturing mechanical springs and wire forms. ASTM A229 steel is made by the open-hearth, basic-oxygen, or electric-furnace process. The wire is drawn to diameter, then hardened and tempered to produce the desired mechanical properties. In addition to tensile requirements, the wire is subjected to a wrap test. Wire 4.10 mm (0.160 in.) and smaller in diameter is to be wound on itself as an arbor without breakage. Wire up to and including 7.90 mm (0.310 in.) diameter is to be wound on a mandrel twice the diameter without breakage. Wrap test on wire larger than 7.90 mm (0.310 in.) is not applicable. The values stated in English units are to be regarded as the standard.

Mechanical Properties

ASTM A229: Relaxation curves for steel helical compression springs after 72 h at temperature. Source: Ref 1, p 298

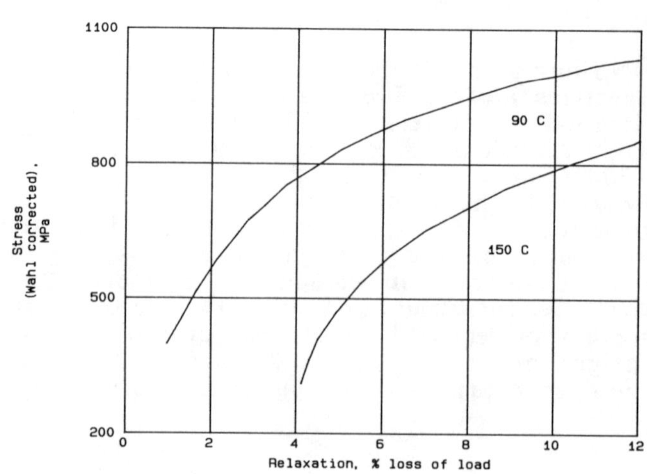

ASTM A231

UNS number: K15048
Specification: ASTM A231
Product forms: Rod, wire
Description: ASTM A231 describes round, chromium-vanadium, alloy steel, spring wire with properties intended for manufacturing springs used at moderately elevated temperatures. ASTM A231 wire is supplied in either the annealed and cold drawn or oil-tempered condition, as specified. The surface of the wire is furnished free of imperfections, such as seams, pits, die marks, scratches, and other imperfections that tend to impair the fatigue value of the springs. Transverse sections of the valve-spring-quality wire, when properly mounted, polished, and etched, shall not show completely decarburized (carbon-free) areas when examined at a magnification of 100 diameters. Partial decarburization shall not exceed a depth of 0.025 mm (0.001 in.) on wire 4.88 mm (0.192 in.) and smaller in diameter, or 0.0381 mm (0.0015 in.) on wire larger than 4.88 mm (0.192 in.) in diameter. The values stated in English units are to be regarded as the standard.

Mechanical Properties

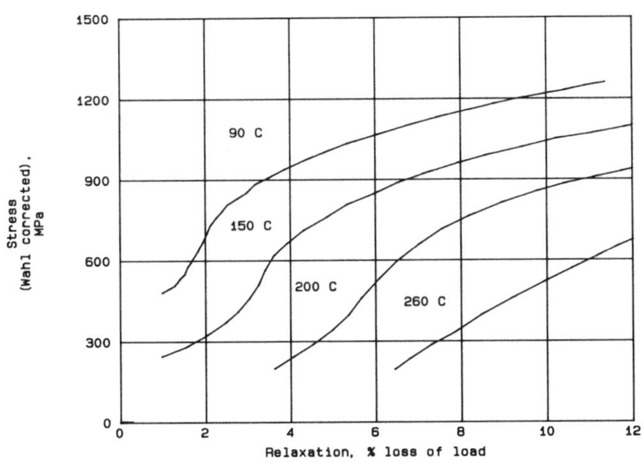

ASTM A231: Relaxation curves for steel helical compression springs after 72 h at temperature. Source: Ref 1, p 298

ASTM A401

UNS number: K15590
Specifications: ASTM A401, FED QQ-W-412 (I)
Nominal composition: 0.510 to 0.590% C, 0.600 to 0.800% Cr, 0.600 to 0.800% Mn, 0.035% P max, 0.040% S max, 1.200 to 1.600% Si
Product forms: Rod, wire
Description: ASTM A401 describes round, chromium-silicon, alloy steel spring wire with properties intended for manufacturing springs that are resistant to set when used at moderately elevated temperatures. This wire is furnished in either the annealed and cold drawn or oil tempered condition. The surface of ASTM A401 wire as received shall be free of rust and excessive scale. No serious die marks, scratches, or seams may be present. Based upon examination of an etched transverse specimen, seams are not to exceed 3.5% of the wire diameter, or 0.250 mm (0.010 in.), whichever is smaller, as measured on a transverse section. The values stated in English units are to be regarded as the standard. ASTM A401 steel may be made by the open-hearth, basic-oxygen, or electric-furnace process.

Mechanical Properties

ASTM A401: Relaxation curves for steel helical compression springs after 72 h at temperature. Source: Ref 1, p 298

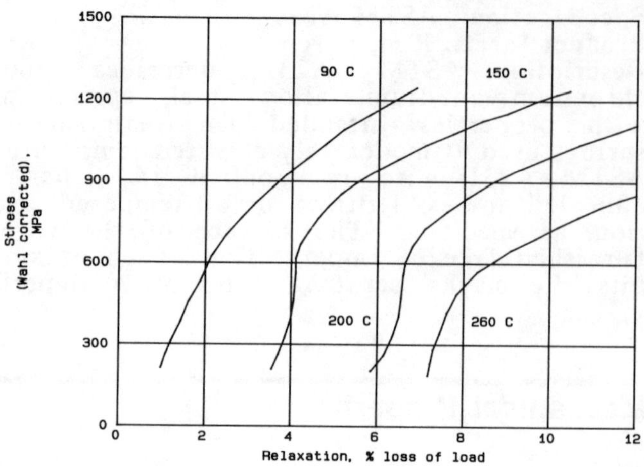

ASTM A441

UNS number: K12211
Nominal composition: 0.220% C max, 0.200% Cu min, 0.850-1.250% Mn, 0.040% P max, 0.050% S max, 0.300% Si max, 0.020% V min
Product forms: Bar, plate, wire, structural shapes

Description: ASTM A441 is a vanadium-modified version of A440 for welded construction. The high-strength low-alloy steel has a minimum yield strength range of 275 to 345 MPa (40 to 50 ksi.

Mechanical Properties

ASTM A441: Ratio of yield stress at elevated temperatures to room-temperature values

TEMP C	TEMP F	YS
27	80	1.00
93	200	0.93
204	400	0.86
316	600	0.82
427	800	0.77
538	1000	0.60
649	1200	0.37
760	1400	0.16
871	1600	0.09
1038	1900	0.03

Source: Ref 21, p 5

ASTM A441: Ratio of tensile strength at elevated temperatures to room-temperature values

TEMP C	TEMP F	UTS
27	80	1.00
93	200	0.93
204	400	0.93
316	600	0.95
427	800	0.81
538	1000	0.54
649	1200	0.30
760	1400	0.14
871	1600	0.11
1038	1900	0.05

Source: Ref 21, p 5

ASTM A537

UNS number: K02400
Specifications: ASTM A537
Nominal composition: 0.240% C max, 0.700 to 1.600% Mn, 0.035% P max, 0.040% S max, 0.150 to 0.300% Si
Product form: Plate

Description: ASTM A537 is available as pressure vessel quality, high-strength low-alloy steel plate that combines weldability with toughness at low temperatures.
Typical applications: Offshore platforms, ships, process tanks, pressure vessels, and storage tanks.

Mechanical Properties

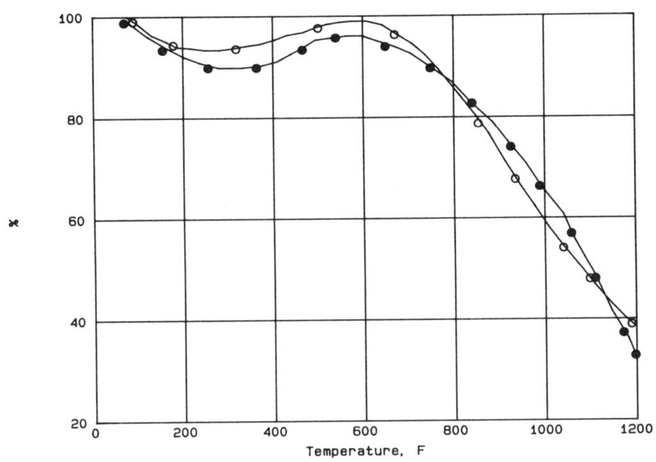

ASTM A537: Short-time tensile strength reported as percentage of room-temperature value. Open circle: 0.667-in. thick, 70 to 85 ksi room-temperature tensile strength, 50 ksi minimum room-temperature yield strength. Solid circle: 0.75-in. thick, 80 to 100 ksi room-temperature tensile strength, 60 ksi minimum room-temperature yield strength. Source: Ref 21, p 3

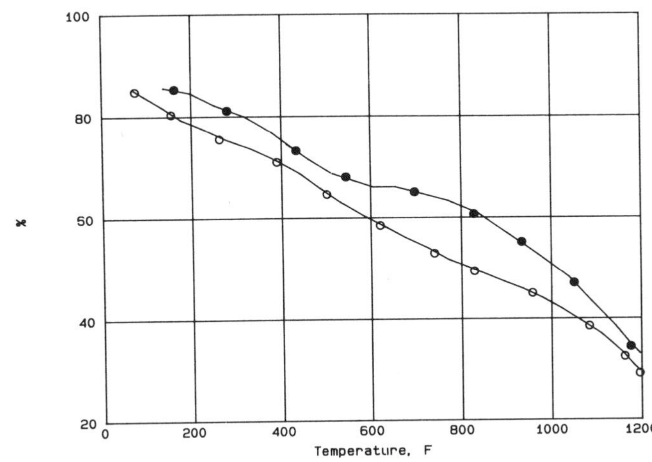

ASTM A537: Short-time yield strength reported as percentage of room-temperature values. Open circle: 0.667 in.-thick, 70 to 85 ksi room-temperature tensile strength, 50 ksi minimum room-temperature yield strength. Solid circle: 0.75 in.- thick, 80 to 100 ksi room-temperature tensile strength, 60 ksi minimum room-temperature yield strength. Source: Ref 21, p 3

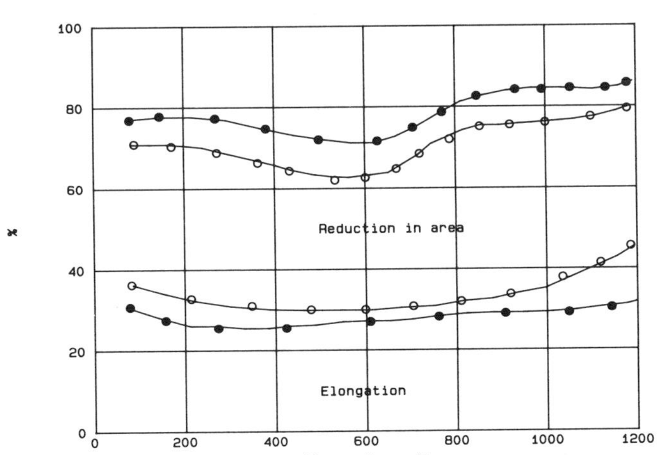

ASTM A537: Short-time reduction in area and elongation. Open circle: 0.667-in., 70 to 85 ksi room-temperature tensile strength, 50 ksi minimum room-temperature yield strength. Solid circle: 0.75-in. thick, 80 to 100 ksi room-temperature tensile strength, 60 ksi minimum room-temperature yield strength. Source: Ref 21, p 3

5. *Elevated-Temperature Service Steels*

Elevated-Temperature Service Steels

Low-alloy constructional steels are used extensively in a variety of applications that involve exposure to elevated temperatures. The term "low-alloy" designates steels that generally contain no more than 10% total alloy content. "Elevated temperature" generally indicates the range 370 to 650 C (700 to 1200 F), although properties at temperatures of 800 to 1200 C (1500 to 2200 F) are very important in mill processing and metal forming.

The materials covered in this section include C-Si steels (ASTM A106 pipe grades, for example); C-Mo steels (A204 pressure vessel plates, for example); Cr-Mo steels (2-1/4Cr-1Mo, for example); Cr-Mo-V steels (AISI 601, for example); and Ni-Cr-Mo/Ni-Cr-Mo-V steels.

In general, carbon is usually limited to 0.20% in C-Mo steels intended for elevated temperature service. In chromium-bearing steels, the maximum is usually 0.15%. Carbon increases short-term elevated temperature tensile strength, but does not add appreciably to creep resistance at temperatures above 540 C (1000 F), because carbides eventually spheroidize.

Manganese prevents hot shortness by forming dispersed manganese sulfide inclusions, whereas phosphorus and sulfur are considered undesirable, because they reduce elevated-temperature ductility (and thus stress-rupture life and thermal fatigue life). Silicon increases elevated-temperature strength and increases the scaling resistance of low-chromium steels.

Molybdenum, even in small amounts (0.1 to 0.5%), enhances resistance to deformation at elevated temperatures. Vanadium may be added to higher carbon grades to provide added resistance to tempering. Chromium in small amounts (0.5%) is a carbide former and stabilizer; in larger amounts (up to 9% or more), it increases resistance to scaling in air. Chromium is most effective as a strengthener in Cr-Mo steels at levels of 1 to 2%. It makes its greatest contribution to creep resistance at a level of about 2.25%

Short-term elevated-temperature mechanical properties of primary interest in the design of structures for high-temperature service are ultimate tensile strength, yield strength, and elastic modulus. Reduction in area during tensile testing and compressive yield strength sometimes are useful, particularly in analysis of hot forming operations. Long-term elevated-temperature mechanical properties of primary interest are creep rate, rupture life, and relaxation behavior.

Physical Properties

Low-alloy steels for elevated-temperature service

UNS No.	Composition	C %	Mn %	Si %	P %	S %	Cr %	Ni %	Mo %
K03006	C-Si	0.30 max	0.29-1.06	0.10 min	0.048 max	0.058 max	...	...	...
K11820	C-1/2Mo	0.18 max	0.90 max	0.15-0.30	0.035 max	0.040 max	...	...	0.45-0.60
K11562	1Cr-1/2Mo	0.15 max	0.30-0.61	0.50 max	0.045 max	0.045 max	0.50-1.25	...	0.44-0.65
K21590	2-1/4Cr-1Mo	0.15 max	0.30-0.60	0.50 max	0.035 max	0.035 max	2.0-2.5	...	0.90-1.10
J82090	9Cr-1Mo	0.02 max	0.35-0.65	1.00 max	0.04 max	0.45 max	8.0-10.0	...	0.90-1.20

Source: Ref 30, p 4.74

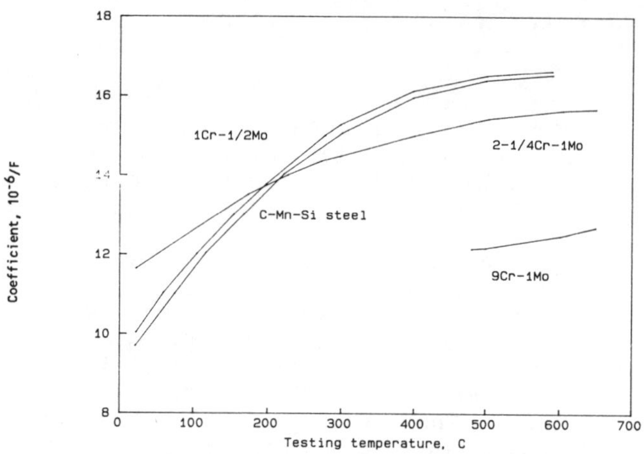

Instantaneous coefficient of linear thermal expansion for several elevated-temperature service steels at indicated temperatures. These are not mean values of the coefficient over a range of temperatures. Source: Ref 1, p 653

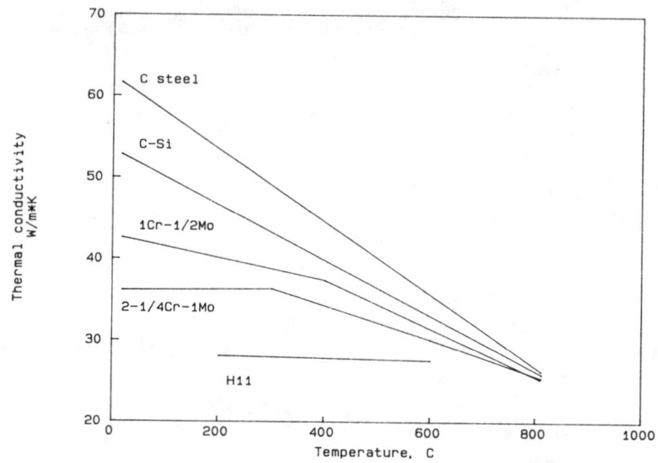

Thermal conductivity of carbon and elevated-temperature service steels. Source: Ref 1, p 652

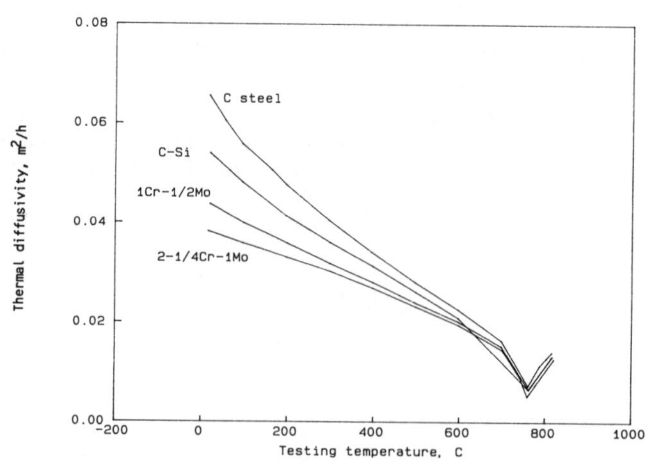

Thermal diffusivity of carbon and several elevated-temperature service steels. Thermal conductivity is divided by product of density and specific heat. Source: Ref 1, p 652

ASME Boiler Code Specifications for 2-1/4Cr-1Mo steel

ASME No.	Grade	Product form
SA-182	F22	Pipe flanges, fillings and valves
SA-199	T22	Seamless cold drawn tubes
SA-213	T22	Seamless ferritic alloy steel tubes
SA-217	WC9	Alloy steel castings
SA-333	P22	Welded and seamless pipe
SA-336	F22	Alloy steel forgings
	F22a	Alloy steel forgings
SA-369	FP22	Ferritic alloy steel forged and bored pipe
SA-387	GR22(a)	Cr-Mo PV plate
	GR22(b)	Cr-Mo PV plate
SA-426	CP22	Centrifugally cast ferritic alloy steel pipe
SA-542	(a)	Cr-Mo alloy steel plate
	(b)	Cr-Mo alloy steel plate

(a) Class 1. (b) Class 2
Source: Ref 1, p 654

Low-alloy steels: Compositions of select steels for elevated-temperature service

UNS No.	C %	Mn %	Si %	Cr %	Mo %	V %
K14675	0.46	0.60	0.26	1.00	0.50	0.30
K23015	0.30	0.55	0.65	1.25	0.50	0.25
K22770	0.27	0.75	0.65	1.25	0.50	0.85

Source: Ref 30, p 4.74

"2-1/4Cr-1Mo Steel"

UNS number: K21590
Specifications: ASTM A182 (F22), ASTM A199 (T22), ASTM A200 (T22), ASTM A213 (T22), ASTM A234 (WP22), ASTM A335 (P22), ASTM A336 (F22, F22a), ASTM A369 (F22), ASTM A387 (22), ASTM A541 (6, 6A), ASTM A542, MIL S-18410
Nominal composition: 0.150% C max, 1.900 to 2.600% Cr, 0.300 to 0.600% Mn, 0.870 to 1.130% Mo, 0.030% P, 0.030% S max, 0.500% Si max

Product forms: Wrought, bar, plate, sheet, strip, tube, pipe, billets, structural shapes

Description: The elevated temperature behavior of 2-1/4Cr-1Mo steel has been studied very thoroughly. The steel has an excellent service record in both conventional and nuclear power plant applications. Ferritic 2-1/4Cr-1Mo has become a reference against which the performance of other steels is measured.

Physical Properties

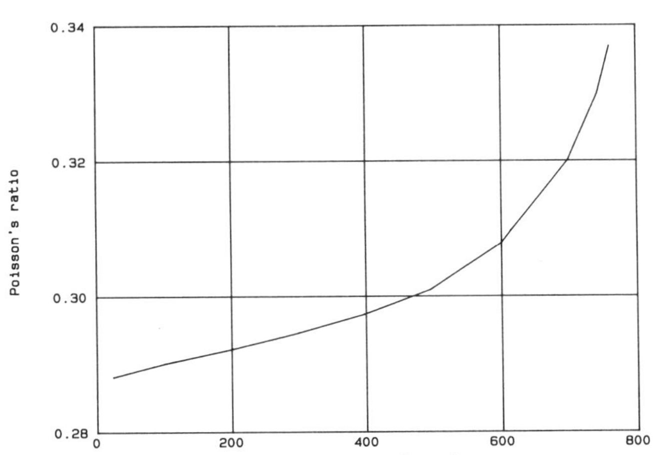

2-1/4Cr-1Mo steel: Poisson's ratio. Source: Ref 1, p 641

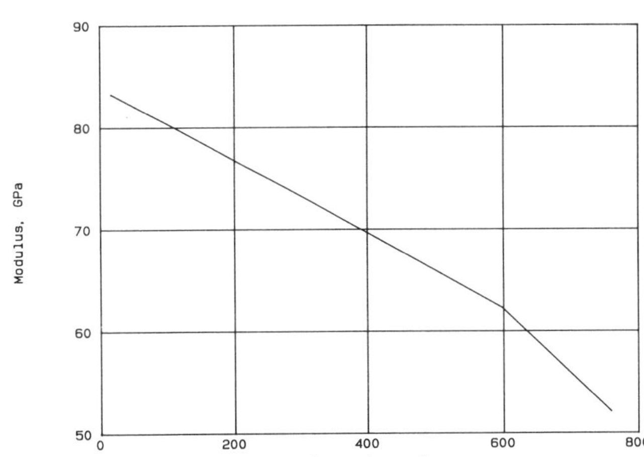

2-1/4Cr-1Mo steel: Modulus of rigidity. Source: Ref 1, p 641

Mechanical Properties

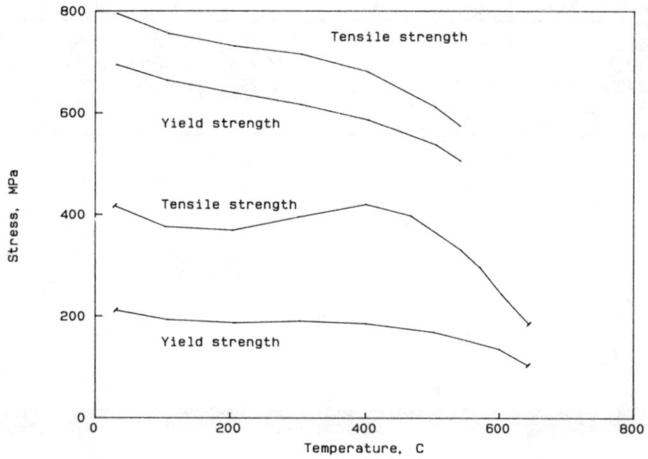

2-1/4Cr-1Mo steel: Tensile and yield strength of annealed specimens (bottom curves) and hardened and tempered specimens (top curves). Source: Ref 1, p 655

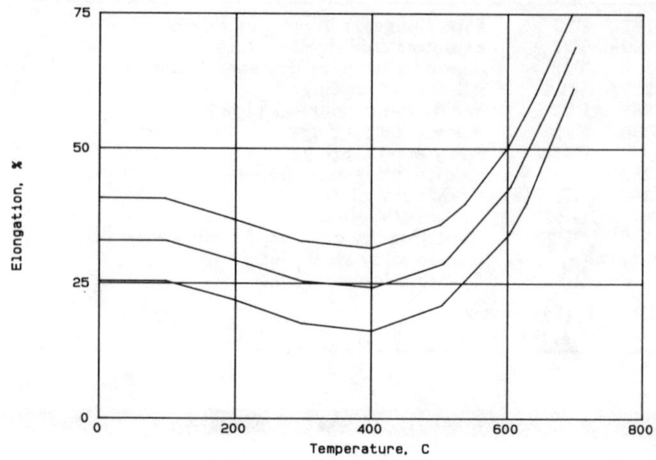

2-1/4Cr-1Mo steel: Scatterband for tensile elongation of annealed specimens. Area between the curves includes +/- two standard deviations from the mean. Source: Ref 1, p 655

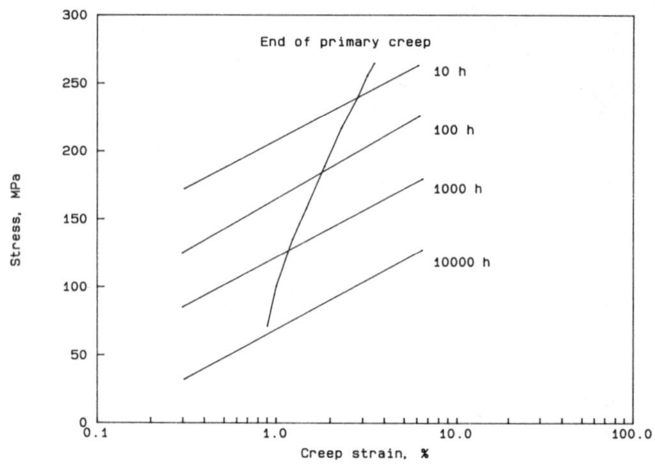

2-1/4Cr-1Mo steel: Creep behavior at 540 C. Stress-creep strain plot. Source: Ref 1, p 644

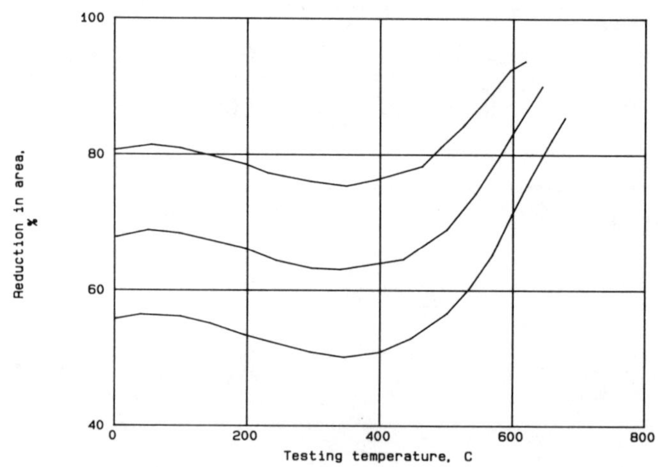

2-1/4Cr-1Mo steel: Scatterband for tensile reduction in area of annealed specimens. Area between the curves includes +/- two standard deviations from the mean. Source: Ref 1, p 655

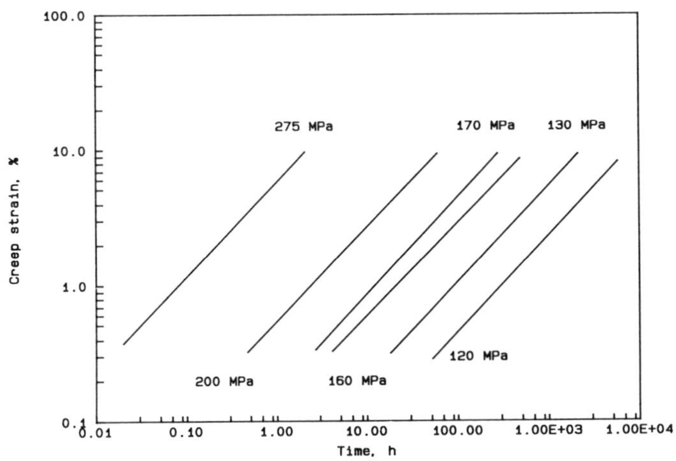

2-1/4Cr-1Mo steel: Creep behavior at 540 C. Creep strain-time plot; constant-stress lines have been drawn parallel. Source: Ref 1, p 644

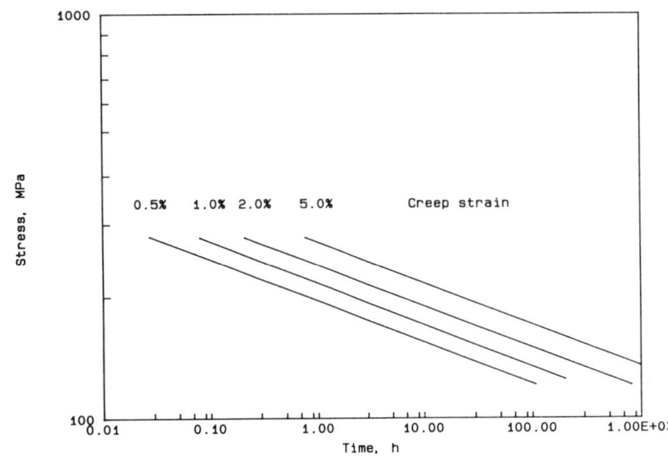

2-1/4Cr-1Mo steel: Creep behavior at 540 C. Stress-time plot; constant-stress lines have been drawn parallel. Source: Ref 1, p 645

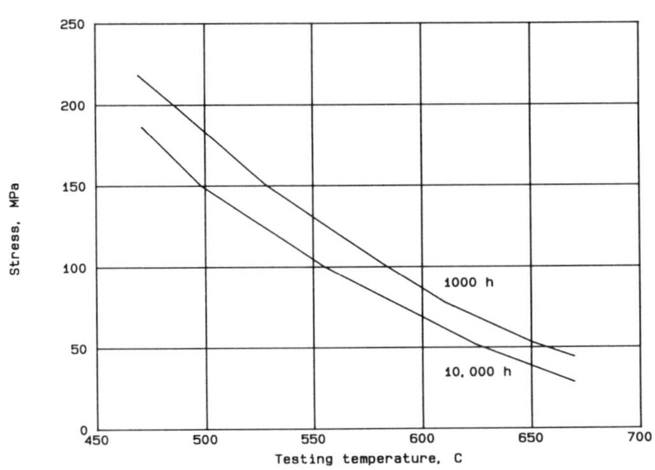

2-1/4Cr-1Mo steel: Effect of prior exposure at test temperature for 10,000 h (without stress) on stress required to cause rupture in 1000 and 10,000 h for normalized and tempered specimens. Source: Ref 1, p 658

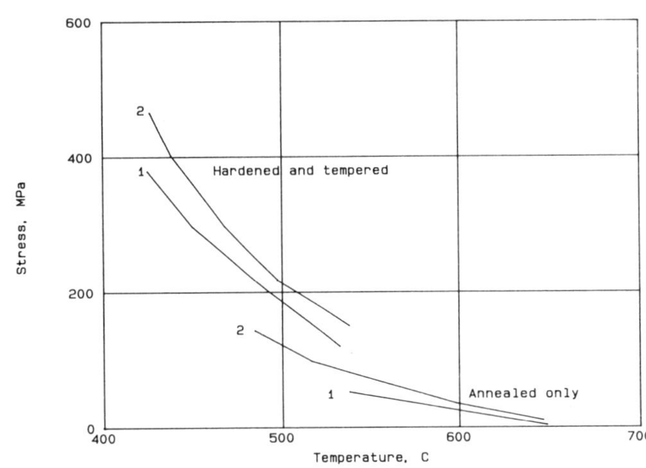

2-1/4Cr-1Mo steel: Stress required to produce a minimum creep rate of 0.001%/h (1) and rupture (2) in 100,000 h for specimens with two different heat treatments. Source: Ref 1, p 655

2-1/4Cr-1Mo Steel: Allowable design stress at elevated temperatures

TEMP C	TEMP F	Stress MPa	Stress 10^3 ksi
480	900	90	13.1
510	950	76	11.0
540	1000	54	7.8
565	1050	40	5.8
595	1100	29	4.2
620	1150	21	3.0
650	1200	14	2.0

Source: Ref 9

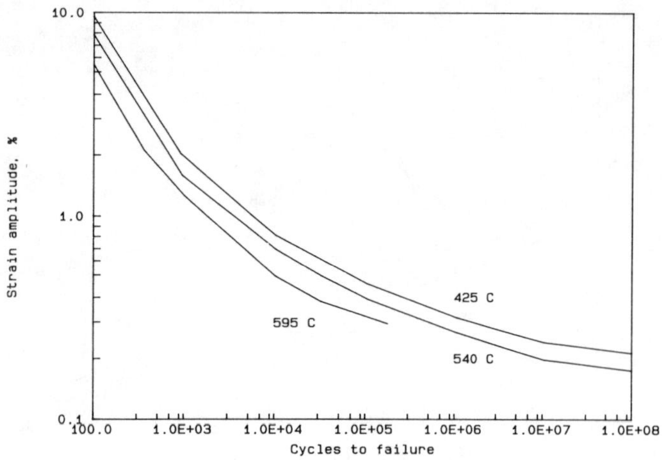

2-1/4Cr-1Mo steel: Strain-controlled fatigue behavior of annealed steel. Strain rate was greater than 4 mm/m*s. Source: Ref 1, p 659

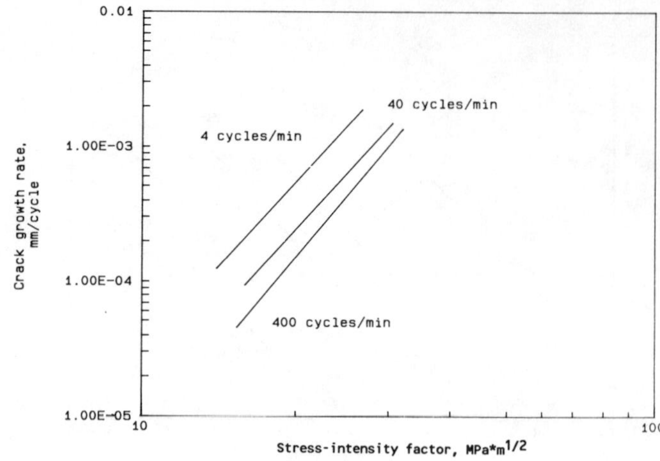

2-1/4Cr-1Mo steel: Effect of cyclic frequency on fatigue crack growth rate. Stress ratio was 0.05, tested in air at 595 C. Source: Ref 1, p 661

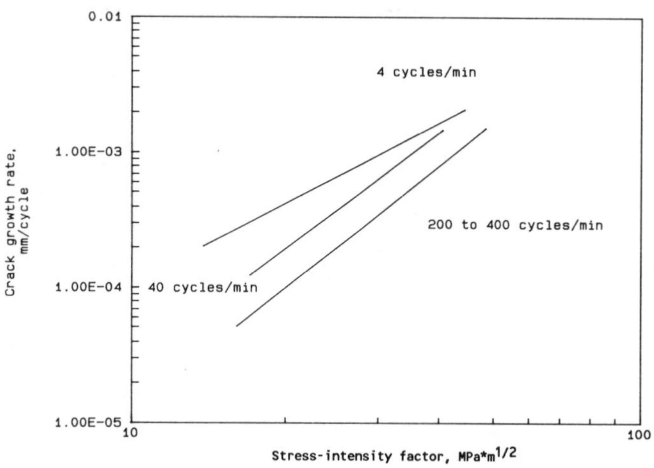

2-1/4Cr-1Mo steel: Effect of cyclic frequency on fatigue crack growth rate. Stress ratio was 0.05, tested in air at 510 C. Source: Ref 1, p 661

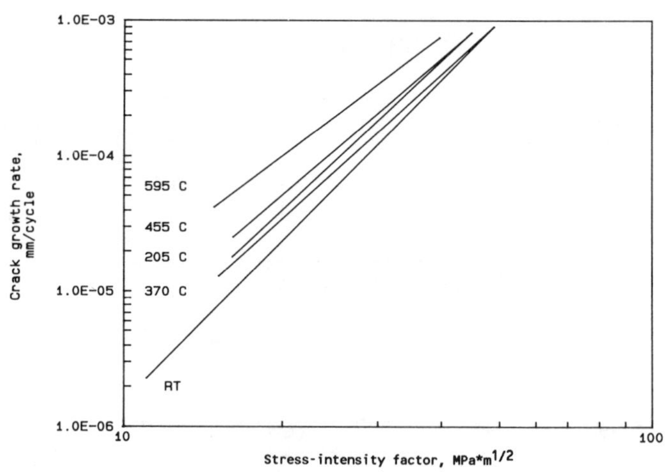

2-1/4Cr-1Mo steel: Effect of temperature on fatigue crack growth rate in specimens tested in air. Stress ratio was 0.05. Test frequency was 400/min. Source: Ref 1, p 660

"9Cr-1Mo Steel"

UNS number: S50400
Specification: ASTM A387 Grade 9
Nominal composition: 0.150% C max, 8.00 to 10.00% Cr, 0.300 to 0.600% Mn, 0.90 to 1.10% Mo, 0.030% P max, 0.030% S max, 1.00% Si max
Product forms: Wrought, bar, plate, sheet, strip, tube, pipe, billets, structural shapes

Mechanical Properties

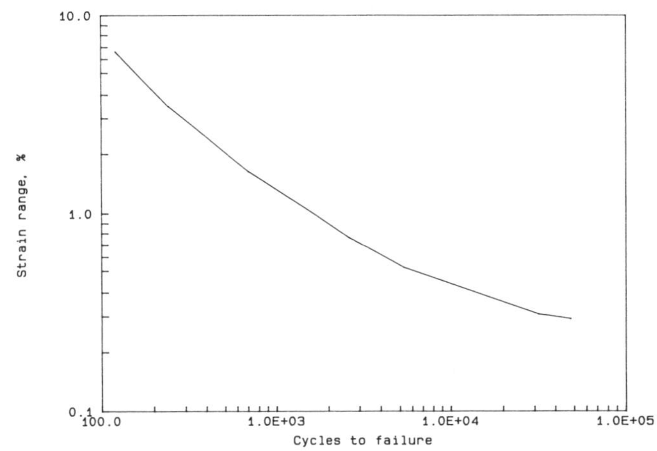

9Cr-1Mo steel: Creep fatigue of specimens at 525 C with or without tensile hold times to 0.5 h dwell time. Source: Ref 23, p 194

"1/2 Cr-Mo-V Steel"

Product forms: Wrought, bar, plate, sheet, strip, tube, pipe, billets, structural shapes

Mechanical Properties

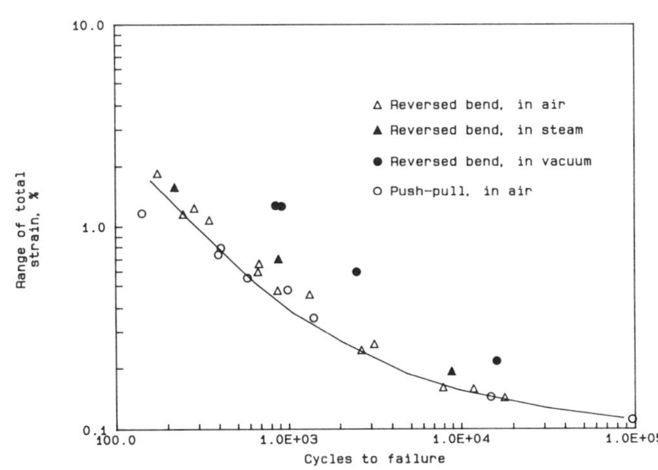

1/2Cr-Mo-V steel: Fatigue endurance behavior of cast specimens at 550 C in air, steam, and vacuum (no dwell period). Source: Ref 23, p 145

17-22A

Alternate designations: 1Cr-1/2Mo-V, AISI 601
UNS number: K14675
Specifications: AMS 6304, MIL S-24502
Nominal composition: 0.460% C, 1.000% Cr, 0.600% Mn, 0.500% Mo, 0.260% Si, 0.300% V

Product forms: Wrought, bar, plate, sheet, strip, tube, pipe, billets, forgings
Description: AISI 601, also known as Alloy 17-22A, is a Cr-Mo-V alloy steel that is conventionally used oil quenched and tempered to a tensile strength of 724 to 965 MPa (105 to 140 ksi). High strength is obtained upon tempering via precipitation of cementite, Mo_2C and VC, the last two of which are responsible for its high-temperature strength. The steel is used primarily for bolting and structural applications that operate at temperatures up to 540 C (1000 F). Short-time exposure at temperatures as high as 595 C (1100 F) can also be tolerated. In terms of corrosion and oxidation resistance, alloy 601 is comparable to other 1Cr-1/2 Mo steels such as those used in the power, boiler, and refinery industries. The vanadium in the steel neither aids nor impairs corrosion or oxidation resistance.

Physical Properties

17-22A: Mean coefficients of linear thermal expansion

TEMP C	TEMP F	THEXP 10^{-6}/C	THEXP 10^{-6}/F
21-427	70-800	13.79	7.66
21-482	70-900	13.93	7.74
21-538	70-1000	14.04	7.80
21-593	70-1100	14.17	7.87
21-649	70-1200	14.27	7.93

Source: Ref 28, p 11

17-22A: Thermal conductivity

TEMP C	TEMP F	THCON W/m*K	THCON Btu/h*ft*F
93	200	33.98	19.63
316	600	36.00	20.80
243	800	34.99	20.21
538	1000	32.69	18.88

Source: Ref 28, p 11

17-22A: Modulus of elasticity

TEMP C	TEMP F	E GPa	E 10^3 ksi
21	70	212.4	30.8
93	200	208.9	30.3
204	400	202.7	29.4
260	500	199.3	28.9
316	600	191.7	27.8
371	700	191.7	27.8
427	800	187.5	27.2
482	900	182.7	26.5
538	1000	177.9	25.8
593	1100	171.7	24.9
649	1200	165.5	24.0

Note: Data was obtained on steel containing 1-1/4% Cr and 1/2% Mo
Source: Ref 28, p 10

17-22A: Modulus of rigidity

TEMP C	TEMP F	G GPa	G 10^3 ksi
21	70	82.1	11.9
93	200	80.7	11.7
204	400	78.6	11.4
260	500	77.2	11.2
316	600	75.2	10.9
371	700	73.8	10.7
427	800	71.7	10.4
482	900	69.6	10.1
538	1000	67.6	9.8
593	1100	64.8	9.4
649	1200	62.7	9.1

Note: Data was obtained on steel containing 1-1/4% Cr and 1/2% Mo
(a) Calculated from modulus of elasticity
Source: Ref 28, p 11

Mechanical Properties

17-22A: Tensile property data

TEMP C	TEMP F	YS MPa	YS ksi	UTS MPa	UTS ksi	EL %	RA %
21	70	710.2	103.0	855.0	124.0	29	61
204	400	...	...	882.6	128.0	19	58
316	600	434.4	63.0	864.6	125.4	21	60
371	700	...	...	755.7	109.6	24	66
427	800	...	...	703.3	102.0	24	68
482	900	372.3	54.0	648.1	94.0	23	71
538	1000	206.9	30.0	532.3	77.2	29	81
593	1100	...	...	491.6	71.3	29	85
649	1200	...	...	267.5	38.8	53	93

Note: The initial strength is 855 MPa (124 ksi) at 21 C (70 F). Specimens were taken in 75-mm (3-in.) rounds, heated to 940 C (1725 F) for 4 h, oil quenched, reheated to 677 C (1250 F) for 8 h, and air cooled. Specimens were held 30 min at temperature before testing
Source: Ref 28, p 11

17-22A: Stress-to-rupture property data

TEMP C	TEMP F	RPSTR 100 h MPa	RPSTR 100 h ksi	RPSTR 1000 h MPa	RPSTR 1000 h ksi	RPSTR 10000 h MPa	RPSTR 10000 h ksi
427	800	579.2	84	496.4	72	441.3	64.0
482	900	462.0	67	386.1	56	310.3	45.0
538	1000	337.9	49	248.2	36	172.4	25.0
593	1100	206.9	30	124.1	18	65.5	9.5

Note: Strength level was 827 to 931 MPa (120 to 135 ksi) tensile strength at 21 C (70 F)
Source: Ref 28, p 11

17-22A: Effect of temperature on Charpy V-notch impact values

TEMP C	TEMP F	IS(a) J	IS(a) ft-lb	IS(b) J	IS(b) ft-lb
24	75	81	60	99	73
66	150	95	70	102	75
121	250	99	73	102	75
177	350	103	76	119	88
232	450	107	79	118	87
288	550	104	77	111	82

Note: Specimens were held 20 min at temperature before testing. The transition zone of -6.67 C (20 F) to 15.56 C (60 F) is based on 50% (+ or - 10%) fibrous fracture
(a) Test 1. (b) Test 2
Source: Ref 28, p 11

17-22AS

Alternate designations: 1/4Cr-1/2Mo-V, AISI 602
UNS number: K23015
Specifications: AMS 6302, AMS 6385, AMS 6458
Nominal composition: 0.300% C, 1.250% Cr, 0.550% Mn, 0.500% Mo, 0.650% Si, 0.250% V
Product forms: Wrought, bar, plate, sheet, strip, tube, pipe, billets, forgings
Description: AISI 602, also known as Alloy 17-22AS, is a low-alloy, high-temperature constructional steel that possesses useful strength after heat treatment at temperatures up to 540 C (1000 F). For best high-temperature properties, normalizing from 940 C (1725 F) and then tempering to the desired hardness is recommended. Oil quenching is chosen instead of normalizing for large sections. Alloy 602 is used for high-temperature bolting and for structural parts such as chain pins in elevators for petroleum catalytic units. The alloy also resists thermal cracking and heat checking and has sufficient strength to resist hot upsetting for short times under rapidly applied stresses at high temperatures. These characteristics make it suitable for hot die applications and for aircraft brakes, both disc and drum type. Recommended coatings for high-temperature service include hot dip aluminizing and nickel electroplating.

Physical Properties

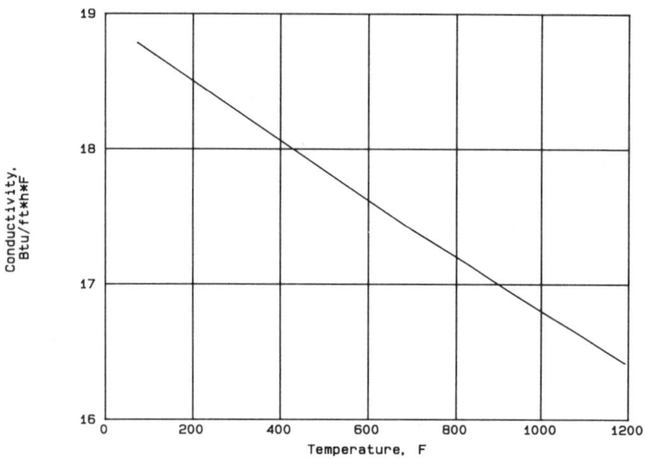

17-22AS: Thermal conductivity. Source: Ref 22, p 5

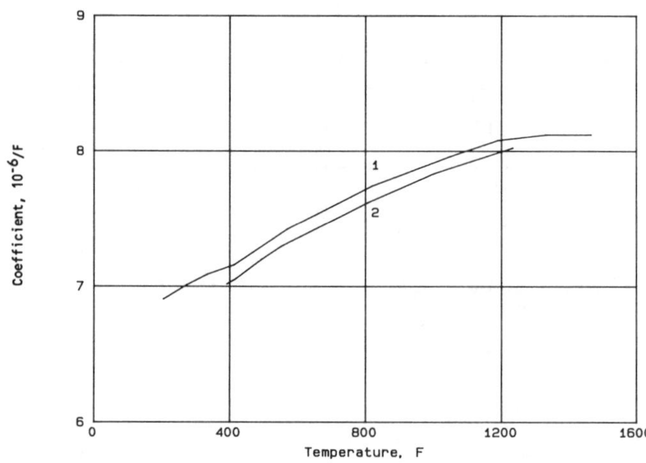

17-22AS: Mean coefficient of linear thermal expansion from room temperature to indicated temperature. Curve 1: heated to 1725 F, air cooled, held at 1200 F for 6 h. Curve 2: heated to 1650 F, oil quenched, held at 1200 F for 6 h. Source: Ref 22, p 5

17-22AS: Mean coefficients of linear thermal expansion

TEMP C	TEMP F	THEXP(a) 10^{-6}/C	THEXP(a) 10^{-6}/F	THEXP(b) 10^{-6}/C	THEXP(b) 10^{-6}/F
21-204	70-400	12.62	7.01	12.38	6.88
21-316	70-600	13.32	7.40	13.18	7.32
21-427	70-800	13.70	7.61	13.63	7.57
21-482	70-900	13.93	7.74	13.81	7.67
21-538	70-1000	14.13	7.85	13.99	7.77
21-593	70-1100	14.29	7.94	14.15	7.86
21-649	70-1200	14.42	8.01	14.31	7.95

(a) Normalized at 940 C (1725 F), tempered at 649 C (1200 F) for 6 h. (b) Oil quenched at 899 C (1650 F), tempered at 649 C (1200 F) for 6 h
Source: Ref 28, p 12

17-22AS: Modulus of elasticity

TEMP C	TEMP F	E GPa	E 10^3 ksi
21	70	209.6-213.7	30.4-31.0
93	200	202.7-209.6	29.4-30.4
149	300	202.7-205.5	29.4-29.8
204	400	198.6-202.0	28.8-29.3
260	500	194.4-197.2	28.2-28.6
316	600	188.9-193.1	27.4-28.0
371	700	184.1-188.2	26.7-27.3
427	800	179.3-184.1	26.0-26.7
482	900	175.8-179.3	25.5-26.0
538	1000	170.3-173.8	24.7-25.2
593	1100	165.5-168.9	24.0-24.5
649	1200	157.2-162.7	22.8-23.6
704	1300	147.6-153.1	21.4-22.2

Source: Ref 28, p 12

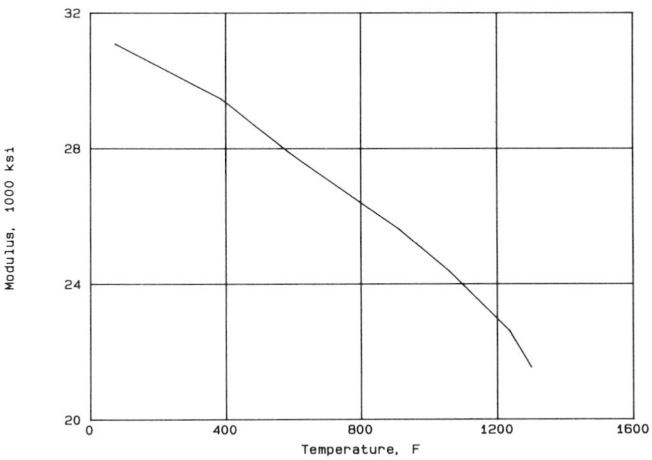

17-22AS: Dynamic modulus of elasticity. Source: Ref 22, p 17

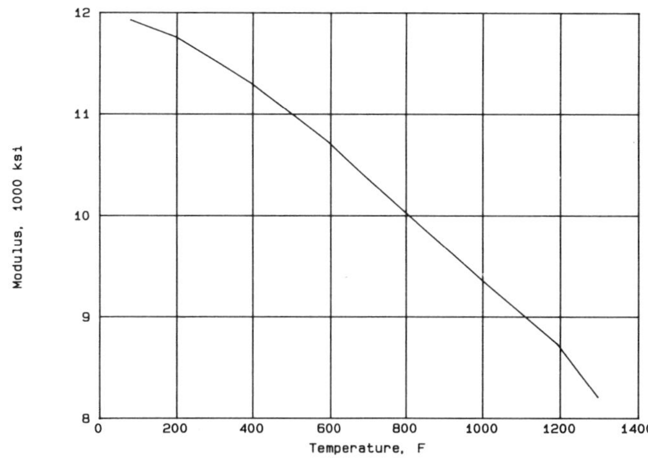

17-22AS: Modulus of rigidity. Source: Ref 22, p 17

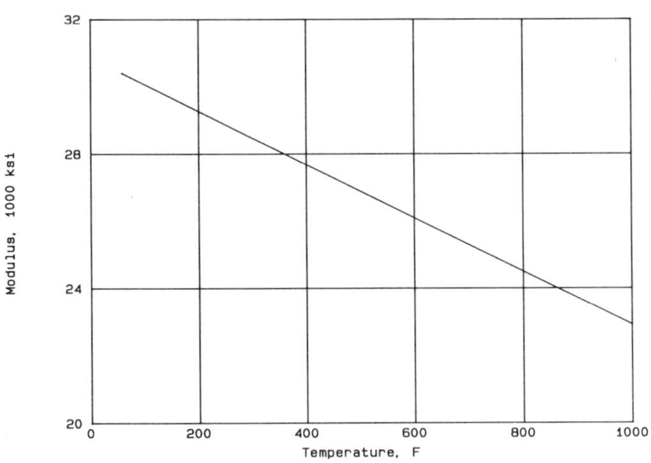

17-22AS: Static modulus of elasticity of 0.062-in. sheet, heat treated at 1750 F for 0.25 h, oil quenched, and tempered at 1300 F for 1 h. Source: Ref 22, p 17

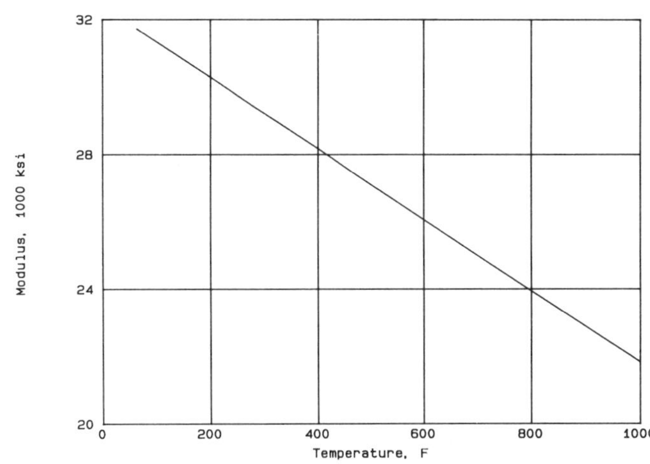

17-22AS: Static modulus of elasticity in compression of 0.062-in. sheet, heat treated at 1750 F for 0.25 h, oil quenched, and tempered at 1300 F for 1 h. Source: Ref 22, p 17

17-22AS: Modulus of rigidity

TEMP C	TEMP F	G GPa	G 10^3 ksi
21	70	80.7-83.4	11.7-12.1
93	200	80.0-81.4	11.6-11.8
149	300	78.6-80.0	11.4-11.6
204	400	77.2-78.6	11.2-11.4
260	500	75.2-76.5	10.9-11.1
316	600	73.1-74.5	10.6-10.8
371	700	71.0-72.4	10.3-10.5
427	800	75.2-70.3	10.9-10.2
482	900	66.9-68.3	9.7-9.9
538	1000	64.8-66.2	9.4-9.6
593	1100	62.7-64.1	9.1-9.3
649	1200	60.0-60.7	8.7-8.8
704	1300	56.5-57.9	8.2-8.4

Source: Ref 28, p 12

17-22AS: Poisson's ratio

TEMP C	TEMP F	PR
21	70	0.285-0.296
93	200	0.287-0.296
149	300	0.288-0.296
204	400	0.290-0.298
260	500	0.292-0.300
316	600	0.294-0.302
371	700	0.297-0.304
427	800	0.298-0.306
482	900	0.301-0.310
538	1000	0.304-0.313
593	1100	0.306-0.313
649	1200	0.311-0.322
704	1300	0.317-0.327

Source: Ref 28, p 12

17-22AS: Poisson's ratio. Source: Ref 22, p 17

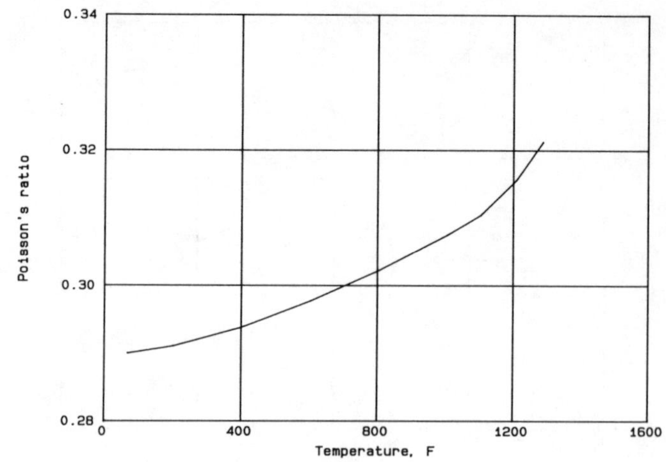

Mechanical Properties

17-22AS: Room-temperature tensile properties of bar previously exposed to stress

EXP TEMP C	EXP TEMP F	STREX MPa	STREX ksi	EXP TIME h	UTS MPa	UTS ksi	YS MPa	YS ksi	EL %	RA %
Heated to 940 C (1725 F), air cooled and tempered at 677 C (1250 F), held 6 h										
...	...	...	...	...	882.6	128	744.7	108	21	61
482	900	224.1	32.5	1096	910.1	132	765.3	111	19	58
510	950	137.9	20.0	1174	813.6	118	634.3	92	20	60
538	1000	103.4	15.0	1270	779.1	113	586.1	85	22	60
Heated to 940 C (1725 F), air cooled and tempered at 663 C (1225 F), held 6 h										
...	...	...	...	...	1020.5	148	882.6	128	16.5	56
482	900	224.1	32.5	1280	1034.3	150	896.4	130	15.5	54
510	950	258.6	37.5	1104	1006.7	146	896.4	130	17	54
538	1000	158.6	23.0	1273	992.9	144	868.8	126	16	54
Heated to 940 C (1725 F), air cooled and tempered at 649 C (1200 F), held 6 h										
...	...	...	...	...	1068.7	155	979.1	142	16	51
482	900	413.7	60.0	2012	1041.1	151	937.7	136	16	50
593	1100	34.5	5.0	1314	899.5	129	765.3	111	19	55
Heated to 899 C (1650 F), oil quenched and tempered at 690 C (1275 F), held 6 h										
...	...	...	...	...	930.8	135	910.1	132	21	65
482	900	172.4	25.0	1010	930.8	135	841.2	122	21	61
510	950	120.7	17.5	1053	951.5	138	848.1	123	19	61
538	1000	69.0	10.0	1356	910.1	132	813.6	118	21	61

Source: Ref 22, p 7

17-22AS: Effect of heat treatment on high-temperature tensile properties

TEMP C	TEMP F	LIMIT(a) ksi	YS(b) MPa	YS(b) ksi	YS MPa	YS ksi	UTS MPa	UTS ksi	EL %	RA %
Normalized and tempered, 235 HB										
27	80	105.0	741.2	107.5	741.2	107.5	882.6	128.0	21	61
260	500	60.0	465.4	67.5	496.4	72.0	739.49	107.25	20	57
371	700	37.5	424.0	61.5	458.5	66.5	729.15	105.75	24	65
427	800	35.0	444.7	64.5	468.9	68.0	658.5	95.5	24	70
482	900	32.5	424.0	61.5	458.5	66.5	610.2	88.5	23	71
538	1000	30.0	396.5	57.5	444.7	64.5	558.5	81.0	28	83
Normalized and tempered, 285 HB										
27	80	110.0	861.9	125.0	875.7	127.0	1032.5	149.75	16	56
260	500	67.5	765.3	111.0	799.8	116.0	961.9	139.5	16	55
371	700	62.5	692.9	100.5	730.9	106.0	879.1	127.5	18	61
427	800	60.0	627.4	91.0	696.4	101.0	830.8	120.5	18	64
482	900	55.0	593.0	86.0	620.6	90.0	761.9	110.5	20	70
538	1000	40.0	510.2	74.0	534.4	77.5	627.4	91.0	18	72
593	1100	...	358.5	52.0	399.9	58.0	496.4	72.0	27	80
649	1200	...	213.7	31.0	241.3	35.0	303.4	44.0	27	80
704	1300	...	96.5	14.0	117.2	17.0	151.7	22.0	52	77
Normalized and tempered, 315 to 324 HB										
27	80	100.0	903.2	131.0	927.4	134.5	1058.4	153.5	18	53
149	300	87.5	827.4	120.0	858.4	124.5	992.9	144.0	17	53
260	500	77.5	792.9	115.0	837.7	121.5	979.1	142.0	17	53
399	750	70.0	724.0	105.0	775.7	112.5	917.0	133.0	19	60
482	900	67.5	655.0	95.0	699.8	101.5	813.6	118.0	19	68
538	1000	45.0	582.6	84.5	637.8	92.5	748.1	108.5	21	70
593	1100	...	482.7	70.0	527.5	76.5	644.7	93.5	22	76
649	1200	...	331.0	48.0	386.1	56.0	520.6	75.5	23	72
704	1300	...	165.5	24.0	200.0	29.0	251.7	36.5	24	50
760	1400	...	82.7	12.0	96.5	14.0	131.0	19.0	32	60
Oil quenched and tempered, 262 HB										
27	80	130.0	910.1	132.0	910.1	132.0	958.4	139.0	20	63
260	500	77.5	675.7	98.0	689.5	100.0	851.5	123.5	21	62
371	700	57.5	644.7	93.5	672.3	97.5	844.6	122.5	21	66
427	800	52.5	575.7	83.5	613.7	89.0	758.5	110.0	21	70
482	900	45.0	558.5	81.0	586.1	85.0	706.7	102.5	22	73
538	1000	37.5	486.1	70.5	517.1	75.0	589.5	85.5	23	79
Oil quenched, tempered at 427 C (800 F), 444 HB										
27	80	...	1310.1	190.0	1375.6	199.5	1506.6	218.5	12	48
316	600	...	1044.6	151.5	1149.7	166.75	1408.3	204.25	16	58
Oil quenched, tempered at 538 C (1000 F), 415 to 429 HB										
27	80	...	1241.1	180.0	1310.1	190.0	1430.7	207.5	15	51
427	800	...	951.5	138.0	1027.4	149.0	1175.6	170.5	18	65
482	900	...	868.8	126.0	944.6	137.0	1068.7	155.0	18	64

Note: Specimens were held 1 h at test temperature prior to testing. Tempering time was 6 h in all cases
(a) Proportional limit. (b) 0.1% offset
Source: Ref 28, p 13

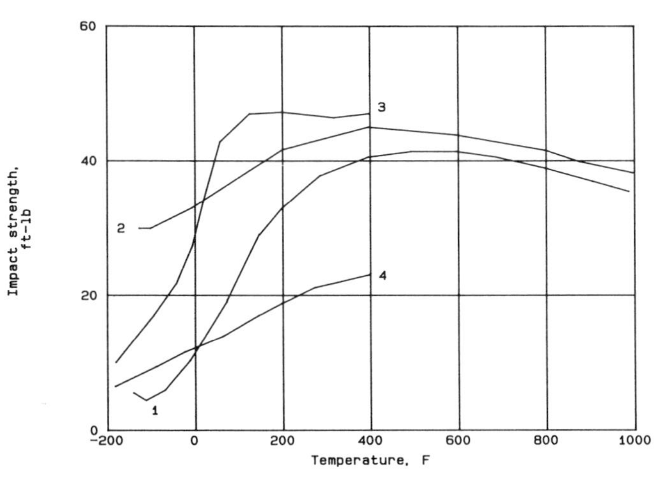

17-22AS: Charpy V-notch impact strength of bar. Curve 1: 1750 F, air cooled, held at 1225 F for 6 h, V-notch. Curve 2: 1650 F, oil quenched, held at 1225 F for 6 h, V-notch. Curve 3: 1650 F, oil quenched, held at 1100 F for 8 h, keyhole notch. Curve 4: 1650 F, oil quenched, held at 800 F for 4 h, keyhole notch. Source: Ref 22, p 17

17-22AS: Stress to produce a minimum creep rate of 0.01%/1000 h

Ttpr C	Tnrm C	HARD HB	CRSTR at 482 C (900 F) MPa	CRSTR at 482 C (900 F) ksi	CRSTR at 510 C (950 F) MPa	CRSTR at 510 C (950 F) ksi	CRSTR at 538 C (1000 F) MPa	CRSTR at 538 C (1000 F) ksi	CRSTR at 593 C (1100 F) MPa	CRSTR at 593 C (1100 F) ksi
940	663	285	303.4	44.0	172.4	25.0	96.5	14.0	...	...
940	677	235	220.6	32.0	96.5	14.0	59.3	8.6	...	...
940	649	315	275.8	40.0	...	...	120.7	17.5	20.7	3.0
899(a)	690	262	137.9	20.0	82.7	12.0	53.8	7.8	...	...

Note: Tempering time was 6 h in all cases
(a) Oil quenched
Source: Ref 28, p 14

17-22AS: Room-temperature mechanical properties before and after creep tests

TEMP C	TEMP F	CRSTR MPa	CRSTR ksi	TIME h	YS(a) MPa	YS(a) ksi	YS MPa	YS ksi	UTS MPa	UTS ksi	EL %	RA %	HARD HB	IS(b) J	IS(b) ft-lb
Normalized at 940 C (1725 F), tempered at 677 C (1250 F); 235 HB															
21	70	...	...	...	741.2	107.5	741.2	107.5	879.1	127.5	21	61	235	57, 71	42, 52
482	900	224.1	32.5	1096	727.4	105.5	765.3	111.0	912.9	132.4	19	58	...	...	...
482	900	275.8	40	1152	...	...	...	...	...	...	...	...	241	52, 54	38, 40
510	950	137.9	20	1174	613.7	89.0	637.8	92.5	812.2	117.8	20	60	...	...	...
510	950	172.4	25	1053	...	...	...	...	...	...	...	...	285	81, 89	60, 66
538	1000	103.4	15	1270	568.8	82.5	586.1	85.0	776.4	112.6	22	60	...	...	...
538	1000	137.9	20	1147	...	...	...	...	...	...	...	...	277	83, 83	61, 61
Normalized at 940 C (1725 F), tempered at 677 C (1250 F); 285 HB															
21	70	...	...	...	861.9	125.0	879.1	127.5	1032.5	149.75	16.5	56	285	34, 46	25, 34
482	900	224.1	32.5	1280	875.7	127.0	896.4	130.0	1030.8	149.5	15.5	54	...	...	...
482	900	310.3	45.0	1122	...	...	...	...	...	...	...	...	311	30, 33	22, 24
510	950	258.6	37.5	1104	875.7	127.0	896.4	130.0	1006.7	146.0	17	54	...	...	...
510	950	344.8	50.0	1053	...	...	...	...	...	...	...	...	302	41, 22	30, 16
538	1000	158.6	23.0	1273	861.9	125.0	868.8	126.0	991.2	143.75	16	54	...	...	...
538	1000	206.9	30.0	1215	...	...	...	...	...	...	...	...	255	71, 72	52, 53
Oil quenched at 899 C (1650 F), tempered at 690 C (1275 F); 262 HB															
21	70	...	...	...	910.1	132.0	910.1	132.0	930.8	135.0	21	65	262	85, 98	63, 72
482	900	172.4	25.0	1010	827.4	120.0	837.7	121.5	932.2	135.2	21	61	...	...	...
482	900	206.9	30.0	1204	...	...	...	...	...	...	...	...	277	81, 84	60, 62
510	950	120.7	17.5	1053	841.2	122.0	848.1	123.0	953.2	138.25	19	61	...	...	...
510	950	137.9	20.0	1057	...	...	...	...	...	...	...	...	277	88, 95	65, 70
538	1000	69.0	10.0	1356	810.2	117.5	813.6	118.0	912.9	132.4	21	61	...	...	...
538	1000	82.7	12.0	1122	...	...	...	...	...	...	...	...	269	79, 89	58, 66
Normalized at 940 C (1725 F), tempered at 649 C (1200 F); 311 to 331 HB															
21	70	...	...	...	972.2	141.0	979.1	142.0	1070.8	155.3	16	51	311/331	31, 37	23, 27
482	900	413.7	60.0	2012	928.8	134.7	937.7	136.0	1042.5	151.2	16	50	...	...	...
593	1100	34.5	5.0	1314	764.0	110.8	765.3	111.0	889.5	129.0	19	55	...	...	...
593	1100	55.2	8.0	1274	...	...	...	...	...	...	...	...	257	33, 34	24, 25

(a) 0.1% offset. (b) Izod impact: undersize specimens were 9.04-mm (0.356-in.) square with a 1.27-mm (0.050-in.) deep V-notch.
Source: Ref 28, p 14

17-22AS: Creep strength at 538 C (1000 F)

CRSTN %	CRSTR 100 h MPa	CRSTR 100 h ksi	CRSTR 1000 h MPa	CRSTR 1000 h ksi
1.0	475.8	69.0	393.0	57.0
0.5	344.8	50.0	289.6	42.0
0.2	234.4	34.0	206.9	30.0
0.1	186.2	27.0	158.6	23.0

Note: The above values are representative for normalized and tempered 25-mm (1-in.) round bars
Source: Ref 28, p 15

17-22AS: Stress to produce 0.1 to 0.2% creep in 1000 h

TEMP C	TEMP F	CRSTN %	CRSTR MPa	CRSTR ksi
371	700	0.1	620.6	90.0
371	700	0.2	744.7	108.0
427	800	0.1	551.6	80.0
427	800	0.2	634.3	92.0
482	900	0.1	365.4	53.0
482	900	0.2	482.7	70.0
538	1000	0.1	220.6	32.0
538	1000	0.2	282.7	41.0

Note: The above values were determined on 25-mm (1-in.) round bars normalized and tempered to 311 to 331 HB
Source: Ref 28, p 15

17-22AS: Stress-to-rupture property data

TEMP C	TEMP F	RPSTR 100 h MPa	RPSTR 100 h ksi	RPSTR 500 h MPa	RPSTR 500 h ksi	RPSTR 1000 h MPa	RPSTR 1000 h ksi	RPSTR 10000 h MPa	RPSTR 10000 h ksi
Normalized at 940 C (1725 F) and tempered at 649 C (1200 F), 311 to 331 HB									
538	1000	517.1	75.0	455.1	66.0	413.7	60.0	196.5	28.5
566	1050	393.0	57.0	289.6	42.0	227.5	33.0	103.4	15.0
593	1100	202.7	41.0	172.4	25.0	141.3	20.5	66.2	9.6
Normalized at 940 C (1725 F) and tempered at 704 C (1300 F), 241 to 255 HB									
538	1000	386.1	56.0	331.0	48.0	303.4	44.0	220.6	32.0
593	1100	268.9	39.0	186.2	27.0	158.6	23.0	93.1	13.5
Normalized at 940 C (1725 F), tempered at 732 C (1350 F), and austenitized at 871 C (1600 F), 197 to 229 HB									
538	1000	331.0	48.0	289.6	42.0	258.6	37.5	189.6	27.5
538	1000	255.1	37.0	206.9	30.0	179.3	26.0	113.8	16.5
Normalized at 940 C (1725 F), oil quenched at 1066 C (1950 F), and tempered at 699 C (1290 F), 321 HB									
538	1000	455.1	66.0	386.1	56.0	296.5	43.0	122.7	17.8
593	1100	241.3	35.0	151.7	22.0	131.0	19.0	69.0	10.0
Oil quenched at 954 C (1750 F) and tempered at 649 C (1200 F), 285 to 302 HB									
538	1000	413.7	60.0	331.0	48.0	300.0	43.5	151.7	22.0

Source: Ref 28, p 14

17-22AS: Stress relaxation

CRIS MPa	CRIS ksi	Limiting Deform(a) in./in.	CRSRD 1000 h MPa	CRSRD 1000 h ksi	CRSRD 10000 h MPa	CRSRD 10000 h ksi
At 482 C (900 F)						
281.7	40.85	0.00176	187.2	27.15	136.2	19.75
At 538 C (1000 F)						
246.8	35.8	0.0016	187.2	27.15	104.8	15.2
232.0	33.65	0.0015	124.8	18.1	88.9	12.9
201.3	29.2	0.0015	106.2	15.4	73.8	10.7

Note: The above values were determined by the Robinson Down Step Method on 25-mm (1-in.) round normalized and tempered bars
(a) Limiting deformation on loading
Source: Ref 28, p 15

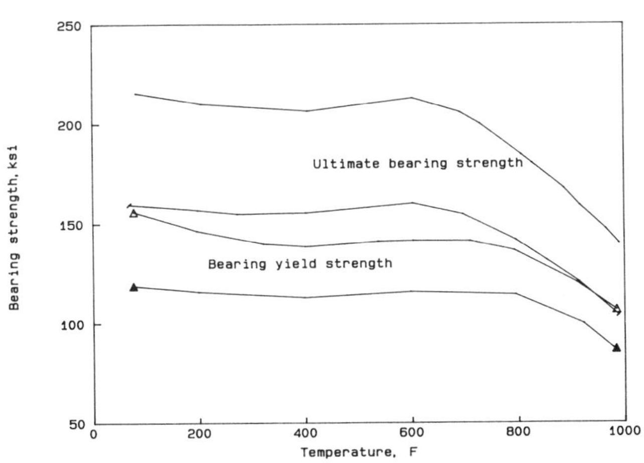

17-22AS: Bearing strength of 0.062-in. sheet. Heat treatment: 1750 F for 0.25 h, oil quenched and tempered at 1300 F for 1 h. Solid symbols: e/D = 1.5. Open symbols: e/D = 2.0. Exposure time: 0.5 to 1000 h. Source: Ref 22, p 12

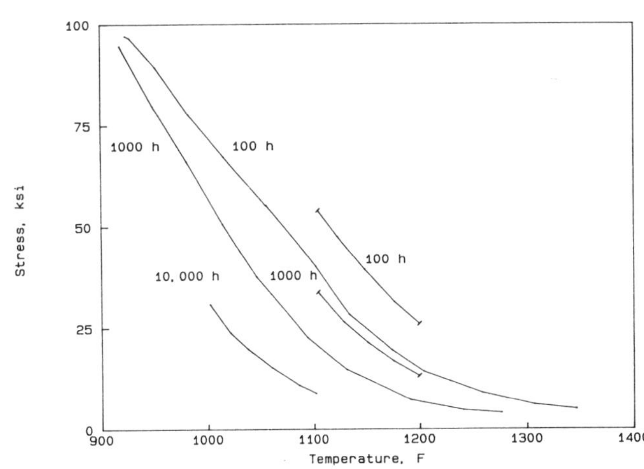

17-22AS: Creep-rupture strength of bar heat treated at 1725 F for 0.5 h, air cooled, tempered at 1200 for 6 h (open circles) or heat treated at 1850 F, air cooled, and tempered at 1200 F for 6 h (solid circles). Source: Ref 22, p 15

17-22AS: Stress-rupture of specimens notched (K = 10), normalized at 950 C, and tempered at 650 C for 6 h. Source: Ref 1, p 651

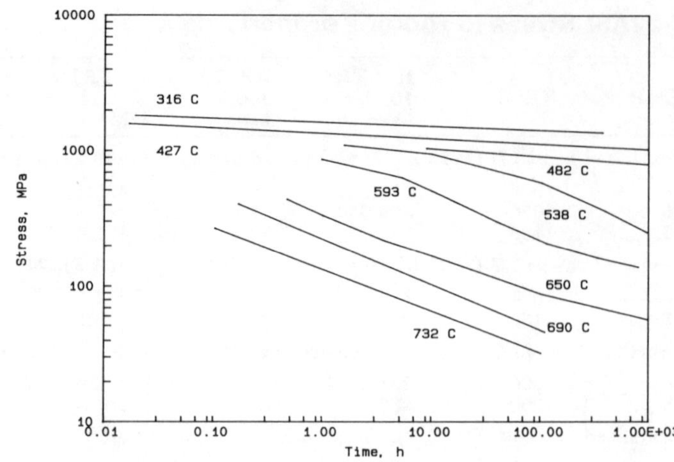

17-22AV

Alternate designations: 1-1/4Cr-1/2Mo-V, AISI 603
UNS number: K22770
Specifications: AMS 6303, AMS 6436
Nominal composition: 0.270% C, 1.250% Cr, 0.750% Mn, 0.500% Mo, 0.650% Si, 0.850% V
Product forms: Wrought, bar, plate, sheet, strip, tube, pipe, billets, forgings, structural shapes
Description: AISI 603, also known as alloy 17-22AV, is a low-alloy, high-temperature, constructional steel that has useful strength after heat treatment at temperatures up to 565 C (1050 F). Optimum high temperature properties are obtained by normalizing from 995 C (1825 F) and then tempering to the desired hardness. For larger sections, oil quenching is typically used instead of normalizing.
Typical applications: Turbine rotors, jet engine and missile components, and high-temperature bolting and mechanical connectors for service at temperatures as high as 565 C (1050 F). Alloy 603 is also used for hot-work dies and aircraft brake discs, applications that take advantage of its resistance to hot upsetting and thermal cracking. Alloy 603 is not particularly resistant to either corrosion or oxidation. Thus, if rusting or scaling poses a problem, exposed surfaces should be nickel plated or aluminized.

Physical Properties

17-22AV: Mean coefficients of linear thermal expansion

TEMP C	TEMP F	THEXP 10^{-6}/C	THEXP 10^{-6}/F
21-38	70-100	11.70	6.50
21-204	70-400	12.01	6.67
21-316	70-600	12.67	7.04
21-427	70-800	12.94	7.19
21-538	70-1000	13.30	7.39
21-593	70-1100	13.39	7.44
21-649	70-1200	13.50	7.50

Note: These values were determined on bar stock normalized at 982 C (1800 F) and tempered for 6 h at 649 C (1200 F)
Source: Ref 28, p 16

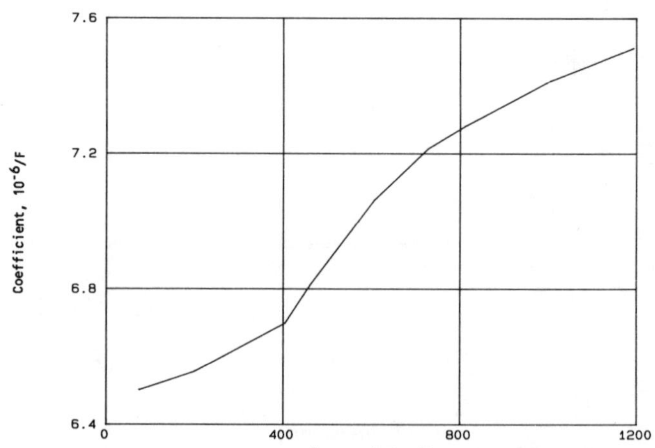

17-22AV: Mean coefficient of linear thermal expansion from room temperature to the temperature indicated. Heat treatment: 1725 F, air cooled, held at 1200 F for 6 h. Source: Ref 22, p 5

17-22AV: Modulus of elasticity

TEMP C	TEMP F	E GPa	E 10^3 ksi
21	70	209.6-213.7	30.4-31.0
93	200	202.7-209.6	29.4-30.4
149	300	209.4-205.5	29.4-29.8
204	400	198.6-202.0	28.8-29.3
260	500	194.4-197.2	28.2-28.6
316	600	188.9-193.1	27.4-28.0
371	700	184.1-188.2	26.7-27.3
427	800	179.3-184.1	26.0-26.7
482	900	175.8-179.3	25.5-26.0
538	1000	170.3-173.8	24.7-25.2
593	1100	165.5-168.9	24.0-24.5
649	1200	157.2-162.7	22.8-23.6
704	1300	147.6-153.1	21.4-22.2

Source: Ref 28, p 16

17-22AV: Poisson's ratio

TEMP C	TEMP F	PR
21	70	0.285-0.296
93	200	0.287-0.296
149	300	0.288-0.296
204	400	0.290-0.298
260	500	0.292-0.300
316	600	0.294-0.302
371	700	0.297-0.304
427	800	0.298-0.306
482	900	0.301-0.310
538	1000	0.304-0.313
593	1100	0.306-0.316
649	1200	0.311-0.322
704	1300	0.317-0.327

Source: Ref 28, p 16

17-22AV: Modulus of rigidity

TEMP C	TEMP F	G GPa	G 10^3 ksi
21	70	80.7-84.4	11.7-12.1
93	200	80.0-81.4	11.6-11.8
149	300	78.6-80.0	11.4-11.6
204	400	77.2-78.6	11.2-11.4
260	500	75.2-76.5	10.9-11.1
316	600	73.1-74.5	10.6-10.8
371	700	71.0-72.4	10.3-10.5
427	800	75.2-70.3	10.9-10.2
482	900	66.9-68.3	9.7-9.9
538	1000	64.8-66.2	9.4-9.6
593	1100	62.7-64.1	9.1-9.3
649	1200	60.0-60.7	8.7-8.8
704	1300	56.5-57.9	8.2-8.4

Source: Ref 28, p 16

Mechanical Properties

17-22AV: Tensile property data

TEMP C	TEMP F	YS(a) MPa	YS(a) ksi	YS MPa	YS ksi	UTS MPa	UTS ksi	EL %	RA %
Normalized at 982 C (1800 F), tempered at 649 C (1200 F), 341 HB									
27	80	965.3	140.0	999.8	145.0	1103.2	160.0	17	52
260	500	896.4	130.0	930.0	135.0	1034.3	150.0	16	55
371	700	827.4	120.0	861.9	125.0	965.3	140.0	16	57
427	800	758.5	110.0	792.9	115.0	896.4	130.0	17	61
482	900	713.6	103.5	69.0	10.0	803.3	116.5	19	66
538	1000	586.1	85.0	637.8	92.5	692.9	100.5	21	66
593	1100	441.3	64.0	503.3	73.0	586.1	85.0	23	72
649	1200	324.1	47.0	379.2	55.0	455.1	66.0	24	66
704	1300	172.4	25.0	210.3	30.5	260.9	39.0	33	72
Oil quenched at 954 C (1750 F), tempered at 538 C (1000 F), 444 HB									
27	80	1241.1	180.0	1330.7	193.0	1510.0	219.0	14	45
427	800	989.4	143.5	1092.9	158.5	1251.4	181.5	15	50
482	900	917.0	133.0	1006.7	146.0	1151.5	167.0	14	40
Oil quenched at 954 C (1750 F), tempered at 427 C (800 F), 444 HB									
27	80	1137.7	165.0	1282.5	186.0	1468.6	213.0	13	44
316	600	979.1	142.0	1120.4	162.5	1461.7	212.0	11	32

Note: Specimens were held 1 h at test temperature prior to testing. Tempering time was 6 h in all cases
(a) 0.1% offset
Source: Ref 28, p 17

17-22AV: Room-temperature tensile properties of bar previously exposed to stress

EXP TEMP C	EXP TEMP F	EXP STRESS MPa	EXP STRESS ksi	EXP TIME h	YS MPa	YS ksi	UTS MPa	UTS ksi	EL %	RA %
...	...	...	...	...	1006.7	146.0	1082.5	157.0	17	54
538	1000	158.6	23.0	2197	992.9	144.0	1041.1	151.0	17	53
538	1000	137.9	20.0	4003	910.1	132.0	992.9	144.0	17	53
649	1100	48.3	7.0	2470	730.9	106.0	813.6	118.0	20	60
649	1100	82.7	12.0	2198	772.2	112.0	827.4	120.0	19	61

Note: These values were determined on bar stock normalized at 982 C (1800 F), air cooled and tempered at 649 C (1200 F)
Source: Ref 21, p 7

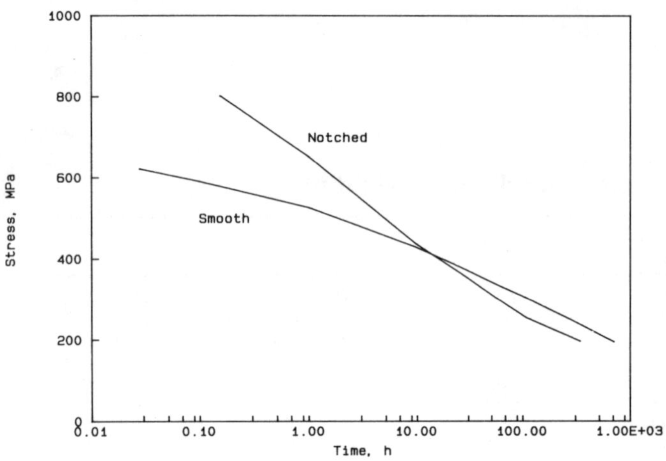

17-22AV: Effect of notch on stress-rupture behavior. Stress-rupture behavior of smooth (K = 1.0) and notched (K = 10.0) specimens tested at 595 C. All specimens were normalized at 980 C and tempered for 6 h at 650 C. Source: Ref 1, p 642

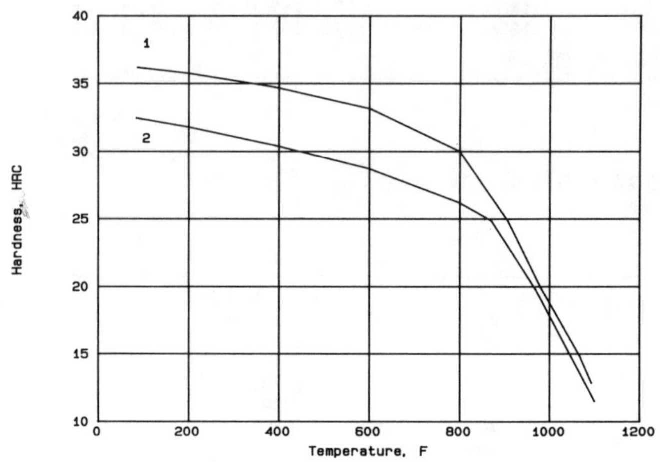

17-22AV: Rockwell C hardness. Curve 1: 17-22AV heated to 1850 F, air cooled, and tempered at 1200 F. Curve 2: 17-22AS heated to 1725 F, air cooled, and tempered at 1200 F. Source: Ref 22, p 4

17-22AV: Creep strength for 0.01% creep in 1000 h

TEMP C	TEMP F	CRSTR MPa	CRSTR ksi
538	1000	137.9	20.0
593	1100	34.5	5.0

Note: Samples were normalized at 982 C (1800 F) and tempered for 6 h at 649 C (1200 F)
Source: Ref 28, p 17

"AISI 604"

Alternate designation: 1Cr-1Mo-V

Nominal composition: 0.200% C, 1.000% Cr, 0.500% Mn, 1.000% Mo, 0.750% Si, 0.100% V

Product forms: Wrought, bar, plate, sheet, strip, tube, pipe, billets, forgings

Description: AISI 604 is a low-carbon, Cr-Mo-V alloy steel used almost exclusively for welded structures in high-temperature service. Tensile strengths up to 896 MPa (130 ksi) can be obtained via heat treatment, but 80 to 85% of this maximum is suggested for design purposes. High-temperature properties are obtained via precipitation of Mo_2C and VC. Alloy 604 is relatively insensitive to thermal shock and does not overtemper at 540 C (1000 F), making it suitable for jet engine frames, casings, and other welded assemblies that must operate at up to 540 C (1000 F). The alloy can also withstand short-time service at temperatures as high as 595 C (1100 F). Alloy 604 does not excel in oxidation or corrosion resistance. In fact, oxidation-resistant coatings are recommended in elevated-temperature applications.

Physical Properties

AISI 604: Mean coefficients of linear thermal expansion

TEMP C	TEMP F	THEXP 10^{-6}/C	THEXP 10^{-6}/F
25-100	77-212	12.60	7.00
25-204	77-400	12.87	7.15
25-316	77-600	13.14	7.30
25-427	77-800	13.59	7.55
25-538	77-1000	14.22	7.90
25-649	77-1200	14.94	8.30

Source: Ref 28, p 19

AISI 604: Modulus of elasticity

TEMP C	TEMP F	E GPa	E 10^3 ksi
21	70	218.6	31.7
93	200	212.4	30.8
204	400	204.1	29.6
316	600	195.1	28.3
427	800	186.9	27.1
538	1000	174.4	25.3
649	1200	160.0	23.2
677	1250	156.5	22.7

Note: The above elastic constants were determined by ultrasonic pulse technique
Source: Ref 28, p 19

AISI 604: Modulus of rigidity

TEMP C	TEMP F	G GPa	G 10^3 ksi
21	70	84.8	12.3
93	200	82.7	12.0
204	400	79.3	11.5
316	600	75.2	10.9
427	800	71.0	10.3
538	1000	66.2	9.6
649	1200	60.7	8.8
677	1250	59.3	8.6

Note: The above elastic constants were determined by ultrasonic pulse technique
Source: Ref 28, p 19

AISI 604: Poisson's ratio

TEMP C	TEMP F	PR
21	70	0.280
93	200	0.283
204	400	0.290
316	600	0.298
427	800	0.305
538	1000	0.312
649	1200	0.320
677	1250	0.321

Source; Ref 28, p 19

AISI 604: Stress-to-rupture property data

TEMP C	TEMP F	RPSTR 10 h MPa	RPSTR 10 h ksi	RPSTR 100 h MPa	RPSTR 100 h ksi	RPSTR 200 h MPa	RPSTR 200 h ksi	RPSTR 1000 h MPa	RPSTR 1000 h ksi
482	900	765.3	111	717.1	104	689.5	100	599.9	87
538	1000	655.0	95	517.1	75	475.8	69	351.6	51
593	1100	441.3	64	275.8	40	234.4	34	144.8	21

Source: Ref 28, p 20

AISI 604: Tensile property data for bar

TEMP C	TEMP F	YS 0.02% MPa	YS 0.02% ksi	YS 0.2% MPa	YS 0.2% ksi	UTS MPa	UTS ksi
21	70	744.7	108	806.7	117	951.5	138
93	200	717.1	104	792.9	115	951.5	138
204	400	682.6	99	758.5	110	923.9	134
316	600	641.2	93	717.1	104	889.5	129
427	800	599.9	87	661.9	96	820.5	119
538	1000	524.0	76	586.1	85	758.5	110
649	1200	344.8	50	413.7	60	524.0	76

Note: Specimens were taken from hot rolled bar stock, heat treated at 927-954 C (1700-1750 F) for 1 h, air cooled, and tempered at 649 C (1200 F) for 2 h
Source: Ref 28, p 20

6. Ultrahigh Strength Steels

Ultrahigh Strength Steels

Structural steels with very high strength levels are often referred to as ultrahigh strength steels. The designation "ultrahigh strength" is arbitrary because no universally accepted strength level for use of the term has been set. Also, as structural steels with greater and greater strength have been developed, the strength range for which the term may be applied has gradually increased.

The ultrahigh strength structural steels are those alloys capable of attaining a minimum yield strength of at least 1380 MPa (200 ksi). Their applications tend to concentrate in the aerospace and defense industries.

9Ni Steel

Alternate designation: 9% Ni
UNS number: K81340
Specifications: ASTM A333 (8), ASTM 1334 (8), ASTM A353M, ASTM A522 (I), ASTM A553M (I)
Nominal composition: 0.130% C max, 0.900% Mn max, 8.500 to 9.500% Ni, 0.035% P max, 0.040% S max, 0.150 to 0.300% Si

Product forms: Wrought, bar, wire, plate, strip, billets, structural shapes
Description: This ferritic alloy steel is used primarily for structural applications at sub-zero temperatures, such as equipment for containing and transporting liquefied hydrocarbon and elemental gases.

Physical Properties

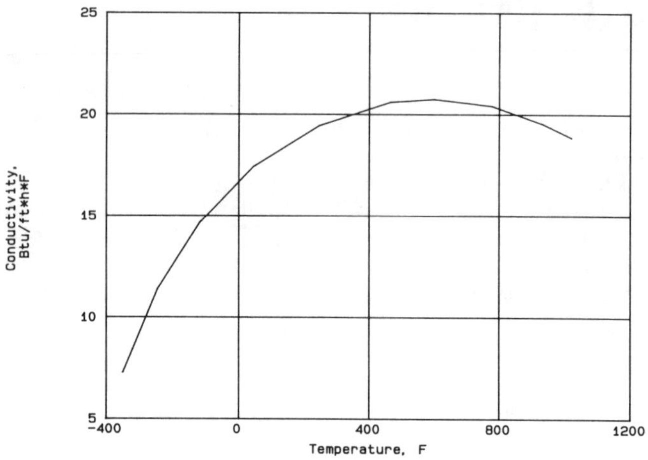

9Ni steel: Thermal conductivity of specimens heat treated at 1650 F, air cooled, held at 1450 F, air cooled, aged at 1000 F, and air cooled. Source: Ref 22, p 27

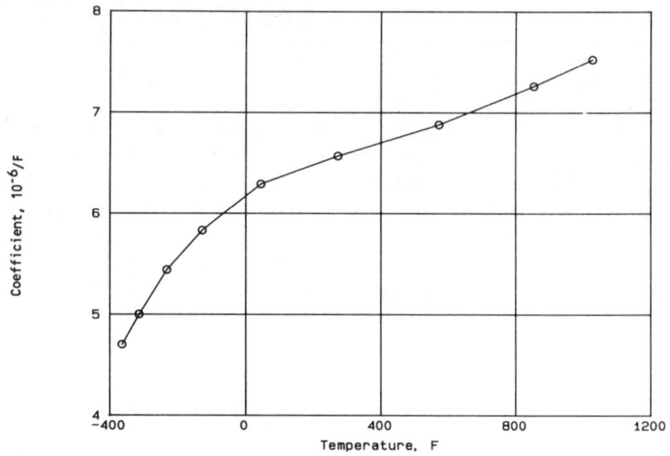

9Ni steel: Coefficient of linear thermal expansion from room temperature to temperature indicated for specimens treated at 1650 F, air cooled, held at 1450 F, air cooled, held at 1100 F, water quenched. Source: Ref 22, p 27

9Ni steel: Specific heat

TEMP C	TEMP F	SHTCP J/kg*k	SHTCP Btu/lb*F
27-371	80-700	498.253	0.119
27-538	80-1000	548.497	0.131
371-538	700-1000	653.172	0.156

Note: Specimens were heated to 899 C (1650 F), air cooled and heated to 788 C (1450 F), air cooled and heated to 593 C (1100 F), and air cooled
Source: Ref 29, Dec 1979

Mechanical Properties

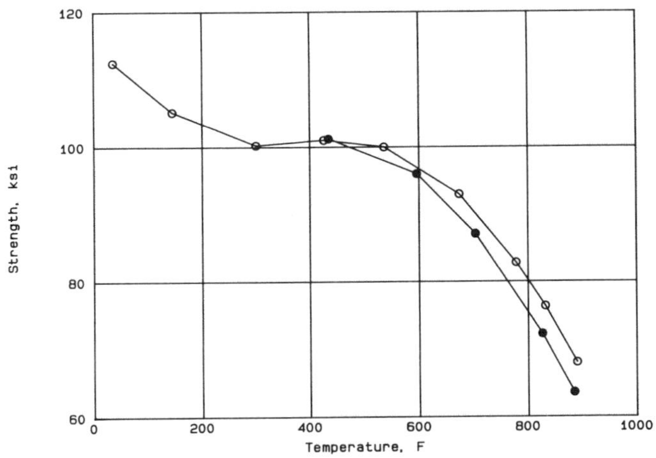

9Ni steel: Tensile strength of 0.5-in. plate. Open circle: heat treated at 1475 F for 0.5 h, water quenched, held at 1050 F for 0.5 h, water quenched. Solid circle: heat treated at 1650 F for 0.5 h, air cooled, held at 1450 F for 0.5 h, air cooled, held at 1050 F for 0.5 h, and air cooled. Source: Ref 22, p 27

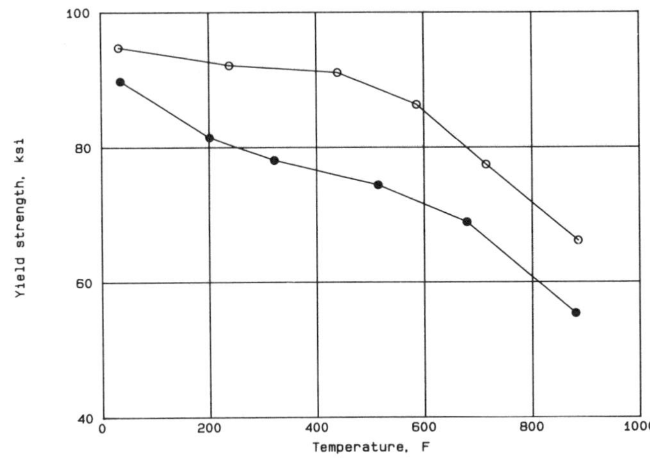

9Ni steel: Yield strength of 0.5-in. plate. Open circle: heat treated at 1475 F for 0.5 h, water quenched, held at 1050 F for 0.5 h, and water quenched. Solid circle: heat treated at 1650 F for 0.5 h , air cooled, held at 1450 F for 0.5 h, air cooled, held at 1050 F for 0.5 h, and air cooled. Source: Ref 22, p 27

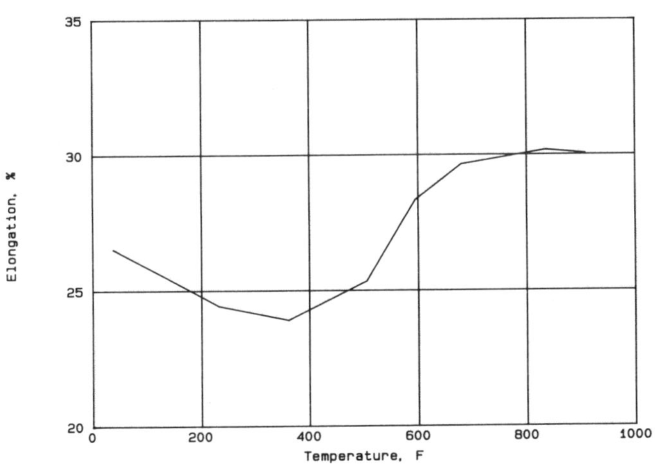

9Ni steel: Tensile elongation of 0.5-in. plate, heat treated at 1650 F, held for 0.5 h, air cooled, aged at 1450 F, held for 0.5 h, air cooled, aged at 1050 F, 0.5 h, air cooled. Source: Ref 22, p 27

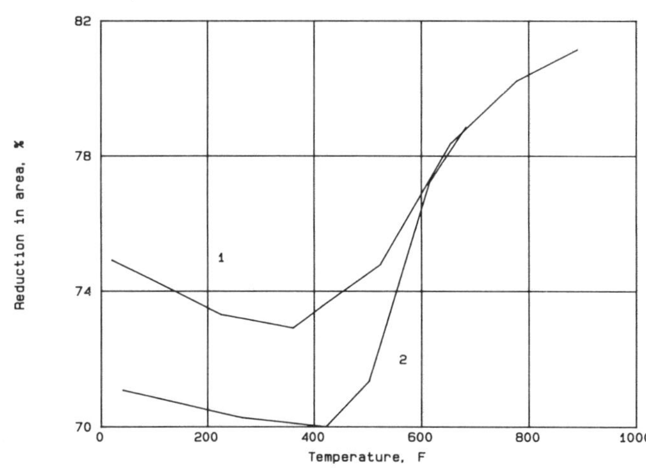

9Ni steel: Reduction in area of 0.5-in. plate. Curve 1: heat treated at 1475 F for 0.5 h, water quenched, held at 1050 F for 0.5 h, and water quenched. Curve 2: heat treated at 1650 F for 0.5 h , air cooled, held at 1450 F for 0.5 h, air cooled, held at 1050 F for 0.5 h, and air cooled. Source: Ref 22, p 27

300M

UNS number: K44220
Specifications: AMS 6419, ASTM A579 (32), MIL S-8844 (3)
Nominal composition: 0.400 to 0.460% C, 0.700 to 0.950% Cr, 0.650 to 0.900% Mn, 0.300 to 0.450% Mo, 1.540 to 2.000% Ni, 1.450 to 1.800% Si, 0.050% V min
Product forms: Wrought, bar, rod, wire, plate, sheet, strip, billets, structural shapes, cast, sand, investment

Description: Alloy 300M is basically a silicon modified AISI 4340 steel with slightly higher carbon and molybdenum contents and an addition of vanadium. Many of its properties are similar to those of 4340, except that the added silicon provides deeper hardenability, increased solid solution strengthening, and better resistance to softening at elevated temperatures.
Typical applications: Aircraft landing gear, airframe parts, fasteners, and pressure vessels.

Mechanical Properties

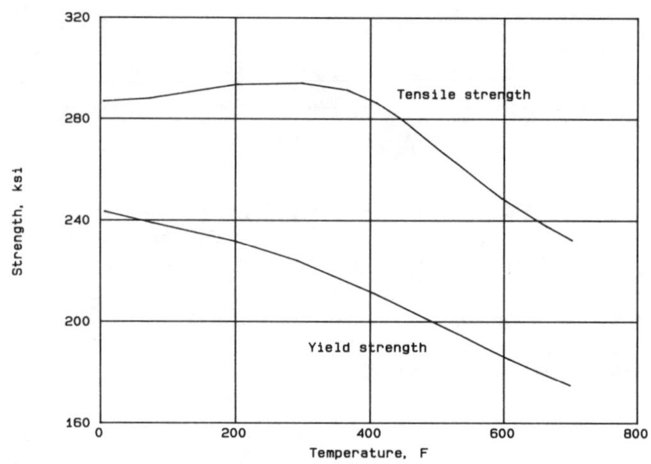

300M: Tensile and yield strength of bar heat treated at 1600 F, oil quenched, and held at 600 F for 2 h. Source: Ref 22, p 9

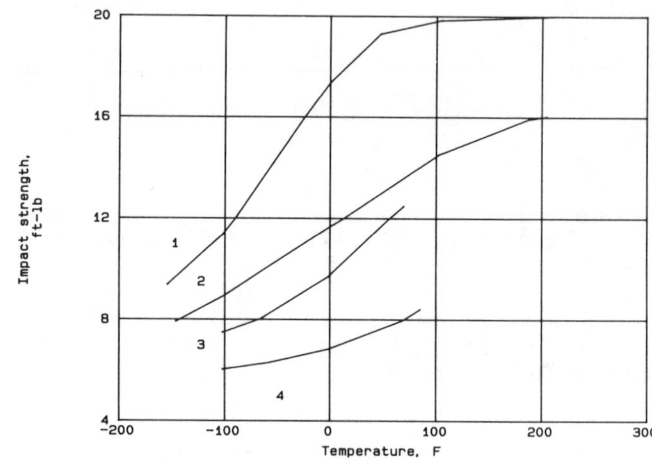

300M: Charpy impact strength of bar heat treated at 1600 to 1650 F, oil quenched, held at 600 F. Curve 1: 4.25-in, square longitudinal test bar. Curve 2: Same as 1, transverse. Curve 3: 5.75-in. diam longitudinal midway test bar. Curve 4: 15-in. square midway test bar. Source: Ref 22, p 10

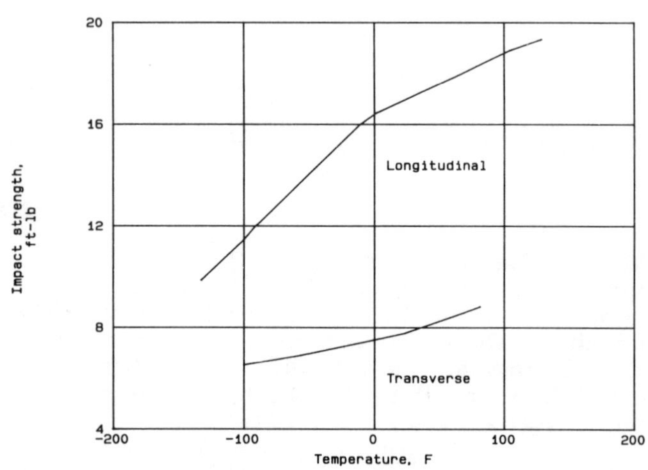

300M: Charpy impact strength of specimens heat treated as tube with 12-in. outside diameter, 9-in. inside diameter; heat treated at 1600 to 1650 F, oil quenched, and held at 600 h. Source: Ref 22, p 10

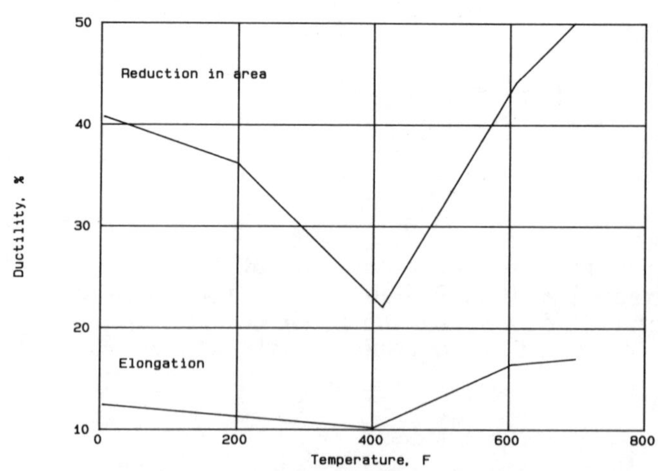

300M: Tensile elongation and reduction in area of bar heat treated at 1600 F, oil quenched and held at 600 F for 2 h. Source: Ref 22, p 9

AISI 613

Nominal composition: 0.810% C max, 4.080% Cr, 0.300% Mn max, 4.250% Mo max, 0.200% Si max, 1.000% V max
Product forms: Wrought, bar, wire, plate, sheet, strip, billets, structural shapes
Description: AISI 613 is a low-alloy, martensitic high speed steel developed primarily for high-strength, high-load applications at elevated temperatures. The alloy combines inherently high hardness and high compressive strength at temperatures up to 540 C (1000 F) with long-time resistance to softening at elevated temperatures.
Typical applications: Aircraft and missile structural components such as bearings, races, gears, cams, and constant speed drive parts. Oxidation resistance is generally satisfactory up to 370 C (700 F).

Physical Properties

AISI 613: Mean coefficients of linear thermal expansion

TEMP C	TEMP F	THEXP 10^{-6}/C	THEXP 10^{-6}/F
21-93	70-200	11.21	6.23
21-149	70-300	11.50	6.39
21-204	70-400	11.84	6.58
21-260	70-500	12.10	6.72
21-316	70-600	12.29	6.83
21-371	70-700	12.51	6.95
21-427	70-800	12.69	7.05
21-482	70-900	12.96	7.20
21-538	70-1000	13.28	7.38

Source: Ref 28, p 29

Mechanical Properties

AISI 613: Tensile property data

TEMP C	TEMP F	UTS MPa	UTS ksi	YS MPa	YS ksi	EL(a) %	RA %	HARD HRC
21	70	2837.3	411.5	2334.0	338.5	2	2	64
316	600	2520.1	365.5	2071.9	300.5	4	12	...
427	800	2378.8	345.0	1809.9	262.5	6	17	...
538	1000	2130.6	309.0	1723.8	250.0	6	12	...

Note: Specimens were air quenched and tempered at 550 C (1025 F)
(a) In 1.4 in.
Source: Ref 29, Feb 1965

AISI 613: Dimensional change after thermal exposure

TEMP(a) C	TEMP(a) F	Change (b) micro-in./in.
204	400	+22
316	600	+12
427	800	+27
538	1000	-50

Note: Heated 20 min at 1149 C (2100 F) and double tempered at 566 C (1050 F)
(a) Exposed 1000 h at indicated temperature. (b) Length change
Source: Ref 29, Feb 1965

AISI 613: Hot hardness data

TEMP C	TEMP F	HARD(a) HRC	HARD(b) HRC
21	70	64	64
204	400	62	61
316	600	59	57
427	800	57	55
538	1000	52	46

(a) After 0.5-h exposure at temperature. (b) After 1000-h exposure at temperature
Source: Ref 29, Feb 1965

AISI 618

Nominal composition: 1.050% C, 14.500% Cr, 0.500% Mn max, 4.000% Mo max, 0.300% Si
Product forms: Wrought, bar, wire, plate, sheet, strip, billets, structural shapes
Description: AISI 618 is a general-purpose, high-temperature alloy that can be heat treated to high strength and hardness levels. For elevated-temperature applications, the alloy is usually tempered in the secondary hardening range of 480 to 540 C (900 to 1000 F) and then refrigerated. The combination produces a room temperature hardness of 62 to 64 HRC. AISI 618 also offers good corrosion and oxidation resistance due to its high chromium and molybdenum contents.

Physical Properties

AISI 618: Mean coefficients of linear thermal expansion

TEMP C	TEMP F	THEXP 10^{-6}/C	THEXP 10^{-6}/F
21-38	70-100	7.33	4.07
21-316	70-600	10.90	6.06
21-538	70-1000	12.01	6.67

Source: Ref 29, Jan 1965

Mechanical Properties

AISI 618: Effect of exposure time at 480 C (900 F) on room-temperature hardness

TIME h	HARD HRC
1	62.5
20	62.5
45	62.5
70	62.0
130	61.0
200	60.0
300	59.0
400	58.0

Note: Specimens were austenitized at 1093 C (2000 F), oil quenched, tempered at 510 C (950 F), cold treated at -73 C (-100 F), and retempered at 510 C (950 F)
Source: Ref 29, Jan 1965

AISI 618: Short-time hot hardness

TEMP C	TEMP F	HARD HRC
22	72	62.0
93	200	61.0
149	300	60.5
204	400	59.0
260	500	58.5
316	600	57.0
371	700	56.5
427	800	55.0
482	900	53.0
538	1000	51.0

Note: Specimens were austenitized at 1121 C (2050 F), oil quenched, refrigerated at -73 C (-100 F), double tempered at 538 C (1000 F). Samples were held at each temperature for 30 min prior to testing
Source: Ref 29, Jan 1965

D-6a/D-6ac

UNS number: K24728

Specifications: AMS 5027, AMS 5028, AMS1 6431, AMS 6438, AMS 6439, ASTM A355 (A), ASTM A579 (23), MIL S-8949, MIL S-47036, MIL S-47264

Nominal composition: 0.450 to 0.500% C, 0.900 to 1.200% Cr, 0.600 to 0.900% Mn, 0.900 to 1.100% Mo, 0.400 to 0.700% Ni, 0.010% P max, 0.05 to 0.10% V, 0.15 to 0.30% Si

Product forms: Wrought, bar, plate, sheet, strip, billets, forgings, structural shapes

Description: D-6a is a low-alloy, ultrahigh strength steel developed for aircraft and missile structural applications such as landing gear and rocket motor cases. Other uses include shafts, gears, springs, dies, dummy blocks, and backer blocks. The alloy was designed primarily for use at room temperature tensile strengths of 1800 to 2000 MPa (260 to 290 ksi). It does not exhibit temper embrittlement, and it offers strength retention at elevated temperatures. D-6a is made by air melting in an electric furnace. D-6ac is the same alloy produced by air melting plus vacuum arc remelting.

Physical Properties

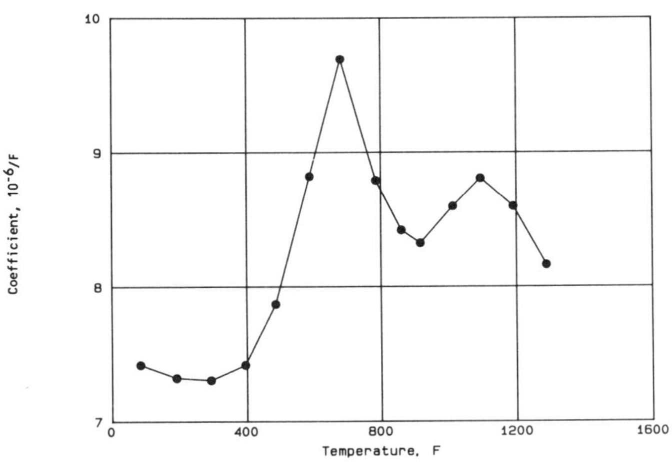

D-6a/D-6ac: Mean coefficient of linear thermal expansion. Source: Ref 22, p 5

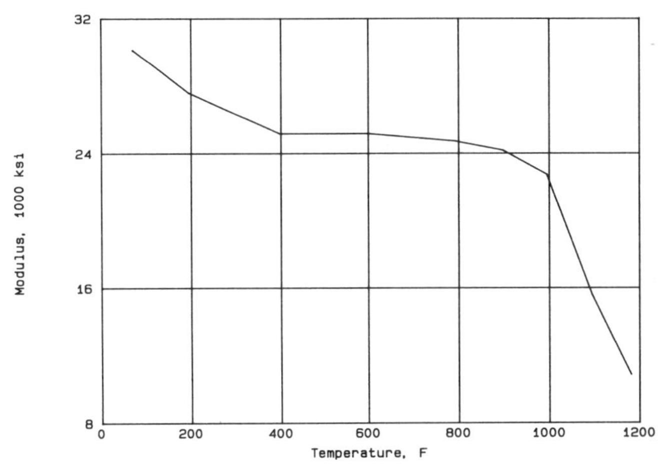

D-6a/D-6ac: Modulus of elasticity. Source: Ref 22, p 5

Mechanical Properties

D-6a/D-6ac: Tensile property data

TEMP C	TEMP F	UTS MPa	UTS ksi	YS MPa	YS ksi	EL %	RA %
260	500	1839.6	266.8	1249.9	187.8	15.2	55.0
315	600	1815.5	236.3	1256.3	182.2	15.8	57.4
370	700	1430.0	207.4	1165.6	168.9	14.7	55.0
425	800	1277.6	185.3	1096.3	159.0	15.2	55.0
480	900	1158.4	168.0	976.3	141.6	16.2	57.5
540	1000	957.7	138.9	831.5	120.6	19.8	64.5
595	1100	497.1	72.1	395.8	57.4	41.7	86.5

Note: Specimens were normalized at 900 C (1650 F), oil quenched from 845 C (1550 F), and tempered at 28 C (50 F) above indicated test temperature
Source: Ref 1, p 431

6.8 High-Temperature Property Data: Ferrous Alloys

D-6a/D-6ac: Room-temperature tensile property data after exposure at 540 C (1000 F)

EXP h	UTS MPa	UTS ksi	YS MPa	YS ksi	EL %	RA %
0	1410	204	1340	195	14.8	50.5
10	1410	204	1330	193	14.5	52.0
100	1330	193	1260	183	14.8	51.0

Note: The steel specimens were normalized at 900 C (1650 F), oil quenched from 845 C (1550 F) and tempered at 565 C (1050 F)
Source: Ref 1, p 431

D-6a/D-6ac: Effect of test temperature on smooth and notched fatigue properties of bar

TEMP C	TEMP F	FATST 10^2 cyc MPa	FATST 10^2 cyc ksi	FATST 10^3 cyc MPa	FATST 10^3 cyc ksi	FATST 10^4 cyc MPa	FATST 10^4 cyc ksi	FATST 10^5 cyc MPa	FATST 10^5 cyc ksi	FATST 10^6 cyc MPa	FATST 10^6 cyc ksi	FATST 10^7 cyc MPa	FATST 10^7 cyc ksi
Smooth													
21	70	...	...	1861.7	270	1585.9	230	1172.2	170	1034.3	150	1034.3	150
		1516.9	220	1241.1	180	1103.2	160	896.4	130	689.5	100	689.5	100
232	450	...		1792.7	260	1516.9	220	1172.2	170	1034.3	150	930.8	135
		1516.9	220	1241.1	180	1103.2	160	896.4	130	689.5	100	620.6	90
288	550	...		1654.8	240	1482.4	215	1172.2	170	1034.3	150	896.4	130
		1516.9	220	1241.1	180	965.3	140	758.5	110	586.1	85	517.1	75
Notched													
21	70	2068.5	300	1310.1	190	827.4	120	551.6	80	551.6	80	551.6	80
		1379.0	200	930.8	135	448.2	65	344.8	50	344.8	50	344.8	50
232	450	...	...	...	...	689.5	100	586.1	85	586.1	85	413.7	60
		...	...	...	...	...	...	310.3	45	310.3	45	275.8	40
288	550	1930.6	280	965.3	140	689.5	100	551.6	80	551.6	80	379.2	55
		1172.2	170	689.5	100	448.2	65	344.8	50	344.8	50	275.8	40

Note: A 12.7-mm (0.5-in.) bar was heated to 816 C (1500 F) for 15 min, oil quenched and heated to 260 C (500 F) for 2 h
Source: Ref 29, Jun 1974

D-6a/D-6ac: Fatigue limits

TEMP C	TEMP F	Tension(a) MPa	Tension(a) ksi	Tension(b) MPa	Tension(b) ksi
24	75	1035	150	690	100
232	450	930	135	550	80
288	550	930	135	575	75

Note: Smooth specimens were heat treated to a tensile strength of 1860 MPa (270 ksi); test speed, 186kHz
(a) Mean stress equal to alternating stress (R=0). (b) Mean stress equal to zero (R=-1)
Source: Ref 1, p 432

D-6a/D-6ac: Charpy V-notch impact strength. Specimens were normalized at 1650 F, aged at 1550 F, oil quenched, and held at 950 F. Source: Ref 22, p 20

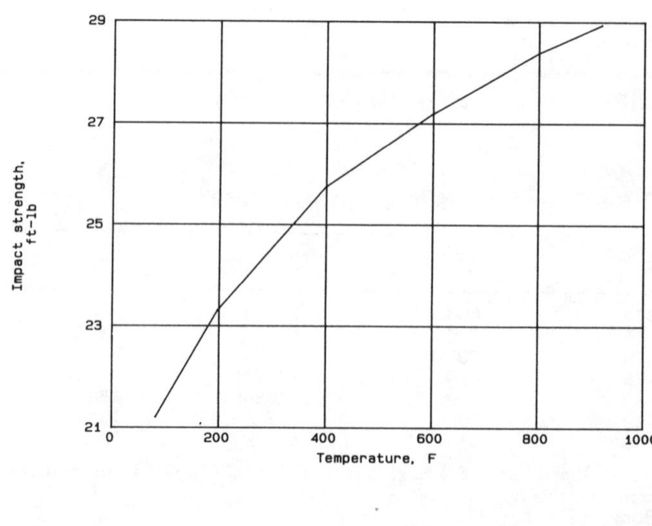

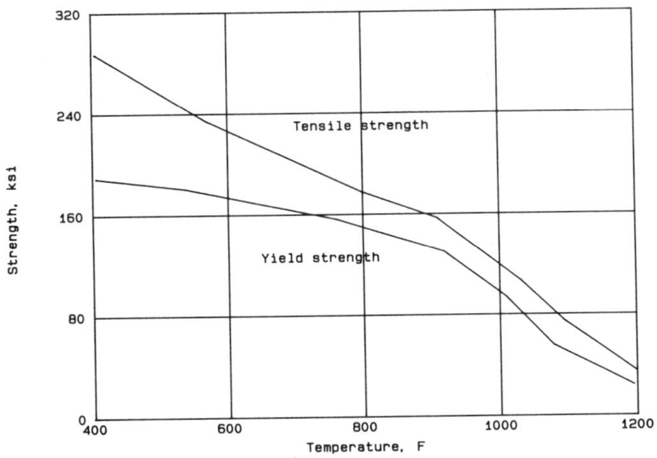

D-6a/D-6ac: Tensile and yield strength of bar, normalized at 1650 F, held at 1550 F, oil quenched, and tempered 50 F above test temperature. Source: Ref 22, p 19

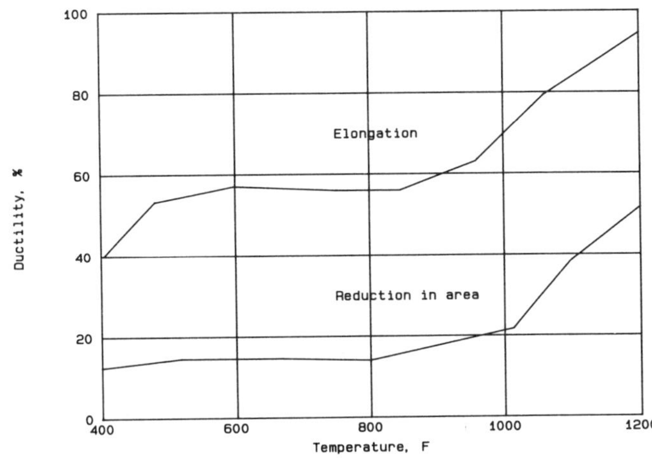

D-6a/D-6ac: Tensile elongation and reduction in area of bar, normalized at 1650 F, aged at 1550 F, oil quenched and tempered 50 F above test temperature. Source: Ref 22, p 19

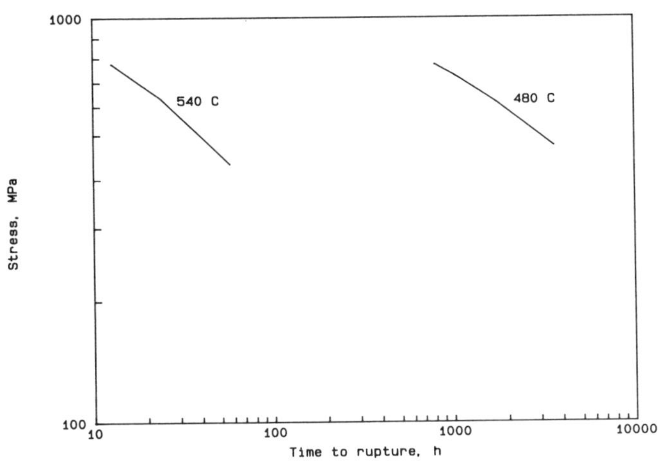

D-6a/D-6ac: Stress-rupture life determined on specimens heat treated to a room-temperature tensile strength of 1380 to 1520 MPa. Source: Ref 1, p 432

HP-9-4-20

Nominal composition: 0.160 to 0.230% C, 4.250 to 4.750% Co, 0.650 to 0.850% Cr, 0.200 to 0.400% Mn, 0.900 to 1.100% Mo, 8.500 to 9.500% Ni, 0.200% Si max, 0.060 to 0.120% V
Product forms: Wrought, bar, rod, plate, sheet, strip, billets, structural shapes
Description: HP-9-4-20 has high fracture toughness when heat treated to very high strength levels. The HP-9-4-20 usually is electric furnace melted and vacuum arc remelted. It has good thermal stability, which makes it suitable for long-term service at temperatures up to at least 370 C (700 F). Its oxidation resistance is superior to that of AISI 4340.
Typical applications: Aircraft structural components, pressure vessels, rotor shafts for metal forming equipment, drop hammer rods, and high-strength, shock-absorbing automotive parts.

Physical Properties

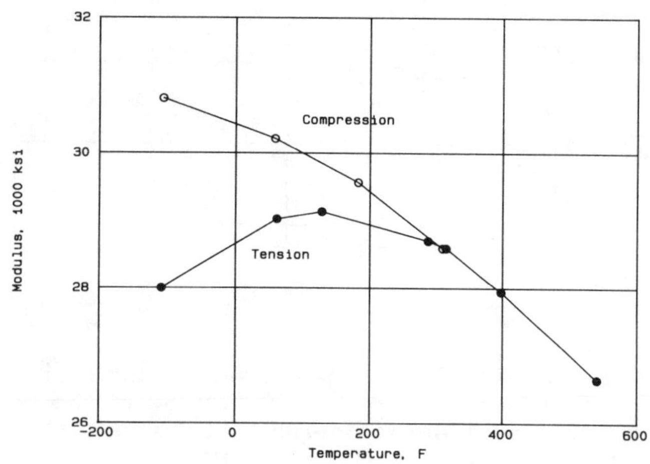

HP-9-4-20: Static modulus of elasticity for 0.30% C grade vacuum melted forgings, 3 by 9 by 24 in., with a room-temperature tensile strength of 230 ksi. Source: Ref 22, p 28

Mechanical Properties

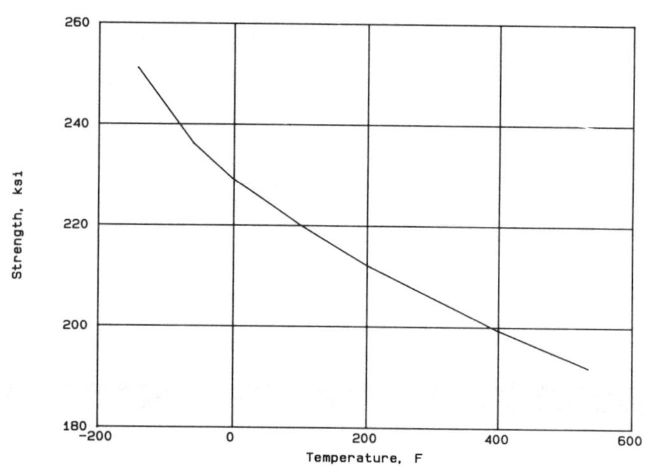

HP-9-4-20: Compressive yield strength of 0.30% C grade vacuum melted forgings. Specimens were heat treated at 1650 F for 1 h, air cooled, held at 1550 F for 1 h, cooled to -100 F for 2 h, double tempered at 950 F for 2 h. Hardness was 212 HRC. Source: Ref 22, p 38

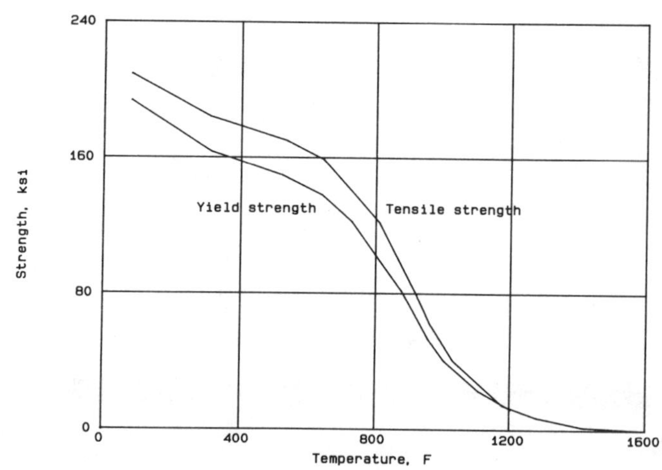

HP-9-4-20: Tensile and yield strength of 0.20% C grade vacuum melted 1-in. plate heat treated at 1700 F for 1 h, air cooled, held at 1500 F for 1 h, water quenched, and held at 1050 F for 2 h. Source: Ref 22, p 33

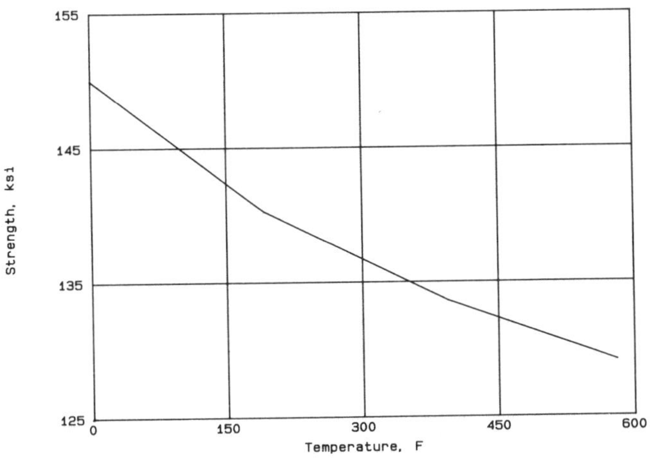

HP-9-4-20: Shear strength of 0.30% C grade vacuum melted forgings. Specimens were heat treated at 1650 F for 1 h, air cooled, held at 1550 F for 1 h, cooled to -100F for 2 h, and double tempered at 950 F for 2 h. Source: Ref 22, p 41

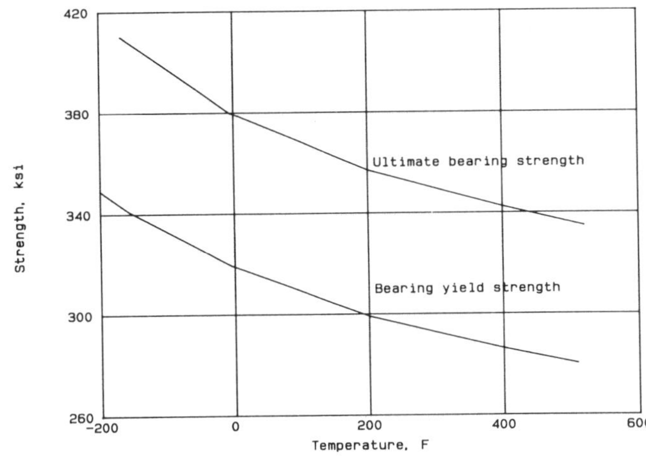

HP-9-4-20: Ultimate bearing and yield strength of 0.30% C grade vacuum melted forgings. Specimens were heat treated at 1650 F for 1 h, air cooled, held at 1550 F for 1 h, cooled to -100 F for 2 h, and double tempered at 950 F for 2 h. e/D = 1.5. Source: Ref 22, p 42

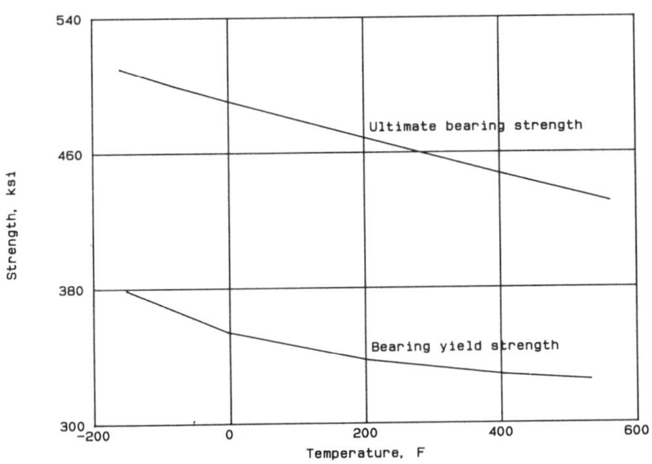

HP-9-4-20: Ultimate bearing and yield strength of 0.30% C grade vacuum melted forgings. Specimens were heat treated at 1650 F for 1 h, air cooled, held at 1550 F for 1 h, cooled to -100 F for 2 h, and double tempered at 950 F for 2 h. e/D = 2. Source: Ref 22, p 42

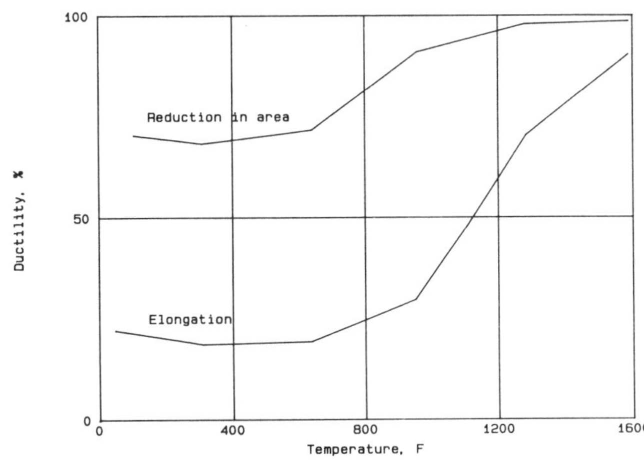

HP-9-4-20: Tensile elongation and reduction in area of 0.20% C grade vacuum melted 1-in. plate, heat treated at 1700 F for 1 h, held at 1500 F for 1 h, water quenched, and held at 1050 F for 2 h. Source: Ref 22, p 33

HP-9-4-20: Crack strength of 0.45% C grade plate of several thicknesses. Specimens were heat treated at 1450 F for 0.5 h, solution quenched at 450 F for 6 h, double tempered at 700 F for 2 h. Source: Ref 22, p 43

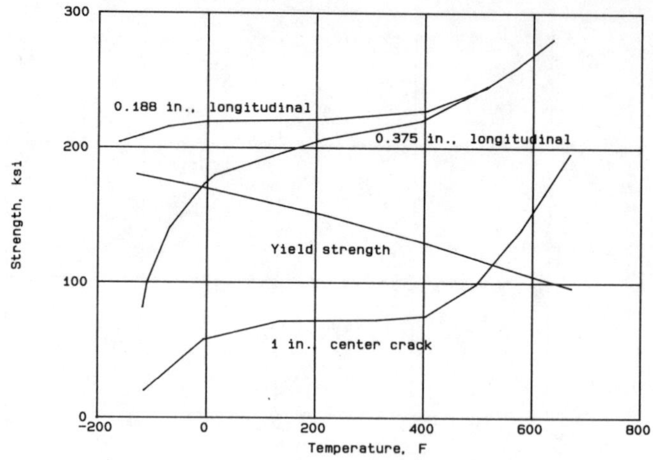

HP-9-4-30

Nominal composition: 0.290 to 0.340% C, 4.250 to 4.750% Co, 0.900 to 1.100% Cr, 0.100 to 0.350% Mn, 0.900 to 1.100% Mo, 7.000 to 8.000% Ni, 0.200% Si max, 0.060 to 0.120% V
Product forms: Wrought, bar, rod, plate, sheet, strip, billets, structural shapes
Description: HP-9-4-30 has high fracture toughness when heat treated to very high strength levels. HP-9-4-20 and HP-9-4-30 (0.20 and 0.30% C, respectively) are produced in significant quantities. It is usually electric furnace melted and vacuum arc remelted. It has good thermal stability, which makes it suitable for long-term service at temperatures up to at least 370 C (700 F). Its oxidation resistance is superior to that of AISI 4340.
Typical applications: Aircraft structural components, pressure vessels, rotor shafts for metal forming equipment, drop hammer rods, and high-strength, shock-absorbing automotive parts.

Physical Properties

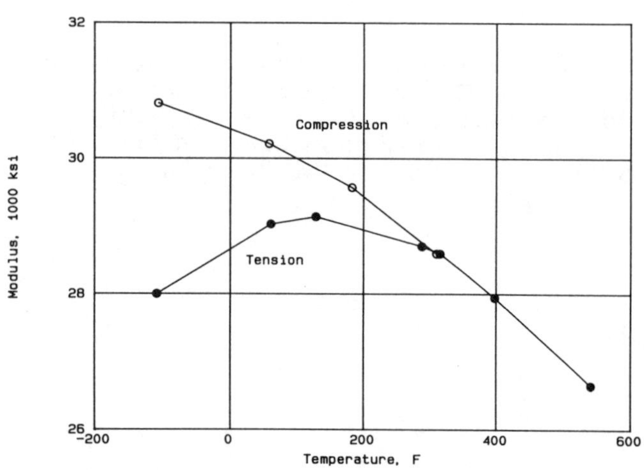

HP-9-4-30: Static modulus of elasticity for 0.30% C grade vacuum melted forgings, 3 by 9 by 24 in., with a room-temperature tensile strength of 230 ksi. Source: Ref 22, p 28

Mechanical Properties

HP 9-4-30: Room-temperature mechanical properties after 1000 h at elevated temperatures

TEMP C	TEMP F	UTS MPa	UTS ksi	YS MPa	YS ksi	EL(a) %	RA %	IS(b) J	IS(b) ft-lb
...	...	1650	239	1350	196	14	52	39	29
205	400	1585	230	1405	204	16	60	41	30
345	650	1585	230	1440	209	15	56	38	28
425	800	1650	239	1400	203	14	50	34	25
480	900	1565	227	1395	202	15	51	26	19

Note: All specimens were austenitized, quenched to martensite, and tempered at 540 C (1000 F)
(a) In 25 mm or 1 in. (b) Charpy V-notch impact energy
Source: Ref 1, p 442

HP-9-4-30: Tensile property data at three test temperatures

TEMP C	TEMP F	UTS MPa	UTS ksi	YS MPa	YS ksi	EL(a) %	FRACTURE(b) MPa*m$^{1/2}$	FRACTURE(b) ksi*in.$^{1/2}$
24	75	1530	222	1275	185	16	109.3	99.5
260	500	1380	200	1215	176	16	106.3	96.7
345	650	1325	192	1110	161	17	103.8	94.5

Note: Specimens were normalized at 900 C (1650 F), oil quenched from 845 C (1550 F) and double tempered, 2 + 2 h at 550 C (1025 F)
(a) in 4D. (b) Plane-strain fracture toughness has WR orientation. An average of two tests were taken at each temperature
Source: Ref 1, p 442

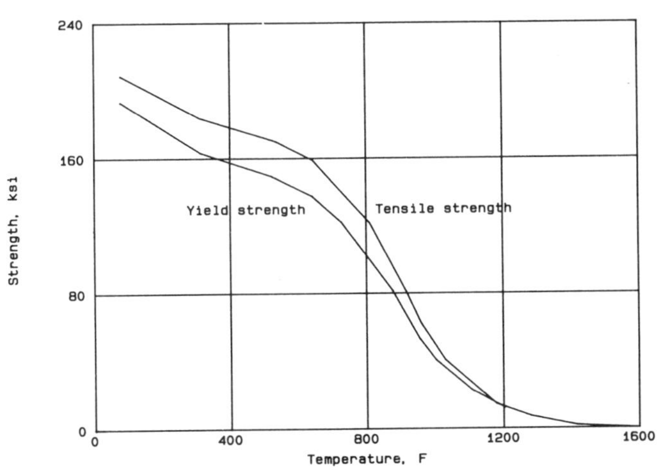

HP-9-4-30: Tensile and yield strength of 0.20% C grade vacuum melted 1-in. plate heat treated at 1700 F for 1 h, air cooled, held at 1500 F for 1 h, water quenched, and held at 1050 F for 2 h. Source: Ref 22, p 33

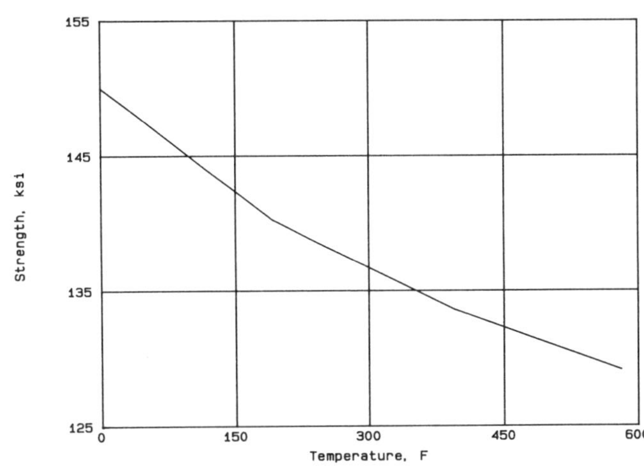

HP-9-4-30: Compressive yield strength of 0.30% C grade vacuum melted forgings. Specimens were heat treated at 1650 F for 1 h, air cooled, held at 1550 F for 1 h, cooled to -100 F for 2 h, double tempered at 950 F for 2 h. Hardness was 212 HRC. Source: Ref 22, p 38

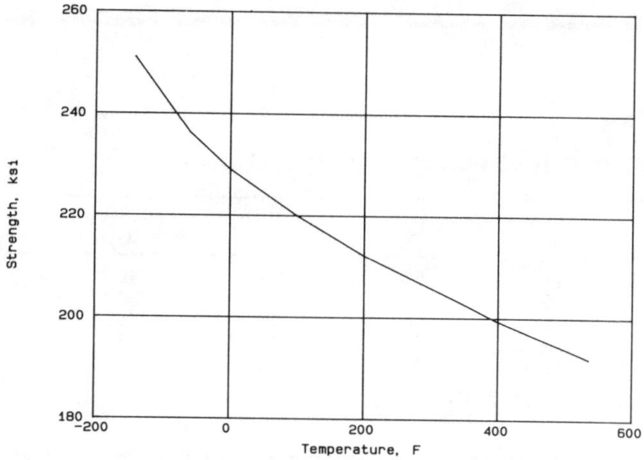

HP-9-4-30: Shear strength of 0.30% C grade vacuum melted forgings. Specimens were heat treated at 1650 F for 1 h, air cooled, held at 1550 F for 1 h, cooled to -100F for 2 h, and double tempered at 950 F for 2 h. Source: Ref 22, p 33

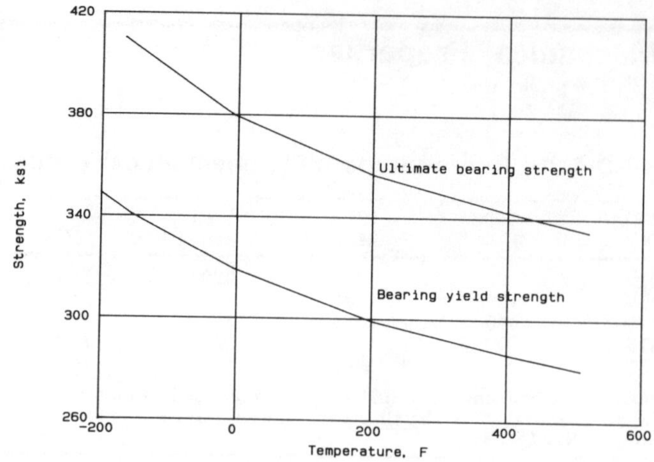

HP-9-4-30: Ultimate bearing and yield strength of 0.30% C grade vacuum melted forgings. Specimens were heat treated at 1650 F for 1 h, air cooled, held at 1550 F for 1 h, cooled to -100 F for 2 h, and double tempered at 950 F for 2 h. e/D = 1.5. Source: Ref 22, p 42

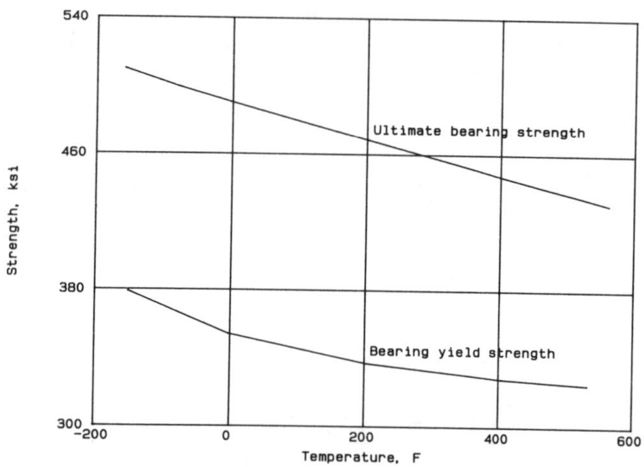

HP-9-4-30: Ultimate bearing and yield strength of 0.30% C grade vacuum melted forgings. Specimens were heat treated at 1650 F for 1 h, air cooled, held at 1550 F for 1 h, cooled to -100 F for 2 h, and double tempered at 950 F for 2 h. e/D = 2. Source: Ref 22, p 42

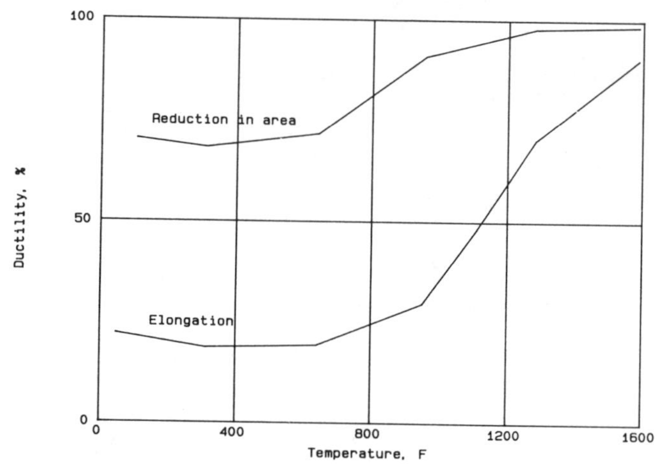

HP-9-4-30: Tensile elongation and reduction in area of 0.20% C grade vacuum melted 1-in. plate, heat treated at 1700 F for 1 h, held at 1500 F for 1 h, water quenched, and held at 1050 F for 2 h. Source: Ref 22, p 33

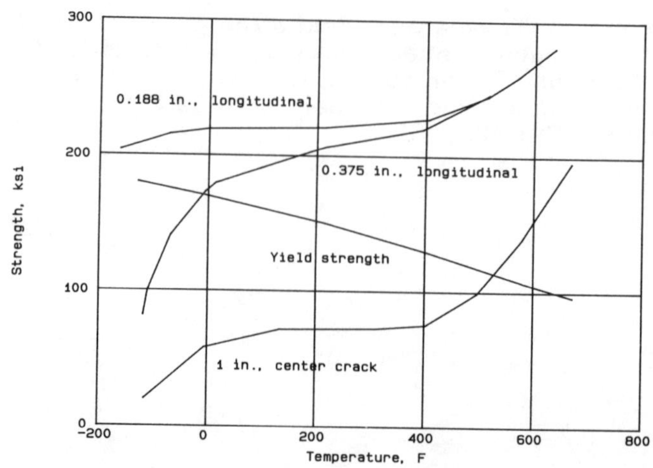

HP-9-4-30: Crack strength of 0.45% C grade plate of several thicknesses. Specimens were heat treated at 1450 F for 0.5 h, solution quenched at 450 F for 6 h, double tempered at 700 F for 2 h. Source: Ref 22, p 43

HY-TUF

UNS number: K32550
Specifications: AMS 6418, ASTM A579 (31), MIL S-7108
Nominal composition: 0.230 to 0.280% C, 0.200 to 0.400% Cr, 1.200 to 1.500% Mn, 0.350 to 0.450% Mo, 1.650 to 2.000% Ni, 0.025% P max, 0.025% S max, 1.300 to 1.700% Si

Product forms: Wrought, bar, rod, wire, plate, sheet, strip, billets, structural shapes
Description: This silicon modified steel (similar to 300M) exhibits excellent toughness at strengths up to at least 1380 MPa (200 ksi). Its toughness is roughly equivalent to that of a maraging steel in this strength range.

Physical Properties

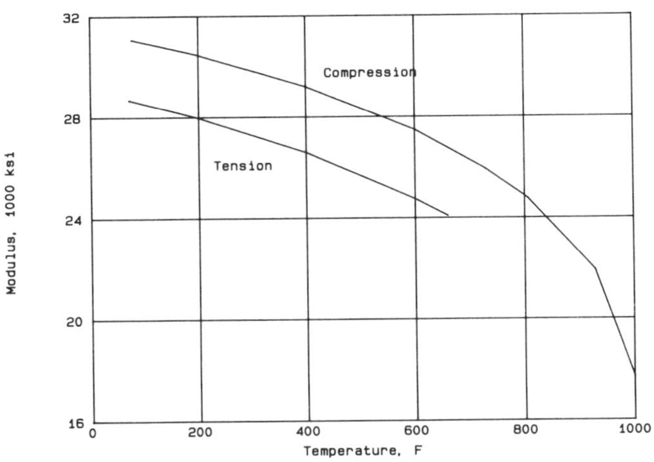

HY-TUF: Static modulus of elasticity in tension and compression for 0.250-in. plate. Specimens were heat treated at 1600 F, held for 25 min, oil quenched, and held at 600 F for 30 min. Tensile strength was 220 ksi. Source: Ref 22, p 5

Mechanical Properties

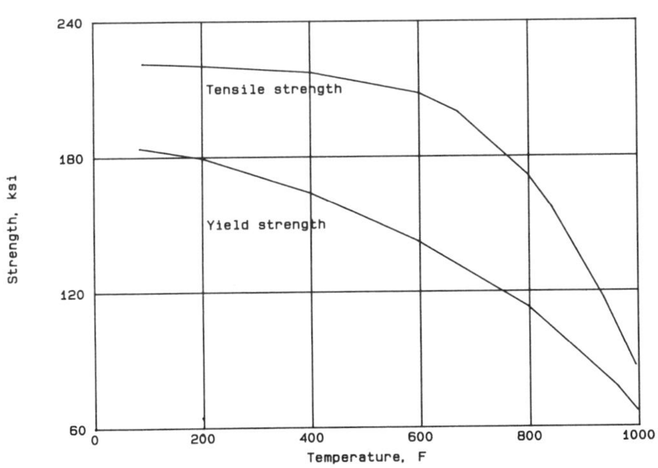

HY-TUF: Tensile and yield strength of 0.250-in. plate. Specimens were heat treated at 1600 F, held for 25 min, oil quenched, and held at 600 F for 30 min. Source: Ref 22, p 4

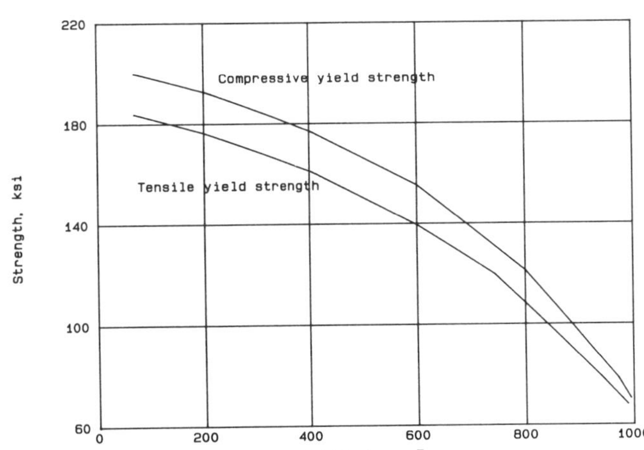

HY-TUF: Tensile and compressive yield strength of 0.250-in. plate. Specimens were heat treated at 1600 F, held 25 min., oil quenched, and held at 600 F for 30 min. Tensile strength was 220 ksi. Source: Ref 22, p 5

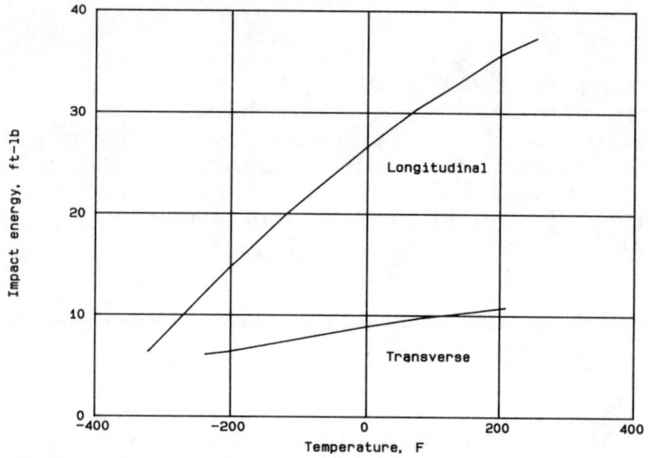

HY-TUF: Charpy V-notch impact energy of 3-in. diam bar. Hardness ranged from 46.5 to 47 HRC. Source: Ref 22, p 5

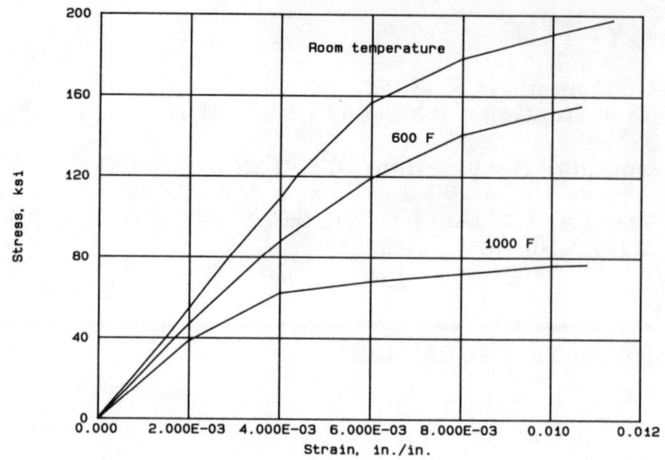

HY-TUF: Stress-strain curves for 0.250-in. plate. Specimens were treated at 1600 F, held 25 min, oil quenched and held at 600 F for 30 min. Tensile strength was 220 ksi. Source: Ref 22, p 4

7. Tool Steels

Tool Steels

Although tool steels are indeed used to make tools for cutting and forming materials, many alloy grades also are used for machinery components and structural applications such as high-temperature springs, ultrahigh strength fasteners, special-purpose valves, and bearings for elevated-temperature service. Thus, tool steels might be best defined as high-quality specialized steels.

As a class, tool steels are generally heat treated, and they tend to have excess carbides in their microstructures, thus making them harder and more wear resistant than the standard AISI-SAE carbon and alloy steels. Commonly used tool steels are classified by the American Iron and Steel Institute (AISI) into ten categories, each identified by a capital letter:

High speed steels
- Molybdenum types, M
- Tungsten types, T

Hot work steels
- Chromium types, H10 to H19
- Tungsten types, H21 to H26
- Molybdenum types, H42

Cold work steels
- Air-hardening, medium alloy types, A
- High-carbon, high-chromium types, D
- Oil-hardening types, O

Shock-resisting steels, S

Special-purpose tool steels
- Low-alloy types, L
- Carbon-tungsten types, F

Mold steels
- Low-carbon types, P1 to P19
- Other types, P20 to P39

Water-hardening tool steels, W

Any given tool steel application typically requires an optimum combination of toughness, resistance to softening at elevated temperatures, and wear resistance. How well a tool steel resists softening at elevated temperatures is related both to its ability to develop secondary hardening and to the amount of special phases (excess alloy carbides, for instance) in its microstructure.

M and T Grades. The high speed steels were developed primarily for high speed cutting tool applications. Among the molybdenum (M) grades, M2 has unusually high resistance to softening at elevated temperatures, because of its high alloy content. The tungsten (T) grades exhibit high red hardness; that is, they resist softening at temperatures so high that the steel emits radiation in the red portion of the visible spectrum. T types are also highly wear resistant. Applications for M and T high speed steels include cutting tools of all types, dies, punches, and high-load, high-temperature structural components such as aircraft bearings and pump parts.

H Grades. The hot work steels were developed to withstand the heat, pressure, and abrasion associated with such metalworking operations as punching, shearing, or forming at high temperatures. Chromium hot work steels (H10 to H19) have good resistance to heat softening. High tungsten and molybdenum contents increase hot strength, but slightly reduce toughness. Vanadium enhances resistance to "washing" (erosive wear) at high temperatures. Silicon improves oxidation resistance at temperatures up to 800 C (1475 F).

These steels are particularly well suited to hot die work. A slightly higher carbon version of H11 is designed for highly stressed aircraft structurals. This steel, called H11 Mod or AISI 610, combines moderate toughness and ductility at a room temperature tensile strength of 1720 to 2070 MPa (250 to 300 ksi) with resistance to softening during continuous exposure to temperatures as high as 540 C (1000 F). Vacuum-arc remelted and electroslag remelted H11 and H11 Mod have improved transverse ductility and toughness; particularly in thick sections.

Tungsten hot work steels (H21 to 26) are

more resistant to high-temperature softening and washing than H11 and H13, but they are also more prone to brittleness at normal working hardnesses. Applications include hot forging dies and high-temperature extrusion dies and mandrels.

The molybdenum hot work steel H42 is less expensive than tungsten hot work grades and is also more resistant to heat checking. Applications are essentially the same.

A, D, and O Grades. Because these grades generally lack sufficient amounts of the alloying elements needed for high heat resistance, cold work tool steels normally are restricted to applications that do not involve prolonged or repeated exposure to temperatures above 200 to 260 C (400 to 500 F). Typical uses for the air-hardening, medium alloy (A) grades include knives, punches, dies, and gages. Types A2, A3, and A7 to A9 contain enough chromium (5%) to give them moderate resistance to elevated temperature softening.

The D grades with 12% Cr offer even better resistance to softening at elevated temperatures. Typical applications for these high carbon, high chromium steels include long-run dies, burnishing tools, rolls, and shear and slitter knives.

The most important service-related property of the oil-hardening (O) steels is high wear resistance at ambient temperature. The O grades have low resistance to softening at elevated temperatures.

S Grades. Shock-resisting steels are used primarily for chisels, rivet sets, punches, driver bits, and other applications requiring toughness and resistance to shock loading. Heat resistance of S1 and S7 is sufficiently high to make them suitable for hot punching and shearing tools.

L and P Grades. The low-alloy, special-purpose (L) steels typically are used for machine parts such as arbors, cams, chucks, and collets, and for other applications requiring good strength and toughness.

The P mold steels are used in low-temperature die casting dies and in molds for injection or compression molding of plastics. P4 and P21 offer medium resistance to softening at elevated temperatures; all the other P grades have low resistance.

W Grades. The water-hardening types are high carbon tool steels that typically contain small amounts of vanadium and chromium. They are used for cold heading tools, woodworking tools, taps and reamers, cutlery, and wear-resistant machine tool components. Their resistance to elevated-temperature softening is low.

Mechanical Properties

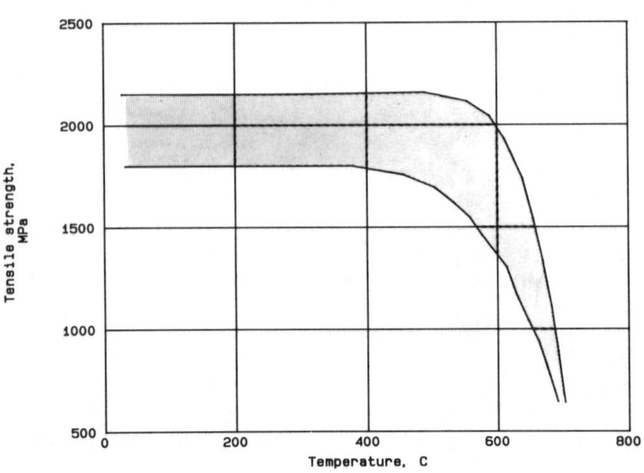

Tool steels: Scatterband of tensile strength for hot work steels. Source: Ref 9, p 46

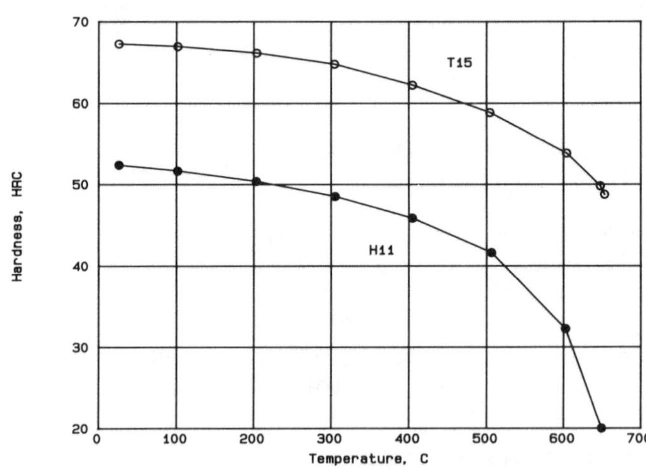

Tool steels: Rockwell C hardness for tool steels with a high resistance to softening. H11 was treated at 1010 C, air cooled, and double tempered at 565 C for 2 h. T15 was heat treated at 1230 C, oil quenched, and double tempered at 550 C for 2 h. After hot hardness testing at 650 C, T15 had a room-temperature hardness of 63.4 HRC. Source: Ref 3, p 442

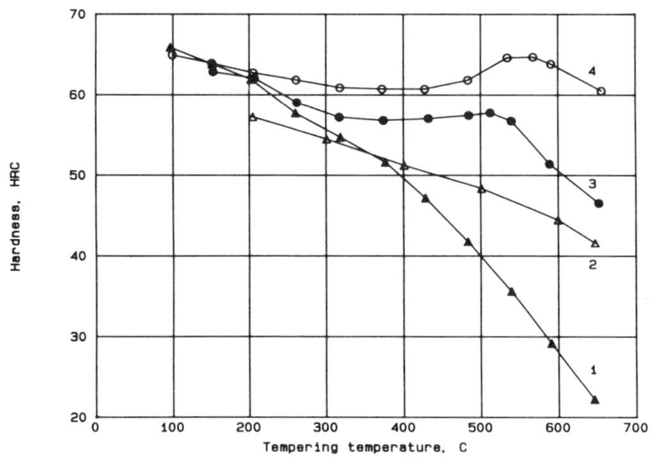

Tool steels: Variation of hardness with tempering for 1 h. Curve 1: low resistance to softening, as in groups W and O. Curve 2: medium resistance to softening, as in type S1. Curves 3 and 4: high and very high resistance to softening, respectively, typical of secondary-hardening tool steels such as A2 and M2. Source: Ref 30, p 18-2

A2

UNS number: T30102
Specifications: ASTM A681 (A2), FED QQ-T-570 (A2), SAE J437 (A2), SAE J438 (A2)
Nominal composition: 0.950 to 1.050% C, 4.750 to 5.500% Cr, 1.000% Mn max, 0.300% Ni max, 0.500% Si max, 0.900 to 1.400% W, 0.150 to 0.500% V
Product forms: Wrought, bar, rod, wire, plate, sheet, strip, billets, forgings, ring, cast, sand, investment, powder

Description: A2 is a medium-alloy, air-hardening, cold work steel. It exhibits a high resistance to softening at elevated temperatures.
Typical applications: Thread rolling dies, extrusion dies, trimming, dies, blanking dies, coining dies, mandrels, shear blades, slitters, spinning rolls, forming rolls, gages, beading dies, burnishing tools, ceramic tools, embossing dies, plastic molds, stamping dies, bushings, punches, and liners for brick molds.

Physical Properties

A2: Mean coefficients of linear thermal expansion

TEMP C	TEMP F	THEXP 10^{-6}/C	THEXP 10^{-6}/F
21-100	70-212	10.7	5.96
21-200	70-390	12.0	6.64
21-300	70-570	12.7	7.05
21-400	70-750	13.2	7.36
21-500	70-930	13.7	7.6
21-600	70-1110	14.0	7.75
21-700	70-1290	14.3	7.92
21-750	70-1380	14.4	7.98

Source: Ref 15, p 445

Mechanical Properties

A2: Hardness and Charpy V-notch property data

TEMP C	TEMP F	HARD HRC	IS J	IS ft-lb
(a)	(a)	63-65	...	...
205	400	60-62	42	31
260	500	59-61	56	41
315	600	58-60	50	37
370	700	57-59	45	33
425	800	57-59	42	31
480	900	57-59	39	29
540	1000	56-58	56	41
595	1100	50-52	...	...
650	1200	42-44	...	...

(a) Air hardened
Source: Ref 15, p 446

A6

UNS number: T30106
Specifications: ASTM A681, FED QQ T-570, DGN B82, COPANT 337, BS 4659, GOST 7ChG2WM

Nominal composition: 0.650 to 0.750% C, 1.800 to 2.500% Mn, 0.500% Si max, 0.900 to 1.200% Cr, 0.300% Ni max, 0.900 to 1.400% Mo
Product forms: Wrought, bar, rod, wire, plate, sheet, strip, billets, forgings, ring, cast, sand, investment, powder
Description: A6 is a medium-alloy, air-hardening, cold work tool steel. It exhibits medium resistance to softening at elevated temperatures.
Typical applications: Blanking dies, forming dies, coining dies, trimming dies, punches, shear blades, spindles, master hobs, retaining rings, mandrels, and plastic dies.

Physical Properties

A6: Mean coefficients of linear thermal expansion

TEMP C	TEMP F	THEXP $10^{-6}/C$	THEXP $10^{-6}/F$
21-100	70-212	11.8	6.57
21-200	70-390	12.4	6.91
21-300	70-570	13.2	7.32
21-400	70-750	13.7	7.59
21-500	70-930	14.1	7.86
21-600	70-1110	14.4	8.01
21-700	70-1290	14.7	8.19

Source: Ref 15, p 446

A8

UNS number: T30108
Specifications: ASTM A681, FED QQ T-570, ABNT EH12, NF 3432 Z38CDWV5, UNI 2955, JIS G 4404, DGN B82, UNE F.537, GOST 6Ch6W3MFS, DIN 1.2606

Nominal composition: 0.500 to 0.600% C, 0.500% Mn max, 0.750 to 1.100% Si, 4.750 to 5.500% Cr, 0.300% Ni max, 1.150 to 1.650% Mo, 1.000 to 1.500% W
Product forms: Wrought, bar, rod, wire, plate, sheet, strip, billets, forgings, ring, cast, sand, investment, powder
Description: A8 is a medium-alloy, air-hardening, cold work tool steel. It exhibits high resistance to softening at elevated temperatures.
Typical applications: Cold slitters, shear blades, hot pressing dies, blanking dies, beading tools, cold forming dies, punches, coining dies, trimming dies, master hobs, rolls, forging die inserts, compression molds, notching dies, and slitter knives.

Physical Properties

A8: Mean coefficients of linear thermal expansion

TEMP C	TEMP F	THEXP $10^{-6}/C$	THEXP $10^{-6}/F$
21-427	70-800	12.0	6.7
21-538	70-1000	12.2	6.8
21-649	70-1200	12.6	7.0

Source: Ref 29, Nov 1981

Mechanical Properties

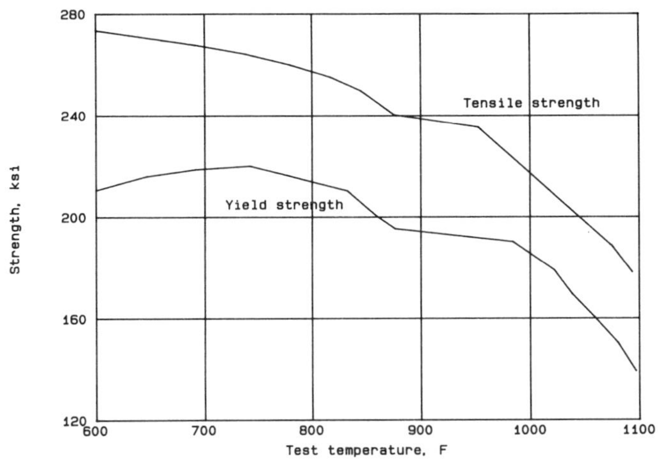

A8: Tensile and yield strength of standard 0.505-in. diameter specimens. Specimens were quenched in still air from 1850 F and double tempered at 1060 and 1035 F. Source: Ref 29, Nov 1981

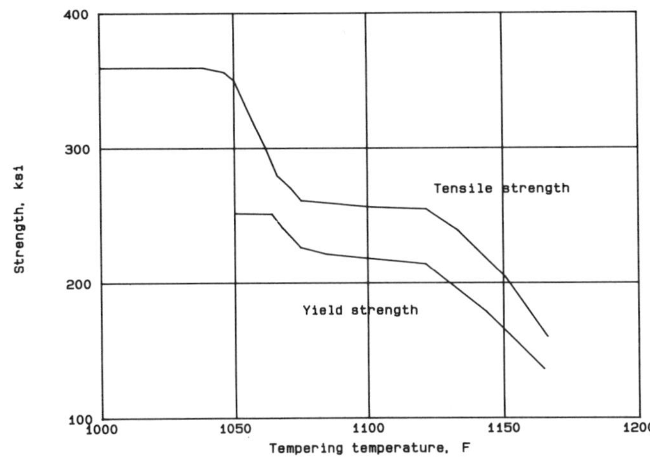

A8: Effect of tempering temperature on the room-temperature tensile and yield strength of standard 0.505-in. diam specimens, quenched in still air from 1850 F. Source: Ref 29, Nov 1981

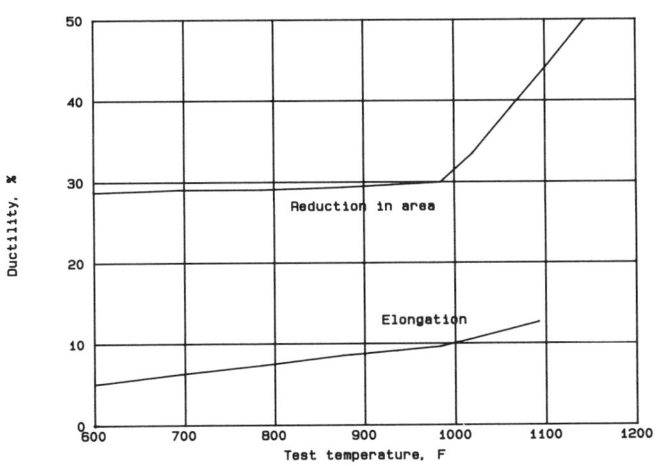

A8: Tensile elongation and reduction in area of standard 0.505-in. diameter specimens. Specimens were quenched in still air from 1850 F and double tempered at 1060 and 1035 F. Source: Ref 29, Nov 1981

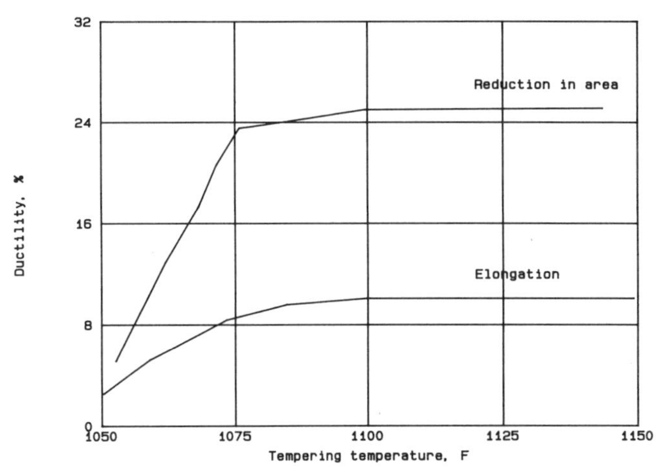

A8: Effect of tempering temperature on tensile elongation and reduction in area of standard 0.505-in. diameter specimens, quenched in still air from 1850 F. Source: Ref 29, Nov 1981

A8: Effect of tempering temperature on Rockwell C hardness. Specimens were austenitized at 1800 F and double tempered. Source: Ref 29, Nov 1981

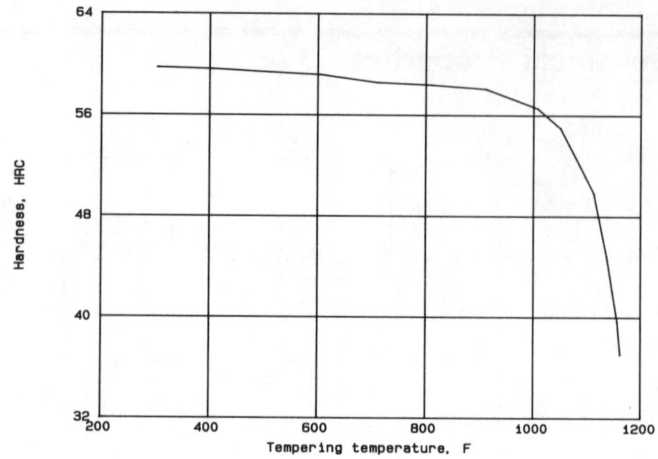

A9

UNS number: T30109
Specifications: ASTM A681, FED QQ T-570, DGN B82
Nominal composition: 0.450 to 0.550% C, 0.500% Mn max, 0.950 to 1.150% Si, 4.750 to 5.500% Cr, 1.250 to 1.750% Ni, 1.300 to 1.800% Mo, 0.800 to 1.400% V
Product forms: Wrought, bar, rod, wire, plate, sheet, strip, billets, forgings, ring, cast, sand, investment, powder

Description: A9 is a medium-alloy, air-hardening type cold work tool steel. It exhibits a high resistance to softening at elevated temperatures.
Typical applications: Solid cold heading dies, die inserts, heading hammers, coining dies, forming dies and rolls, die casings, and gripper dies. Hot work applications include punches, piercing tools, mandrels, extrusion tooling, forging dies, gripper dies, die casings, heading dies, hammers, and coining and forming dies.

Mechanical Properties

A9: Tensile property data

TEMP C	TEMP F	UTS MPa	UTS ksi	YS MPa	YS ksi	EL %	RA %
21	70	1696	246	1517	220	10.5	41.8
427	800	1403	203.5	1193	173	13.2	49.0
482	900	1296	188	1100	159.5	15.1	54.8
538	1000	1145	166	969	140.5	19.5	60.4
593	1100	876	127	738	107	20.4	70.1
649	1200	545	79	452	65.5	33.7	88.5

Note: Specimens were air cooled from 1010 C (1850 F), tempered at 565 C (1050 F), 3+3 h to 48.5 HRC
Source: Ref 14, p 258

A9: Hot hardness data

TEMP C	TEMP F	HARD(a) HRC	HARD(b) HRC
Room-temperature hardness after heat treatment			
21	70	57.0	48.0
After 0.5 h at temperature			
316	600	53.5	44.0
427	800	51.0	39.5
538	1000	41.5	32.5
593	1100	35.5	20.5
649	1200	21.5	15.5

(a) Test 1. (b) Test 2
Source: Ref 14, p 258

D2

UNS number: T30402
Specifications: ASTM A681, FED QQ T-570, SAE J437, SAE J438, DIN 1.2379, NF 2235 Z160CDV12, BS 4659 BD2, UNI X 150CrMo12KD, STAS VMoC120, COPANT 337, DGN B82, SIS 14 23 10

Nominal composition: 1.400 to 1.600% C, 0.600% Mn max, 0.600% Si max, 11.000 to 13.000% Cr, 0.300% Ni max, 0.700 to 1.200% Mo, 1.100% V max, 1.000% Co max

Product forms: Wrought, bar, plate, sheet, strip, billets, forgings, ring, cast, sand, investment
Description: D2 is a high-carbon, high-chromium cold work tool steel. It exhibits high resistance to softening at elevated temperatures.
Typical applications: Blanking dies, cold forming dies, drawing dies, lamination dies, thread rolling dies, shear blades, slitter knives, forming rolls, burnishing tools, punches, gages, knurling tools, lathe centers, broaches, cold extrusion dies, mandrels, swaging dies, and cutlery.

Physical Properties

D2: Mean coefficients of linear thermal expansion

TEMP C	TEMP F	THEXP $10^{-6}/C$	THEXP $10^{-6}/F$
21-100	70-212	11.8	6.57
21-200	70-390	12.4	6.91
21-300	70-570	13.2	7.32
21-400	70-750	13.7	7.59
21-500	70-930	14.1	7.86
21-600	70-1110	14.4	8.01
21-700	70-1290	14.7	8.19

Source: Ref 15, p 449

D3

UNS number: T30403
Specifications: ASTM A681, FED QQ T-570, SAE J347, SAE J438, DIN 1.2080, NF 2234 Z200C12, BS 4659 BD3, UNI X 205Cr12KU, JIS G 4404 SKD1, GOST Ch12, UNE F.5212
Nominal composition: 2.000 to 2.350% C, 0.600% Mn max, 0.600% Si max, 11.000 to 13.500% Cr, 0.300% Ni max, 1.000% W max, 1.000% V max
Product forms: Wrought, bar, plate, sheet, strip, billets, forgings, ring, cast, sand, investment
Description: D3 is a high-carbon, high-chromium cold work tool steel. It exhibits a high resistance to softening at elevated temperatures.
Typical applications: Blanking dies, cold forming dies, drawing dies, lamination dies, thread rolling dies, shear blades, slitter knives, forming rolls, seaming rolls, burnishing tools, punches, gages, crimping dies, and swaging dies.

Physical Properties

D3: Mean coefficients of linear thermal expansion

TEMP C	TEMP F	THEXP 10^{-6}/C	THEXP 10^{-6}/F
21-100	70-212	10.7	5.93
21-200	70-390	11.6	6.45
21-300	70-570	12.1	6.73
21-400	70-750	12.4	6.89
21-500	70-930	12.8	7.09
21-600	70-1110	12.9	7.14
21-700	70-1290	13.0	7.21
21-750	70-1380	13.1	7.26

Source: Ref 15, p 449

D5

UNS number: T30405
Specifications: ASTM A681, FED QQ T-570, SAE J437, SAE J438
Nominal composition: 1.400 to 1.600% C, 0.600% Mn max, 0.600% Si max, 11.000 to 13.000% Cr, 0.300% Ni max, 0.700 to 1.200% Mo, 1.000% V max, 2.500 to 3.500% Co
Product forms: Wrought, bar, plate, sheet, strip, billets, forgings, ring, cast, sand, investment
Description: D5 is a high-carbon, high-chromium, cold work steel. It exhibits high resistance to softening at elevated temperatures. Its high red hardness makes it preferable to other high-carbon, high-chromium steels for intermittent hot work applications, such as dies for trimming hot forging flash.
Typical applications: Cold forming dies, thread rolling dies, blanking dies, coining dies, trimming dies, draw dies, shear blades, punches, quality cutlery, and rolls.

Physical Properties

D5: Mean coefficients of linear thermal expansion

TEMP C	TEMP F	THEXP 10^{-6}/C	THEXP 10^{-6}/F
27-260	80-500	9.9	5.5
27-427	80-800	10.9	6.1
27-538	80-1000	12.1	6.7
27-649	80-1200	12.6	7.0

Source: Ref 29, Nov 1980

Mechanical Properties

D5: Hot hardness data

Ttpr C	Ttpr F	HARD(a) HRC	TEMP C	TEMP F	HARD HRC
427	800	60.0	427	800	56.0
482	900	60.5	482	900	54.5
538	1000	59.0	538	1000	48.5
593	1100	50.0	593	1100	29.0

Note: Specimens were hardened by air cooling from 1010 C (1850 F) and tempered 2 h. Specimens, indentor, and anvil were at test temperature 30 min before testing
(a) At room temperature
Source: Ref 29, Nov 1980

Mechanical Properties

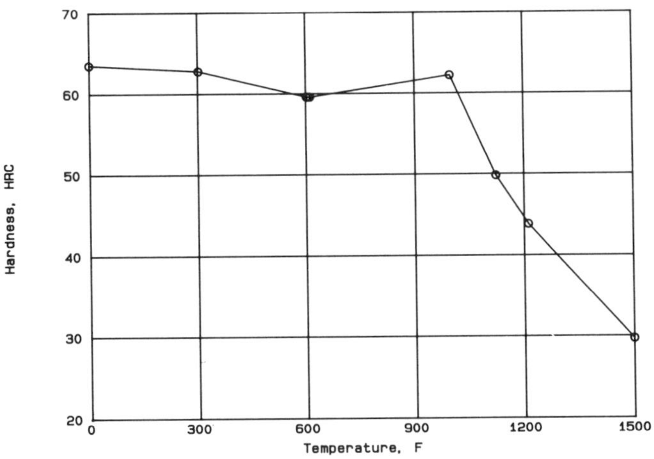

D5: Effect of tempering temperature on hardness. Note pronounced secondary hardening at 980 F. Source: Ref 29, Nov 1980

H10

UNS number: T20810
Specifications: ASTM A600, FED QQ T-570, DIN 1.2365, NF 13451 30DCV28, BS 4659 BH10, UNI 2955, GOST 3Ch3M3F, UNE F.5313, COPANT 37H10, UNE F.5200, DGN B82
Nominal composition: 0.350 to 0.450% C, 0.250 to 0.700% Mn, 0.800 to 1.200% Si, 3.000 to 3.750% Cr, 0.300% Ni max, 2.000 to 3.000% Mo, 0.250 to 0.750% V
Product forms: Wrought, bar, rod, wire, plate, sheet, strip, billets, forgings, ring, cast, sand, investment, powder
Description: H10 is a chromium hot work tool steel. It exhibits high resistance to softening at elevated temperatures. Compared with most other high-alloy tool steels, AISI H10 offers greater resistance to softening at no sacrifice in wear resistance. It can be used at temperatures as high as 650 C (1200 F) while retaining a hardness of 41 to 43 HRC.
Typical applications: Mandrels, extrusion and forging dies, die holders, dummy blocks, punches, die inserts, gripper and header dies, hot shears, aluminum die casting dies, inserts for forging dies and upsetters, and shell piercing tools.

Physical Properties

H10: Mean coefficients of linear thermal expansion

TEMP C	TEMP F	THEXP 10^{-6}/C	THEXP 10^{-6}/F
27-204	80-400	13.3	7.4
27-621	80-1150	14.2	7.9
260-621	500-1150	14.9	8.3

Note: Sample hardened to 41 HRC
Source: Ref 29, Jul 1964

Mechanical Properties

H10: Tensile property data

TEMP C	TEMP F	UTS MPa	UTS ksi	YS MPa	YS ksi	EL %	RA %
21	70	1679	243.5	1451	210.5	11.4	35.0
427	800	1348	195.5	1214	176	14.0	58.0
482	900	1258	182.5	1145	166	16.6	63.2
538	1000	1134	164.5	1048	152	17.4	63.2
593	1100	941	136.5	810	117.5	21.3	68.9
649	1200	596	86.5	527	76.5	22.1	79.4

Note: Specimens were air cooled from 1010 C (1850 F), tempered at 606 C (1125 F), 3+3 h, to 48 HRC
Source: Ref 14, p 261

H10: Effect of hardness and temperature on Charpy V-notch impact values

HARD HRC	IS(a) J	IS(a) ft-lb	IS(b) J	IS(b) ft-lb	IS(c) J	IS(c) ft-lb	IS(d) J	IS(d) ft-lb
40	11	8	18	13	22	16	27	20
45	9	7	12	9	15	11	23	17
50	8	6	11	8	14	10	20	15
55	8	6	11	8	14	10	20	15

Note: Specimens from a 25-mm (1-in.) diam bar were austenitized at 1038 C (1900 F) for 1 h, air cooled and tempered to indicated hardness
(a) 27 C (80 F). (b) 149 C (300 F). (c) 260 C (500 F). (d) 427 C (800 F)
Source: Ref 29, Jul 1964

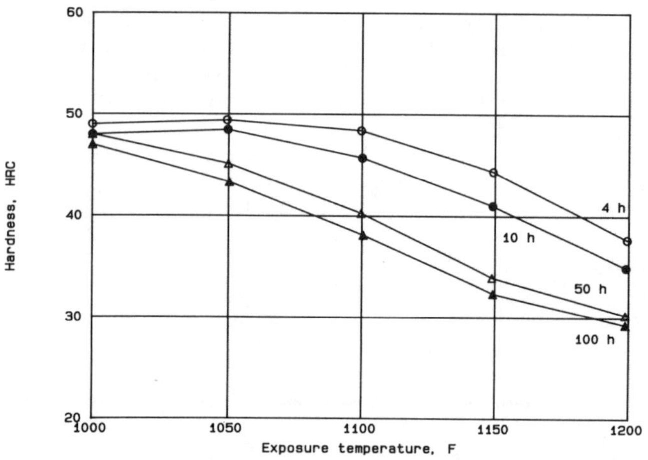

H10: Rockwell hardness values after exposure at temperature for different times. Initial hardness: 49 HRC. Specimens were heat treated at 1850 F for 0.5 h, air cooled, double tempered at 1125 F for 3 h, and air cooled. Source: Ref 14, p 262

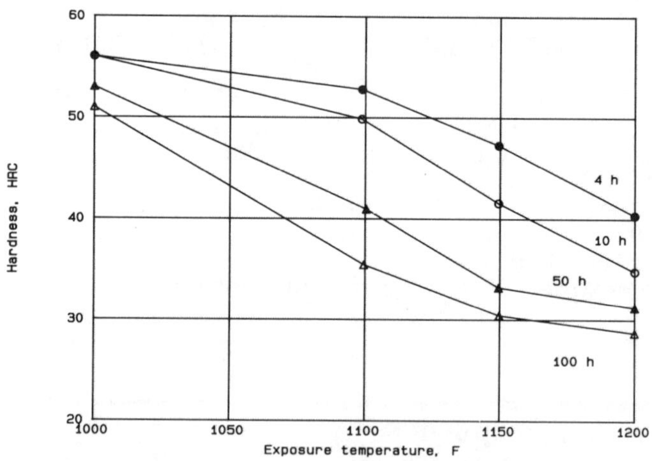

H10: Rockwell hardness values after exposure to temperature for different times. Initial hardness: 56.5 HRC. Specimens were heat treated at 1850 F for 0.5 h, air cooled, double tempered at 1025 F for 3 h, and air cooled. Source: Ref 14, p 262

H10: Hot hardness data

TEMP C	TEMP F	HARD(a) HRC	HARD(b) HRC
Room-temperature hardness after heat treatment			
21	70	55.5	46
After 0.5 h at temperature			
316	600	50.5	42
427	800	48	40
538	1000	42	32
593	1100	32	19
649	1200	21	15

(a) Test 1. (b) Test 2
Source: Ref 14, p 261

H10: Hot hardness data

TEMP C	TEMP F	HARD(a) HRC	HARD(b) HRC
21	70	57	47
121	250	55	45
204	400	55	45
316	600	54	42
427	800	51	41
538	1000	48	32
566	1050	...	25

Note: Specimens were oil hardened and tempered
(a) Tempered at 538 C (1000 F) for 2+2 h, 57 HRC.
(b) Tempered at 621 C (1150 F) for 2+2 h, 47 HRC
Source: Ref 29, Jul 1964

H11

UNS number: T20811
Specifications: AMS 6437, AMS 6485, AMS 6488, ASTM A579, ASTM A681, MIL S-47086, MIL S-47262, FED QQ T-570, SAE J437, SAE J438, SAE J467, DIN 1.2343, NF Z 38CDV5, BS BH11, UNI X 37CrMoV51KU, JIS S KD6, GOST 4Ch5MFS, UNE F.5317

Nominal composition: 0.330 to 0.430% C, 0.200 to 0.500% Mn, 0.800 to 1.200% Si, 4.750 to 5.500% Cr, 0.300% Ni max, 1.100 to 1.600% Mo, 0.300 to 0.600% V

Product forms: Wrought, bar, rod, wire, plate, sheet, strip, billets, forgings, ring, cast, sand, investment, powder
Description: AISI H11 is an air-hardening chromium steel that combines a room-temperature tensile strength of 1515 to 2070 MPa (220 to 300 ksi) with stable elevated temperature properties up to 480 to 540 C (900 to 1000 F).
Typical applications: Ultrahigh-strength landing gear and air frame components, solid fuel rocket cases, aerospace fasteners, and a variety of hot work tooling components.

Physical Properties

H11: Thermal conductivity

TEMP C	TEMP F	THCON W/m*K	THCON Btu/ft*h*F
100	200	42.2	24.4
260	500	36.3	21.0
400	750	33.4	19.3
540	1000	31.5	18.2
675	1250	30.1	17.4
315	1500	28.6	16.5

Source: Ref 3, p 442

H11: Mean coefficients of linear thermal expansion

TEMP F	TEMP C	THEXP 10^{-6}/C	THEXP 10^{-6}/F
200	93	10.9	6.1
400	204	11.5	6.4
800	427	12.2	6.8
1000	538	12.4	6.9

Source: Ref 1, p 653

H11: Instantaneous coefficient of linear thermal expansion. These are not mean values of the coefficient over a range of temperatures. Source: Ref 1, p 653

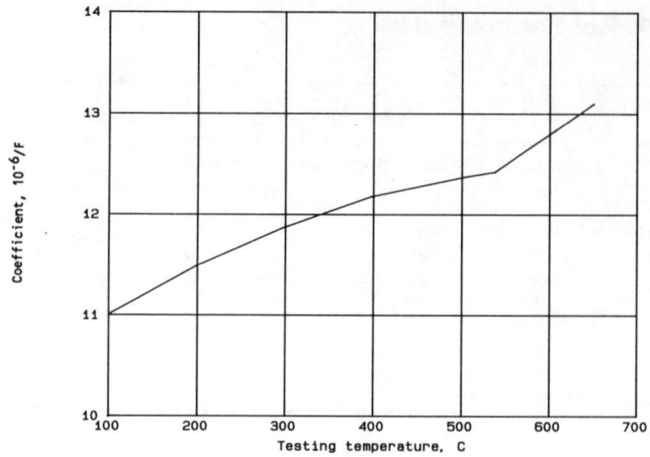

Mechanical Properties

H11: Tensile property data

TEMP C	TEMP F	UTS MPa	UTS ksi	YS MPa	YS ksi	EL %	RA %
21	70	1520	220.5	1358	197	14.7	52.3
427	800	1189	172.5	1027	149	15.6	61.3
482	900	1124	163	955	138.5	18.8	61.2
538	1000	1051	152.5	872	126.5	22.5	69.0
593	1100	855	124	741	107.5	25.6	76.4
649	1200	593	86	514	74.5	26.1	81.4

Note: Specimens were air cooled from 1010 C (1850 F), tempered at 593 C (1100 F) for 3+3 h, 46 HRC
Source: Ref 14, p 263

H11: Tensile property data

TEMP C	TEMP F	UTS MPa	UTS ksi	YS MPa	YS ksi	EL %	RA %	HARD HB
510	950	1531	222	1276	185	11	45	440
538	1000	1365	198	1151	167	12	46	380
566	1050	1172	170	965	140	15	51	330
593	1100	910	132	793	115	19	60	275
621	1150	531	77	483	70	30	78	140

Note: Specimens were austenitized for 30 min at 1010 C (1850 F), air cooled, tempered for 3 h at temperature and tested at the corresponding temperature. The specimens were standard 12.8-mm (0.505-in.) diameter for hot tensile testing. Elevated temperature hardness was determined by the mutual indentation method using cylinders 10 mm (0.394 in.) in diameter by 10 mm (0.394 in.) long
Source: Ref 29, Aug 1964

H11: Tensile property data

TEMP C	TEMP F	UTS MPa	UTS ksi	YS MPa	YS ksi	EL %	RA %	HARD HB
510	950	1034	150	938	136	18	66	265
538	1000	945	137	862	125	20	69	250
566	1050	848	123	779	113	23	73	230
593	1100	800	116	662	96	27	77	205
621	1150	572	83	510	74	28	80	170

Note: Specimens were austenitized for 30 min at 1010 C (1850 F), air cooled, tempered 3 h at 621 C (1150 F), air cooled and tested at lower temperatures
Source: Ref 29, Aug 1964

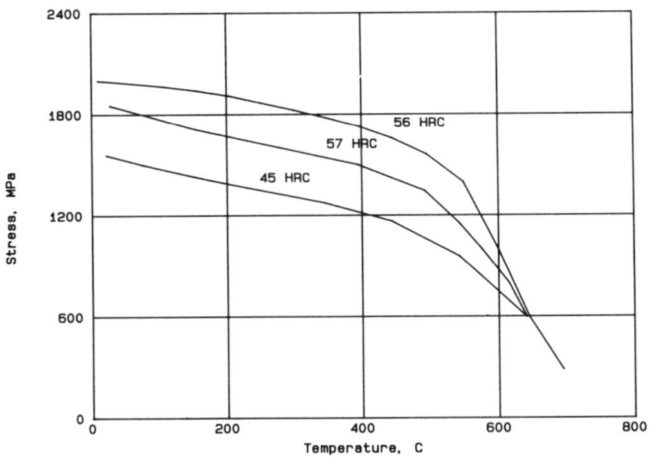

H11: Tensile strength of specimens with indicated hardness. Hardness was measured using the Rockwell C hardness test. Source: Ref 15, p 460

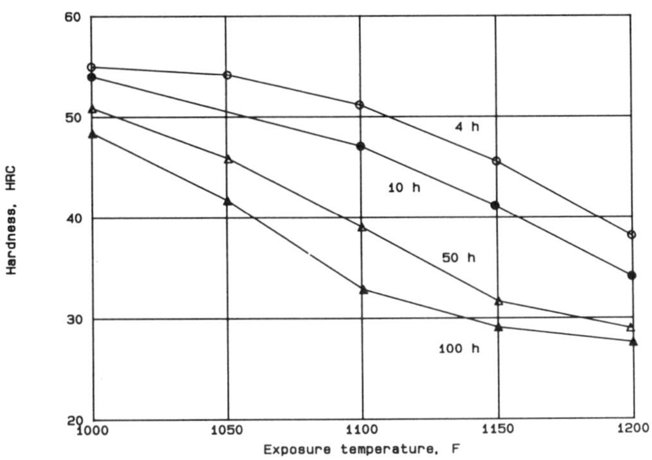

H11: Hardness after exposure to temperature for different times. Initial hardness: 55 HRC. Specimens were heat treated to 1850 F for 0.5 h, air cooled, double tempered at 1000 F for 3 h. Source: Ref 14, p 266

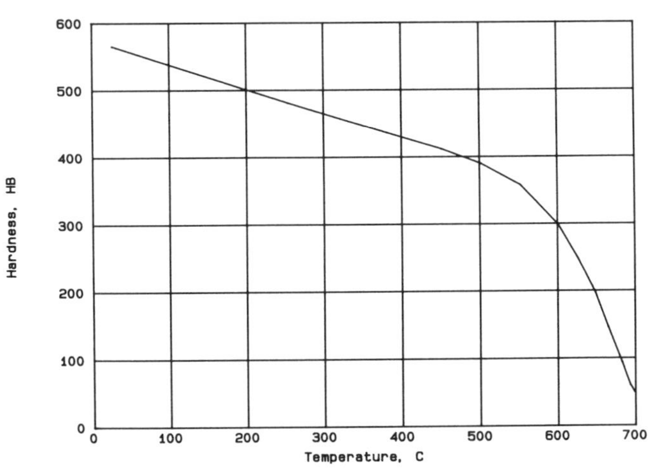

H11: Brinell hardness. Source: Ref 15, p 460

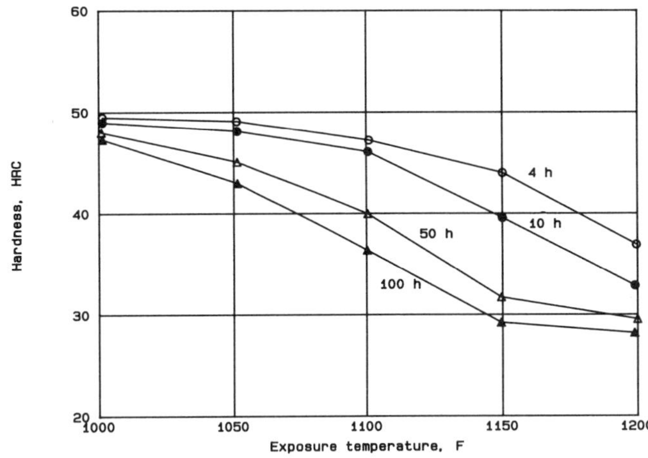

H11: Hardness after exposure to temperature for different times. Initial hardness: 49 HRC. Specimens were heat treated at 1850 F for 0.5 h, air cooled, double tempered at 1100 F for 3 h. Source: Ref 14, p 266

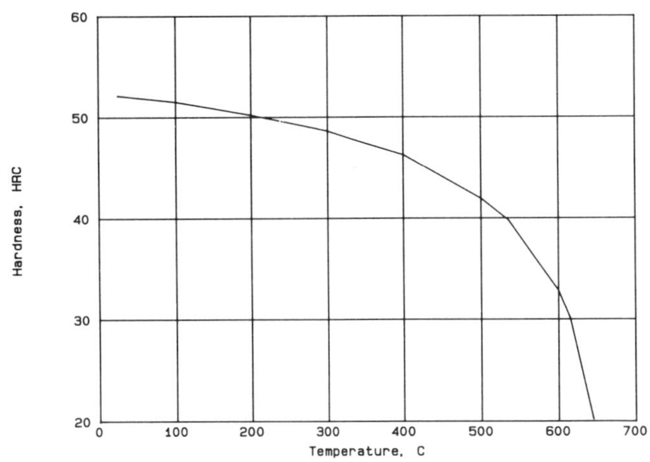

H11: Rockwell C hardness. H11 has high resistance to softening at elevated temperatures. For these tests, H11 was air cooled from 1010 C and double tempered at 565 C for 2 h. Source: Ref 3, p 442

H11: Hot hardness data

TEMP C	TEMP F	HARD(a) HRC	HARD(b) HRC
Room-temperature hardness after heat treatment			
21	70	57.0	48.5
After 0.5 h at temperature			
316	600	52.5	42.0
427	800	51.0	39.5
538	1000	42.0	33.0
593	1100	33.5	20.5
649	1200	23.5	17.0

(a) Test 1. (b) Test 2
Source: Ref 14, p 263

H11 Mod (AISI 610)

Specifications: AMS 6437, AMS 6485, AMS 6488, ASTM A579, ASTM A681, MIL S-47086, MIL S-47262, FED QQ T-570, SAE J437, SAE J438, SAE J467, DIN 1.2343, NF Z 38CDV5, BS BH11, UNI X 37CRMoV51KU, JIS S KD6, GOST 4Ch5MFS, UNE F.5317
Nominal composition: 0.400% C, 0.300% Mn, 0.900% Si, 5.000% Cr, 1.300% Mo, 0.500% V
Product forms: Wrought, bar, rod, wire, plate, sheet, strip, billets, forgings, ring, cast, sand, investment, powder

Description: This modification of AISI H11 is an air-hardening chromium steel that combines a room temperature tensile strength of 1515 to 2070 MPa (220 to 300 ksi) with stable elevated temperature properties up to 480 to 540 C (900 to 1000 F).

Typical applications: Ultrahigh strength landing gear and air frame components, solid fuel rocket cases, aerospace fasteners, and a variety of hot work tooling components.

Physical Properties

H11 Mod: Thermal conductivity

TEMP C	TEMP F	THCON W/m*K	THCON Btu/ft*h*F
216	420	29	16.5
349	660	28	16.4
477	890	28	16.3
604	1120	29	16.6

Note: Samples were annealed
Source: Carpenter Technology

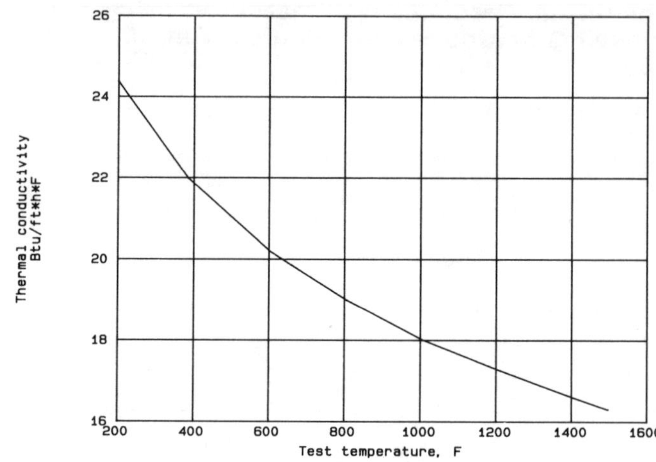

H11 Mod: Thermal conductivity. Source: Ref 22, p 7

H11 Mod: Mean coefficients of linear thermal expansion

TEMP C	TEMP F	THEXP 10^{-6}/C	THEXP 10^{-6}/F
Annealed			
21-100	70-212	11.9	6.6
21-200	70-392	12.4	6.9
21-300	70-572	12.8	7.1
21-400	70-752	13.0	7.2
21-500	70-932	13.1	7.3
21-600	70-1112	13.3	7.4
21-700	70-1292	13.6	7.5
21-800	70-1472	13.8	7.7
Hardened			
21-100	70-212	11.9	6.6
21-200	70-392	12.4	6.9
21-300	70-572	12.7	7.0
21-400	70-752	12.9	7.2
21-500	70-932	13.3	7.4
21-600	70-1112	13.6	7.5

Note: Test samples were heat treated to 1517 MPa (220 ksi)
Source: Ref 28, p 22

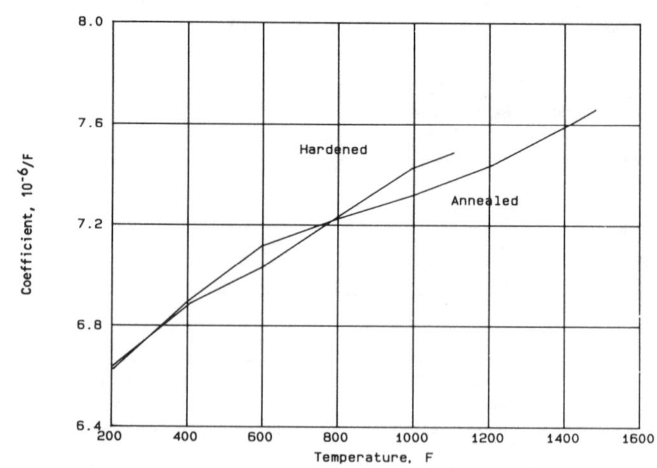

H11 Mod: Mean coefficient of linear thermal expansion from room temperature to indicated temperature. Source: Ref 22, p 7

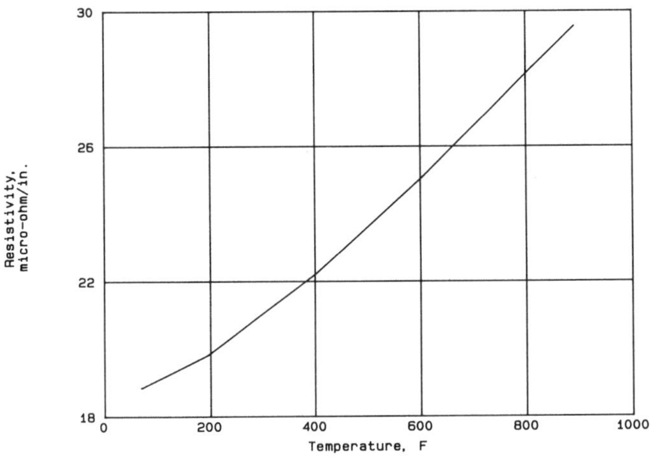

H11 Mod: Electrical resistivity of stress relieved and annealed bar. Source: Ref 22, p 7

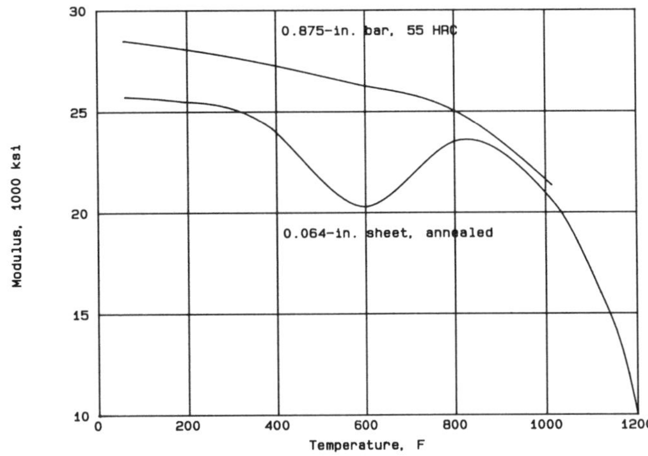

H11 Mod: Modulus of elasticity. Source: Ref 22, p 46

H11 Mod: Modulus of elasticity

TEMP C	TEMP F	E GPa	E 10^6 psi
21	70	210	30.5
93	200	197	28.5
204	400	186	27
260	500	183	26.5
316	600	190	27.5
371	700	193	28
427	800	186	27
482	900	186	27
538	1000	159	23
593	1100	145	21
649	1200	110	16

Note: Test samples were heat treated to 1792 MPa (260 ksi) tensile strength
Source: Ref 28, p 23

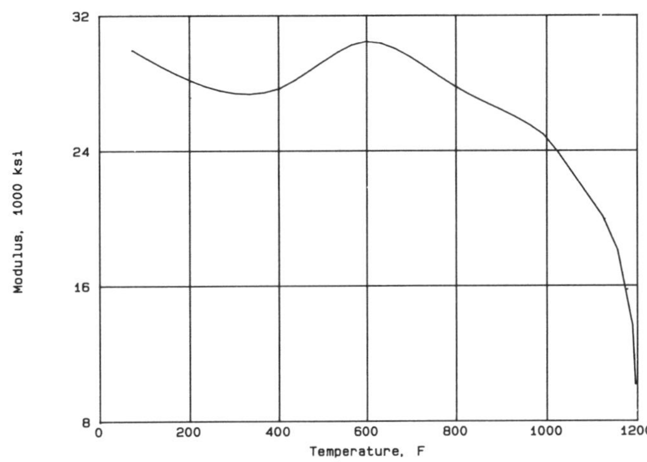

H11 Mod: Modulus of elasticity in compression for 0.064-in. annealed sheet. Source: Ref 22, p 47

H11 Mod: Modulus of rigidity

TEMP C	TEMP F	G GPa	G 10^6 psi
21	70	83	12
93	200	76	11
204	400	69	10
260	500	69	10
316	600	76	11
371	700	76	11
427	800	76	11
482	900	69	10
538	1000	62	9
593	1100	55	8
649	1200	41	6

Note: Samples were heat treated to 1792 MPa (260 ksi) tensile strength
Source: Ref 28, p 23

Mechanical Properties

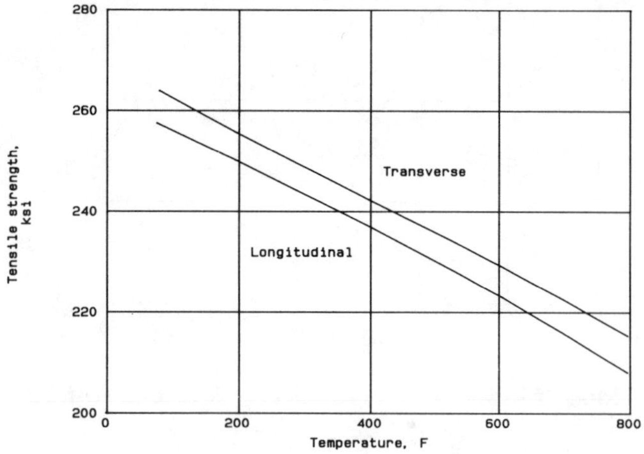

H11 Mod: Tensile strength of sheet heat treated to 260 to 280 ksi tensile strength. Sheet thickness: 0.062 and 0.078 in. Heat treatment: 1150 F for 0.5 h, 1850 F for 0.5 h, air cooled and double tempered for 3 h. Source: Ref 22, p 28

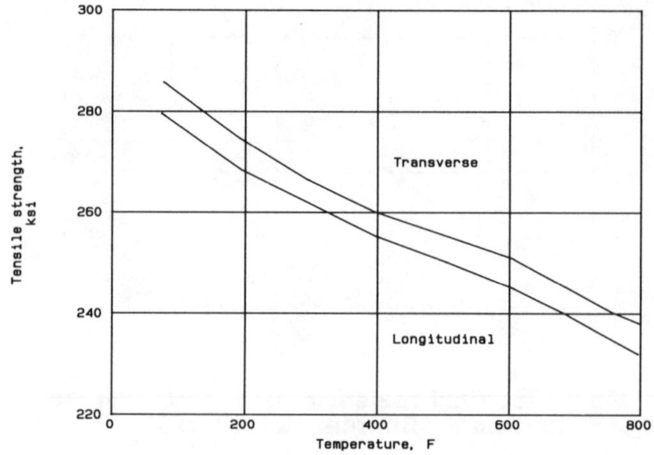

H11 Mod: Tensile strength of 0.062- and 0.078-in. sheet, heat treated to 280 to 300 ksi tensile strength. Heat treatment: 1150 F for 0.5 h, 1850 F for 0.5 h, air cooled, and double tempered for 3 h. Source: Ref 22, p 28

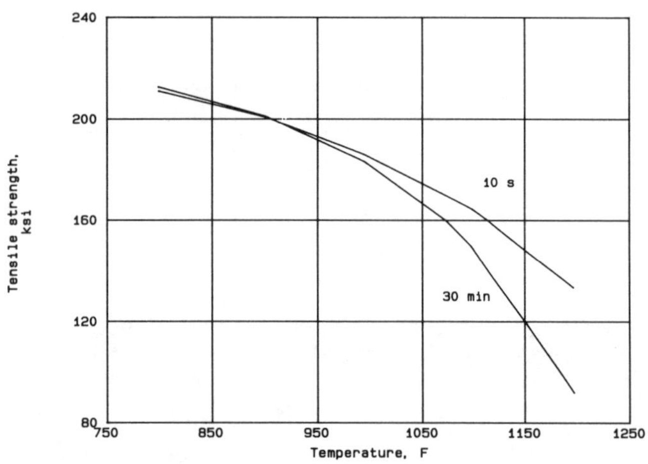

H11 Mod: Effect of test temperature and holding time on tensile strength at a strain rate of 0.0003 in./in.*min. Heat treatment of 0.063-in. sheet: 1850 F, air cooled, and double tempered for 3 h at 1000 F. Source: Ref 22, p 29

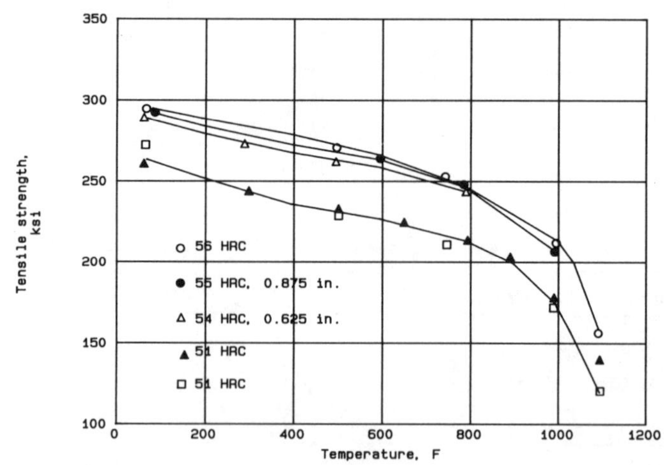

H11 Mod: Tensile strength of bar heat treated to 260 to 290 ksi room-temperature tensile strength. Source: Ref 22, p 29

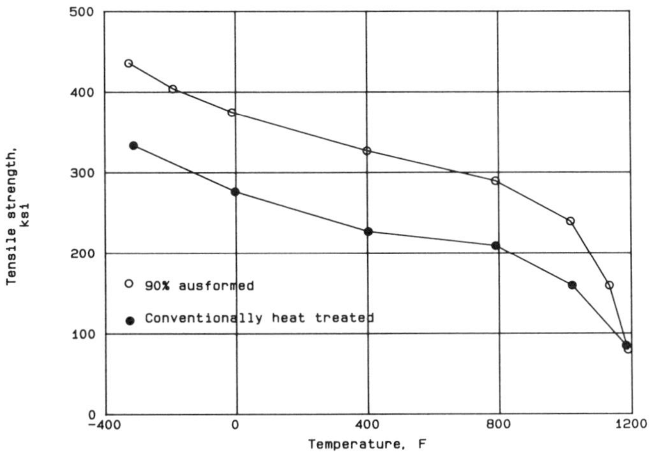

H11 Mod: Tensile strength of 90% ausformed steel tempered at 1050 F and conventionally heat treated steel tempered at 1060 F. Source: Ref 22, p 32

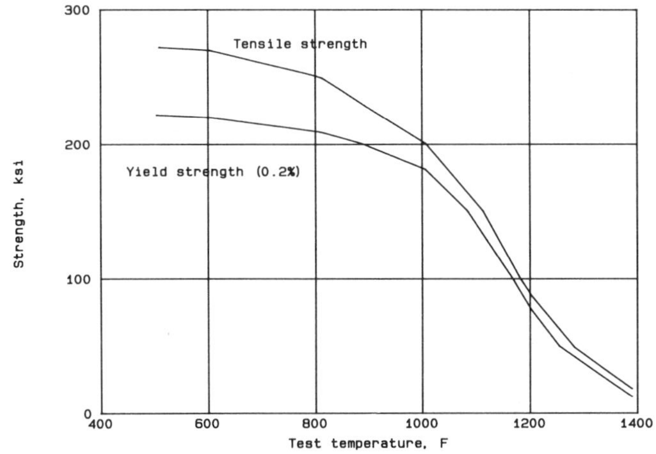

H11 Mod: Tensile and yield strength of longitudinal specimens, heat treated in air at 1850 F for 1 h and double tempered at 1000 F for 2 h. Source: Carpenter Technology

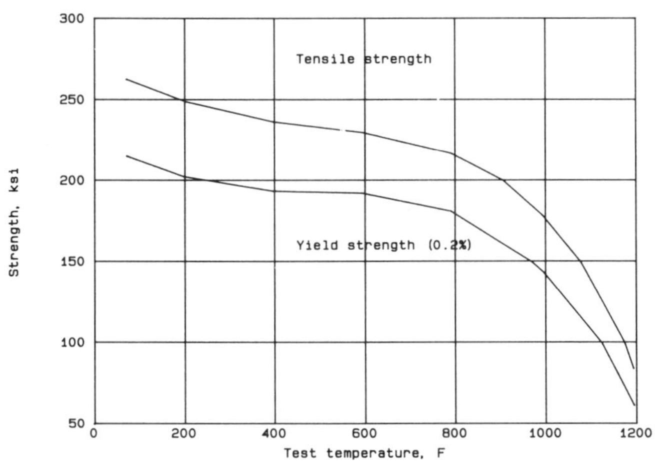

H11 Mod: Tensile and yield strength of longitudinal specimens, heat treated in air at 1850 F for 1 h and double tempered at 1050 F for 2 h. Source: Carpenter Technology

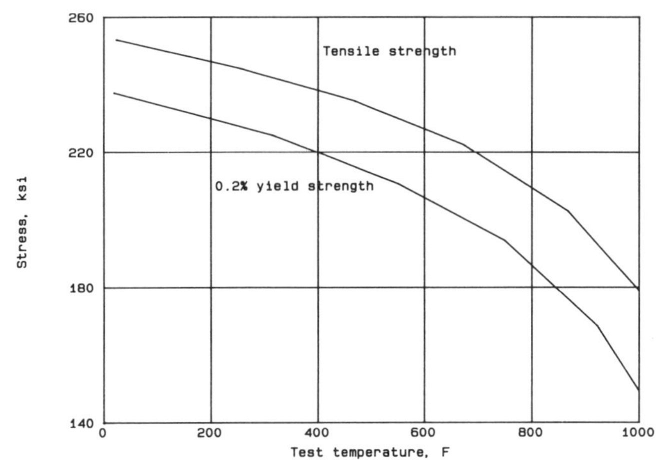

H11 Mod: Tensile and yield strength of cast, vacuum degassed specimens, homogenized at 2000 F for 3 h, air cooled, austenitized at 1850 F for 2 h, air cooled, double tempered at 1050 F for 2 h, and air cooled. Source: Ref 22, p 30

H11 Mod: Short-time longitudinal tensile property data

TEMP C	TEMP F	YS 0.02% MPa	YS 0.02% ksi	YS 0.20% MPa	YS 0.20% ksi	UTS MPa	UTS ksi	EL(a) %	RA %
21	70	1089	158	1482	215	1806	262	10	36
149	300	1000	145	1358	197	1689	245	10	36
260	500	1007	146	1345	195	1600	232	10	35
343	650	931	135	1317	191	1579	229	10	35
427	800	800	116	1289	187	1510	219	11	39
482	900	683	99	1145	166	1427	207	12	39
538	1000	655	95	965	140	1241	180	11	35
593	1100	434	63	724	105	979	142	13	46
649	1200	234	34	434	63	586	85	19	67

Note: Test samples were rolled bars under 25 mm (1 in.) round with an initial tensile strength of 1792 MPa (260 ksi) at 21 C (70 F)
(a) Test specimen had a diameter of 44.5 mm (1.75 in.) round
Source: Ref 28, p 23

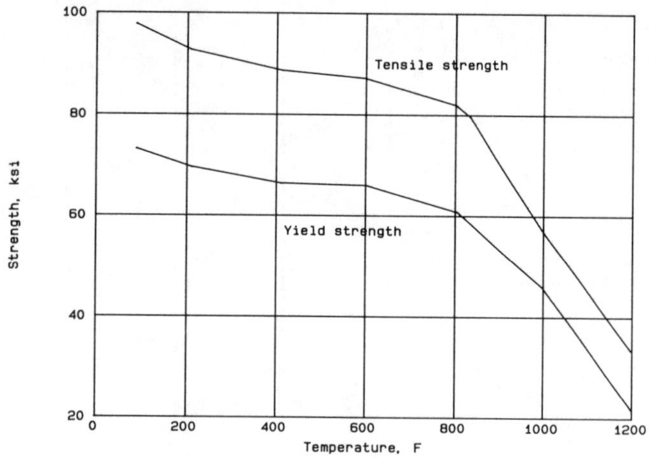

H11 Mod: Tensile and yield strength of annealed 0.064-in. sheet after 0.5 h at indicated temperature. Source: Ref 22, p 31

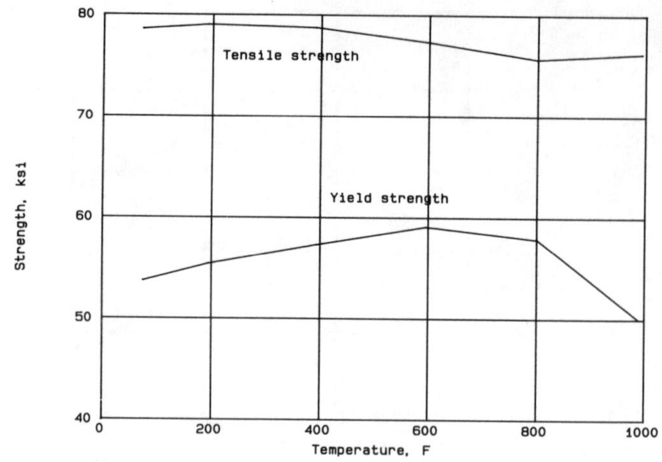

H11 Mod: Effect of 100-h exposure at elevated temperature on room-temperature tensile and yield strength of 0.064-in. annealed sheet. Source: Ref 22, p 15

H11 Mod: Short-time tensile property data and Charpy V-notch impact strength

TEMP C	TEMP F	UTS MPa	UTS ksi	YS MPa	YS ksi	EL %	RA %	IS J	IS ft-lb
Tempered at 540 C (1000 F)									
260	500	1860	270	1520	220	9.9	33.2	...	...
315	600	1840	267	1490	216	10.3	34.5	...	...
425	800	1670	242	1440	209	12.0	42.6	...	...
480	900	1580	229	1365	198	12.3	46.1	...	...
540	1000	1480	215	1250	182	13.7	48.2	...	...
650	1200	610	88	582.6	84.5	24.8	95.2	...	...
Tempered at 565 C (1050 F)									
21	70	1810	262	1480	215	9.8	35.4	...	...
150	300	1700	246	1365	198	10.1	36.1	29.4	21.7
260	500	1610	233	1340	195	10.2	35.8	41.2	30.4
315	600	1600	231.5	1330	193	10.3	36.0	42.7	31.5
425	800	1500	217	1270	184	11.4	38.8	40.0	29.5
480	900	1420	206	1140	166	12.2	39.3	39.7	29.3
540	1000	1240	180	970	141	12.2	41.3	41.4	30.5
595	1100	980	142	720	105	12.8	46.8	45.0	33.2
650	1200	590	85	440	64	19.0	66.8	80.0	59
Tempered at 595 C (1100 F)									
260	500	1340	195	1130	164	10	45	44	33
315	600	1310	190	1100	160	10	48.1	...	...
425	800	1230	178	1010	146	12.4	52.2	41	30
480	900	1130	164	900	131	13.5	56	...	..
540	1000	980	142	790	115	15.5	62	...	...

Note: Longitudinal specimens taken from bar stock, air cooled from 1010 C (1850 F), and double tempered, 2+2 h at indicated tempering temperature
Source: Ref 1, p 436

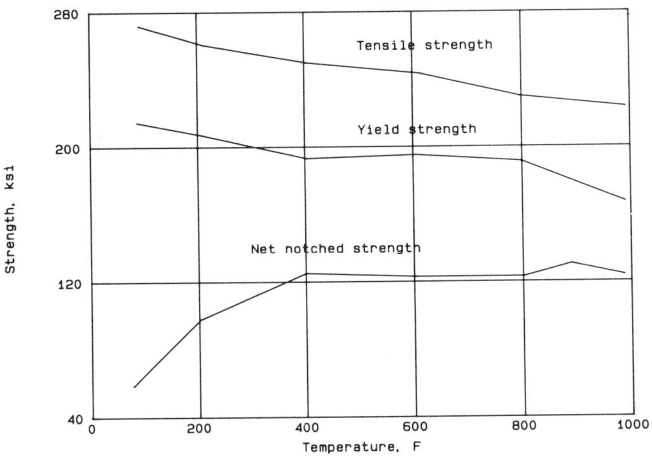

H11 Mod: Effect of test temperature on tensile properties of 0.080-in. sheet. Heat treatment: austenitized at 1850 F in salt, air cooled, and double tempered at 1000 F for 2 h. Source: Ref 22, p 37

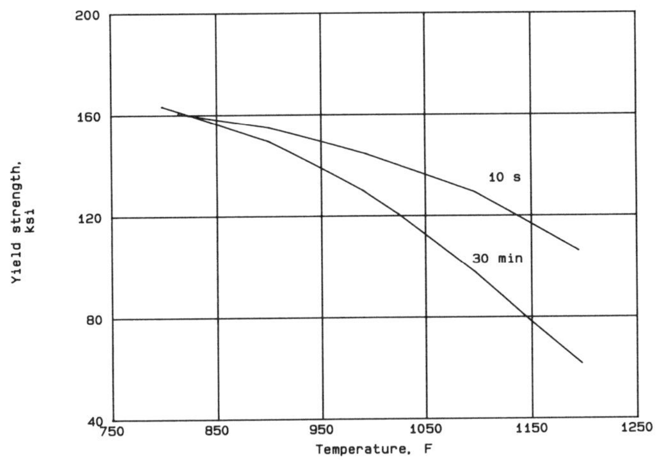

H11 Mod: Effect of test temperature and holding time on yield strength at a strain rate of 0.0003 in./in.*min for 0.063-in sheet, heat treated at 1850 F, air cooled, and double tempered at 1000 F for 3 h. Source: Ref 22, p 29

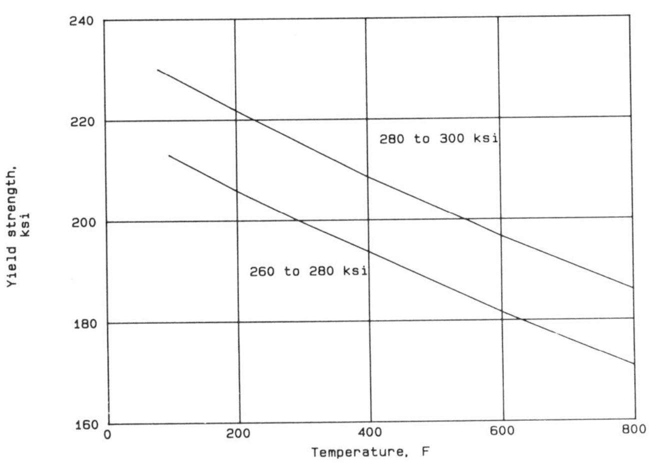

H11 Mod: Yield strength of 0.062- to 0.078-in. sheet, heat treated at 1150 F for 0.5 h, held at 1850 F for 0.5 h, air cooled, and double tempered for 3 h. Source: Ref 22, p 28

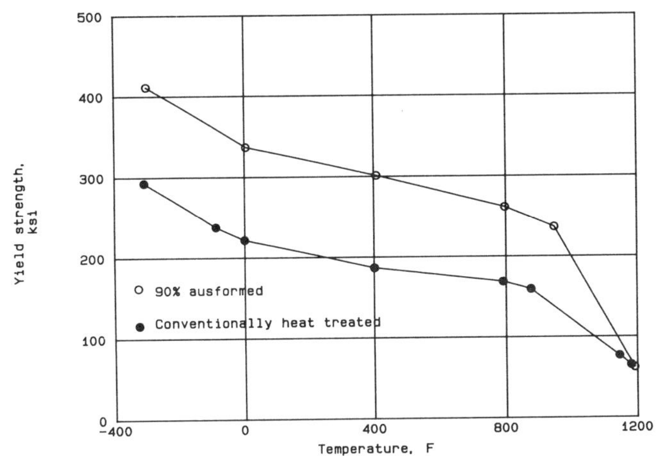

H11 Mod: Yield strength of ausformed and conventionally heat treated steel. Source: Ref 22, p 32

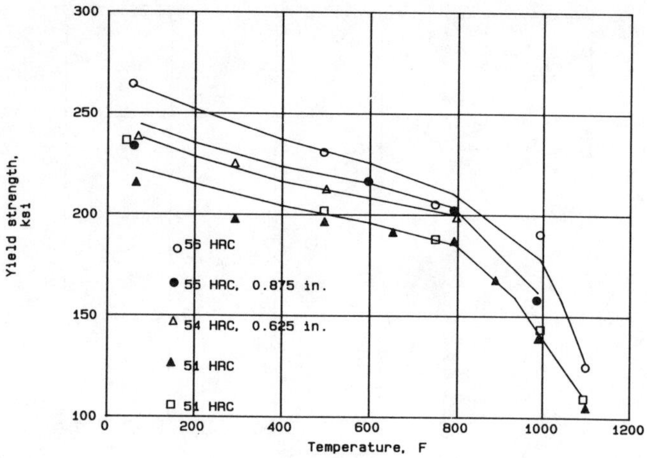

H11 Mod: Effect of test temperature on yield strength of bar heat treated to 260 to 290 ksi tensile strength. Source: Ref 22, p 29

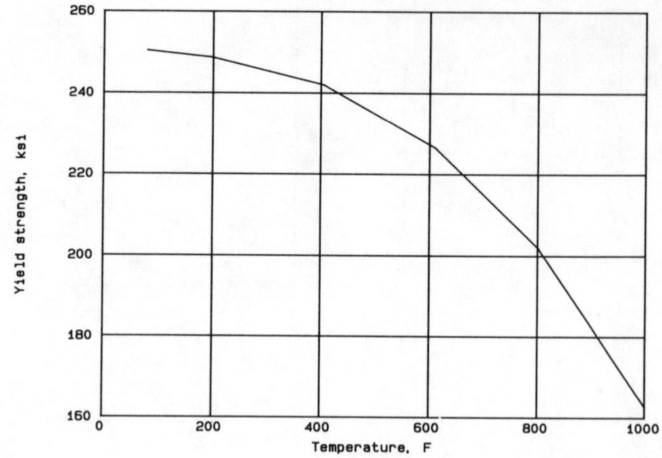

H11 Mod: Compressive yield strength of 0.033-in. sheet, heat treated at 1900 F for 30 s, air cooled, and double tempered at 1025 F for 2 h. Source: Ref 22, p 28

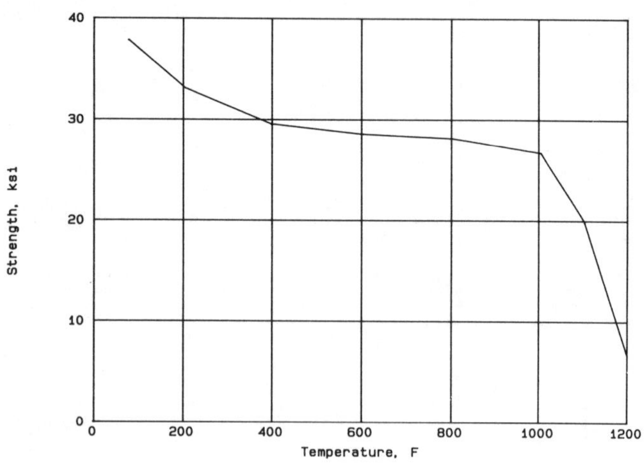

H11 Mod: Compressive yield strength of annealed 0.064-in. sheet. Source: Ref 22, p 28

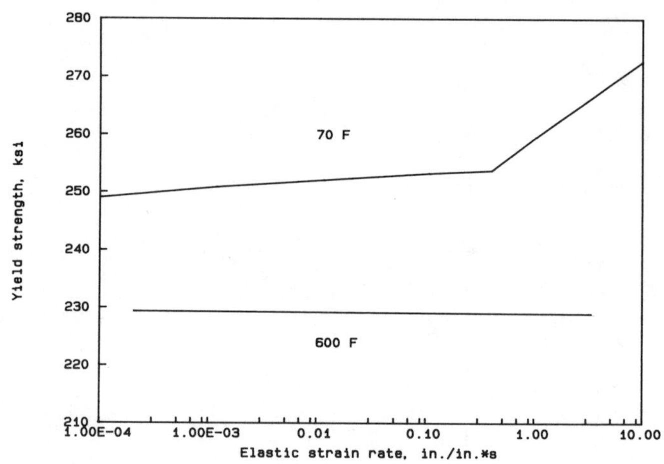

H11 Mod: Effect of elastic strain rate on yield strength of 1-in. diam bar at room temperature and 600 F. Heat treatment: 1850 F for 1 h, air cooled, double tempered at 1050 F for 1 h, and air cooled. Source: Ref 22, p 28

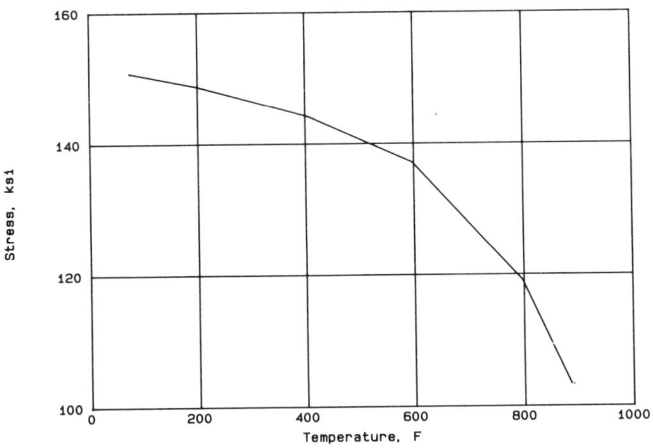

H11 Mod: Shear strength of heat treated bolts. Minimum tensile strength, 220 ksi. Source: Ref 22, p 33

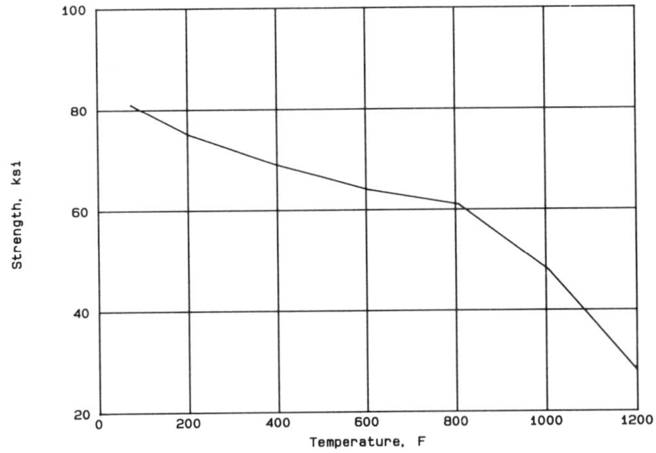

H11 Mod: Shear strength of 0.064-in. annealed sheet. Source: Ref 22, p 33

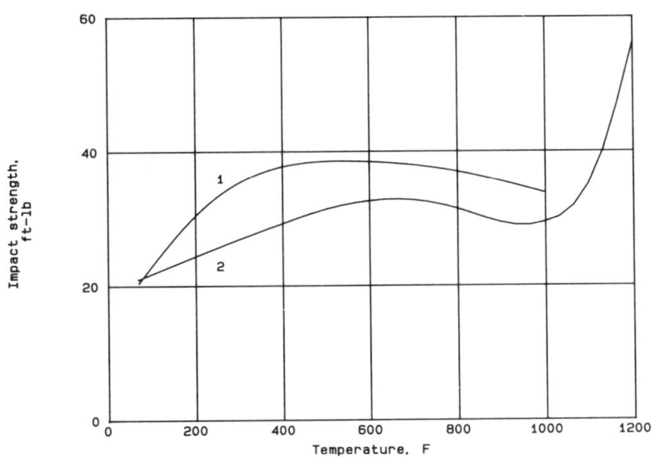

H11 Mod: Impact strength of specimens heat treated to 51 HRC. Curve 1: heat treated at 1850 F, air cooled, and double tempered at 1025 F for 2 h. Curve 2: heat treated at 1850 F, air cooled, and double tempered at 1060 F for 2.5 h. Source: Ref 22, p 33

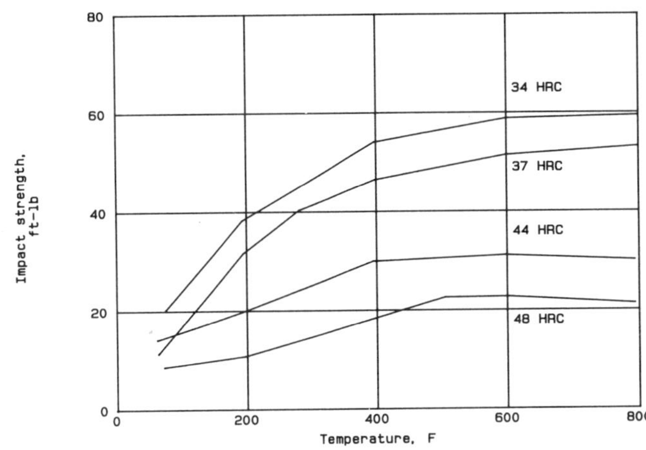

H11 Mod: Effect of test temperature on Charpy V-notch impact strength of bar at various hardness levels. Source: Ref 22, p 33

H11 Mod: Longitudinal Charpy V-notch impact properties for indicated tempering temperature

TEST TEMP C	TEST TEMP F	IS(a) J	IS(a) ft-lb	IS(b) J	IS(b) ft-lb	IS(c) J	IS(c) ft-lb	IS(d) J	IS(d) ft-lb
149	300	29	21.7	34	25	56	41	66	49
204	400	33	24	...	...	...	...	...	...
260	500	41	30.4	45	33	65	48	76	56
316	600	43	31.5	...	...	...	...	...	...
343	650	43	32	...	...	...	...	...	...
427	800	40	29.5	41	30	72	53	81	60

Note: Specimens were air treated at 1010 C (1850 F), 1 h at heat, double tempered 2+2 h, and tested at the indicated temperature
(a) 566 C (1050 F). (b) 593 C (1100 F). (c) 621 C (1150 F). (d) 649 C (1200 F)
Source: Carpenter Technology

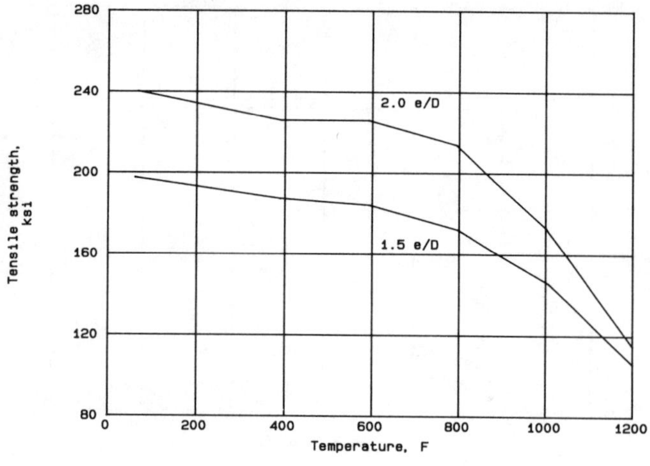

H11 Mod: Ultimate bearing tensile strength of 0.064-in. sheet. Source: Ref 22, p 34

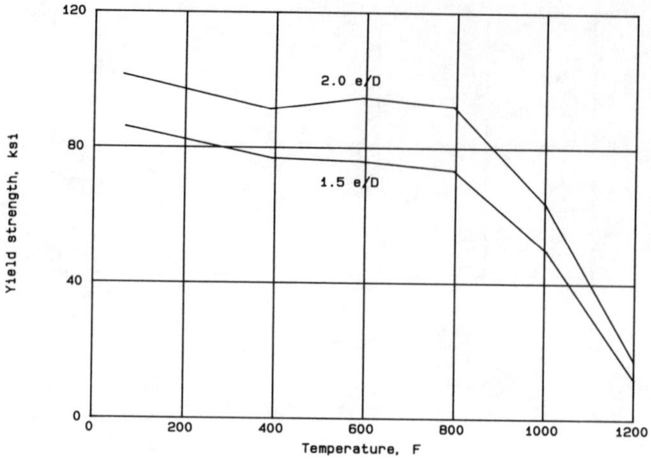

H11 Mod: Bearing yield strength of 0.064-in. annealed sheet. Source: Ref 22, p 34

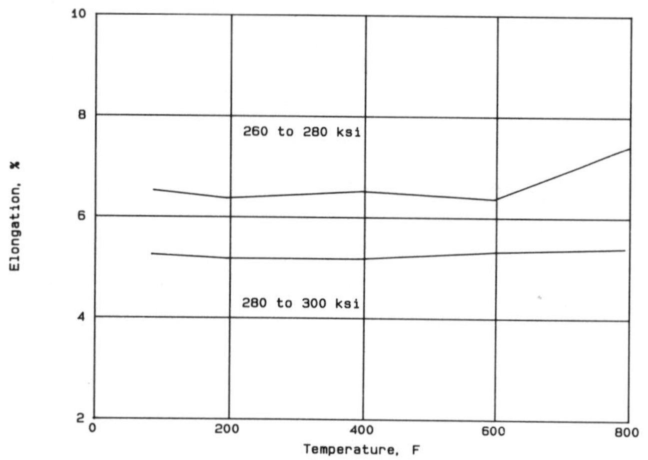

H11 Mod: Tensile elongation of 0.062- to 0.078-in sheet, heat treated at 1150 F for 0.5 h, held at 1850 F for 0.5 h, air cooled, and double tempered for 3 h. Source: Ref 22, p 28

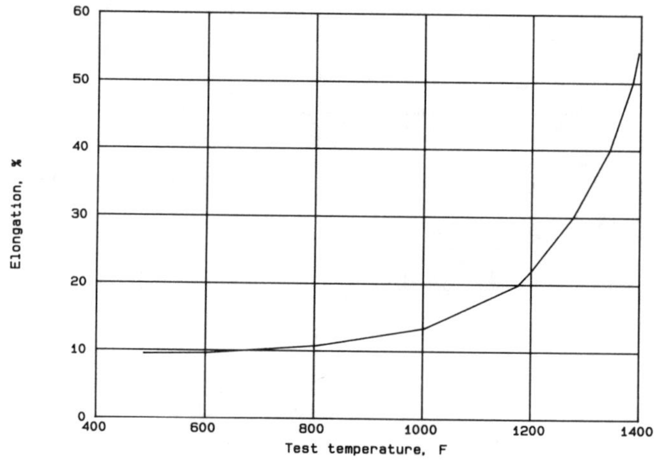

H11 Mod: Tensile elongation of longitudinal specimens, heat treated at 1850 F for 1 h and double tempered at 1000 F for 2 h. Source: Carpenter Technology

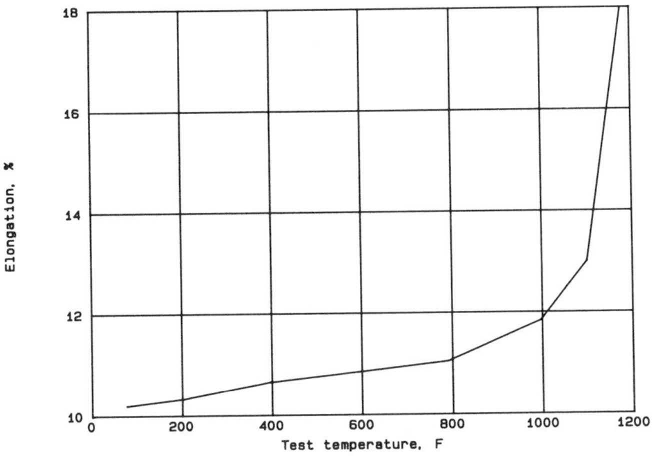

H11 Mod: Tensile elongation of longitudinal specimens, heat treated at 1850 F for 1 h and double tempered at 1050 F for 2 h. Source: Carpenter Technology

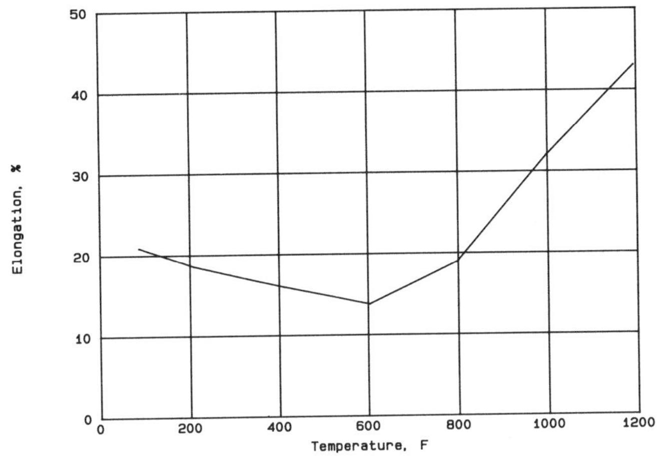

H11 Mod: Tensile elongation of 0.064-in. annealed sheet. Source: Ref 22, p 31

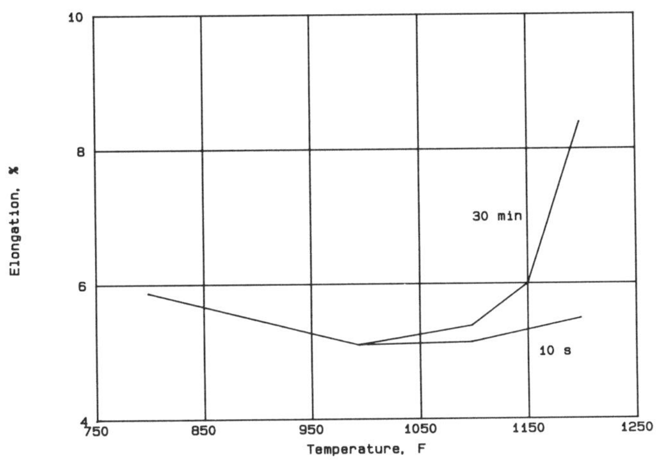

H11 Mod: Effect of test temperature and holding time on tensile elongation at a strain rate of 0.0003 in./in.*min. Heat treatment of 0.063-in. sheet: 1850 F, air cooled, double tempered at 1000 F for 3 h. Source: Ref 22, p 29

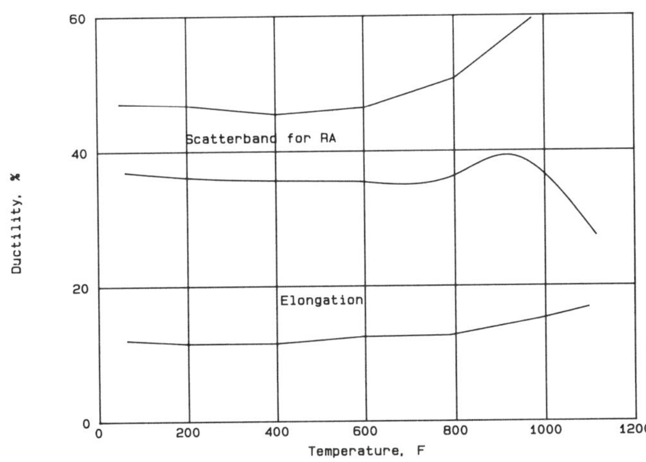

H11 Mod: Effect of test temperature on tensile elongation and reduction in area of bar heat treated to 260 to 290 ksi tensile strength and 51 to 56 HRC. Source: Ref 22, p 29

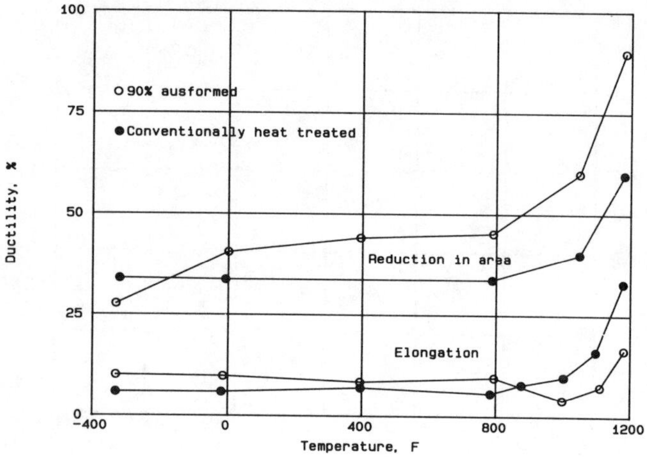

H11 Mod: Tensile elongation and reduction in area. Effect of temperature on 90% ausformed steel (tempered at 1050 F) and conventionally heat treated steel (tempered at 1060 F). Source: Ref 22, p 33

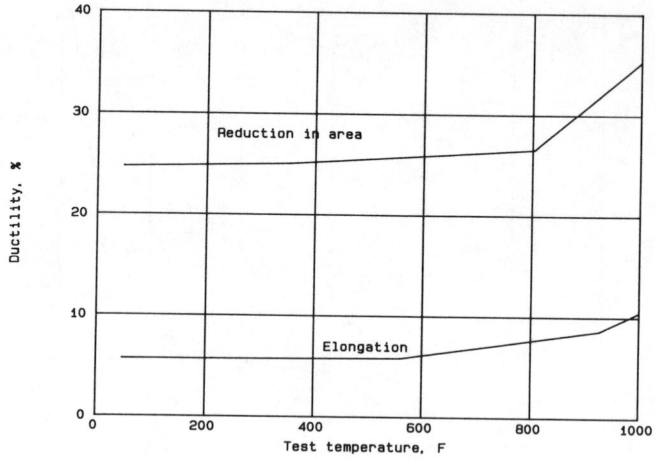

H11 Mod: Tensile elongation and reduction in area of cast, vacuum degassed specimens. Treatment: homogenized at 2000 F for 3 h, air cooled, austenitized at 1850 F for 2 h, air cooled, double tempered at 1050 F for 2 h, and air cooled. Source: Ref 22, p 30

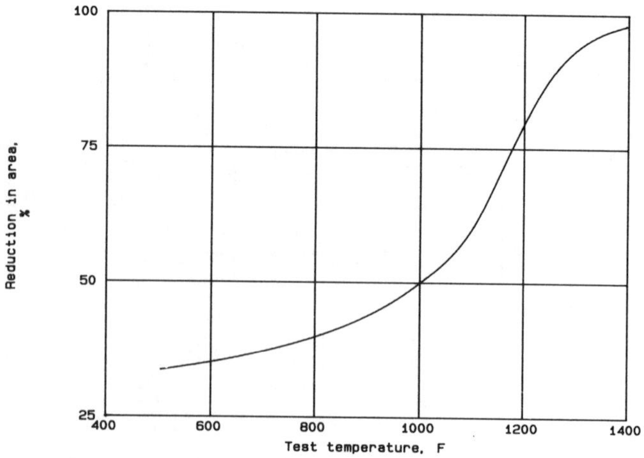

H11 Mod: Reduction in area of longitudinal specimens heat treated in air at 1850 F for 1 h and double tempered at 1000 F for 2 h. Source: Carpenter Technology

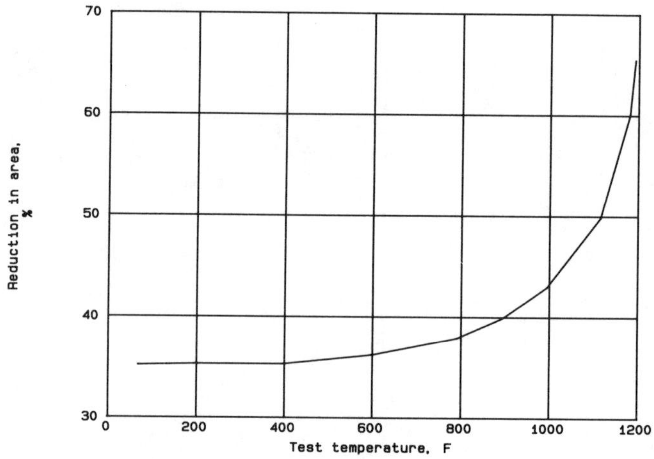

H11 Mod: Reduction in area of longitudinal specimens heat treated in air at 1850 F for 1 h and double tempered at 1050 F for 2 h. Source: Carpenter Technology

H11 Mod: Creep data

Ttpr C	Ttpr F	Stress MPa	Stress ksi	Time for 0.2% creep h	Time for 0.5% creep h	Time for 1% creep h
Air treated at 1010 C (1850 F), tempered 1 h at 566 C (1050 F)						
427	800	1379	200	...	1.2	308
427	800	1344	195	0.1	5	757
427	800	896	130	302	...	...
482	900	1368	200	...	...	...
482	900	1172	170	0.5	9	97
482	900	862	125	24	263	...
482	900	689	100	172	...	...
538	1000	1069	155	...	0.5	4.5
538	1000	758	110	...	7	27
538	1000	517	75	7	43	116
538	1000	483	70	9.8	52	134
538	1000	345	50	24	113	...
Air treated at 1010 C (1850 F), tempered at 649 C (1200 F), 1 h at heat						
538	1000	655	95	...	0.2	0.7
538	1000	483	70	0.2	0.8	2.6
538	1000	276	40	11	80	158
593	1100	414	60	0.1	0.3	0.6
593	1100	276	40	2.5	10	18
593	1100	138	20	6.7	22	46
649	1200	103	15	0.7	2.2	49
649	1200	69	10	1.6	5.8	16

Source: Ref 29, Jan 1968

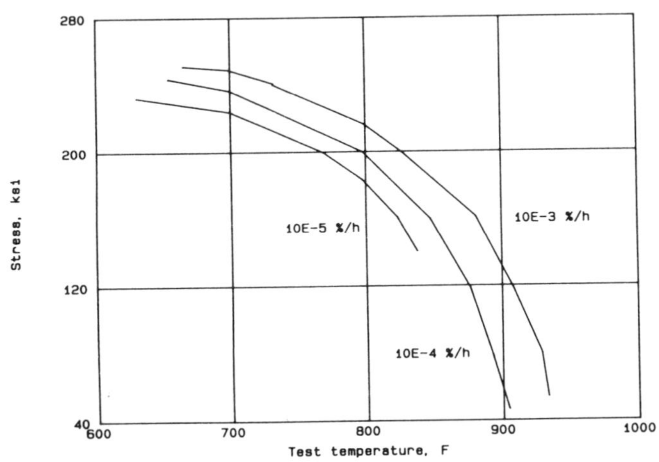

H11 Mod: Stress to produce creep rates of 10^{-3} to 10^{-5} %/h at 700 to 900 F in forged 2-in. bar. Heat treatment: 1850 F for 0.5 h, air cooled, tempered 3 times at 1000 F for 2 h, at 1020 F for 2 h, and air cooled. Source: Ref 22, p 40

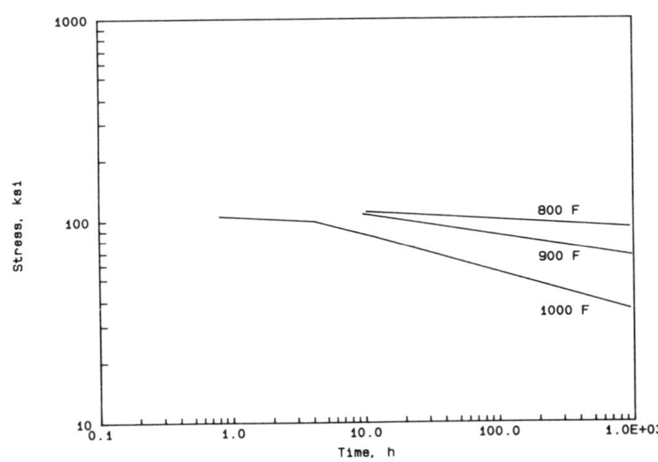

H11 Mod: Stress-rupture life for specimens heat treated to a tensile strength of 175 ksi. Source: Ref 22, p 40

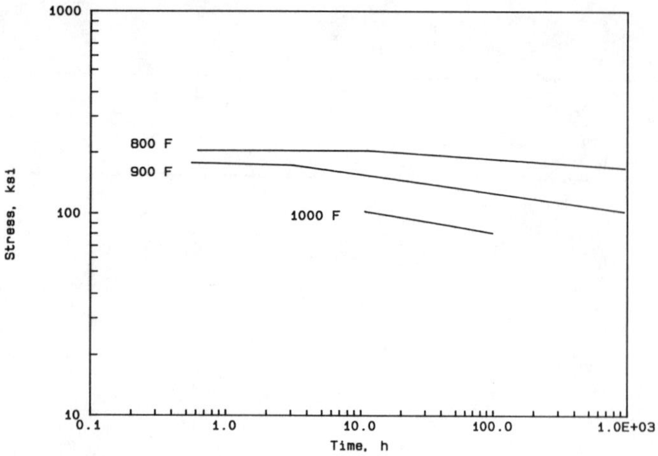

H11 Mod: Stress-rupture life for specimens heat treated to a tensile strength of 220 ksi. Source: Ref 22, p 40

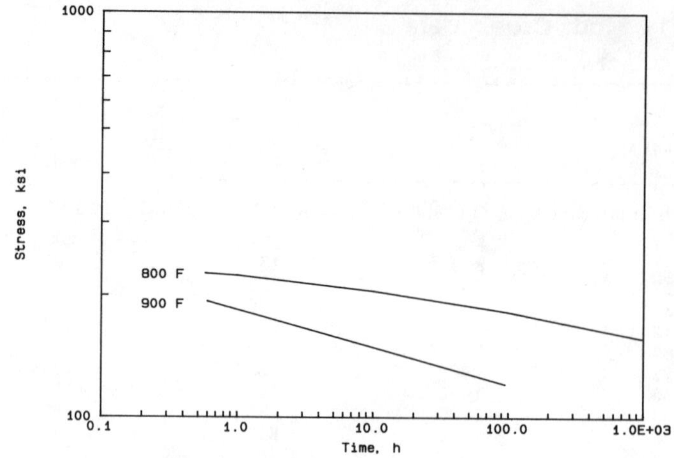

H11 Mod: Stress-rupture life for specimens heat treated to a tensile strength of 260 ksi. Source: Ref 22, p 40

H11 Mod: Stress-rupture property data

TEMP C	TEMP F	RPSTR MPa	RPSTR ksi	RPT h	EL(a) %	RA(a) %
Tempered at 565 C (1050 F)						
480	900	1380	200	10.4	9.9	33.7
480	900	1170	170	257.3	5.7	33.7
540	1000	1070	155	11.1	9.9	20.9
540	1000	760	110	58	12.4	24.9
540	1000	520	75	254	13.3	30.3
540	1000	480	70	318	17.3	35.4
Tempered at 650 C (1200 F)						
540	1000	655	95	6.9	24.5	32.3
540	1000	480	70	31.4	25.4	46.7
540	1000	275	40	435.4	31.3	68.7
595	1100	410	60	3.3	25.7	60.4
595	1100	205	30	73.1	49.3	64.6
595	1100	140	20	444.3	55.2	71.3
650	1200	205	30	2.3	28.1	75
650	1200	100	15	70.5	69.9	84.8

Note: Specimens were air cooled from 1010 C (1850 F) and tempered 1 h at temperature
(a) At rupture
Source: Ref 1, p 437

H11 Mod: Effect of exposure temperature on time-to-rupture for specimens heat treated in air at 1850 F for 1 h and tempered at 1050 F for 1 h. Source: Carpenter Technology

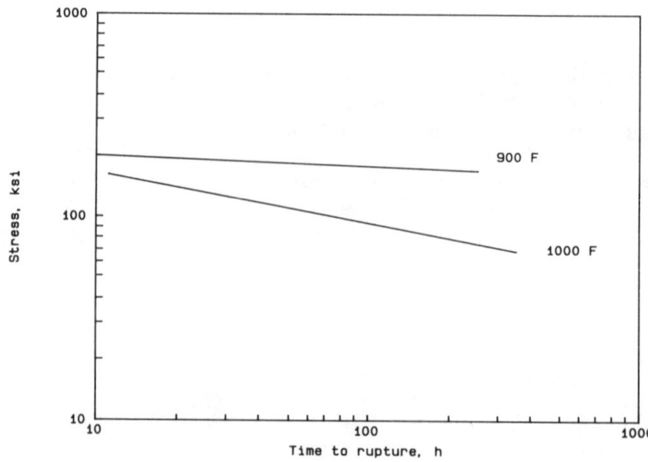

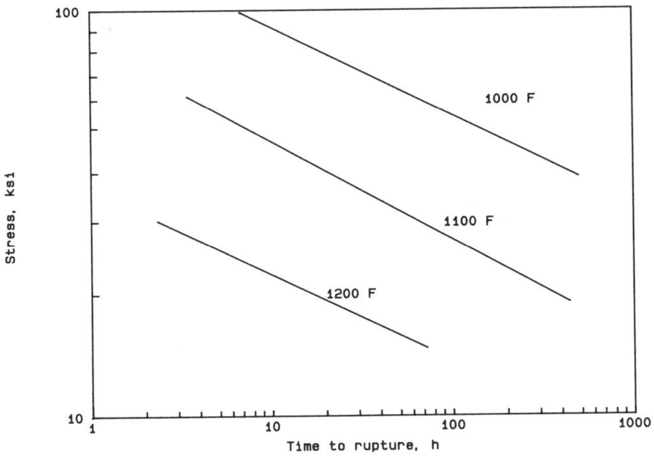

H11 Mod: Effect of exposure temperature on time-to-rupture for specimens heat treated in air at 1850 F for 1 h and tempered at 1200 F for 1 h. Source: Carpenter Technology

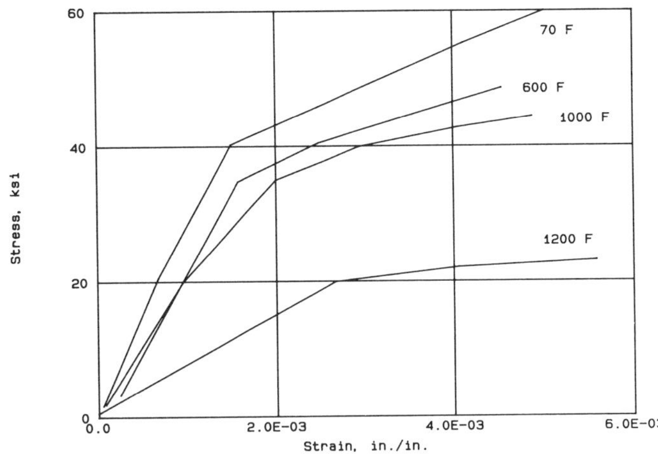

H11 Mod: Stress-strain curves in compression at room and elevated temperatures for 0.064-in. annealed sheet. Source: Ref 22, p 32

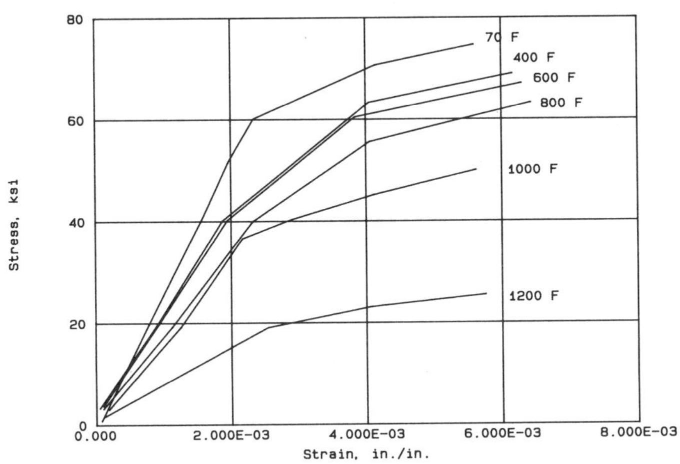

H11 Mod: Stress-strain curves for 0.064-in. annealed sheet. Source: Ref 22, p 26

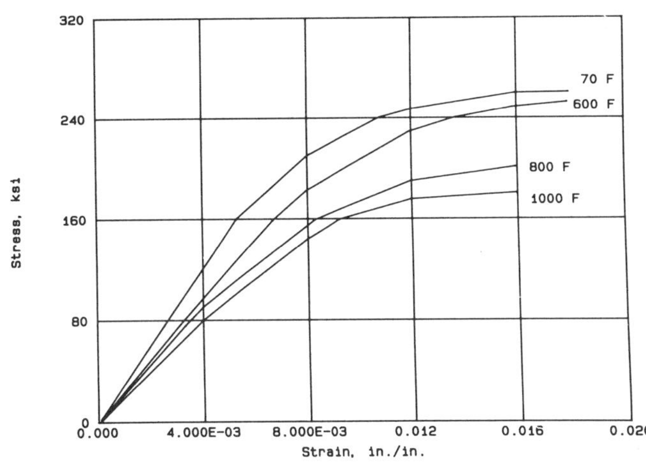

H11 Mod: Stress-strain curves for 0.090-in. sheet heat treated to tensile strength of 280 ksi. Heat treatment: 1850 F, air cooled, and double tempered at 1000 F for 3 h. Specimens tested at 1000 F were double tempered at 1025 F. Source: Ref 22, p 26

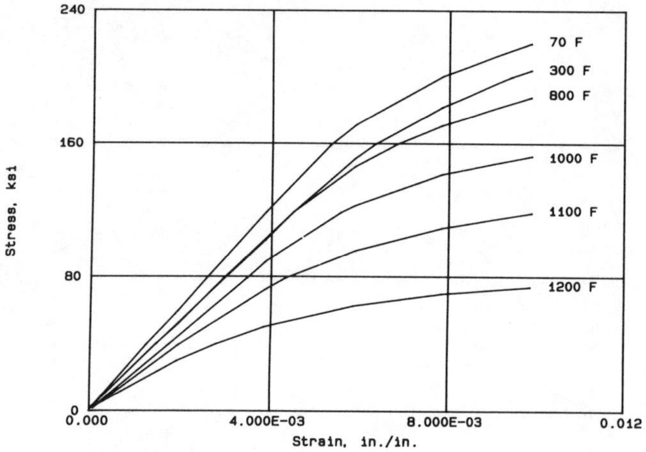

H11 Mod: Stress-strain curves at room and elevated temperatures for material heat treated to a tensile strength of 260 ksi and a hardness of 50 HRC. Source: Ref 22, p 26

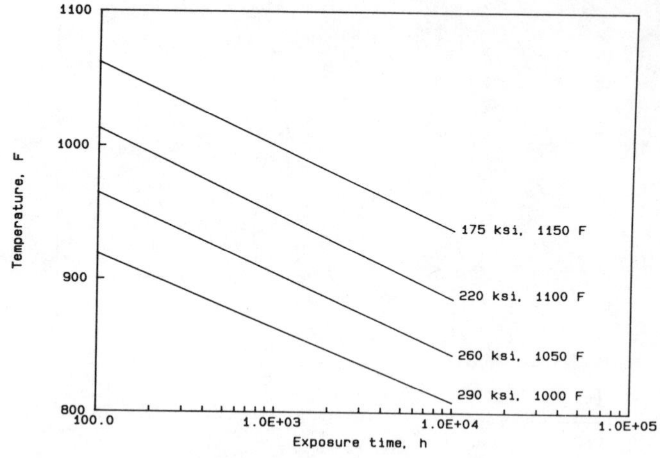

H11 Mod: Maximum temperatures for stability of mechanical properties for various tempering temperatures and exposure times at indicated tensile strength. Temper: 2 x 3 h at indicated time and temperature. Source: Ref 22, p 15

H12

UNS number: T20812
Specifications: ASTM A681, FED QQ T-570, SAE J437, SAE J438, SAE J467, DIN 1.2606, NF Z 35CWDV5, BS BH12, JIS SKD62
Nominal composition: 0.300 to 0.400% C, 0.200 to 0.500% Mn, 0.800 to 1.200% Si, 4.750 to 5.500% Cr, 0.300% Ni max, 1.250 to 1.750% Mo, 1.000 to 1.700% W, 0.500% V max
Product forms: Wrought, bar, rod, wire, plate, sheet, strip, billets, forgings, ring, cast, sand, investment, powder
Description: H12 is a chromium hot work tool steel. It exhibits high resistance to softening at elevated temperatures.
Typical applications: Extrusion dies, dummy blocks, holders, gripper and header dies, forging die inserts, punches, mandrels, and sleeves for cold heading dies.

Physical Properties

H12: Thermal conductivity

TEMP C	TEMP F	THCON W/m*K	THCON Btu/ft*h*F
27	80	17.4	10.1
204	400	23.0	13.3
427	800	25.1	14.5
649	1200	26.5	15.3

Source: Ref 29, Apr 1962

Mechanical Properties

H12: Tensile property data

TEMP C	TEMP F	UTS MPa	UTS ksi	YS MPa	YS ksi	EL %	RA %	HARD HB
510	950	972	141	896	130	19.0	61.0	275
538	1000	896	130	827	120	21.5	65.5	260
566	1050	810	117.5	745	108	24.5	70.0	230
593	1100	703	102	638	92.5	27.5	75.5	200
621	1150	552	80	469	68	29.0	80.0	165

Note: Specimens were austenitized 30 min at 1010 C (1850 F), air cooled, tempered 3 h at 621 C (1150 F), air cooled, and tested at lower temperatures
Source: Ref 29, Apr 1962

H12: Tensile property data

TEMP C	TEMP F	UTS MPa	UTS ksi	YS MPa	YS ksi	EL %	RA %
21	70	1613	234	1455	211	13.4	50.4
427	800	1272	184.5	1120	162.5	13.3	52.5
482	900	1179	171	1027	149	16.2	64.5
538	1000	1014	147	893	129.5	17.5	68.6
593	1100	896	130	807	117	17.3	70.4
649	1200	517	75	455	66	22.3	80.7

Note: Specimens were air cooled from 1010 C (1850 F), tempered at 582 C (1080 F) for 3+3 h, 48 HRC
Source: Ref 29, p 265

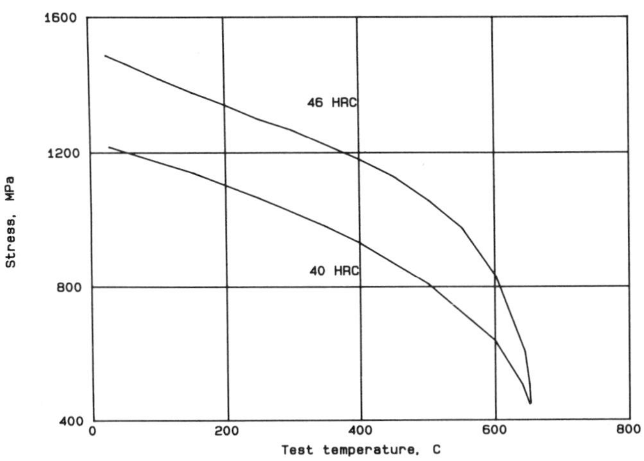

H12: Effect of elevated-temperature exposure on tensile strength. Hardness was measured using Rockwell C hardness test. Source: Ref 15, p 460

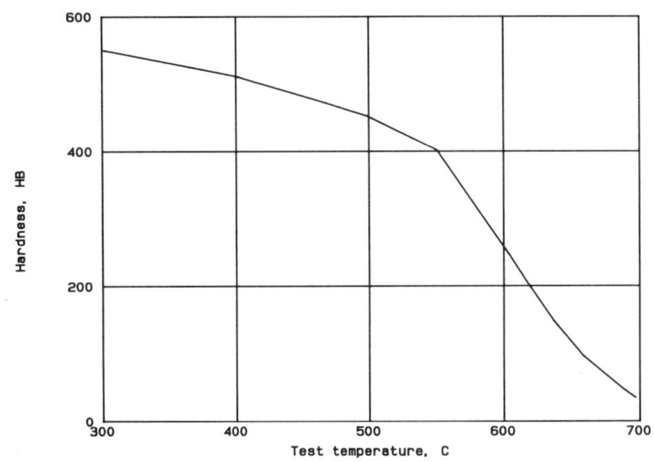

H12: Effect of temperature on Brinell hardness. Specimens were air quenched from 1010 C. Rockwell C hardness in the as-quenched condition was 52 to 54 HRC. Source: Ref 15, p 462

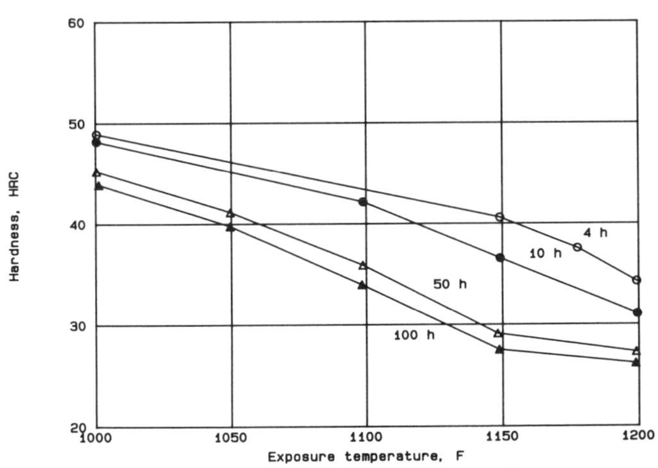

H12: Resistance to softening during elevated-temperature exposure based on room-temperature hardness. Initial hardness: 49 HRC. Prior treatment: 1850 F for 0.5 h, air cooled, tempered 3 times at 1060 F, air cooled, heated to 1073 F for 1 h, and air cooled. Source: Ref 14, p 266

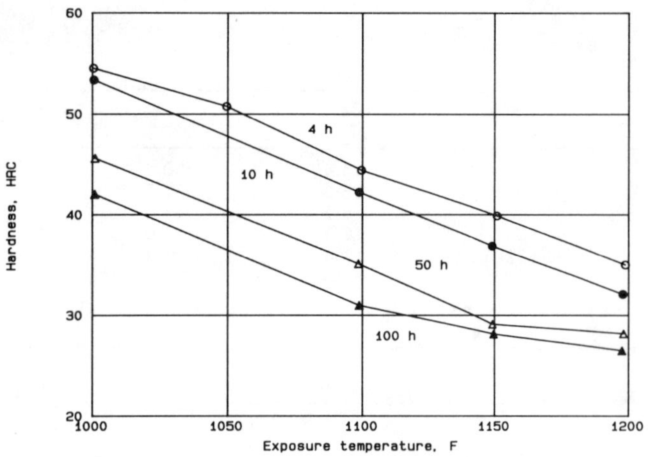

H12: Resistance to softening during elevated-temperature exposure based on room-temperature hardness. Initial hardness: 55.5 HRC. Prior heat treatment: 1850 F for 0.5 h, air cooled, tempered 3 times at 1000 F for 3 h, and air cooled. Source: Ref 14, p 266

H12: Hot hardness data

TEMP C	TEMP F	HARD(a) HRC	HARD(b) HRC
Room-temperature hardness after heat treatment			
21	70	56.0	46.5
After 0.5 h at temperature			
316	600	51.0	41.5
427	800	49.0	40.0
538	1000	41.0	32.5
593	1100	33.5	20.5
649	1200	23.5	15.0

(a) Test 1. (b) Test 2
Source: Ref 14, p 265

H13

UNS number: T20813
Specifications: ASTM A681, FED QQ T-570, SAE J437, SAE J438, SAE J467, DIN 1.2344, NF Z 40CVD5, BS BH13, UNI X 40CrMoV511KU, JIS S KD61, SS 2242, GOST 4Ch5MF1S, UNE F.5318

Nominal composition: 0.320 to 0.450% C, 0.200 to 0.500% Mn, 0.800 to 1.200% Si, 4.750 to 5.500% Cr, 0.300% Ni max, 1.100 to 1.750% Mo, 0.800 to 1.200% V

Product forms: Wrought, bar, rod, wire, plate, sheet, strip, billets, forgings, ring, cast, sand, investment, powder
Description: H13 is a chromium hot work tool steel. It exhibits a high resistance to softening at elevated temperatures.
Typical applications: Die casting dies and inserts, dummy blocks, ejector pins, plungers, sleeves, slides, extrusion dies, and forging dies and inserts.

Physical Properties

H13: Thermal conductivity

TEMP C	TEMP F	THCON W/m*K	THCON Btu/ft*h*F
215	420	28.6	16.5
350	660	28.4	16.4
475	890	28.4	16.4
605	1120	28.7	16.6

Source: Ref 3, p 442

H13: Mean coefficients of linear thermal expansion

TEMP C	TEMP F	THEXP 10^{-6}/C	THEXP 10^{-6}/F
27-93	80-200	10.4	5.8
27-427	80-800	12.4	6.9
27-816	80-1500	13.5	7.5

Source: Ref 29, Apr 1965

H13: Modulus of elasticity

TEMP C	TEMP F	E GPa	E 10^3 ksi
22	72	207	30
316	600	197	28.5
538	1000	159	23

Source: Ref 29, Apr 1965

Mechanical Properties

H13: Tensile property data

TEMP C	TEMP F	UTS MPa	UTS ksi	YS MPa	YS ksi	EL %	RA %	HARD HRC
21	70	1407	204	1196	173.5	15.7	51.2	45
427	800	1158	168	972	141	16.2	56.4	38
482	900	1079	156.5	900	130.5	18.2	59.3	35
538	1000	965	140	858	124.5	21.9	70.4	29
593	1100	858	124.5	752	109	22.5	72.3	20
649	1200	638	92.5	565	82	29.2	84.1	...

Note: Specimens were air cooled from 1010 C (1850 F), tempered at 606 C (1125 F), 3+3 h to 44 HRC
Source: Ref 14, p 267

H13: Short-time tensile property data

TEMP C	TEMP F	UTS MPa	UTS ksi	YS MPa	YS ksi	EL %	RA %	HARD HRC
27	80	1482	215	1269	184	13	50	45
427	800	1165	169	931	135	17	56	38
482	900	1069	155	862	125	20	65	35
538	1000	986	143	710	103	19	65	29
593	1100	745	108	593	86	21	71	20

Source: Ref 29, Apr 1965

H13: Longitudinal short-time tensile property data for steel bar

TEMP C	TEMP F	UTS MPa	UTS ksi	YS MPa	YS ksi	EL %	RA %
Tempered 2+2 h at 527 C (980 F) to 52 HRC							
425	800	1620	235	1240	180	13.7	50.6
540	1000	1305	189	1000	145	13.9	54
595	1100	1020	148	825	120	17.5	65.4
650	1200	450	65	340	49	28.9	88.9
Tempered 2+2 h at 575 C (1065 F) to 48 HRC							
425	800	1400	203	1150	167	15	59.9
540	1000	1160	168	960	139	17.1	62.4
595	1100	940	136	750	109	18	68.5
650	1200	455	66	350	51	33.6	89
Tempered 2+2 h at 605 C (1120 F) to 44 HRC							
425	800	1200	174	1005	146	17	64.1
540	1000	995	144	820	119	20.6	70
595	1100	827	120	690	100	22.6	74
650	1200	450	65	350	51	28.4	87.6

Note: Bars were oil quenched from 1010 C (1850 F) and double tempered to indicated room temperature hardness
Source: Ref 1, p 438

H13: Longitudinal Charpy V-notch impact property data for tempered bar

Ttpr C	Ttpr F	HARD(a) HRC	IS(a) J	IS(a) ft-lb	IS(b) J	IS(b) ft-lb	IS(c) J	IS(c) ft-lb	IS(d) J	IS(d) ft-lb
524	975	54	14	10	27	20	31	23	...	...
565	1050	52	14	10	30	22	34	25	34 (e)	25 (e)
607	1125	47	24	18	41	30	45	33	43	32
615	1140	43	24	18	52	38	60	44	57	42

Note: Specimens were air cooled from 1010 C (1850 F) and double tempered, 2+2 h at indicated temperature
(a) At 21 C (70 F). (b) At 260 C (500 F). (c) At 540 C (1000 F). (d) 595 C (1100 F). (e) At 565 C (1050 F)
Source: Ref 1, p 438

H13: Stress-rupture property data at 480 C (900 F)

TIME h	RPSTR MPa	RPSTR ksi
25	1207	175
50	1172	170
100	1138	165
200	1076	156
400	1034	150
1000	965	140

Source: Ref 29, Apr 1965

H13: Resistance to softening

TEMP C	TEMP F	HARD(a) HRC	HARD(b) HRC
...	...	50.2	41.7
480	900	48.7	38.6
540	1000	45.3	39.3
600	1100	29.0	27.7
650	1200	22.7	23.7
700	1300	20.1	20.2
760	1400	13.9	13.2

(a) Original hardness. (b) After 100 h at indicated temperature
Source: Ref 3, p 426

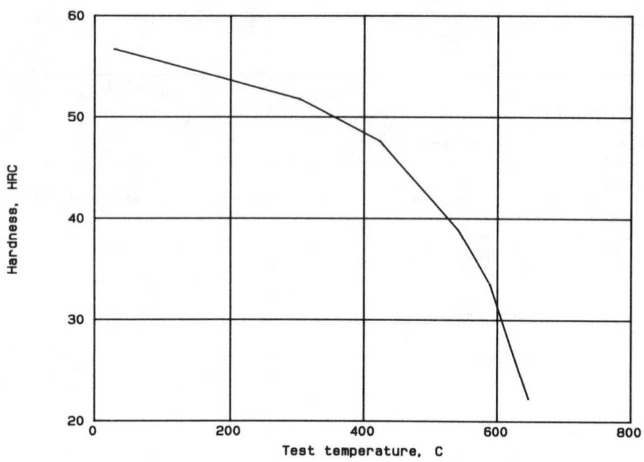

H13: Rockwell C hardness. Source: Ref 7, p 796

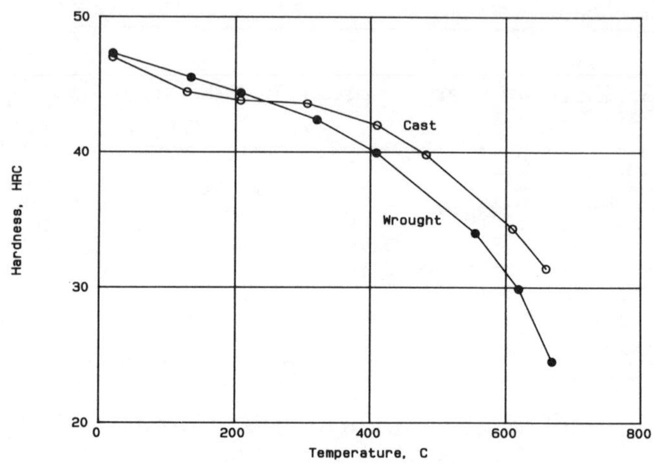

H13: Comparison of hardness for cast and wrought specimens. Source: Ref 3, p 446

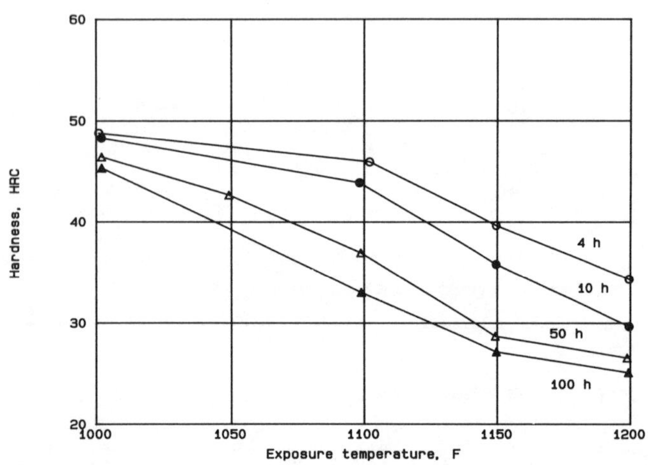

H13: Resistance to softening during elevated-temperature exposure based on room-temperature hardness. Initial hardness: 49 HRC. Prior heat treatment: 1850 F for 0.5 h, air cooled, tempered 3 times at 1100 F for 3 h, and air cooled. Source: Ref 14, p 268

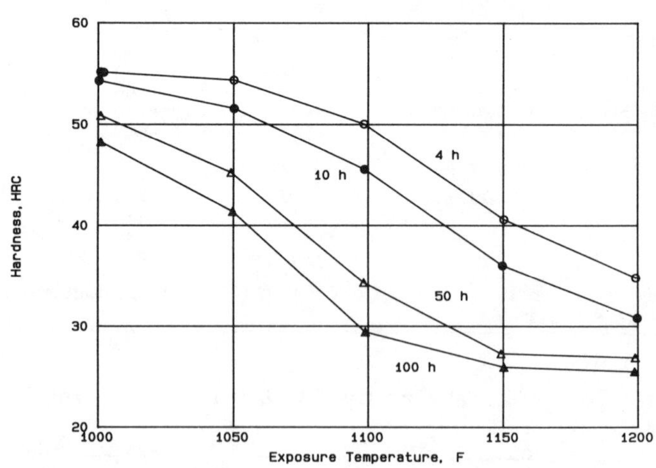

H13: Resistance to softening during elevated-temperature exposure based on room-temperature hardness. Initial hardness: 55 HRC. Prior heat treatment: 1850 F for 0.5 h, air cooled, tempered 3 times at 1025 F for 3 h, and air cooled. Source: Ref 14, p 268

H13: Hot hardness data

TEMP C	TEMP F	HARD(a) HRC	HARD(b) HRC
Room-temperature hardness after heat treatment			
21	70	55.5	46.5
After 0.5 h at temperature			
316	600	51.0	41.0
482	900	48.0	39.0
538	1000	39.0	33.0
593	1100	33.5	21.5
649	1200	23.5	17.0

(a) Test 1. (b) Test 2
Source: Ref 29, Apr 1965

H19

UNS number: T20819
Specifications: ASTM A681, FED QQ T-570, DGN B82, BS 4659, DIN 1.2678
Nominal composition: 0.320 to 0.450% C, 0.200 to 0.500% Mn, 4.000 to 4.750% Cr, 0.300% Ni max, 0.200 to 0.500% Si, 0.300 to 0.550% Mo, 3.750 to 4.500% W, 1.750 to 2.200% V, 4.000 to 4.500% Co
Product forms: Wrought, bar, rod, wire, plate, sheet, strip, billets, forgings, ring, cast, sand investment, powder

Description: H19 is a chromium hot work tool steel. It exhibits high resistance to softening at elevated temperatures. Mechanical properties at elevated temperatures are superior to those of 9% W hot work grades. H19 is also more resistant to heat checking than any of the standard tungsten hot work die steels.
Typical applications: Extrusion dies and die inserts, dummy blocks, punches, forging dies and die inserts, mandrels, and hot punch tools.

Mechanical Properties

H19: Short-time tensile property data

TEMP C	TEMP F	UTS MPa	UTS ksi	YS MPa	YS ksi	EL %	RA %	HARD HRC
Preheated at 815 C (1500 F), austenitized at 1205 C (2200 F), air cooled, tempered at 660 C (1225 F) for 2 h to 45 HRC								
427	800	1227	178	...	...	9	36	44.5
538	1000	1014	147	...	...	12	42	44.5
649	1200	758	110	...	...	16	37	44
Oil quenched from 1175 C (2150 F), tempered at 635 C (1175 F), 2+2 h to 48 HRC								
21	70	1610	233.5	1500	217.5	5.6	17.1	...
427	800	1282	186	1162	168.5	7.8	23.6	...
538	1000	1141	165.5	1043	151.2	8.3	27.3	...
593	1100	976	141.5	893	129.5	8.3	30	...
649	1200	834	121	741	107.5	10.1	36	...

Source: Ref 29, Jul 1982 and Ref 14, p 269

H19: Charpy V-notch impact properties

TEMP C	TEMP F	IS J	IS ft-lb	HARD HRC
21	70	7	5	54
204	400	11	8	53
316	600	8	6	53
427	800	16	12	54
538	1000	12	9	53
649	1200	12	9	50

Note: Specimens were austenitized at 1215 C (2200 F), air cooled and tempered at 620 C (1120 F)
Source: Ref 29, Jul 1982

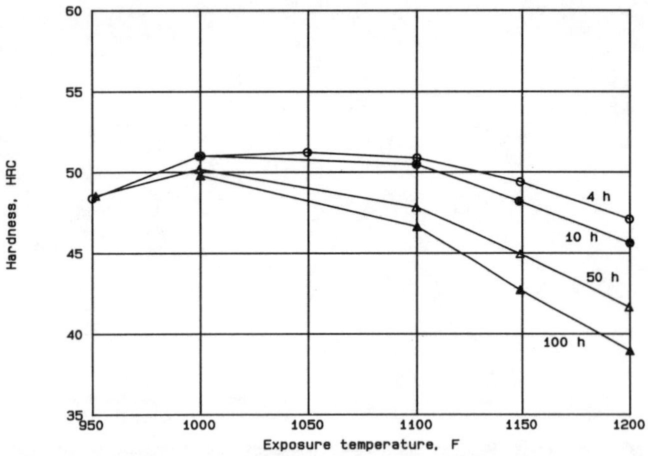

H19: Resistance to softening during elevated-temperature exposure based on room-temperature hardness. Initial hardness: 49.5 HRC. Prior heat treatment: 2175 F for 5 min, oil quenched, double tempered at 1200 F for 2 h, and air cooled. Source: Ref 14, p 270

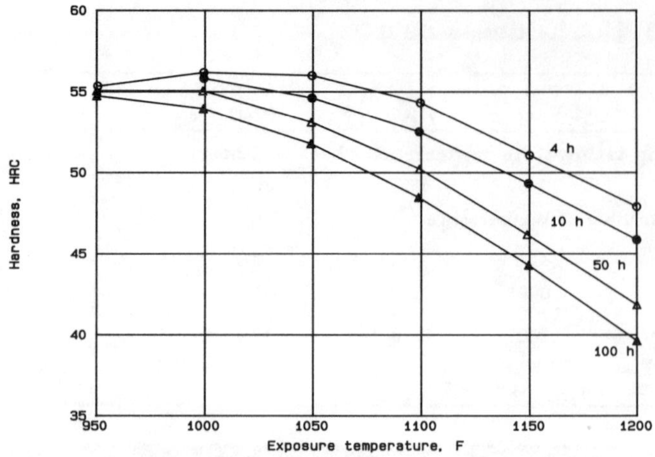

H19: Resistance to softening during elevated-temperature exposure based on room-temperature hardness. Initial hardness: 55 HRC. Prior heat treatment: 2175 F for 5 min, oil quenched, double tempered at 1050 F for 2 h, and air cooled. Source: Ref 14, p 270

H19: Hot hardness data

TEMP C	TEMP F	HARD(a) HRC	HARD(b) HRC
Room-temperature hardness after heat treatment			
21	70	56.5	47.5
After 0.5 h at temperature			
316	600	50.5	43.5
427	800	49	40
538	1000	44.5	35
593	1100	39	22.5
649	1200	31	19

(a) Test 1. (b) Test 2
Source: Ref 14, p 269

H21

UNS number: T20821
Specifications: ASTM A681, FED QQ T-570, SAE J437, SAE J438, DIN 1.2581, NF Z 30WCV9, BS 4659 BH21, UNI X-30WCrV93KU, JIS SKD5, GOST 3Ch2W8F, UNE F.5323
Nominal composition: 0.260 to 0.360% C, 0.150 to 0.400% Mn, 0.150 to 0.500% Si, 3.000 to 3.750% Cr, 0.300% Ni max, 8.500 to 10.000% W, 0.300 to 0.600% V
Product forms: Wrought, bar, rod, wire, plate, sheet, strip, billets, forgings, ring, cast, sand, investment, powder

Description: H21 is a tungsten hot work tool steel. It exhibits high resistance to softening at elevated temperatures. H21 shows little reduction in hardness in hot applications up to 595 C (1100 F) and will withstand operating temperatures as high as 540 C (1000 F) for extended periods.
Typical applications: Mandrels, hot blanking dies, hot punches, blades for flying shear, hot trimming dies, extrusion and die casting dies for brass, dummy blocks, piercer points, gripper dies, hot nut tools, and hot headers.

Physical Properties

H21: Thermal conductivity

TEMP C	TEMP F	THCON W/m*K	THCON Btu/ft*h*F
100	200	27.0	15.6
260	500	29.8	17.2
400	750	29.8	17.2
540	1000	29.4	17.0
675	1250	29.1	16.8

Source: Ref 3, 442

H21: Mean coefficients of linear thermal expansion

TEMP C	TEMP F	THEXP 10^{-6}/C	THEXP 10^{-6}/F
27-427	80-800	12.4	6.8
27-788	80-1450	13.7	7.6
27-788	80-1450	14.6	8.1
27-788	80-1450	15.1	8.4

Source: Ref 29, Aug 1964

Mechanical Properties

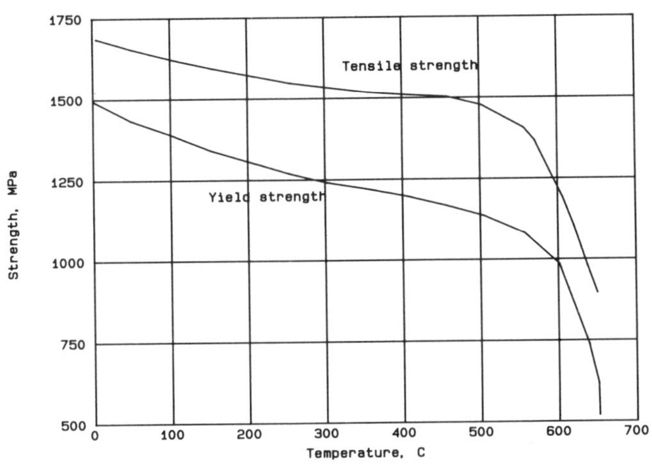

H21: Tensile and yield strength of specimens oil quenched from 1175 C and tempered to 50 HRC. Source: Ref 15, p 462

H21: Tensile property data

TEMP C	TEMP F	UTS MPa	UTS ksi	YS MPa	YS ksi	EL %	RA %	HARD(a) HRC
Quenched in oil from 1175 C (2150 F), tempered at 625 C (1170 F) to 50 HRC								
21	70	1675	243	1482	215	12	37	...
204	400	1579	229	1310	190	12	42	...
260	600	1531	222	1241	180	11	42	...
427	800	1517	220	1207	175	10	40	...
538	1000	1448	210	1124	163	9	32	...
649	1200	896	130	586	85	5	15	...
Short-time properties of specimens hardened and tempered to 44 HRC								
27	80	1379	200	1193	173	8	30	43
427	800	1138	165	952	138	11	23	42
482	900	1048	152	896	130	14	43	42
538	1000	889	129	765	111	11	29	41
593	1100	938	136	627	91	6	12	44
649	1200	745	108	586	85	4	9	43
Oil quenched from 1175 C (2150 F), tempered 620 C (1150 F), 2+2 h to 49 HRC								
21	70	1672	242.5	1520	220.5	7.4	27	...
427	800	1365	198	1245	180.5	7.9	27.3	...
538	1000	1182	171.5	1117	162	7.5	28.3	...
593	1100	1048	152	969	140.5	8.8	34.4	...
649	1200	879	127.5	789	114.5	10.5	38.2	...

(a) At room temperature
Source: Ref 14, p 271 and Ref 29, Aug 1964

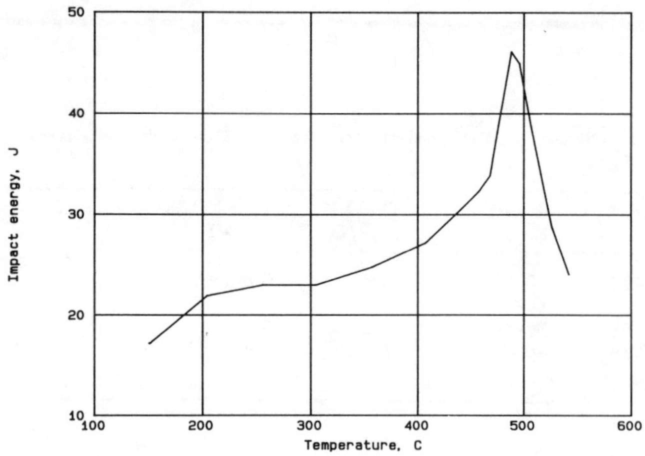

H21: Impact energy. Specimens were oil quenched from 1175 C and tempered to 50 HRC. Source: Ref 15, p 462

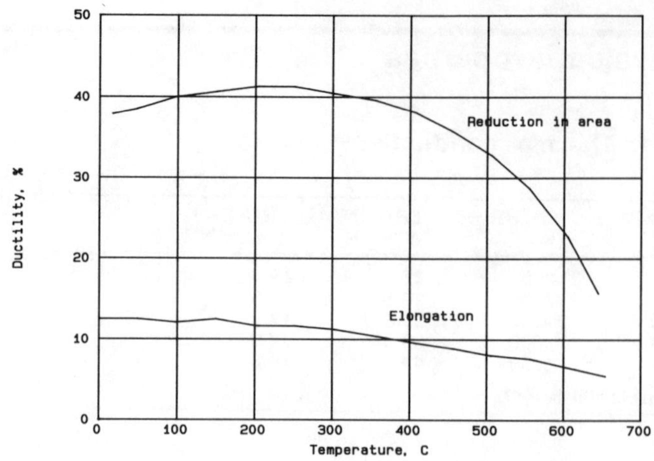

H21: Tensile elongation and reduction in area. Test specimens were oil quenched from 1175 C and tempered to 50 HRC. Source: Ref 15, p 462

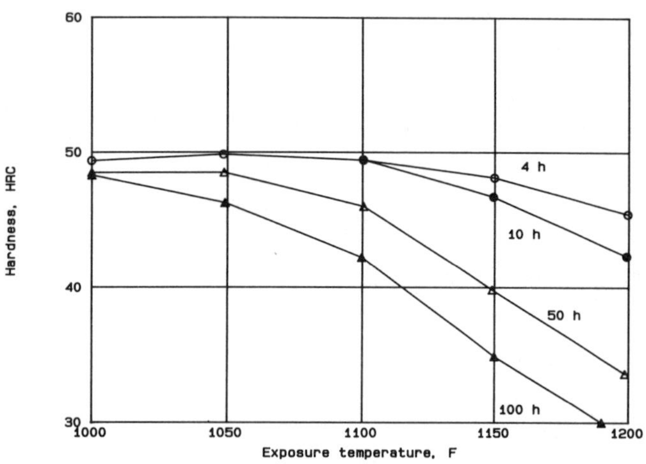

H21: Resistance to softening during elevated-temperature exposure as measured by room-temperature hardness. Initial hardness: 47 HRC. Heat treated at 2150 F for 5 min, oil quenched, double tempered at 1200 F for 2 h, air cooled, and heated to 1225 F for 1 h. Source: Ref 14, p 272

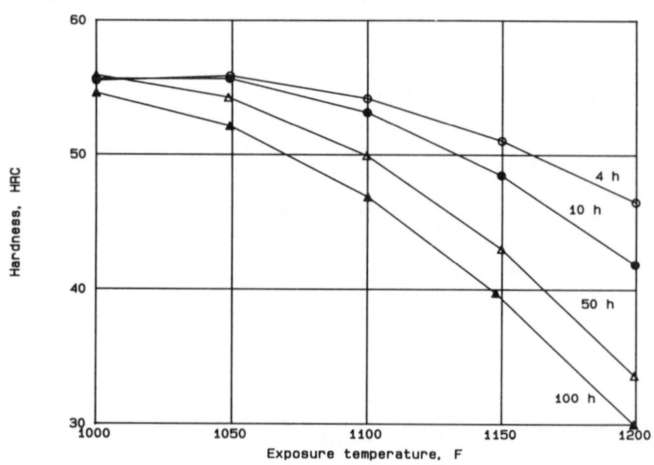

H21: Resistance to softening during elevated-temperature exposure as measured by room-temperature hardness. Initial hardness: 54 HRC. Prior treatment: 2150 F for 5 min, oil quenched, double tempered at 1050 F for 2 h, and air cooled. Source: Ref 14, p 272

H21: Resistance to softening

TEMP C	TEMP F	HARD 6 min HRC	HARD 1 h HRC	HARD 10 h HRC	HARD 100 h HRC
538	1000	43	43	43	43
593	1100	43	43	43	35
677	1250	43	37	29	23
732	1350	37	29	21	18
788	1450	29	22	18	15

Note: Heat treated and tempered to 45 HRC and reheated to temperature for time indicated
Source: Ref 29, Aug 1964

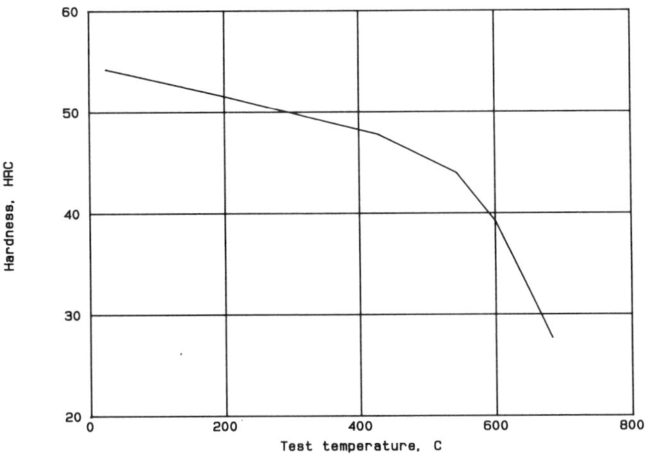

H21: Rockwell C hardness. Source: Ref 7, p 796

H21: Hot hardness data

TEMP C	TEMP F	HARD(a) HRC	HARD(b) HRC
Room-temperature hardness after heat treatment			
21	70	54.5	49.5
After 0.5 h at temperature			
316	600	49.5	45.0
427	800	48.0	40.5
538	1000	44.0	34.5
593	1100	36.0	25.5
649	1200	32.5	19.0

(a) Test 1. (b) Test 2
Source: Ref 14, p 271

H22

UNS number: T20822
Specifications: ASTM A681, FED QQ T-570, MSZ 4352, UNI 2955, JIS G D4404 SKD5, DGN B82, STAS VCW85
Nominal composition: 0.300 to 0.400% C, 0.150 to 0.400% Mn, 0.150 to 0.400% Si, 1.750 to 3.750% Cr, 0.300% Ni max, 10.000 to 11.750% W, 0.250 to 0.500% V
Product forms: Wrought, bar, rod, wire, plate, sheet, strip, tube, billets, forgings, ring, cast, sand, investment, powder
Description: H22 is a tungsten hot work tool steel. It exhibits high resistance to softening at elevated temperatures.
Typical applications: Mandrels, hot blanking dies, hot punches, blades for flying shear, hot trim dies, extrusion dies, dummy blocks, piercer points, and gripper dies.

Physical Properties

H22: Thermal conductivity

TEMP C	TEMP F	THCON W/m*K	THCON Btu/h*ft*F
27	80	25.1	14.5
204	400	29.7	17.2
427	800	29.7	17.2
649	1200	28.8	16.7

Note: Values were converted from electrical resistance measurements
Source: Ref 29, Nov 1963

H22: Mean coefficients of linear thermal expansion

TEMP C	TEMP F	THEXP 10^{-6}/C	THEXP 10^{-6}/F
27-93	80-200	10.9	6.1
27-427	80-800	12.2	6.8
27-649	80-1200	13.1	7.3
27-788	80-1450	13.7	7.6
260-649	500-1200	14.0	7.8
260-788	500-1450	14.6	8.1

Source: Ref 29, Nov 1963

H22: Tensile property data

TEMP C	TEMP F	UTS MPa	UTS ksi	YS MPa	YS ksi	EL %	RA %	HARD(a) HRC
Hardened and tempered to indicated hardness(b)								
27	80	1310	190	1103	160	9	21	42
427	800	1138	165	986	143	8	17	43
482	900	1055	153	896	130	8	22	43
538	1000	938	136	807	117	6	14	43
593	1100	827	120	731	106	5	6	43
649	1200	641	93	552	80	4	10	42
Oil quenched from 1175 C (2150 F), tempered 648 C (1200 F), 2+2 h to 46.5 HRC								
21	70	1551	225	1382	200.5	8.8	29.9	...
427	800	1265	183.5	1151	167	9.1	32.1	...
538	1000	1100	159.5	1007	146	9.7	32.5	...
593	1100	917	133	838	121.5	9.3	34.7	...
649	1200	696	101	631	91.5	10.3	36.2	...

(a) Room-temperature values. (b) Loaded at a rate of 0.05 in./min crosshead speed
Source: Ref 29, Nov 1963

H22: Effect of hardness and temperature on Charpy V-notch impact strength

HARD HRC	IS(a) J	IS(a) ft-lb	IS(b) J	IS(b) ft-lb	IS(c) J	IS(c) ft-lb	IS(d) J	IS(d) ft-lb
36	7	5	11	8	23	17	28	21
39	7	5	9	7	12	9	20	15
45	4	3	8	6	9	7	11	8
49	5	4	7	5	8	6	11	8

(a) At 27 C (80 F). (b) At 149 C (300 F). (c) At 260 C (500 F). (d) At 427 C (800 F)
Source: Ref 29, Nov 1963

H22: Brinell hardness data

HARD(a) HB	HARD(b) HB	HARD(c) HB	HARD(d) HB	HARD(e) HB	HARD(f) HB	HARD(g) HB
415	390	370	350	320	300	285
355	330	300	280	260	245	240

(a) As heat treated. (b) At 150 C (300 F). (c) At 260 C (500 F). (d) At 370 C (700 F). (e) At 480 C (900 F). (f) At 595 C (1100 F). (g) At 650 C (1200 F)
Source: Ref 29, Nov 1963

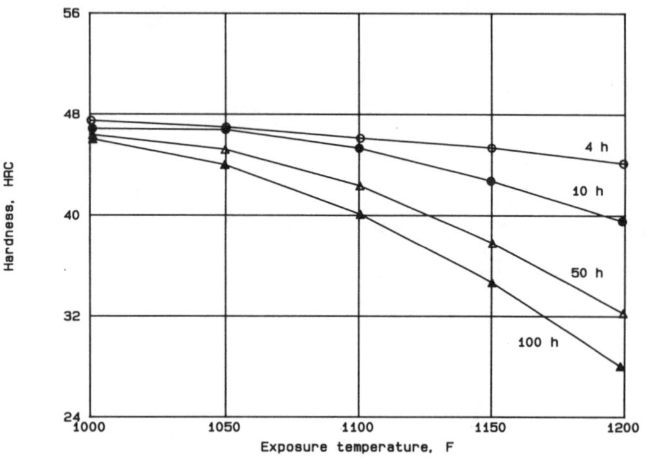

H22: Resistance to softening during elevated-temperature exposure as measured by room-temperature hardness. Initial hardness: 46.5 HRC. Prior treatment: 2150 F for 5 min, oil quenched, double tempered at 1175 F for 2 h, and air cooled. Source: Ref 14, p 274

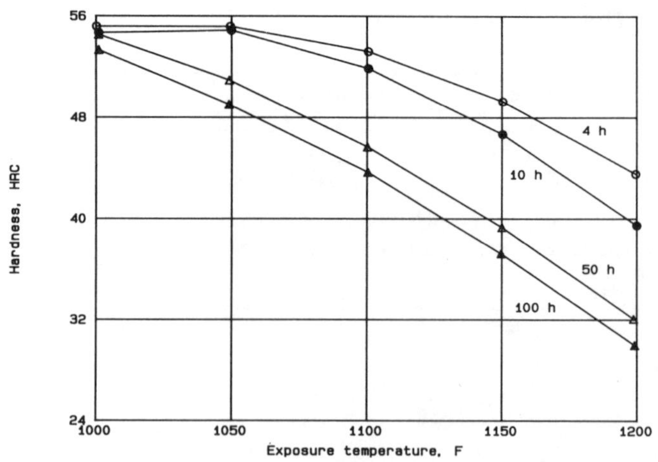

H22: Resistance to softening during elevated-temperature exposure as measured by room-temperature hardness. Initial hardness: 53.5 HRC. Prior heat treatment: 2150 F for 5 min, oil quenched, double tempered at 1050 F for 2 h, and air cooled. Source: Ref 14, p 274

H22: Resistance to softening

TEMP C	TEMP F	HARD 6 min HRC	HARD 1 h HRC	HARD 10 h HRC	HARD 100 h HRC
538	1000	44	44	44	42
593	1100	44	44	44	37
677	1250	44	38	30	23
727	1350	38	30	22	19
788	1450	30	23	19	16

Note: Specimens heat treated to 44 HRC and reheated to temperatures for times indicated
Source: Ref 29, Nov 1963

H23

UNS number: T20823
Specifications: ASTM A681, FED QQ T-570, DGN B82, DIN 1.2625
Nominal composition: 0.250 to 0.350% C, 0.150 to 0.400% Mn, 0.150 to 0.600% Si, 11.000 to 12.750% Cr, 0.300% Ni max, 11.000 to 12.750% W, 0.750 to 1.250% V
Product forms: Wrought, bar, rod, wire, plate, sheet, strip, billets, forgings, ring, cast, sand, investment, powder
Description: H23 is a tungsten hot work tool steel. It exhibits very high resistance to softening at elevated temperatures.
Typical applications: Extrusion and die casting dies for brass, and brass and bronze permanent molds.

Mechanical Properties

H23: Tensile property data

TEMP C	TEMP F	UTS MPa	UTS ksi	YS MPa	YS ksi	EL %	RA %
427	800	1265	183.5	848	123	7.0	11.4
482	900	1241	180	979	142	6.9	12.4
538	1000	1155	167.5	934	135.5	7.4	13.5
593	1100	965	140	683	99	8.0	16.5
649	1200	758	110	617	89.5	9.4	18.2

Note: Specimens were oil quenched from 1260 C (2300 F), tempered at 595 C (1100 F), 2+2 h to 46.5 HRC
Source: Ref 14, p 275

H23: Resistance to softening

TEMP C	TEMP F	HARD(a) 100 h HRC
Original hardness: 40.8 HRC		
480	900	40.0
540	1000	40.6
600	1100	40.8
650	1200	38.6
700	1300	33.2
760	1400	25.8
Original hardness: 38.9 HRC		
480	900	38.9
540	1000	38.0
600	1100	38.0
650	1200	37.1
700	1300	32.5
760	1400	25.6

(a) Room temperature values after exposure at temperature for time indicated
Source: Ref 3, p 426

H23: Hot hardness data

TEMP C	TEMP F	HARD HRC
21	70	46.0(a)
316	600	41.0
427	800	37.0
538	1000	33.5
593	1100	25.0
649	1200	20.0

Note: Values are for 30 min at temperature, unless otherwise indicated
(a) Room temperature after heat treatment
Source: Ref 14, p 275

H23: Resistance to softening during elevated-temperature exposure as measured by room-temperature hardness. Inital hardness: 49 HRC. Prior treatment: 2300 F for 5 min, oil quenched, triple tempered at 1200 F for 2 h, and air cooled. Source: Ref 14, p 276

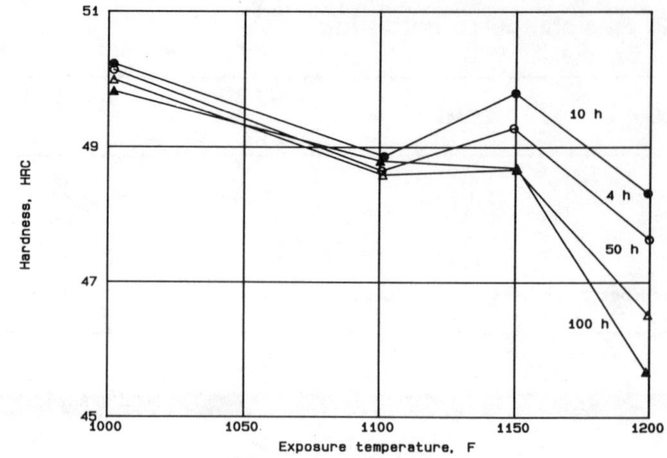

H26

UNS number: T20826
Specifications: ASTM A681, FED QQ T-570, DGN B82, COPANT 337, BS 4659 BH26
Nominal composition: 0.450 to 0.550% C, 0.150 to 0.400% Mn, 0.150 to 0.400% Si, 3.750 to 4.500% Cr, 0.300% Ni max, 17.250 to 19.000% W, 0.750 to 1.250% V
Product forms: Wrought, bar, rod, wire, plate, sheet, strip, billets, forgings, ring, cast, sand, investment, powder

Description: H26 is a tungsten hot work tool steel. It exhibits a very high resistance to softening at elevated temperatures.
Typical applications: Mandrels, hot blanking dies, hot punches, blades for flying shear, hot trimming dies, extrusion dies, dummy blocks, piercer points, gripper dies, pipe threading dies, nut chisels, forging press inserts, and extrusion dies for brass and copper.

Mechanical Properties

H26: Tensile property data

TEMP C	TEMP F	UTS MPa	UTS ksi	YS MPa	YS ksi	EL %	RA %
Oil quenched from 1175 C (2150 F), tempered 675 C (1250 F), 2+2 h to 42 HRC							
21	70	1320	191.5	1100	159.5	4.9	12.8
427	800	1041	151	931	135	5.2	15.1
538	1000	872	126.5	789	114.5	7.4	23.1
593	1100	714	103.5	645	93.5	8.4	24.9
649	1200	593	86	531	77	9.1	26.4
Oil quenched from 1205 C (2200 F), tempered 635 C (1175 F), 2+2 h to 51 HRC							
21	70	1779	258	1624	235.5	3.6	6.1
427	800	1631	236.5	1358	197	3.9	9.5
538	1000	1317	191	1238	179.5	4.5	11.6
593	1100	1124	163	1083	157	4.5	11.6
649	1200	989	143.5	924	134	5.7	13.7

Source: Ref 14, p 277

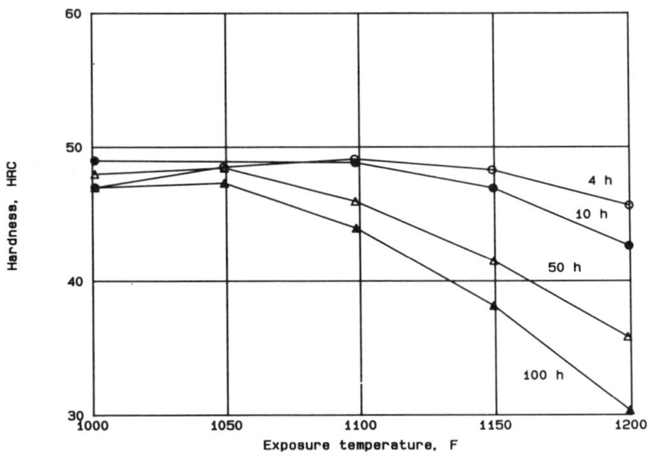

H26: Resistance to softening during elevated-temperature exposure as measured by room-temperature hardness. Initial hardness: 46.5 HRC. Prior treatment: 2150 F for 5 min, oil quenched, double tempered at 1250 F for 2 h, and air cooled. Source: Ref 14, p 278

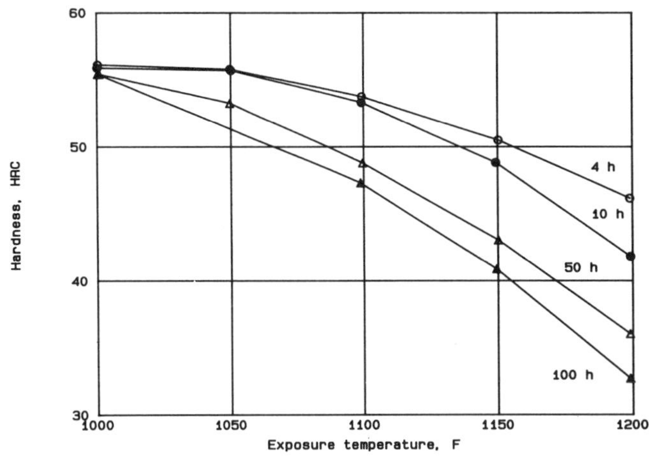

H26: Resistance to softening during elevated-temperature exposure as measured by room-temperature hardness. Initial hardness: 54.5 HRC. Prior treatment: 2150 F for 5 min, oil quenched, double tempered at 1150 F for 2 h, and air cooled. Source: Ref 14, p 278

H26: Resistance to softening

TEMP C	TEMP F	HARD(a) 100 h HRC
Original hardness: 51.0 HRC		
480	900	50.6
540	1000	50.3
600	1100	47.1
650	1200	38.4
700	1300	26.9
760	1400	21.3
Original hardness: 42.9 HRC		
480	900	42.4
540	1000	42.3
600	1100	41.3
650	1200	34.9
700	1300	26.4
760	1400	21.1

(a) Room temperature values after exposure at temperature for time indicated
Source: Ref 3, p 426

H26: Hot hardness data

TEMP C	TEMP F	HARD(a) HRC
Test 1		
21(b)	70(b)	57.0
316	600	53.5
427	800	49.5
538	1000	46.0
593	1100	44.0
649	1200	35.0
Test 2		
21(b)	70(b)	46.5
316	600	42.0
427	800	40.0
538	1000	32.5
593	1100	24.5
649	1200	20.0

(a) After 30 min at temperature. (b) After heat treatment
Source: Ref 14, p 277

M1

UNS number: T11301
Specifications: ASTM A600, SAE J437, SAE J438, DIN 1.3346, NF Z 85DCWV 08040201, BS 4659 BM1, GOST H41, ABNT 4441 2-8-2, MSZ 4351 R10, COPANT 337 M1, UNE F.5607
Nominal composition: 0.780 to 0.880% C, 0.150 to 0.400% Mn, 0.200 to 0.500% Si, 3.500 to 4.000% Cr, 0.300% Ni max, 8.200 to 9.200% Mo, 1.400 to 2.100% W, 1.000 to 1.350% V

Product forms: Wrought, bar, rod, wire, plate, sheet, strip, billets, forgings, ring, powder
Description: M1 is a molybdenum high speed tool steel. It exhibits very high resistance to softening at elevated temperatures.
Typical applications: Drills, taps, end mills, reamers, milling cutters, hobs, punches, lathe and planer tools, form tools, saws, chasers, broaches, routers, and woodworking tools.

Physical Properties

M1: Mean coefficients of linear thermal expansion

TEMP C	TEMP F	THEXP $10^{-6}/C$	THEXP $10^{-6}/F$
38-260	100-500	10.01	5.61
38-427	100-800	11.38	6.32
38-538	100-1000	11.99	6.66
38-649	100-1200	12.35	6.86
38-816	100-1500	12.73	7.07

Source: Ref 29, May 1980

Mechanical Properties

M1: Hot hardness data

TEMP C	TEMP F	HARD(a) HRC
21	70	65.0(b)
316	600	59.5
427	800	57.0
538	1000	54.0
593	1100	47.0
649	1200	37.5

(a) After 30 min at temperature. (b) After heat treatment
Source: Ref 14, p 279

M1: Hardness after treating for maximum hardness before testing

TEMP C	TEMP F	HARD
21	70	66.8 HRC
595	1100	475 HB
650	1200	308 HB

Source: Ref 15, p 467

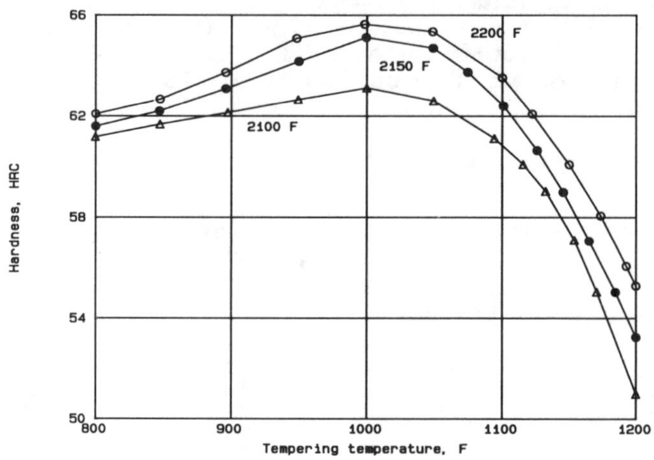

M1: Effect of tempering temperature on Rockwell C hardness for specimens austenitized at 3 different temperatures. Source: Ref 29, May 1980

M2

Alternate designation: AISI 611
UNS number: T11302
Specifications: ASTM A600, SAE J437, SAE J438, DIN 1.3343, NF Z 85WDCV 06050402, BS BM2, UNI HS 652, SS 2722, GOST R6AM5, UNE F.5063
Nominal composition: 0.780 to 0.880% C, 0.150 to 0.400% Mn, 0.200 to 0.450% Si, 3.750 to 4.500% Cr, 0.300% Ni max, 4.500 to 5.500% Mo, 5.500 to 6.750% W, 1.750 to 2.200% V
Product forms: Wrought, bar, rod, wire, plate, sheet, strip, billets, forgings, ring, powder

Description: M2 is a molybdenum high speed tool steel. It exhibits very high resistance to softening at elevated temperatures. Oxidation resistance is satisfactory up to 370 C (700 F). M2 is also known as AISI 611, a heat resistant material for high-strength structural components that operate at temperatures up to 540 C (1000 F). It is usually vacuum melted for critical applications.

Typical applications: Bearings, races, rollers, cams, gears, and pump bodies, pistons, and vanes.

Physical Properties

M2: Thermal conductivity

TEMP C	TEMP F	THCON W/m*K	THCON Btu/ft*h*F
100	200	21.3	12.3
200	500	23.5	13.6
400	750	25.6	14.8
540	1000	27.0	15.6
675	1250	28.9	16.7

Source: Ref 3, p 442

M2: Mean coefficients of linear thermal expansion

TEMP C	TEMP F	THEXP(a) 10^{-6}/C	THEXP(a) 10^{-6}/F	THEXP(b) 10^{-6}/C	THEXP(b) 10^{-6}/F
21-93	70-200	10.24	5.69	11.25	6.25
21-149	70-300	10.64	5.91	11.7	6.5
21-204	70-400	10.96	6.09	12.08	6.71
21-260	70-500	11.30	6.28	12.20	6.78
21-316	70-600	11.56	6.42	12.28	6.82
21-371	70-700	11.74	6.52	12.33	6.85
21-427	70-800	12.01	6.67	12.6	7.0
21-482	70-900	12.28	6.82	12.78	7.1
21-538	70-1000	12.55	6.97	12.89	7.11

(a) Heat treated. (b) Annealed
Source: Ref 28, p 25

Mechanical Properties

M2: Tensile property data

TEMP C	TEMP F	UTS MPa	UTS ksi	YS MPa	YS ksi	EL %	RA %	HARD HRC
538	1000	2006	291	1827	265	3	4	65

Note: Specimens were heat treated to 1205 C (2200 F), air cooled , and double tempered at 565 C (1050 F) for 2 h
Source: Ref 14, p 280

M2: Short-time compression property data

TEMP C	TEMP F	YS 0.1% MPa	YS 0.1% ksi	YS 0.2% MPa	YS 0.2% ksi	HARD HRC
21	70	...	...	...	...	66
204	400	2648	384	2703	392	62
316	600	2427	352	2530	367	61
427	800	1979	287	2069	300	59
538	1000	1544	224	1710	248	55

Note: Values from a single heat. Specimens were soaked 30 min before testing
Source: Ref 28, p 25

M2: Hot hardness data

TEMP C	TEMP F	HARD HB
21	70	745
595	1100	465
650	1200	342

Note: Values were obtained after treating for maximum hardness using the mutual indentation method before testing
Source: Ref 15, p 467

M2: Hot hardness data

TEMP C	TEMP F	HARD HRC
21	70	65.3(a)
538	1000	58.0
565	1050	57.0
595	1100	55.9

(a) Original room temperature hardness value before tests
Source: Ref 15, p 467

M2: Hot hardness data

TEMP C	TEMP F	HARD 30 min HRC	HARD 1000 h HRC
21	70	66	66
204	400	63	62
316	600	61	60
427	800	58	57
538	1000	51	49
649	1200	34	...

Note: Actual HRC values after exposure for indicated time at temperature
Source: Ref 28, p 25

M3, class 1

UNS number: T11313
Specifications: ASTM A600, SAE J437, SAE J438, NF Z 90 WDCV 060502, DIN 1.3342
Nominal composition: 1.000 to 1.100% C, 0.150 to 0.400% Mn, 0.200 to 0.450% Si, 3.750 to 4.500% Cr, 0.300% Ni max, 4.750 to 6.500% Mo, 5.000 to 6.750% W, 2.250 to 2.750% V
Product forms: Wrought, bar, rod, wire, plate, sheet, strip, billets, forgings, ring, powder
Description: M3, class 1 is a molybdenum high speed tool steel. It exhibits very high resistance to softening at elevated temperatures.

Mechanical Properties

M3, class 1: Hot hardness data

TEMP C	TEMP F	HARD(a) HRC
21	70	66.0(b)
316	600	61.5
427	800	59.0
538	1000	55.0
593	1100	52.0
649	1200	43.0

(a) After exposure for 0.5 h at temperature. (b) After heat treatment
Source: Ref 14, p 281

M3, class 2

UNS number: T11323
Specifications: ASTM A600, SAE J437, SAE J438, DIN 1.3344, NF Z 130WDCV, JIS G 4403 SKH53, COPANT 337, UNE F.5605
Nominal composition: 1.150 to 1.250% C, 0.150 to 0.400% Mn, 0.200 to 0.450% Si, 3.750 to 4.500% Cr, 0.300% Ni max, 4.750 to 6.500% Mo, 5.000 to 6.750% W, 2.750 to 3.750% V
Product forms: Wrought, bar, rod, wire, plate, sheet, strip, billets, forgings, ring, powder
Description: M3, class 2 is a molybdenum high speed tool steel. It exhibits very high resistance to softening at elevated temperatures.
Typical applications: Drills, taps, end mills, reamers and counterbores, broaches, hobs, form tools, lathe and planer tools, cheeking tools, slitting saws, punches, drawing dies, and woodworking tools.

Physical Properties

M3, class 2: Mean coefficients of linear thermal expansion

TEMP C	TEMP F	THEXP 10^{-6}/C	THEXP 10^{-6}/F
21-260	100-500	9.83	5.46
21-427	100-800	11.50	6.39
21-538	100-1000	12.0	6.67
21-649	100-1200	12.69	7.05
21-816	100-1500	12.91	7.17

Source: Ref 29, Nov 1982

Mechanical Properties

M3, class 2: Hot hardness data

TEMP C	TEMP F	HARD(a) HRC
21	70	66(b)
316	600	61.5
427	800	58.5
538	1000	55.0
593	1100	51.0
649	1200	41.0

(a) After exposure for 0.5 h at temperature. (b) After heat treatment
Source: Ref 14, p 282

M4

UNS number: T11304
Specifications: ASTM A600, SAE J437, SAE J438, NF 4361 6-5-4, JIS G 4403 SKH54, DGN B82, COPANT 337, BS 4659 BM4
Nominal composition: 1.250 to 1.400% C, 0.150 to 0.400% Mn, 0.200 to 0.450% Si, 3.750 to 4.750% Cr, 0.300% Ni max, 4.250 to 5.500% Mo, 5.250 to 6.500% W, 3.750 to 4.500% V
Product forms: Wrought, bar, rod, wire, plate, sheet, strip, billets, forgings, ring, powder
Description: M4 is a molybdenum high speed tool steel. It exhibits very high resistance to softening at elevated temperatures.
Typical applications: Broaches, reamers, milling cutters, chasers, form tools, lathe and planer tools, cheeking tools, blanking dies and punches for abrasive materials, and swaging dies.

Mechanical Properties

M4: Hot hardness data

TEMP C	TEMP F	HARD HRC	HARD(a) HRC
21	70	65.4(b)	66.5
316	600	...	61.5
427	800	...	58.5
540	1000	59.0	54.5
565	1050	57.5	...
595	1100	56.7	51.0
649	1200	...	40.0

(a) After 30 min at temperature. (b) Original hardness before tests
Source: Ref 15, p 467 and Ref 14, p 283

M7

UNS number: T11307
Specifications: ASTM A600, DIN 1.3348, NF Z 100DCWV 09040202, UNI HS 292, SS 2782, UNE F.5607
Nominal composition: 0.970 to 1.050% C, 0.150 to 0.400% Mn, 0.200 to 0.550% Si, 3.500 to 4.000% Cr, 0.300% Ni max, 8.200 to 9.200% Mo, 1.400 to 2.100% W, 1.750-2.250% V
Product forms: Wrought, bar, rod, wire, plate, sheet, strip, tube, billets, forgings, ring, powder
Description: M7 is a molybdenum high speed tool steel. It exhibits very high resistance to softening at elevated temperatures.
Typical applications: Drills, taps, end mills, reamers, routers, saws, milling cutters, lathe and planer tools, chasers, borers, woodworking tools, hobs, form tools, and punches.

Mechanical Properties

M7: Hot hardness data

TEMP C	TEMP F	HARD(a) HRC
21	70	66.5(b)
316	600	61.0
427	800	58.0
538	1000	55.0
593	1100	50.5
649	1200	38.0

(a) After exposure for 0.5 h at temperature. (b) Room temperature after heat treatment
Source: Ref 14, p 284

M10

Alternate Designation: AISI 612
UNS number: T11310
Specifications: ASTM A600, NF 4441 2-8-2, DGN B82
Nominal composition: 0.840 to 0.940% C, 0.100 to 0.400% Mn, 0.200 to 0.450% Si, 3.750 to 4.500% Cr, 0.300% Ni max, 7.750 to 8.500% Mo, 1.800 to 2.200% V
Product forms: Wrought, bar, rod, wire, plate, sheet, strip, tube, billets, forgings, ring, powder
Description: M10 is a molybdenum high speed tool steel. It exhibits very high resistance to softening at elevated temperatures. Oxidation resistance is satisfactory up to 370 C (700 F). M10 is also known as AISI 612, a heat resistant material for high-strength aircraft and missile components that operate at temperatures up to 540 C (1000 F). It is usually vacuum melted for critical applications.
Typical applications: Drills, taps, reamers, chasers, end mills, lathe and planer tools, woodworking tools, routers, saws, milling cutters, hobs, form tools, punches, and broaches. Other uses include bearings, races, rollers, cams, gears, and pump bodies, pistons, and vanes.

Mechanical Properties

M10: Tensile property data

TEMP C	TEMP F	UTS MPa	UTS ksi	YS MPa	YS ksi	EL %	RA %	HARD HRC
538	1000	1934	280.5	1738	252	1.5	2.5	64

Note: Heat treated at 1190 C (2175 F), air cooled, double tempered at 565 C (1050 F) for 2 h
Source: Ref 14, p 285

M10: Hot hardness data

TEMP C	TEMP F	HARD 30 min HRC	HARD 1000 h HRC
21	70	66	66
93	200	64	...
204	400	62	61
316	600	61	61
427	800	58	57
538	1000	54	45

Source: Ref 14, p 27

M10: Hot hardness data

TEMP C	TEMP F	HARD(a) HRC
21	70	65.5(b)
316	600	60.5
427	800	57.5
538	1000	54.0
593	1100	50.0
649	1200	40.0

(a) After exposure for 0.5 h at temperature. (b) After heat treatment
Source: Ref 14, p 285

M10: Hot hardness data

TEMP C	TEMP F	HARD HRC
21	70	65.3(a)
538	1000	58.0
565	1050	57.0
593	1100	55.5

(a) Hardness before tests
Source: Ref 15, p 467

M10: Hardness after treating for maximum hardness before testing

TEMP C	TEMP F	HARD HRC
21	70	745
595	1100	458
650	1200	313

Note: Mutual indentation method
Source: Ref 15, p 467

M30

UNS number: T11330
Specifications: ASTM A600, DGN B82, UNE F.5611, BS 4659 BM34, DIN 1.3249
Nominal composition: 0.750 to 0.850% C, 0.150 to 0.400% Mn, 0.200 to 0.450% Si, 3.500 to 4.250% Cr, 0.300% Ni max, 7.750 to 9.000% Mo, 1.300 to 2.300% W, 1.000 to 1.400% V, 4.500 to 5.500% Co
Product forms: Wrought, bar, rod, wire, plate, sheet, strip, billets, forgings, ring, powder
Description: M30 is a molybdenum high speed tool steel. It exhibits an extremely high resistance to softening at elevated temperatures.
Typical applications: Broaches, chasers, cheeking tools, drills, end mills, hobs, lathe tools, milling cutters, planer tools, punches, reamers, routers, saws, taps, and woodworking tools.

Mechanical Properties

M30: Hot hardness data

TEMP C	TEMP F	HARD(a) HRC
21	70	66.0(b)
316	600	61.5
427	800	59.5
538	1000	57.0
593	1100	51.0
649	1200	42.0

(a) After exposure for 0.5 h at temperature. (b) After heat treatment
Source: Ref 14, p 286

M33

UNS number: T11333
Specifications: ASTM A600, DIN 1.3249, BS BM42, UNE F.5611
Nominal composition: 0.850 to 0.920% C, 0.150 to 0.400% Mn, 0.150 to 0.500% Si, 3.500 to 4.000% Cr, 0.300% Ni max, 9.000 to 10.000% Mo, 1.300 to 2.100% W, 1.000 to 1.350% V, 7.750 to 8.750% Co
Product forms: Wrought, bar, rod, wire, plate, sheet, strip, billets, forgings, ring, powder

Description: M33 is a molybdenum high speed tool steel. It exhibits an extremely high resistance to softening at elevated temperatures.

Typical applications: Drills, taps, end mills, lathe tools, milling cutters, form tools, and chasers.

Physical Properties

M33: Mean coefficients of linear thermal expansion

TEMP C	TEMP F	THEXP 10^{-6}/C	THEXP 10^{-6}/F
38-260	100-500	9.5	5.28
38-427	100-800	11.1	6.14
38-538	100-1000	11.7	6.5
38-649	100-1200	12.08	6.71
38-816	100-1500	12.49	6.94

Source: Ref 29, Aug 1982

Mechanical Properties

M33: Hot hardness data

TEMP C	TEMP F	HARD(a) HRC
21	70	66.0(b)
316	600	62.0
427	800	59.0
538	1000	56.5
593	1100	51.5
649	1200	43.0

(a) After exposure for 0.5 h at temperature. (b) After heat treatment
Source: Ref 14, p 287

M34

UNS number: T11334
Specifications: ASTM A600 (M34), FED QQ-T-590 (M34)
Nominal composition: 0.850 to 0.920% C, 0.150 to 0.400% Mn, 0.200 to 0.450% Si, 3.500 to 4.000% Cr, 0.300% Ni max, 7.750 to 9.200% Mo, 1.400 to 2.100% W, 1.900 to 2.300% V, 7.750 to 8.750% Co
Description: M34 is a molybdenum high speed tool steel. It exhibits an extremely high resistance to softening at elevated temperatures.
Typical applications: Drills, taps, end mills, lathe tools, milling cutters, form tools, and chasers.

Mechanical Properties

M34: Hot hardness data

TEMP C	TEMP F	HARD(a) HRC
21	70	65.5(b)
316	600	61.5
427	800	58.5
538	1000	55.5
593	1100	51.0
649	1200	43.0

(a) After exposure for 0.5 h at temperature. (b) After heat treatment
Source: Ref 14, p 288

M36

UNS number: T11336
Specifications: ASTM A600, NF 4371 6-5-2-5, UNI 2955, JIS G 4403 SKH55, DGN B82, COPANT 337, UNE F.5613, SIS 2723, GOST R6M5K5, DIN 1.3243
Nominal composition: 0.800 to 0.900% C, 0.150 to 0.400% Mn, 0.200 to 0.450% Si, 3.750 to 4.500% Cr, 0.300% Ni max, 4.500 to 5.500% Mo, 5.500 to 6.500% W, 1.750 to 2.250% V, 7.750 to 8.750% Co

Product forms: Wrought, bar, rod, wire, plate, sheet, strip, billets, forgings, ring, powder

Description: M36 is a molybdenum high speed tool steel. It exhibits an extremely high resistance to softening at elevated temperatures.

Typical applications: Heavy-duty lathe and planer tools, boring tools, milling cutters, drills, cutoff tools, and tool holder bits.

Mechanical Properties

M36: Hot hardness data obtained after treating for maximum hardness before testing

TEMP C	TEMP F	HARD HB
21	70	757
595	1100	500
650	1200	390

Note: Mutual indentation method
Source: Ref 15, p 467

M50

UNS number: T11350
Specifications: AMS 6490, AMS 6491
Nominal composition: 0.770 to 0.850% C, 0.200 to 0.350% Mn, 0.100 to 0.250% Si, 3.750 to 4.250% Cr, 0.100% Ni max, 4.000 to 4.500% Mo, 0.250% W max, 0.900 to 1.100% V
Product forms: Wrought, bar, rod, wire, plate, sheet, strip, billets, forgings, ring, powder
Description: M50 is a vacuum melted, intermediate high speed tool steel. It exhibits excellent resistance to wear and softening at temperatures up to 425 C (800 F), as well as relatively good oxidation resistance.
Typical applications: M50 is widely used for highly loaded bearings in gas turbine engines and for other critical aircraft and missile components. It is also specified for premium quality chasers, drills, hydraulic pump components, punches, router bits, taps, and woodworking tools.

Physical Properties

M50: Mean coefficients of linear thermal expansion

TEMP C	TEMP F	THEXP 10^{-6}/C	THEXP 10^{-6}/F
21-93	70-200	10.22	5.68
21-204	70-400	11.32	6.29
21-371	70-700	12.44	6.91
21-538	70-1000	12.99	7.22
21-649	70-1200	13.39	7.44
21-760	70-1400	14.15	7.86

Source: Ref 29, Sep 1983

Mechanical Properties

M50: Short-time tensile property data

TEMP C	TEMP F	UTS MPa	UTS ksi	YS MPa	YS ksi	EL(a) %	RA %	E GPa	E 10^3 ksi
Annealed to 92.5 HRB									
21	70	681	98.8	302	43.8	26.8	54.2	205	29.7
Heat treated to 64 HRC									
21	70	2837	411.5	2334	338.5	1.6	1.7	202	29.3
316	600	2520	365.5	2072	300.5	4.2	12.0	174	25.3
427	800	2382	345.5	1810	262.5	6.3	17.0	164	23.8
538	1000	2131	309.0	1724	250.0	6.1	11.7	162	23.5

Note: Specimens were preheated to 815 C (1500 F), austenitized at 1120 C (2050 F) for 4 min, air cooled and tempered at 550 C (1050 F) for 2 h before final grinding
(a) 4D
Source: Ref 29, Sep 1983

M50: Hot hardness data

TEMP C	TEMP F	HARD HRC	HARD HB
Tempered at 556 C (1025 F)			
316	600	62.5	600
371	700	62.5	578
427	800	62.5	555
482	900	62.5	555
538	1000	62.5	512
Tempered at 593 C (1100 F)			
593	1100	60.0	430
Tempered at 649 C (1200 F)			
649	1200	49.0	241

Note: Specimens were austenitized at 1120 C (2050 F), oil quenched and tempered 2 h at 550 C (1025 F), or to the tempering temperature if higher
Source: Ref 29, Sep 1983

M50: Oxidation resistance

Ttpr C	Ttpr F	EXP TEMP C	EXP TEMP F	WTGN 50 h mg/cm^2	WTGN 150 h mg/cm^2
538	1000	538	1000	0.281	0.481
593	1100	593	1100	1.81	2.36

Note: Specimens were preheated, austenitized, oil quenched and tempered at the indicated temperatures for 2 h. These specimens were then weighed, exposed to air for 50 and 150 h and reweighed.
Source: Ref 29, Sep 1983

O1

UNS number: T31501
Specifications: ASTM A681, FED QQ T-570, SAE J437, SAE J438, DIN 1.2510, BS BO1, UNI 95 MnWCr5KU, UNE F.5220,
Nominal composition: 0.850 to 1.000% C, 1.000 to 1.400% Mn, 0.500% Si max, 0.300% Ni max, 0.400 to 0.600% Cr, 0.400 to 0.600% W, 0.300% V max
Product forms: Wrought, bar, rod, wire, plate, sheet, strip, billets, forgings, ring, cast, investment, powder
Description: O1 is an oil-hardening cold work tool steel. It exhibits low resistance to softening at elevated temperatures.
Typical applications: Blanking dies, plastic mold dies, drawing dies, trim dies, paper knives, shear blades, taps, reamers, tools, gages, bending and forming dies, bushings, and punches.

Physical Properties

O1: Mean coefficients of linear thermal expansion

TEMP C	TEMP F	THEXP 10^{-6}/C	THEXP 10^{-6}/F
38-260	100-500	10.8	6.0
38-425	100-800	12.8	7.1
38-540	100-1000	14.0	7.8
38-650	100-1200	14.4	8.0

Source: Ref 15, p 441

O2

UNS number: T31502
Specifications: ASTM A681, FED QQ T-570, SAE J437, SAE J438, DIN 1.2842, NF 90 MV8, BS BO2, UNI 90 MnVCr8KU,
Nominal composition: 0.850 to 0.950% C, 1.400 to 1.800% Mn, 0.500% Si max, 0.350% Cr max, 0.300% Ni max, 0.300% Mo max, 0.300% V max
Product forms: Wrought, bar, rod, wire, plate, sheet, strip, billets, forgings, ring, cast, investment, powder
Description: O2 is an oil-hardening cold work tool steel. It exhibits low resistance to softening at elevated temperatures.
Typical applications: Blanking, stamping, trimming, cold forming dies and punches, cold forming tools, threading dies and taps, reamers, gages, plugs and master tools, broaches, circular cutters and saws, thread roller dies, bushings, and plastic molding dies.

Physical Properties

O2: Mean coefficients of linear thermal expansion

TEMP C	TEMP F	THEXP 10^{-6}/C	THEXP 10^{-6}/F
25-40	77-100	10.3	5.7
25-95	77-200	11.7	6.5
25-150	77-300	12.6	7.0
25-205	77-400	12.6	7.0
25-260	77-500	13.0	7.2
25-315	77-600	13.3	7.4
25-370	77-700	13.9	7.7
25-425	77-800	13.9	7.7
25-480	77-900	14.4	8.0
25-540	77-1000	14.6	8.1
25-595	77-1100	14.9	8.3
25-650	77-1200	15.1	8.4

Source: Ref 15, p 441

O6

UNS number: T31506
Specifications: ASTM A681, FED QQ T-570, SAE J437, SAE J438, NF 2121 14aSMD4, DGN B82, COPANT 337
Nominal composition: 1.250 to 1.550% C, 0.300 to 1.100% Mn, 0.550 to 1.500% Si, 0.300% Cr max, 0.300% Ni max, 0.200 to 0.300% Mo
Product forms: Wrought, bar, rod, wire, plate, sheet, strip, billets, forgings, ring, cast, investment, powder
Description: O6 is an oil-hardening cold work tool steel. It exhibits a low resistance to softening at elevated temperatures.
Typical applications: Blanking dies, forming dies, mandrels, punches, cams, brake dies, deep drawing dies, cold forming rollers, bushings, gages, blanking and forming punches, piercing and perforating dies, taps, paper cutting dies, wear plates, tool shanks, jigs, machine spindles, arbors, guides in grinders and straighteners.

Physical Properties

O6: Mean coefficients of linear thermal expansion

TEMP C	TEMP F	THEXP 10^{-6}/C	THEXP 10^{-6}/F
25-250	77-480	12.48	6.9
25-300	77-570	12.86	7.1
25-350	77-660	13.17	7.3
25-400	77-750	13.43	7.5
25-450	77-840	13.65	7.6
25-500	77-930	13.86	7.7

Source: Ref 15, p 441

S2

UNS number: T41902
Specifications: ASTM A681, FED QQ T-570, SAE J437, SAE J438
Nominal composition: 0.400 to 0.550% C, 0.300 to 0.500% Mn, 0.900 to 1.200% Si, 0.300% Ni max, 0.300 to 0.600% Mo, 0.500% V max
Product forms: Wrought, bar, rod, wire, plate, sheet, strip, billets, forgings, ring, cast, investment, powder
Description: S2 is a shock-resisting tool steel that exhibits low resistance to softening at elevated temperatures.
Typical applications: Hand and pneumatic chisels, drift pins, forming tools, knockout pins, mandrels, nail sets, pipe cutters, rivet sets, screw driver bits, shear blades, spindles, stamps, tool shanks, and track tools.

Physical Properties

S2: Mean coefficients of linear thermal expansion

TEMP C	TEMP F	THEXP 10^{-6}/C	THEXP 10^{-6}/F
21-100	70-212	10.9	6.04
21-200	70-390	11.9	6.63
21-300	70-570	13.0	7.24
21-400	70-750	13.6	7.54
21-500	70-930	14.1	7.83
21-600	70-1110	14.4	8.01
21-700	70-1290	14.6	8.12
21-750	70-1380	14.8	8.22

Source: Ref 15, p 437

S7

UNS number: T41907
Specifications: ASTM A681, FED QQ T-570, DGN B82, COPANT 337
Nominal composition: 0.450 to 0.550% C, 0.200 to 0.800% Mn, 0.200 to 1.000% Si, 3.000 to 3.500% Cr, 1.300 to 1.800% Mo, 0.200-0.300% V
Product forms: Wrought, bar, rod, wire, plate, sheet, strip, billets, forgings, ring, cast, investment, powder
Description: S7 is a shock-resisting tool steel that exhibits high resistance to softening at elevated temperatures.
Typical applications: Shear blades, punches, slitters, chisels, forming dies, hot header dies, blanking dies, rivet sets, gripper dies, engraving dies, plastic molds, die casting dies, master hobs, beading tools, caulking tools, chuck jaws, clutches, pipe cutters, and swaging dies.

Physical Properties

S7: Mean coefficients of linear thermal expansion

TEMP C	TEMP F	THEXP 10^{-6}/C	THEXP 10^{-6}/F
25-200	77-390	12.59	6.99
25-300	77-570	12.99	7.22
25-400	77-570	13.33	7.41
25-500	77-930	13.68	7.6
25-600	77-1110	14.01	7.78
25-700	77-1290	14.27	7.93

Source: Ref 15, p 437

T1

UNS number: T12001
Specifications: AMS 5626, ASTM A600, SAE J437, SAE J438, DIN 1.3355, NF Z 80WCV 180401, BS BT1, UNI HS 1801, JIS SKH2, GOST R18, UNE F.5520
Nominal composition: 0.650 to 0.800% C, 0.100 to 0.400% Mn, 0.200 to 0.400% Si, 3.750 to 4.000% Cr, 0.300% Ni max, 17.250 to 18.750% W, 0.900 to 1.300% V
Product forms: Wrought, bar, rod, wire, plate, sheet, strip, tube, billets, forgings, ring, powder

Description: T1 is a tungsten high speed tool steel that exhibits very high resistance to softening at elevated temperatures.

Typical applications: Drills, taps, reamers, hobs, lathe and planer tools, broaches, crowners, burnishing dies, cold extrusion dies, cold heading die inserts, lamination dies, chasers, cutters, taps, end mills, and milling cutters.

Physical Properties

T1: Thermal conductivity

TEMP C	TEMP F	THCON W/m*K	THCON Btu/ft*h*F
100	200	19.9	11.5
260	500	21.6	12.5
400	750	23.2	13.4
540	1000	24.7	14.3

Source: Ref 3, p 442

Mechanical Properties

T1: Hot hardness data

TEMP C	TEMP F	HARD HB
21	70	750
595	1100	480
650	1200	337

Source: Ref 15, p 463

T4

UNS number: T12004
Specifications: ASTM A600 (T4), FED QQ-T-590 (T4), SAE J437 (T4), SAE J438 (T4)
Nominal composition: 0.700 to 0.800% C, 0.100 to 0.400% Mn, 0.200 to 0.400% Si, 3.750 to 4.500% Cr, 0.300% Ni max, 0.400 to 1.000% Mo, 17.500 to 19.000% W, 0.800 to 1.200% V, 4.250 to 5.750% Co
Description: T4 is a tungsten high speed tool steel that exhibits an extremely high resistance to softening at elevated temperatures.
Typical applications: Lathe and planer tools, drills, boring tools, broaches, roll turning tools, milling cutters, shaper tools, form tools, hobs, and single point cutting tools.

Mechanical Properties

T4: Hot hardness data

TEMP C	TEMP F	HARD HRC
21	70	66.2(a)
540	1000	60.2
565	1050	59.5
595	1100	58.7

(a) Rockwell C hardness before tests
Source: Ref 15, p 464

T5

UNS number: T12005
Specifications: ASTM A600, SAE J437, SAE J438, DIN 1.3265, BS BT5, UNI HS 180110, JIS SKH4A, UNE F.5540
Nominal composition: 0.750 to 0.850% C, 0.200 to 0.400% Mn, 0.200 to 0.400% Si, 3.750 to 5.000% Cr, 0.300% Ni max, 0.500 to 1.250% Mo, 17.500 to 19.000% W, 1.800 to 2.400% V, 7.000 to 9.500% Co
Product forms: Wrought, bar, rod, wire, plate, sheet, strip, billets, forgings, ring, powder
Description: T5 is a tungsten high speed tool steel that exhibits an extremely high resistance to softening at elevated temperatures. It also exhibits unusually high red hardness.

Physical Properties

T5: Mean coefficients of linear thermal expansion

TEMP C	TEMP F	THEXP $10^{-6}/C$	THEXP $10^{-6}/F$
21-93	70-200	11.07	6.15
21-204	70-400	10.91	6.06
21-316	70-600	11.21	6.23
21-427	70-800	11.34	6.30
21-538	70-1000	11.81	6.56
21-649	70-1200	12.11	6.73
21-760	70-1400	12.46	6.92
21-816	70-1500	12.47	6.93

Source: Ref 29, Oct 1981

Mechanical Properties

T5: Hot hardness data obtained after treatment for maximum hardness before testing

TEMP C	TEMP F	HARD HB
21	70	745
595	1100	520
650	1200	400

Note: Mutual indentation method
Source: Ref 15, p 463

T5: Hot hardness data

TEMP C	TEMP F	HARD HB 538 C (1000 F)	HARD HB 593 C (1100 F)	HARD HB 649 C (1200 F)
21	70	712	680	590
93	200	705	665	570
149	300	697	660	557
204	400	690	654	548
260	500	680	643	537
316	600	670	632	523
371	700	657	617	506
427	800	640	600	485
482	900	629	577	459
538	1000	593	543	428
593	1100	...	490	381
649	1200	...	...	310

Note: Specimens were austenitized at 1285 C (2350 F), oil quenched, and double tempered at the indicated temperature
Source: Ref 29, Oct 1981

T6

UNS number: T12006
Specifications: ASTM A600, JIS G 4403 SKH4B, DGN B82, BS 4659 BT6, DIN 1.3257
Nominal composition: 0.750 to 0.850% C, 0.200 to 0.400% Mn, 0.200 to 0.400% Si, 4.000 to 4.750% Cr, 0.300% Ni max, 0.400 to 1.000% Mo, 18.500 to 21.000% W, 1.500 to 2.100% V, 11.000 to 13.000% Co
Product forms: Wrought, bar, rod, wire, plate, sheet, strip, billets, forgings, ring, powder
Description: T6 is a tungsten high speed tool steel that exhibits an extremely high resistance to softening at elevated temperatures.
Typical applications: Heavy duty lathe and planer tools, drills, checking tools, cutoff tools, milling cutters, and hobs.

Mechanical Properties

T6: Hot hardness data

TEMP C	TEMP F	HARD HRC
21	70	67.2(a)
540	1000	62.0
565	1050	61.4
595	1100	59.6

(a) Rockwell C hardness before testing
Source: Ref 15, p 464

T15

UNS number: T12015
Specifications: ASTM A600, DIN 1.3202, BS BT15, UNE F.5563
Nominal composition: 1.500 to 1.600% C, 0.150 to 0.400% Mn, 0.150 to 0.400% Si, 3.750 to 5.000% Cr, 0.300% Ni max, 1.000% Mo max, 11.750 to 13.000% W, 4.500 to 5.250% V, 4.750 to 5.250% Co
Product forms: Wrought, bar, rod, wire, plate, sheet, strip, billets, forgings, ring, powder
Description: T15 is a tungsten high speed tool steel that has an extremely high resistance to softening at elevated temperatures.
Typical applications: Form tools, lathe and planer tools, broaches, milling cutters, blanking dies, punches, and heavy duty tools requiring good wear resistance.

Physical Properties

T15: Average thermal conductivity

TEMP C	TEMP F	THCON W/m*K	THCON Btu/ft*h*F
100	200	20.9	12.1
200	500	24.1	13.9
400	750	25.4	14.7
500	1000	26.3	15.2

Source: Ref 3, p 442

Mechanical Properties

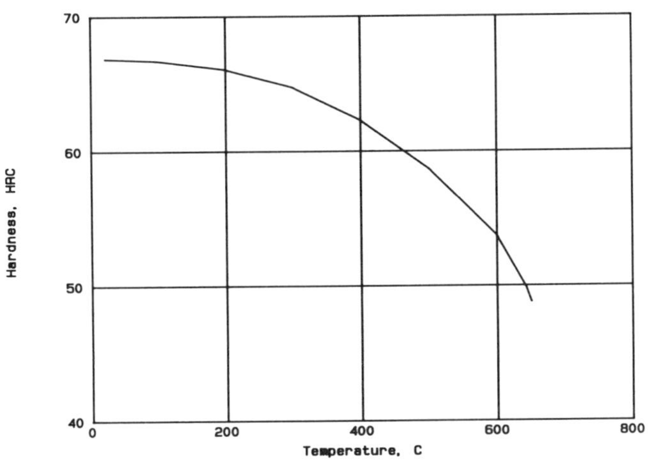

T15: Rockwell C hardness data. T15 exhibits high resistance to softening. For these tests, specimens were oil quenched from 1230 C and double tempered at 550 C for 2 h. After hot hardness testing at 650 C, specimens had a room-temperature hardness of 63.4 HRC. Source: Ref 3, p 442

W1

UNS number: T72301
Specifications: ASTM A686, SAE J437, SAE J438, ABNT W1k, TGL 4392 C80W2, NF 1101 Y120, MSZ 4354 S71, JIS G 4401 SK1, DGN B82, COPANT 337, PN 85020 N5, STAS OSC7, UNE F.510A, SIS 14 18 80, BS 4659 BW1A, GOST U7, DIN 1.1525, DIN 1.1545, DIN 1.1560
Nominal composition: 0.700 to 1.500% C, 0.100 to 0.400% Mn, 0.100 to 0.400% Si, 0.150% Cr max, 0.200% Ni max, 0.100% Mo max, 0.150% W max, 0.100% V max
Product forms: Wrought, bar, rod, wire, plate, sheet, strip, tube, billets, forgings, ring, cast, sand, investment, powder
Description: W1 is a water-hardening tool steel with low resistance to softening at elevated temperatures.
Typical applications: Low-carbon blacksmith tools, blanking tools, caulking tools, cold chisels, forging dies, rammers, rivet sets, shear blades, punches, and sledges. Medium carbon arbors, beading tools, blanking dies, reamers, bushings, cold heading dies, chisels, coining dies, counter-sinks, drills, forming dies, jeweler dies, mandrels, punches, shear blades, and woodworking tools. High-carbon cutters, jeweler dies, lathe tools, reamers, taps and dies, twist drills, and woodworking tools.

Physical Properties

W1: Thermal conductivity

TEMP C	TEMP F	THCON W/m*K	THCON Btu/ft*h*F
100	200	48.3	27.9
260	500	41.5	24.0
400	750	38.1	22.0
540	1000	34.6	20.0
675	1250	29.4	17.0
815	1500	24.2	14.0

Source: Ref 3, p 442

W1: Mean coefficients of linear thermal expansion

TEMP C	TEMP F	THEXP 10^{-6}/C	THEXP 10^{-6}/F
21-100	70-212	10.4	5.76
21-200	70-390	11.0	6.13
21-300	70-570	12.2	6.8
21-400	70-750	13.1	7.28
21-500	70-930	13.8	7.64
21-600	70-1110	14.2	7.9
21-700	70-1290	14.6	8.11

Source: Ref 15, p 432

8. Maraging Steels

Maraging Steels

The high-strength maraging steels are hardened by a reaction that does not involve carbon. Instead, they are strengthened by the precipitation at 480 C (900 F) of intermetallic compounds in a low-carbon martensite matrix. Indeed, carbon is considered an impurity in these alloys and is kept as low as commercially feasible (about 0.03% max). These alloys feature yield strengths near their tensile strengths and good machinability and formability in the annealed condition. Upon subsequent aging, however, strength and hardness increase with low attendant distortion.

Maraging steels contain about 18% Ni and varying amounts of other alloying elements such as cobalt, molybdenum, aluminum, and titanium. The most common grades are the four 18Ni steels: 18Ni(200), 18Ni(250), 18Ni(300), and 18Ni(350). The number in parentheses designates the yield strength in ksi. The most widely used is 18Ni(250).

The maraging grades can be used for prolonged service at temperatures up to 400 C (750 F); yield and tensile strengths at that temperature are about 80% of their room temperature values. Long-time stress-rupture failures can occur at 400 C (750 F), but rupture stresses are fairly high. At higher temperatures, reversion of the martensite matrix to austenite accelerates , and load-carrying capacity decays fairly rapidly.

Applications for the maraging steels are found in the aerospace industry, where their mechanical properties and weldability play key roles, and for tools and dies, where mechanical properties and fabricability (particularly low distortion during aging) are of primary importance. Specific examples include missile cases, aircraft forgings, cannon recoil springs, Belleville springs, bearings, transmission shafts, couplings, hydraulic hoses, bolts, and punches and dies.

18Ni(200)

UNS number: K92810
Specifications: ASTM A538/A538M (A), MIL S-46850, SPEC MIL S-47139
Nominal composition: 0.050 to 0.150% Al, 0.030% C max, 7.000 to 8.500% Co, 0.100% Mn max, 4.000 to 4.500% Mo, 17.000 to 19.000% Ni, 0.010% P max, 0.010% S max, 0.100% Si max, 0.100 to 0.250% Ti
Product forms: Wrought, bar, plate, sheet, forgings

Physical Properties

18Ni(200): Thermal conductivity

TEMP C	TEMP F	THCON W/m*K	THCON Btu/ft*h*F
21	70	19.6	11.3
50	120	20.1	11.6
100	212	21.2	12.1

Source: Ref 15, p 472

Mechanical Properties

18Ni(200): Mechanical property data

TEMP C	TEMP F	UTS MPa	UTS ksi	YS MPa	YS ksi	EL(a) %	RA %
315	600	1217	176.5	1141	165.5	12.5	60.0
425	800	1154	167.4	1059	153.6	14.0	61.0
480	900	1044	151.4	977	141.7	18.0	66.3
510	950	953	138.2	876	127.1	18.5	69.6
540	1000	840	121.9	743	107.7	24.0	73.2

Note: Specimens were solution annealed at 815 C (1500 F) for 1 h, air cooled and aged at 480 C (900 F) for 3 h
(a) In gage length of 4.5 times the square root of the area
Source: Ref 15, p 474

18Ni(200): Charpy V-notch impact energy of bar with small cross section, solution annealed at 815 C for 0.5 h, air cooled, and aged at 480 C for 3 h. Source: Ref 15, p 473

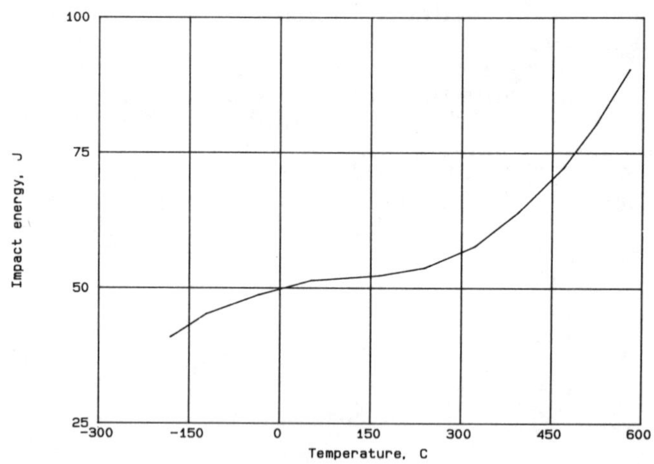

18Ni(250)

UNS number: K92890
Specifications: AMS 6512, AMS 6520, ASTM A538/A538M (B), SPEC MIL S-46850, SPEC MIL S-47139, SAE J1099 (A538B)
Nominal composition: 0.050 to 0.150% Al, 0.030% C max, 7.000 to 8.500% Co, 0.100% Mn max, 4.600 to 5.100% Mo, 17.000 to 19.000% Ni, 0.010% P max, 0.010% S max, 0.100% Si max, 0.300 to 0.500% Ti

Product forms: Wrought, bar, plate, sheet, strip, forgings

Physical Properties

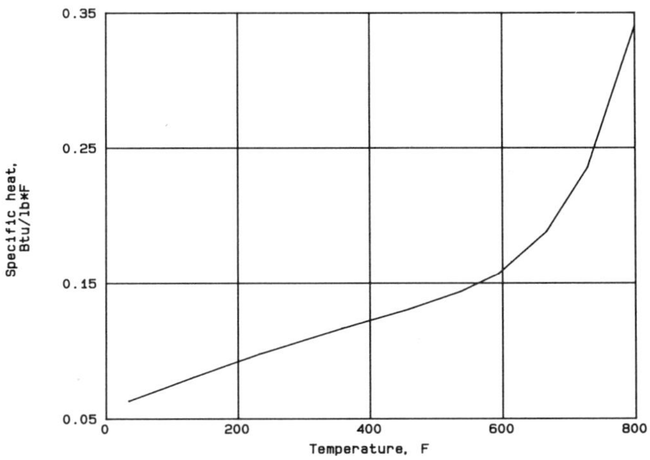

18Ni(250): Specific heat at constant pressure of 1 atm. Source: Ref 22, p 13

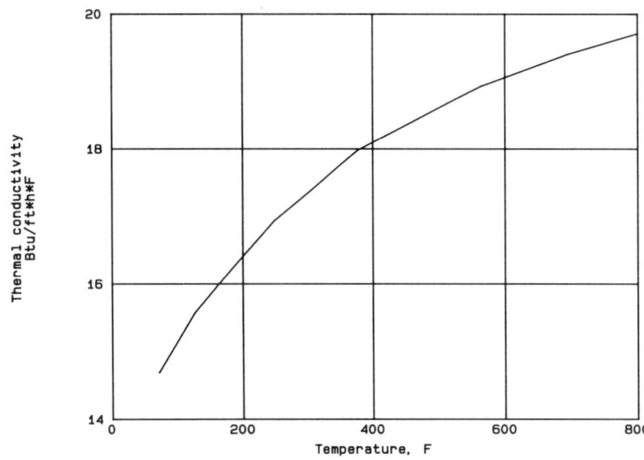

18Ni(250): Thermal conductivity. Source: Ref 22, p 13

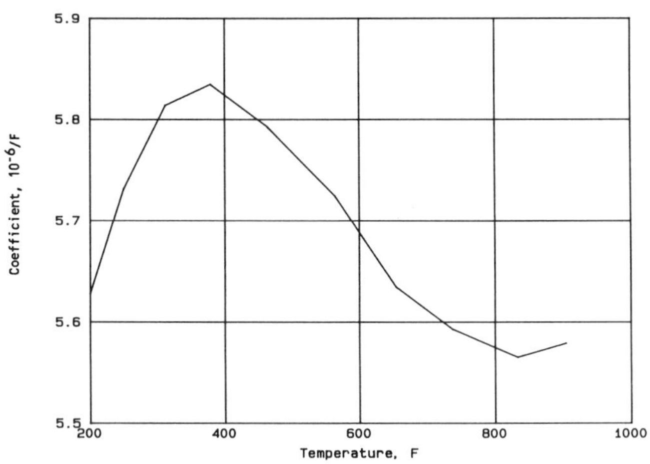

18Ni(250): Mean coefficient of linear thermal expansion from 75 F. Source: Ref 22, p 13

Mechanical Properties

18Ni(250): Tensile property data

TEMP C	TEMP F	UTS MPa	UTS ksi	YS MPa	YS ksi	EL(a) %	RA %
315	600	1609	233.4	1548	224.5	11.5	56.0
425	800	1524	221.0	1453	210.8	12.0	56.1
480	900	1379	200.0	1276	185.1	16.5	64.6
540	1000	1029	149.2	890	129.1	23.0	79.2

Note: Specimens were solution annealed at 815 C (1500 F) for 1 h, air cooled, and aged at 480 C (900 F) for 3 h
(a) In gage length of 4.5 times the square root of the area
Source: Ref 15, p 475

18Ni(250): Tensile property data for solution annealed and aged specimens

TEMP C	TEMP F	UTS MPa	UTS ksi	YS MPa	YS ksi	EL %	RA %
316	600	1765	256	1689	245	7	48
427	800	1655	240	1565	227	8	56
482	900	1448	210	1338	194	12	62
538	1000	1158	168	1055	153	13	66

Source: Ref 29, May 1965

18Ni(250): Effect of 1000-h elevated-temperature exposure on the tensile properties of sheet

TEST TEMP C	TEST TEMP F	Direction	UTS MPa	UTS ksi	YS MPa	YS ksi	EL %
Exposure temperature: 315 C (600 F)							
21	70	T	1965	285	1937	281	3
316	600	T	1675	243	1613	234	3
Exposure temperature: 425 C (800 F)							
21	70	L	1910	277	1868	271	3
		T	1979	287	1937	281	3
427	800	L	1503	218	1365	198	5
		T	1565	227	1427	207	5
Exposure temperature: 540 C (1000 F)							
21	70	T	1289	187	1213	176	8
538	1000	T	896	130	669	97	12

Note: Specimens were vacuum melted, 1.8-mm (0.07-in.) sheet, mill annealed at 815 C (1500 F), aged at 480 C (900 F) for 3 h. Room-temperature longitudinal properties: UTS 1820 MPa (264 ksi), YS 1785 MPa (259 ksi). Room-temperature transverse properties: UTS 1855 MPa (269 ksi), YS 1827 MPa (265 ksi). Nominal composition: Fe-18Ni-7.5 Co-5Mo-Ti-Al
Source: Ref 22, p 7

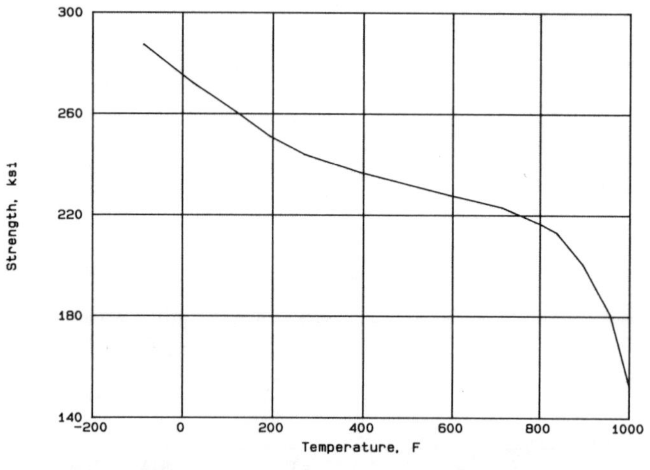

18Ni(250): Tensile strength of vacuum melted 0.07-in. sheet, mill annealed at 1500 F and aged at 900 F for 3 h. Exposure time at test temperature was 0.5 h. Both transverse and longitudinal sections were tested. Source: Ref 22, p 27

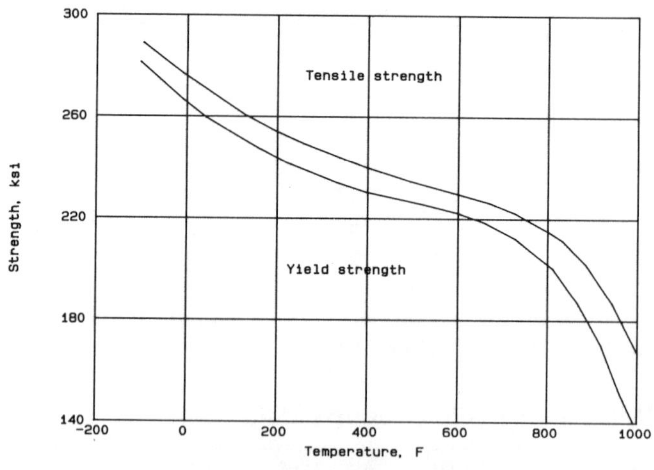

18Ni(250): Tensile and yield strength of 0.875-in. diam vacuum melted longitudinal bar, mill annealed at 1500 F and aged at 900 F for 3 h. Exposure time at test temperature was 0.5 h. Source: Ref 22, p 28

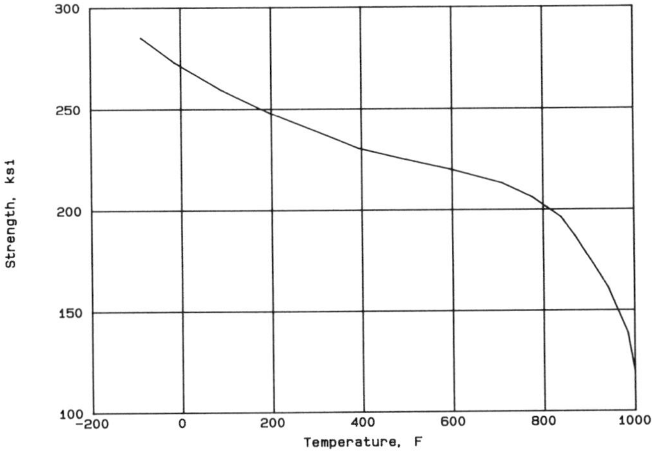

18Ni(250): Yield strength of vacuum melted 0.07-in. sheet, mill annealed at 1500 F and aged at 900 F for 3 h. Exposure time at test temperature was 0.5 h. Both transverse and longitudinal sections were tested. Source: Ref 22, p 27

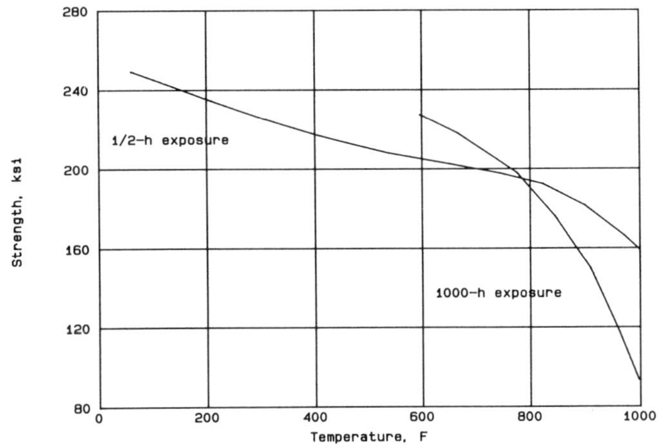

18Ni(250): Compressive yield strength of vacuum melted 0.07-in. sheet, mill annealed at 1500 F and aged at 900 F for 3 h. Transverse and longitudinal sections were tested. Source: Ref 22, p 29

18Ni(250): Effect of 1000-h exposure at elevated temperatures on the compressive properties of sheet

TEMP C	TEMP F	Direction	YSC MPa	YSC ksi
21	70	L	1781	255
		T	1697	243
316	600	T	1809	259
427	800	L	1774	254
		T	1760	252
538	1000	T	1132	162

Note: Specimens were vacuum melted 1.8-mm (0.07-in.) sheet, mill annealed at 815 C (1500 F) and aged at 480 C (900 F) for 3 h. Nominal composition: Fe-18-Ni-7.5Co-Ti-Al
Source: Ref 22, p 7

18Ni(250): Effect of 1000-h elevated-temperature exposure on the shear strength of sheet at room temperature

TEMP C	TEMP F	Direction	USS MPa	USS ksi
21	70	L	1069	155
		T	1089	158
317	600	T	1165	169
427	800	L	1207	175
		T	1172	170
538	1000	T	834	121

Note: Specimens were 1.8-mm (0.07-in.) vacuum melted sheet, mill annealed at 815 C (1500 F) and aged at 480 C (900 F) 3 h. Nominal composition: Fe-18Ni-7.5Co-5Mo-Ti-Al
Source: Ref 29, Sep 1970

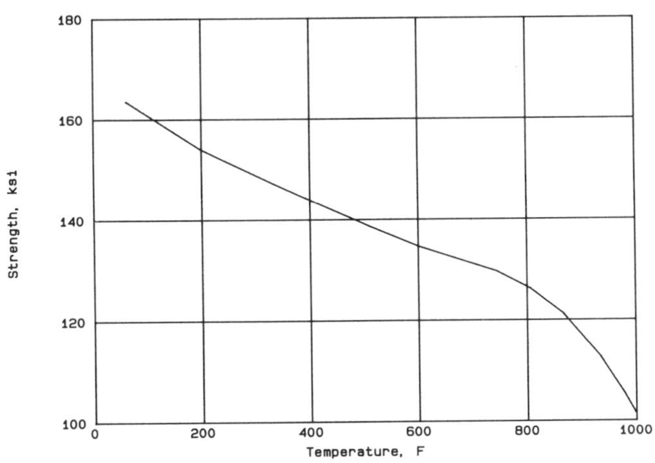

18Ni(250): Shear strength of vacuum melted 0.875-in. diam bar, mill annealed at 1500 F and aged at 900 F for 3 h. Exposure time at test temperature was 0.5 h. Source: Ref 22, p 32

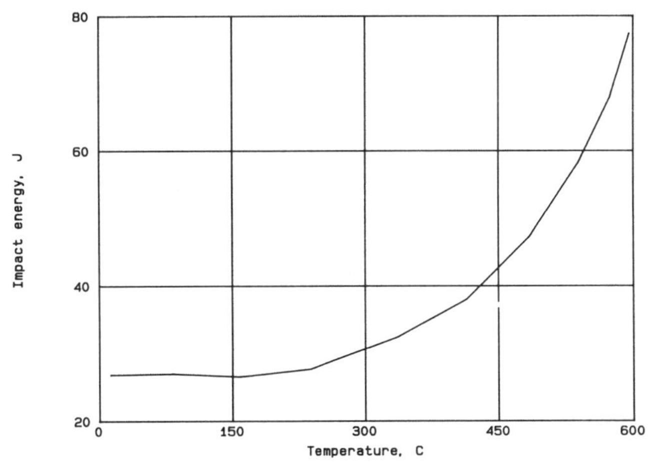

18Ni(250): Charpy V-notch impact energy for bar of small cross section, solution annealed at 815 C for 0.5 h, air cooled, and aged at 480 C for 3 h. Source: Ref 15, p 475

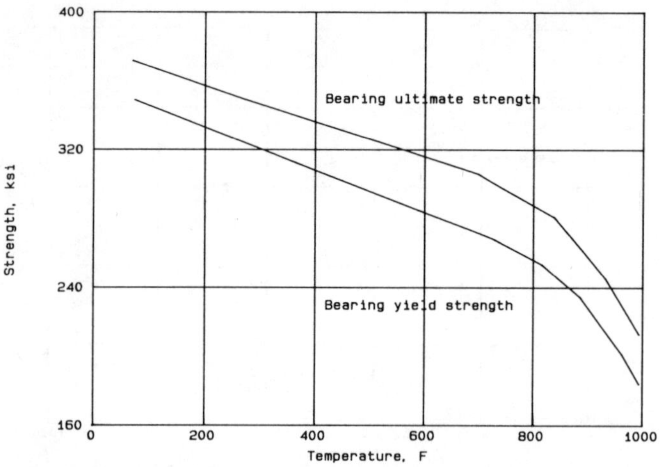

18Ni(250): Bearing ultimate and yield strength (e/D = 1.5) of vacuum melted 0.25-in. plate, tested in transverse section, mill annealed at 1500 F and aged at 900 F for 3 h. Exposure time at test temperature was 0.5 h. Source: Ref 22, p 32

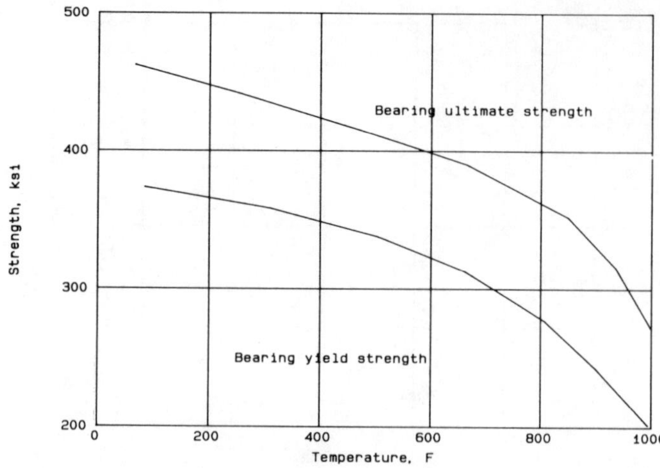

18Ni(250): Bearing ultimate and yield strength (e/D = 2) of vacuum melted 0.25-in. plate, tested in transverse section, mill annealed at 1500 F and aged at 900 F for 3 h. Exposure time at test temperature was 0.5 h. Source: Ref 22, p 32

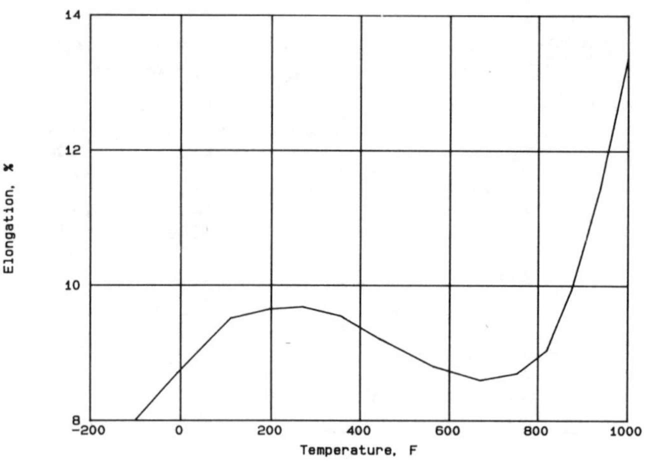

18Ni(250): Tensile elongation of 0.875-in. diam vacuum melted longitudinal bar, mill annealed at 1500 F and aged at 900 F for 3 h. Exposure time at test temperature was 0.5 h. Source: Ref 22, p 28

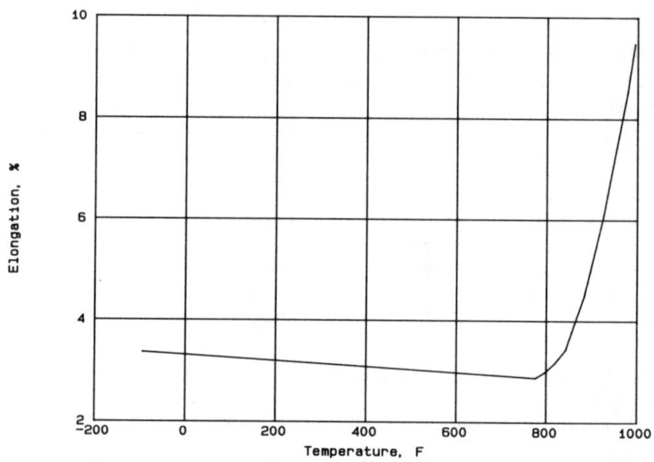

18Ni(250): Tensile elongation of vacuum melted 0.07-in. sheet, mill annealed at 1500 F and aged at 900 F for 3 h. Exposure time at test temperature was 0.5 h. Both transverse and longitudinal sections were tested. Source: Ref 22, p 27

18Ni(250): Creep property data for 1.8-mm (0.07-in.) sheet

TIME h	0.05% CRSTR MPa	0.05% CRSTR ksi	0.10% CRSTR MPa	0.10% CRSTR ksi	0.50% CRSTR MPa	0.50% CRSTR ksi	1.0% CRSTR MPa	1.0% CRSTR ksi
At 315 C (600 F)								
10	1413	205	1482	215	1606	233	1627	236
100	1344	195	1413	205	1586	230	1606	233
1000	...	...	1344	195	...	...	1551	225
At 425 C (800 F)								
10	793	115	896	130	1138	165	1207	175
100	...	...	724	105	965	140	1034	150
1000	...	...	...	...	758	110	793	115

Note: Specimens were mill annealed at 815 C (1500 F) and aged at 480 C (900 F) for 3 h. YS: 1827 MPa (265 ksi). Nominal composition: Fe-18Ni-7.5 Co-5Mo-Ti-Al
Source: Ref 22, p 8

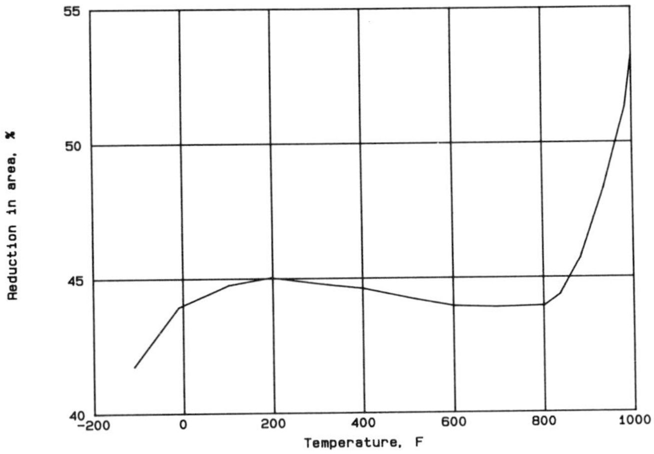

18Ni(250): Reduction in area of 0.875-in. diam vacuum melted longitudinal bar, mill annealed at 1500 F and aged at 900 F for 3 h. Exposure time at test temperature was 0.5 h. Source: Ref 22, p 28

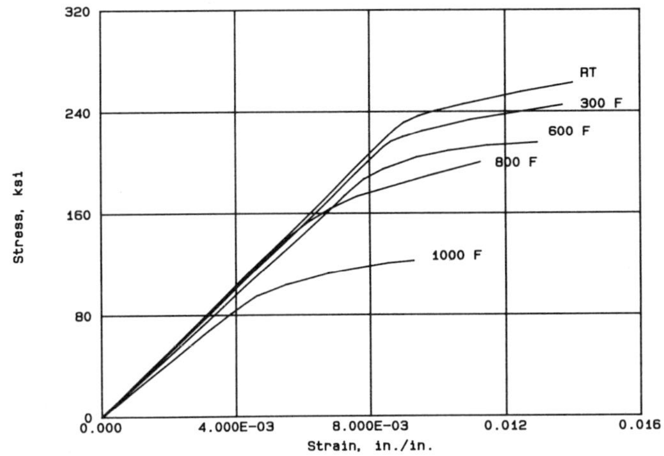

18Ni(250): Stress-strain curves in tension for vacuum melted bar, mill annealed at 1500 F and aged at 900 F for 3 h. Exposure time at test temperature was 0.5 h. Source: Ref 22, p 26

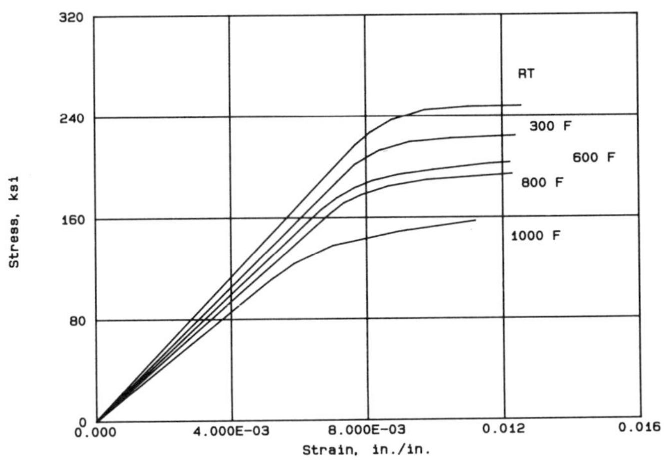

18Ni(250): Stress-strain curves in compression for vacuum melted bar, mill annealed at 1500 F and aged at 900 F for 3 h. Exposure time at test temperature was 0.5 h. Source: Ref 22, p 29

18Ni(250): Creep-rupture property data for air-melted bar

RPT h	RPSTR MPa	RPSTR ksi	EL %	RA %
38	1207	175	13	51
561	1034	150	13	59
7	1034	150	17	63
38	862	125	24	70
5	689	100	32	71
48	517	75	31	80

Note: Specimens were held at 815 C (1500 F) for 1 h, held at 480 C (900 F) for 3 h. Nominal composition: Fe-18 Ni-7.5Co-5Mo-Ti-Al
Source: Ref 29, Sep 1970

18Ni(250): Notch tensile strength of vacuum-melted bar

TEMP C	TEMP F	YS MPa	YS ksi	NTS(a) MPa	NTS(a) ksi
21	70	1765	256	1565	227
316	600	1538	223	1393	202
427	800	1407	204	1269	184

Note: Test specimens were 40-mm (1.5-in.) diameter bar tested in the longitudinal mode. Condition: mill annealed 480 C (900 F), 3 h. Nominal composition: Fe-18Ni-7.5Co-5Mo-Ti-Al
(a) 22-mm (7/8-in.) diameter specimen
Source: Ref 29, Sep 1970

18Ni(250): Hot hardness data

TEMP C	TEMP F	HARD HB
316	600	418
343	650	418
371	700	418
399	750	401
427	800	370
454	850	370
482	900	356
510	950	343
538	1000	318
566	1050	287
593	1100	250

Note: Specimens were solution treated at 815 C (1500 F) for 30 min, aged at 510 C (950 F) for 2 h. Specimens were 30 min at heat before testing
Source: Ref 29, May 1965

18Ni(300)

UNS number: K93120
Specifications: ASM 6514, ASM 6521, ASTM A538/538M (C), SPEC MIL S-46850, SPEC MIL S-47139, SAW J1099 (A538C)
Nominal composition: 0.050 to 0.150% Al, 0.030% C max, 8.000 to 9.500% Co, 0.100% Mn max, 4.600 to 5.200% Mo, 17.000 to 19.000% Ni, 0.010% P max, 0.010% S max, 0.100% Si max, 0.550 to 0.800% Ti

Product forms: Wrought, bar, plate, sheet, forgings

Physical Properties

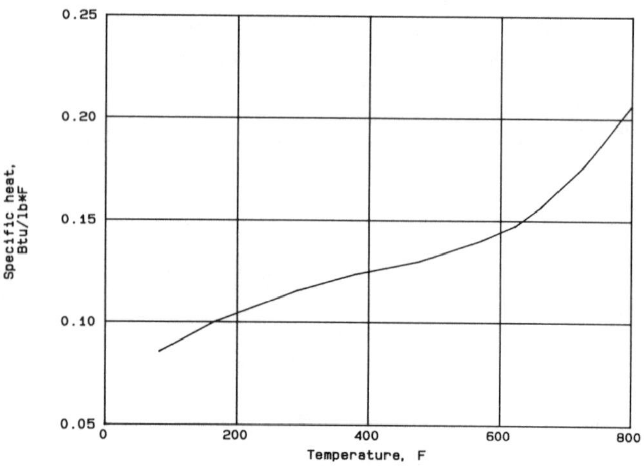

18Ni(300): Specific heat at 1 atm constant pressure. Source: Ref 22, p 14

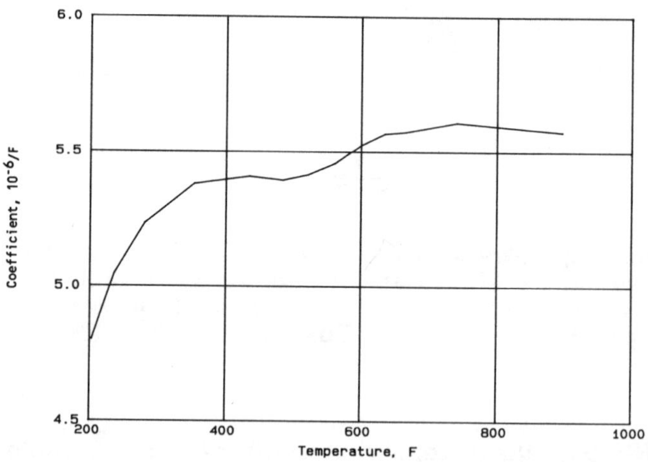

18Ni(300): Mean coefficient of linear thermal expansion from 75 F. Source: Ref 22, p 14

Mechanical Properties

18Ni(300): Tensile property data

TEMP C	TEMP F	UTS MPa	UTS ksi	YS MPa	YS ksi	EL(a) %	RA %
315	600	1771	256.8	1693	245.6	12.0	61.8
425	800	1655	240.1	1570	227.7	14.0	61.3
480	900	1454	210.9	1343	194.8	17.3	68.4
510	950	1304	189.1	1192	172.9	22.0	76.5
540	1000	1158	168	1056	153.2	24.0	77.2

(a) In gage length of 4.5 times the square root of the area
Source: Ref 15, p 477

18Ni(300): Effect of 1000-h elevated-temperature exposure on the room-temperature tensile properties of sheet

TEMP C	TEMP F	Direction	UTS MPa	UTS ksi	YS MPa	YS ksi	EL %
21	70	L	1931	280	1875	272	4.0
		T	1958	284	1875	272	4.0
315	600	T	2020	293	1972	286	4.0
430	800	L	2137	310	2082	302	4.0
		T	2144	311	2103	305	3.0
540	1000	T	1400	203	1234	179	9.0

Note: Specimens were 1.8-mm (0.07-in.) vacuum melted sheet, mill annealed 815 C (1500 F) and aged at 480 C (900 F), 3 h. Nominal composition: Fe-18Ni-9Co-5Mo-Ti-Al
Source: Ref 29, Sep 1968

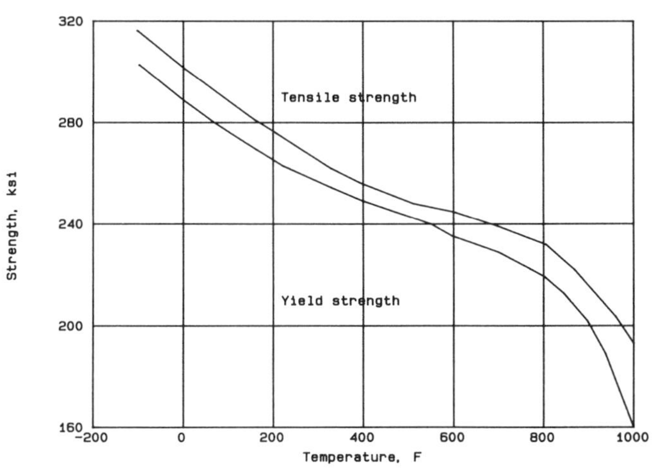

18Ni(300): Tensile and yield strength of longitudinal vacuum melted 0.875-in. bar, mill annealed at 1500 F and aged at 900 F for 3 h. Exposure time at test temperature was 0.5 h. Source: Ref 22, p 29

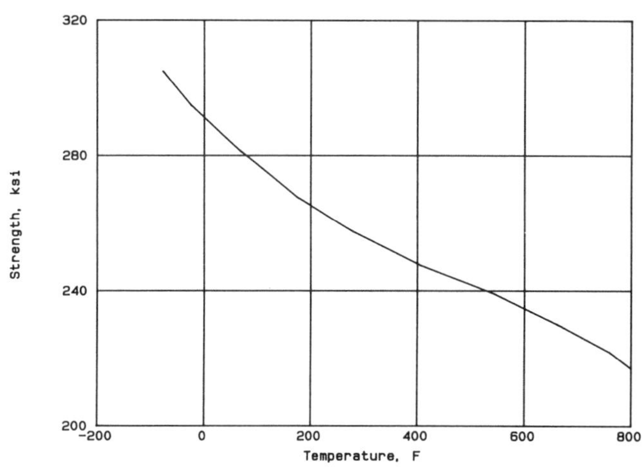

18Ni(300): Yield strength of vacuum melted longitudinal 1.5-in. bar, mill annealed at 1500 F and aged at 900 F for 3 h. Specimens were notched and precracked after annealing. Source: Ref 22, p 36

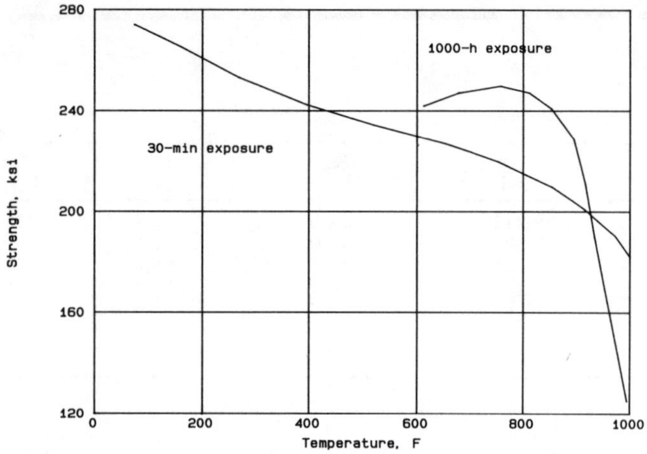

18Ni(300): Compressive strength of vacuum melted 0.07-in. sheet, mill annealed at 1500 F and aged at 900 F for 3 h. Transverse and longitudinal sections were tested. Source: Ref 22, p 32

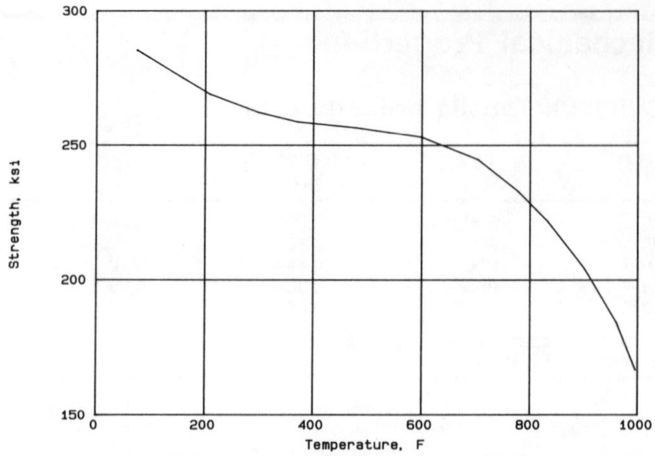

18Ni(300): Compressive yield strength of transverse vacuum melted 0.25-in. bar, mill annealed at 1500 F and aged at 900 F for 3 h. Exposure time was 0.5 h. Source: Ref 22, p 32

18Ni(300): Effect of 1000-h elevated-temperature exposure on the compressive properties of sheet at room temperature

TEMP C	TEMP F	Direction	YSC MPa	YSC ksi
21	70	L	1875	272
		T	1903	276
315	600	T	1917	278
427	800	L	2062	299
		T	2179	316
540	1000	T	1200	174

Note: Specimens were 1.8-mm (0.07-in.) vacuum melted sheet, mill annealed at 815 C (1500 F) and aged at 480 C (900 F), 3 h.
Nominal composition: Fe-18Ni-9Co-5Mo-Ti-Al
Source: Ref 29, Sep 1969

18Ni(300): Effect of 1000-h elevated temperature exposure on the ultimate shear strength of sheet at room temperature

TEMP C	TEMP F	Direction	USS MPa	USS ksi
21	70	L	1186	172
		T	1207	175
315	600	T	1179	171
427	800	L	1282	186
		T	1289	187
540	1000	T	855	124

Note: Specimens were 1.8-mm (0.07-in.) vacuum melted sheet, mill annealed at 815 C (1500 F) and aged at 480 C (900 F) 3 h.
Nominal composition: Fe-18Ni-9Co-5Mo-Ti-Al
Source: Ref 29, Sep 1968

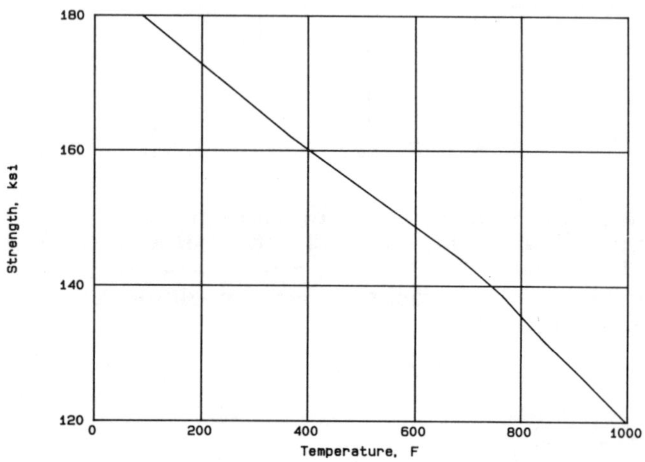

18Ni(300): Shear strength of longitudinal vacuum melted 0.875-in.diam bar, mill annealed at 1500 F and aged at 900 F for 3 h. Source: Ref 22, p 33

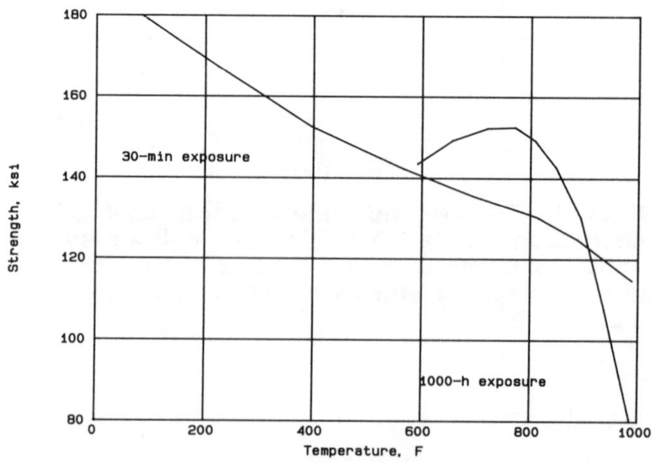

18Ni(300): Shear strength of vacuum melted 0.07-in. sheet, mill annealed at 1500 F and aged at 900 F for 3 h. Transverse and longitudinal sections were tested. Source: Ref 22, p 33

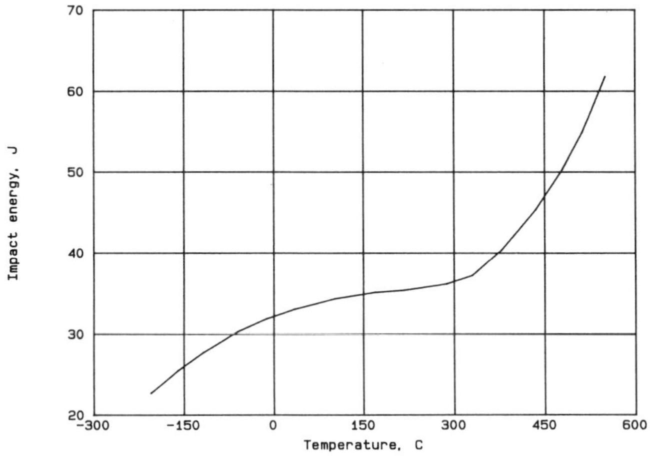

18Ni(300): Charpy V-notch impact energy of bar with a small cross section, solution annealed at 815 C for 0.5 h, air cooled, and aged at 480 C for 3 h. Source: Ref 15, p 477

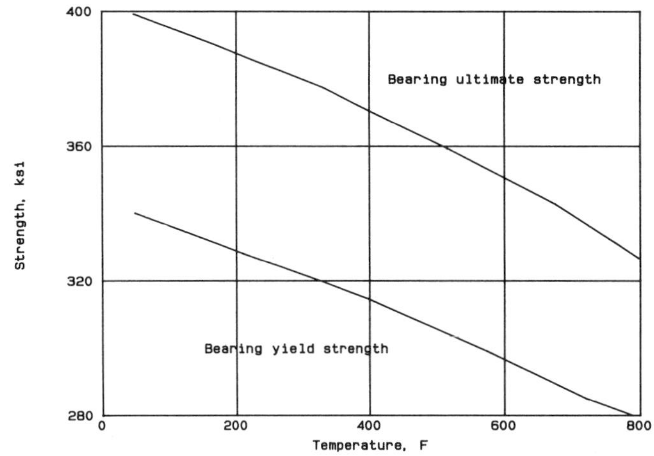

18Ni(300): Bearing ultimate strength and yield strength of bar. e/D = 1.5. Source: Ref 22, p 34

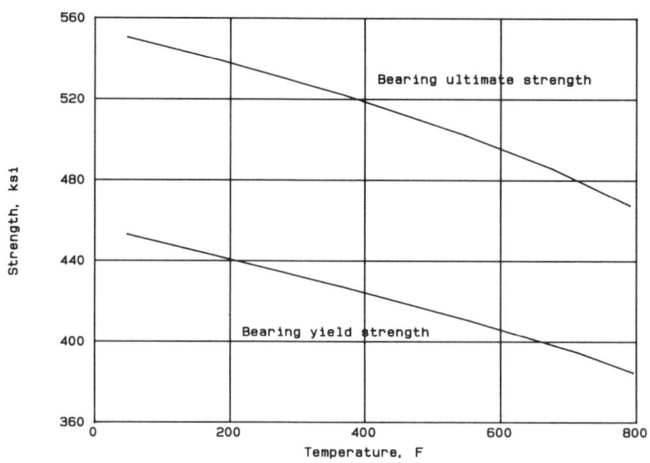

18Ni(300): Bearing ultimate strength and yield strength of bar. e/D = 2.0. Source: Ref 22, p 34

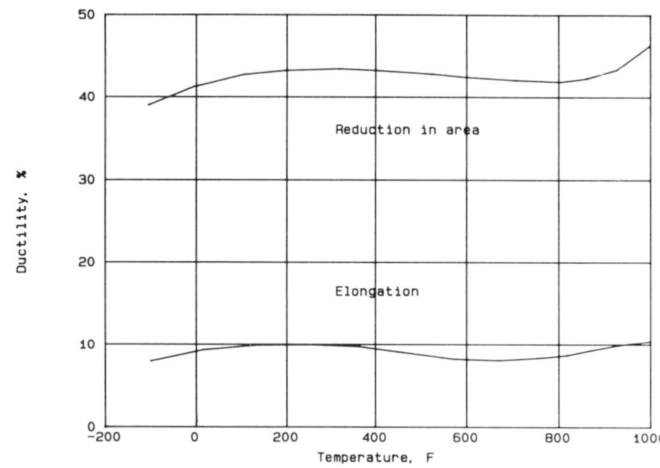

18Ni(300): Tensile elongation and reduction in area of longitudinal vacuum melted 0.875-in. bar, mill annealed at 1500 F and aged at 900 F for 3 h. Exposure time at test temperature was 0.5 h. Source: Ref 22, p 29

18Ni(300): Creep property data for sheet

TIME h	0.05% RPSTR MPa	0.05% RPSTR ksi	0.10% RPSTR MPa	0.10% RPSTR ksi	0.50% RPSTR MPa	0.50% RPSTR ksi	1.0% RPSTR MPa	1.0% RPSTR ksi
At 315 C (600 F)								
10	1586	230	1655	240	...	...	...	...
100	1517	220	1586	230	...	...	...	...
1000	1448	210	1517	220	...	...	...	...
At 425 C (800 F)								
10	896	130	1069	155	1379	200	1448	210
100	...	...	862	125	1138	165	1344	195
1000	...	...	655	95	862	125	931	135

Note: Specimens were 1.8-mm (0.07-in.) vacuum melted sheet mill annealed 815 C (1500 F) and aged at 480 C (900 F) 3 h. Nominal composition: Fe-18Ni-9Co-5Mo-Ti-Al
Source: Ref 29, Sep 1969

18Ni(300): Creep property data for bar

TIME h	0.10% RPSTR MPa	0.10% RPSTR ksi	0.20% RPSTR MPa	0.20% RPSTR ksi	1.0% RPSTR MPa	1.0% RPSTR ksi
At 345 C (650 F)						
10	1517	220	...	...	...	...
100	1407	204	...	...	...	...
1000	1276	185	...	...	...	...
At 425 C (800 F)						
10	1034	150	1207	175	...	...
100	758	110	931	135	1241	180
1000	...	...	655	95	758	110

Note: Specimens were heat treated at 1500 F for 1 h, air cooled, heated to 900 F, and held for 3 h. Yield strength was 272 ksi.
Nominal composition: Fe-18Ni-9Co-5Mo-Ti-Al
Source: Ref 29, Sep 1969

18Ni(300): Creep strength data for vacuum melted forgings

Direction	CRSTR MPa	CRSTR ksi	CRSTN 100 h %
At 425 C (800 F)			
L	896	130	1
T	896	130	0
At 480 C (900 F)			
L	689	100	4
T	689	100	3

Note: Specimens were heat treated at 1500 F for 1 h, water quenched, heated to 900 F, and held 3 h. Nominal composition: Fe-18Ni-9Co-5Mo-Ti-Al
Source: Ref 29, Sep 1969

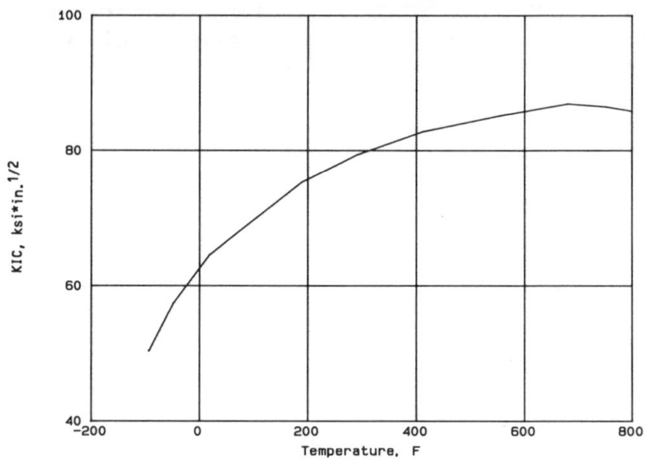

18Ni(300): Fracture toughness of longitudinal 1.5-in. vacuum melted bar, mill annealed at 1500 F and aged at 900 F for 3 h. Specimens were notched and precracked after annealing. Source: Ref 22, p 31

18Ni(350)

Nominal composition: 0.050 to 0.150% Al, 0.030% C max, 11.500 to 13.000% Co, 0.100% Mn max, 4.000 to 5.000% Mo, 17.000 to 19.000% Ni, 0.010% P max, 0.010% S max, 0.100% Si max, 1.200 to 1.800% Ti max

Product forms: Wrought, bar, plate, sheet, forgings

Physical Properties

18Ni(350): Thermal conductivity

TEMP C	TEMP F	THCON W/m*K	THCON Btu/ft*h*F
21	70	25.5	14.6
50	120	25.8	14.9
100	210	27.0	15.6

Source: Ref 15, p 479

18Ni(350): Mean coefficient of linear thermal expansion

TEMP C	TEMP F	THEXP 10^{-6}/C	THEXP 10^{-6}/F
21-480	70-900	11.3	6.3

Source: Ref 15, p 479

Mechanical Properties

18Ni(350): Tensile property data

TEMP C	TEMP F	UTS MPa	UTS ksi	YS MPa	YS ksi	EL(a) %	RA %
315	600	2139	310.2	2037	295.4	12.3	54.9
425	800	1988	288.4	1912	277.3	15.6	57.6
480	900	1864	270.4	1737	251.9	17.4	60.3
540	1000	1611	233.6	1460	211.8	20	70.9

Note: Specimens were solution annealed at 815 C (1500 F) for 1 h, air cooled, and aged at 480 C (900 F) for 6 h
(a) In gage length of 4.5 times the square root of the area
Source: Ref 15, p 481

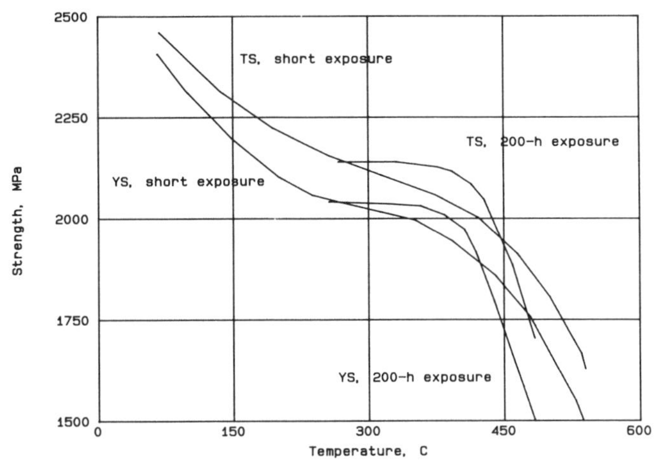

18Ni(350): Tensile and yield strength at two exposure times for bar of small cross section, solution annealed at 800 C for 1 h, air cooled, and aged at 510 C for 3 h. Source: Ref 15, p 480

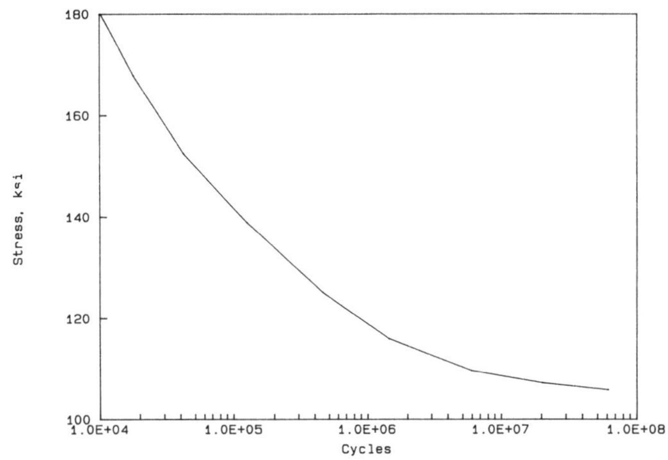

18Ni(350): Charpy V-notch impact strength of bar of small cross section, solution annealed at 1500 F for 0.5 h, air cooled, and aged at 900 F for 3 h. Source: Ref 15, p 480

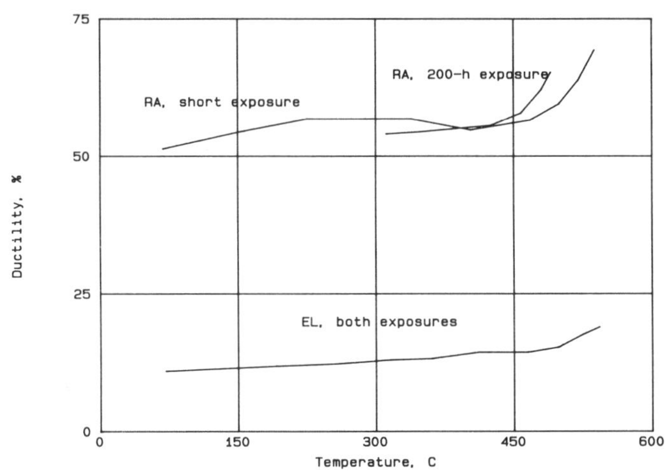

18Ni(350): Reduction in area and tensile elongation at two exposure times for bar of small cross section, solution annealed at 800 C for 1 h, air cooled, and aged at 510 C for 3 h. Elongation in gage length of 4.5 times the square root of the area. Source: Ref 15, p 480

9. Wrought Stainless Steels

Wrought Stainless Steels

Many stainless steels, particularly certain austenitic and precipitation-hardening types, are used extensively for elevated-temperature applications. Strength and hardness of austenitic stainlesses can be raised by cold working. Precipitation-hardening and martensitic stainlesses respond to heat treatment. Ferritic types can be strengthened somewhat by cold working, and certain ferritic steels can be hardened slightly by heat treating. Duplex (austenitic-ferritic) stainless steels can be hardened by cold working.

Ferritic Stainless Steels. Most of these steels are completely ferritic at all temperatures. However, AISI 405, 409, 430, 434, and 446 may form a certain amount of austenite when heated to high temperatures. Chromium in amounts ranging from 11 to 29% is added for corrosion and oxidation resistance. Molybdenum also increases corrosion resistance, and aluminum and silicon enhance oxidation resistance. Titanium is used to tie up carbon and nitrogen for structure control and resistance to intergranular corrosion.

Room-temperature yield strength of ferritic stainlesses is nearly equivalent to that of austenitic steels. At higher temperatures, the austenitic grades have a decided edge. High-temperature creep and rupture strengths of ferritic types are also relatively low compared with those of austenitic stainless steels.

Oxidation resistance of the ferritic steels, however, is comparable to that of the austenitic grades. Type 430 (16% Cr), for instance, can be used without excessive scaling to 840 C (1550 F). Type 446 (25% Cr) can handle temperatures up to 1090 C (2000 F). Thus, stress conditions permitting, a ferritic stainless steel can be the material of choice for a high-temperature application because of its lower alloy content and lower cost. Other factors, such as component repairability and tolerance for long-term thermal embrittlement, also must be considered however.

All ferritic grades with chromium in excess of 13% are susceptible to precipitation of alpha prime, which is a chromium-rich phase, on exposure to temperatures of 370 to 540 C (700 to 1000 F). This "475 C (885 F) embrittlement" results in an increase in hardness and a drastic drop in room-temperature toughness. Embrittling sigma phase is also encountered in certain higher chromium alloys at temperatures above 565 C (1050 F), and chi phase will form in certain ferritic stainless steels with more than 0.5% Ti.

Type 409 is used for automotive exhaust systems, catalytic converters, and exhaust ducting and silencers in gas turbines; type 405 is used in stationary vanes and spacers in steam turbines and in various furnace components. Types 430 and 439 typically are used for heat exchangers, hot water tanks, condensers, and furnace parts, and type 446 is specified for industrial oven applications, blowers, exhaust systems, annealing boxes, kiln liners, and pyrometer tubes.

Martensitic Stainless Steels. Quenched and tempered martensitic stainless steels offer usable strength up to 590 C (1100 F) and relatively good corrosion resistance. They can be grouped according to increasing strength and heat resistance:

- Lowest strength and heat resistance: types 403, 410, and 416
- Greek Ascoly and type 431
- Highest strength and heat resistance: type 422

Type 403, for example, is used extensively for steam turbine rotor blades and gas turbine compressor blades operating at up to 480 C (895 F). Type 410 is selected for steam valves, pump shafts, and bolts needing corrosion resistance and moderate strength up to 540 C (1000 F). At 650 C (1200 F), the tensile strength of type 422 is equal to that of type 403 at 590 C (1100 F). Its 540 C (1000 F) rupture strength is also considerably higher than those of other quenched and tempered martensitic stainless steels.

Precipitation-Hardening Stainless Steels. Precipitation-hardening martensitic (maraging) stainless steels (12 to 16% Cr) are intermediate in heat resistance between the chromium-free 18% Ni maraging steels and the 12% Cr, low-nickel, quenched and tempered martensitic stainless steels such as type 422. Steels included in this category include Custom 450, 17-4 PH, and 15-5 PH (lower strength) and Custom 455 and PH 13-8 Mo (higher strength). Typically, tensile and yield strengths drop rapidly above 425 C (800 F), and tensile elongation is higher than 10% at all temperatures from ambient to 540 C (1000 F). A wide variety of useful properties can be produced in a given steel by varying the aging temperature.

The precipitation-hardening martensitic stainless steels are used where corrosion resistance and high mechanical properties are needed at temperatures up to 425 C (800 F). Examples include valve parts, ball bearings, turbine blades, mandrels, conveyor chains, and a variety of aircraft mechanical and structural components.

Precipitation-hardening semiaustenitic stainlesses are modifications of standard 18Cr-8Ni austenitic grades. Nickel contents are lower, and elements such as aluminum, copper, molybdenum, and niobium are added. As with the precipitation-hardening martensitic steels, different hardening heat treatments may produce a variety of properties in the same alloy.

These precipitation-hardening steels are also used at temperatures up to 425 C (800 F). Typical applications include landing gear hooks, poppet valves, fuel tanks, hydraulic lines and fittings, compressor casings, and aircraft structural components.

Austenitic stainless steels contain 16 to 25% Cr and up to 20% Ni. Some types have as much as 18% Mn. The austenitic grades are not hardenable by heat treatment, but can be strengthened by cold work. The effect of cold work on elevated-temperature mechanical properties depends on the recrystallization temperature, the amount of residual stress resulting from the cold work, and the duration of thermal exposure.

Austenitic stainless steels are noted for high strength and exceptional toughness, ductility, and formability. As a class, they are more corrosion and oxidation resistant than martensitic or ferritic steels, and they combine excellent strength with oxidation resistance at elevated temperatures.

Room-temperature tensile properties of annealed 300 series alloys are similar. At higher temperatures, 425 and 650 C (800 and 1200 F), types 321, 347, and 309 have yield strengths higher than those of types 304, 310, and 316. Types 309, 310, and 316 have the highest tensile strengths at 650 C (1200 F).

Type 316 has the highest stress-rupture capability of all the 300 series alloys. Type 347 is somewhat weaker. Types 304, 309, and 310 are inferior to both types 316 and 347 in stress-rupture capability.

The Cr-Ni-Mn alloys generally have higher stress-rupture capabilities than 300 series stainlesses. However, type 316 bests Nitronic 40, and both types 316 and 347 are stronger than Nitronic 60.

Austenitic stainless steels are cooled rapidly from the solution heat treating temperature of 955 to 1180 C (1750 to 2150 F). Carbides dissolved in the matrix may subsequently precipitate at grain boundaries when the alloy is exposed to temperatures from 425 to 870 C (800 to 1600 F). This "sensitizes" the steel to intergranular corrosion by creating chromium-depleted grain boundary regions. Sensitization (precipitation of chromium carbides) can be controlled by reducing carbon content (as in types 304L and 316L) or by adding stronger carbide formers such as titanium and niobium (as in types 321 and 347).

Furnace parts, heat exchanger tubing, steam lines, exhaust systems in reciprocating engines and gas turbines, and afterburner parts are typical of the many components that take advantage of the heat-resisting capabilities of austenitic stainless steels.

Type 304 has good atmospheric corrosion and oxidation resistance. Types 309 and 310 have properties superior to those for type 304 due to higher chromium and nickel contents. Type 310 is preferred where intermittent heating and cooling are encountered, because its oxide scale is more adherent. Both types 309 and 310 are used for firebox sheets, furnace linings, boiler baffles, and thermocouple wells.

Stabilized types 321 and 347 are used primarily where solution treatment after welding is not practical. Steam lines, superheater tubes, and exhaust systems that operate in the sensitization range of 425 to 870 C (800 to 1600 F) are typical examples. Types 304L and 316L have similar applications, but they are more susceptible to intergranular attack during long exposures to high temperatures.

Duplex Stainless Steels. In general, duplex stainless steels are tougher than the ferritic grades and fill the gap between the ferritic and austenitic grades in terms of resistance to chloride stress-corrosion cracking. Duplex grades containing nitrogen can be cold worked to strengths that are not obtainable in ferritic and most austenitic types. Compared with conventional duplex stainlesses, their as-welded corrosion resistance is improved because of reduced segregation of alloying elements between the austenite and ferrite phases.

Mechanical Properties

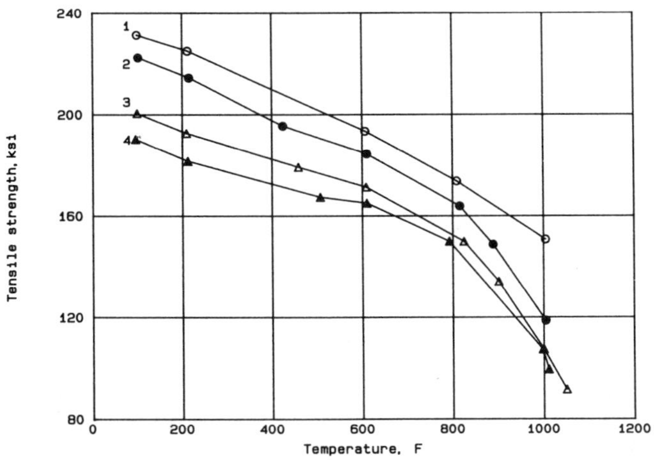

Stainless steels: Tensile strength. Curve 1: Custom 455, heat treated at 1500 F for 0.5 h, water quenched, held at 950 F for 4 h, and air cooled. Curve 2: PH 13-8 Mo, oil quenched from 1700 F, held at 1000 F for 4 h, and air cooled. Curve 3: 17-4 PH, heat treated at 1900 F for 0.5 h, oil quenched, held at 900 F for 4 h, and air cooled. Curve 4: Custom 450, heat treated at 1900 F for 1 h, water quenched, held at 900 F for 4 h, and air cooled. Source: Ref 3, p 200

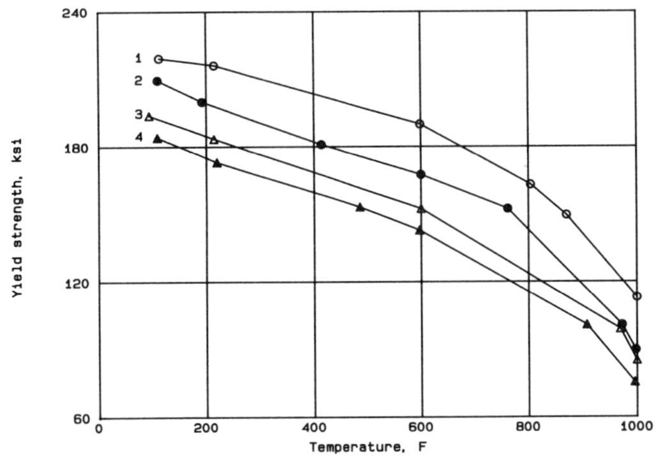

Stainless steels: Yield strength. Curve 1: Custom 455, heat treated at 1500 F for 0.5 h, water quenched, held at 950 F for 4 h, and air cooled. Curve 2: PH 13-8 Mo, oil quenched from 1700 F, held at 1000 F for 4 h, and air cooled. Curve 3: 17-4 PH, heat treated at 1900 F for 0.5 h, oil quenched, held at 900 F for 4 h, and air cooled. Curve 4: Custom 450, heat treated at 1900 F for 1 h, water quenched, held at 900 F for 4 h, and air cooled. Source: Ref 3, p 200

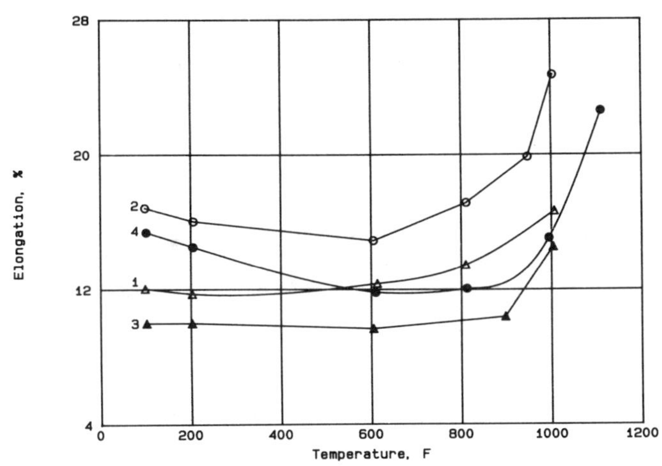

Stainless steels: Tensile elongation. Curve 1: Custom 455, heat treated at 1500 F for 0.5 h, water quenched, held at 950 F for 4 h, and air cooled. Curve 2: PH 13-8 Mo, oil quenched from 1700 F, held at 1000 F for 4 h, and air cooled. Curve 3: 17-4 PH, heat treated at 1900 F for 0.5 h, oil quenched, held at 900 F for 4 h, and air cooled. Curve 4: Custom 450, heat treated at 1900 F for 1 h, water quenched, held at 900 F for 4 h, and air cooled. Source: Ref 3, p 200

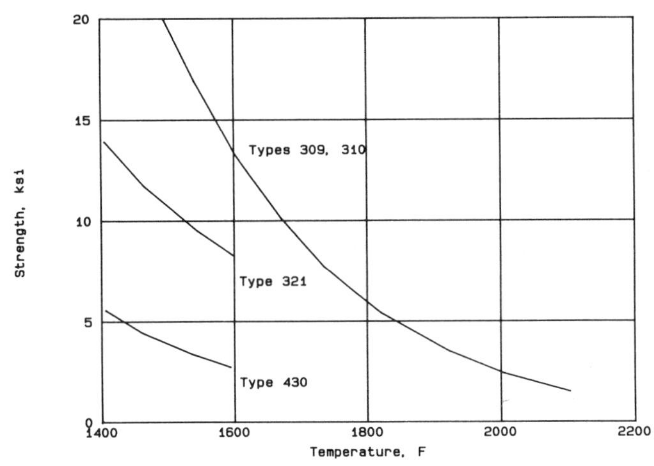

Stainless steels: Yield strength (0.2% offset) of annealed specimens. Source: Ref 21, p 7

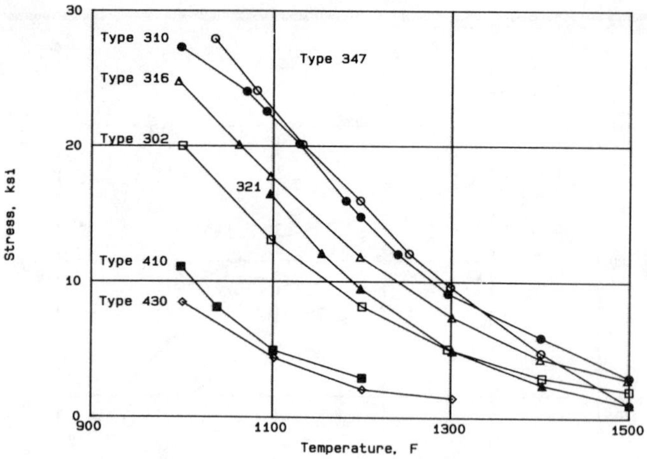

Stainless steels: Stress for secondary creep rate of 0.0001%/h (extrapolated data). Source: Ref 21, p 9

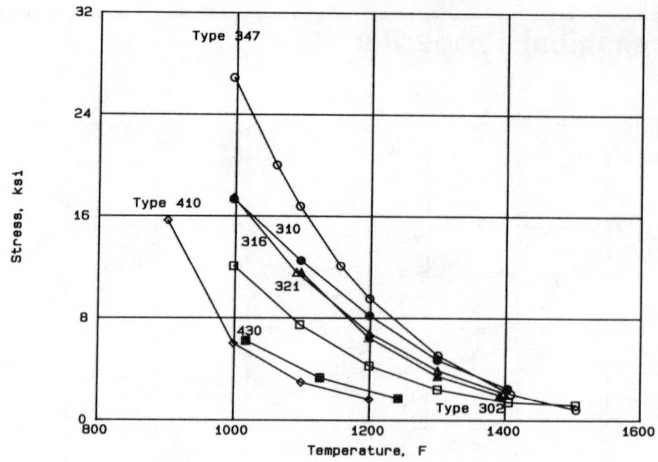

Stainless steels: Stress for secondary creep rate of 0.00001%/h (extrapolated data). Source: Ref 21, p 9

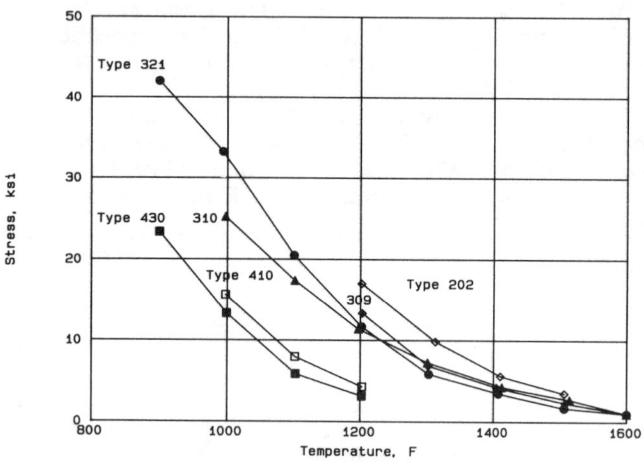

Stainless steels: Comparison of 10,000 h stress-rupture strength. Type 410 was heat treated at 1800 F for 0.5 h, oil quenched at 1200 F for 2 h, and air cooled. All others were annealed. Source: Ref 21, p 9

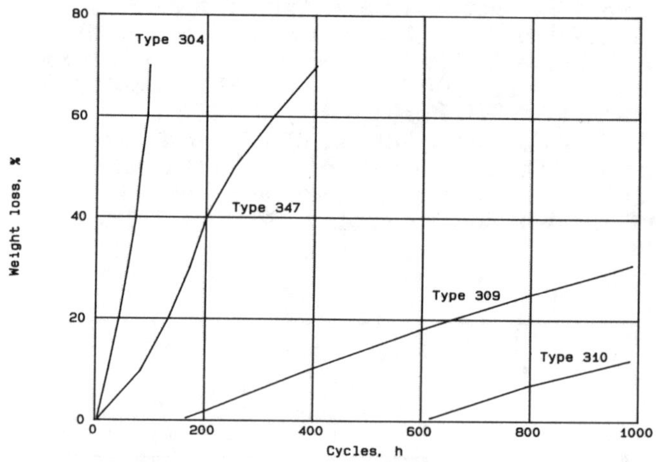

Stainless steels: Resistance of austenitic specimens to cyclic conditions at 1800 F. One cycle consisted of 15 min at temperature, followed by cooling in air for 5 min. Source: Ref 9, p 202

AISI 201

UNS number: S20100
Specifications: ASME SA412, ASTM A412, ASTM A429, ASTM A666, FED QQ S-766, SAE J405, JIS G 4303
Nominal composition: 0.150% C max, 16.000 to 18.000% Cr, 5.500 to 7.500% Mn, 0.250% N max, 3.500 to 5.500% Ni, 0.060% P max, 0.030% S max, 1.000% Si max
Product forms: Wrought, bar, wire, plate, sheet, strip, tube, billets, forgings

Description: AISI 201 is an austenitic Cr-Ni-Mn steel that was developed to conserve nickel. It is a satisfactory alternative to Type 310 for many applications. Type 201 is used primarily in sheet and strip form. Scaling temperatures in air are: 845 C (1550 F) max for continuous service and 790 C (1450 F) max for intermittent service.

Typical applications: Flatware, automobile wheel covers, and trim.

Physical Properties

AISI 201: Mean coefficients of linear thermal expansion

TEMP C	TEMP F	THEXP $10^{-6}/C$	THEXP $10^{-6}/F$
0-100	32-212	15.7	8.7
0-205	32-400	16.4	9.1
0-315	32-600	17.5	9.7
0-425	32-800	18.0	10.0
0-480	32-900	18.2	10.1
0-540	32-1000	18.4	10.2
0-650	32-1200	18.9	10.5
0-870	32-1600	20.3	11.3

Source: Ref 15, p 246

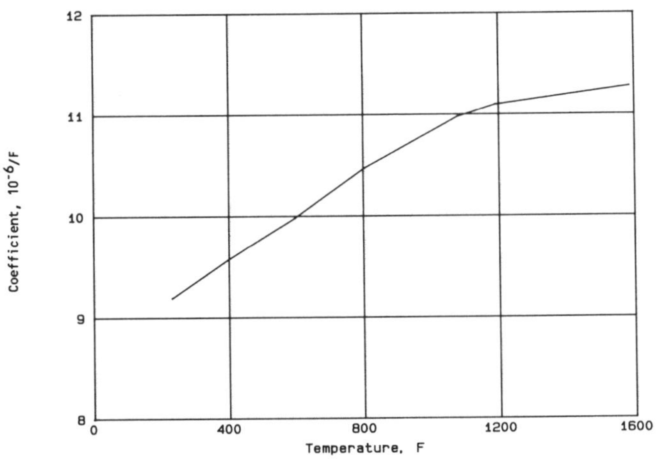

AISI 201: Thermal conductivity

TEMP C	TEMP F	THCON W/m*K	THCON Btu/ft*h*F
100	212	16.3	9.4
500	930	21.5	12.4

Source: Ref 15, p 246

AISI 201: Mean coefficient of linear thermal expansion from room temperature to indicated temperature for annealed specimens. Source: Ref 21, p 2

Mechanical Properties

AISI 201: Tensile property data

TEMP C	TEMP F	UTS MPa	UTS ksi	YS MPa	YS ksi	EL %
21	70	809	117.3	365	53.0	56.5
95	200	673	97.6	268	38.9	62.5
205	400	561	81.4	208	30.1	46.5
315	600	545	79.0	189	27.4	44.0
425	800	527	76.5	181	26.2	45.5
540	1000	479	69.5	161	23.4	33.0
650	1200	328	47.6	141	20.5	28.5
760	1400	188	27.2	125	18.2	27.5
870	1600	130	18.9	97	14.0	55.0

Source: Ref 15, p 247

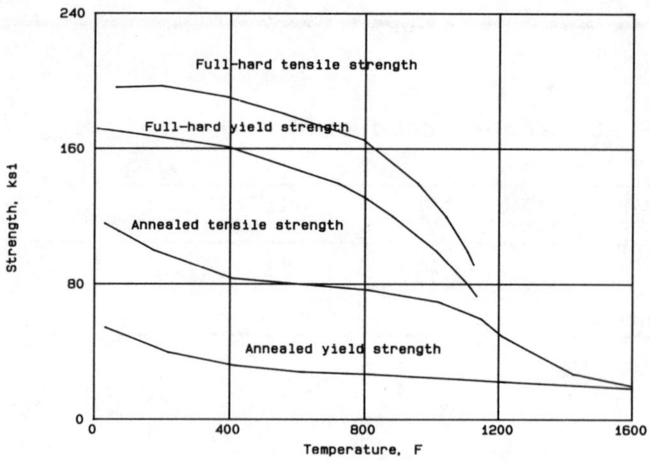

AISI 201: Tensile and yield strength of annealed and longitudinal full-hard specimens. Source: Ref 21, p 9

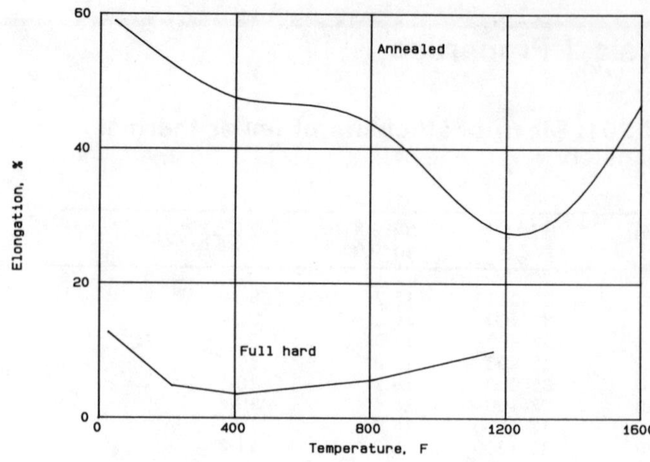

AISI 201: Tensile elongation of annealed and longitudinal full-hard specimens. Source: Ref 21, p 9

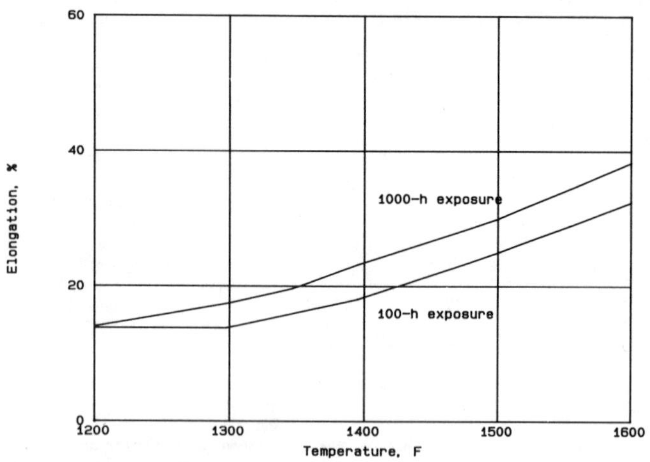

AISI 201: Effect of exposure time on tensile elongation. Source: Ref 21, p 11

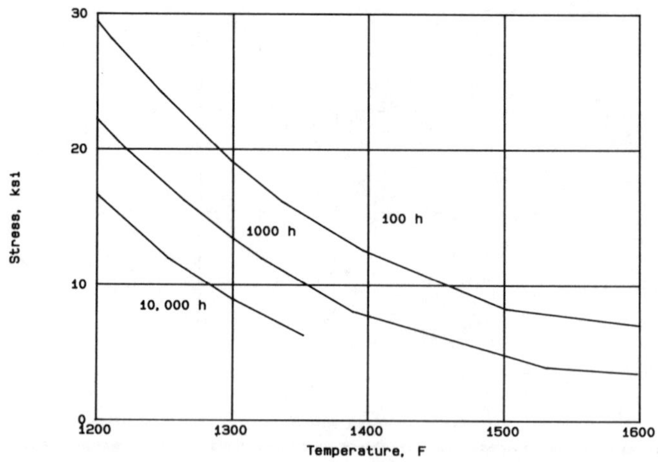

AISI 201: Stress-rupture at indicated time. Values for failure at 10,000 h were extrapolated. Source: Ref 21, p 11

AISI 201: Stress-rupture property data

TEMP C	TEMP F	RPSTR 100 h MPa	RPSTR 100 h ksi	RPSTR(a) 1000 h MPa	RPSTR(a) 1000 h ksi	RPSTR 10000 h MPa	RPSTR 10000 h ksi	EL, % 100 h	EL, % 1000 h
650	1200	200	29.0	150	22	114	16.5	14	14
730	1350	103	15.0	69	10	46	6.6	16	22
815	1500	51	7.4	30	4	...	...	35	42

(a) Extrapolated
Source: Ref 15, p 247

AISI 202

UNS number: S20200
Specifications: ASTM A314, ASTM A412, ASTM A429, ASTM A473, ASTM A666, FED QQ S-763, FED QQ S-766, SAE J405, JIS G 4303, SIS 14 23 57, BS 970 284S16, DIN 1.4371, GOST Kh17AG9N4, GOST Kh17G9AN4, GOST 1Kh18N4G4,
Nominal composition: 17.000 to 19.000% Cr, 4.000 to 6.000% Ni, 0.150% C max, 7.500 to 10.000% Mn, 1.000% Si max, 0.060% P max, 0.030% S max, 0.250% N max
Product forms: Wrought, bar, wire, plate, sheet, strip, tube, billets, forgings
Description: AISI 202 is a general purpose austenitic Cr-Ni-Mn steel that was developed to conserve nickel. It represents an attractive alternative to Type 302 steel.
Applications: Sinks, cooking utensils, hospital ware, dairy equipment, cabinet tops, antennas, beer barrels, chemical handling equipment, refrigerator trays, springs, textile dyeing equipment, tubular furniture, wire mesh, and hub caps.

Physical Properties

AISI 202: Thermal conductivity

TEMP C	TEMP F	THCON W/m*K	THCON Btu/ft*h*F
100	212	16.3	9.4
500	930	21.6	12.5

Source: Ref 15, p 252

AISI 202: Mean coefficients of linear thermal expansion

TEMP C	TEMP F	THEXP 10^{-6}/C	THEXP 10^{-6}/F
0-100	32-212	17.5	9.7
0-315	32-600	18.4	10.2
0-425	32-800	18.9	10.5
0-480	32-900	19.1	10.6
0-540	32-1000	19.3	10.7
0-650	32-1200	19.8	11.0
0-760	32-1400	20.3	11.3

Source: Ref 15, p 252

Mechanical Properties

AISI 202: Tensile property data for annealed strip

TEMP C	TEMP F	UTS MPa	UTS ksi	YS MPa	YS ksi	EL %
20	70	745	108	355	52	55
95	200	660	96	300	43	61
205	400	595	86	235	34	45
315	600	560	81	205	30	45
425	800	540	78	195	28	45
540	1000	495	72	170	25	36
650	1200	370	54	150	22	28
760	1400	215	31	130	19	31
870	1600	125	18	105	15	48

Source: Ref 15, p 252 and Ref 29, Aug 1964

AISI 202: Tensile property data for bar

TEMP C	TEMP F	UTS MPa	UTS ksi	YS MPa	YS ksi	EL %
650	1200	347	50.4	146	21.2	36
760	1400	255	37.0	132	19.1	38

Source: Ref 29, Aug 1964

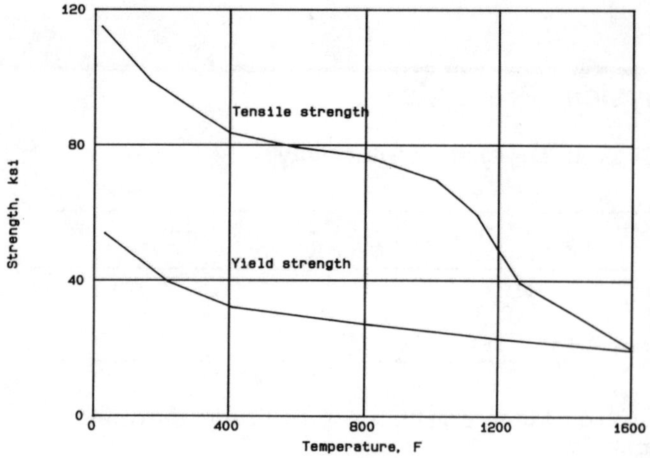

AISI 202: Tensile and yield strength of annealed specimens. Source: Ref 21, p 9

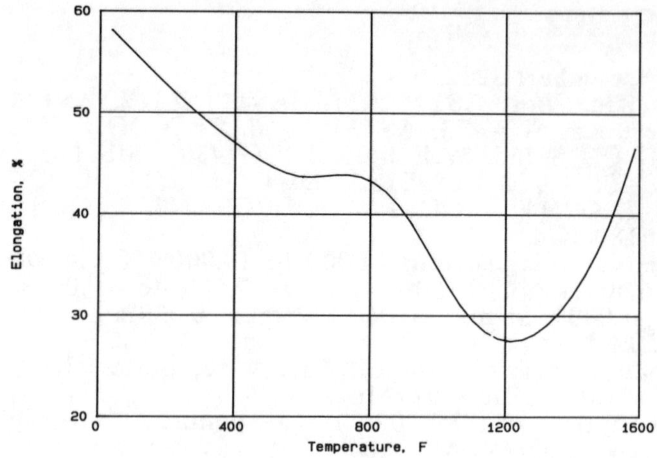

AISI 202: Tensile elongation of annealed specimens. Source: Ref 21, p 9

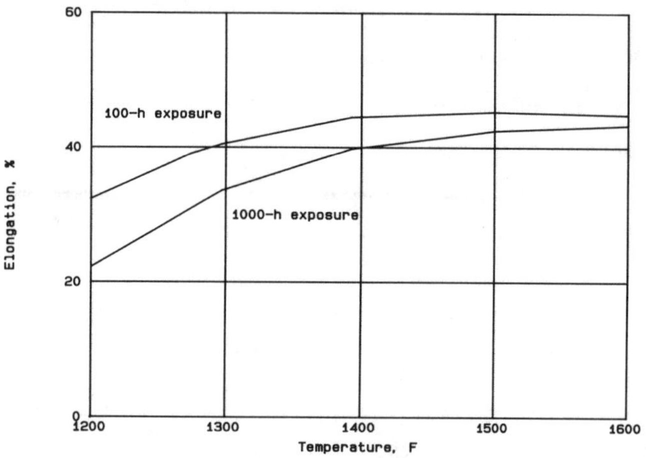

AISI 202: Effect of exposure time on tensile elongation. Source: Ref 21, p 11

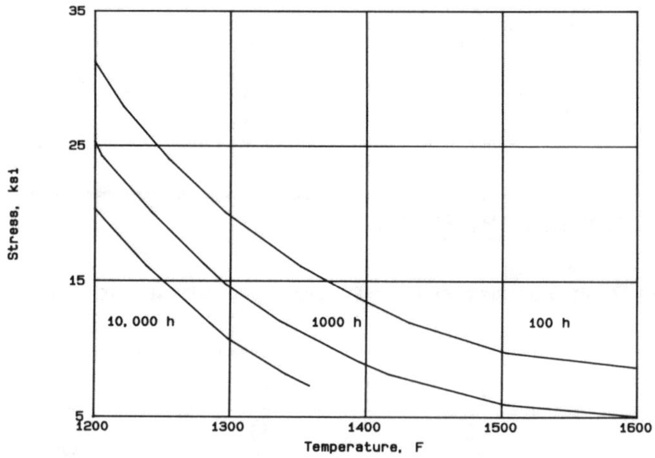

AISI 202: Stress-rupture at indicated time. Values for failure at 10,000 h were extrapolated. Source: Ref 21, p 11

AISI 301

UNS number: S30100

Specifications: AMS 5517, AMS 5518, AMS 5519, ASTM A167, ASTM A177, ASTM A554, ASTM A666, FED QQ S-766, MIL S-5059, SAE J405, NF Z12CN17-08, UNI X12CrNi17-07, JIS G 4303-39 SUS39B, SIS 14 23 31, BS 801 Grade A, GOST Kh17N8, DIN 1.4310 X12CrNi177

Nominal composition: 0.150% C max, 16.000 to 18.000% Cr, 2.000% Mn max, 6.000 to 8.000% Ni, 0.045% P max, 0.030% S max, 1.000% Si max

Product forms: Wrought, bar, wire, plate, sheet, strip, tube, billets, forgings

Description: AISI 301 is an austenitic Cr-Ni steel capable of attaining high tensile strength and ductility via moderate or severe cold working. Most popular product forms are cold rolled or cold drawn sheet, strip, and wire.

Typical applications: Structurals requiring high strength and ductility. Railroad cars, trailer bodies, aircraft structurals, fasteners, automobile wheel covers, auto trim, and pole line hardware.

Physical Properties

AISI 301: Density of annealed specimens

TEMP C	TEMP F	DENS g/cm3	DENS lb/in.3
20	68	7.91	0.2859
93	200	7.88	0.2848
204	400	7.84	0.2832
316	600	7.79	0.2814
427	800	7.74	0.2797
538	1000	7.69	0.2779
650	1200	7.64	0.2760
760	1400	7.59	0.2741
871	1600	7.54	0.2723
982	1800	7.49(a)	0.2705(a)

(a) Extrapolated
Source: Ref 21, p 4

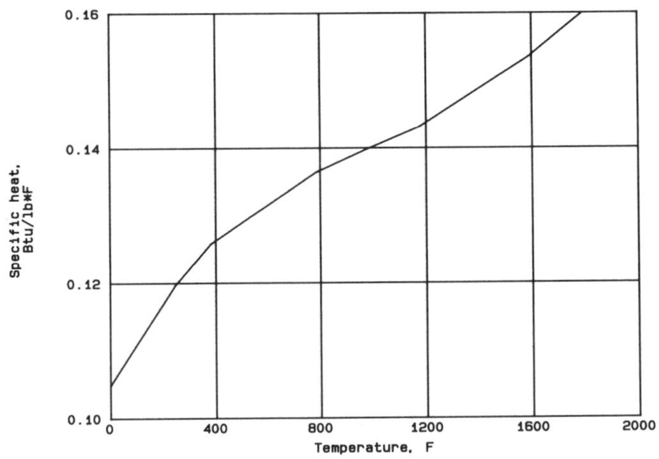

AISI 301: Specific heat of annealed specimens. Source: Ref 22, p 7

AISI 301: Thermal conductivity

TEMP C	TEMP F	THCON W/m*K	THCON Btu/ft*h*F
100	212	16.3	9.4
500	930	21.5	12.4

Source: Ref 15, p 254

AISI 301: Mean coefficients of linear thermal expansion for extra-hard stress relieved sheet

TEMP C	TEMP F	THEXP 10^{-6}/C	THEXP 10^{-6}/F
21-100	70-212	14.0	7.8
21-200	70-392	14.7	8.2
21-303	70-572	15.1	8.4
21-403	70-752	15.4	8.6
21-504	70-932	15.6	8.7

Note: Composition: 0.11% C, 0.57% Mn, 0.5% Si, 17.25% Cr, 7.00% Ni. Specimens were 0.71 mm (0.028 in.) thick, cold rolled 65%, stress relieved 8 h at 400 C (750 F), and air cooled
Source: Ref 21, p 4

AISI 301: Mean apparent specific heat of annealed specimens

TEMP C	TEMP F	SHTCP J/kg*K	SHTCP Btu/lb*F
20	68	456	0.109
94	200	490	0.117
204	400	532	0.127
316	600	557	0.133
427	800	574	0.137
538	1000	586	0.140
649	1200	599	0.143
760	1400	620	0.148
871	1600	645	0.154
982	1800	678	0.162(a)

(a) Extrapolated
Source Ref 21, p 5

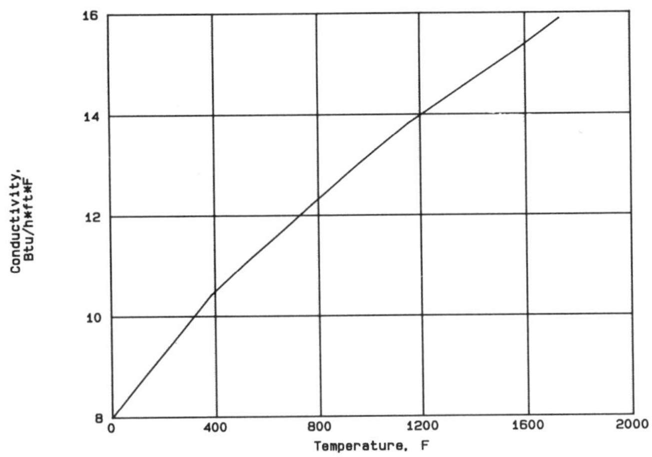

AISI 301: Thermal conductivity of annealed specimens. Source: Ref 22, p 7

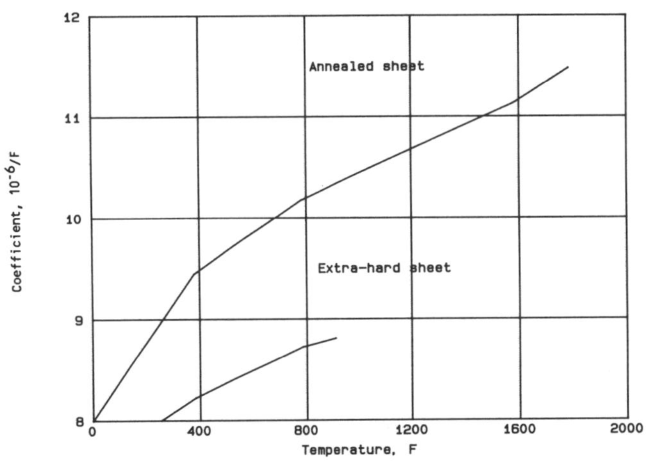

AISI 301: Mean coefficient of linear thermal expansion from room temperature to indicated temperature. Extra-hard sheet was cold rolled 65% and aged at 750 F for 8 h. Source: Ref 21, p 4

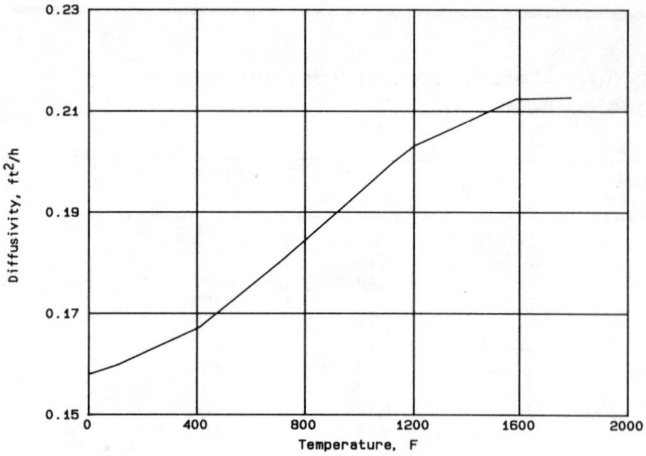

AISI 301: Thermal diffusivity of annealed specimens. Source: Ref 21, p 5

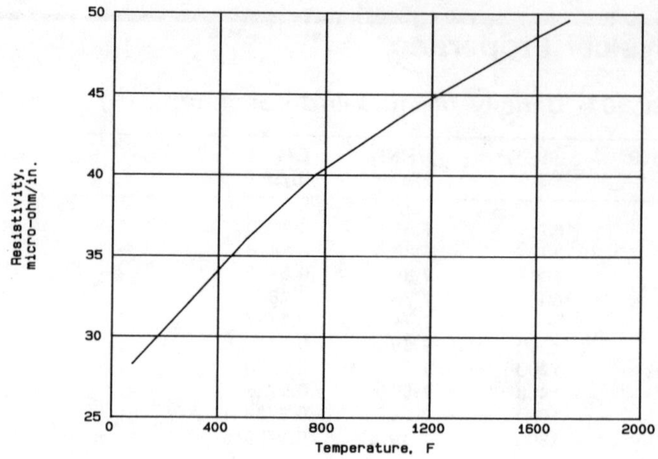

AISI 301: Electrical resistivity of annealed specimens. Source: Ref 22, p 7

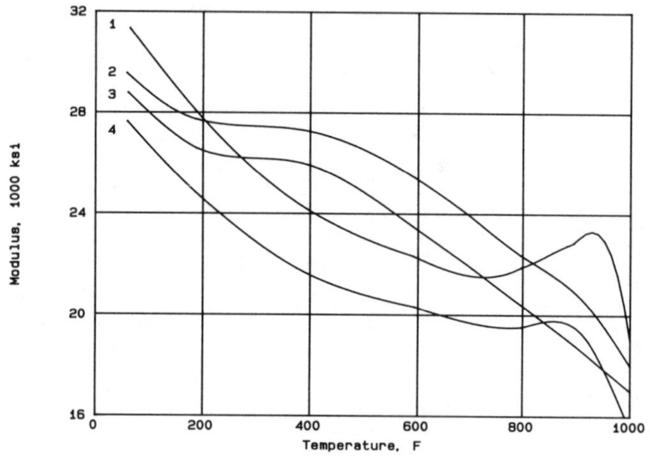

AISI 301: Modulus of elasticity of full-hard specimens. Curve 1: transverse, as rolled. Curve 2: transverse, 800 F for 8 h. Curve 3: longitudinal, 800 F for 8 h. Curve 4: longitudinal, as rolled. Source: Ref 21, p 32

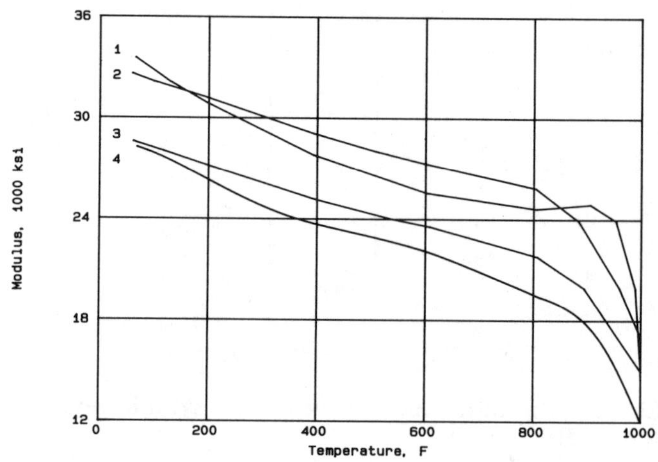

AISI 301: Modulus of elasticity of extra-hard specimens. Curve 1: transverse, as rolled. Curve 2: transverse, 750 F for 8 h. Curve 3: longitudinal, 750 F for 8 h. Curve 4: longitudinal, as rolled. Source: Ref 21, p 32

AISI 301: Mean coefficients of linear thermal expansion for annealed specimens

TEMP C	TEMP F	THEXP 10^{-6}/C	THEXP 10^{-6}/F
21-93	70-200	16.6	9.2
21-149	70-300	17.0	9.4
21-204	70-400	17.1	9.5
21-260	70-500	17.3	9.6
21-316	70-600	17.5	9.7
21-371	70-700	17.6	9.8
21-427	700-800	17.8	9.9
21-482	700-900	18.0	10.0
21-538	700-1000	18.2	10.1

Source: Ref 21, p 4

AISI 301: Electrical resistivity

TEMP C	TEMP F	ELRES n{o}*m
21	70	720
100	212	780
200	390	860
400	750	1000
600	1110	1110
800	1470	1210
900	1650	1260

Source: Ref: 15, p 254

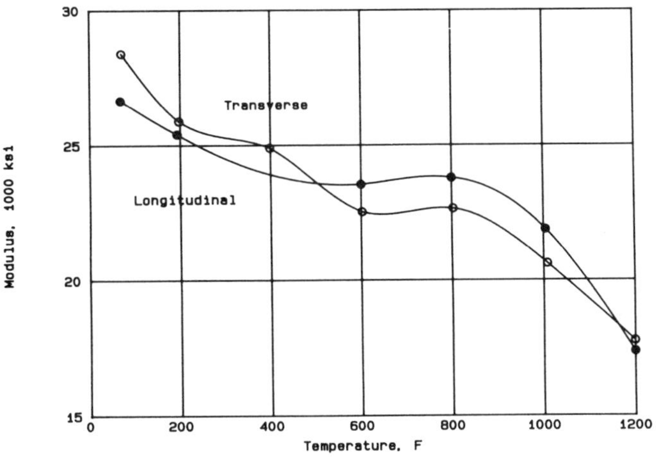

AISI 301: Modulus of elasticity in compression for half-hard sheet. Open circles: transverse specimens. Closed circles: longitudinal specimens. Source: Ref 22, p 34

Mechanical Properties

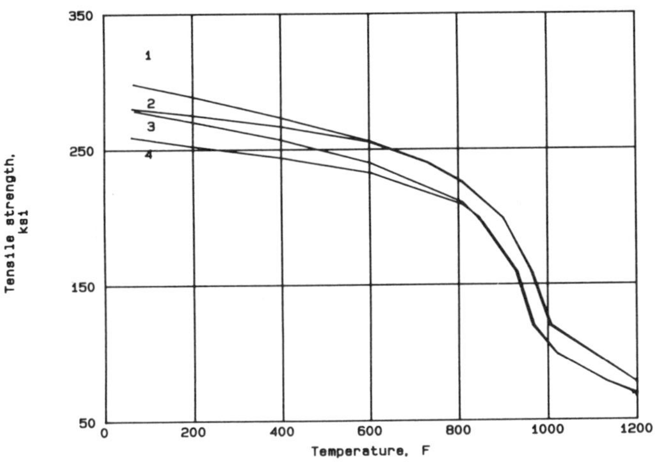

AISI 301: Tensile strength of extra-hard sheet, 0.020 and 0.027 in. Curve 1: transverse, rolled 65% and aged 8 h at 750 F. Curve 2: transverse, as rolled 65%. Curve 3: longitudinal, rolled 65% and aged 8 h at 750 F. Curve 4: longitudinal, as rolled 65%. Source: Ref 21, p 24

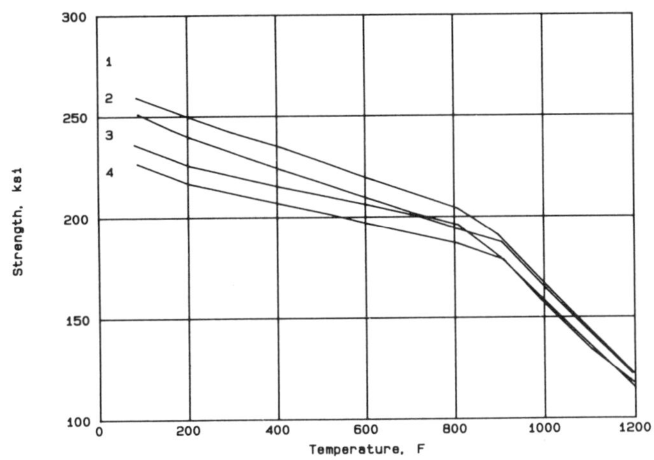

AISI 301: Tensile strength of 0.020- to 0.060-in. sheet. Curve 1: transverse, rolled 40% and aged at 800 F for 8 h. Curve 2: longitudinal, rolled 40% and aged at 800 F for 8 h. Curve 3: longitudinal, as rolled 40%. Curve 4: transverse, as rolled 40%. Source: Ref 21, p 22

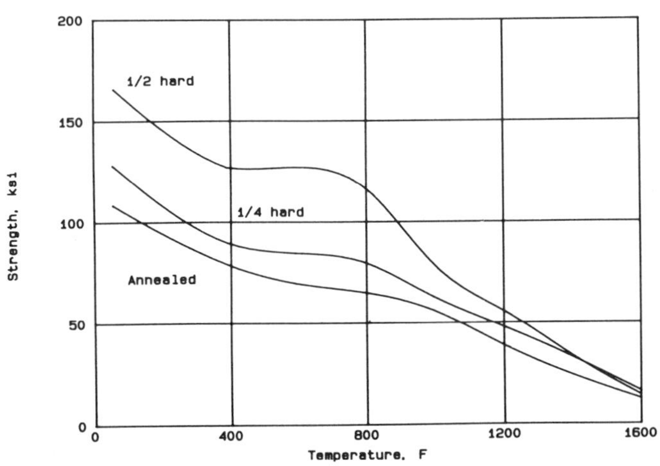

AISI 301: Effect of cold work on tensile strength. Source: Ref 9, p 179

AISI 301: Tensile property data

TEMP C	TEMP F	UTS MPa	UTS ksi	YS MPa	YS ksi	EL %
Annealed						
21	70	724	105	276	40	55
205	400	552	80	152	22	46
315	600	485	70.4	134	19.4	40
425	800	463	67.2	134	19.5	39
540	1000	401	58.2	126	18.3	34
650	1200	282	40.9	106	15.4	36
760	1400	204	29.6	99	14.4	30
870	1600	109	15.8	66	9.5	29
25% hard						
21	70	889	129	503	73	43.5
205	400	625	90.6	424	61.5	23
315	600	594	86.2	412	59.8	20
425	800	506	73.4	377	54.7	17.5
540	1000	447	64.9	353	51.2	16.5
650	1200	352	51	276	40	20
760	1400	248	36	186	27	17
870	1600	134	19.4	106	15.4	15
50% hard						
21	70	1138	165	772	112	28.5
205	400	876	127	731	106	9
315	600	846	122.7	656	95.2	6.5
425	800	734	105.1	590	85.5	7
540	1000	538	78	464	67.3	7
650	1200	396	57.5	331	48	10
760	1400	241	35	214	31	10
870	1600	113	16.4	96	13.9	12.5

Source: Ref 15, p 255

AISI 301: Tensile property data for aged, extra-hard specimens tested in the longitudinal direction

TEMP C	TEMP F	YS MPa	YS ksi	UTS MPa	UTS ksi	EL %
Cold rolled 65% at 400 C (750 F) 2 h, air cooled						
427	800	1265	183.5	1484	215.2	5.0
482	900	1020	147.9	1287	186.7	7.0
Cold rolled 65% at 480 C (900 F) 2 h, air cooled						
427	800	1200	174.1	1420	205.9	6.0
482	900	894	129.6	1157	167.8	6.5

Note: Composition: 0.11% C, 0.57% Mn, 0.50% Si, 17.25% Cr, 7.00% Ni
Source: Ref 21, p 24

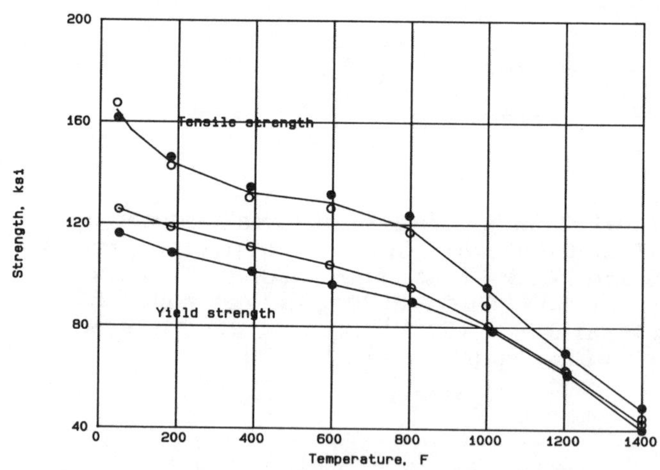

AISI 301: Tensile and yield strength of 0.052-in. half-hard sheet. Open circles: longitudinal specimens. Solid circles: transverse specimens. Source: Ref 22, p 18

AISI 301: Tensile, compressive, and bearing property data for half-hard specimens

EXP h(a)	YS MPa	YS ksi	YSC 0.2% MPa	YSC 0.2% ksi	YSB 0.2% MPa	YSB 0.2% ksi	UTS MPa	UTS ksi	USB MPa	USB ksi
At 25 C (78 F)										
...	730	105.84	518	75.2	1134	164.4	1070	155.16	1593	231
At 205 C (400 F)										
0.5	591	85.7	412	59.8	1027	148.9	748	108.5	1229	178.3
2	572	83	433	62.8	956	138.6	756	109.6	1094	158.6
10	548	79.5	428	62.1	942	136.6	749	108.7	1200	174.1
100	543	78.7	439	63.7	934	135.5	775	112.4	1141	165.5
At 315 C (600 F)										
0.5	557	80.8	485	70.3	979	142	747	108.3	1107	160.6
2	550	79.8	498	72.3	1017	147.5	720	104.4	1159	168.1
10	562	81.5	477	69.2	981	142.3	731	106.0	1107	160.6
100	573	83.1	525	76.1	1006	145.9	745	108.0	1176	170.5
At 425 C (800 F)										
0.5	557	80.8	476	69.1	905	131.2	693	100.5	1114	161.6
2	543	78.8	494	71.7	905	131.2	717	104	1054	152.9
10	516	74.9	530	76.9	887	128.6	689	99.9	1086	157.5
100	516	74.9	544	78.9	845	122.5	641	93	1024	148.5
At 540 C (1000 F)										
0.5	483	70	442	64.1	776	112.6	592	85.9	919	133.3
2	447	64.8	451	65.4	727	105.5	565	81.9	885	128.4
10	437	63.4	439	63.6	749	108.6	556	80.7	866	125.6
100	440	63.8	439	63.6	689	100	541	78.5	824	119.5
At 650 C (1200 F)										
0.5	350	50.8	...	...	538	78.0	453	65.7	681	98.8
2	347	50.4	...	...	597	86.6	440	63.8	694	100.7
10	367	53.2	...	...	565	82.0	460	66.7	658	95.4
100	352	51	...	...	568	82.4	426	61.8	674	97.7

Note: Composition: 0.10% C, 1.14% Mn, 0.39% Si, 17.38% Cr, 7.46% Ni, 0.1% Mo
(a) Time held prior to testing at indicated temperature
Source: Ref 21, p 40

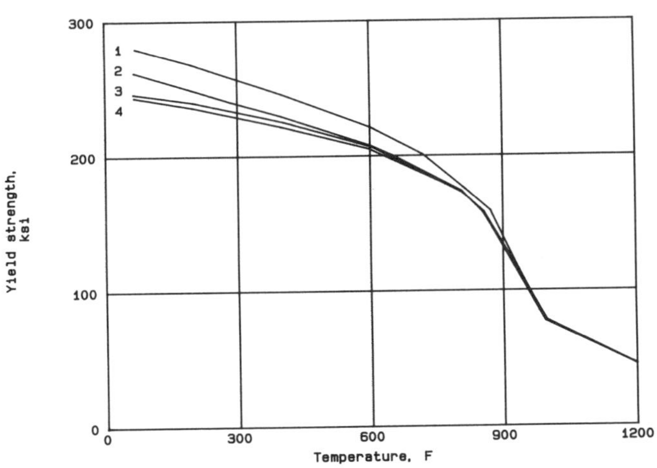

AISI 301: Yield strength of 0.020- and 0.027-in. extra-hard sheet. Curve 1: transverse, rolled 65% and aged 8 h at 750 F. Curve 2: longitudinal, rolled 65% and aged 8 h at 750 F. Curve 3: transverse, as rolled 65%. Curve 4: longitudinal, as rolled 65%. Source: Ref 21, p 24

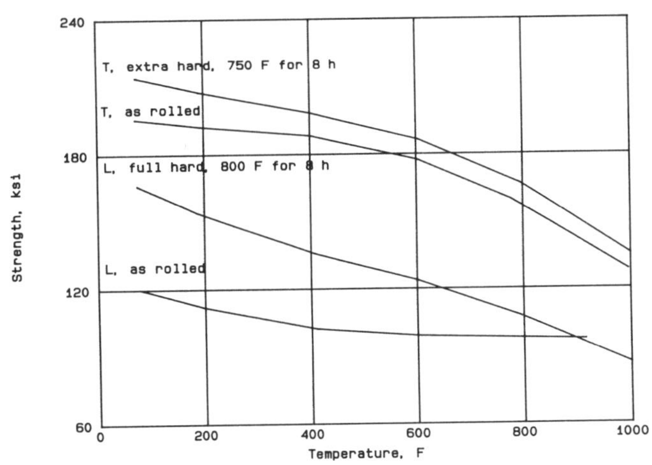

AISI 301: Compressive yield strength of 0.020- to 0.062-in. full-hard sheet, cold rolled 40%. Source: Ref 22, p 22

AISI 301: Compressive property data for extra hard specimens

TEMP C	TEMP F	YSC(a) MPa	YSC(a) ksi	Ec(a) GPa	Ec(a) 10^3 ksi	YSC(b) MPa	YSC(b) ksi	Ec(b) GPa	Ec(b) 10^3 ksi
Cold rolled 65%, 50 HRC									
21	70	1246	180.7	159	23.1	1968	285.5	205	29.8
204	400	1187	172.1	140	20.3	1820	264.0	190	27.6
316	600	1166	169.1	125	18.2	1708	247.7	181	26.2
427	800	1058	153.4	123	17.8	1515	219.7	160	23.2
482	900	999	144.9	119	17.3	1391	201.8	139	20.1
538	1000	659	95.6	108	15.7	974	141.2	126	18.3
Cold rolled 65% + 8 h at 400 C (750 F), 52 HRC									
21	70	1624	235.5	190	27.5	2230	323.4	219	31.8
204	400	1408	204.2	179	26.0	1968	285.5	203	29.4
316	600	1195	173.3	161	23.4	1778	257.9	191	27.7
427	800	1009	146.3	146	21.2	1601	232.2	170	24.6
482	900	962	139.5	139	20.1	1326	192.3	152	22.1
538	1000	598	86.8	114	16.6	807	117.0	125	18.1

Note: Composition: 0.11% C, 0.57% Mn, 0.50% Si, 17.25% Cr, 7.00% Ni
(a) Longitudinal specimens. (b) Transverse specimens
Source: Ref: 21, p 36

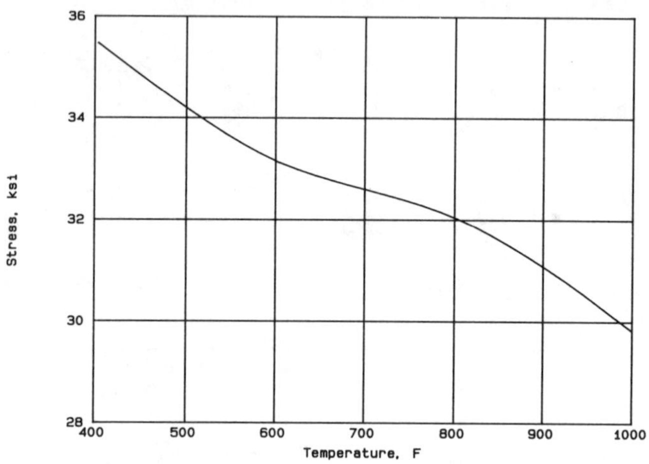

AISI 301: Compressive yield strength of 0.063-in. annealed sheet. Source: Ref 22, p 22

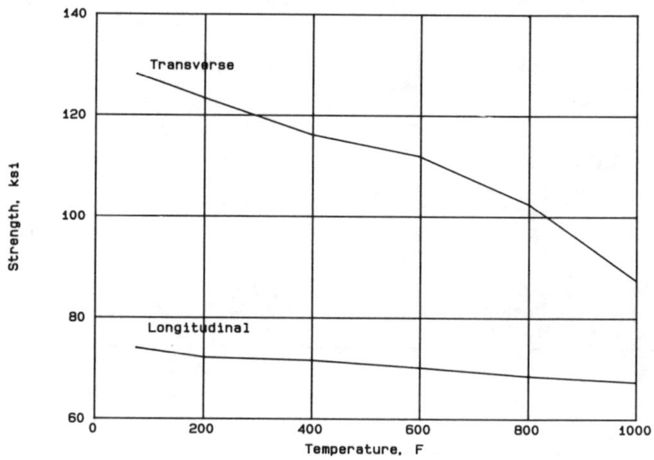

AISI 301: Compressive yield strength of 0.052-in. half-hard sheet. Source: Ref 21, p 36

AISI 301: Compressive property data for half-hard specimens

TEMP C	TEMP F	Direction	YS MPa	YS ksi	E GPa	E 10^3 ksi
21	70	L	512	74.3	184.8	26.8
		T	873	126.6	193.0	28.7
94	200	L	476	69.1	175.9	25.5
		T	836	121.3	172.4	25.7
206	400	L	527	76.4	169.6	24.6
		T	793	115.0	169.6	24.6
318	600	L	487	70.7	162.8	23.6
		T	785	113.8	155.9	22.6
430	800	L	472	68.5	163.4	23.7
		T	687	99.7	156.6	22.7
542	1000	L	472	68.4	151.0	21.9
		T	616	89.3	140.7	20.4

Note: Composition: 0.10% C, 0.60% Mn, 0.027% P, 0.019% S, 0.50% Si, 19.32% Cr, 7.04% Ni. L=longitudinal and T=transverse
Source: Ref 21, p 35

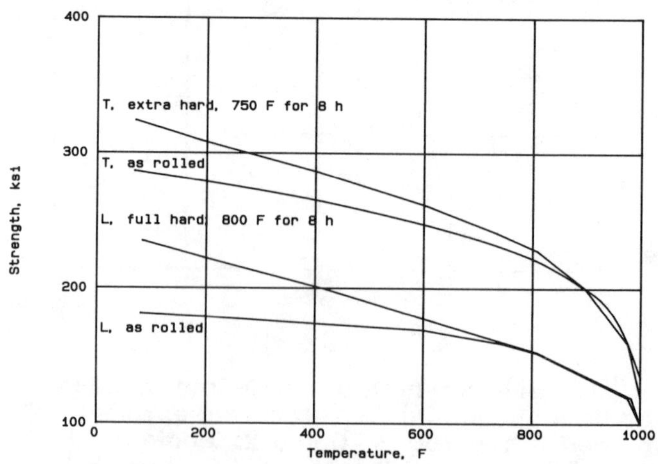

AISI 301: Compressive yield strength of 0.020- to 0.027-in. extra-hard sheet, cold rolled 65%. Source: Ref 22, p 22

AISI 301: Shear strength of full-hard specimens

TEMP C	TEMP F	USS (a) MPa	USS (a) ksi	USS:UTS (a)	USS (b) MPa	USS (b) ksi	USS:UTS (b)
As rolled 0.53-mm (0.021-in.) thick							
21	70	938	136.0	0.69	938	136.0	0.70
204	400	736	106.7	0.60	719	104.3	0.63
316	600	628	91.1	0.55	608	88.2	0.55
427	800	632	91.6	0.58	589	85.4	0.58
538	1000	532	77.2	0.64	547	79.3	0.71
As rolled 427 C (800 F) 8 h							
21	70	1009	146.3	0.74	1009	146.3	0.76
204	400	756	109.6	0.61	756	109.6	0.64
316	600	708	102.7	0.59	690	100.1	0.61
427	800	654	94.9	0.59	645	93.5	0.62
538	1000	552	80.0	0.65	553	80.2	0.72

Note: Composition: 0.11% C, 0.58% Mn, 0.38% Si, 17.99% Cr, 6.92% Ni
(a) Longitudinal specimens. (b) Transverse specimens
Source: Ref 21, p 41

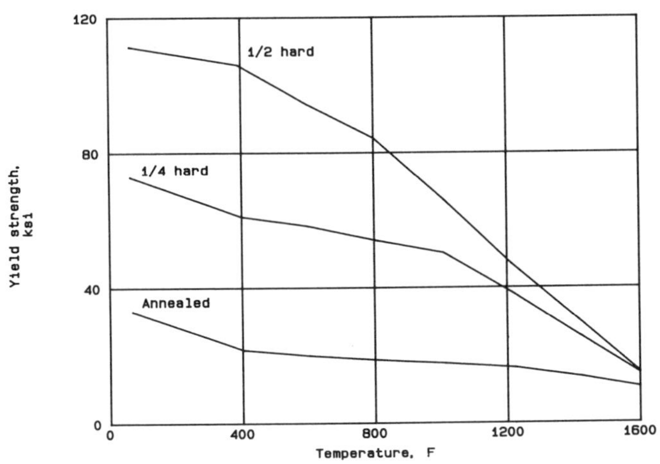

AISI 301: Effect of cold work on short-time 0.2% offset yield strength. Source: Ref 9, p 179

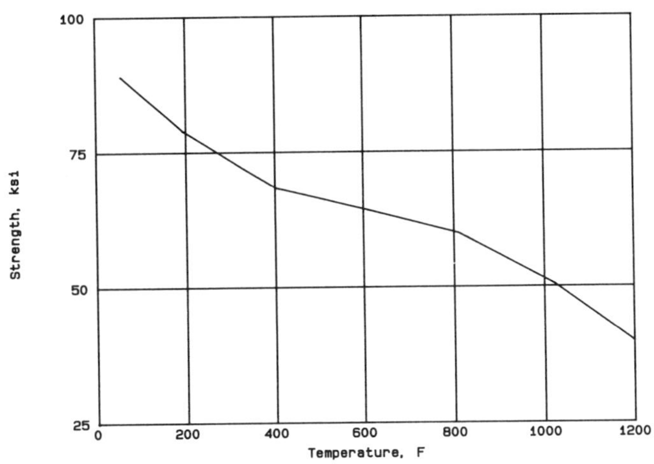

AISI 301: Shear strength of half-hard sheet. Source: Ref 22, p 23

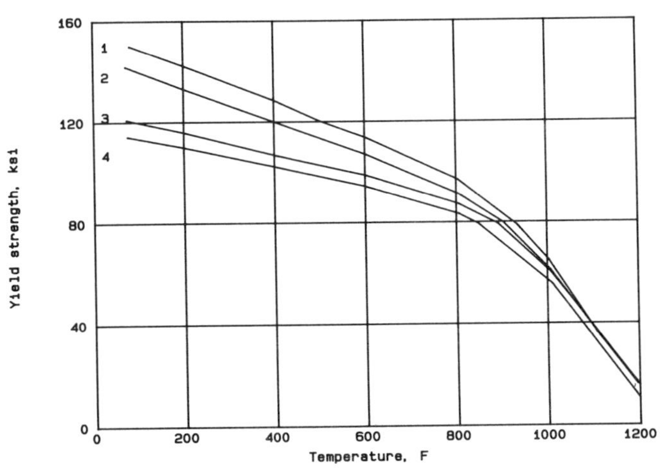

AISI 301: Yield strength of 0.020- to 0.060-in. sheet. Curve 1: transverse, rolled 40% and aged at 800 F for 8 h. Curve 2: longitudinal, rolled 40% and aged at 800 F for 8 h. Curve 3: longitudinal, as rolled 40%. Curve 4: transverse, as rolled 40%. Source: Ref 21, p 22

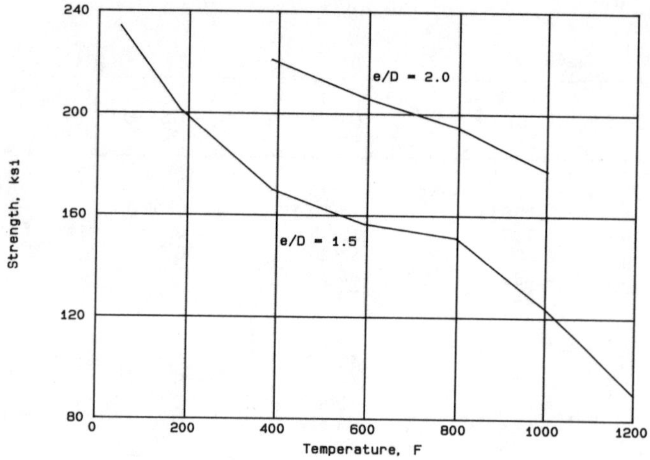

AISI 301: Ultimate bearing strength of half-hard sheet. Source: Ref 22, p 23

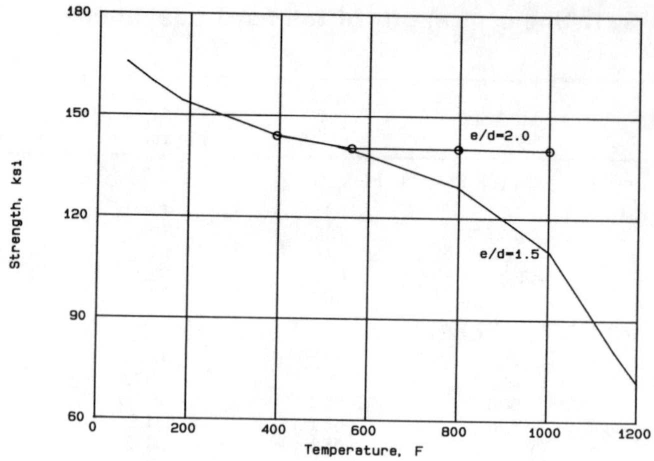

AISI 301: Bearing yield strength of half-hard sheet. Source: Ref 22, p 23

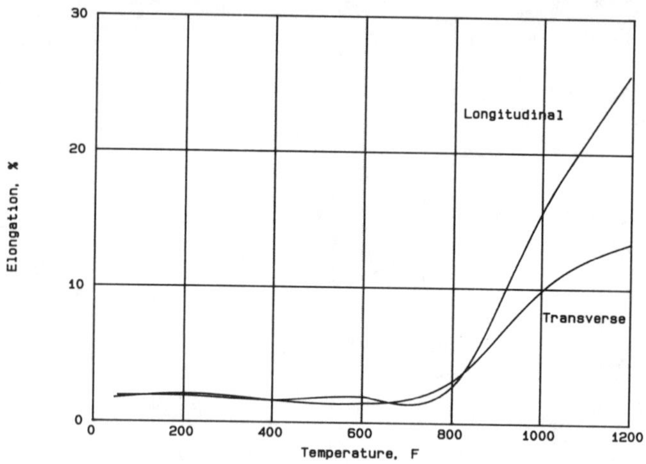

AISI 301: Tensile elongation of 0.020- and 0.027-in. extra-hard sheet, using specimens both as rolled 65%, and as rolled 65% and stress relieved at 750 F for 8 h. Source: Ref 21, p 24

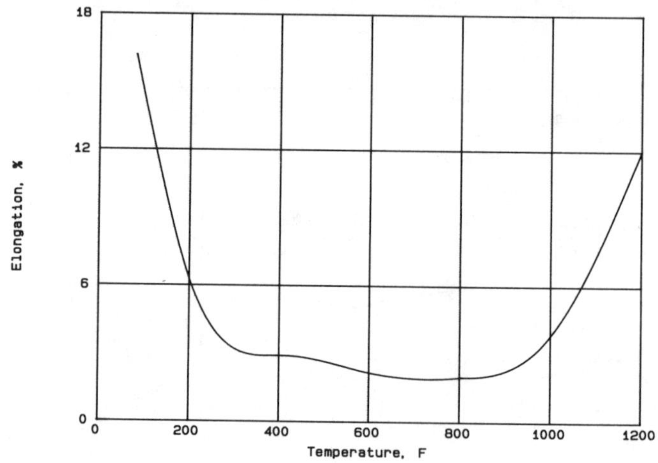

AISI 301: Tensile elongation of 0.020- to 0.060-in. full-hard sheet, transverse and longitudinal specimens, as rolled 40% and aged at 800 F for 8 h. Source: Ref 21, p 22

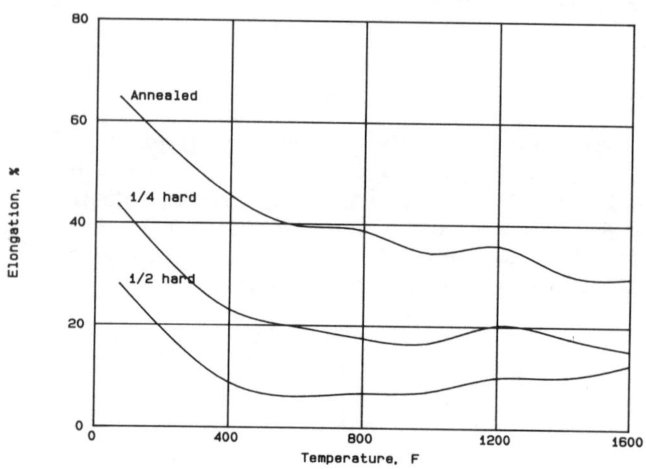

AISI 301: Effect of cold work on tensile elongation. Source: Ref 9, p 179

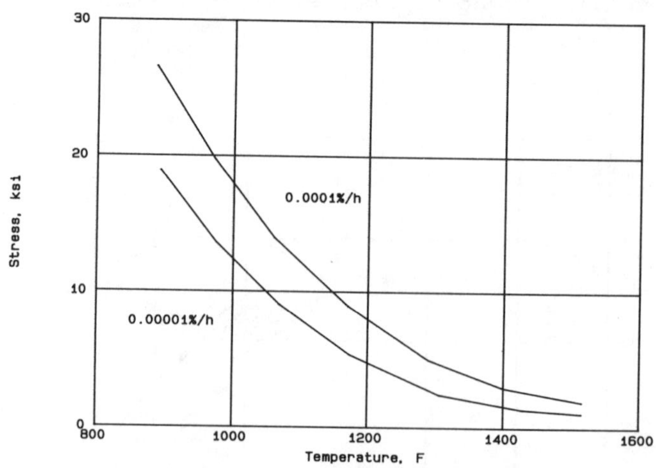

AISI 301: Creep rates in annealed sheet as a function of stress and temperature. Source: Ref 21, p 51

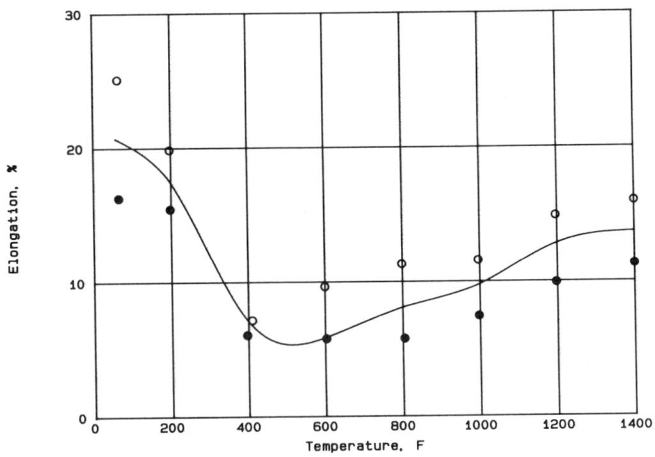

AISI 301: Elongation of 0.052-in. half-hard sheet. Open circles: longitudinal specimens. Solid circles: transverse specimens. Source: Ref 22, p 18

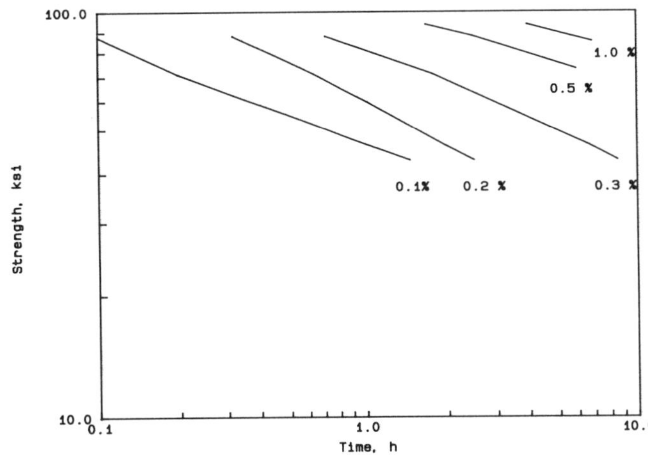

AISI 301: Creep lives of 0.040-in. extra fully hard sheet (60% reduction) under cyclic load at 900 F. The cyclic load was 15 min at stress and 15 min unstressed. Source: Ref 22, p 31

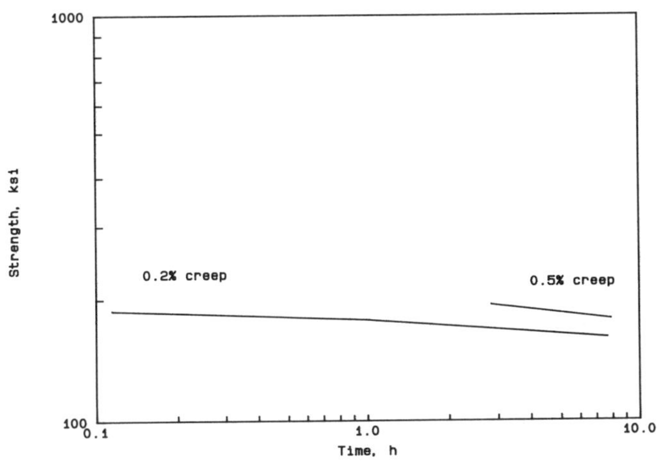

AISI 301: Creep lives of 0.040-in. extra fully hard sheet (60% reduction) at 600 F. Source: Ref 22, p 31

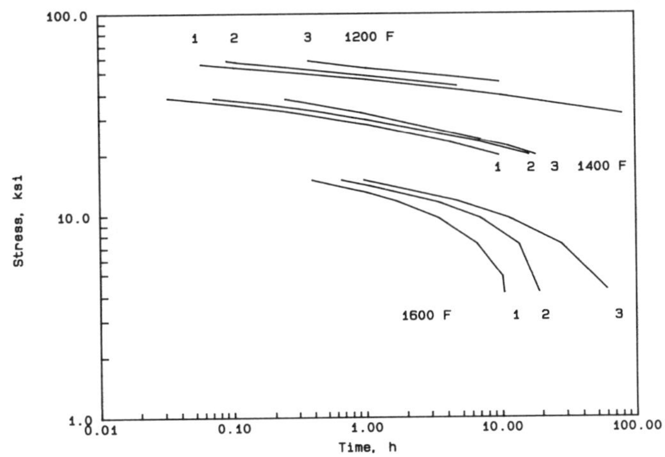

AISI 301: Creep and creep-rupture lives of 0.050-in. half-hard sheet. Curve 1: 1.0% creep. Curve 2: 2.0% creep. Curve 3: rupture. Source: Ref 22, p 30

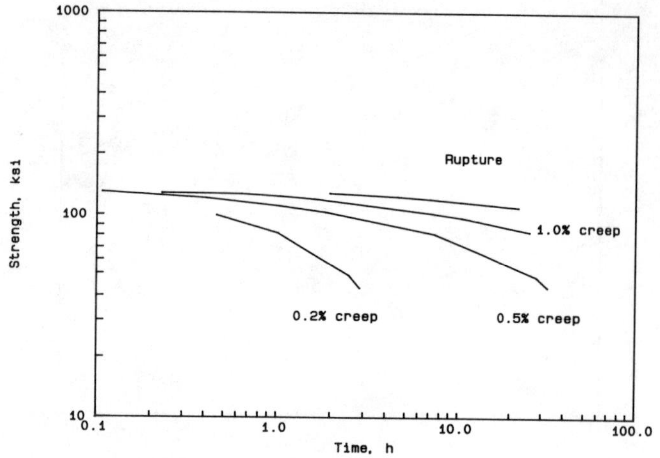

AISI 301: Creep and creep-rupture lives of 0.040-in. extra fully hard sheet (60% reduction) at 900 F. Source: Ref 22, p 31

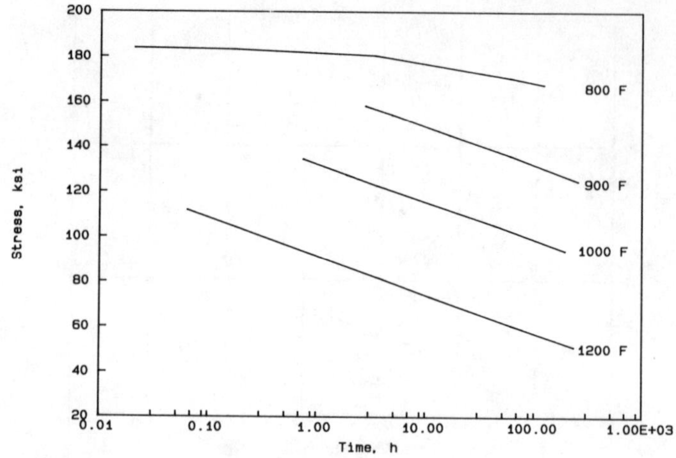

AISI 301: Creep-rupture lives of full-hard 0.032-in. sheet, stress relieved at 800 F for 8 h. Source: Ref 22, p 31

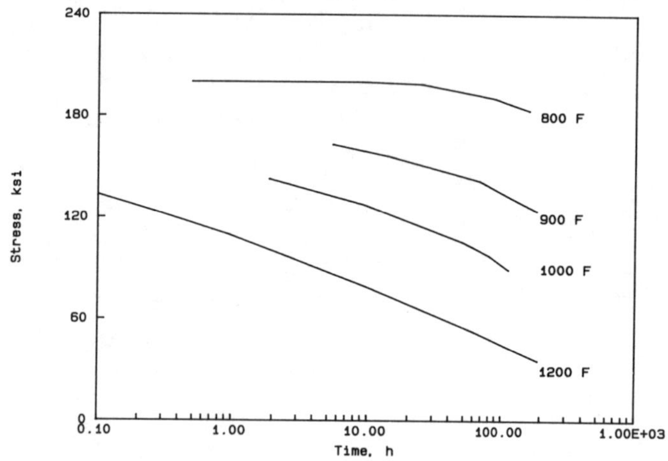

AISI 301: Creep-rupture lives of extra-hard 0.0285-in. sheet, stress relieved at 750 F for 8 h. Source: Ref 22, p 31

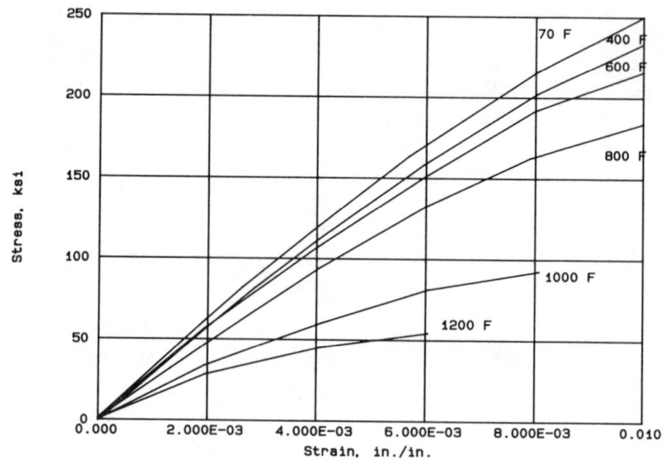

AISI 301: Stress-strain curves for 0.020-in. extra hard sheet, stress relieved at 750 F for 8 h. Source: Ref 22, p 11

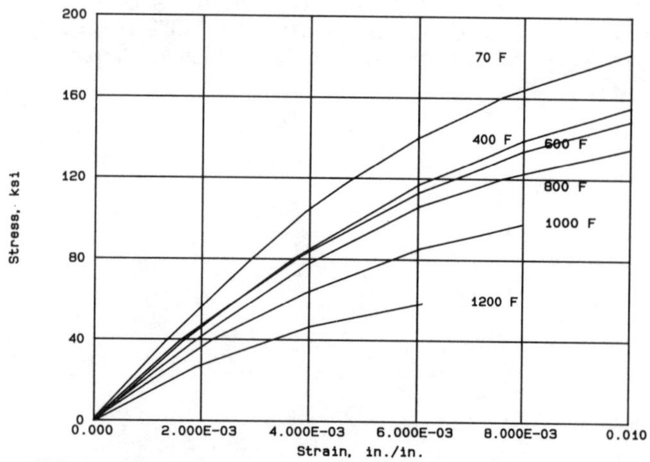

AISI 301: Stress-strain curves for 0.020-in. full-hard sheet (40% reduction). Source: Ref 22, p 10

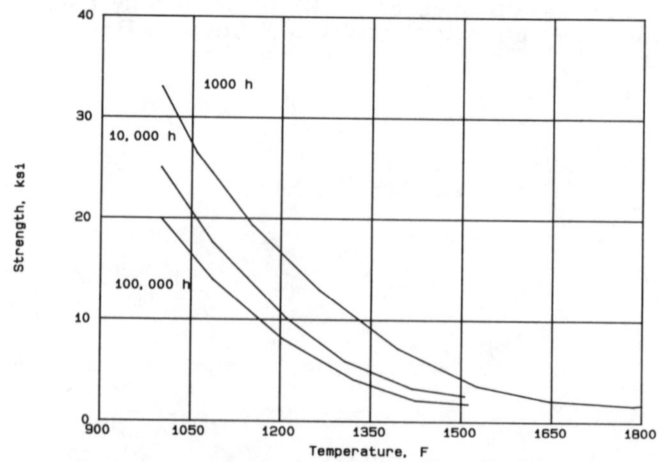

AISI 301: Rupture strength of annealed sheet. Source: Ref 21, p 51

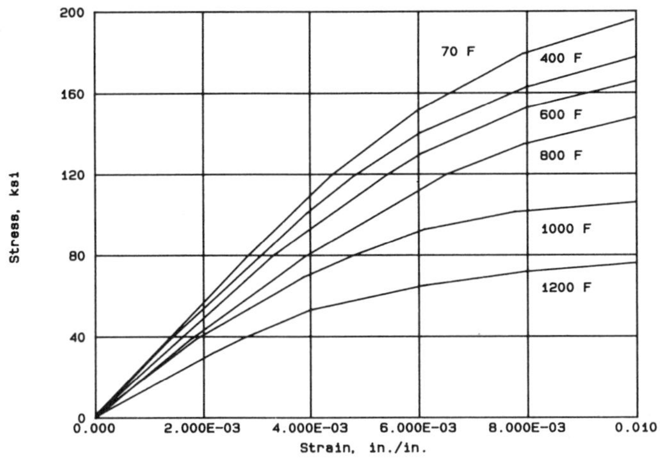

AISI 301: Stress-strain curves for 0.032-in. full-hard sheet, stress relieved at 800 F for 8 h. Source: Ref 22, p 10

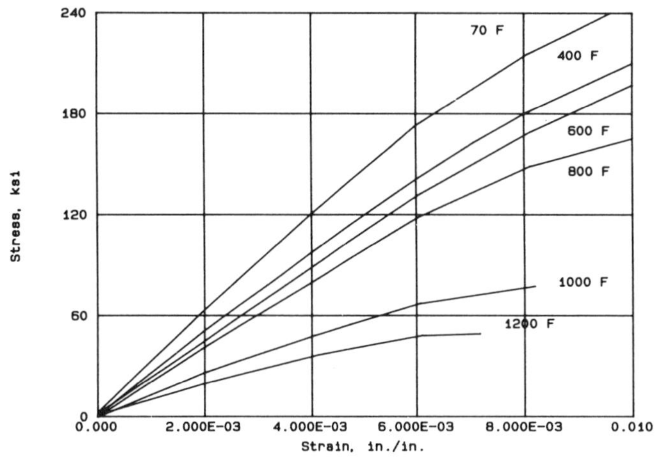

AISI 301: Stress-strain curves for 0.020-in. extra-hard sheet (65% reduction). Source: Ref 22, p 11

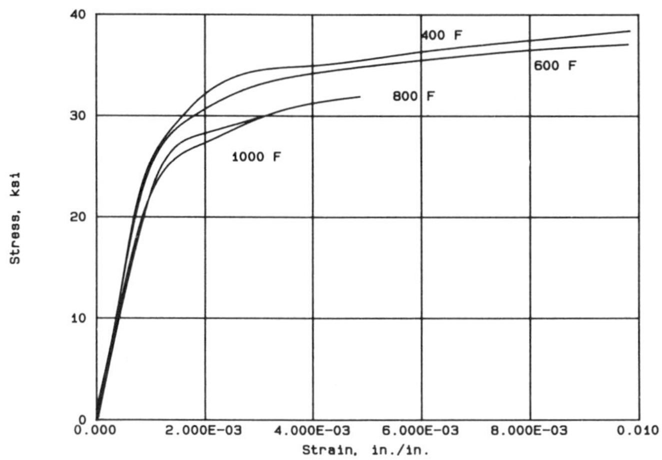

AISI 301: Stress-strain curves in compression for 0.063-in. annealed sheet. Source: Ref 22, p 20

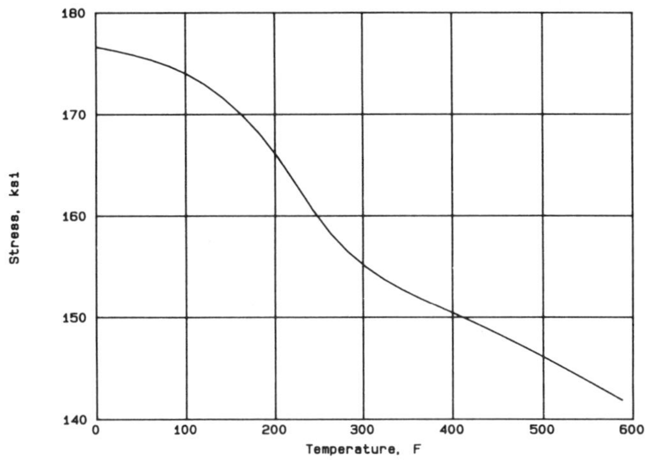

AISI 301: Net fracture stress of full-hard 0.046-in. sheet. Source: Ref 21, p 48

AISI 302

UNS number: S30200
Specifications: AMS 5515, AMS 5516, AMS 5600, ASME SA240, ASME SA479, ASTM A167, ASTM A240, ASTM A276, ASTM A313, FED QQ S-763, FED QQ S-766, FED QQ W-423, MIL S-862, MIL S-5059, MIL S-7720, SAE J230, SAE J405, NF Z10CN18-09, BS 970 302S25 En58A, DIN 1.4300 X12CrNi188
Nominal composition: 0.150% C max, 17.000 to 19.000% Cr, 2.000% Mn max, 8.000 to 10.000% Ni, 0.045% P max, 0.030% S max, 1.000% Si max, 0.100% N max
Product forms: Wrought, bar, wire, plate, sheet, strip, tube, billets, forgings

Description: AISI 302 is the original "18-8" austenitic Cr-Ni stainless steel. Its corrosion resistance is somewhat superior to that of Type 301, but its ductility at high tensile strength levels is slightly lower. Scaling temperatures in air are: 900 C (1650 F) max for continuous service and 815 C (1500 F) max for intermittent service.

Typical applications: This general-purpose stainless steel is used for trim, food-handling equipment, aircraft cowlings, antennas, springs, cookware, building exteriors, tanks, household appliances, jewelry, oil-refining equipment, and signs.

Physical Properties

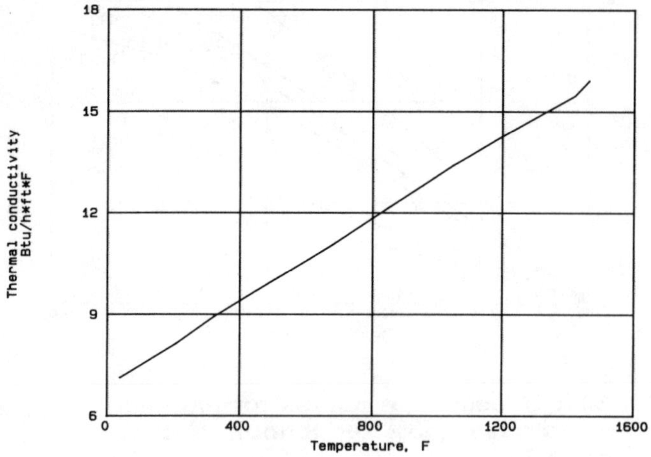

AISI 302: Thermal conductivity of annealed specimens. Source: Ref 22, p 7

AISI 302: Thermal conductivity

TEMP C	TEMP F	THCON W/m*K	THCON Btu/ft*h*F
100	212	16.3	9.4
500	930	21.5	12.4

Source: Ref 15, p 258

AISI 302: Mean coefficients of linear thermal expansion

TEMP C	TEMP F	THEXP 10^{-6}/C	THEXP 10^{-6}/F
0-100	32-212	17.3	9.6
0-315	32-600	17.8	9.9
0-540	32-1000	18.4	10.2
0-650	32-1200	18.7	10.4

Source: Ref 15, p 258

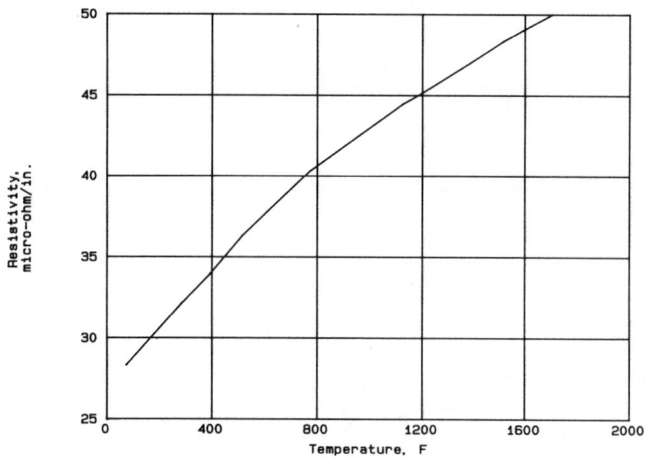

AISI 302: Electrical resistivity of annealed specimens. Source: Ref 22, p 7

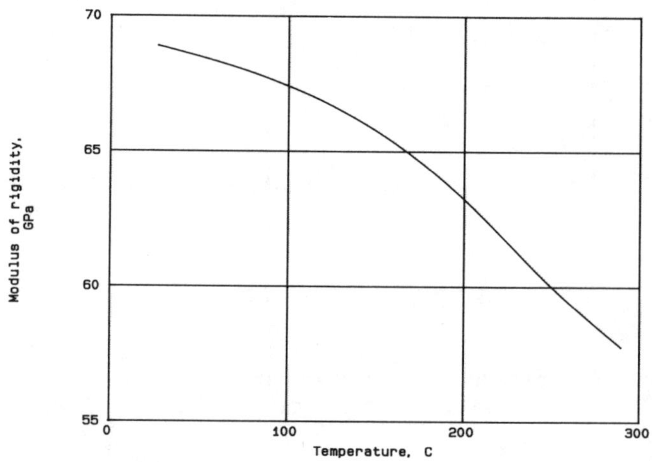

AISI 302: Modulus of rigidity. Source: Ref 1, p 300

AISI 302: Electrical resistivity

TEMP C	TEMP F	ELRES n{o}*m
21	70	720
100	212	780
200	390	860
400	750	1000
600	1110	1110
800	1470	1210
900	1650	1260

Source: Ref: 15, p 258

Mechanical Properties

AISI 302: Tensile property data

TEMP C	TEMP F	UTS MPa	UTS ksi	YS MPa	YS ksi	EL %	RA %
21	70	585	85	255	37	57	70
205	400	495	72	160	23	51	73
315	600	470	68	135	19.5	45	71
425	800	440	64	115	16.5	40	69
540	1000	385	56	97	14	36	66
650	1200	305	44	86	12.5	34	58
790	1400	200	29	76	11	36	46
870	1600	110	16	...	...	40	40

Source: Ref: 15, p 259

AISI 302: Effect of temperature on tensile, compressive, and bearing properties of annealed specimens

EXP(a) h	YS 0.2% MPa	YS 0.2% ksi	YSC 0.2% MPa	YSC 0.2% ksi	YSB 0.2% MPa	YSB 0.2% ksi	UTS MPa	UTS ksi	USB MPa	USB ksi
At 26 C (78 F)										
...	310	45	317	46	503	73	614	89	1158	168
At 204 C (400 F)										
0.5	228	33	248	36	427	62	476	69	765	111
2	214	31	248	36	386	56	462	67	758	110
10	221	32	255	37	393	57	469	68	765	111
100	221	32	255	37	379	55	462	67	758	110
At 316 C (600 F)										
0.5	214	31	241	35	393	57	469	68	752	109
2	214	31	234	34	372	54	469	68	731	106
10	214	31	221	32	372	54	455	66	752	109
100	207	30	228	33	400	58	462	67	752	109
At 427 C (800 F)										
0.5	200	29	214	31	379	55	448	65	717	104
2	193	28	214	31	365	53	434	63	724	105
10	193	28	207	30	365	53	455	66	779	113
100	200	29	214	31	345	50	448	65	689	100
At 538 C (1000 F)										
0.5	172	25	200	29	324	47	414	60	641	93
2	165	24	200	29	345	50	421	61	648	94
10	172	25	193	28	331	48	434	63	641	93
100	165	24	200	29	310	45	427	62	648	94
At 648 C (1200 F)										
0.5	159	23	...	...	303	44	372	54	558	81
2	159	23	...	...	303	44	372	54	558	81
10	165	24	...	...	296	43	372	54	552	80
100	152	22	...	...	290	42	359	52	538	78

(a) Exposure to temperature before test
Source: Ref 9, p 187

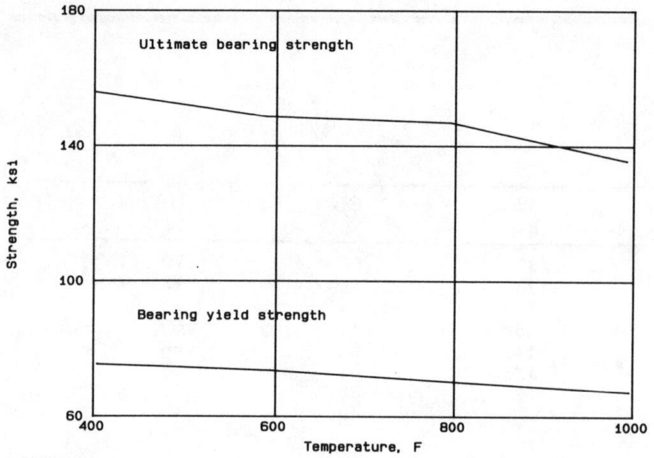

AISI 302: Ultimate bearing and yield strength of 0.063-in. annealed sheet. Source: Ref 22, p 23

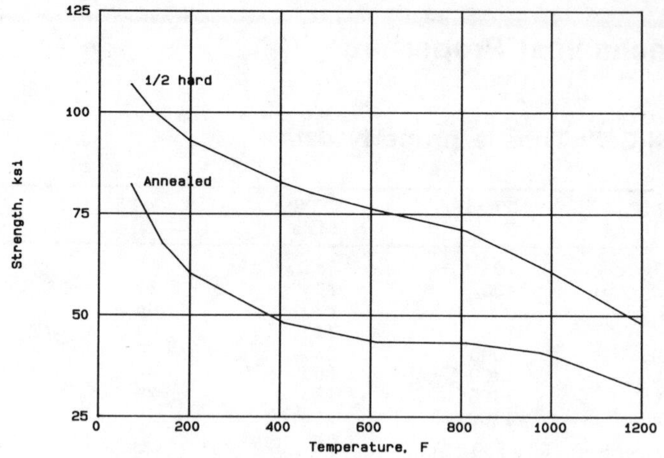

AISI 302: Shear strength of annealed and half-hard specimens. Source: Ref 22, p 23

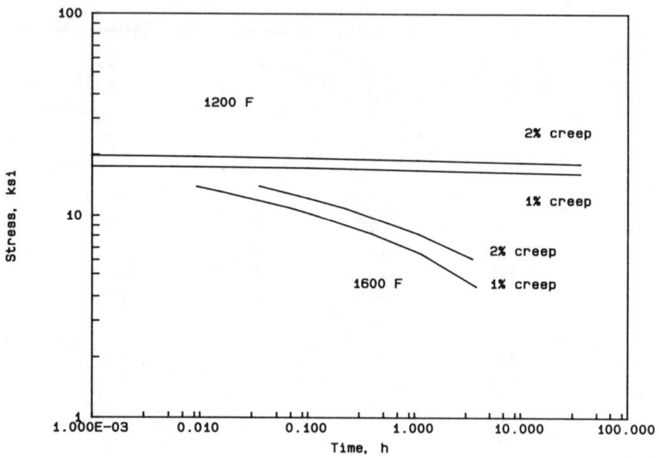

AISI 302: Creep lives of 0.050-in. annealed sheet. Source: Ref 22, p 30

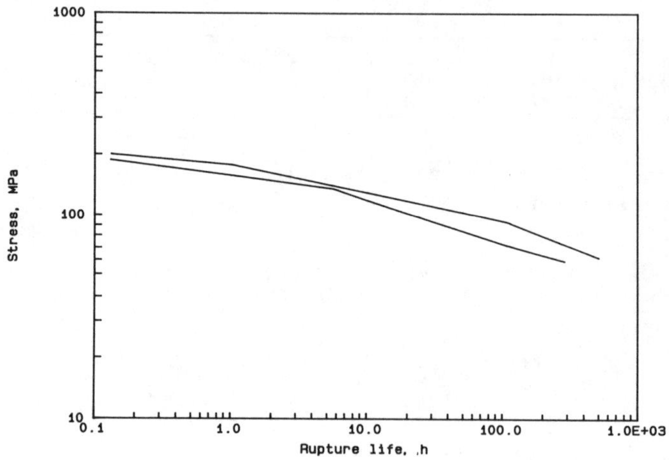

AISI 302: Stress-rupture life at 700 C. Source: Ref 3, p 229

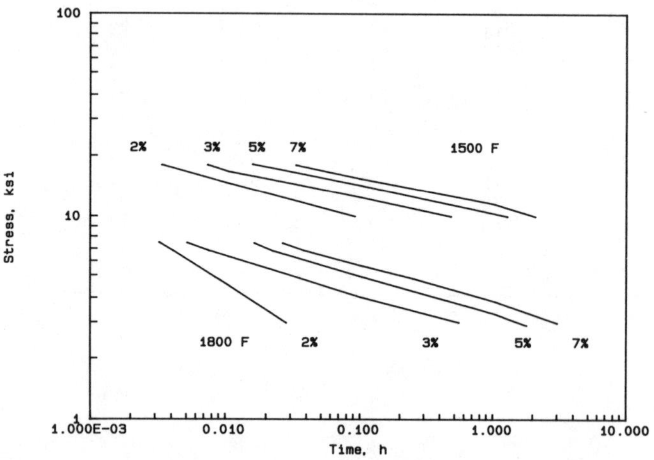

AISI 302: Short-time creep for annealed specimens. Thermal expansion included: 1200 F = 1.17%; 1500 F = 1.54%; 1800 F = 1.91%. Source: Ref 22, p 30

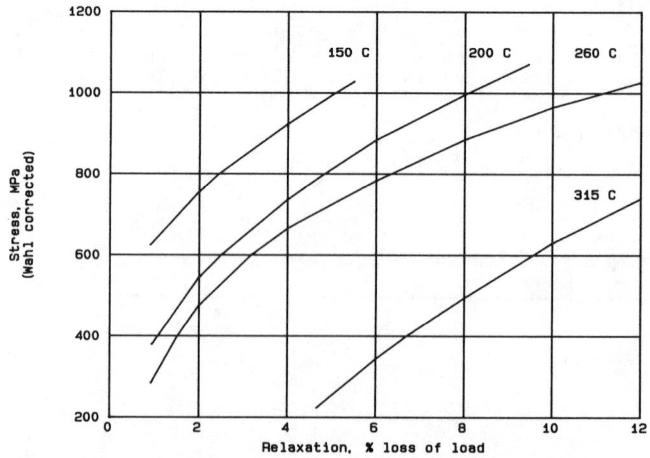

AISI 302: Relaxation curves for steel helical compression springs after 72 h at indicated temperature. Source: Ref 1, p 298

AISI 302: Stress-rupture and creep property data

TEMP C	RPSTR (a) 1000 h MPa	RPSTR (a) 1000 h ksi	RPSTR (a) 10000 h MPa	RPSTR (a) 10000 h ksi	RPSTR (a) 100000 h MPa	RPSTR (a) 100000 h ksi	CRSTR 1%/10000 h MPa	CRSTR 1%/10000 h ksi	CRSTR 1%/100000 h MPa	CRSTR 1%/100000 h ksi
480	338	49	262	38	...	...	...	...	...	...
540	241	35	183	26.5	145	21	138	20	83	12
595	152	22	117	17	90	13	90	13	52	7.5
650	97	14	69	10	48	7	57	8.3	30	4.3
705	59	8.5	41	6	28	4	34	5	17	2.5
760	41	6	26	3.7	17	2.5	21	3	10	1.5
815	26	3.7	19	2.7	10	1.5	17	2.4	8	1.2
870	19	2.7	...	...	...	...	...	...	...	...

(a) Extrapolated
Source: Ref 15, p 259

AISI 302B

UNS number: S30215
Specifications: ASTM A167, ASTM A276, ASTM A314, ASTM A473, ASTM A580, SAE J405, DGN B83, GOST EI543, GOST EI95, GOST 0Kh18N952
Nominal composition: 0.150% C max, 17.000 to 19.000% Cr, 2.000% Mn max, 8.000 to 10.000% Ni, 0.045% P max, 0.030% S max, 2.000 to 3.000% Si
Product forms: Wrought, bar, wire, plate, sheet, strip, tube, billets, forgings

Description: This high-silicon version of Type 302 boasts increased resistance to scaling at elevated temperatures. Scaling temperatures in air are: 955 C (1750 F) max for continuous service and 870 C (1600 F) max for intermittent service.
Typical applications: Components exposed to high temperatures, such as furnace parts, sill liners, heating elements, annealing covers, and burner sections.

Physical Properties

AISI 302B: Mean coefficients of linear thermal expansion

TEMP C	TEMP F	THEXP 10^{-6}/C	THEXP 10^{-6}/F
0-100	32-212	16.2	9.0
0-315	32-600	18.0	10.0
0-540	32-1000	19.4	10.8
0-650	32-1200	20.2	11.2

Source: Ref 15, p 262

AISI 302B: Thermal conductivity

TEMP C	TEMP F	THCON W/m*K	THCON Btu/ft*h*F
100	212	15.9	9.2
500	930	21.6	12.5

Source: Ref 15, p 262

Mechanical Properties

AISI 302B: Stress for 1% creep in 10,000 h

TEMP C	TEMP F	CRSTR MPa	CRSTR ksi
650	1200	48	7
705	1300	31	4.5
815	1500	7	1

Source: Ref 19, p 25

AISI 303

UNS number: S30300
Specifications: AMS 5640 (TYPE 1), ASME SA194, ASME SA320, ASTM A194, ASTM A314, ASTM A320, ASTM A473, ASTM A473, ASTM A581, ASTM A582, MIL S-862, SAE J405, ABNT V303, NF Z10CNF18-09, UNI X10CrNiS18-09, JIS G 4303 8 SUS303, BS 970 303S21 En58M, DIN 1.4305 X12CrNiS188
Nominal composition: 0.150% C max, 17.000 to 19.000% Cr, 2.000% Mn max, 0.600% Mo max, 8.000 to 10.000% Ni, 0.200% P max, 0.150% S max, 1.000% Si max
Product forms: Wrought, bar, wire, billets, forgings

Description: AISI 303 is a free-machining modification of Type 302. It is the most readily machinable of the austenitic Cr-Ni steels. Type 303 contains added sulfur and is used for heavier cuts. It is suitable for use in automatic screw machines, and is widely specified to minimize seizing or galling. Scaling temperatures in air are: 900 C (1650 F) max for continuous service and 760 C max for intermittent service.

Typical applications: Screws, rivets, bolts, nuts, studs, automobile trim, appliance parts, forged and swaged parts, aircraft fittings, and valve parts.

Physical Properties

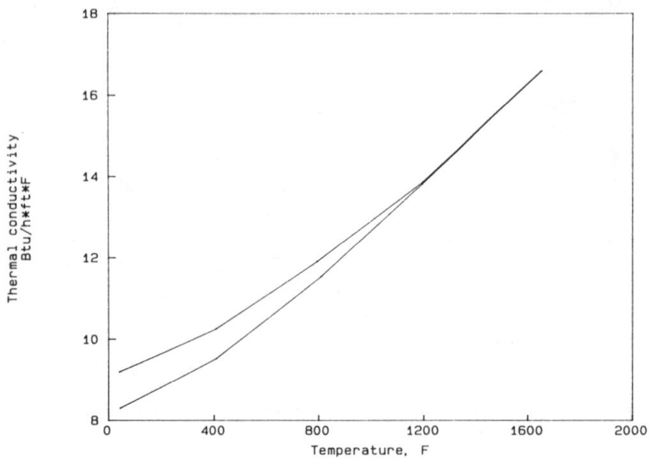

AISI 303: Thermal conductivity. Data from several reports are shown. Source: Ref 22 p 3

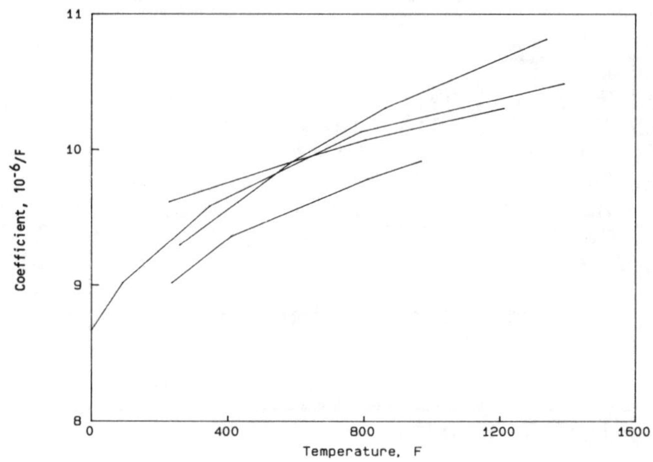

AISI 303: Mean coefficient of thermal expansion. Results of several studies are shown. Source: Ref 22, p 3

AISI 303: Thermal conductivity

TEMP C	TEMP F	THCON W/m*K	THCON Btu/ft*h*F
100	212	16.3	9.4
500	930	21.5	12.4

Source: Ref 15, p 266

AISI 303: Mean coefficients of linear thermal expansion

TEMP C	TEMP F	THEXP 10^{-6}/C	THEXP 10^{-6}/F
0-100	32-212	17.3	9.6
0-315	32-600	17.8	9.9
0-540	32-1000	18.4	10.2
0-650	32-1200	18.7	10.4

Source: Ref 15, p 266

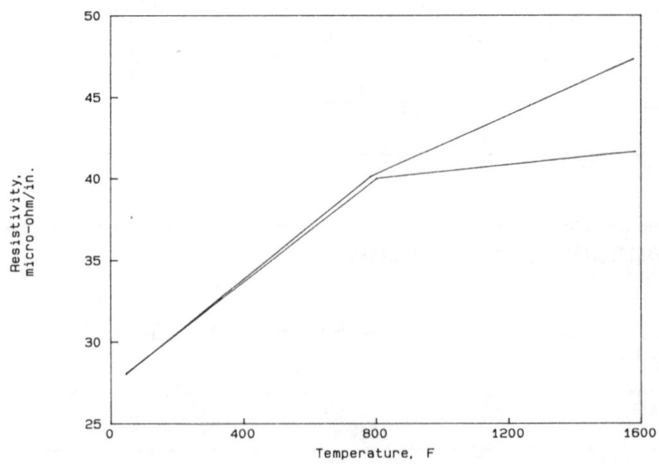

AISI 303: Electrical resistivity. Source: Ref 22, p 3

AISI 303: Electrical resistivity

TEMP C	TEMP F	ELRES n{o}*m
93	200	860
204	400	1000
427	800	1210

Source: Ref 29, Nov 1978

Mechanical Properties

AISI 303: Tensile and Izod impact property data

TEMP C	TEMP F	UTS MPa	UTS ksi	YS MPa	YS ksi	EL %	RA %	CRSTR 1%/10000 h MPa	CRSTR 1%/10000 h ksi	HARD HB	IS ft-lb (J)
21	70	620	90	240	35	50	55	...	...	160	80 (108)
425	800	420	61	240	35	35	51	...	...	...	...
540	1000	380	55	235	34	34	55	115	17	...	...
650	1200	310	45	205	30	30	54	50	7	...	...
760	1400	205	30	145	21	31	45	5	2	...	...
870	1600	140	20	70	10	34	43	...	...	...	...

Note: Test specimens were annealed and tested in short-time tensile tests.
Source: Ref: 15, p 266

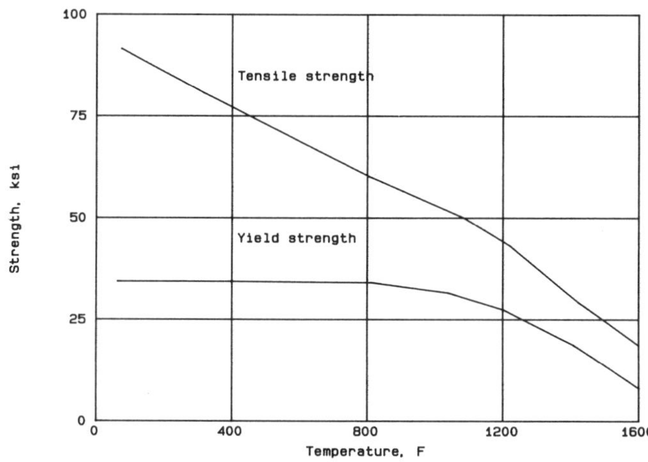

AISI 303: Tensile and yield strength of 1-in. diam round bar, annealed at 1900 F. Source: Ref 22, p 6

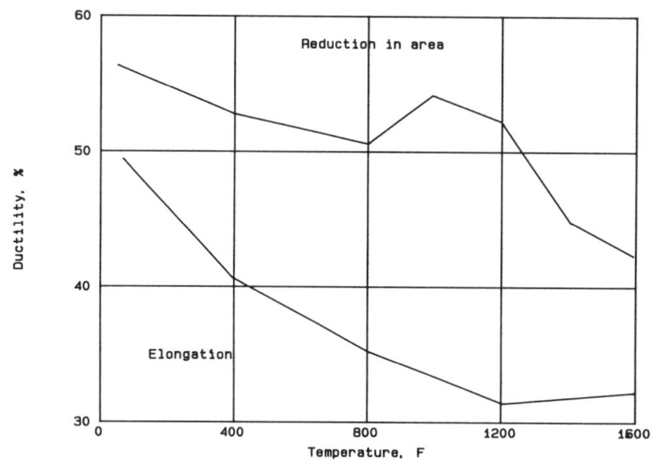

AISI 303: Tensile elongation and reduction in area of 1-in. diam round bar, annealed at 1900 F. Source: Ref 22, p 6

AISI 303Se

UNS number: S30323

Specifications: AMS 5640 (TYPE 2), AMS 5641, AMS 5738, ASME SA194, ASME SA340, ASTM A194, ASTM A314, ASTM A320, ASTM A473, ASTM A581, ASTM A582, MIL S-862, SAE J405, JIS G 4303 8 SUS303Se, DGN B83, BS 970 303541 En58M

Nominal composition: 0.150% C max, 17.000 to 19.000% Cr, 2.000% Mn max, 8.000 to 10.000% Ni, 0.200% P max, 0.060% S max, 0.150% Se min, 1.000% Si max

Product forms: Wrought, bar, wire, billets, forgings

Description: AISI 303Se is a free-machining modification of Type 302. It is the most readily machinable of the austenitic Cr-Ni steels. Type 303Se contains selenium and is used for lighter cuts. It is also suitable for use in automatic screw machines and is widely specified to minimize seizing or galling. Scaling temperatures in air are: 900 C (650 F) max for continuous service and 760 C (1400 F) max for intermittent service.

Typical applications: Screws, rivets, bolts, nuts, studs, automobile trim, appliance parts, forged and swaged parts, aircraft fittings, and valve parts.

Physical Properties

AISI 303Se: Thermal conductivity

TEMP C	TEMP F	THCON W/m*K	THCON Btu/ft*h*F
100	212	16.3	9.4
500	930	21.5	12.4

Source: Ref 15, p 266

AISI 303Se: Mean coefficients of linear thermal expansion

TEMP C	TEMP F	THEXP 10^{-6}/C	THEXP 10^{-6}/F
0-100	32-212	17.3	9.6
0-315	32-600	17.8	9.9
0-540	32-1000	18.4	10.2
0-650	32-1200	18.7	10.4

Source: Ref 15, p 266

Mechanical Properties

AISI 303Se: Tensile and Izod impact property data

TEMP C	TEMP F	UTS MPa	UTS ksi	YS MPa	YS ksi	EL %	RA %	CRSTR 1%/10000 h MPa	CRSTR 1%/10000 h ksi	HARD HB	IS ft-lb (J)
21	70	620	90	240	35	50	55	...	...	160	80 (108)
425	800	420	61	240	35	35	51	...	...	...	...
540	1000	380	55	235	34	34	55	115	17	...	...
650	1200	310	45	205	30	30	54	50	7	...	...
760	1400	205	30	145	21	31	45	5	2	...	...
870	1600	140	20	70	10	34	43	...	...	...	...

Note: Test specimens were annealed and tested in short-time tensile tests.
Source: Ref: 15, p 266

AISI 304

UNS number: S30400
Specifications: AMS 5501, AMS 5513, ASME SA182, ASME SA194, ASTM A167, ASTM A182, ASTM A193, FED QQ S-763, MIL S-862, SAE J405, NF Z6CN18-09, UNI X5CrNi18-10, JIS G 4303-27 SUS27HP, BS 970 304515 En58E, GOST 08Ch18N10, DIN 1.4301 X5CrNi189
Nominal composition: 0.080% C max, 18.000 to 20.000% Cr, 2.000% Mn max, 8.000 to 10.500% Ni, 0.045% P max, 0.030% S max, 1.000% Si max
Product forms: Wrought, bar, wire, plate, sheet, strip, tube, billets, forgings
Description: Type 304 is the most widely used of the stainless and heat-resisting steels. Its carbon content is lower than Type 302 to restrict carbide precipitation during welding. Corrosion resistance of Type 304 is somewhat higher than that of 302. Scaling temperatures in air are: 900 C (1650 F) max for continuous service; 845 C (1550 F) max for intermittent service.
Typical applications: Chemical and food-processing equipment, brewing equipment, cryogenic vessels, gutters, downspouts, and flashings.

Physical Properties

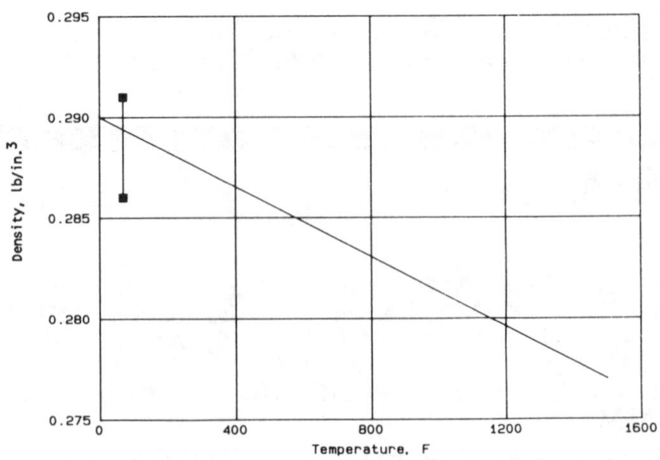

AISI 304: Density, showing room-temperature range typical for all conditions. Source: Ref 21, p 5

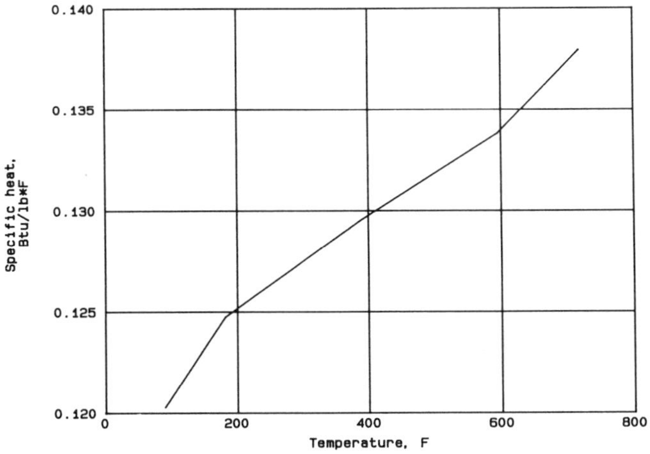

AISI 304: Specific heat. Source: Ref 22, p 6

AISI 304: Mean coefficients of linear thermal expansion

TEMP C	TEMP F	THEXP 10^{-6}/C	THEXP 10^{-6}/F
0-100	32-212	17.3	9.6
0-315	32-600	17.8	9.9
0-540	32-1000	18.4	10.2
0-650	32-1200	18.7	10.4

Source: Ref 15, p 273

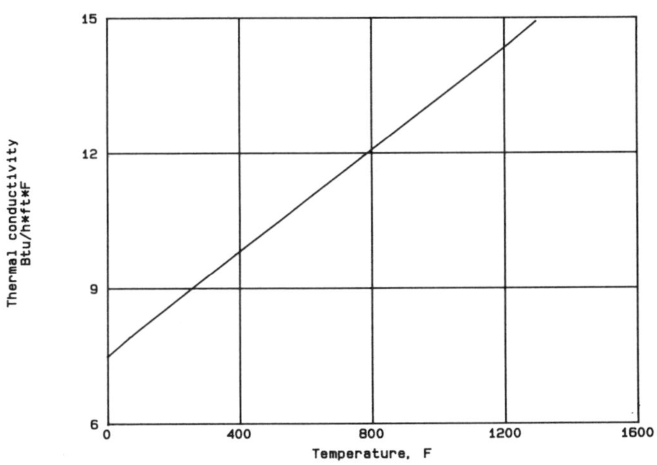

AISI 304: Thermal conductivity. Source: Ref 22, p 5

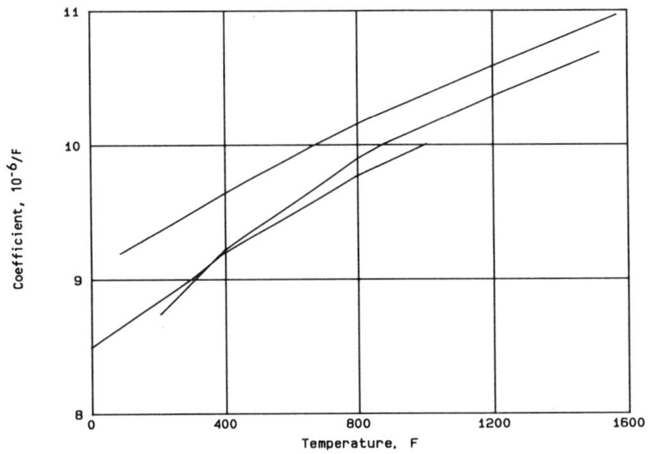

AISI 304: Mean coefficient of linear thermal expansion from room temperature to temperature indicated. Data from three tests are shown. Source: Ref 22, p 5

AISI 304: Thermal conductivity

TEMP C	TEMP F	THCON W/m*K	THCON Btu/ft*h*F
100	212	16.3	9.4
500	932	21.5	12.4

Source: Ref: 15, p 273

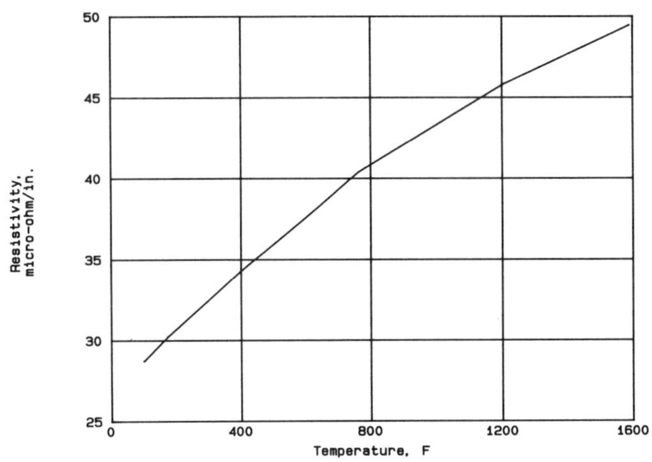

AISI 304: Electrical resistivity: Source: Ref 22, p 6

AISI 304: Modulus of elasticity

TEMP C	TEMP F	E GPa	E 10^3 ksi
93	200	192	27.9
149	300	187	27.1
204	400	183	26.6
260	500	179	26.0
316	600	177	25.6
371	700	170	24.7
427	800	166	24.1
482	900	160	23.2
538	1000	155	22.5
593	1100	150	21.8
649	1200	145	21.1
704	1300	141	20.4
760	1400	134	19.4
816	1500	125	18.1

Source: Ref 21, p 70

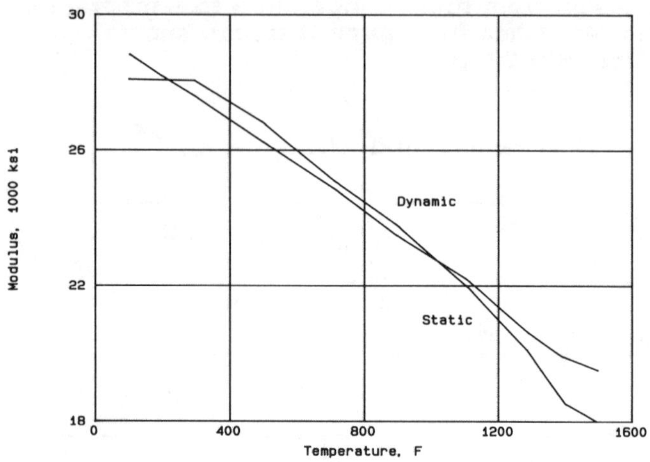

AISI 304: Modulus of elasticity of annealed bar. Source: Ref 21, p 69

AISI 304: Modulus of rigidity

TEMP C	TEMP F	G GPa	G 10^3 ksi
93	200	77	11.1
149	300	74	10.8
204	400	72	10.5
260	500	70	10.2
316	600	68	9.9
371	700	67	9.7
427	800	66	9.5
482	900	63	9.2
538	1000	61	8.9
593	1100	59	8.6
649	1200	57	8.3
704	1300	55	8.0
760	1400	53	7.7
816	1500	51	7.4

Source: Ref 21, p 70

AISI 304: Electrical resistivity

TEMP C	TEMP F	ELRES n{o}*m
21	70	720
93	200	780
204	400	860
316	600	950
427	800	1020
538	1000	1080
649	1200	1140
760	1400	1180
871	1600	1250

Source: Ref 21, p 4

AISI 304: Poisson's ratio

TEMP C	TEMP F	PR
149	300	0.28
260	500	0.30
371	700	0.32
482	900	0.28
593	1100	0.29
704	1300	0.28
816	1500	0.25

Source: Ref 21, p 71

Mechanical Properties

AISI 304: Tensile property data

TEMP C	TEMP F	UTS MPa	UTS ksi	YS MPa	YS ksi	EL %	RA %
21	70	585	85	330	37	57	70
205	400	495	72	275	23	51	73
315	600	470	68	255	19.5	45	71
425	800	440	64	160	16.5	40	69
540	1000	385	56	135	14	36	66
650	1200	305	44	115	12.5	34	58
760	1400	200	29	76	11	36	46
870	1600	110	16	...	...	40	40

Source: Ref 15, p 274

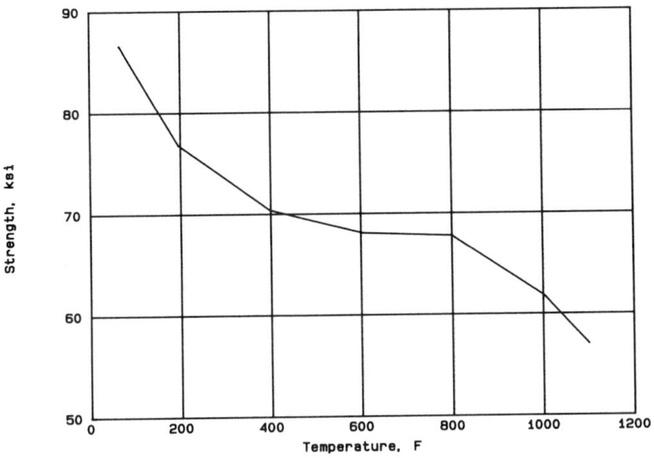

AISI 304: Average tensile strength. Source: Ref 21, p 50

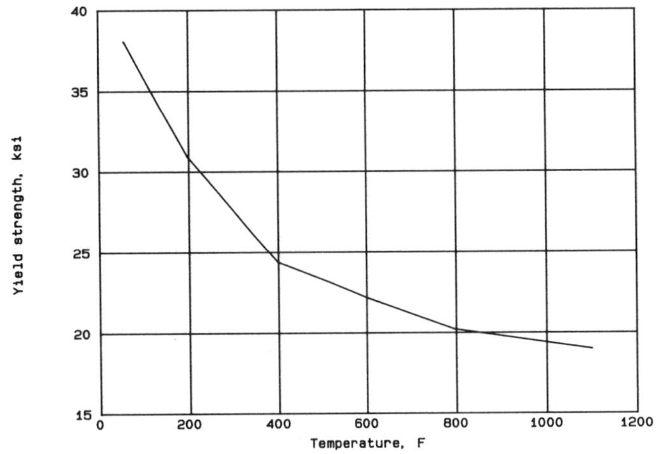

AISI 304: Average yield strength. Source: Ref 21, p 50

AISI 304: Tensile property data

TEMP C	TEMP F	UTS MPa	UTS ksi	YS MPa	YS ksi	EL(a) %	RA %
21	70	669	97.0	410	59.5	66.5	81.5
205	400	483	70.0	331	48.0	36.5	72.5
425	800	472	68.5	290	42.0	35.5	62.0
540	1000	427	62.0	265	38.5	34.5	68.0
650	1200	324	47.0	214	31.0	35.0	54.5
760	1400	207	30.0	162	23.5	44.5	63.0
870	1600	119	17.2	112	16.3	58.5	61.5
980	1800	68	9.9	66	9.6	75.5	89.5

Note: Test specimens were 16-mm (0.625-in) diam bar annealed and cold worked to hardness of 95 HRB
(a) In 4D
Source: Ref 15, p 274

AISI 304: Tensile property data at 595 C (1100 F)

	YS(a) MPa	YS(a) ksi	YS(b) MPa	YS(b) ksi	UTS(a) MPa	UTS(a) ksi	UTS(b) MPa	UTS(b) ksi	FRSTN %(a)	FRSTN %(b)	RA(a) %	RA(b) %
Plate												
0.375 in.	162	23.5	72	10.4	356	51.7	337	48.9	42.3	49.6	69.8	68.5
0.50 in.	...	...	76	11.0	...	...	341	49.4	...	47.7	...	65.1
0.75 in.	...	...	72	10.5	...	...	341	49.5	...	50.2	...	67.2
1.0 in.	...	...	77	11.1	...	...	321	46.5	...	37.5	...	51.5
2.0 in.	...	...	74	10.8	...	...	314	45.5	...	41.7	...	37.8
Bar												
1.75 in.	...	...	90	13.0	...	...	336	48.7	...	49.3	...	66.7
2.0 in.	...	...	66	9.5	...	...	336	48.8	...	48.0	...	67.5
4.5 in.	102	14.8	74	10.7	324	47	325	47.2	43.3	48.2	61.6	65.4
Pipe												
8 in., sched 80	110	15.9	77	11.2	341	49.5	339	49.2	50.6	53.9	65.9	68.9
4 in., sched 160	114	16.6	81	11.7	341	49.5	341	49.5	40.8	55.3	47.4	68.6
Tube												
2.5 in. OD x 0.625 in. hot reduced	245	35.5	654	94.9	325	47.1	337	48.9	42.9	51.3	56.9	67.5
2.5 in. OD x 0.625 in. cold drawn	107	15.5	75	10.9	359	52	350	50.7	44.4	46.0	66.9	68.5

(a) As received. (b) Annealed 1095 C (2000 F), 30 min in argon
Source: Ref: 21, p 52

AISI 304: Tensile characteristics for annealed specimens

TEMP C	TEMP F	TSRA 10^{-3}/s	SHX	FRSTN
21	68	4	0.568	0.009
430	806	4	0.392	0.0058
650	1202	4	0.314	0.0034
816	1489	4	0.166	0.002

Source: Ref 21, p 55

AISI 304: Tensile property data

TEMP C	TEMP F	YS MPa	YS ksi	UTS MPa	UTS ksi
21	70	214-238	31.0-34.5	555-593	80.5-86.0
93	200	197-228	28.5-33.1	476-497	69.1-72.1
204	400	162-193	23.5-28.0	427-448	61.9-65.0
316	600	145-166	21.0-24.1	421-445	61.0-64.5
427	800	128-152	18.5-22.1	407-441	59.0-64.0
538	1000	125-136	18.1-19.7	374-442	54.2-64.1
649	1200	121-125	17.5-18.1	295-352	42.8-51.1

Source: Ref 22, p 17

AISI 304: Effect of thermal exposure on Charpy keyhole notch impact properties of plate

EXP TEMP C	EXP TEMP F	IS(a) J	IS(a) ft-lb	IS 1000 h J	IS 1000 h ft-lb	IS 5000 h J	IS 5000 h ft-lb	IS 10000 h J	IS 10000 h ft-lb
677	1250	149	110	61	45	68	50	58	43
732	1350	...	...	57	42	61	45	57	42
788	1450	...	...	58	43	61	45	...	...

(a) Unexposed
Note: Composition: 0.057% C, 0.58% Mn, 0.21% Si, 10.41% Ni, 18.64% Cr, 0.27% Mo
Source: Ref 21, p 80

AISI 304: Stress-rupture property data at 595 C (1100 F)

Product	RPT at 241 MPa (35 ksi) h	CRTMN at 241 MPa (35 ksi) %/h	CRSTN at 241 MPa (35 ksi) %	RPT at 207 MPa (30 ksi) h	CRTMN at 207 MPa (30 ksi) %/h	CRSTN at 207 MPa (30 ksi) %	RPT at 172 MPa (25 ksi) h	CRTMN at 172 MPa (25 ksi) %/h	CRSTN at 172 MPa (25 ksi) %
Plate									
0.375 in.	14.9	0.34	14.9	150.7	0.055	15.6	527.3	0.0130	10.2
0.50 in.	36.3	0.45	25.4	137.5	0.011	22.3	759.3	0.0130	19.8
0.75 in.	27.0	0.29	12.9	123.9	0.043	9.4	769.6	0.0065	11.8
1.0 in.	23.5	0.41	...	107.1	0.100	17.4	440.1	0.0110	16.0
2.0 in.	22.2	0.37	16.2	92.8	0.078	11.4	357.1	0.0090	5.9
Bar									
0.675 in.	43.7	0.35	...	168.8	0.061	...	...	...	...
1.75 in.	33.9	0.31	19.8	142.5	0.066	14.7	600.7	0.0110	14.0
2.0 in.	33.1	0.30	15.9	169.0	0.058	17.1	614.4	0.0084	12.8
4.5 in.	40.2	0.33	19.6	159.5	0.095	23.4	611.8	0.0140	15.3
Pipe									
8 in., sched 20	32.2	0.29	15.4	165.7	0.049	13.4	630.4	0.011	13.6
4 in., sched 160	37.8	0.56	24.6	160.9	0.067	22.1	758.9	0.0074	13.0
Tube									
2.5 in. OD x 0.675 in. hot reduced	23.3	0.38	15.0	142.9	0.045	10.7	690.7	0.0076	11.1
2.5 in. OD x 0.675 in. cold drawn	38.5	0.26	15.2	179.1	0.045	13.0	758.4	0.0064	12.2

Note: All material annealed 30 min at 1093 C (2000 F). Data are average of up to three specimens
Source: Ref 21, p 117

AISI 304: Stress-rupture and creep property data

TEMP C	RPSTR 1000 h MPa	RPSTR 1000 h ksi	RPSTR(a) 10000 h MPa	RPSTR(a) 10000 h ksi	RPSTR(a) 100000 h MPa	RPSTR(a) 100000 h ksi	Stress(b) 0.0001%/h MPa	Stress(b) 0.0001%/h ksi	Stress(b) 0.00001%/h MPa	Stress(b) 0.00001%/h ksi
480	338	49.0	262	38.0	...	...	...	...	...	...
540	241	35.0	183	26.5	145	21.0	138	20.0	83	12.0
595	152	22.0	117	17.0	90	13.0	90	13.0	52	7.5
650	97	14.0	69	10.0	48	7.0	57	8.3	30	4.3
705	59	8.5	41	6.0	28	4.0	34	5.0	17	2.5
760	41	6.0	26	3.7	17	2.5	21	3.0	10	1.5
815	26	3.7	19	2.7	10	1.5	17	2.4	8	1.2
870	19	2.7	...	...	...	...	...	...	...	...

(a) Extrapolated (b) Stress for secondary creep rate of extrapolated data
Source: Ref 15, p 275

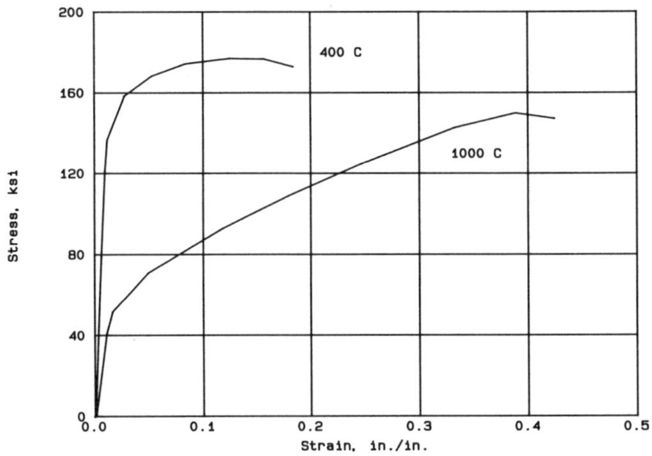

AISI 304: Compressive stress-strain curves. Strain rate is 0.01/s at room temperature. Source: Ref 21, p 72

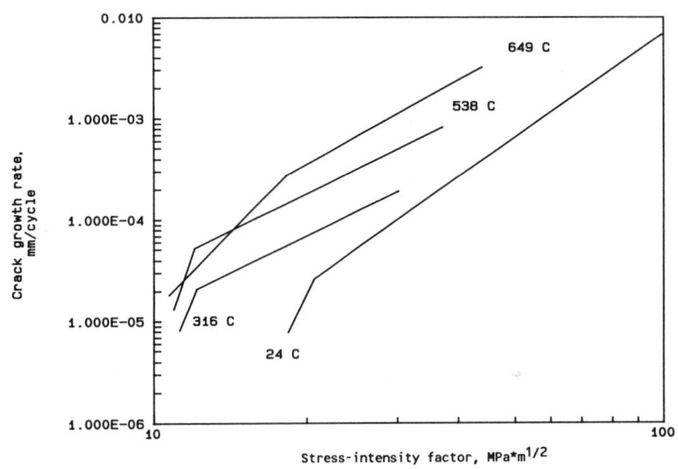

AISI 304: Fatigue-crack growth rates of annealed specimens tested in air at a frequency of 0.066 Hz and R = 0 to 0.05. Source: Ref 23, p 203

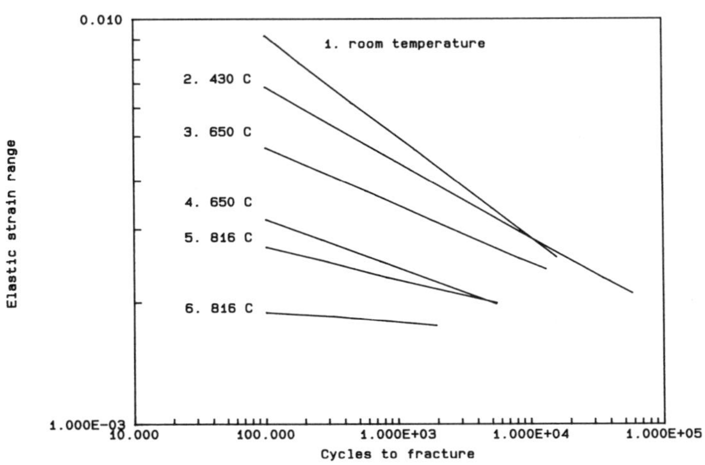

AISI 304: Low-cycle fatigue of annealed sheet at a strain rate of 4 X 10^{-3}/s (Curves 1, 2, 3), 4 X 10^{-5}/s (Curves 4, 6), and 4 X 10^{-3}/s (Curve 5). Source: Ref 21, p 96

AISI 304: Fatigue data for tension/compression cycle tests at 650 C (1200 F)

Tension min	Compression min	TSRGX %	FAT.n(a) cycles	FAT.n(b) cycles
0	0	1.98	592	546
30	30	1.98	380	416
30	0	2.08	146	...
30	0	2.02	...	165
30	3	1.98	308	...
30	3	2.00	...	336

Note: Testing consisted of a 30-min hold period in tension plus a short hold period in compression. Tested in air at a strain rate of 4 x 10^{-3}/s.
(a) Test 1. (b) Test 2
Source: Ref 8, p 350

AISI 304: Effect of hold-time in tension-hold-only testing on fatigue resistance at 650 C (1200 F)

Time min	FAT.n cycles(a)	FAT.n cycles(b)
0	592	546
0.1	570	545
1.0	329	331
10.0	193	201
30.0	146	165
60.0	144	158
180.0	150	120

Note: Strain rate of 4 x 10^{-3}/s for a strain range of about 2.0%
(a) Test 1. (b) Test 2
Source: Ref 8, p 349

AISI 304: Comparison of monotonic and cyclic stress-strain properties of annealed specimens

TEMP C	TEMP F	TSR 1/s	TSRGX	YS(a) MPa	YS(a) ksi	YS(b) MPa	YS(b) ksi	YS ratio(c)	CSHX(a)	CSHX(b)
21	70	4x10^{-3}	0.03	216	31.4	457	66.3	2.11	0.152	0.163
427	800	4x10^{-3}	0.03	121	17.5	333	48.3	2.76	0.126	0.116
538	1000	4x10^{-3}	0.03	101	14.6	300	43.5	2.98	0.164	0.097
538	1000	4x10^{-3}	0.01	101	14.6	236	34.3	2.35	0.293	0.143
538	1000	4x10^{-5}	0.01	101	14.6	276	40.0	2.74	0.288	0.141
649	1200	4x10^{-3}	0.03	98	14.2	252	36.6	2.58	0.172	0.101
649	1200	4x10^{-3}	0.01	98	14.2	223	32.4	2.28	0.182	0.120
649	1200	4x10^{-5}	0.01	91	13.2	170	24.6	1.86	0.183	0.115

Note: 0.252-in. diameter specimens
(a) Monotonic. (b) Cyclic. (c) Ratio of monotonic yield strength to cyclic yield strength
Source: Ref 21, p 130

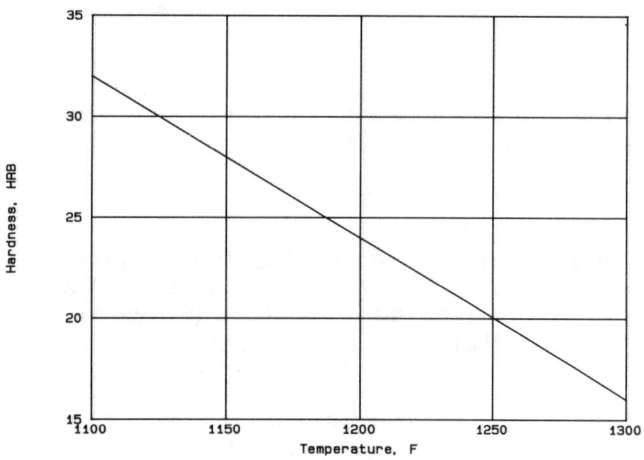

AISI 304: Effect of temperature on hot hardness of annealed specimens. At room temperature, HRB = 70. Source: Ref 21, p 10

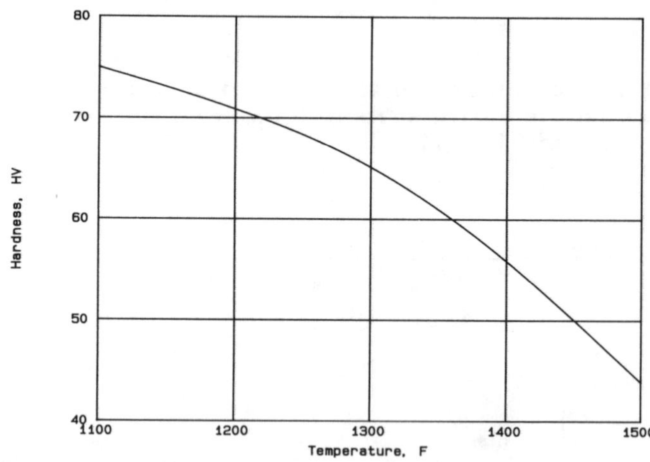

AISI 304: Vickers hardness of annealed specimens. At room temperature, HV = 130. Source: Ref 21, p 10

AISI 304H

UNS number: S30409
Specifications: ASME SA182, ASME SA213, ASME SA240, ASME A249, ASME A312, ASME A376, ASME A403, ASME A430, ASME A479, ASTM 182, ASTM 213, ASTM 240, ASTM 249, ASTM 271, ASTM 312, ASTM A358, ASTM A376, ASTM 403, ASTM A430, ASTM A479
Nominal composition: 0.040 to 0.100% C, 18.000 to 20.000% Cr, 2.000% Mn max, 8.000 to 10.500% Ni, 0.045% P max, 0.030% S max, 1.000% Si max
Description: This higher carbon modification of Type 304 has improved elevated-temperature strength. Scaling temperatures in air are: 900 C (1650 F) max for continuous service and 845 C (1550 F) max for intermittent service.
Typical applications: AISI 304H is used for Type 304 applications that require slightly higher elevated-temperature strength.

Mechanical Properties

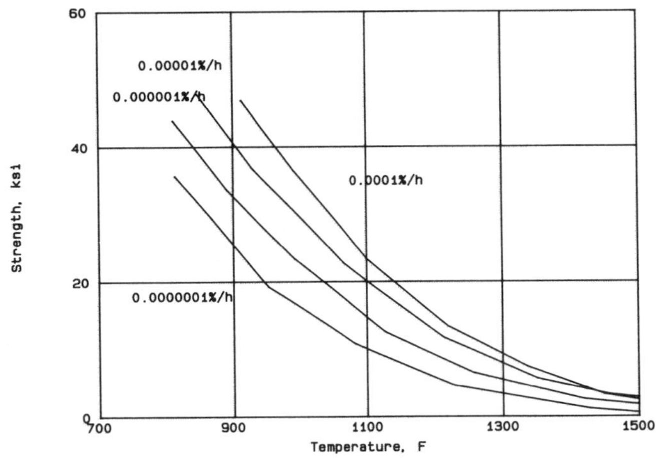

AISI 304H: Creep rates for annealed specimens as a function of stress and temperature. Source: Ref 21, p 121

AISI 304L

UNS number: S30403
Specifications: AMS 5511, AMS 5647, ASME SA182, ASME SA213, ASTM A167, ASTM A213, FED QQ S-763, MIL S-862, SAE J405, NF Z2CN18-10, UNI X2CrNi18-10, JIS G 403-28 SUS 28B, DGN B83, BS 1501 801 Grade C, GOST 03CH18N11, DIN 1.4306 X2CrNi189
Nominal composition: 0.030% C max, 18.000 to 20.000% Cr, 2.000% Mn max, 8.000 to 12.000% Ni, 0.045% P max, 0.030% S max, 1.000% Si max, 0.100% N max
Product forms: Wrought, bar, wire, plate, sheet, strip, tube, billets, forgings
Description: AISI 304L is a very low carbon modification of Type 304 with similar general corrosion resistance, but superior resistance to intergranular corrosion after welding or stress relieving. It is recommended for weldments that cannot be subsequently annealed. Maximum service temperature is about 425 C (800 F).
Typical applications: Coal hopper linings, and tanks for liquid fertilizer and tomato paste.

Physical Properties

AISI 304L: Mean coefficients of linear thermal expansion

TEMP C	TEMP F	THEXP 10^{-6}/C	THEXP 10^{-6}/F
0-100	32-212	17.28	9.6
0-316	32-600	17.82	9.9
0-538	32-1000	18.36	10.2
0-649	32-1200	18.72	10.4

Source: Ref 14, p 206

AISI 304L: Thermal conductivity

TEMP C	TEMP F	THCON W/m*K	THCON Btu/h*ft F
100	212	16.27	9.4
500	932	21.46	12.4

Source: Ref 14, p 206

Mechanical Properties

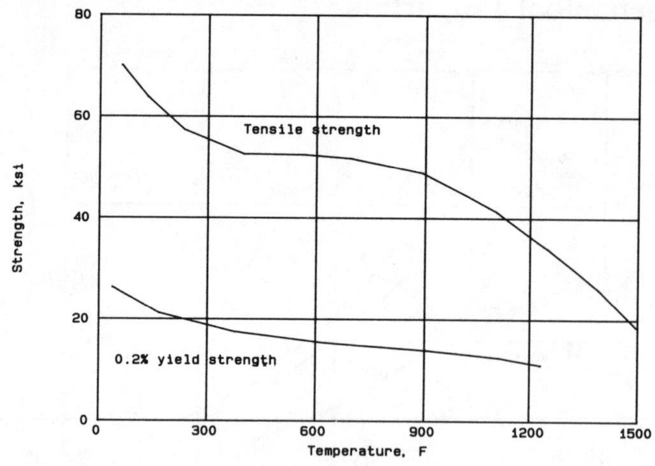

AISI 304L: Tensile and yield strength of annealed specimens. Source: Ref 21, p 121

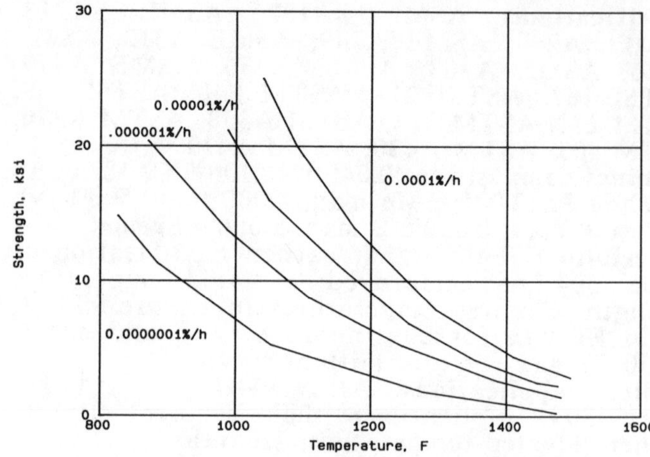

AISI 304L: Creep strength of annealed specimens. Source: Ref 21, p 121

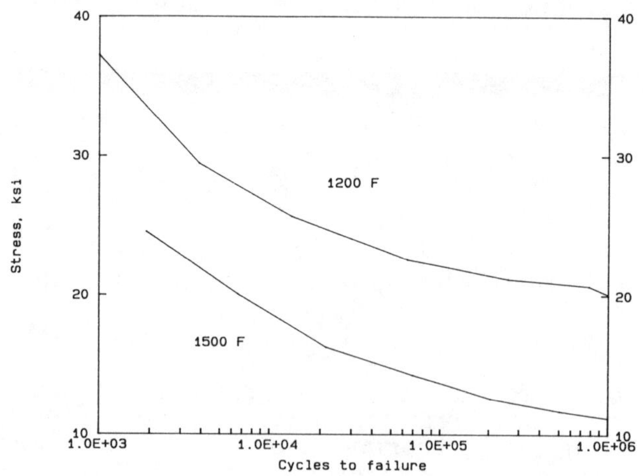

AISI 304L: Axial fatigue for 0.75-in. diam annealed bar. Source: Ref 21, p 97

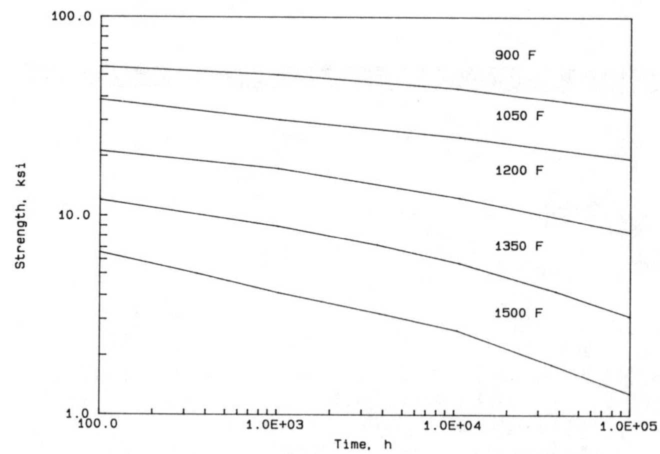

AISI 304L: Stress-rupture strength of bar. Specimens were treated at 1950 F for 0.5 h and water quenched. Source: Ref 22, p 23

AISI 304N

UNS number: S30451
Specifications: AMS 5511, AMS 5647, ASME SA182, ASME SA213, ASME SA240, ASME SA312, ASME SA358, ASME SA376, ASME SA430, ASME SA479, ASTM A182, ASTM A213, ASTM A240, ASTM A249, ASTM A312, ASTM A358, ASTM A376, ASTM A403, ASTM A430
Nominal composition: 0.080% C max, 18.000 to 20.000% Cr, 2.000% Mn max, 0.100 to 0.160% N, 8.000 to 10.500% Ni, 0.045% P max, 0.030% S max, 1.000% Si max

Product forms: Wrought, bar, wire, plate, sheet, strip, tube, billets, forgings
Description: AISI 304N is a modification of Type 304 with added nitrogen to increase strength with minimum effect on ductility and corrosion resistance. Type 304N is also more resistant to increased magnetic permeability on cold working than Type 304.
Typical applications: Type 304N is used for Type 304 applications that require increased strength.

Mechanical Properties

AISI 304N: Stress-rupture strength of bar

TEMP C	TEMP F	RPSTR 10000 h MPa	RPSTR 10000 h ksi	RPSTR 100000 h MPa	RPSTR 100000 h ksi
566	1050	234	34.0	186	27.0
650	1200	124	18.0	86	12.5
732	1350	61	8.8	41	6.0
816	1500	23	3.3	28	4.0

Note: Alloy contained 0.14% N
Source: Ref 22, p 24

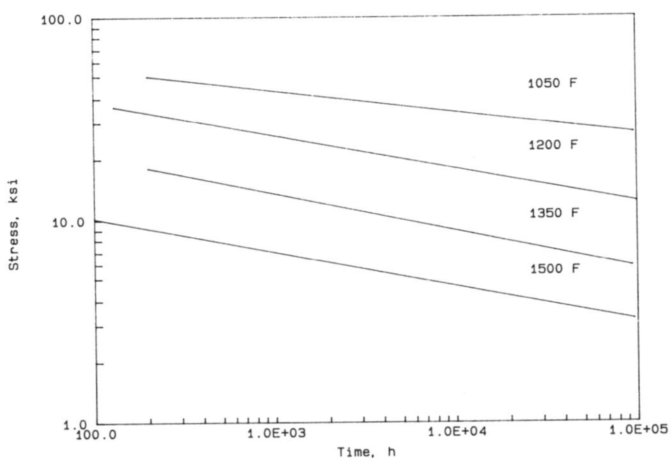

AISI 304N: Stress-rupture strength of bar with a nitrogen content of 0.14%. Source: Ref 22, p 24

AISI 304N: Tensile property data

TEMP C	TEMP F	YS MPa	YS ksi	UTS MPa	UTS ksi
21	70	214-238	31.0-34.5	555-593	80.5-86.0
93	200	197-228	28.5-33.1	476-497	69.1-72.1
204	400	162-193	23.5-28.0	427-448	61.9-65.0
316	600	145-166	21.0-24.1	421-445	61.0-64.5
427	800	128-152	18.5-22.1	407-441	59.0-64.0
538	1000	125-136	18.1-19.7	374-442	54.2-64.1
649	1200	121-125	17.5-18.1	295-352	42.8-51.1

Note: Specimens were taken from 3 in. x 3 in. bar, were annealed at 1080 C (1975 F) for 1 h, and water quenched. Alloy contained 0.12 to 0.16% N
Source: Ref 22, p 17

AISI 305

UNS number: S30500
Specifications: AMS 5514, AMS 5685, ASME SA193, ASME SA194, ASME SA240, ASTM A167, ASTM A240, ASTM A249, FED QQ S-763, SAE J405, NF Z8CN18-12, UNI X8CrNi18-12, JIS G 4304 SUS62HP, DGN B83, SIS 14 23 31, BS 801 Grade A, GOST 10Ch18N9L, DIN 1.4303, DIN 1.4312
Nominal composition: 0.120% C max, 17.000 to 19.000% Cr, 2.000% Mn max, 10.500 to 13.000% Ni, 0.045% P max, 0.030% S max, 1.000% Si max
Product forms: Wrought, bar, wire, plate, sheet, strip, tube, billets, forgings
Description: AISI 305 is a low alloy, work-hardening stainless steel used for spin forming, severe drawing, and cold heading. Scaling temperature for continuous service in air is 900 C (1650 F) max.
Typical applications: Coffee urn tops, mixing bowls, and reflectors.

Physical Properties

AISI 305: Thermal conductivity

TEMP C	TEMP F	THCON W/m*K	THCON Btu/ft*h*F
100	212	16.3	9.4
500	930	21.5	12.4

Source: Ref 15, p 282

AISI 305: Mean coefficients of linear thermal expansion

TEMP C	TEMP F	THEXP 10^{-6}/C	THEXP 10^{-6}/F
0-100	32-212	17.3	9.6
0-315	32-600	17.8	9.9
0-540	32-1000	18.4	10.2
0-650	32-1200	18.7	10.4

Source: Ref 15, p 282

AISI 308

UNS number: S30800
Specifications: ASTM A167, ASTM A276, ASTM A314, ASTM A473, ASTM A580, SAE G 4303 SUS308, DGN B83
Nominal composition: 0.080% C max, 19.000 to 21.000% Cr, 2.000% Mn max, 10.000 to 12.000% Ni, 0.045% P max, 0.030% S max, 1.000% Si max
Product forms: Wrought, bar, wire, plate, sheet, strip

Description: AISI 308 is a higher alloy austenitic steel combining corrosion and heat resistance. Scaling temperatures in air are: 925 C (1700 F) max for continuous service and 845 C (1550 F) max for intermittent service.

Typical applications: Filler metals that compensate for alloy loss in welding and industrial furnace components.

Physical Properties

AISI 308: Mean coefficients of linear thermal expansion

TEMP C	TEMP F	THEXP 10^{-6}/C	THEXP 10^{-6}/F
0-100	32-212	17.3	9.6
0-315	32-600	17.8	9.9
0-540	32-1000	18.4	10.2
0-650	32-1200	18.7	10.4

Source: Ref 15, p 284

AISI 308: Thermal conductivity

TEMP C	TEMP F	THCON W/m*K	THCON Btu/ft*h*F
100	212	15.2	8.8
500	930	21.6	12.5

Source: Ref 15, p 284

AISI 308: Electrical resistivity

TEMP C	TEMP F	ELRES n{o}*m
21	70	720
650	1200	1160

Source: Ref 15, p 284

AISI 309

UNS number: S30900
Specifications: ASME SA249, ASME SA312, ASTM 167, ASTM A249, ASTM A276, ASTM A312, FED QQ S-763, FED QQ S-766, MIL S-862, SAE J405, NF Z12CNS25-13, UNI X16CrNi23-14, DGN B83, DGN B171, BS 970 309S24, GOST Kh21N15, DIN 1.4830
Nominal composition: 0.200% C max, 22.000 to 24.000% Cr, 2.000% Mn max, 12.000 to 15.000% Ni, 0.045% P max, 0.030% S max, 1.000% Si max
Product forms: Wrought, bar, wire, plate, sheet, strip
Description: AISI 309 is an austenitic Cr-Ni steel combining high-temperature strength with resistance to scaling. Scaling temperatures in air are: 165 C (1950 F) max for continuous service and 1010 C (1850 F) for intermittent service.
Typical applications: Aircraft heaters, heat treating trays, annealing covers, furnace parts, heat exchangers, oven linings, pump parts, air preheater baffles, and sulfite liquor-handling equipment.

Physical Properties

AISI 309: Thermal conductivity

TEMP C	TEMP F	THCON W/m*K	THCON Btu/ft*h*F
100	212	15.6	9.0
500	930	18.7	10.8

Source: Ref 15, p 286

AISI 309: Mean coefficients of linear thermal expansion

TEMP C	TEMP F	THEXP 10^{-6}/C	THEXP 10^{-6}/F
0-100	32-212	14.9	8.3
0-315	32-600	16.7	9.3
0-540	32-1000	17.3	9.6
0-650	32-1200	18.0	10.0
0-980	32-1800	20.7	11.5

Source: Ref 15, p 286

Mechanical Properties

AISI 309: Tensile property data

TEMP C	TEMP F	UTS MPa	UTS ksi	YS MPa	YS ksi	EL %	RA %
21	70	620	90	290	42	50	77
150	300	560	81	255	37	47	77
205	400	550	80	240	35	46	75
260	500	530	77	220	32	45	75
315	600	515	75	205	30	44	73
370	700	510	74	195	28	43	72
425	800	495	72	185	27	40	67
480	900	475	69	170	25	39	66
540	1000	455	66	165	24	36	58
595	1100	405	59	160	23	35	54
650	1200	380	55	150	22	35	50
705	1300	305	44	145	21	37	41
760	1400	250	36	140	20	40	40
815	1500	185	27	130	19	46	37
870	1600	145	21	125	18	50	48
925	1700	105	15	...	...	59	47
980	1800	76	11	...	...	65	66

Source: Ref 15, p 287

AISI 309: Stress-rupture property data

TEMP C	TEMP F	RPSTR 1000 h MPa	RPSTR 1000 h ksi	RPSTR 10000 h MPa	RPSTR 10000 h ksi	RPSTR 100000 h MPa	RPSTR 100000 h ksi
650	1200	141	20.5	103	15.0	78	11.3
705	1300	81	11.7	48	7.0	26	3.7
760	1400	42	7.5	31	4.5	21	3.0
815	1500	34	5.0	28	4.0	10	1.5
870	1600	22	3.2	10	1.5	6	0.9
980	1800	7	1.0	6	0.8	3	0.5

Source: Ref 15, p 286

AISI 309: Minimum creep rate data for annealed bar

TEMP C	TEMP F	CRSTR 0.0001%/h MPa	CRSTR 0.0001%/h ksi
538	1000	110	15.9
593	1100	80	11.6
649	1200	55	8.0
704	1300	31	4.5
816	1500	7	1.0

Note: Specimens were taken from 25-mm (1-in.) diam bars
Source: Ref 29, Apr 1982

AISI 310

UNS number: S31000
Specifications: ASME SA182, ASME SA213, ASTM A167, ASTM A182, ASTM A213, ASTM A249, FED QQ S-763, FED QQ S-766, MIL S-862, SAE J405, NF Z12CN25-20, UNI X16Cr-NiSi25-20. JIS G 3311-33 SUH33B, DGN B83, SIS 14 23 61, GOST 20Ch25N2052, DIN 1.4841
Nominal composition: 0.250% C max, 24.000 to 26.000% Cr, 2.000% Mn max, 19.000 to 22.000% Ni, 0.045% P max, 0.030% S max, 1.500% Si max
Product forms: Wrought, bar, wire, plate, sheet, strip, billets, forgings

Description: AISI 310 is an austenitic Cr-Ni steel with higher elevated-temperature strength and scale resistance than Type 309. Scaling temperatures in air are: 1120 C (2050 F) max for continuous service and 1040 C (1900 F) max for intermittent service.

Typical applications: Heat exchangers, furnace parts, combustion chambers, welding filler metals, gas turbine parts, incinerators, recuperators, and rolls for roller hearth furnaces.

Physical Properties

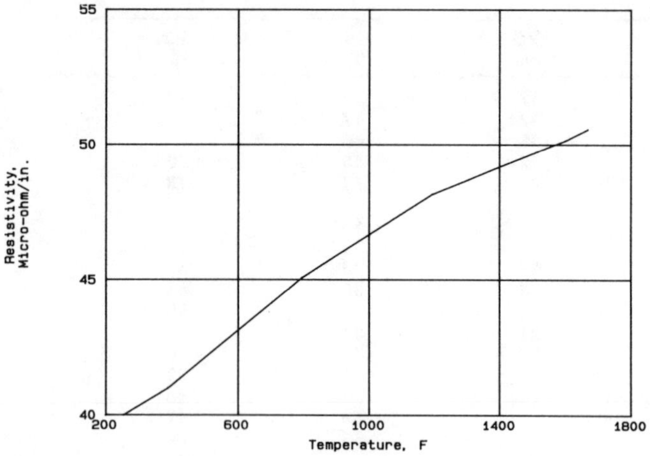

AISI 310: Electrical resistivity of wrought specimens. Source: Ref 22, p 5

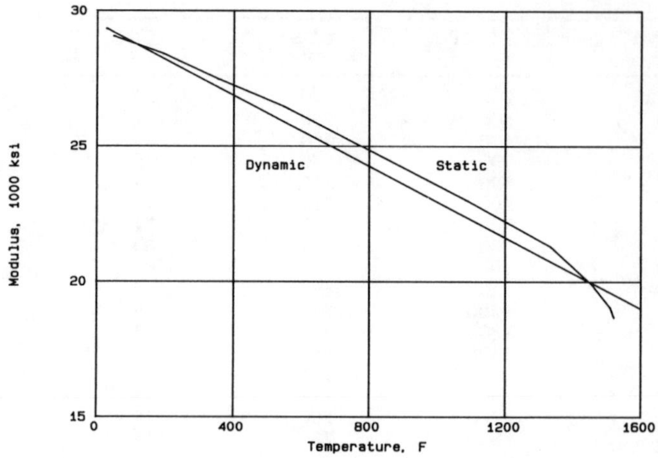

AISI 310: Modulus of elasticity. Source: Ref 22, p 20

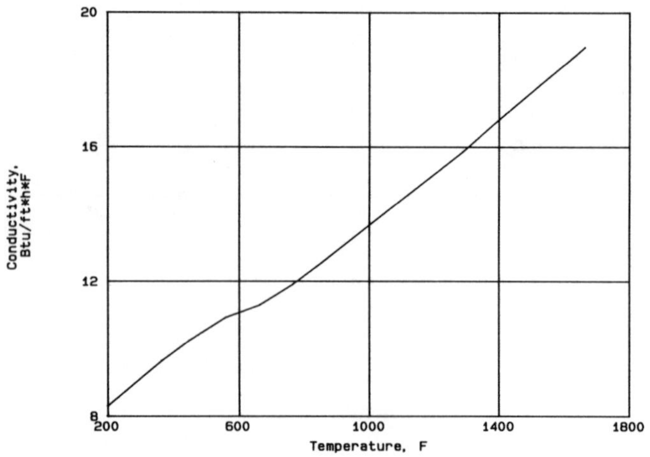

AISI 310: Thermal conductivity. Source: Ref 22, p 5

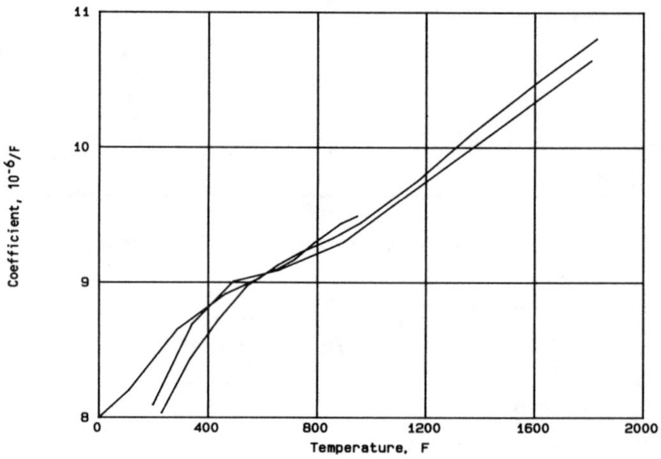

AISI 310: Mean coefficient of linear thermal expansion obtained from several different studies. Source: Ref 22, p 5

AISI 310: Thermal conductivity

TEMP C	TEMP F	THCON W/m*K	THCON Btu/h*ft*F
100	212	14.19	8.2
500	932	18.69	10.8

Source: Ref 27, p 49

AISI 310: Mean coefficients of linear thermal expansion

TEMP C	TEMP F	THEXP 10^{-6}/C	THEXP 10^{-6}/F
0-100	32-212	15.9	8.8
0-315	32-600	16.2	9.0
0-538	32-1000	17.0	9.4
0-649	32-1200	17.5	9.7
0-982	32-1800	19.1	10.6

Source: Ref 27, p 49

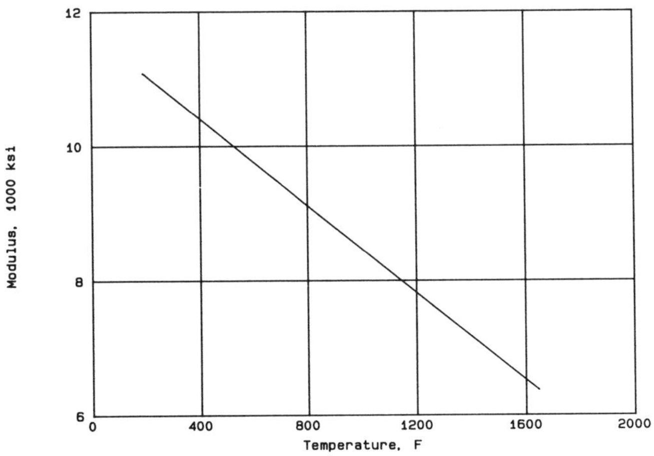

AISI 310: Modulus of rigidity of annealed specimens. Source: Ref 22, p 21

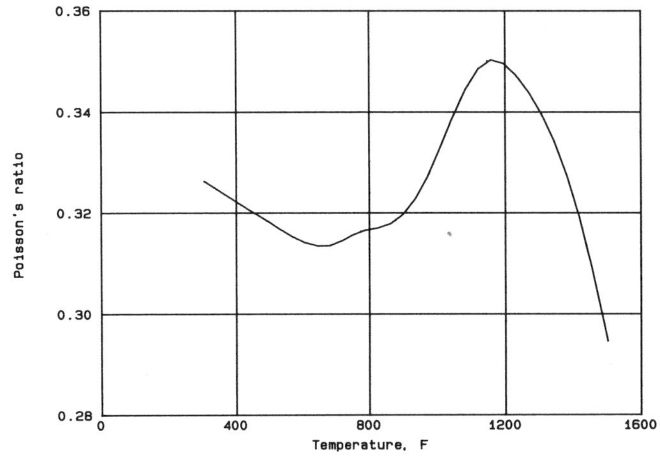

AISI 310: Poisson's ratio. Source: Ref 22, p 19

Mechanical Properties

AISI 310: Tensile property data

TEMP C	TEMP F	UTS MPa	UTS ksi	YS MPa	YS ksi	EL %	RA %
21	70	625	91	290	42	47	70
150	300	600	87	240	35	39	70
205	400	585	85	230	33	38	69
260	500	570	83	220	32	37	69
315	600	565	82	205	30	37	69
370	700	560	81	200	29	37	67
425	800	525	76	185	27	36	63
480	900	515	75	180	26	34	61
540	1000	495	72	165	24	33	55
595	1100	450	65	150	22	33	50
650	1200	385	56	145	21	35	45
705	1300	340	49	140	20	36	40
760	1400	285	41	130	19	37	39
815	1500	235	34	125	18	50	38
870	1600	185	27	110	16	45	37
925	1700	215	31	...	...	50	40
980	1800	90	13	...	...	57	40

Source: Ref 15, p 289

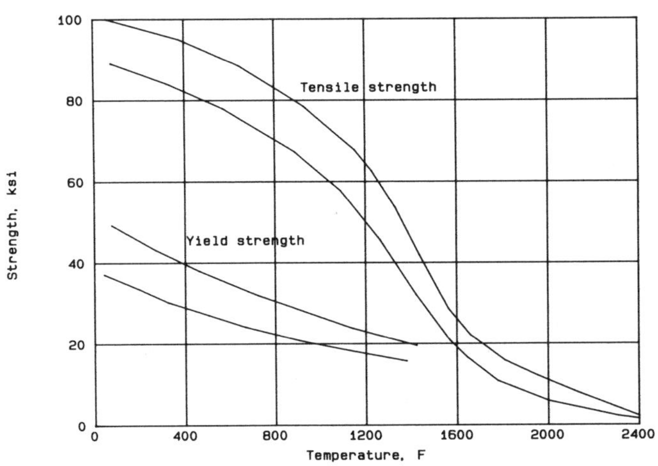

AISI 310: Scatterbands for tensile and yield strength of annealed bar. Source: Ref 22, p 14

AISI 310: Stress-rupture property data and creep strength for annealed bar

TEMP C	RPSTR 1000 h MPa	RPSTR 1000 h ksi	RPSTR 10000 h MPa	RPSTR 10000 h ksi	RPSTR 100000 h MPa	RPSTR 10000 h ksi	CRSTR 1%/10000 h MPa	CRSTR 1%/10000 h ksi	CRSTR 1%/100000 h MPa	CRSTR 1%/100000h ksi
540	220	32	170	25	125	18	140	20	90	13
595	160	23	110	16	83	12	105	15	55	8
650	110	16	69	10	55	8	55	8	40	6
705	69	10	40	6	35	5	35	5	30	4
760	50	7	30	4	20	3	15	2	15	2
815	35	5	15	2	7	1	7	1	...	...
870	20	3	7	1	...	...	...	...	...	...
980	7	1	...	...	...	...	...	...	...	...

Source: Ref 15, p 289

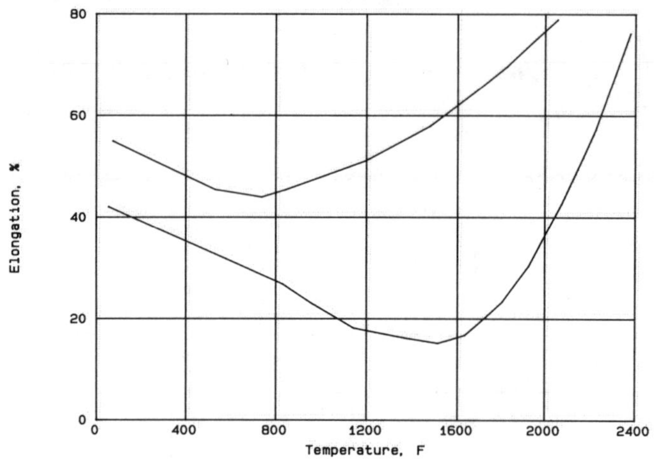

AISI 310: Scatterband for tensile elongation of annealed bar. Source: Ref 22, p 14

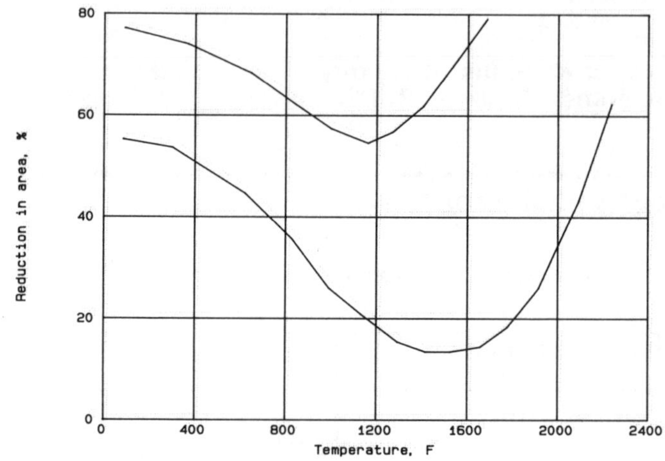

AISI 310: Scatterband for tensile reduction in area of annealed bar. Source: Ref 22, p 14

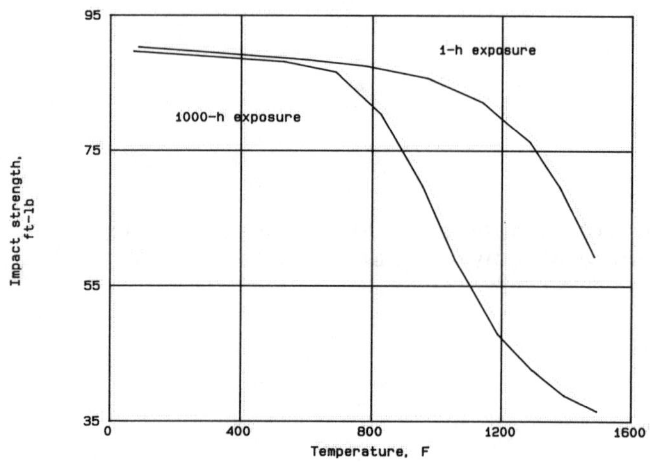

AISI 310: Effect of exposure time and temperature on Charpy V-notch impact strength of 1-in. bar. Specimens were heat treated at 2150 F and water quenched, with an ASTM grain size of 0.67. Source: Ref 22, p 15

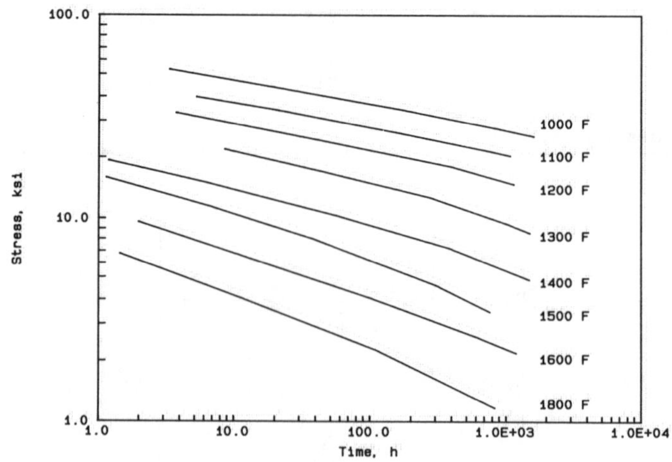

AISI 310: Stress-rupture strength of 1-in. annealed bar. Specimens were treated at 2150 F and water quenched, with an ASTM grain size of 1 to 4. Source: Ref 22, p 17

AISI 310S

UNS number: S31008
Specifications: AMS 5521, AMS 5572, AMS 5577, AMS 5651, ASME SA240, ASME SA479, ASTM A167, ASTM A240, ASTM A276, ASTM A314, FED QQ S-763, SAE J405, UNI CrNi2520, JIS G 4303-42 SUS 42B, DGN B83, DGN B171, SIS 14 23 61, DIN 1.4335
Nominal composition: 0.080% C max, 24.000 to 26.000% Cr, 2.000% Mn max, 19.000 to 22.000% Ni, 0.045% P max, 0.030% S max, 1.500% Si max
Product forms: Wrought, bar, wire, plate, sheet, strip, billets, forgings

Description: AISI 310S is an austenitic Cr-Ni steel with higher elevated-temperature strength and scale resistance than Type 309. Type 310S is the low-carbon modification for welded construction. Scaling temperatures in air are: 1120 C (2050 F) max for continuous service and 1040 C (1900 F) max for intermittent service.
Typical applications: Heat exchangers, furnace parts, combustion chambers, welding filler metals, gas turbine parts, jet engine rings, incinerators, recuperators, and rolls for roller hearth furnaces.

Physical Properties

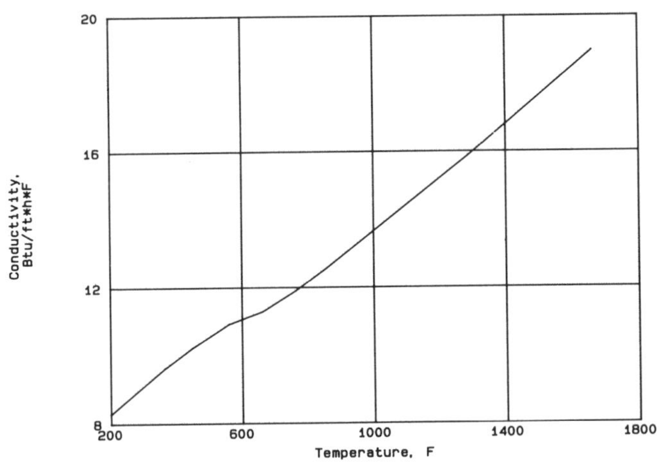

AISI 310S: Thermal conductivity. Source: Ref 22, p 5

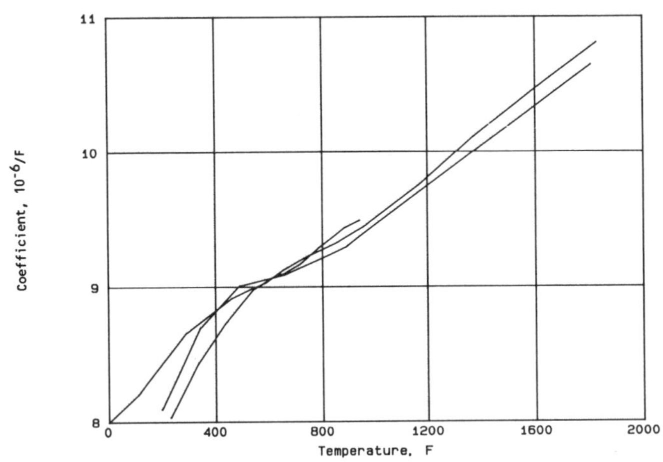

AISI 310S: Mean coefficient of linear thermal expansion obtained from several different studies. Source: Ref 22, p 5

AISI 310S: Thermal conductivity

TEMP C	TEMP F	THCON W/m*K	THCON Btu/h*ft*F
100	212	14.19	8.2
500	932	18.69	10.8

Source: Ref 27, p 49

AISI 310S: Mean coefficients of linear thermal expansion

TEMP C	TEMP F	THEXP 10^{-6}/C	THEXP 10^{-6}/F
0-100	32-212	15.9	8.8
0-315	32-600	16.2	9.0
0-538	32-1000	17.0	9.4
0-649	32-1200	17.5	9.7
0-982	32-1800	19.1	10.6

Source: Ref 27, 49

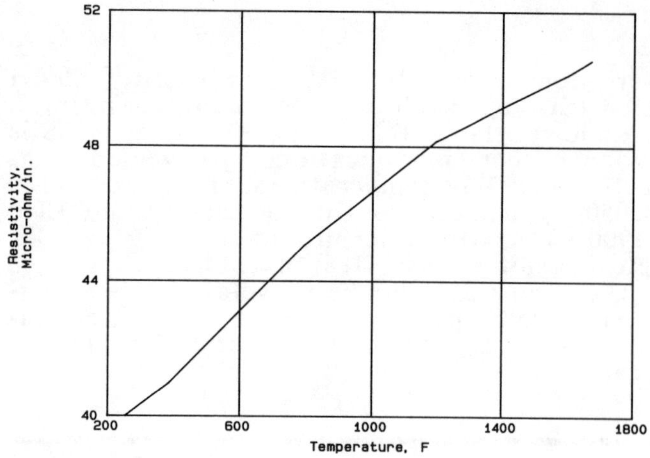

AISI 310S: Electrical resistivity of wrought specimens. Source: Ref 22, p 5

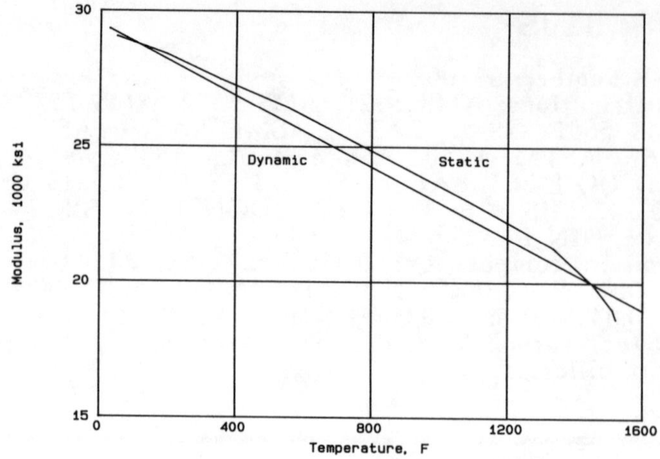

AISI 310S: Modulus of elasticity. Source: Ref 22, p 20

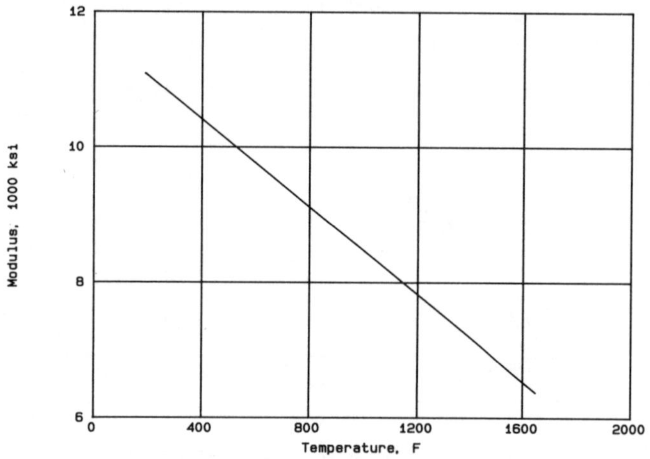

AISI 310S: Modulus of rigidity of annealed specimens. Source: Ref 22, p 21

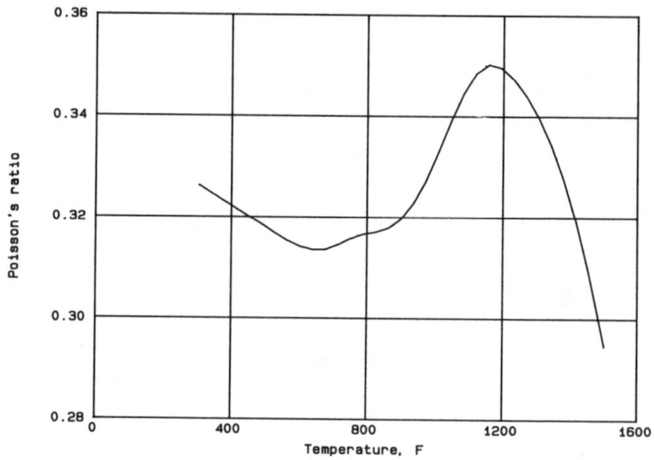

AISI 310S: Poisson's ratio. Source: Ref 22, p 19

Mechanical Properties

AISI 310S: Tensile property data

TEMP C	TEMP F	UTS MPa	UTS ksi	YS MPa	YS ksi	EL %	RA %
21	70	625	91	290	42	47	70
150	300	600	87	240	35	39	70
205	400	585	85	230	33	38	69
260	500	570	83	220	32	37	69
315	600	565	82	205	30	37	69
370	700	560	81	200	29	37	67
425	800	525	76	185	27	36	63
480	900	515	75	180	26	34	61
540	1000	495	72	165	24	33	55
595	1100	450	65	150	22	33	50
650	1200	385	56	145	21	35	45
705	1300	340	49	140	20	36	40
760	1400	285	41	130	19	37	39
815	1500	235	34	125	18	50	38
870	1600	185	27	110	16	45	37
925	1700	215	31	...	...	50	40
980	1800	90	13	...	...	57	40

Source: Ref 15, p 289

AISI 310S: Stress-rupture property data and creep strength for annealed bar

TEMP C	RPSTR 1000 h MPa	RPSTR 1000 h ksi	RPSTR 10000 h MPa	RPSTR 10000 h ksi	RPSTR 100000 h MPa	RPSTR 10000 h ksi	CRSTR 1%/10000 h MPa	CRSTR 1%/10000 h ksi	CRSTR 1%/100000 h MPa	CRSTR 1%/100000 h ksi
540	220	32	170	25	125	18	140	20	90	13
595	160	23	110	16	83	12	105	15	55	8
650	110	16	69	10	55	8	55	8	40	6
705	69	10	40	6	35	5	35	5	30	4
760	50	7	30	4	20	3	15	2	15	2
815	35	5	15	2	7	1	7	1	...	...
870	20	3	7	1	...	...	...	...	...	...
980	7	1	...	...	...	...	...	...	...	...

Source: Ref 15, p 289

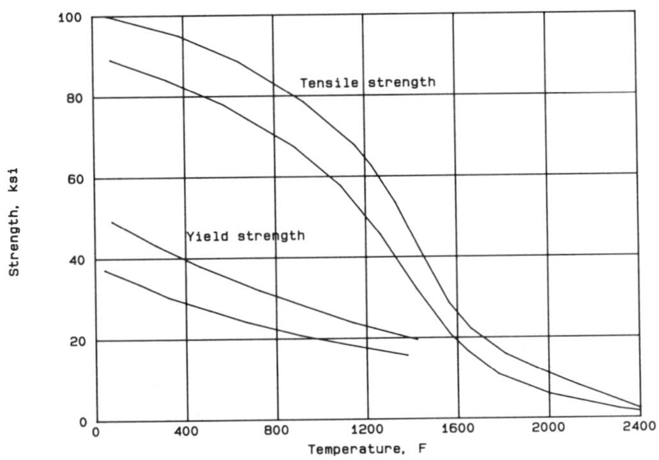

AISI 310S: Scatterbands for tensile and yield strength of annealed bar. Source: Ref 22, p 14

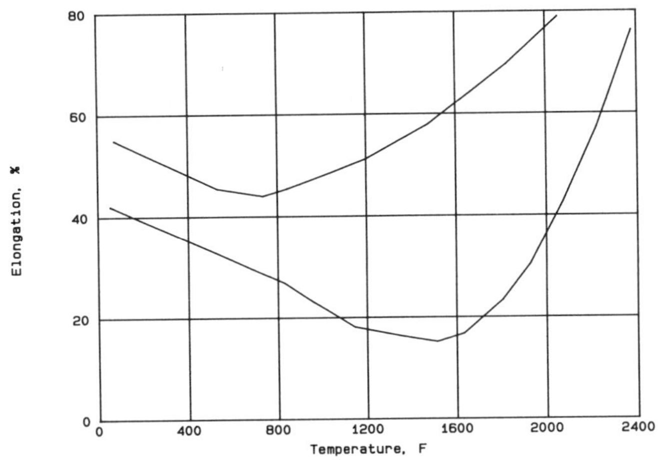

AISI 310S: Scatterband for tensile elongation of annealed bar. Source: Ref 22, p 14

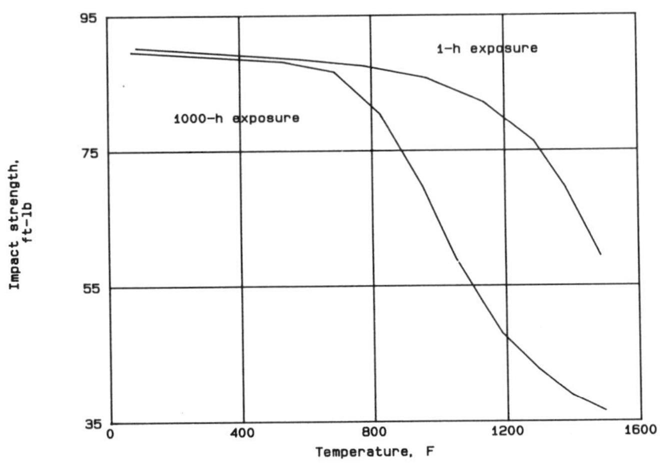

AISI 310S: Effect of exposure time and temperature on Charpy V-notch impact strength of 1-in. bar. Specimens were heat treated at 2150 F and water quenched, with an ASTM grain size of 0.67. Source: Ref 22, p 15

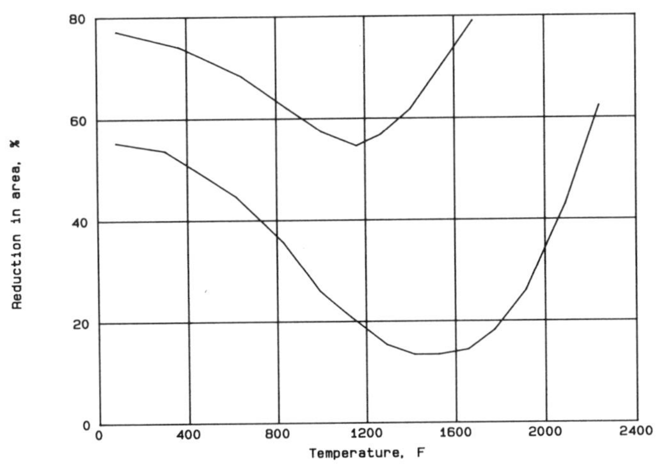

AISI 310S: Scatterband for tensile reduction in area of annealed bar. Source: Ref 22, p 14

AISI 310S: Stress-rupture curves for 1-in. annealed bar. Specimens were treated at 2150 F and water quenched, with an ASTM grain size of 1 to 4. Source: Ref 22, p 17

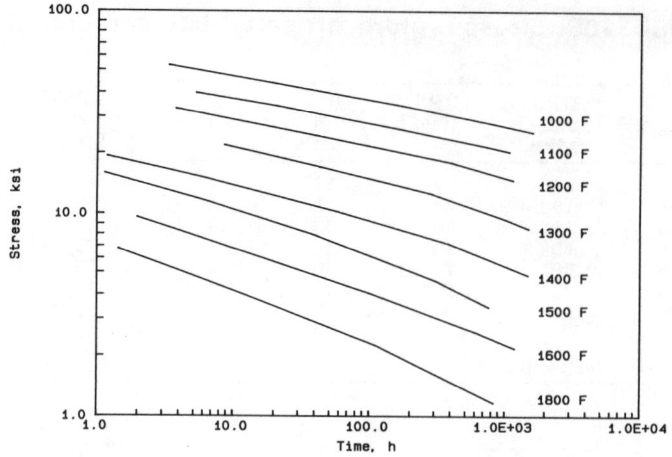

AISI 314

UNS number: S31400
Specifications: AMS 5522, AMS 5652, ASTM A276, ASTM A314, ASTM A473, ASTM A580, SAE J405, UNI X16CrNiSi2520, DGN B83, BS 970 310S24, GOST 20Ch25N20S2, DIN 1.4841, ABNT V310, CSN 17255, PN 86022 H25N2052
Nominal composition: 0.250% C max, 23.000 to 26.000% Cr, 2.000% Mn max, 19.000 to 22.000% Ni, 0.045% P max, 0.030% S max, 1.500 to 3.000% Si

Product forms: Wrought, bar, wire, plate, sheet, strip, tube, billets, forgings

Description: AISI 314 is an austenitic stainless steel with better heat resistance than any of the Cr-Ni types.

Typical applications: Furnace parts, aircraft and jet engine parts, oil burner parts, and chemical processing equipment.

Physical Properties

AISI 314: Thermal conductivity

TEMP C	TEMP F	THCON W/m*K	THCON Btu/ft*h*F
100	212	17.5	10.1
500	930	21.1	12.2

Source: Ref 15, p 291

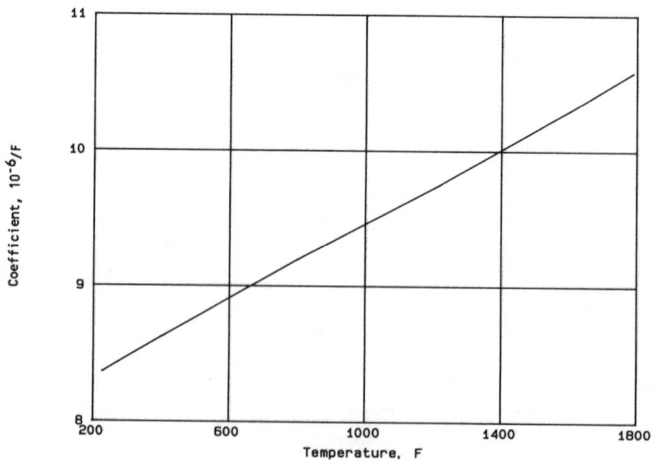

AISI 314: Mean coefficient of linear thermal expansion from room temperature to temperature indicated. Source: Ref 21, p 2

Mechanical Properties

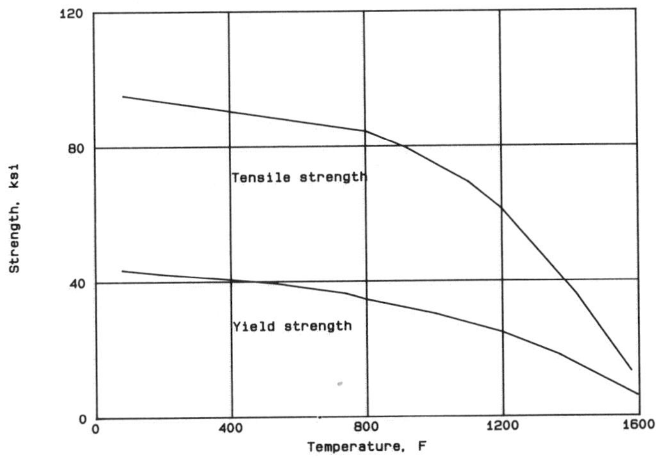

AISI 314: Tensile and yield strength of wrought bar, annealed at 1950 F, held 0.5 h, and water quenched. Source: Ref 22, p 3

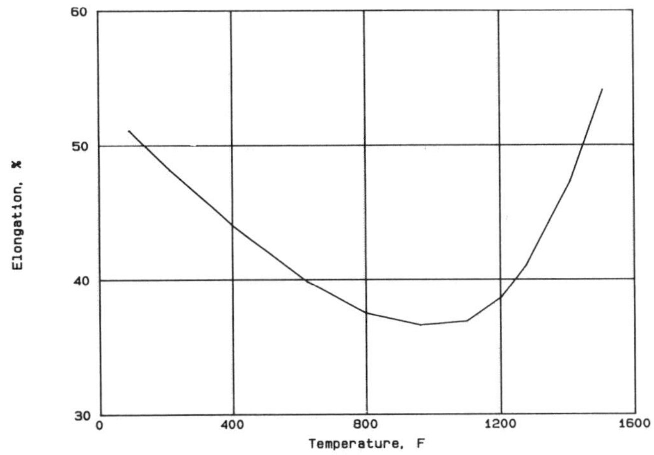

AISI 314: Tensile elongation of wrought bar annealed at 1950 F, held 0.5 h, and water quenched. Source: Ref 22, p 3

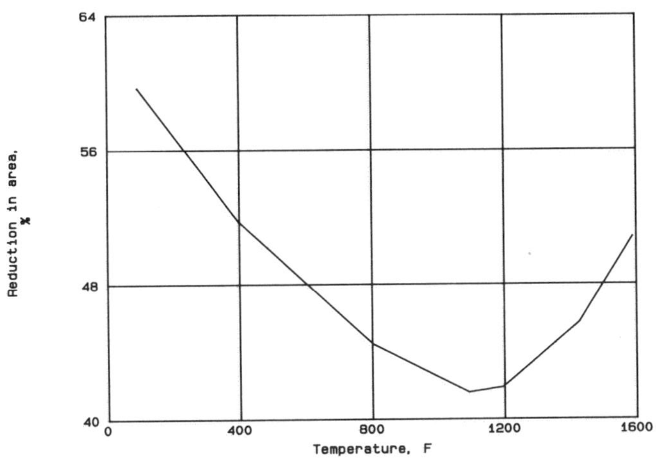

AISI 314: Tensile reduction in area of wrought bar, annealed at 1950 F, held for 0.5 h, and water quenched. Source: Ref 22, p 3

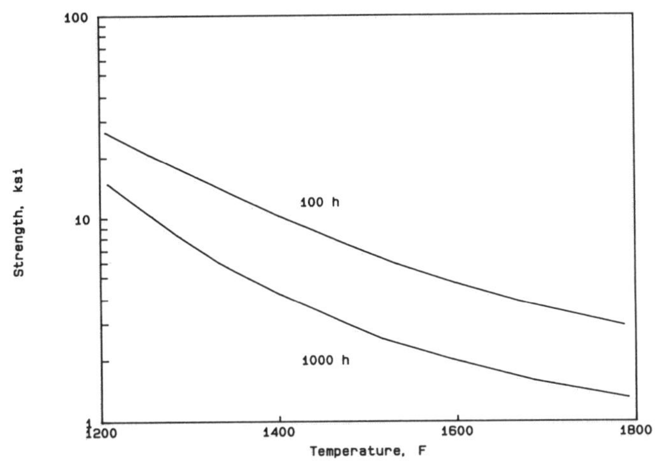

AISI 314: Creep-rupture strength of annealed 1-in. diam specimens, heat treated to 2100 F, and water quenched. Source: Ref 22, p 3

AISI 316

UNS number: S31600

Specifications: AMS 5524, AMS 5573, AMS 5648, AMS 5690, AMS 5691, ASME SA182, ASME-SA193, ASME SA194, ASME SA213, ASME SA240, ASME SA249, ASME SA312, ASME SA320, ASME SA358, ASME SA376, ASME SA403, ASME SA409, ASME SA430, ASME SA479, ASME SA688

Nominal composition: 0.080% C max, 16.000 to 18.000% Cr, 2.000% Mn max, 2.000 to 3.000% Mo, 0.100% N max, 10.000 to 14.000% Ni, 0.045% P max, 0.030% S max, 1.000% Si max

Product forms: Wrought, bar, wire, plate, sheet, strip, tube, billets, forgings

Description: AISI 316 is an austenitic Cr-Ni stainless and heat-resisting steel with better corrosion resistance than Types 302 and 304 when exposed to many kinds of chemical mediums, as well as to marine atmospheres. The molybdenum addition is largely responsible for the enhanced corrosion resistance. Type 316 also exhibits superior creep strength at elevated temperatures. Scaling temperatures in air are 900 C (1650 F) max for continuous service and 845 C (1550 F) max for intermittent service.

Typical applications: Chemical and pump handling equipment, photographic equipment, brandy vats, fertilizer machine parts, catsup cooking kettles, and yeast tubs.

Physical Properties

AISI 316: Physical property data for annealed specimens

TEMP C	TEMP F	DENS Mg/m^3	DENS lb/in.^{3}	SHTCP J/kg*K	SHTCP Btu/lb*F	THDIF ft^2/h	ELRES n{o}*m
21	70	7.958	0.2873	452	0.108	0.143	740
93	200	7.925	0.2861	486	0.116	0.144	790
204	400	7.883	0.2846	528	0.126	0.148	870
316	600	7.836	0.2829	548	0.131	0.158	930
427	800	7.792	0.2813	565	0.135	0.167	990
538	1000	7.745	0.2796	574	0.137	0.173	1040
649	1200	7.698	0.2779	586	0.14	0.188	1100
760	1400	7.645	0.276	615	0.147	0.191	...
871	1600	7.593	0.2741	649	0.155	0.194	...
982	1800	7.543	0.2723	...	...	...	...

Source: Ref 21, p 5

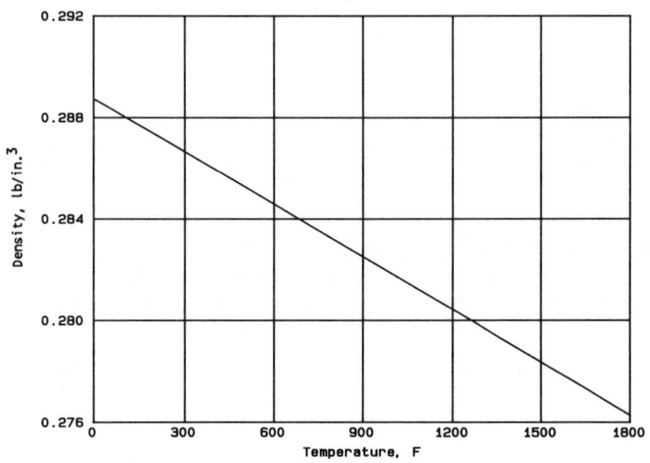

AISI 316: Density of annealed specimens. Source: Ref 22, p 5

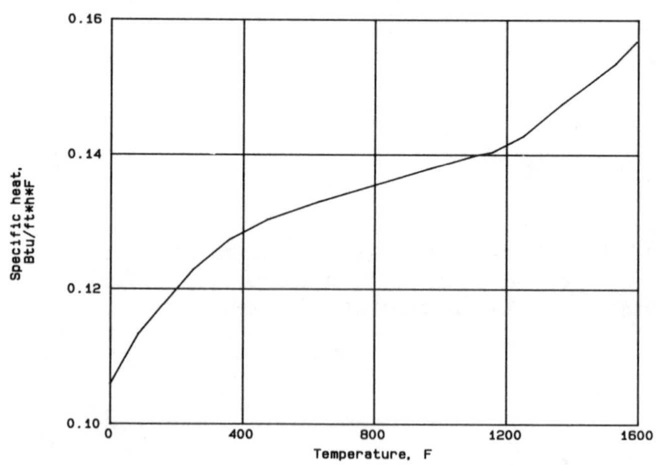

AISI 316: Specific heat of annealed specimens. Source: Ref 22, p 5

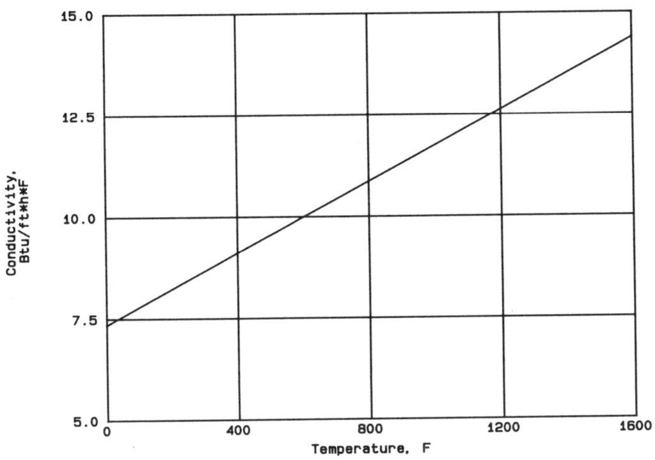

AISI 316: Thermal conductivity of annealed specimens. Source: Ref 22, p 4

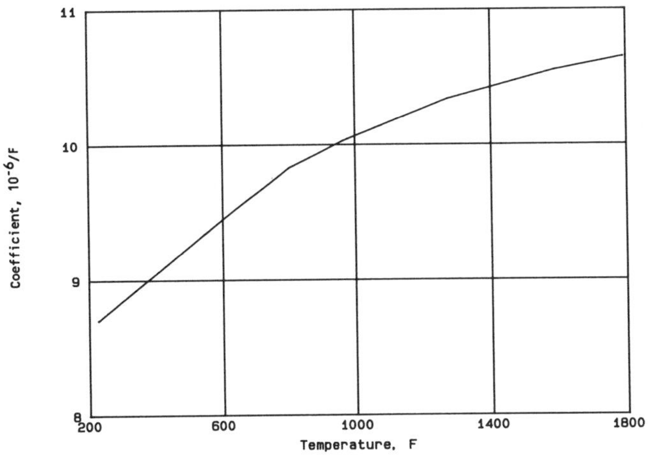

AISI 316: Mean coefficient of thermal expansion of annealed specimens. Source: Ref 22, p 4

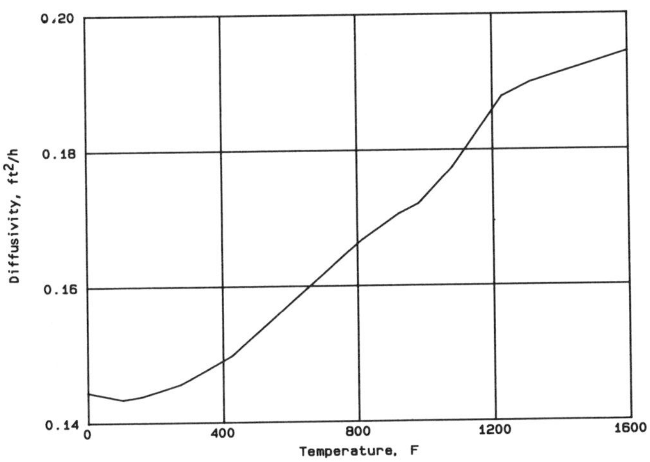

AISI 316: Thermal diffusivity of annealed specimens. Source: Ref 22, p 5

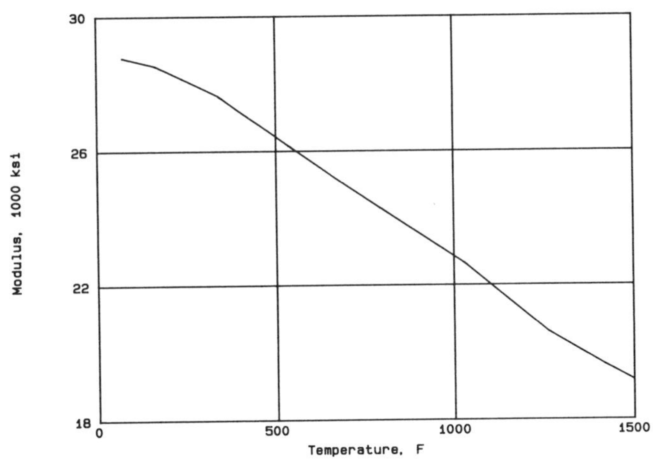

AISI 316: Static modulus of elasticity of wrought and annealed specimens. Source: Ref 22, p 15

AISI 316: Mean coefficients of linear thermal expansion

TEMP C	TEMP F	THEXP 10^{-6}/C	THEXP 10^{-6}/F
0-100	32-212	16.0	8.9
0-315	32-600	16.2	9.0
0-540	32-1000	17.5	9.7
0-650	32-1200	18.5	10.3
0-815	32-1500	20.0	11.1

Source: Ref 15, p 293

AISI 316: Thermal conductivity

TEMP C	TEMP F	THCON W/m*K	THCON Btu/ft*h*F
100	212	16.3	9.4
500	930	21.5	12.4

Source: Ref 15, p 293

AISI 316: Modulus of elasticity

TEMP C	TEMP F	E GPa	E 10^3 ksi
93	200	194	28.1
149	300	190	27.5
204	400	185	26.9
260	500	181	26.3
316	600	177	25.6
371	700	172	24.9
427	800	167	24.2
482	900	162	23.5
538	1000	157	22.8
593	1100	153	22.2
649	1200	148	21.5
704	1300	143	20.8
760	1400	138	20.0
816	1500	131	19.0

Source: Ref 21, p 35

AISI 316: Modulus of rigidity

TEMP C	TEMP F	G GPa	G 10^3 ksi
93	200	76	11.0
149	300	73	10.6
204	400	71	10.3
260	500	69	10.0
316	600	67	9.7
371	700	65	9.4
427	800	63	9.1
482	900	61	8.8
538	1000	59	8.5
593	1100	57	8.3
649	1200	56	8.1
704	1300	54	7.9
760	1400	53	7.7
816	1500	52	7.5

Source: Ref 21, p 35

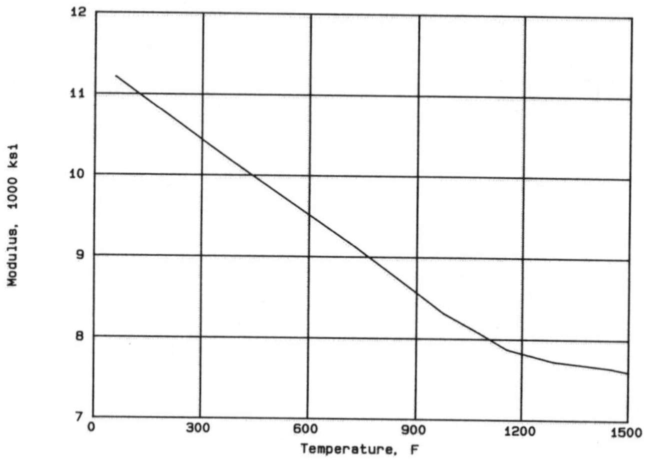

AISI 316: Modulus of rigidity of wrought and annealed specimens. Source: Ref 22, p 15

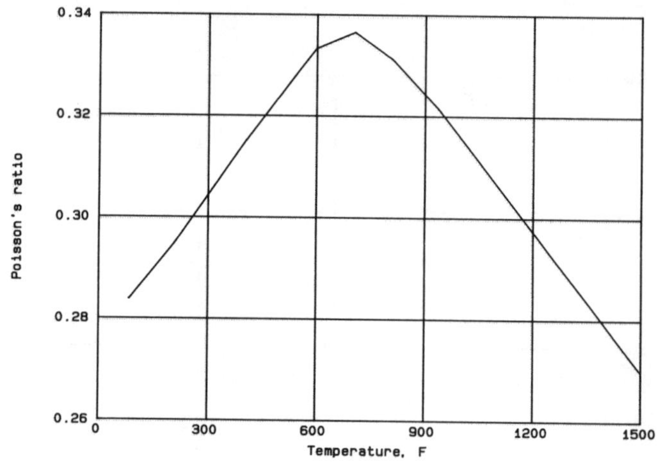

AISI 316: Poisson's ratio of wrought and annealed specimens. Source: Ref 22, p 15

AISI 316: Poisson's ratio

TEMP C	TEMP F	PR
149	300	0.26
260	500	0.29
371	700	0.34
482	900	0.30
593	1100	0.32
704	1300	0.31
816	1500	0.24

Source: Ref 21, p 35

Mechanical Properties

AISI 316: Effect of thermal exposure on tensile and impact property data

Material	EXP TEMP C	EXP TEMP F	EXP h	UTS MPa	UTS ksi	YS MPa	YS ksi	EL %	RA %	IS(b) J	IS(b) ft-lb
Forged											
Material A	20	70	...	607	88	276	40	62(a)	65	108.5	80
	482	900	10000	614	89	310	45	62(a)	66	118.0	87
	566	1050	10000	634	92	276	40	60(a)	60	66.4	49
	649	1200	10000	690	100	296	43	56(a)	55	43.4	32
Plate											
Material B	20	70	...	531	77	221	32	60	72	203.4	150(c)
	649	1200	10000	634	92	331	48	41	48	94.9	70(c)
	732	1350	10000	648	94	290	42	33	40	54.2	40(c)
	816	1500	10000	607	88	276	40	40	45	54.2	40(c)
Material C	20	70	...	621	90	296	43	52	72	257.6	190(c)
	649	1200	10000	641	93	276	40	41	55	88.1	65(c)
	732	1350	10000	621	90	255	37	39	50	67.8	50(c)
	816	1500	10000	558	81	228	33	42	41	101.7	75(c)

Note: Specimens were annealed. Material A: 0.07C-1.32Mn-0.32Si-10.59Ni-18.27Cr-2.16Mo-0.009Al-0.030N-0.16Cu-0.003Ti. Material B: 0.06C-1.52Mn-0.43Si-12.06Ni-17.35Cr-2.08Mo. Material C: 0.062C-1.65Mn-0.35Si-13.16Ni-16.65Cr-2.30Mo
(a) Elongation in 25 mm (1 in.). (b) Charpy impact keyhole notch. (c) Charpy impact V-notch
Source: Ref 21, p 11

AISI 316: Tensile property data

TEMP C	TEMP F	UTS MPa	UTS ksi	YS MPa	YS ksi	EL %	RA %
205	400	560	81	240	35	51	76
315	600	540	78	215	31	48	72
425	800	525	76	195	28	47	66
540	1000	485	70	165	24	44	60
650	1200	395	57	145	21	40	53
760	1400	240	35	125	18	37	46
870	1600	165	24	110	16	39	44

Source: Ref 15, p 294

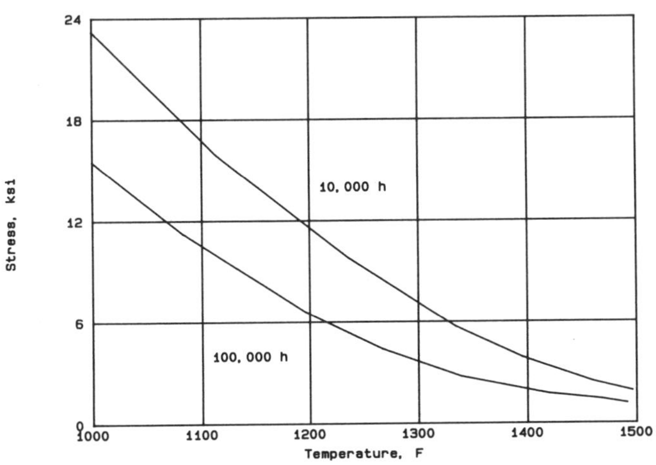

AISI 316: Creep rate (1%) of annealed specimens. Source: Ref 15, p 296

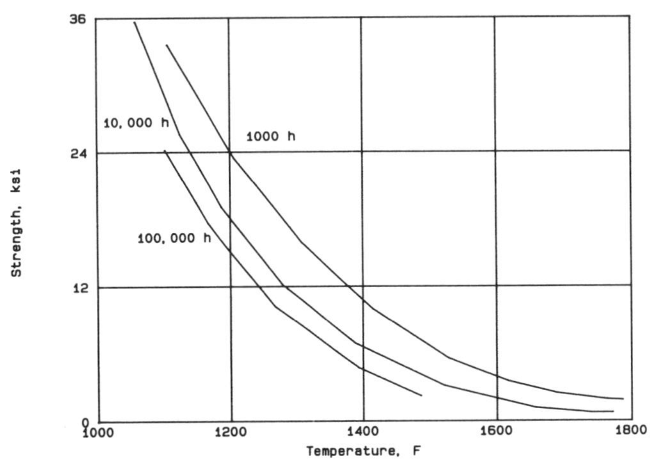

AISI 316: Rupture strength of annealed specimens. Source: Ref 15, p 295

AISI 316: Stress-rupture and creep property data

TEMP C	RPSTR 1000 h MPa	RPSTR 1000 h ksi	RPSTR(a) 10000 h MPa	RPSTR(a) 10000 h ksi	RPSTR(a) 100000 h MPa	RPSTR(a) 100000 h ksi	CRSTR 1%/10000 h MPa	CRSTR 1%/10000 h ksi	CRSTR 1%/100000 h MPa	CRSTR 1%/100000 h ksi
595	245	36	195	28	170	25	124	18	79	11.5
650	165	24	115	17	97	14	83	12	48	7
705	110	16	69	10	50	7	52	7.5	28	4
760	69	10	40	6	30	4	31	4.5	15	2.2
815	40	6	30	4	15	2	21	3	...	...
870	30	4	...	...	...	...	...	...	...	...

(a) Extrapolated
Source: Ref 15, p 294

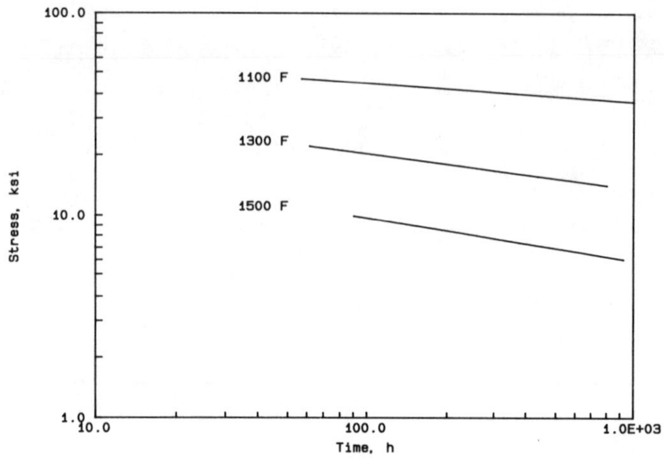

AISI 316: Stress-rupture strength of 0.75-in. diam wrought bar. Source: Ref 21, p 37

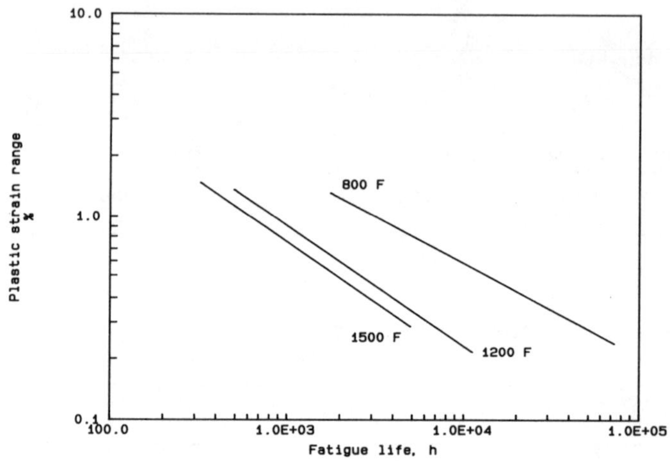

AISI 316: Effect of temperature on fatigue life. Source: Ref 16, p 181

AISI 316: Fatigue crack propagation rate for 20% cold worked specimens. Curves are averages for L-T and T-L specimens tested in air; 3 Hz at 75 F, 0.67 Hz as higher temperatures; R=0.05. Source: Ref 23, p 220

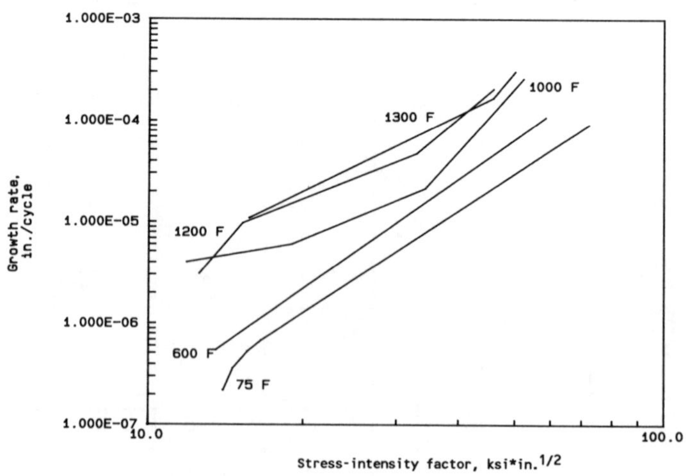

AISI 316F

UNS number: S31620
Specification: AMS 5649
Nominal composition: 0.080% C max, 17.000 to 19.000% Cr, 2.000% Mn max, 1.750 to 2.500% Mo, 10.000 to 14.000% Ni max, 0.200% P max, 0.100% S min, 1.000% Si max
Product forms: Wrought, bar, plate, sheet, strip, billets, forgings
Description: AISI 316F is an austenitic Cr-Ni stainless steel with higher phosphorus and sulfur than Type 316. It has improved machining and nonseizing characteristics.
Typical applications: Screw machine parts.

Physical Properties

AISI 316F: Mean coefficients of linear thermal expansion

TEMP C	TEMP F	THEXP 10^{-6}/C	THEXP 10^{-6}/F
0-100	32-212	16.6	9.2
0-315	32-600	17.5	9.7
0-540	32-1000	18.2	10.1

Source: Ref 15, p 299

AISI 316H

UNS number: S31609

Specifications: ASME SA182, ASME SA213, ASME SA240, ASME SA249, ASME SA312, ASME A358, ASME A376, ASME A403, ASME A430, ASME A479

Nominal composition: 0.040 to 0.100% C, 16.000 to 18.000% Cr, 2.000% Mn max, 2.000 to 3.000% Mo, 10.000 to 14.000% Ni, 0.045% P max, 0.030% S max, 1.000% Si max

Description: AISI 316H is a higher carbon modification of Type 316 with improved elevated-temperature strength. Scaling temperatures in air are: 900 C (1650 F) max for continuous service and 845 C (1550 F) for intermittent service.
Typical applications: Chemical and pump-handling equipment, photographic equipment, brandy vats, fertilizer machine parts, catsup-cooking kettles, and yeast tubs.

Mechanical Properties

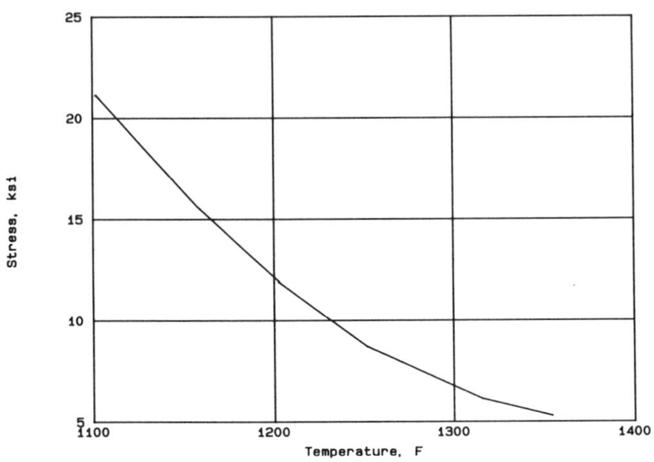

AISI 316H: Stress-rupture strength of tubing with a 6.625-in. outside diameter and 1.313-in. wall. Specimens were water quenched at 1900 F and hot finished. Source: Ref 21, p 43

AISI 316N

UNS number: S31651
Specifications: ASME SA182, ASME SA213, ASME SA240, ASME SA249, ASME SA312, ASME SA358, ASME SA376, ASME SA403, ASME SA430, ASME SA479, ASTM A182, ASTM A213, ASTM A240, ASTM A249, ASTM A276, ASTM A312, ASTM A358, ASTM A376, ASTM A403, ASTM A430
Nominal composition: 0.080% C max, 16.000 to 18.000% Cr, 2.000% Mn max, 2.000 to 3.000% Mo, 0.100 to 0.160% N, 10.000 to 14.000% Ni, 0.045% P max, 0.030% S max, 1.000% Si max
Product forms: Wrought, bar, wire, plate, sheet, strip, tube, billets, forgings
Description: AISI 316N is a high nitrogen modification of Type 316. It has increased strength with minimum effect on ductility and corrosion resistance.
Typical applications: AISI 316N is suitable for Type 316 applications requiring higher strength.

Physical Properties

AISI 316N: Thermal conductivity

TEMP C	TEMP F	THCON W/m*K	THCON Btu/ft*h*F
95	200	16.1	9.3
540	1000	21.5	12.4

Source: Ref 15, p 300

AISI 316N: Mean coefficients of linear thermal expansion

TEMP C	TEMP F	THEXP 10^{-6}/C	THEXP 10^{-6}/F
0-100	32-212	16.0	8.9
0-315	32-600	16.2	9.0
0-540	32-1000	17.5	9.7
0-650	32-1200	18.5	10.3

Source: Ref 15, p 300

AISI 317

UNS number: S31700
Specifications: ASME SA240, ASME SA249, ASME SA312, ASME SA403, ASME SA409, ASTM A167, ASTM A240, ASTM A249, ASTM A269, ASTM A276, ASTM A312, ASTM A314, ASTM A554, ASTM A580, ASTM A632, FED QQ-S-763, MIL S-862, SAE J405 (30317)
Nominal composition: 0.080% C max, 2.000% Mn max, 1.000% Si max, 18.000 to 20.000% Cr, 11.000 to 15.000% Ni, 0.045% P max, 0.030% S max, 3.000 to 4.000% Mo
Description: AISI 317 is an austenitic Cr-Ni steel with higher corrosion and creep resistance than Type 316. Scaling temperatures in air are: 900 C (1600 F) max for continuous service and 870 C (1600 F) max for intermittent service.
Typical applications: Dye and ink manufacturing equipment.

Physical Properties

AISI 317: Thermal conductivity

TEMP C	TEMP F	THCON W/m*K	THCON Btu/ft*h*F
100	212	16.3	9.4
500	930	21.5	12.4

Source: Ref 15, p 302

AISI 317: Mean coefficients of linear thermal expansion

TEMP C	TEMP F	THEXP 10^{-6}/C	THEXP 10^{-6}/F
0-100	32-212	16.0	8.9
0-315	32-600	16.2	9.0
0-540	32-1000	17.5	9.7
0-650	32-1200	18.5	10.3
0-815	32-1500	20.0	11.1

Source: Ref 15, p 302

Mechanical Properties

AISI 317: Tensile and creep property data

TEMP C	TEMP F	UTS MPa	UTS ksi	CRSTR 1%/10000 h MPa	CRSTR 1%/10000 h ksi
21	70	585	85	...	...
540	1000	505	73	172	25.0
595	1100	485	70	125	18.2
650	1200	460	67	88	12.7
705	1300	450	65	54	7.9
760	1400	350	51	31	4.5
815	1500	275	40	19	2.8

Note: Specimens were 25-mm (1-in.) diam bar, water quenched from 1120 C (2050 F)
Source: Ref 15, p 302

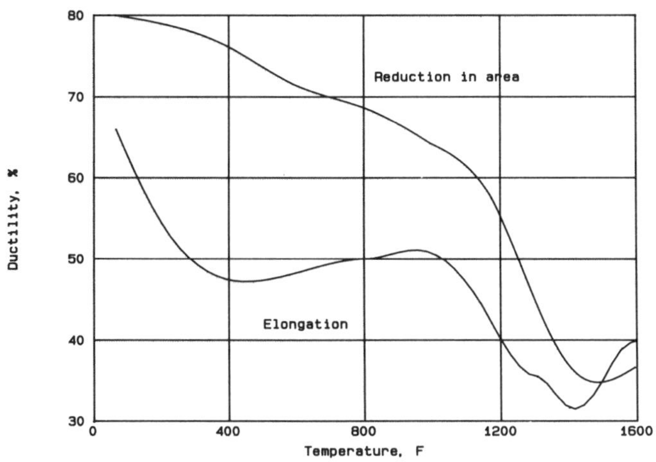

AISI 317: Tensile elongation and reduction in area of wrought, annealed steel. Source: Ref 22, p 9

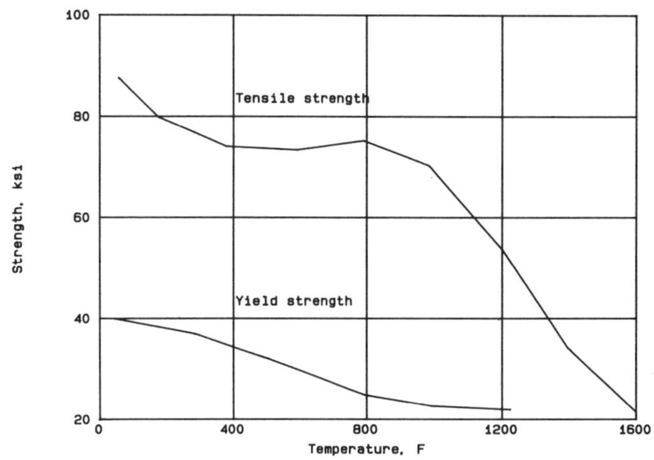

AISI 317: Tensile and yield strength of wrought, annealed steel. Source: Ref 22, p 9

AISI 317L

UNS number: S31703

Specifications: ASME SA240, ASTM A167, ASTM A240, NF Z2CN19-15, JIS G 4303, JIS G 4304, JIS G 4305, SIS 14 23 53, BS 970 317S12, DIN 1.4438 X2CrNiMo1816

Nominal composition: 0.030% C max, 18.000 to 20.000% Cr, 2.000% Mn max, 3.000 to 4.000% Mo, 11.000 to 15.000% Ni, 0.045% P max, 0.030% S max, 1.000% Si max

Product forms: Bar, wire, plate, sheet, strip, tube, billets, forgings

Description: AISI 317L is an extra-low carbon modification of Type 317, with similar general corrosion resistance but superior resistance to intergranular corrosion following welding or stress relieving. Parts made from this alloy are generally limited to service at temperatures no higher than 425 C (800 F).

Typical applications: Welded assemblies that cannot be annealed after fabrication.

Physical Properties

AISI 317L: Thermal conductivity

TEMP C	TEMP F	THCON W/m*K	THCON Btu/ft*h*F
100	212	16.3	9.4
500	930	21.5	12.4

Source: Ref 15, p 304

AISI 317L: Mean coefficients of linear thermal expansion

TEMP C	TEMP F	THEXP 10^{-6}/C	THEXP 10^{-6}/F
0-100	32-212	16.5	9.2
0-538	32-1000	18.1	10.1
0-649	32-1200	18.6	10.3

Source: Ref 27, p 56

AISI 317LM

Nominal composition: 0.030% C max, 18.000 to 20.000% Cr, 2.000% Mn max, 4.000 to 4.500% Mo, 12.000 to 16.000% Ni, 0.045% P max, 0.030% S max, 1.000% Si max

Description: AISI 317LM is a nonstandard, austenitic Cr-Ni stainless steel similar to Type 317L, but with greater strength and corrosion resistance due to a higher nickel content and a molybdenum content of 4% min. It was originally developed for sulfurous acid service. Scaling temperatures in air are 925 C (1700 F) max for continuous service and 815 C (1500 F) max for intermittent service.

Typical applications: Paper pulp-handling equipment, textile and food processing equipment, and equipment for producing inks, rayon, rubber, bleaches, chemicals, and dyestuffs.

Physical Properties

AISI 317LM: Thermal conductivity

TEMP C	TEMP C	THCON W/m*K	THCON Btu/ft*h*F
100	212	16.3	9.4
500	932	21.5	12.4

Source: Ref 29, Apr 1979

AISI 317LM: Mean coefficients of linear thermal expansion

TEMP C	TEMP F	THEXP 10^{-6}/C	THEXP 10^{-6}/F
0-100	32-212	14.4	8.0
0-316	32-600	16.2	9.0
0-538	32-1000	17.5	9.7
0-649	32-1200	18.5	10.3
0-816	32-1500	20.0	11.1

Source: Ref 29, Apr 1979

Mechanical Properties

AISI 317LM: Creep strength

TEMP C	TEMP F	CRSTR 1%/10000 h MPa	CRSTR 1%/10000 h ksi
538	1000	172.4	25.0
593	1100	125.5	18.2
649	1200	87.6	12.7
704	1300	53.8	7.8
760	1400	31.0	4.5
816	1500	19.3	2.8

Note: Specimens were annealed at 1121 to 1149 C (2050 to 2100 F) and air cooled
Source: Ref 29, Apr 1979

AISI 317LM: Ultimate tensile strength of annealed specimens

TEMP C	TEMP F	UTS MPa	UTS ksi
21	70	627.4	91.0
538	1000	503.3	73.0
593	1100	482.7	70.0
649	1200	462.0	67.0
704	1300	448.2	65.0
760	1400	351.6	51.0
816	1500	275.8	40.0

Note: Specimens were annealed at 1121 to 1149 C (2050 to 2100 F) and air cooled
Source: Ref 29, Apr 1979

AISI 321

UNS number: S32100

Specifications: AMS 5510, AMS 5557, ASME SA182, ASME SA193, ASME SA194, ASTM A167, ASTM A182, ASTM A193, FED QQ S-763, FED QQ S-766, MIL S-862, MIL S-27419, SAE J405, NF Z6CN18-10, JIS G 4303, BS 970 321S12 En58B/C, GOST Ch18N10T-WD, GOST Kh18N9T, DIN 1.4541

Nominal composition: 0.080% C max, 17.000 to 19.000% Cr, 2.000% Mn max, 9.000 to 12.000% Ni, 0.045% P max, 0.030% S max, 1.000% Si max, 0.100% N max, % Ti min = 5 X carbon content

Product forms: Wrought, bar, wire, plate, sheet, strip, tube, billets, forgings

Description: AISI 321 is a titanium-stabilized, austenitic Cr-Ni steel for welded assemblies that cannot be subsequently annealed and for parts to be used at temperatures between 425 and 900 C (800 and 1650 F). The alloy is free from harmful sensitization (chromium carbide precipitation) and associated intergranular corrosion after welding or during service in the sensitizing temperature range. Scaling temperatures in air are 900 C (1650 F) for continuous service and 845 C (1550 F) for intermittent service.

Typical applications: Aircraft exhaust manifolds, boiler shells, process equipment, expansion joints, cabin heaters, fire walls, flexible couplings, and pressure vessels.

Physical Properties

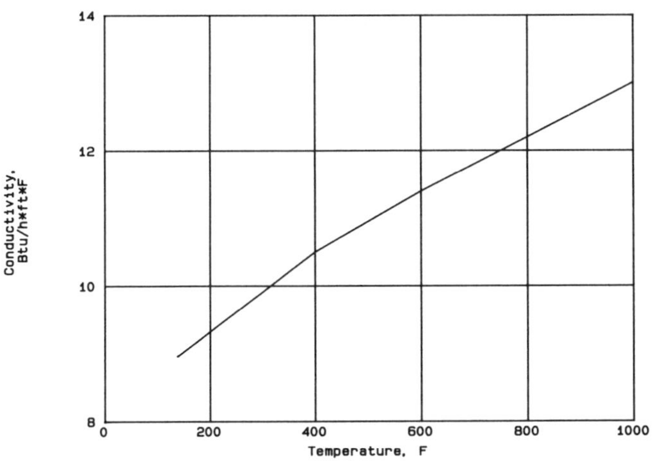

AISI 321: Thermal conductivity. Source: Ref 22, p 5

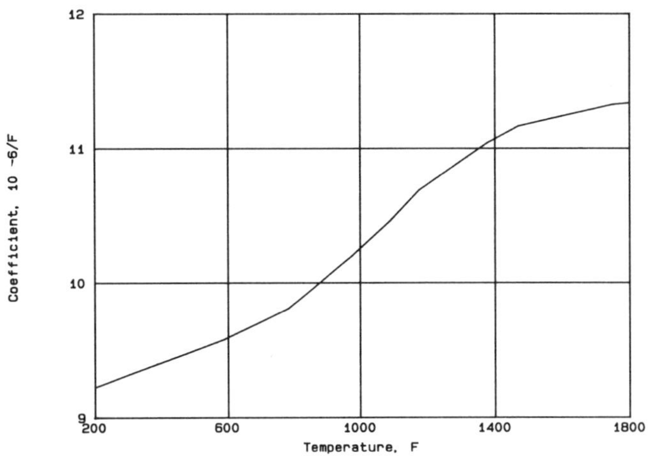

AISI 321: Mean coefficient of linear thermal expansion. Source: Ref 22, p 5

AISI 321: Thermal conductivity

TEMP C	TEMP F	THCON W/m*K	THCON Btu/ft*h*F
100	212	16.1	9.3
500	930	22.1	12.8

Source: Ref 15, p 305

AISI 321: Mean coefficients of linear thermal expansion

TEMP C	TEMP F	THEXP 10^{-6}/C	THEXP 10^{-6}/F
0-100	32-212	16.7	9.3
0-315	32-600	17.1	9.5
0-540	32-1000	18.5	10.3
0-650	32-1200	19.3	10.7
0-815	32-1500	20.2	11.2
21-925	70-1700	20.5	11.4

Source: Ref 15, p 305

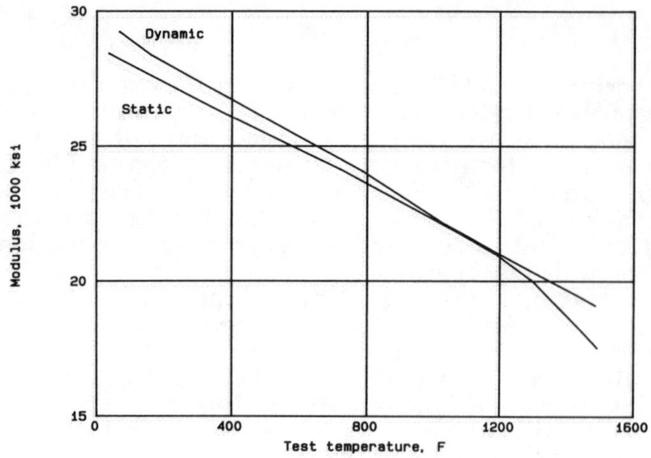

AISI 321: Modulus of elasticity of bar. Source: Ref 22, p 14

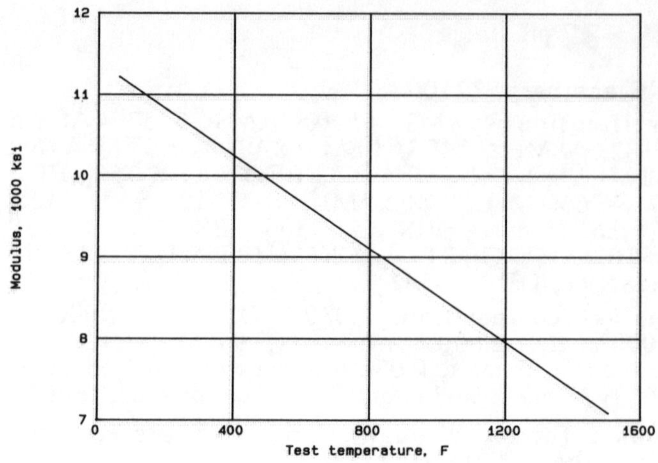

AISI 321: Modulus of rigidity of bar. Source: Ref 22, p 15

AISI 321: Electrical resistivity and conductivity of annealed specimens

TEMP C	TEMP F	ELCON %IACS	ELRES n{o}*m
21	70	2.41	283
100	212	2.22	307
200	392	2.01	338
400	7532	1.73	394
600	1112	1.56	437
800	1472	1.43	476
900	1652	1.37	496

Source: Ref 29, Jun 1976

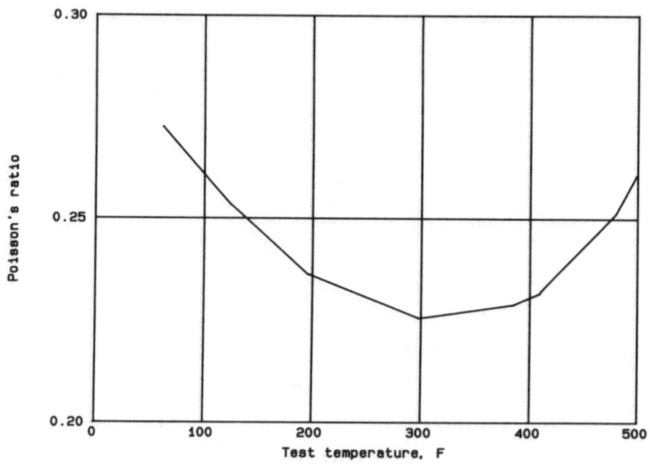

AISI 321: Poisson's ratio for simultaneous bending and torsion. Specimens were annealed at 1950 F for 0.5 h and water quenched. Source: Ref 22, p 14

Mechanical Properties

AISI 321: Tensile property data

TEMP C	TEMP F	UTS MPa	UTS ksi	YS MPa	YS ksi	EL(a) %	RA %
21	70	569	82.5	207	30.0	71.0	79.5
205	400	421	61.0	145	21.0	48.0	76.0
425	800	400	58.0	121	17.5	42.5	71.5
540	1000	379	55.0	124	18.0	42.0	70.5
650	1200	303	44.0	117	17.0	37.0	57.5
760	1400	210	30.5	114	16.5	44.5	53.0
870	1600	117	17.0	93	13.5	57.0	82.0
980	1800	63	9.1	50	7.3	78.0	96.0

Note: Specimens were 16.0-mm (0.625-in.) diam rod, annealed at 1040 C (1900 F) for 30 min, and water quenched. Hardness was 75 HRB
(a) In 4D.
Source: Ref 15, p 305

AISI 321: Tensile property data

TEMP C	TEMP F	UTS MPa	UTS ksi	YS MPa	YS ksi	EL(a) %	RA %
21	70	655	95.0	465	67.5	48.0	73.0
205	400	493	71.5	390	56.5	27.5	74.0
425	800	476	69.0	390	56.5	23.5	63.0
540	1000	448	65.0	365	53.0	22.0	66.0
650	1200	390	56.5	331	48.0	23.0	55.0
760	1400	296	43.0	279	40.5	18.0	32.5
870	1600	169	24.5	162	23.5	50.5	75.5
980	1800	69	10.0	63	9.2	102	93.5

Note: Specimens were 16-mm (0.625-in.) diam rod, annealed and cold worked to a hardness of 96 HRB
(a) In 4D.
Source: Ref 15, p 306

AISI 321: Tensile property data

EXP TEMP C	EXP TEMP F	YS MPa	YS ksi	UTS MPa	UTS ksi	EL %	RA %
Unexposed	Unexposed	207	30.0	533	77.0	61	81
482	900	314	46.0	584	85.0	51	81
566	1050	222	32.0	597	87.0	48	70
649	1200	178	26.0	566	82.0	48	62

Note: Specimens were annealed and exposed for 100000 h at indicated temperature
Source: Ref 29, Jun 1976

AISI 321: Effect of thermal exposure under stress on tensile property data

TEMP C	TEMP F	Stress MPa	Stress ksi	EXP h	YS MPa	YS ksi	UTS MPa	UTS ksi	EL %	RA %
21	70	...	...		462	67.0	641	93.0	49	75
593	1100	86	12.5	1679	421	61.0	665	96.5	48	70
649	1200	48	7.0	1367	412	59.7	664	96.3	42	56
704	1300	34	5.0	1656	376	54.5	665	96.5	39	52

Note: Specimens taken from 25-mm (1-in.) diam bar were annealed at 1038 C (1900 F)
Source: Ref 29, Jun 1976

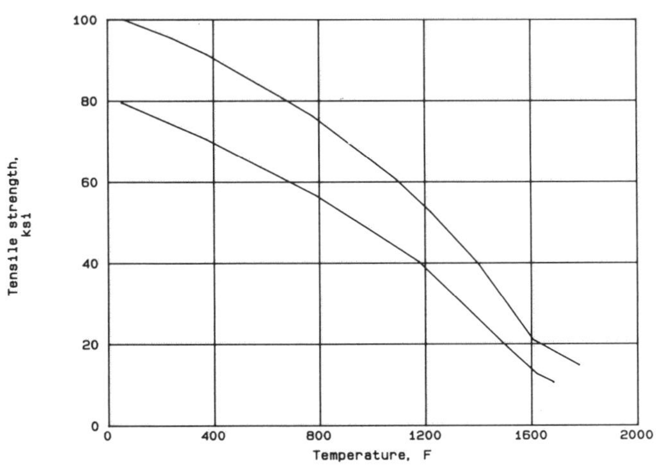

AISI 321: Scatterband for tensile strength of annealed bar. Source: Ref 22, p 8

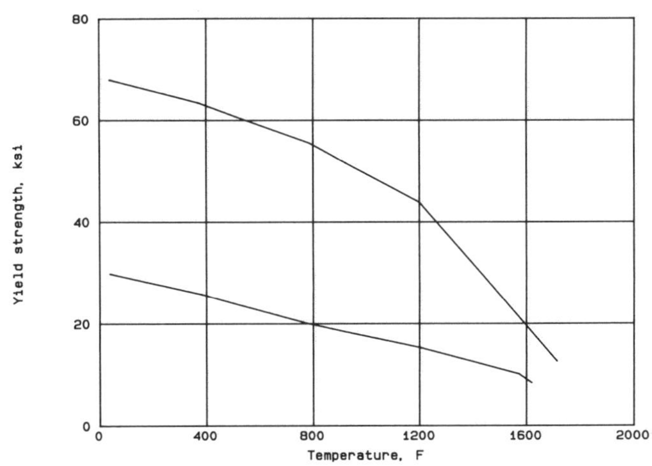

AISI 321: Scatterband for yield strength of annealed bar. Source: Ref 22, p 8

AISI 321: Effect of thermal exposure on room-temperature Charpy keyhole impact values

TEMP C	TEMP F	IS 1000 h J	IS 1000 h ft-lb	IS 10000 h J	IS 10000 h ft-lb
Unexposed	Unexposed	145	107	145	107
482	900	137	101	119	88
621	1050	134	99	98	72
649	1200	94	69	84	62

Source: Ref 29, Jun 1976

AISI 321: Effect of thermal exposure under stress on Izod impact values

TEMP C	TEMP F	Stress MPa	Stress ksi	EXP h	IS J	IS ft-lb
21	70	...	...	...	125	92
593	1100	48	7.0	2750	73	54
649	1200	28	4.0	2753	66	49
704	1300	14	2.0	2612	62	46
816	1500	7	1.0	1601	56	41

Note: Specimens taken from 25-mm (1-in.) diam bar were annealed at 1038 C (1900 F)
Source: Ref 29, Jun 1976

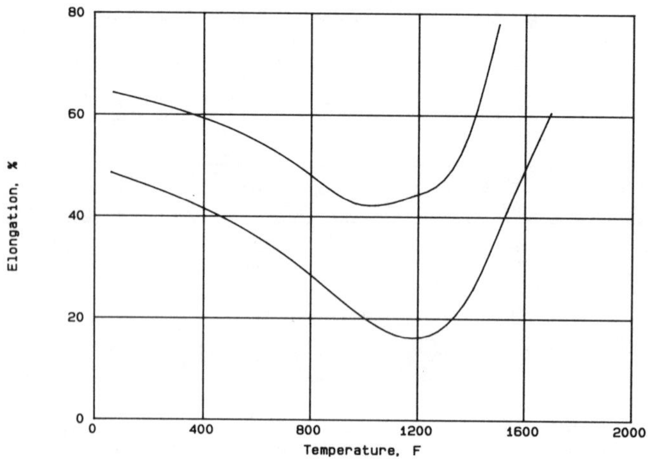

AISI 321: Scatterband for tensile elongation of annealed bar. Source: Ref 22, p 8

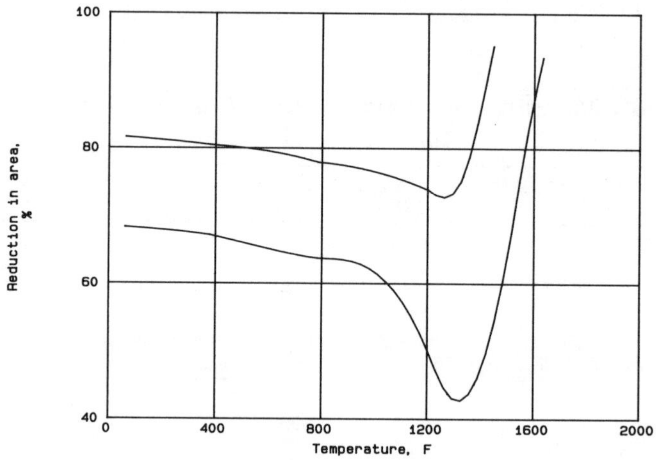

AISI 321: Scatterband for tensile reduction in area of annealed bar. Source: Ref 22, p 8

AISI 321: Stress-rupture and creep property data

TEMP C	RPSTR 1000 h MPa	RPSTR 1000 h ksi	RPSTR 10000 h MPa	RPSTR 10000 h ksi	RPSTR(a) 100000 h MPa	RPSTR(a) 100000 h ksi	CRSTR 1%/10000 h MPa	CRSTR 1%/10000 h ksi	CRSTR 1%/100000 h MPa	CRSTR 1%/100000 h ksi
480	340	49.0	295	43.0	...	...	...	...	...	...
540	290	42.0	235	34.0	...	...	...	...	...	...
595	200	29.0	145	21.0	105	15.0	115	16.5	80	11.5
650	130	18.5	85	12.0	50	7.5	65	9.5	45	6.5
705	75	11.0	40	6.0	30	4.0	35	5.0	25	3.5
760	45	6.5	25	3.5	17	2.5	17	2.5	15	2.0
815	30	4.0	15	2.0	10	1.5	10	1.5	...	...

(a) Extrapolated
Source: Ref 15, p 306

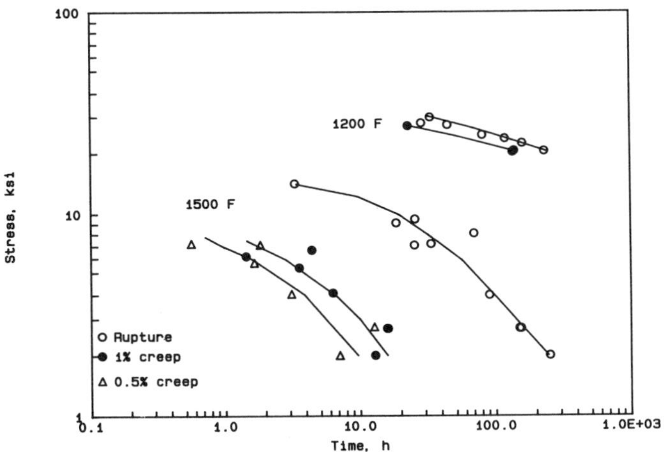

AISI 321: Creep and creep-rupture strength of annealed 0.045-in. sheet at 1200 and 1500 F. Source: Ref 22, p 12

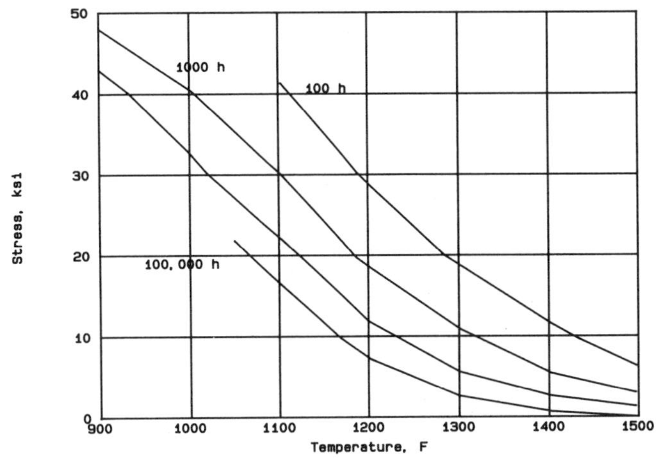

AISI 321: Creep-rupture strength of annealed specimens as a function of temperature. Source: Ref 22, p 12

AISI 321: Hot hardness data at room-temperature after exposure

TEMP C	TEMP F	HARD 1000 h HB	HARD 10000 h HB
Unexposed	Unexposed	168	168
482	900	143	156
566	1050	149	151
649	1200	166	148

Source: Ref 29, Jun 1976

AISI 329

UNS number: S32900
Specifications: ASME SA268, ASTM A268, ASTM A511, DGN B171 Grade MT329, DGN B216 Grade TP329
Nominal composition: 0.100% C max, 25.000 to 30.000% Cr, 2.000% Mn max, 1.000 to 2.000% Mo, 3.000 to 6.000% Ni, 0.0450% P max, 0.0300% S max, 1.000% Si max
Product forms: Wrought, bar, wire, plate, sheet, strip, billets, forgings, welded
Description: AISI 329 is an austenitic-ferritic stainless steel with general corrosion resistance similar to that of Type 316, but with superior resistance to chloride stress corrosion cracking. Type 329 can be age hardened; however, material in this condition is brittle and is therefore restricted to service under compressive loads.
Typical applications: Valves, valve fittings, piping, and pump parts.

Physical Properties

AISI 329: Thermal conductivity

TEMP C	TEMP F	THCON W/m*K	THCON Btu/h*ft*h
38	100	15.3	8.83
93	200	16.2	9.33
149	300	17.0	9.83
204	400	17.9	10.33
260	500	18.7	10.83
316	600	19.6	11.33

Source: Ref 24, p 185

Mechanical Properties

AISI 329: Tensile property data

TEMP C	TEMP F	YS MPa	YS ksi	UTS MPa	UTS ksi	EL %
21	70	565	82	683	99	31
93	200	510	74	627	91	32
149	300	462	67	593	86	28
204	400	434	63	593	86	26
260	500	414	60	593	86	26
316	600	393	57	593	86	25
371	700	393	57	593	86	24

Note: Specimens were annealed to 99 HRB
Source: Ref 24, p 186

AISI 329: Effect of thermal exposure on Charpy V-notch impact values of bar

TEMP C	TEMP F	IS 10 h J	IS 10 h ft-lb	IS 100 h J	IS 100 h ft-lb	IS 500 h J	IS 500 h ft-lb	IS 1000 h J	IS 1000 h ft-lb
260	500	106	78	111	82	115	85	126	93
316	600	92	68	99	73	58	43	35	26
371	700	33	24	12	9	7	5	5	4

Source: Ref 24, p 186

AISI 329: Effect of temperature on Charpy V-notch impact values of bar

TEMP C	TEMP F	IS J	IS ft-lb
21	70	123	91
149	300	168	124
260	500	183	135
371	700	203	150

Source: Ref 24, p 187

AISI 347

UNS number: S34700
Specifications: AMS 5512, AMS 5556, ASME SA182, ASME SA193, ASTM A167, ASTM A182, ASTM A193, FED QQ S-763, FED QQ S-766, MIL S-862, MIL S-23195, MIL S-23196, SAE J405, NF Z6CNN618-11, JIS G 4303, BS A8.Nb, BS 821 Grade Nb, GOST 08Ch18N12B, GOST Kh18N12B, DIN 1.4550
Nominal composition: 0.080% C max, 17.000 to 19.000% Cr, 2.000% Mn max, 9.000 to 13.000% Ni, 0.045% P max, 0.030% S max, 1.000% Si max, % (Nb + Ta) min = 10 X carbon content
Product forms: Wrought, bar, wire, plate, sheet, strip, tube, billets, forgings

Description: This niobium-stabilized, austenitic stainless steel is similar to Type 321, but has higher creep strength. It is recommended for fabricated structures that cannot be annealed after welding, and for parts that operate in the 425 to 900 C (800 to 1650 F) temperature range. Maximum scaling temperature in air is 900 C (1650 F) max for continuous service and 845 C (1550 F) max for intermittent service.

Typical applications: Airplane exhaust stacks, welded tank cars for carrying chemicals, and jet engine parts.

Physical Properties

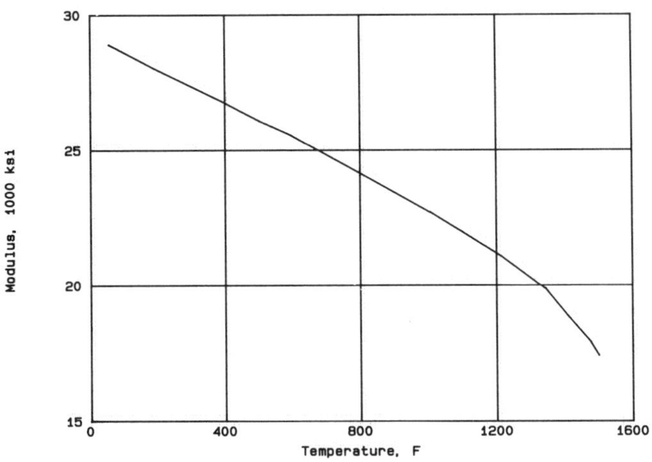

AISI 347: Modulus of elasticity of wrought, annealed steel. Source: Ref 22, p 20

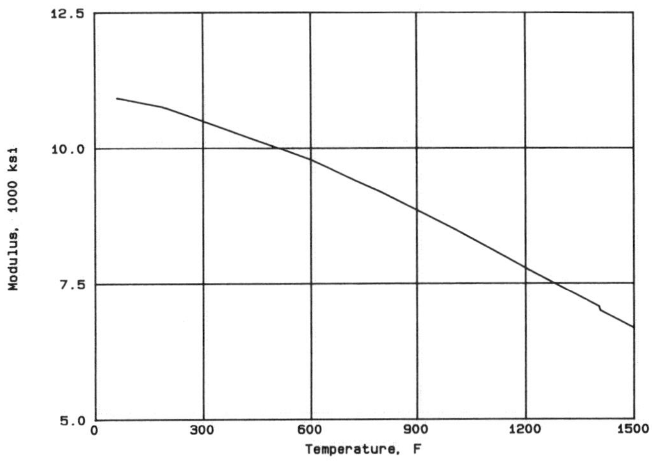

AISI 347: Modulus of rigidity of wrought, annealed steel. Source: Ref 22, p 20

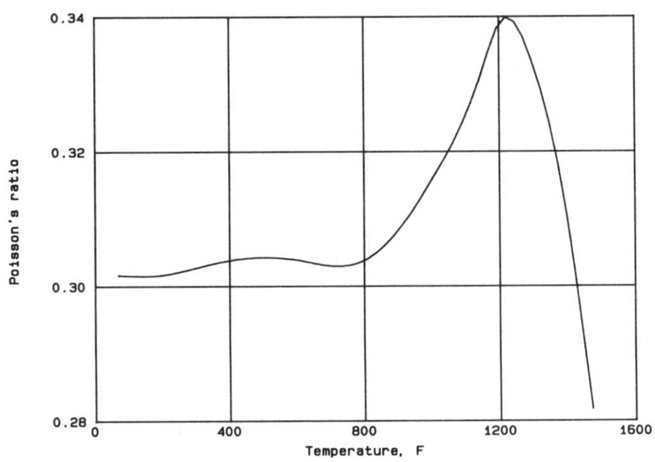

AISI 347: Poisson's ratio of wrought, annealed steel. Source: Ref 22, p 20

AISI 347: Electrical conductivity and resistivity

TEMP C	TEMP F	ELCON %IACS	ELRES n{o}*m
21	70	2.40	719
100	212	2.22	780
144	292	2.01	859
200	392	1.73	1001
400	752	1.56	1110
600	1112	1.43	1209
800	1472	1.37	1260

Source: Ref 29, Dec 1977

AISI 347: Thermal conductivity

TEMP C	TEMP F	THCON W/m*K	THCON Btu/h*ft*F
100	212	16.09	9.3
500	932	22.16	12.8

Source: Ref 27, p 60

AISI 347: Mean coefficients of linear thermal expansion

TEMP C	TEMP F	THEXP 10^{-6}/C	THEXP 10^{-6}/F
0-100	32-212	16.6	9.3
0-315	32-600	18.2	9.5
0-538	32-1000	18.6	10.3
0-649	32-1200	19.1	10.6
0-815	32-1500	20.0	11.1

Source: Ref 27, p 60

Mechanical Properties

AISI 347: Tensile property data for annealed specimens

TEMP C	TEMP F	UTS MPa	UTS ksi	YS MPa	YS ksi	EL %	RA %	CRSTR 1%/10000 h MPa	CRSTR 1%/10000 h ksi
21	70	620	90	255	37	50	65	...	...
95	200	580	84	250	36	45	70	...	...
150	300	540	78	240	35	42	72	...	...
205	400	525	76	205	30	38	74	...	...
260	500	510	74	195	28	36	71	...	...
315	600	495	72	170	25	35	72	...	...
370	700	470	68	160	23	35	72	...	...
425	800	455	66	150	22	35	70	...	...
480	900	435	63	145	21	34	69	...	...
540	1000	405	59	145	21	34	65	115	17
595	1100	380	55	140	20	35	67	90	13
650	1200	330	48	140	20	36	67	50	7
705	1300	275	40	140	20	36	70	30	4.5
760	1400	215	31	130	19	37	70	15	2
815	1500	170	25	115	17	40	70	10	1.5
870	1600	140	20	105	15	40	70	7	1
925	1700	105	15	83	12	45	72	...	...
980	1800	83	12	...	...	50	75	...	...
1040	1900	69	10	...	...	60	77	...	...

Source: Ref 15, p 313

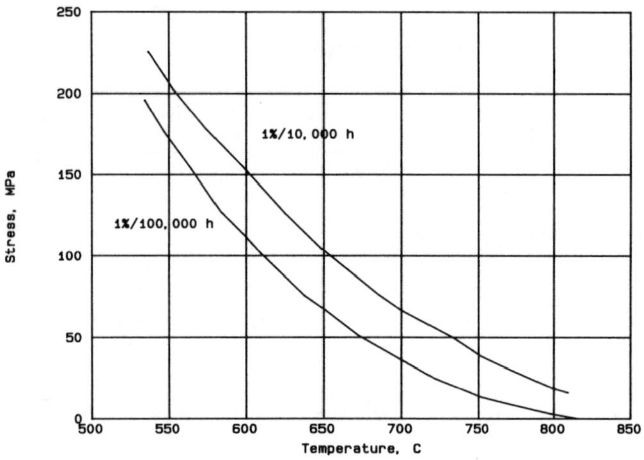

AISI 347: Creep strength of annealed specimens. Source: Ref 15, p 315

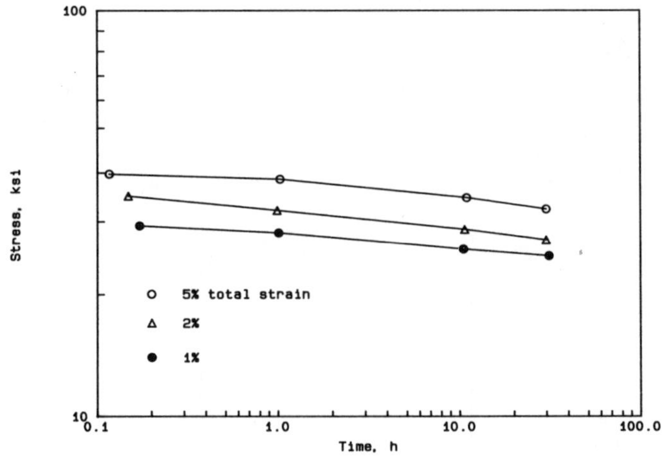

AISI 347: Total creep-strain curves for annealed 0.050-in. sheet at 1200 F. Source: Ref 22, p 18

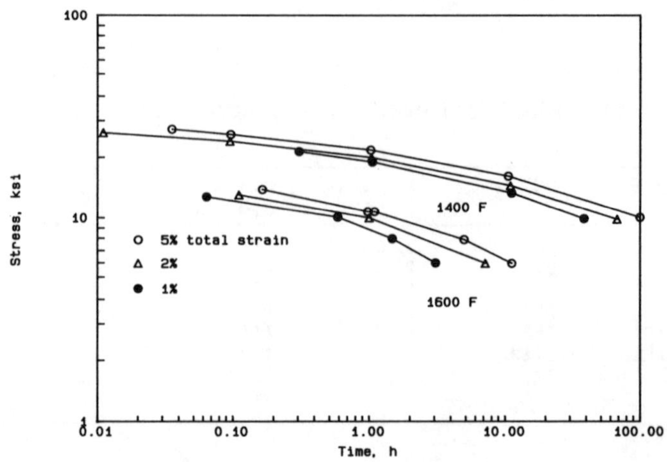

AISI 347: Total creep-strain strength of annealed 0.050-in. sheet at 1400 and 1600 F. Source: Ref 22, p 18

AISI 347: Stress-rupture and creep property data

TEMP C	RPSTR 1000 h MPa	RPSTR 1000 h ksi	RPSTR 10000 h MPa	RPSTR 10000 h ksi	RPSTR(a) 100000 h MPa	RPSTR(a) 100000 h ksi	CRSTR 1%/10000 h MPa	CRSTR 1%/10000 h ksi	CRSTR 1%/100000 h MPa	CRSTR 1%/100000 h ksi
540	340	49.0	260	38.0	230	33.0	220	32.0	185	27.0
595	240	34.5	185	27.0	150	21.5	160	23.0	115	16.5
650	155	22.5	120	17.5	90	13.0	110	16.0	65	9.5
705	97	14.0	72	10.5	50	7.5	60	9.5	35	5.0
760	55	8.0	40	5.5	25	3.5	35	5.0	15	2.0
815	30	4.5	15	2.5	10	1.5	12	1.8	...	...

(a) Extrapolated data
Source: Ref 15, p 312

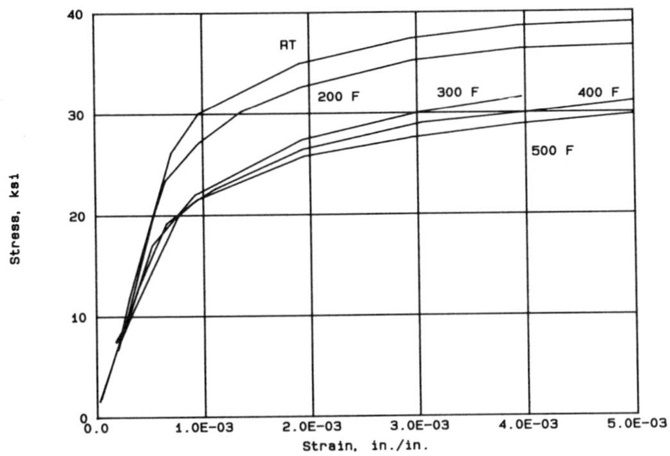

AISI 347: Stress-strain curves for 0.063-in. sheet. Source: Ref 22, p 11

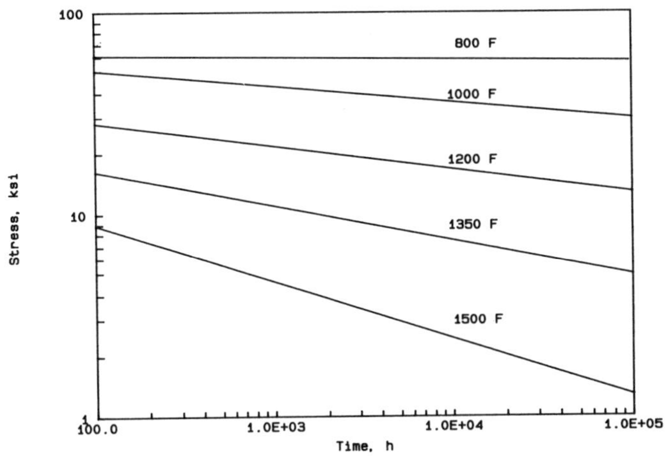

AISI 347: Creep-rupture strength of annealed bar. Source: Ref 22, p 17

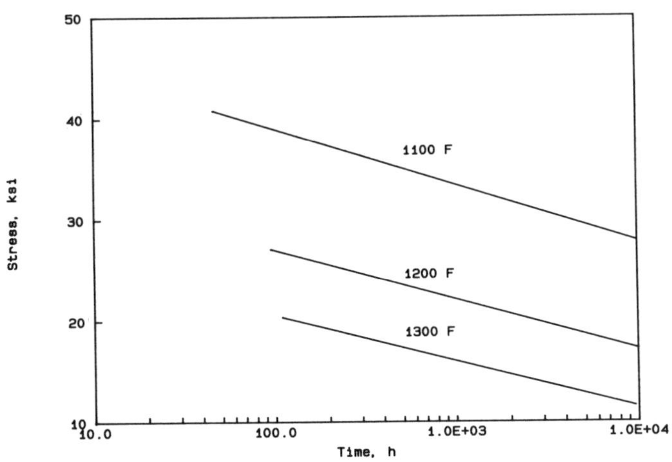

AISI 347: Creep-rupture strength of annealed tube. Source: Ref 22, p 17

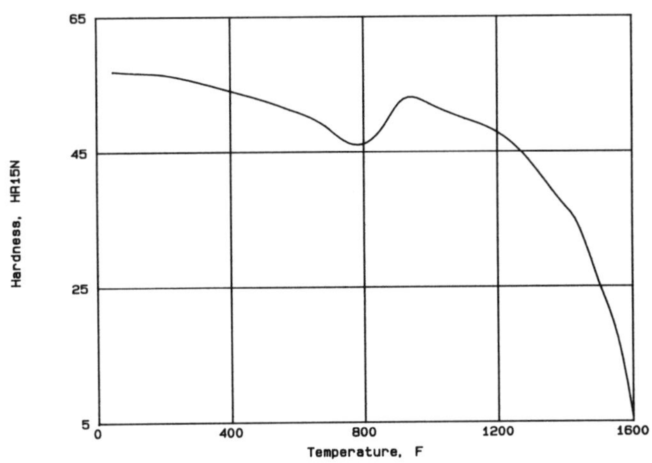

AISI 347: Rockwell 15N hardness of annealed bar. Source: Ref 22, p 4

AISI 348

UNS number: S34800
Specifications: ASME SA182, ASME SA213, ASTM A167, ASTM A182, ASTM A213, ASTM A240, ASTM A249, ASTM A269, ASTM A276, ASTM A312, ASTM A314, FED QQ S-766, MIL S-23195, MIL S-23196, DGN B83, DGN B185 Grade TP348, DGN B185 Grade TP348H, COPANT 513, DIN 1.4878
Nominal composition: 0.080% C max, 0.200% Cu max, 17.000 to 19.000% Cr, 2.000% Mn max, 9.000 to 13.000% Ni, 0.045% P max, 0.030% S max, 1.000% Si max, % (Nb + Ta) min = 10 X carbon content
Product forms: Wrought, bar, wire, plate, sheet, strip, tube, billets, forgings, welded

Description: This niobium-stabilized, austenitic stainless steel is similar to Type 321, but has higher creep strength. It is recommended for fabricated structures that cannot be annealed after welding and for parts that operate in the 425 to 900 C (800 to 1650 F) temperature range. Type 348 is used for nuclear power and other radioactive service applications that require a material with restricted tantalum and cobalt contents. Maximum scaling temperature in air is 900 C (1650 F) max for continuous service and 845 C (1550 F) max for intermittent service.
Typical applications: Tubes and pipes in nuclear energy and other systems that utilize radioactive materials.

Physical Properties

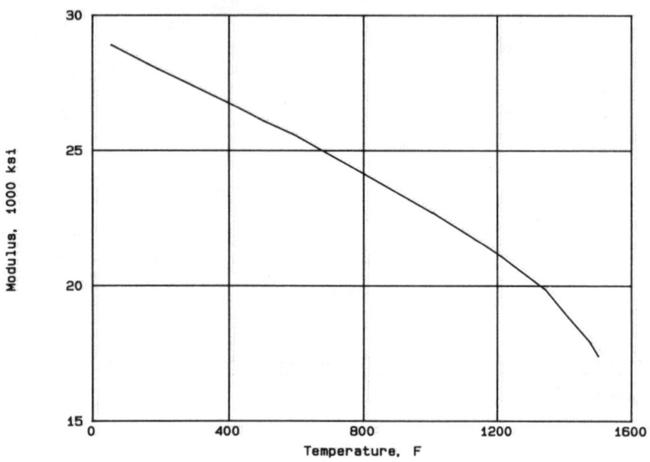

AISI 348: Modulus of elasticity of wrought, annealed steel. Source: Ref 22, p 20

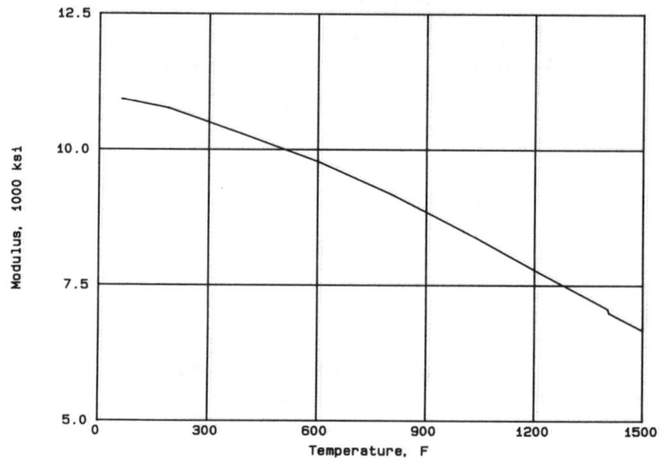

AISI 348: Modulus of rigidity of wrought, annealed steel. Source: Ref 22, p 20

AISI 348: Mean coefficients of linear thermal expansion

TEMP C	TEMP F	THEXP 10^{-6}/C	THEXP 10^{-6}/F
0-100	32-212	16.6	9.3
0-315	32-600	18.2	9.5
0-538	32-1000	18.6	10.3
0-649	32-1200	19.1	10.6
0-815	32-1500	20.0	11.1

Source: Ref 27, p 60

AISI 348: Thermal conductivity

TEMP C	TEMP F	THCON W/m*K	THCON Btu/h*ft*F
100	212	16.09	9.3
500	932	22.16	12.8

Source: Ref 27, p 60

AISI 348: Electrical conductivity and resistivity

TEMP C	TEMP F	ELCON %IACS	ELRES n{o}*m
21	70	2.40	719
100	212	2.22	780
144	292	2.01	859
200	392	1.73	1001
400	752	1.56	1110
600	1112	1.43	1209
800	1472	1.37	1260

Source: Ref 29, Dec 1977

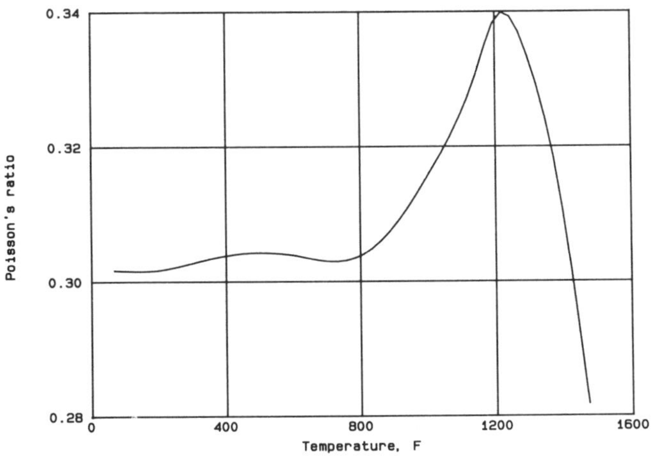

AISI 348: Poisson's ratio of wrought, annealed steel. Source: Ref 22, p 20

Mechanical Properties

AISI 348: Tensile property data for annealed specimens

TEMP C	TEMP F	UTS MPa	UTS ksi	YS MPa	YS ksi	EL %	RA %	CRSTR 1%/10000 h MPa	CRSTR 1%/10000 h ksi
21	70	620	90	255	37	50	65	...	...
95	200	580	84	250	36	45	70	...	...
150	300	540	78	240	35	42	72	...	...
205	400	525	76	205	30	38	74	...	...
260	500	510	74	195	28	36	71	...	...
315	600	495	72	170	25	35	72	...	...
370	700	470	68	160	23	35	72	...	...
425	800	455	66	150	22	35	70	...	...
480	900	435	63	145	21	34	69	...	...
540	1000	405	59	145	21	34	65	115	17
595	1100	380	55	140	20	35	67	90	13
650	1200	330	48	140	20	36	67	50	7
705	1300	275	40	140	20	36	70	30	4.5
760	1400	215	31	130	19	37	70	15	2
815	1500	170	25	115	17	40	70	10	1.5
870	1600	140	20	105	15	40	70	7	1
925	1700	105	15	83	12	45	72	...	...
980	1800	83	12	...	...	50	75	...	...
1040	1900	69	10	...	...	60	77	...	...

Source: Ref 15, p 313

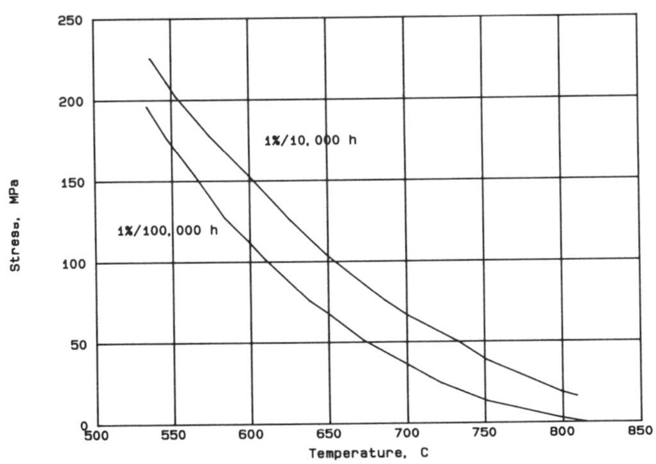

AISI 348: Creep strength of annealed steel. Source: Ref 15, p 315

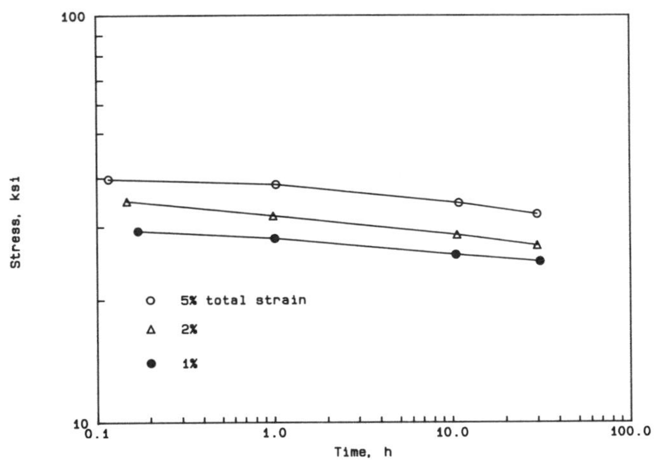

AISI 348: Total creep strength of annealed 0.050-in. sheet at 1200 F. Source: Ref 22, p 18

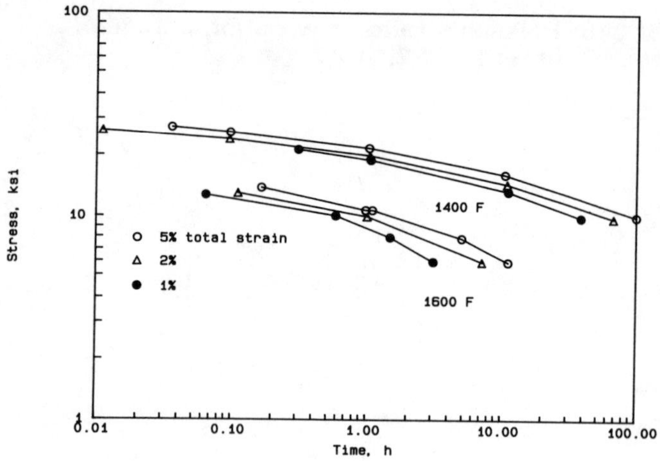

AISI 348: Total creep strength of annealed 0.050-in. sheet at 1400 and 1600 F. Source: Ref 22, p 18

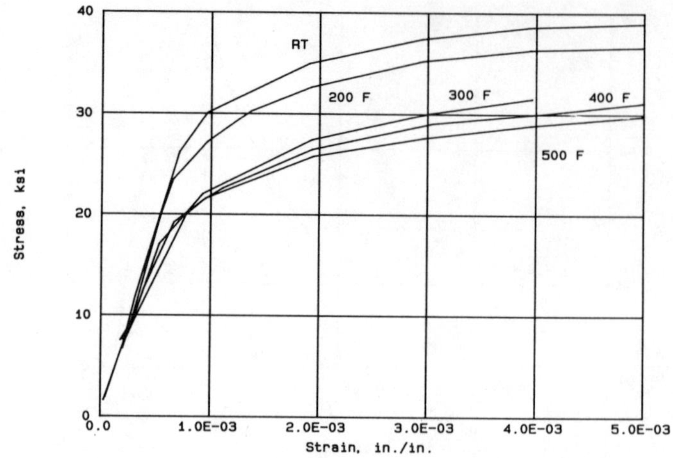

AISI 348: Stress-strain curves for 0.063-in. sheet at room and elevated temperatures. Source: Ref 22, p 11

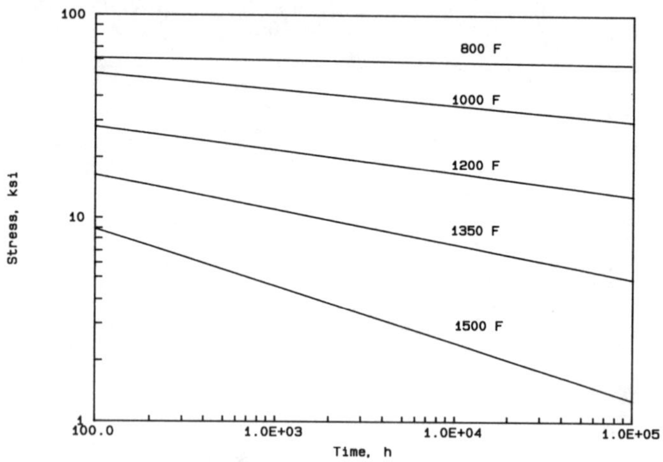

AISI 348: Creep-rupture strength of annealed bar. Source: Ref 22, p 17

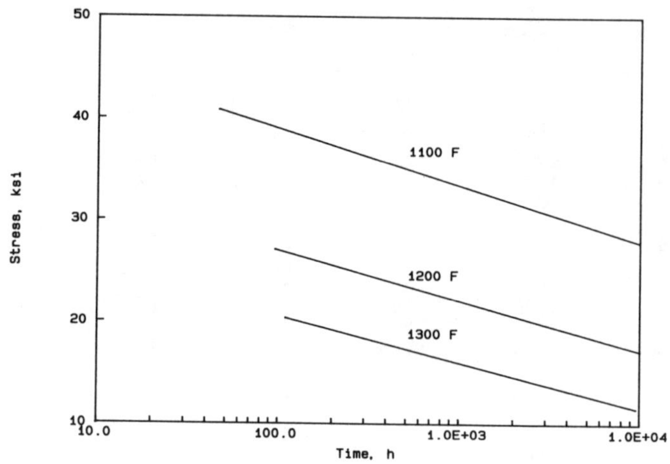

AISI 348: Creep-rupture strength of annealed tube. Source: Ref 22, p 17

AISI 348: Stress-rupture and creep property data

TEMP C	RPSTR 1000 h MPa	RPSTR 1000 h ksi	RPSTR 10000 h MPa	RPSTR 10000 h ksi	RPSTR(a) 100000 h MPa	RPSTR(a) 100000 h ksi	CRSTR 1%/10000 h MPa	CRSTR 1%/10000 h ksi	CRSTR 1%/100000 h MPa	CRSTR 1%/100000 h ksi
540	340	49.0	260	38.0	230	33.0	220	32.0	185	27.0
595	240	34.5	185	27.0	150	21.5	160	23.0	115	16.5
650	155	22.5	120	17.5	90	13.0	110	16.0	65	9.5
705	97	14.0	72	10.5	50	7.5	60	9.5	35	5.0
760	55	8.0	40	5.5	25	3.5	35	5.0	15	2.0
815	30	4.5	15	2.5	10	1.5	12	1.8	...	...

(a) Extrapolated data
Source: Ref 15, p 312

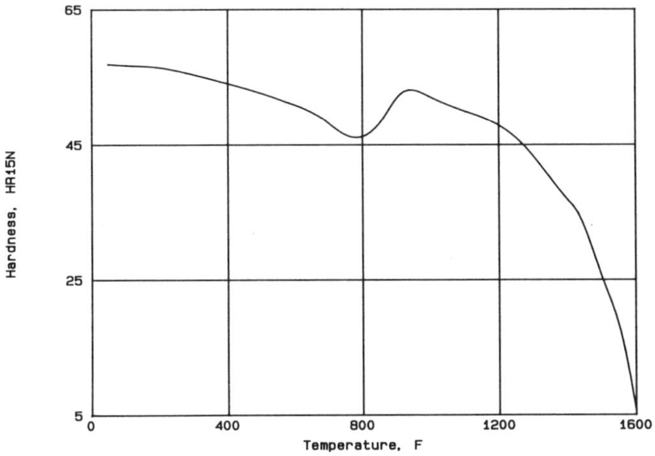

AISI 348: Rockwell 15N hardness of annealed bar. Source: Ref 22, p 4

AISI 403

UNS number: S40300
Specifications: ASTM A176, ASTM A276, ASTM A314, ASTM A473, FED QQ S-763, MIL S-861, MIL S-862, SAE J405, NF Z6-C13, JIS G 4303, JIS G 4305, BS 970 410S21 En56A, GOST 08Ch13, GOST 12Ch13, DIN 1.4000 X7Cr13, DIN 1.4006
Nominal composition: 0.150% C max, 11.500 to 13.000% Cr, 1.000% Mn max, 0.040% P max, 0.030% S max, 0.500% Si max
Product forms: Wrought, bar, wire, sheet, strip, billets, forgings

Description: AISI 403 is a "turbine quality" hardenable martensitic chromium steel that combines both corrosion and heat resistance. Scaling temperatures in air are 705 C (1300 F) max for continuous service and 790 C (1450 F) max for intermittent service.

Typical applications: Steam turbine buckets and blades, jet engine rings, and other highly stressed parts.

Physical Properties

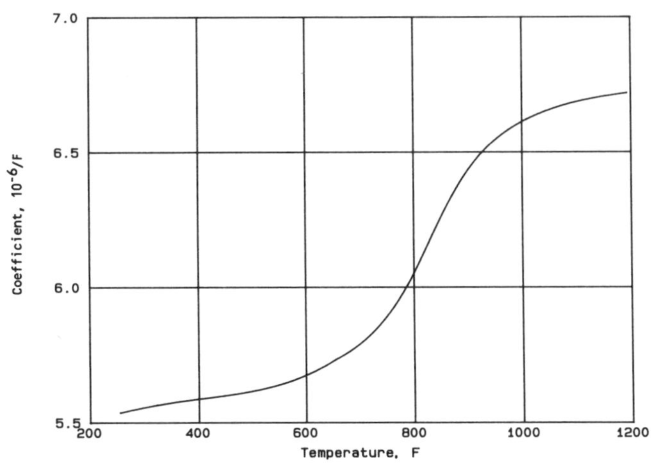

AISI 403: Mean coefficient of linear thermal expansion. Source: Ref 22, p 6

AISI 403: Thermal conductivity

TEMP C	TEMP F	THCON W/m*K	THCON Btu/ft*h*F
100	212	24.9	14.4
500	930	28.7	16.6

Source: Ref 15, p 318

AISI 403: Mean coefficients of linear thermal expansion

TEMP C	TEMP F	THEXP 10^{-6}/C	THEXP 10^{-6}/F
0-100	32-212	9.9	5.5
0-315	32-600	11.3	6.3
0-540	32-1000	11.5	6.4
0-650	32-1200	11.7	6.5

Source: Ref 15, p 318

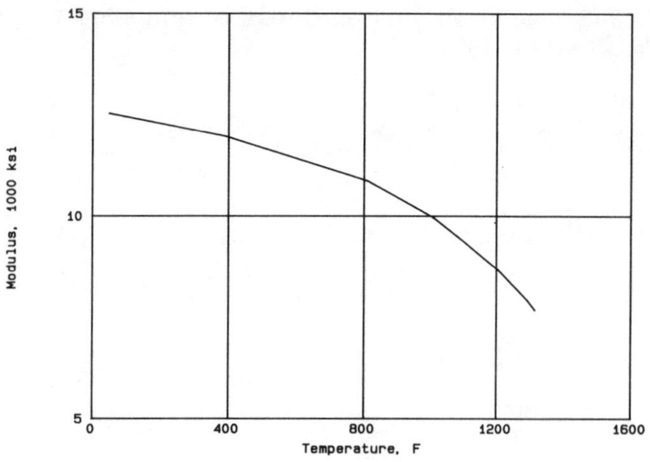

AISI 403: Modulus of rigidity. Source: Ref 22, p 14

AISI 403: Electrical resistivity

TEMP C	TEMP F	ELRES n{o}*m
21	70	570
649	1200	1080

Source: Ref 14, p 209

AISI 403: Modulus of elasticity

TEMP C	TEMP F	E GPa	E 10^3 ksi
21	70	205.06	29.75
260	500	179.27	26.00
482	900	164.10	23.80

Source: Ref 29, Jun 1964

AISI 403: Poisson's ratio

TEMP C	TEMP F	PR
21	70	0.24
482	900	0.19

Source: Ref 29, Jun 1964

Mechanical Properties

AISI 403: Tensile property data

TEMP C	TEMP F	UTS MPa	UTS ksi	YS MPa	YS ksi	EL(a) %	RA %
204	400	1048	152.0	927	134.5	11.0	69.5
316	600	1003	145.5	838	121.5	10.5	65.5
427	800	896	130.0	765	111.0	12.0	70.0
538	1000	700	101.5	645	93.5	16.0	77.5

Note: Specimens were heat treated at 982 C (1800 F) for 30 min, oil quenched, tempered at 538 C (1000 F) for 2 h, and air cooled
(a) In 4D
Source: Ref 29, Jun 1965

AISI 403: Tensile property data

TEMP C	TEMP F	UTS MPa	UTS ksi	YS MPa	YS ksi	EL(a) %	RA %
204	400	817	118.5	727	105.5	17.5	70.5
316	600	772	112.0	690	100.0	16.5	67.5
427	800	724	105.0	652	94.5	17.0	65.5
482	900	627	91.0	586	85.0	18.5	74.5
593	1100	558	81.0	524	76.0	22.0	82.0

Note: Specimens were heat treated at 982 C (1800 F) for 30 min, oil quenched, tempered for 2 h at 607 C (1125 F) and air cooled
(a) In 4D
Source: Ref 29, Jun 1965

AISI 403: Tensile property data

TEMP C	TEMP F	UTS MPa	UTS ksi	YS MPa	YS ksi	EL(a) %	RA %
204	400	741	107.5	652	94.5	18.0	69.0
316	600	696	101.0	617	89.5	16.0	70.0
427	800	634	92.0	572	83.0	17.0	68.0
482	900	583	84.5	527	76.5	18.5	73.5
538	1000	496	72.0	469	68.0	22.5	81.5
593	1100	403	58.5	383	55.5	25.5	88.0

Note: Specimens were heat treated at 982 (1800 F) for 30 min, oil quenched, tempered at 649 C (1200 F) for 2 h, and air cooled
(a) In 4D
Source: Ref 29, Jun 1965

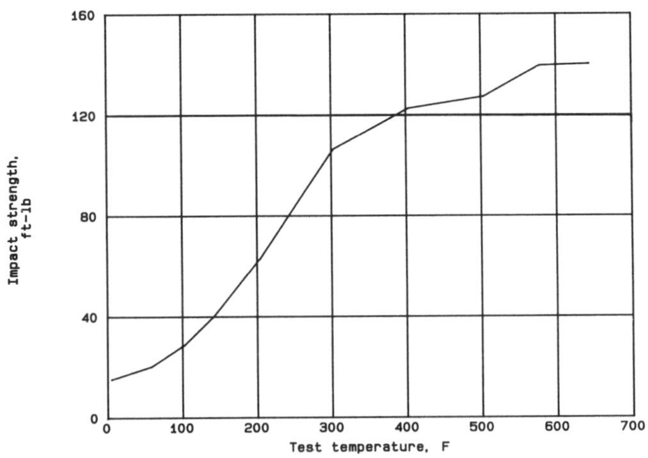

AISI 403: Charpy V-notch impact strength. Specimens were heat treated at 1800 F, brine quenched, and tempered at 1125 F. Source: Ref 22, p 12

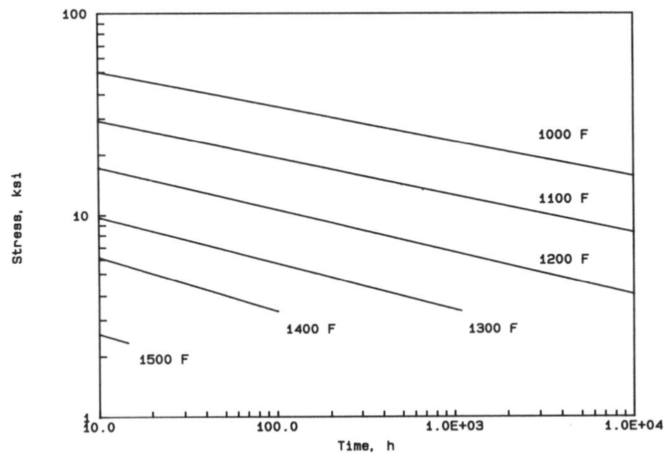

AISI 403: Creep-rupture strength of bar at 1000 to 1500 F. Specimens were heat treated at 1800 F for 0.5 h, oil quenched, and held at 1000 F for 2 h. Source: Ref 22, p 13

AISI 403: Minimum creep rate data

TEMP C	TEMP F	CRSTR 0.0001%/h MPa	CRSTR 0.0001%/h ksi	CRSTR 0.00001%/h MPa	CRSTR 0.00001%/h ksi
482	900	224	32.5	110	16.0
538	1000	76	1.01	41	6.0
593	1100	34	5.0	21	3.0
649	1200	21	3.0	10	1.5

Note: Specimens were oil quenched at 982 C (1800 F), tempered at 621 C (1150 F), and air cooled
Source: Ref 29, Jun 1964

AISI 403: Stress-rupture property data

TEMP C	TEMP F	RPSTR 10 h MPa	RPSTR 10 h ksi	RPSTR 100 h MPa	RPSTR 100 h ksi	RPSTR 1000 h MPa	RPSTR 1000 h ksi	RPSTR 10000 h MPa	RPSTR 10000 h ksi	RPSTR 100000 h MPa	RPSTR 100000 h ksi
538	1000	344.8	50.0	234.4	34.0	158.6	23.0	110.3	16.0	73.8	10.7
593	1100	200.0	29.0	134.5	19.5	86.2	12.5	57.2	8.3	35.9	5.2
649	1200	117.2	17.0	73.8	10.7	45.5	6.6	27.6	4.0	16.5	2.4
704	1300	65.5	9.5	39.3	5.7	22.8	3.3	...	...	...	...
760	1400	35.2	5.1	20.0	2.9	...	...	...	...	...	...
816	1500	17.9	2.6	...	...	...	...	...	...	...	...

Note: Specimens were hardened by heating at 982 C (1800 F) for 30 min, oil quenched, and tempered at 538 C (1000 F) for 2 h
Source: Ref 29, Jun 1964

AISI 403: Stress relaxation

TEMP C	TEMP F	CRIS MPa	CRIS ksi	CRSRD 100 h MPa	CRSRD 100 h ksi	CRSRD 1000 h MPa	CRSRD 1000 h ksi	CRSRD 100000 h MPa	CRSRD 100000 h ksi
427	800	345	50.0	214	31.0	179	26.0	148	21.5
482	900	317	46.0	146	21.2	121	17.5	94	13.6
538	1000	263	38.2	83	12.0	63	9.2	32	4.6
566	1050	280	40.6	81	11.8	53	7.7	31	4.5

Note: Specimens were heat treated at 788 C (1450 F) by hot finishing off mill, air cooled, tempered at 635 C (1175 F), and air cooled
Source: Ref 29, Jun 1964

AISI 403: Fatigue strength

TEMP C	TEMP F	FATST MPa 10^6 cyc	FATST ksi 10^6 cyc	FATST MPa 10^7 cyc	FATST ksi 10^7 cyc	FATST MPa 10^8 cyc	FATST ksi 10^8 cyc
21	70	537.8	78.0	510.2	74.0	482.7	70.0
260	500	441.3	64.0	386.1	56.0	344.8	50.0
371	700	365.4	53.0	341.3	49.5	317.2	46.0
482	900	289.6	42.0	268.9	39.0	248.2	36.0

Note: Specimens were heat treated by hardening at 954 C (1750 F) for 15 min, oil quenched, and tempered at 538 C (1000 F) for 1-1/2 h
Source: Ref 29, Jun 1964

AISI 403: Damping capacity

TEMP C	TEMP F	DMPCP % (a)	DMPCP % (b)	DMPCP % (c)
21	70	0.0060	0.0300	0.0370
149	300	0.0080	0.3100	0.0470
316	600	0.1000	0.2900	0.0300
482	900	0.0092	0.0170	0.0180
649	1200	0.0140	0.0140	0.0140

Note: Specimens were heat treated by hardening at 1010 C (1850 F) for 2 h, oil quenched, tempered at 816 C (1500 F) for 50 h, and air cooled
(a) Logarithmic decrement at a stress of 41.4 MPa (5 ksi).
(b) Logarithmic decrement at a stress of 103.4 MPa (15 ksi).
(c) Logarithmic decrement at a stress of 172.4 MPa (25 ksi)
Source: Ref 29, Jun 1964

AISI 405

UNS number: S40500
Specifications: ASME SA240, ASME SA268, ASME SA479, ASTM A176, ASTM A240, ASTM A268, ASTM A276, ASTM A314, FED QQ S-763, MIL S-861, MIL S-862, NF Z6CA13, JIS G 4303-38 SUS38HP, BS 970 405S17, DIN 1.4002 X7CrA113
Nominal composition: 0.080% C max, 11.500 to 14.500% Cr, 1.000% Mn max, 0.040% P max, 0.030% S max, 1.000% Si max, 0.100 to 0.300% Al
Product forms: Wrought, bar, wire, plate, sheet, strip, billets, forgings
Description: AISI 405 is a ferritic stainless and heat-resisting chromium steel characterized by its ability to be cooled from elevated temperatures without incurring significant hardening. It is ideal for applications for which air-hardening steels such as Types 410 and 403 are not suitable. Scaling temperatures in air are 760 C (1400 F) max for continuous service and 790 C (1450 F) max for intermittent service.

Typical applications: Annealing boxes, quench tanks, oxidation-resistant partitions, heat exchangers, conveyer lines for chemical processing, oil refinery tubing, catalytic crackers, boiler tubing, for hydrogenation and dehydrogenation processes.

Physical Properties

AISI 405: Mean coefficients of linear thermal expansion

TEMP C	TEMP F	THEXP 10^{-6}/C	THEXP 10^{-6}/F
0-100	32-12	10.8	6.0
0-315	32-600	11.5	6.4
0-540	32-1000	12.1	6.7
0-815	32-1500	13.5	7.5

Source: Ref 15, p 324

Mechanical Properties

AISI 405: Tensile property data

TEMP C	TEMP F	UTS MPa	UTS ksi	YS MPa	YS ksi	EL %
21	70	469	68.0	276	40.0	30
95	200	400	58.0	200	29.0	28
205	400	372	54.0	165	24.0	22
315	600	352	51.0	152	22.0	21
425	800	321	46.5	138	20.0	22
540	1000	252	36.5	117	17.0	28
650	1200	159	23.0	93	13.5	34
760	1400	59	8.5	31	4.5	68
870	1600	29	4.2	15	2.2	80

Source: Ref 15, p 325

AISI 405: Creep strength

TEMP C	TEMP F	CRSTR 1%/10000 h MPa	CRSTR 1%/10000 h ksi	CRSTR 1%/100000 h MPa	CRSTR 1%/100000 h ksi
427	800	275.8	40.0	158.6	23.0
482	900	206.8-296.5	30.0-43.0	96.5-120.7	14-17.5
538	1000	55.2-57.9	8.0-8.4	31.0-51.7	4.5-7.5
593	1100	13.8-23.271	2.0-3.375	3.4-19.1	0.5-2.775
649	1200	10.34	1.50	6.723	0.975
704	1300	6.723	0.975	4.1	0.6

Note: These values were taken from published data. Where ranges are shown, they represent two or three sets of data for the particular temperature and time involved
Source: Ref 29, May 1964

AISI 409

UNS number: S40900

Specifications: ASME SA268, ASTM A176, ASTM A268, ASTM A651, SAE J405, DIN 1.4512 X5CrTi12, NF Z6CT12, BS 409S19, UNI X6CrTi12, JIS SUH 409

Nominal composition: 0.080% C max, 10.500 to 11.750% Cr, 1.000% Mn max, 0.500% Ni max, 0.045% P max, 0.045% S max, 1.000% Si max, 0.750% Ti max (6 X carbon content min)

Product forms: Wrought, bar, plate, sheet, strip, billets, forgings

Description: AISI 409 is a ferritic chromium stainless steel for general-purpose applications, in which appearance is secondary to mechanical properties and corrosion resistance. The titanium addition makes Type 409 nonhardenable by heat treatment and prevents hardening during welding. The relatively low cost stainless steel is a higher performance alternative to coated carbon steels.

Typical applications: Automobile exhaust systems, transformer and capacitor cases, dry fertilizer spreaders, and tanks for agricultural sprays.

Physical Properties

AISI 409: Thermal conductivity

TEMP C	TEMP F	THCON W/m*K	THCON Btu/ft*h*F
95	200	24.9	14.4
540	1000	28.6	16.5

Source: Ref 15, p 327

AISI 409: Mean coefficients of linear thermal expansion

TEMP C	TEMP F	THEXP 10^{-6}/C	THEXP 10^{-6}/F
21-100	70-212	11.7	6.5
21-260	70-500	11.9	6.6
21-480	70-900	12.4	6.9
21-650	70-1200	12.9	7.2
21-815	70-1500	13.5	7.5

Source: Ref 15, p 328

Mechanical Properties

AISI 409: Tensile property data

TEMP C	TEMP F	UTS MPa	UTS ksi	YS MPa	YS ksi	EL %
27	80	448	65.0	238	34.5	32.5
93	200	406	58.9	207	30.0	29.5
205	400	365	53.0	172	25.0	25.5
315	600	334	48.5	172	25.0	22.5
425	800	326	47.3	165	24.0	21.0
540	1000	254	36.9	136	19.7	24.0
650	1200	148	21.4	92	13.4	44.0
760	1400	61	8.8	39	5.7	67.5
870	1600	30	4.3	19	2.7	75.5

Source: Ref 15, p 328

AISI 410

UNS number: S41000
Specifications: AMS 5504, AMS 5505, ASME SA194(6), ASME SA240, ASTM A176, ASTM A182, FED QQ S-763, FED QQ S-766, MIL S-861, SAE J405, NF Z10C-13, NF Z12C-13, JIS G 4303-51, JIS G 4304-51, BS 970 420S29 En56B, GOST Zh1, DIN 1.4024 X15Cr13, SIS 14 23 02, UNI X12Cr13
Nominal composition: 0.150% C max, 11.500 to 13.500% Cr, 1.000% Mn max, 0.040% P max, 0.030% S max, 1.000% Si max
Product forms: Wrought, bar, wire, plate, sheet, strip, billets, forgings
Description: AISI 410 is a general-purpose, heat treatable chromium steel. The martensitic stainless combines good corrosion resistance with fair machining properties. Scaling temperatures in air are 705 C (1300 F) max for continuous service and 790 C (1450 F) max for intermittent service.
Typical applications: Pump shafts, bolts, bushings, coal chutes, cutlery, fishing tackle, hardware, jet engine parts, mining machinery, rifle barrels, screws, and valves.

Physical Properties

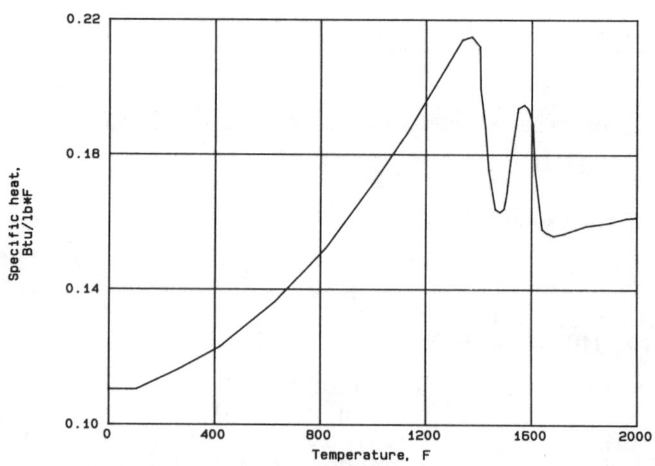

AISI 410: Specific heat. Source: Ref 21, p 3

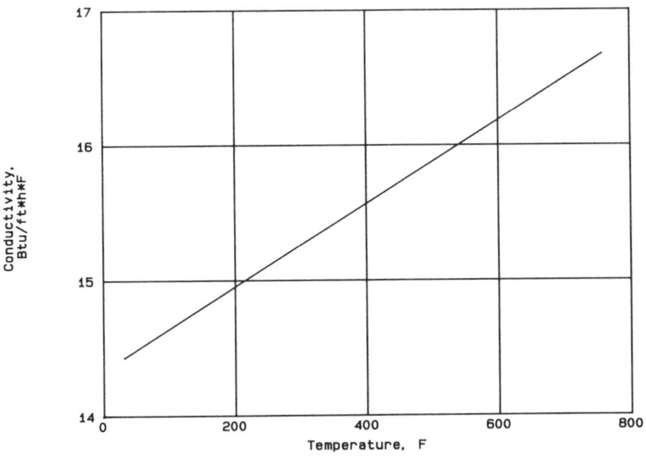

AISI 410: Thermal conductivity. Source: Ref 22, p 6

AISI 410: Thermal conductivity

TEMP C	TEMP F	THCON W/m*k	THCON Btu/h*ft*F
100	212	24.9	14.4
200	392	25.9	15.0
300	572	26.8	15.5
400	752	27.8	16.1
500	932	28.5	16.5

Source: Ref 21, p 3

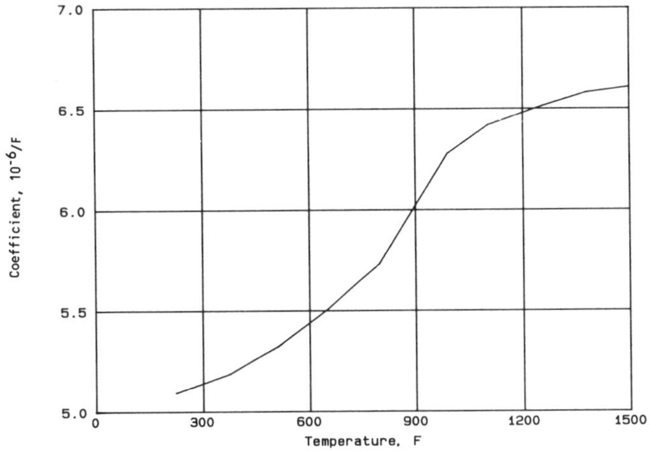

AISI 410: Mean coefficient of linear thermal expansion. Source: Ref 22, p 6

AISI 410: Mean coefficients of linear thermal expansion

TEMP C	TEMP F	THEXP 10^{-6}/C	THEXP 10^{-6}/F
0-100	32-212	9.9	5.5
0-315	32-600	11.0	6.1
0-540	32-1000	11.5	6.4
0-650	32-1200	11.7	6.5

Source: Ref 15, p 329

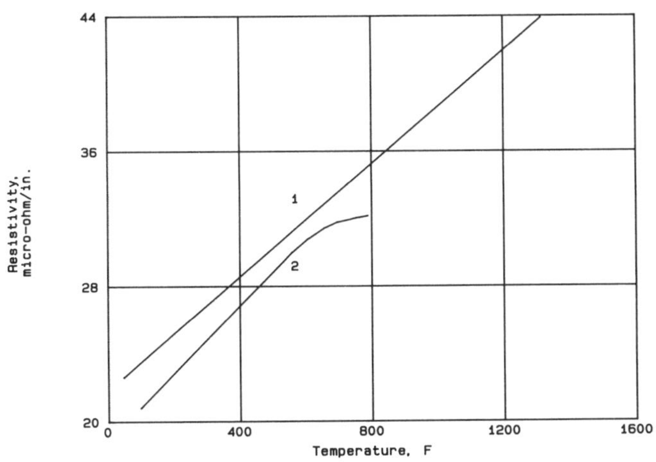

AISI 410: Electrical resistivity. Curve 1: annealed. Curve 2: heat treated to 1750 F, air cooled and heated to 1375 F, and held for 2 h. Source: Ref 21, p 3

AISI 410: Electrical resistivity

TEMP C	TEMP F	ELRES n{o}*m
21	70	570
399	750	880

Source: Ref 21

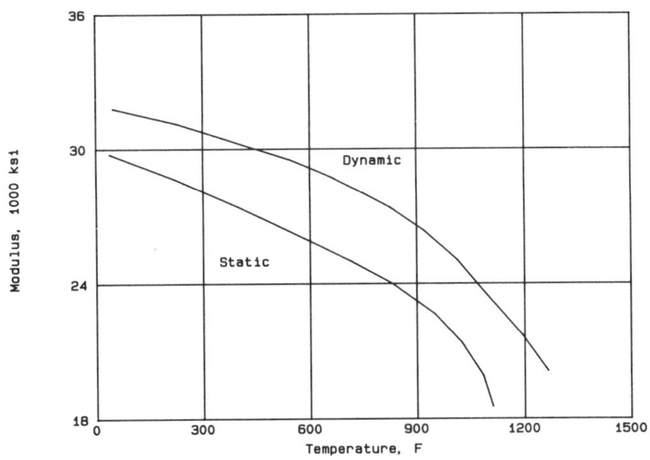

AISI 410: Static and dynamic modulus of elasticity. Source: Ref 21, p 2

AISI 410: Moduli of elasticity and rigidity

TEMP C	TEMP F	E GPa	E 10^3 ksi	G GPa	G 10^3 ksi
21	70	206.9	30	82.7	12
260	500	179.3	26	75.9	11
482	900	158.6	23	69.0	10

Source: Ref 21, p 2

AISI 410: Dynamic modulus of rigidity. Source: Ref 21, p 2

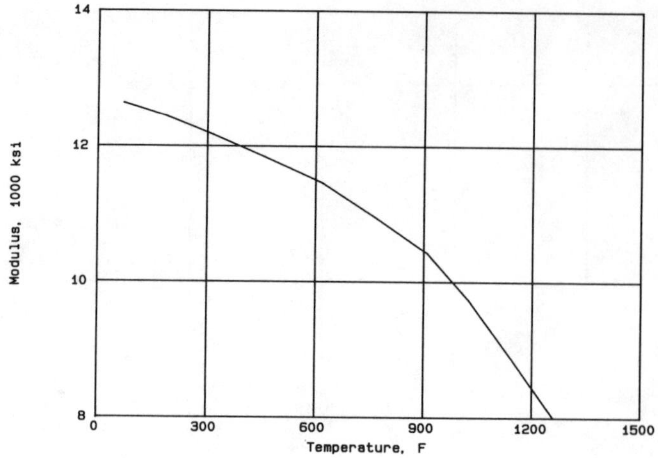

Mechanical Properties

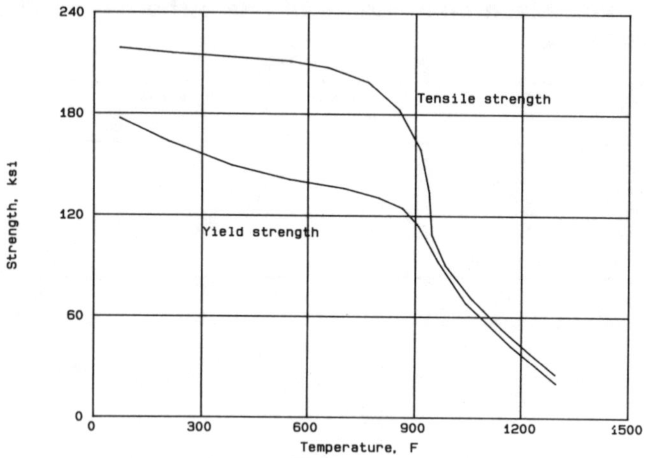

AISI 410: Tensile and yield strength of 0.625-in. bar. Specimens were heat treated at 1800 F for 0.5 h, oil quenched, and tempered for 2 h at 50 F above test temperature. Source: Ref 15, p 331

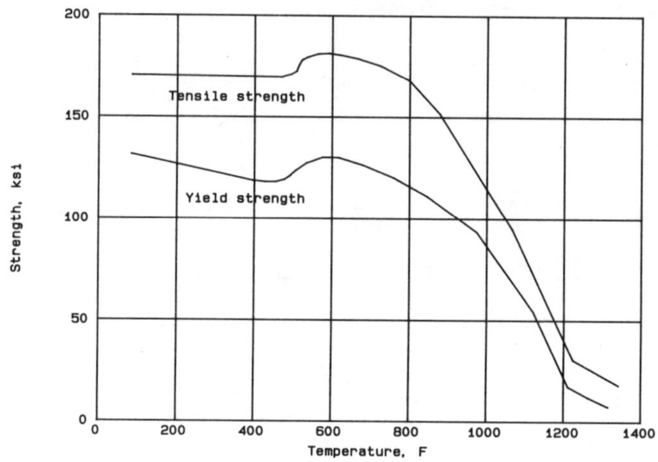

AISI 410: Tensile and yield strength of 0.063-in. sheet. Specimens were heated to 1800 F for 0.5 h, oil quenched, aged at 700 F for 1 h and air cooled. Source: Ref 22, p 10

AISI 410: Tensile property data

TEMP C	TEMP F	YS MPa	YS ksi	UTS MPa	UTS ksi	EL %	RA %
21	70	276	40	517	75	35	70
482	900	241	35	331	48	31	76
538	1000	214	31	283	41	36	79
593	1100	172	25	228	33	41	84
649	1200	124	18	159	23	47	90
704	1300	83	12	110	16	55	94
760	1400	48	7	76	11	66	96

Note: Specimens were annealed
Source: Ref 21, p 16

AISI 410: Tensile property data

TEMP C	TEMP F	UTS MPa	UTS ksi	YS MPa	YS ksi	EL %	RA %	HARD(a) HRC
21	70	1525	221	1225	177.5	14.5	63.5	45
205	400	1475	214	1005	146	11	51	43
315	600	1470	213	961	139.5	18	57	43
425	800	1340	194	920	133.5	18.5	59	43.5
480	900	1150	166.5	835	121	14	57	...
540	1000	605	87.5	565	82	21.5	81.5	...
595	1100	440	64	395	57.5	25.5	87	...
650	1200	300	43.5	270	39.5	29.5	96.5	...
705	1300	195	28.5	165	24	34	91.5	...

Note: Specimens were 16-mm (0.625-in.) diam bar; heated to 980 C (1800 F) for 30 min; oil quenched; tempered for 2 h at 28 C (50 F) above test temperature
(a) Before testing
Source: Ref 15, p 329

AISI 410: Tensile property data

TEMP C	TEMP F	UTS MPa	UTS ksi	YS MPa	YS ksi	EL %	RA %
21	70	1085	157.5	1005	145.5	13	69.5
205	400	1050	152	927	134.5	11	69.5
315	600	1005	145.5	838	121.5	10.5	65.5
425	800	896	130	758	110	12	70
540	1000	700	101.5	645	93.5	16	77.5
650	1200	275	40	260	38	35	91
760	1400	105	15	90	13	54	96
870	1600	96	14	66	9.5	81	77

Note: Specimens were 16-mm (0.625-in.) diam bar; heated to 980 C (1800 F) for 30 min; oil quenched; tempered at 540 C (1000 F) for 2 h; air cooled; hardness of 35.5 HRC
Source: Ref 15, p 330

AISI 410: Tensile property data

TEMP C	TEMP F	UTS MPa	UTS ksi	YS MPa	YS ksi	EL %	RA %
21	70	924	134	807	117	20	68.5
205	400	817	118.5	727	105.5	17.5	70.5
315	600	772	112	689	100	16.5	67.5
425	800	724	105	650	94.5	17	65.5
480	900	625	91	585	85	18.5	74.5
540	1000	560	81	525	76	22	82
595	1100	450	65	415	60	25	87

Note: Specimens were 16-mm (0.625-in.) diam bar; heated to 980 C (1800 F) for 30 min; oil quenched; tempered at 605 C (1125 F) for 2 h; air cooled; hardness of 28.5 HRC
Source: Ref 15, p 330

AISI 410: Tensile property data

TEMP C	TEMP F	UTS MPa	UTS ksi	YS MPa	YS ksi	EL %	RA %
21	70	834	121	721	104.5	21.5	68.5
205	400	741	107.5	650	94.5	18	69
315	600	696	101	615	89.5	16	70
425	800	635	92	570	83	17	68
480	900	585	84.5	525	76.5	18.5	73.5
540	1000	495	72	470	68	22.5	81.5
595	1100	405	58.5	385	55.5	25.5	88
650	1200	305	44	280	40.5	29	90

Note: Specimens were 16-mm (0.625-in.) diam bar; heated to 980 C (1800 F) for 30 min; oil quenched; tempered at 650 C (1200 F) for 2 h; air cooled; hardness of 24.5 HRC
Source: Ref 15, p 329

AISI 410: Tensile property data

TEMP C	TEMP F	UTS MPa	UTS ksi	YS MPa	YS ksi	EL %	RA %
Tempered at 593 C (1100 F)(a)							
21	70	931	135	841	122	19	69
260	500	807	117	738	107	17	70
371	700	758	110	696	101	15	68
482	900	669	97	634	92	18	72
593	1100	510	74	490	71	21	81
Tempered at 649 C (1200 F)(a)							
21	70	834	121	738	107	21	66
260	500	724	105	669	97	20	66
371	700	696	101	621	90	28	64
482	900	634	92	558	81	19	66
593	1100	476	69	427	62	22	79
Tempered at 704 C (1300 F)(a)							
21	70	786	114	648	94	21	71
260	500	627	91	545	79	19	73
371	700	586	85	517	75	18	72
482	900	524	76	469	68	20	74
593	1100	379	55	352	51	25	84

(a) Hardened at 1010 C (1850 F), held 30 min, oil quenched, then tempered for 4 h
Source: Ref 21, p 15

AISI 410: Effect of thermal exposure under stress on room-temperature property data

Stress MPa	Stress ksi	EXP h	UTS MPa	UTS ksi	EL %	HARD HRC
Unexposed						
...	...	...	876	127	20	26
...	...	...	1434	208	...	26
At 482 C (900 F)						
344(a)	50	17229	883	128	18	27
379(b)	55	16818	1538	223	...	27

Note: Specimens were heat treated at 1010 C (1850 F) for 30 min, oil quenched, and tempered at 649 C (1200 F)
(a) Smooth specimen. (b) Notched specimen
Source: Ref 10, p 101

AISI 410: Effect of heat treatment and temperature on Charpy V-notch impact values

TEMP C	TEMP F	IS(a) J	IS(a) ft-lb	IS(b) J	IS(b) ft-lb
21	70	22	16	11	8
21	70	22	16	12	9
316	600	50	37	26	19
316	600	43	31.5	26	19.5
371	700	43	32	25	18.5
371	700	53	39	30	22
427	800	54	40	22	16.5
427	800	58	42.5	25	18.5

(a) Heat treated at 982 C (1800 F) for 1 h, oil quenched, heated to 579 C (1075 F) for 3 h, air cooled, heated to 593 C (1100 F) for 2 h, and air cooled. Hardness was 23 HRC. (b) Heat treated at 982 C (1800 F) for 1 h, oil quenched, heated to 510 C (950 F) for 3 h, air cooled, heated to 523 C (975 F) for 2 h, and air cooled. Hardness was 38 HRC
Source: Ref 21, p 25

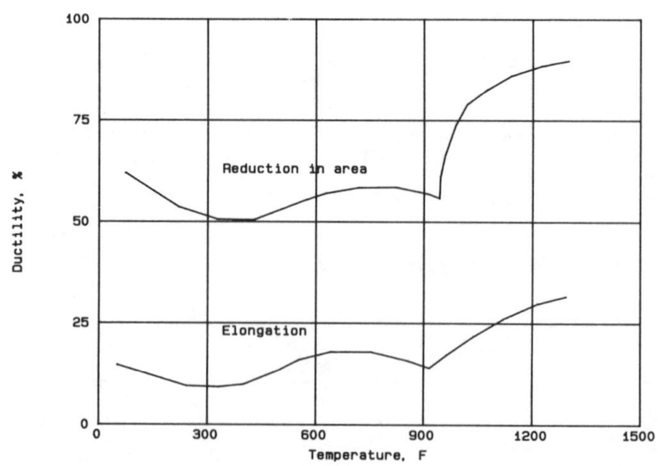

AISI 410: Tensile elongation and reduction in area of 0.625-in. round bar. Specimens were hardened at 1800 F for 0.5 h, oil quenched, and tempered for 2 h at 50 F above test temperature. Source: Ref 15, p 331

AISI 410: Stress-rupture and creep property data

TEMP C	RPSTR 100 h MPa	RPSTR 100 h ksi	RPSTR 1000 h MPa	RPSTR 1000 h ksi	RPSTR(a) 10000 h MPa	RPSTR(a) 10000 h ksi	CRSTR 1%/10000 h MPa	CRSTR 1%/100000 h ksi	CRSTR 1%/100000 h MPa	CRSTR 1%/100000 h ksi
480	...	...	265	38.5	...	...	...	...	110	16.0
540	234	34.0	159	23.0	110	16.0	76	11.0	41	6.0
595	134	19.5	86	12.5	57	8.3	34	5.0	21	3.0
650	74	10.7	46	6.6	28	4.0	21	3.0	10	1.5
705	39	5.7	23	3.3	...	...	...	...	...	...

(a) Extrapolated data
Source; Ref 15, p 331

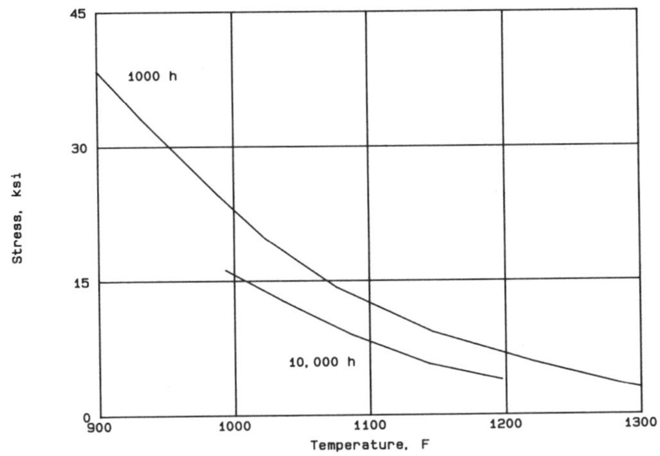

AISI 410: Stress-rupture strength. Specimens were heat treated at 1800 F for 0.5 h, oil quenched, aged at 1200 F for 2 h, and air cooled. Source: Ref 21, p 33

AISI 410: Stress-rupture property data for bar

TEMP C	TEMP F	RPSTR MPa	RPSTR ksi	RPT h
Smooth				
316	600	1000	145	100
316	600	1069	155	100
316	600	1138	165	(a)
316	600	1103	160	(a)
371	700	1034	150	100
371	700	1034	150	500
427	800	1000	145	163
427	800	1000	145	8.7
482	900	931	135	(a)
Notched				
316	600	1138	165	100
316	600	1207	175	500
371	700	1138	165	200
371	700	1172	170	1000
427	800	1069	155	20.3
482	900	1000	145	12.5

Note: Room-temperature material hardness was 38 HRC
(a) Failed on loading
Source: Ref 21, p 32

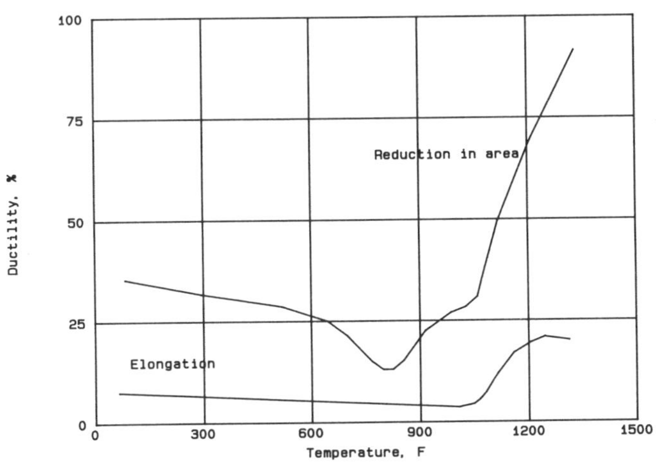

AISI 410: Tensile elongation and reduction in area of 0.063-in. sheet. Specimens were heat treated at 1800 F for 0.5 h, oil quenched and aged at 700 F for 1 h and air cooled. Source: Ref 22, p 10

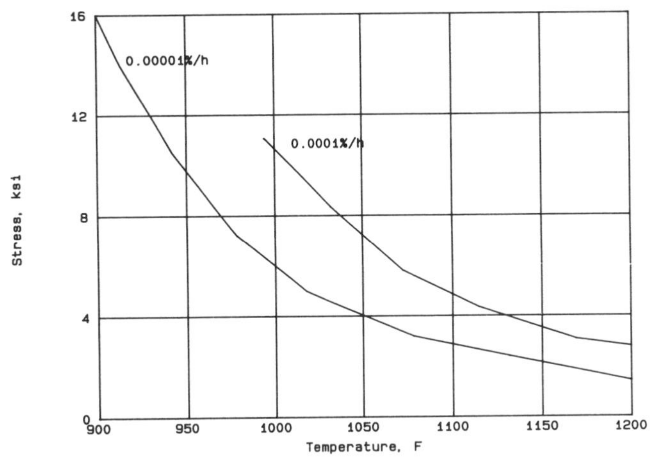

AISI 410: Minimum creep rate data. Source: Ref 21, p 31

AISI 410: Stress-rupture property data

TEMP C	TEMP F	RPSTR 1000 h MPa	RPSTR 1000 h ksi	RPSTR 100000 h MPa	RPSTR 100000 h ksi
482	900	228	33	165	24
538	1000	179	26	138	20

Note: Specimens were heat treated at 1010 C (1850 F), oil quenched, and tempered at 649 C (1200 F)
Source: Ref 10, p 101

AISI 410: Rotating beam fatigue strength comparison

Ttpr C	Ttpr F	UTS MPa	UTS ksi	FATST MPa 10^7 cyc	FATST ksi 10^7 cyc
410 Cb					
566	1050	1014	147	545	79
612	1125	931	135	483	70
649	1200	889	129	469	68
410					
566	1050	924	134	476	69
612	1125	827	120	427	62
649	1200	752	109	386	56

Note: Specimens were hardened at 1010 C (1850 F) for 30 min, oil quenched, then tempered for 4 h
Source: Ref 21, p 26

AISI 410: Fatigue property data

FATST MPa	FATST ksi	HARD HRC	FAT.n 10^6 cyc
At 21 C (70 F)			
655.0	95	31	0.420
634.3	92	35	0.586
620.6	90	37	5.470
613.7	89	35	0.185
606.8	88	28	2.400
599.9	87	28	0.230
593.0	86	28	8.880
586.1	85	33	12.025
At 204 C (400 F)			
586.1	85	25	0.240
572.3	83	28	10.545
572.3	83	...	3.560
558.5	81	35	30.384
551.6	80	37	2.600
544.7	79	31	10.350
537.8	78	31	42.400
530.9	77	38	22.580
524.0	76	28	30.000
At 316 C (600 F)			
565.4	82	23	2.370
551.6	80	32	3.370
537.8	78	32	5.660
524.0	76	28	9.000
517.1	75	...	...
510.2	74	37	7.950
496.4	72	32	...
At 427 C (800 F)			
551.6	80	38	1.270
524.0	76	29	19.300
496.4	72	30	26.300
482.7	70	28	10.647
475.8	69	36	10.172
462.0	67	34	165.000
At 482 C (900 F)			
551.6	80	35	2.143
510.2	74	35	8.121
496.4	72	36	10.801
482.7	70	35	9.936
468.9	68	35	25.235

Source: Ref 21, p 26

AISI 414

UNS number: S41400
Specifications: AMS 5615, ASTM A276, ASTM A314, ASTM A473, ASTM A511, ASTM A580, FED QQ S-763, SAE J405, DGN B83, DGN B171, GOST EI474, GOST 1Kh13N3L
Nominal composition: 0.150% C max, 11.500 to 13.500% Cr, 1.000% Mn max, 1.250 to 2.500% Ni, 0.040% P max, 0.030% S max, 1.000% Si max
Product forms: Wrought, bar, wire, plate, sheet, strip, billets, forgings
Description: AISI 414 is a martensitic stainless steel containing nickel to enhance air-hardening properties, resulting in improved ductility in the hardened condition. Corrosion resistance is somewhat better than that of Type 410, and its mechanical properties in the heat treated condition are slightly higher. Scaling temperatures in air are 705 C (1300 F) max for continuous service and 790 C (1450 F) max for intermittent service.
Typical applications: Springs, tempered rules, bolts, scissors, ship bells, spindles, valve seats, steering wheels, knife blades, and airplane and automobile parts.

Physical Properties

AISI 414: Mean coefficients of linear thermal expansion

TEMP C	TEMP F	THEXP 10^{-6}/C	THEXP 10^{-6}/F
0-100	32-212	10.4	5.8
0-315	32-600	11.0	6.1
0-540	32-1000	12.1	6.7
0-650	32-1200	12.4	6.9

Source: Ref 15, p 332

AISI 414: Thermal conductivity

TEMP C	TEMP F	THCON W/m*K	THCON Btu/ft*h*F
100	212	24.9	14.4
500	930	28.7	16.6

Source: Ref 15, p 332

Mechanical Properties

AISI 414: Tensile property data

TEMP C	TEMP F	UTS MPa	UTS ksi	YS MPa	YS ksi	EL %	RA %	HARD(a) HRC
21	70	1795	260	1013	147	15	49	48
205	400	1500	217.5	1017	147.5	14.5	45.5	45.5
315	600	1430	207.5	960	139.5	18	55.5	44.5
425	800	1385	201	985	142.5	16.5	55.5	47.5
480	900	1195	173.5	885	128	14.5	55	44.5
540	1000	620	90	560	81.5	21	75.5	36.5
595	1100	445	64.5	395	57.5	24.5	84	33
650	1200	275	40	230	33.5	33.5	91	32
705	1300	185	27	145	21	42.5	93.5	29

Note: Specimens were 16-mm (0.625-in.) diam bar; heated to 980 C (1800 F) for 30 min; oil quenched; tempered at 28 C (50 F) above test temperature
(a) Before testing
Source: Ref 15, p 332

AISI 414: Tensile property data

TEMP C	TEMP F	UTS MPa	UTS ksi	YS MPa	YS ksi	EL %	RA %
21	70	1005	145.5	800	116	19	58
205	400	903	131	721	104.5	17	60.5
315	600	879	127.5	680	98.5	15.5	55
425	800	817	118.5	630	91.5	15.5	55.5
480	900	703	102	595	86	17	68
540	1000	550	79.5	495	71.5	21.5	79
595	1100	420	61	385	55.5	25.5	85.5
650	1200	285	41.5	250	36.5	29	90.5

Note: Specimens were 16-mm (0.625-in.) diam bar; heated to 980 C (1800 F) for 30 min; oil quenched; tempered at 650 C (1200 F) for 2 h; air cooled; hardness of 30.5 HRC
Source: Ref 15, p 333

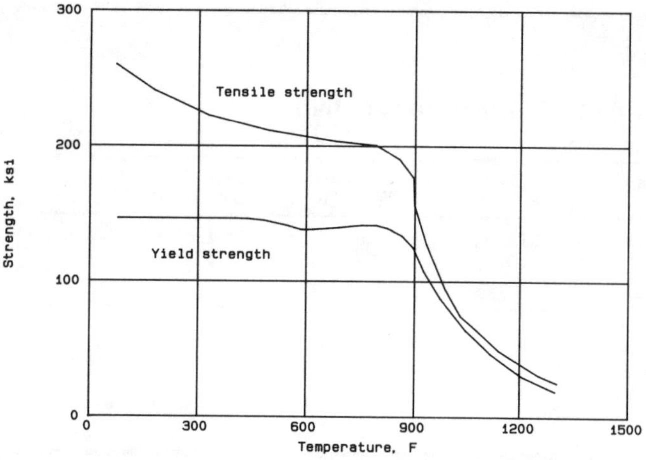

AISI 414: Tensile and yield strength. Specimens were austenitized at 1800 F for 0.5 h, oil quenched and tempered for 2 h at 50 F above test temperature. Source: Ref 14, p 234

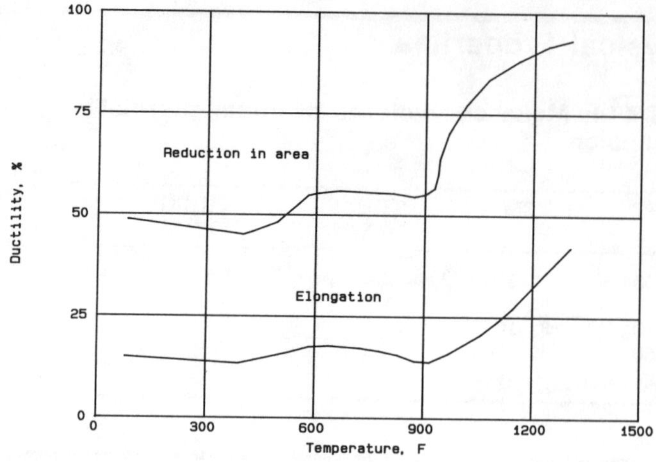

AISI 414: Tensile elongation and reduction in area. Specimens were austenitized at 1800 F for 0.5 h, oil quenched and tempered for 2 h at 50 F above test temperature. Source: Ref 14, p 234

AISI 414: Creep strength

TEMP C	TEMP F	CRSTR 1%/100 h MPa	CRSTR 1%/100 h ksi
482	900	108.9	15.8
593	1100	35.9	5.2
649	1200	27.6	4.0

Source: Ref 29, Jun 1961

AISI 414: Stress-rupture strength

TEMP C	TEMP F	RPSTR 1000 h MPa	RPSTR 1000 h ksi	RPSTR 10000 h MPa	RPSTR 10000 h ksi
482	900	179.3	26.0	132.4	19.2
593	1100	60.0	8.7	44.8	6.5
649	1200	43.4	6.3	27.6	4.0

Source: Ref 29, Jun 1961

AISI 416

UNS number: S41600
Specifications: AMS 5610, ASME SA194, ASTM A194, ASTM A314, ASTM A473, ASTM A581, ASTM A582, MIL S-862, SAE J405, ABNT V416, UNI X12CrS13, JIS G 4030, DGN B83, SIS 14 23 80, DIN 1.4005 X12CrS13, BS 970 416S21 En56AM,
Nominal composition: 0.150% C max, 12.000 to 14.000% Cr, 1.250% Mn max, 0.600% Mo max, 0.060% P max, 0.150% S max, 1.000% Si max
Product forms: Wrought, bar, wire, billets, forgings
Description: AISI 416 is a modification of Type 410, with elements added to boost machining and nonseizing characteristics. These martensitic alloys are the most machinable of all the stainless steels and can be cut using automatic screw machines. Scaling temperatures in air are 675 C (1250 F) max for continuous service and 760 C (1400 F) max for intermittent service.
Typical applications: Aircraft fittings, bolts, nuts, rivets, screws, fire extinguisher inserts, valve trim, pump shafts and rods, gears, golf club heads, valve seats, axles, fuel nozzles, and carburetor needles.

Physical Properties

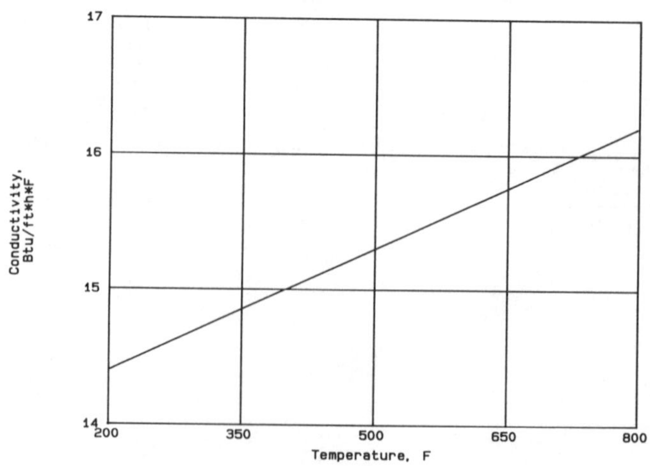

AISI 416: Thermal conductivity. Source: Ref 22, p 6

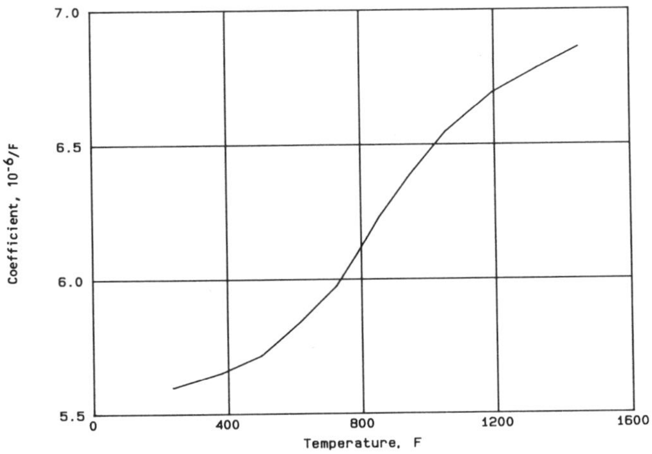

AISI 416: Mean coefficient of linear thermal expansion. Source: Ref 22, p 6

AISI 416: Mean coefficients of linear thermal expansion

TEMP C	TEMP F	THEXP 10^{-6}/C	THEXP 10^{-6}/F
0-100	32-212	9.9	5.5
0-315	32-600	11.0	6.1
0-540	32-1000	11.5	6.4
0-650	32-1200	11.7	6.5

Source: Ref 15, p 338

Mechanical Properties

AISI 416: Tensile property data for annealed specimens

TEMP C	TEMP F	UTS MPa	UTS ksi	YS MPa	YS ksi	EL %	RA %
21	70	517	75	276	40	30.0	60.0
482	900	331	48	241	35	30.0	57.0
538	1000	283	41	214	31	34.0	53.0
593	1100	228	33	172	25	40.0	63.0
649	1200	159	23	124	18	45.0	68.0
704	1300	110	16	83	12	52.0	77.0
760	1400	76	11	52	7.5	60.0	82.0

Source: Ref 29, Nov 1964

AISI 416: Stress-rupture property data

TEMP C	TEMP F	RPSTR 100 h MPa	RPSTR 100 h ksi	RPSTR 1000 h MPa	RPSTR 1000 h ksi	RPSTR 10000 h MPa	RPSTR 10000 h ksi
427	800	414	60	379	55	359	52
482	900	324	47	276	40	228	33
538	1000	221	32	179	26	138	20
593	1100	114	16.5	76	11	48	7
649	1200	55	8	41	6	...	...

Source: Ref 29, Nov 1964

AISI 416: Creep strength of annealed specimens

TEMP C	TEMP F	CRSTR 1%/10000 h MPa	CRSTR 1%/10000 h ksi
538	1000	62-90	9-13
593	1100	28-55	4-8
649	1200	14-17	2-2.5
704	1300	7-10	1-1.5

Source: Ref 29, Nov 1964

AISI 416Se

UNS number: S41623
Specifications: AMS 5610, ASME SA194, ASTM A194, ASTM A314, ASTM A473, ASTM A511, ASTM A581, ASTM A582, FED QQ S-763, MIL S-763, MIL S-862, SAE J405, DGN B83, DGN B171
Nominal composition: 0.150% C max, 12.000 to 14.000% Cr, 1.250% Mn max, 0.060% P max, 0.060% S max, 0.150% Se min, 1.000% Si max
Product forms: Wrought, bar, wire, billets, forgings
Descriptions: AISI 416Se is a modification of Type 410, with elements added to boost machining and nonseizing characteristics. These martensitic alloys are the most machinable of all the stainless steels and can be cut using automatic screw machines. Scaling temperatures in air are 675 C (1250 F) max for continuous service and 760 C (1400 F) max for intermittent service.

Typical applications: Aircraft fittings, bolts, nuts, rivets, screws, fire extinguisher inserts, valve trim, pump shafts and rods, gears, golf club heads, valve seats, axles, fuel nozzles, and carburetor needles.

Physical Properties

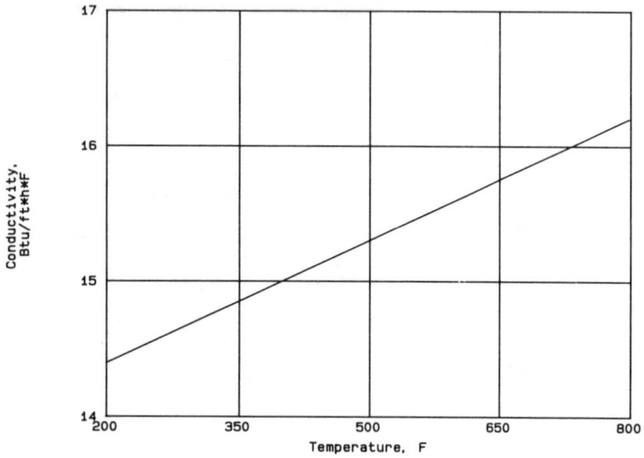

AISI 416Se: Thermal conductivity. Source: Ref 22, p 6

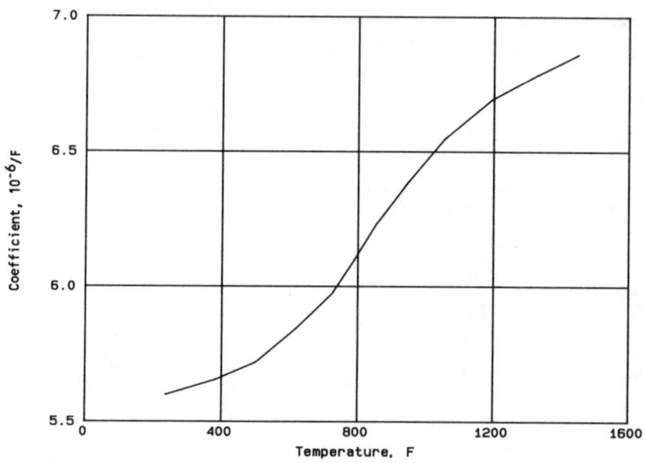

AISI 416Se: Mean coefficient of linear thermal expansion. Source: Ref 22, p 6

AISI 416Se: Mean coefficients of linear thermal expansion

TEMP C	TEMP F	THEXP 10^{-6}/C	THEXP 10^{-6}/F
0-100	32-212	9.9	5.5
0-315	32-600	11.0	6.1
0-540	32-1000	11.5	6.4
0-650	32-1200	11.7	6.5

Source: Ref 15, p 338

Mechanical Properties

AISI 416Se: Tensile property data for annealed specimens

TEMP C	TEMP F	UTS MPa	UTS ksi	YS MPa	YS ksi	EL %	RA %
21	70	517	75	276	40	30.0	60.0
482	900	331	48	241	35	30.0	57.0
538	1000	283	41	214	31	34.0	53.0
593	1100	228	33	172	25	40.0	63.0
649	1200	159	23	124	18	45.0	68.0
704	1300	110	16	83	12	52.0	77.0
760	1400	76	11	52	7.5	60.0	82.0

Source: Ref 29, Nov 1964

AISI 416Se: Stress-rupture property data

TEMP C	TEMP F	RPSTR 100 h MPa	RPSTR 100 h ksi	RPSTR 1000 h MPa	RPSTR 1000 h ksi	RPSTR 10000 h MPa	RPSTR 10000 h ksi
427	800	414	60	379	55	359	52
482	900	324	47	276	40	228	33
538	1000	221	32	179	26	138	20
593	1100	114	16.5	76	11	48	7
649	1200	55	8	41	6	...	...

Source: Ref 29, Nov 1964

AISI 416Se: Creep strength of annealed specimens

TEMP C	TEMP F	CRSTR 1%/10000 h MPa	CRSTR 1%/10000 h ksi
538	1000	62-90	9-13
593	1100	28-55	4-8
649	1200	14-17	2-2.5
704	1300	7-10	1-1.5

Source: Ref 29, Nov 1964

AISI 420

UNS number: S42000

Specifications: AMS 5506, AMS 5621, AMS 7207, ASTM A276, ASTM A314, ASTM A473, ASTM A580, FED QQ S-763, FED QQ S-766, FED QQ W-423, MIL S-862, SAE J405, UNI X20Cr13, JIS G 4303-52 SUS52B, BS 970 420S37 En56C, GOST 20Ch13, GOST 2Kh13, DIN 1.4021 X20Cr13

Nominal composition: 0.150% C min, 12.000 to 14.000% Cr, 1.000% Mn max, 0.040% P max, 0.030% S max, 1.000% Si max

Product forms: Wrought, bar, wire, plate, sheet, strip, billets, forgings

Description: This high-carbon modification of Type 410 can be hardened to about 500 HB. It resists corrosion best in the hardened and tempered condition. Scaling temperatures are 650 C (1200 F) max for continuous service and 760 C (1400 F) max for intermittent service.

Typical applications: Cutlery, surgical instruments, valves, wear parts, glass molds, hand tools, and vegetable choppers.

Physical Properties

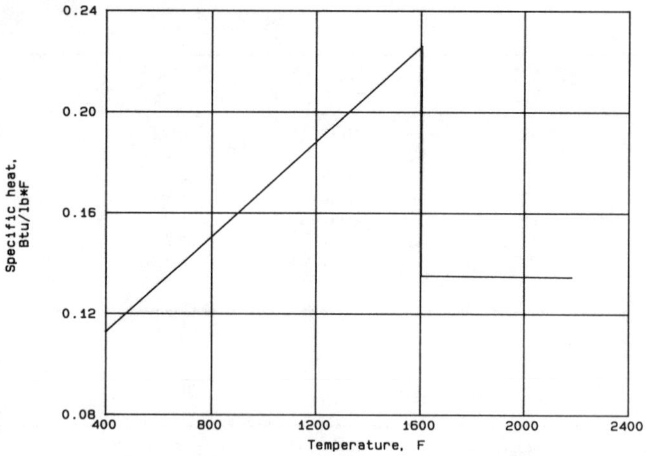

AISI 420: Specific heat. Source: Ref 22, p 5

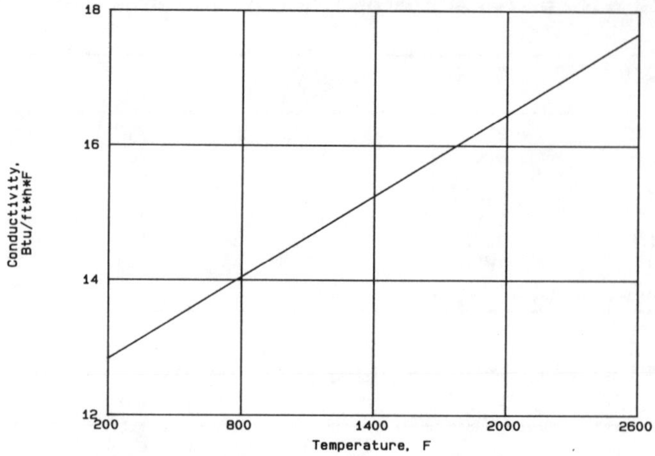

AISI 420: Thermal conductivity. Source: Ref 22, p 4

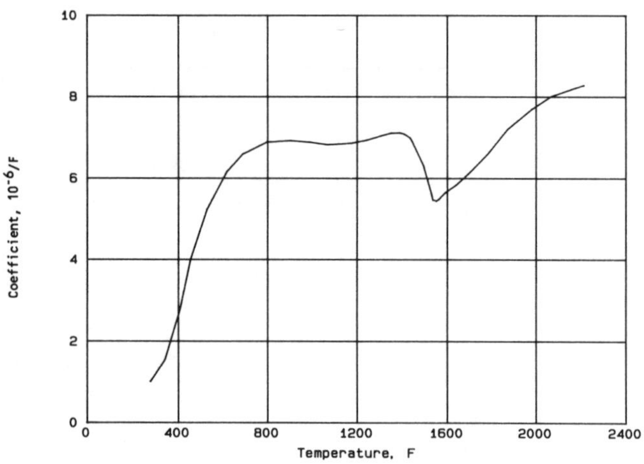

AISI 420: Mean coefficient of thermal expansion. Source: Ref 22, p 4

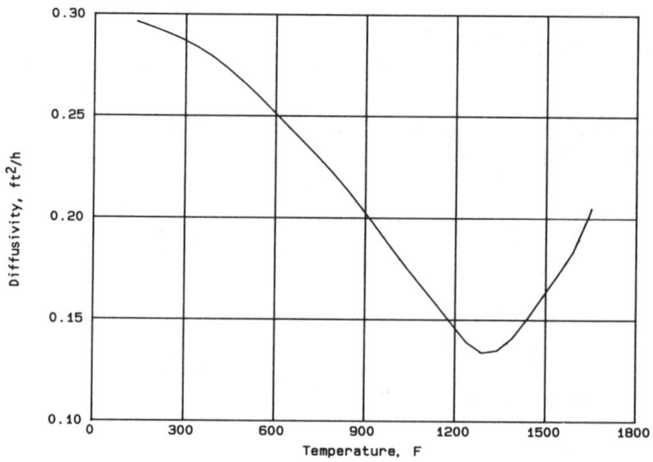

AISI 420: Thermal diffusivity. Source: Ref 22, p 5

AISI 420: Mean coefficients of linear thermal expansion

TEMP C	TEMP F	THEXP 10^{-6}/C	THEXP 10^{-6}/F
0-100	32-212	10.3	5.7
0-315	32-600	10.8	6.0
0-540	32-1000	11.7	6.5
0-650	32-1200	12.2	6.8

Source: Ref 15, p 344

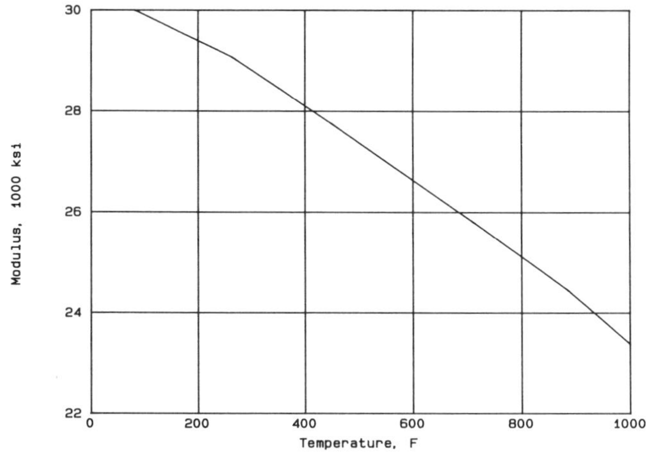

AISI 420: Effect of temperature on static modulus of elasticity of 0.062-in. sheet. Specimens were heat treated for 0.25 h, oil quenched and aged at 900 F for 3 h. Source: Ref 22, p 11

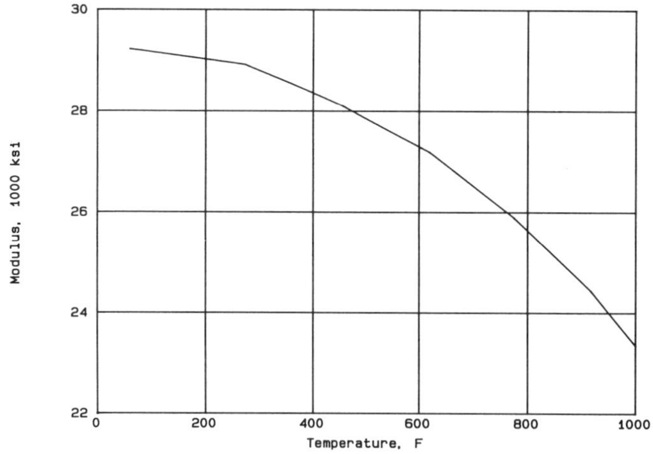

AISI 420: Effect of temperature on compressive modulus of elasticity of 0.62-in. sheet. Specimens were heat treated at 1800 F for 0.25 h, oil quenched, and aged at 900 F for 3 h. Source: Ref 22, p 11

Mechanical Properties

AISI 420: Tensile property data

TEMP C	TEMP F	UTS MPa	UTS ksi	YS MPa	YS ksi	EL %	RA %	HARD(a) HRC	HARD(b) HRC
21	70	2025	294	1360	197	2.5	...	57	...
205	400	1820	264	1085	157.5	11.5	36	51.5	51
315	600	1705	247.5	1040	150.5	13.5	34.5	50.5	49
425	800	1715	248.5	1155	167.5	12.5	35.5	51.5	52
480	900	1415	205	1095	159	9	28.5	52.5	50
540	1000	660	96	585	84.5	20.5	73.5	36	35.5
595	1100	450	65	380	55	26	83	31	30.5
650	1200	290	42	240	35.5	31.5	90	26.5	26
705	1300	170	25	115	16.5	36	93.5	23.5	21

Note: Specimens were 16.0-mm (0.625-in.) diam bar; austenitized at 980 C (1800 F) for 30 min; oil quenched; tempered for 2 h at 28 C (50 F) above test temperature
(a) Before testing. (b) After testing
Source: Ref 15, p 344

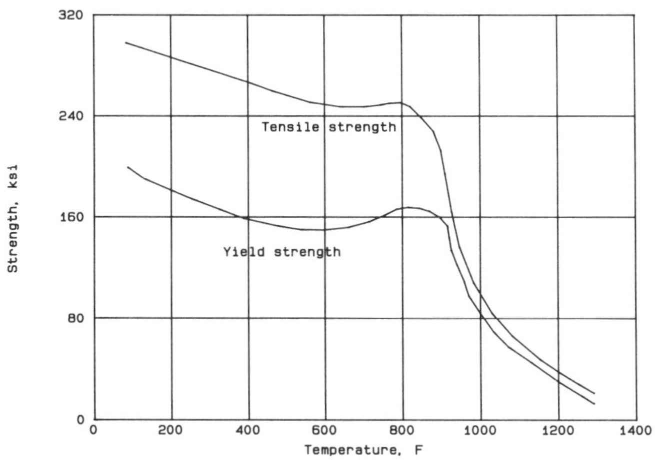

AISI 420: Tensile and yield strength of 0.625-in. round bar. Specimens were heat treated at 1900 F for 0.5 h, oil quenched, and tempered at 50 F above test temperature. Source: Ref 14, p 235

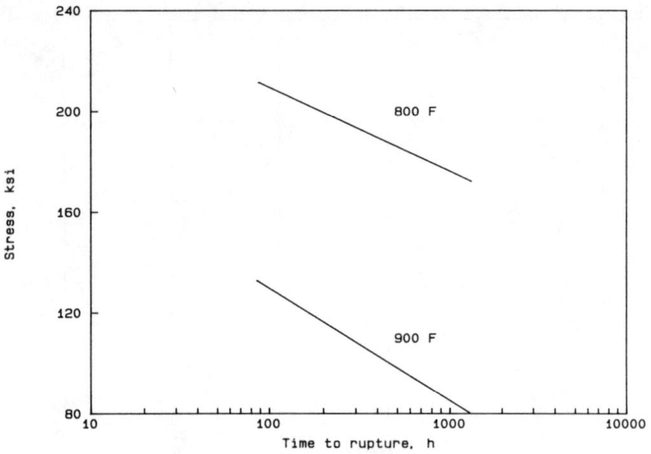

AISI 420: Stress-rupture strength. Specimens were heat treated at 1850 F, oil quenched, and tempered at 50 F above test temperature. Source: Ref 22, p 10

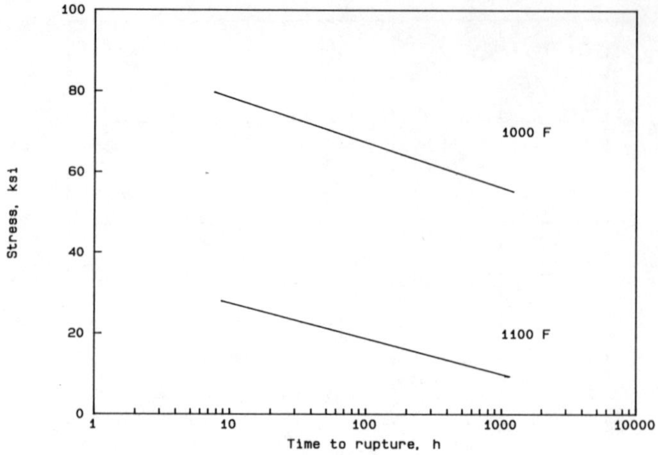

AISI 420: Stress-rupture strength. Specimens were heat treated at 1850 F, oil quenched, and tempered at 50 F above test temperature. Source: Ref 22, p 10

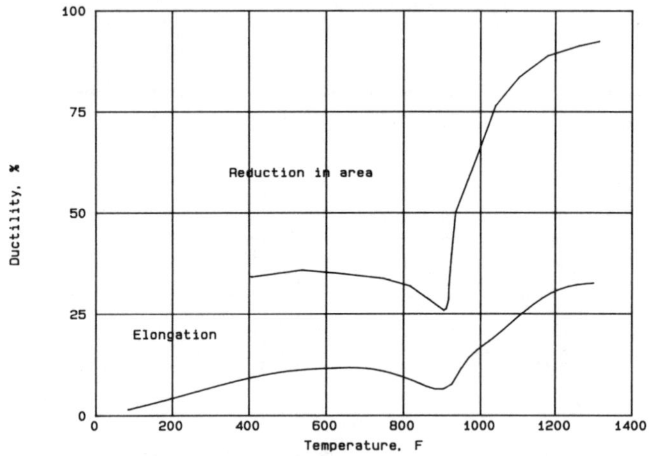

AISI 420: Tensile elongation and reduction in area of 0.625-in. round bar. Specimens were heat treated at 1900 F for 0.5 h, oil quenched, and tempered for 2 h at 50 F above test temperature. Source: Ref 14, p 235

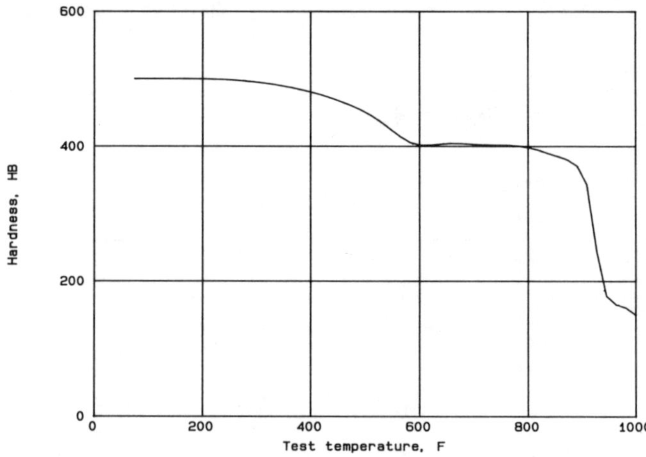

AISI 420: Brinell hardness of bar. Specimens were heat treated at 1850 F for 1 h, oil quenched, and tempered for 1 h at 50 F above test temperature. Source: Ref 22, p 3

AISI 422

UNS number: S42200
Specifications: AMS 5655, ASTM A565, MIL S-861, SAE J467, GOST EI802, GOST KV-7, GOST Kh11V2NMF, GOST Kh12MVFB
Nominal composition: 0.200 to 0.250% C, 11.500 to 12.500% Cr, 0.500 to 1.000% Mn, 0.900 to 1.250% Mo, 0.500 to 1.000% Ni, 0.025% P max, 0.025% S max, 0.500% Si max, 0.200 to 0.300% V, 0.900 to 1.250% W
Product forms: Wrought, bar, wire, plate, sheet, strip, billets, forgings
Description: AISI 422 is a martensitic stainless steel designed for service temperatures up to 650 C (1200 F). It features high strength and toughness and stress-rupture properties in the 315 to 650 C (600 to 1200 F) range that are better than those of other 12% Cr stainless steels. Type 422 has good resistance to destructive scaling or oxidation at continuous service temperatures as high as 790 C (1450 F).

Typical applications: Steam turbine and compressor blades and wheels, rotors, high-temperature bolting, valve trim and other valve applications, and aircraft and missile structural components.

Physical Properties

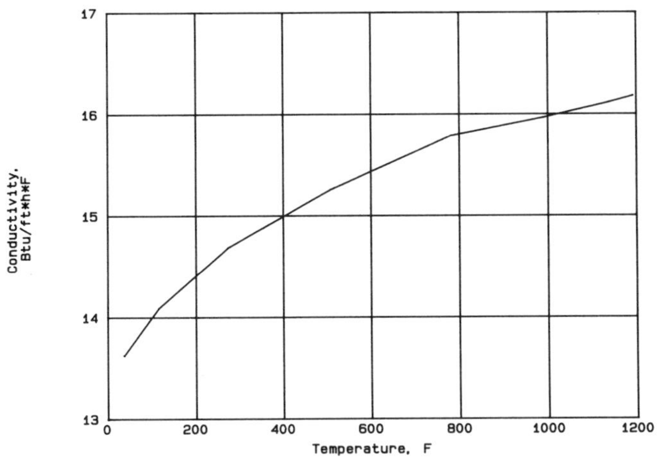

AISI 422: Thermal conductivity. Source: Ref 22, p 5

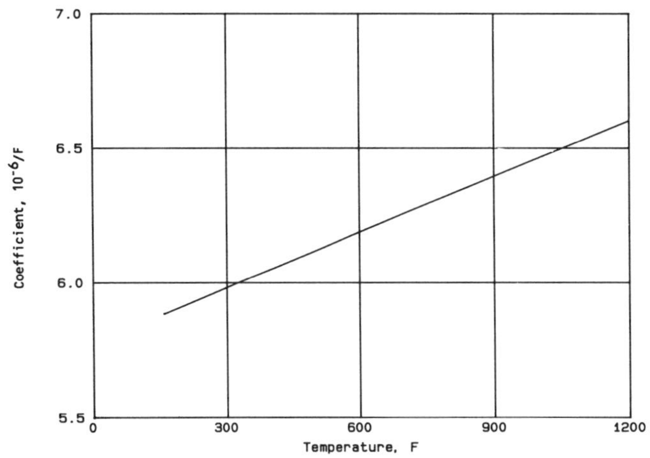

AISI 422: Mean coefficient of thermal expansion. Source: Ref 22, p 5

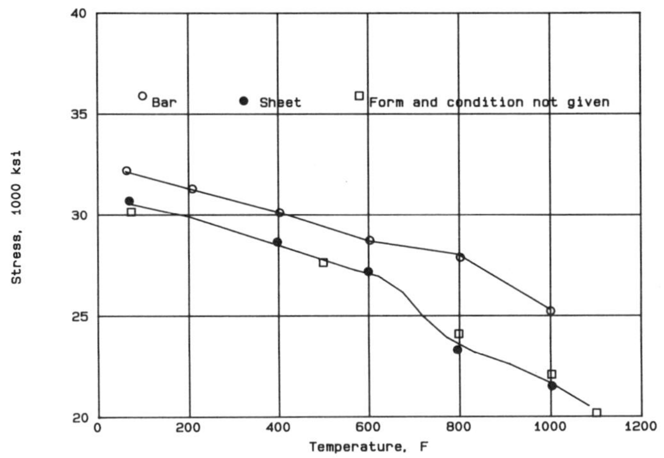

AISI 422: Effect of temperature on static modulus of elasticity of bar, treated at 1850 F for 0.25 h, air cooled, and treated at 900 F for 4 h. Sheet (0.062 in.) was treated at 1900 F for 0.25 h, oil quenched, and treated at 1000 F for 2 h. Source: Ref 22, p 16

AISI 422: Thermal conductivity

TEMP C	TEMP F	THCON W/m*K	THCON Btu/ft*h*F
100	212	23.9	13.8
500	930	27.3	15.8

Source: Ref 15, p 347

AISI 422: Mean coefficients of linear thermal expansion

TEMP C	TEMP F	THEXP 10^{-6}/C	THEXP 10^{-6}/F
0-100	32-212	11.2	6.2
0-315	32-600	11.3	6.3
0-540	32-1000	11.9	6.6
0-650	32-1200	12.1	6.7

Source: Ref 15, p 347

AISI 422: Relationship of electrical properties and tempering temperature

Ttpr C	Ttpr F	ELCON %IACS	ELRES n{o}*m
At 21 C (70 F)			
371	700	2.79	620
427	800	2.83	610
482	900	2.86	605
At 799 C (1470 F)			
371-482	700-900	1.57-1.43	1102-1204

Note: Specimens were quenched and tempered
Source: Ref 22, p 5

AISI 422: Moduli of elasticity, rigidity, and elasticity in compression

TEMP C	TEMP F	E GPa	E 10^3 ksi	G GPa	G 10^3 ksi	E(a) GPa	E(a) 10^3 ksi
21	70	205	29.8	83	12.0	219	31.7
260	500	190	27.5	79	11.5	...	...
316	600	...	...	...	...	214	31.1
427	800	167	24.2	72	10.5	188	27.3
593	1100	141	20.5	62	9.0	...	...

(a) In compression
Source: Ref 28, p 38

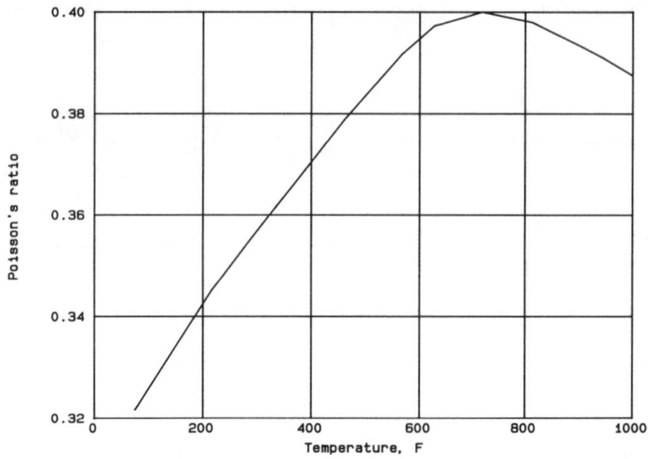

AISI 422: Poisson's ratio of bar. Specimens were heat treated at 1850 F for 0.25 h, air cooled and aged at 900 F for 4 h. Source: Ref 22, p 16

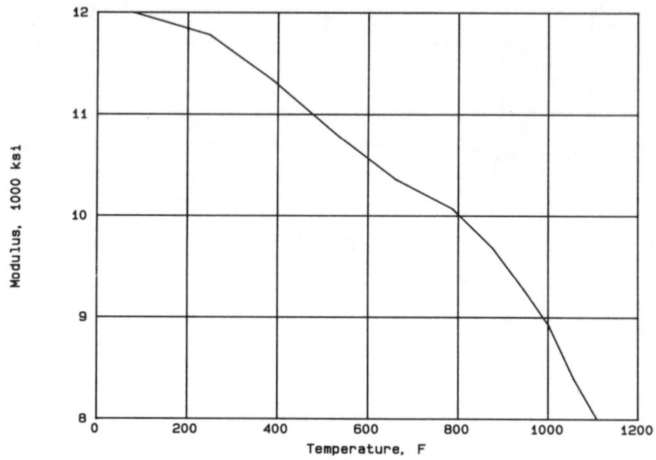

AISI 422: Modulus of rigidity of bar. Specimens were heat treated at 1850 F for 0.25 h, air cooled and aged at 900 F for 4 h. Source: Ref 22, p 17

AISI 422: Poisson's ratio

TEMP C	TEMP F	PR
24	75	0.23
260	500	0.21
427	800	0.20
593	1100	0.16

Source: Ref 28, p 38

Mechanical Properties

AISI 422: Tensile property data

TEMP C	TEMP F	UTS MPa	UTS ksi	YS MPa	YS ksi	EL(a) %	RA %	IS(b) J	IS(b) ft-lb
427	800	814	118	724	105	10	55	...	...
538	1000	662	96	565	82	25	65	54	40
649	1200	359	52	310	45	30	80	68	50

Note: Specimens were austenitized at 1038 C (1900 F), oil quenched, tempered at 649 C (1200 F) for 4 h
(a) In 4D. (b) Charpy V-notch
Source: Ref 29, Feb 1965

AISI 422: Effect of tempering temperature on tensile properties

Ttpr C	Ttpr F	TEMP C	TEMP F	UTS MPa	UTS ksi	YS MPa	YS ksi	EL(a) %	RA %
315	600	21	70	1580	229	1000	145	16	56
425	800	21	70	1635	237	1160	168	16	53
480	900	21	70	1675	243	1220	177	16	50
540	1000	21	70	1610	233	1005	146	16	54
650	1200	21	70	1030	149	862	125	19	52
425	800	315	600	1585	230	1055	153	16	13
425	800	425	800	1490	216	951	138	14	17
455	850	425	800	1490	216	979	142	19	55
525	980	425	800	1510	219	1080	157	16	52
525	980	480	900	1440	209	1040	151	14	46
525	980	540	1000	1220	177	903	131	12	43
650	1200	650	1200	390	57	390	46	21	85

Note: Specimens were austenitized at 1040 C (1900 F) for 1 h, oil quenched, tempered for 2 h at indicated temperature
(a) In 36 mm (1.4 in.)
Source: Ref 15, p 348

AISI 422: Charpy V-notch impact properties

TEMP C	TEMP F	IS J	IS ft-lb
21	70	8	6
		10	7
71	160	14	10
		15	11
100	212	18	13
		20	15
125	260	22	16
		23	17
140	280	28	21
150	300	31	23
		37	27
170	340	38	28
		39	29
205	400	43	32
		52	38

Note: Specimens were austenitized at 1040 C (1900 F) for 1 h, oil quenched, tempered at 525 C (980 F) for 2 h, and air cooled
Source: Ref 15, p 348

AISI 422: Yield strength in compression

TEMP C	TEMP F	YS MPa	YS ksi
22	72	1407	204
316	600	1179	171
427	800	1165	169

Note: All sheet specimens were austenitized at 1038 C (1900 F) for 15 min, in argon atmosphere, oil quenched, tempered at 538 C (1000 F) for 2 h, and air cooled
Source: Ref 28, p 38

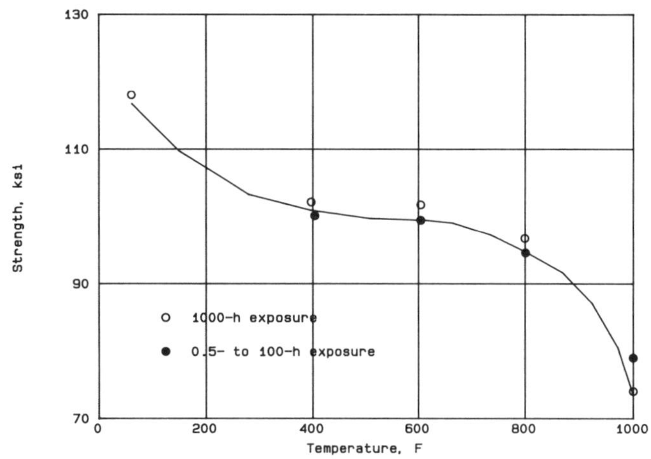

AISI 422: Shear strength of 0.188-in. sheet. Specimens were heat treated at 1900 F for 0.25 h, oil quenched and tempered at 1000 F for 2 h. Source: Ref 22, p 13

AISI 422: Effect of exposure at 538 C (1000 F) on Charpy V-notch impact values

EXP TIME h	IS(a) J	IS(a) ft-lb	IS(b) J	IS(b) ft-lb	IS(c) J	IS(c) ft-lb
Tempered at 649 C (1200 F) for 2 h to 34 HRC						
0	20	15	34	25	58	43
25	19	14	35	26	57	42
100	20	15	34	25	57	42
500	16	12	22	16	38	28
1000	14	10	...	...	28	21
Tempered at 760 C (1400 F) for 2 h to 26 HRC						
0	52	38	92	68	96	71
25	49	36	84	62	87	64
100	54	40	79	58	87	64
300	53	39	88	65	88	65
500	31	23	69	51	75	55
700	28	21	62	46	72	53
1000	23	17	50	37	65	48

Note: Specimens were austenitized at 1038 F (1900 F) for 1 h, oil quenched, tempered at 649 C (1200 F) or 760 C (1400 F) for 2 h
(a) 21 C (70 F). (b) 66 C (150 F). (c) 100 C (212 F)
Source: Ref 28, p 38

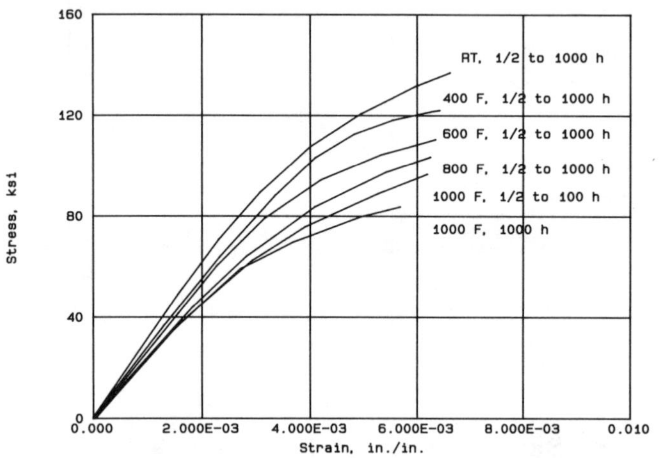

AISI 422: Tensile stress-strain curves for 0.062-in. sheet. Specimens were heat treated at 1900 F for 0.25 h, oil quenched and tempered at 1000 F for 2 h. Source: Ref 22, p 10

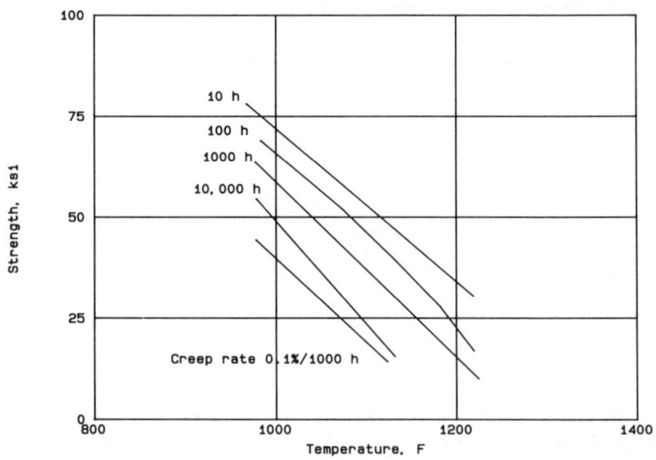

AISI 422: Creep and stress-rupture properties of bar. Specimens were tempered at 1200 F to a room-temperature tensile strength of 150 ksi. Source: Ref 28, p 39

AISI 422: Stress-rupture strength and creep rates

TEMP C	RPSTR 10 h MPa	RPSTR 10 h ksi	RPSTR 100 h MPa	RPSTR 100 h ksi	RPSTR 1000 h MPa	RPSTR 1000 h ksi	RPSTR 10000 h MPa	RPSTR 10000 h ksi	CRSTR 1%/1000 h MPa	CRSTR 1%/1000 h ksi	CRSTR 1%/10000 h MPa	CRSTR 1%/10000 h ksi
540	490	71	435	63	390	57	340	49	360	52	270	39
595	365	53	320	46	255	37	160	23	235	34	140	20
650	385	56	170	25	120	17	...	...	...	...	...	...

Note: Specimens were austenitized at 1040 C (1900 F), oil quenched, tempered at 650 C (1200 F) for 2 h, and air cooled
Source: Ref 15, p 348

AISI 422: Stress-rupture properties

TEMP C	TEMP F	RPSTR 10 h MPa	RPSTR 10 h ksi	RPSTR 100 h MPa	RPSTR 100 h ksi	RPSTR 1000 h MPa	RPSTR 1000 h ksi
538	1000	469	68	434	63	379	55
593	1100	386	56	317	46	234	34
649	1200	262	38	172	25	103	15

Note: Specimens were austenitized at 1038 C (1900 F), oil quenched, tempered at 649 C (1200 F) for 4 h
Source: Ref 29, Feb 1965

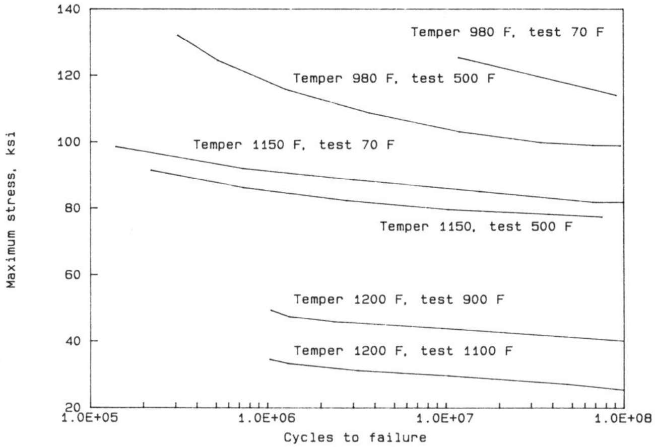

AISI 422: Reverse-bend fatigue properties of bar. Specimens were treated at 1900 F for 1 h, oil quenched, and tempered at 980, 1150, or 1200 F. Source: Ref 22, p 15

AISI 430

UNS number: S43000
Specifications: AMS 5503, AMS 5627, ASME SA182, ASME SA240, ASME SA268, ASTM A176, ASTM A182, ASTM A240, ASTM A276, FED QQ S-763, FED QQ S-766, FED QQ W-423, MIL S-862, SAE J405, NF Z8C-17, JIS G 4303 SUS430, BS 970 430S15 En60, GOST 2Ch17, DIN 1.4016 X8Cr17
Nominal composition: 0.120% C max, 16.000 to 18.000% Cr, 1.000% Mn max, 0.040% P max, 0.030% S max, 1.000% Si max
Product forms: Wrought, bar, wire, plate, sheet, strip

Description: Of the 400 series stainless steels, this nonhardening ferritic type most nearly resembles the 18-8 stainless steels in fabrication and service characteristics. Type 430 combines excellent corrosion resistance with good resistance to elevated temperature scaling. Scaling temperatures in air are 845 C (1550 F) max for continuous service and 900 C (1650 F) max for intermittent service.
Typical applications: Decorative trim, nitric acid tanks, annealing baskets, combustion chambers, dishwashers, heaters, mufflers, range hoods, recuperators, and restaurant equipment.

Physical Properties

AISI 430: Thermal conductivity

TEMP C	TEMP F	THCON W/m*K	THCON Btu/ft*h*F
100	212	26.1	15.1
500	932	26.3	15.2

Source: Ref 15, p 355

AISI 430: Mean coefficients of linear thermal expansion

TEMP C	TEMP F	THEXP 10^{-6}/C	THEXP 10^{-6}/F
0-100	32-212	10.4	5.8
0-315	32-600	11.0	6.1
0-540	32-1000	11.3	6.3
0-650	32-1200	11.9	6.6
0-815	32-1500	12.4	6.9

Source: Ref 15, p 355

Mechanical Properties

AISI 430: Tensile property data

TEMP C	TEMP F	UTS MPa	UTS ksi	YS MPa	YS ksi	EL %	RA %
Annealed at 815 C (1500 F) and cold worked to 89 HRB							
21	70	539	78	441	64	32	77
205	400	459	66.5	390	56.5	28	77.5
425	800	379	55	317	46	28.5	78
540	1000	272	39.5	234	34	30	84.5
650	1200	134	19.5	114	16.5	43.5	95.5
760	1400	66	9.5	52	7.5	70.5	98
870	1600	34	5	28	4	95.5	98.5
Annealed							
21	70	517	75	310	45	32	...
205	400	441	64	265	38.5	30	...
425	800	365	53	231	33.5	29	...
540	1000	269	39	169	24.5	40	...
650	1200	134	19.5	89.6	13	56	...
760	1400	66	9.5	41	6	74	...

Note: Specimens were 16-mm (0.63-in.) diam bar
Source: Ref 15, p 354

AISI 430: Stress-rupture and creep rate property data

TEMP C	TEMP F	RPSTR 1000 h MPa	RPSTR 1000 h ksi	RPSTR(a) 10000 h MPa	RPSTR(a) 10000 h ksi	CRSTR(a) 1% in 10000 h MPa	CRSTR(a) 1% in 10000 h ksi	CRSTR(a) 1% in 100000 h MPa	CRSTR(a) 1% in 100000 h ksi
480	900	205	30	165	24	...	...	...	...
500	1000	120	17	90	13	59	8.5	45	6.5
595	1100	60	9	40	6	30	4.3	22	3.2
650	1200	35	5	20	3	15	2.2	12	1.7
705	1300	...	...	...	...	9	1.3	...	...

(a) Extrapolated data
Source: Ref 15, p 354

AISI 430F

UNS number: UNS S43020
Specifications: ASTM A314, ASTM A473, ASTM A581, ASTM A582, MIL S-862, SAE J405
Nominal composition: 16.000 to 18.000% Cr, 0.12% C max, 1.25% Mn max, 1.0% Si max, 0.060% P max, 0.15% S min, 0.60% Mo max
Product forms: Bar, wire
Description: AISI 430F is a free-machining modification of Type 430. This nonhardenable ferritic steel is designed for use in high-speed, automatic machine operations. Scaling temperatures in air are 815 C (1500 F) max for intermittent service. A selenium containing version of Type 430F Se (UNS S43023) has 0.060% S max, 0.15% Se min, and no molybdenum.
Typical applications: Fasteners, auto trim, appliance parts, forged and swaged parts, aircraft fittings, value parts and hardware.

Physical Properties

AISI 430F: Thermal conductivity

TEMP C	TEMP F	THCON W/m*K	THCON Btu/ft*h*F
100	212	26.14	15.1
500	932	26.31	15.2

Source: Ref 27, p 73

AISI 430F: Electrical resistivity

TEMP C	TEMP F	ELRES n{o}*m
93	200	770
204	400	925
427	800	1150

Source: Ref 29, Dec 1978

AISI 430F: Mean coefficients of linear thermal expansion

TEMP C	TEMP F	THEXP 10^{-6}/C	THEXP 10^{-6}/F
0-100	32-212	10.4	5.8
0-315	32-600	11.0	6.1
0-538	32-1000	11.4	6.3
0-649	32-1200	11.9	6.6
0-815	32-1500	12.4	6.9

Source: Ref 27, p 73

Mechanical Properties

AISI 430F: Tensile property data

TEMP C	TEMP F	UTS MPa	UTS ksi	EL %	RA %
21	70	531	77	32	84
93	200	496	72	30	74
205	400	462	67	27	76
315	600	441	64	26	75
425	800	386	56	29	75
535	1000	248	36	35	84
650	1200	131	19	61	97
760	1400	48	7	70	99

Note: Annealed bar were held at temperature for 1 h and pulled at the rate of 1.3 mm/min (0.05 in./min)
Source: Ref 15, p 349

AISI 431

UNS number: S43100
Specifications: AMS 5628, ASTM A276, ASTM A314, ASTM A473, ASTM A493, ASTM A479, ASTM A579, ASTM A580, MIL S-862, MIL S-8967, MIL S-18732, SAE J405, NF Z15CN16-02, UNI X16CrNi16, JIS G 4303 SUS431, BS 970 431S29 En57, GOST 20Ch17N2, DIN 1.4057 X22CrNi17, SIS 14 23 21
Nominal composition: 0.200% C max, 15.000 to 17.000% Cr, 1.000% Mn max, 1.250 to 2.500% Ni, 0.040% P max, 0.030% S max, 1.000% Si max
Product forms: Wrought, bar, wire, billets, forgings
Description: AISI 431 is a special-purpose hardenable steel used where particularly high mechanical properties are needed. Corrosion resistance of the nickel-bearing, martensitic chromium steel is superior to that of Types 410, 420, 430, and 440. Scaling temperatures in air are 815 C (1500 F) max for continuous service and 870 C (1600 F) max for intermittent service.
Typical applications: Aircraft fittings, beater bars, paper-making machinery, and bolts.

Physical Properties

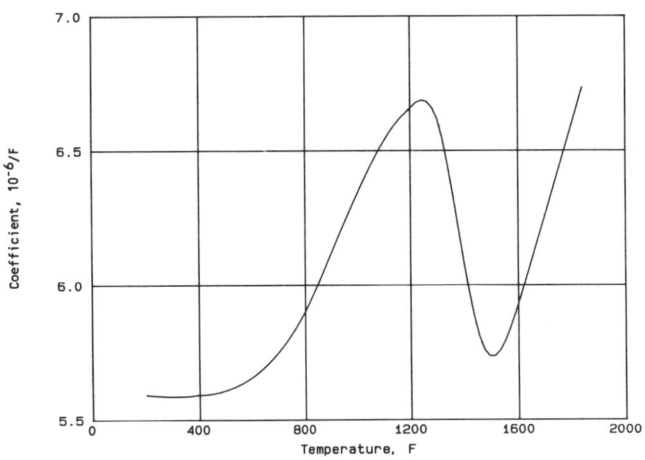

AISI 431: Mean coefficient of linear thermal expansion. Source: Ref 22, p 4

AISI 431: Mean coefficients of linear thermal expansion

TEMP C	TEMP F	THEXP 10^{-6}/C	THEXP 10^{-6}/F
0-100	32-212	10.1	5.6
0-315	32-600	12.1	6.7
0-540	32-1000	11.9	6.6
0-650	32-1200	12.2	6.8

Source: Ref 15, p 356

Mechanical Properties

AISI 431: Tensile property data

TEMP C	TEMP F	UTS MPa	UTS ksi	YS MPa	YS ksi	EL %	RA %
21	70	831	120.5	738	107	20	63.5
205	400	727	105.5	650	94	17.5	66
315	600	685	99	610	88.5	15.5	61.5
425	800	625	91	570	82.5	15.5	61
480	900	560	81.5	510	74	17.5	68.5
540	1000	485	70	470	68	23	77.5
595	1100	385	56	355	51.5	22	83
650	1200	295	42.5	270	39.5	25	86

Note: Specimens were 16-mm (0.625-in.) diam bar; austenitized at 980 C (1800 F) for 30 min; tempered at 650 C (1200 F) for 2 h; air cooled; hardness of 25.5 HRC
Source: Ref 15, p 357

AISI 431: Tensile property data

TEMP C	TEMP F	UTS MPa	UTS ksi	YS MPa	YS ksi	EL %	RA %	HARD(a) HRC
21	70	1435	208	1140	165.5	16.5	59	45
205	400	1350	196	865	125.5	14	44.5	40.5
315	600	1365	198	995	144	15	44.5	41.5
425	800	1280	185.5	915	133	17	55	42.5
480	900	1035	150	760	110.5	13.5	53.5	42.5
540	1000	565	82	525	76.5	20.5	75	31.5
595	1100	415	60.5	360	52.5	23.5	82	28
650	1200	285	41	260	38	26	86.5	24.5
705	1300	170	25	150	21.5	29	91	23.5

Note: Specimens were 16-mm (0.625-in.) diam bar, austenitized at 980 C (1800 F) for 30 min, oil quenched, tempered at 28 C (50 F) above test temperature for 2 h
(a) Before testing
Source: Ref 15, p 357

AISI 431: Effect of test temperature on tensile strength of 0.625-in. square bar. Specimens were heat treated at 1800 F for 0.5 h, oil quenched, held at 1225 F for 4 h, and oil quenched. Source: Ref 22, p 7

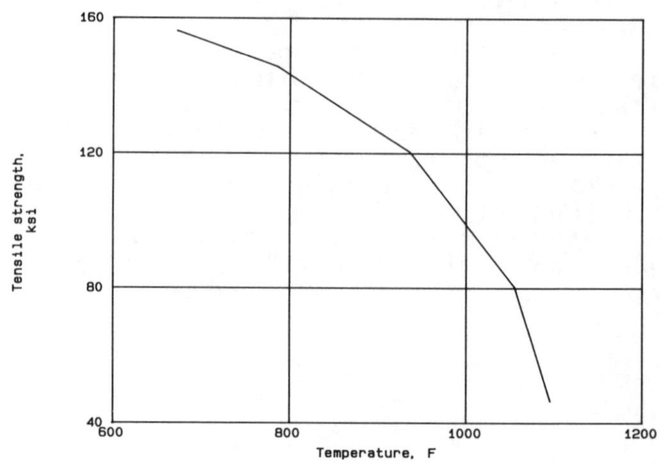

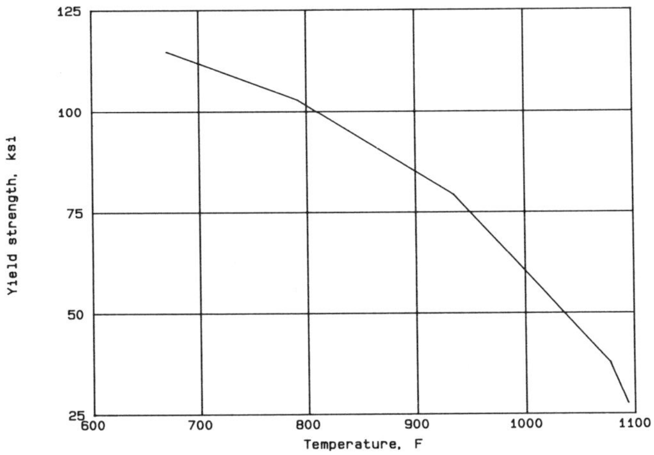

AISI 431: Effect of test temperature on yield strength of 0.625-in. square bar. Specimens were heat treated at 1800 F for 0.5 h, oil quenched, held at 1225 F for 4 h, and oil quenched. Source: Ref 22, p 7

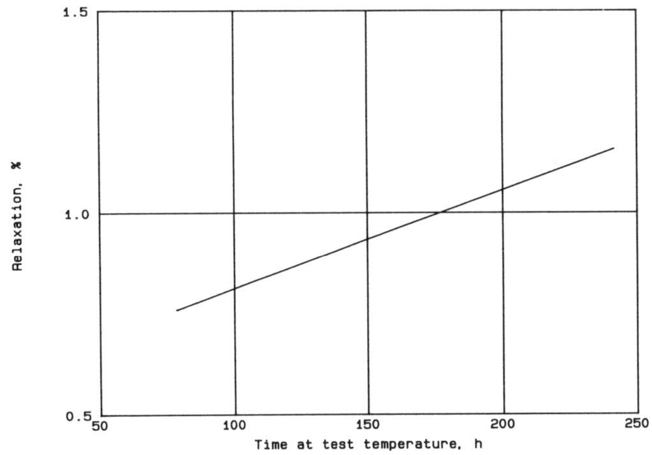

AISI 431: Relaxation of 10-turn helical springs. Wire diameter: 2.69 mm. Spring diameter: 25 mm. Free length: 75 mm. 690 MPa stress corrected by Wahl factor. Spring hardness: 35 HRC. Source: Ref 1, p 299

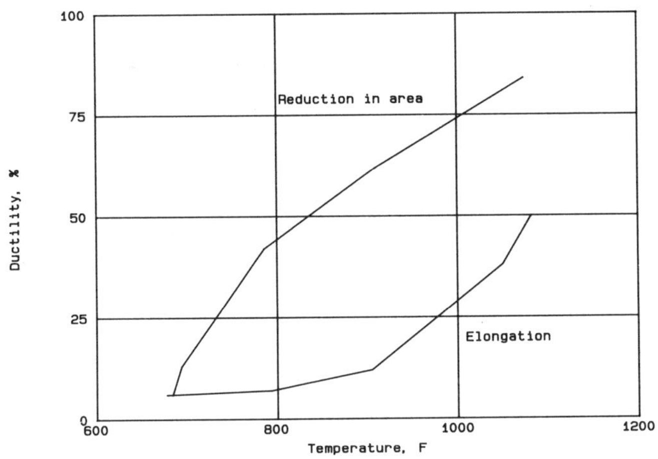

AISI 431: Effect of test temperature on tensile elongation and reduction in area of 0.625-in. square bar. Specimens were heat treated at 1800 F, held for 0.5 h, oil quenched, held at 1225 F for 4 h, and oil quenched. Source: Ref 22, p 7

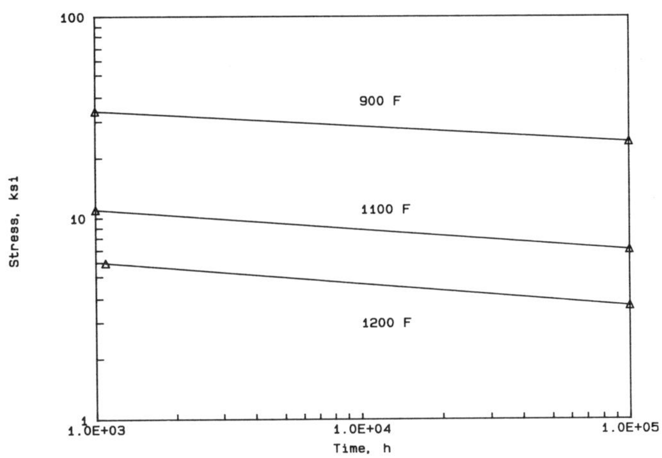

AISI 431: Stress-rupture strength. Specimens were oil quenched and tempered at 1200 F. Source: Ref 22, p 7

AISI 439

UNS number: S43035
Specifications: ASME SA240 (XM-8), ASME SA268, ASME SA479, ASTM A240 (XM-8), ASTM A268, ASTM A479, ASTM A651
Nominal composition: 0.150% Al max, 0.070% C max, 17.000 to 19.000% Cr, 1.000% Mn max, 0.500% Ni max, 0.040% P max, 0.030% S
Description: AISI 439 is a ferritic chromium-titanium stainless steel also known as HWT and Aqualloy.

Physical Properties

AISI 439: Mean coefficients of linear thermal expansion

TEMP C	TEMP F	THEXP 10^{-6}/C	THEXP 10^{-6}/F
21-100	70-212	10.1	5.6
21-500	70-930	11.5	6.4
21-800	70-1470	12.4	6.9

Source: Ref 15, p 352

Mechanical Properties

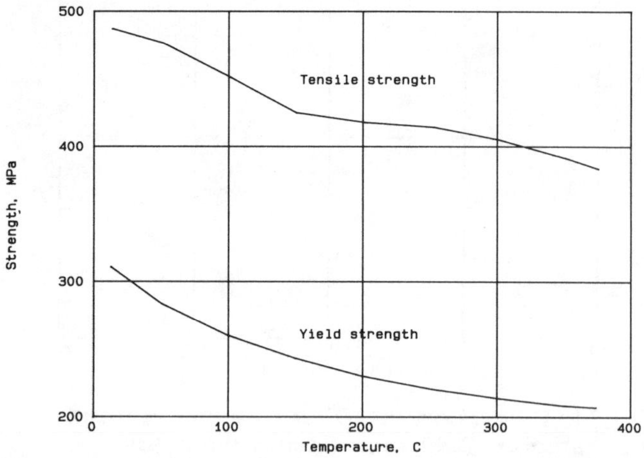

AISI 439: Tensile and yield strength. Source: Ref 15, p 353

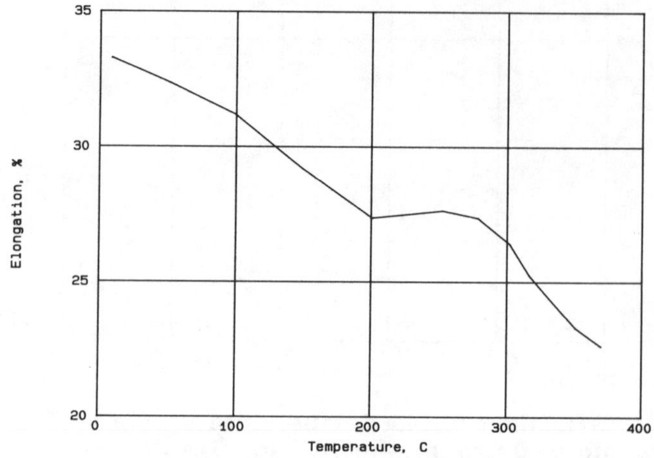

AISI 439: Tensile elongation. Source: Ref 15, p 353

AISI 440A

UNS number: S44002
Specifications: AMS 5631, AMS 5632, ASTM A276, ASTM A314, ASTM A473, ASTM A511, ASTM A580, FED QQ S-763, MIL S-862, SAE J405, NF Z50CD14, UNI X40Cr13, JIS G 4303,5,6,7, DGN B83, DIN 1.4109 X65CrMo14, DIN 1.4110
Nominal composition: 0.600 to 0.750% C, 16.000 to 18.000% Cr, 1.000% Mn max, 0.750% Mo max, 0.040% P max, 0.030% S max, 1.000% Si max
Product forms: Wrought, bar, wire, plate, strip, billets, forgings
Description: AISI 440A is a hardenable chromium steel that attains optimum corrosion resistance in the hardened and tempered condition. Type 440A can be made harder than type 420 and tougher than types 440B and 440C. Scaling temperatures in air are 760 C (1400 F) max for continuous service and 815 C (1500 F) max for intermittent service.
Typical applications: Cutlery, bearings, and surgical tools.

Physical Properties

AISI 440A: Mean coefficients of linear thermal expansion

TEMP C	TEMP F	THEXP 10^{-6}/C	THEXP 10^{-6}/F
0-100	32-212	10.3	5.7
0-316	32-600	10.8	6.0
0-538	32-1000	11.7	6.5
0-649	32-1200	12.2	6.8

Source: Ref 22, p 3

Mechanical Properties

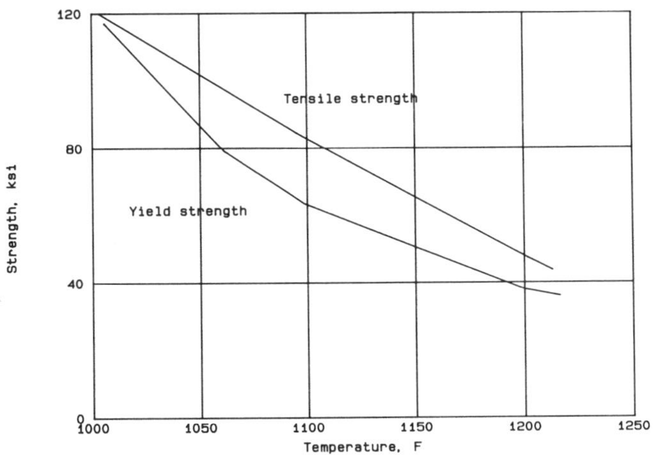

AISI 440A: Tensile and yield strength. Specimens were austenitized at 1900 F, tempered at 50 F above test temperature, and held for 1 h. Source: Ref 22, p 4

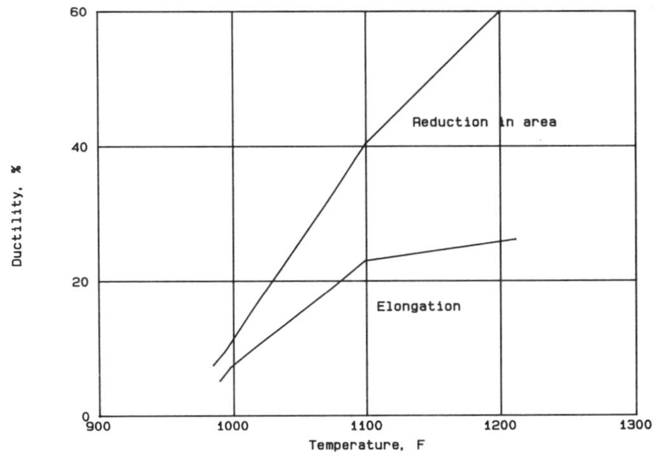

AISI 440A: Tensile elongation and reduction in area. Specimens were austenitized at 1900 F, tempered at 50 F above test temperature, and held for 1 h. Source: Ref 22, p 4

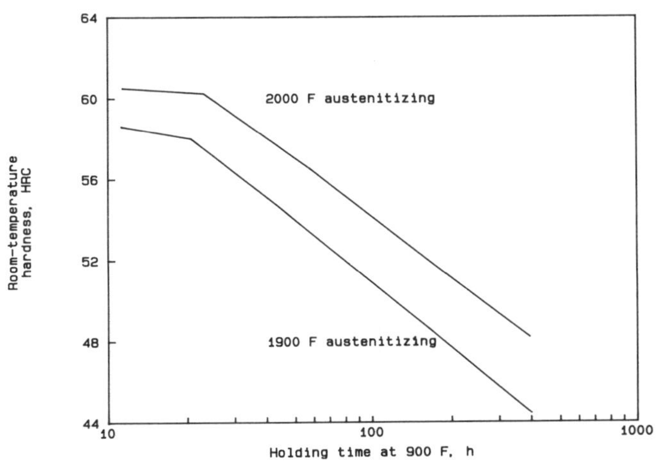

AISI 440A: Softening after extended holding at 900 F. Specimens were double tempered at 1000 F. Source: Ref 9, p 187

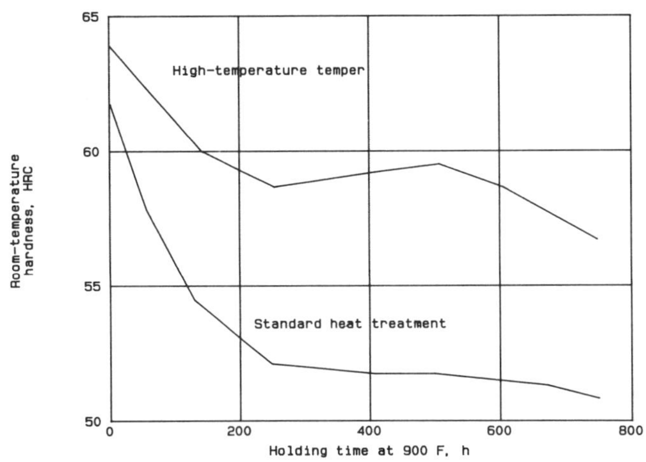

AISI 440A: Softening after extended holding at 900 F. Source: Ref 9, p 187

AISI 440B

UNS number: S44003
Specifications: AMS 7445, ASTM A276, ASTM A314, ASTM A473, ASTM A580, FED QQ S-763, MIL S-862, SAE J405, ABNT E440B, JIS G 4303 SUS440B, DGN B83, DIN 1.4112 X90CrMoV18
Nominal composition: 0.750 to 0.950% C, 16.000 to 18.000% Cr, 1.000% Mn max, 0.750% Mo max, 0.040% P max, 0.030% S max, 1.000% Si max
Product forms: Wrought, bar, wire, strip, billet, forgings
Description: AISI 440B is a hardenable chromium steel that attains optimum corrosion resistance in the hardened and tempered condition. Type 440B can be made harder than 440A and tougher than 440C. Scaling temperatures in air are 760 C (1400 F) max for continuous service and 815 C (1500 F) max for intermittent service.
Typical applications: Cutlery, valve parts, and instrument bearings.

Physical Properties

AISI 440B: Mean coefficients of linear thermal expansion

TEMP C	TEMP F	THEXP 10^{-6}/C	THEXP 10^{-6}/F
0-100	32-212	10.3	5.7
0-316	32-600	11.0	6.1
0-538	32-1000	11.9	6.6
0-649	32-1200	12.2	6.8

Source: Ref 22, p 3

Mechanical Properties

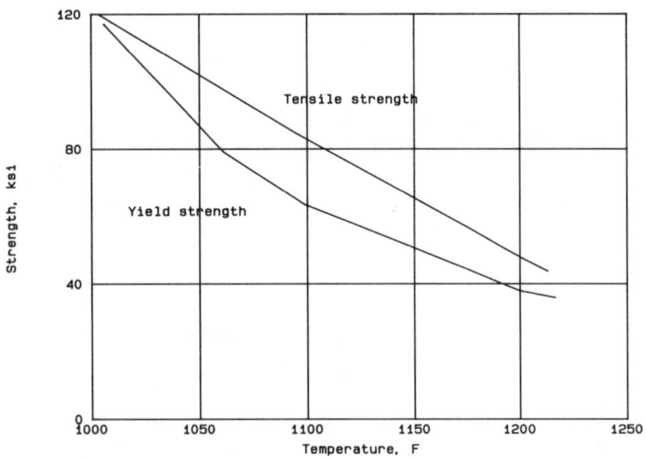

AISI 440B: Tensile and yield strength. Specimens were austenitized at 1900 F, tempered at 50 F above test temperature, and held for 1 h. Source: Ref 22, p 4

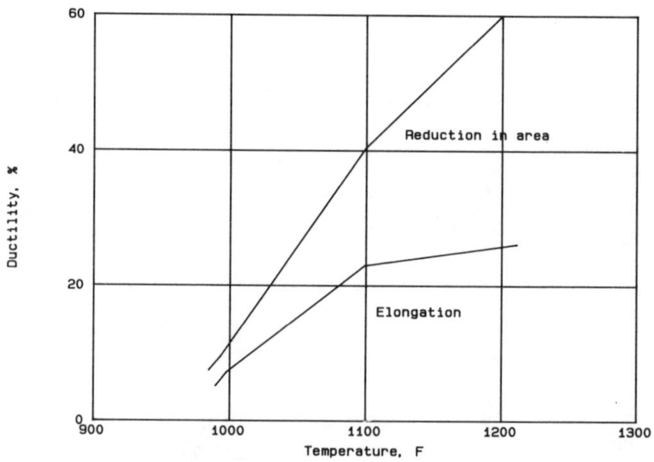

AISI 440B: Tensile elongation and reduction in area. Specimens were austenitized at 1900 F, tempered at 50 F above test temperature, and held for 1 h. Source: Ref 22, p 4

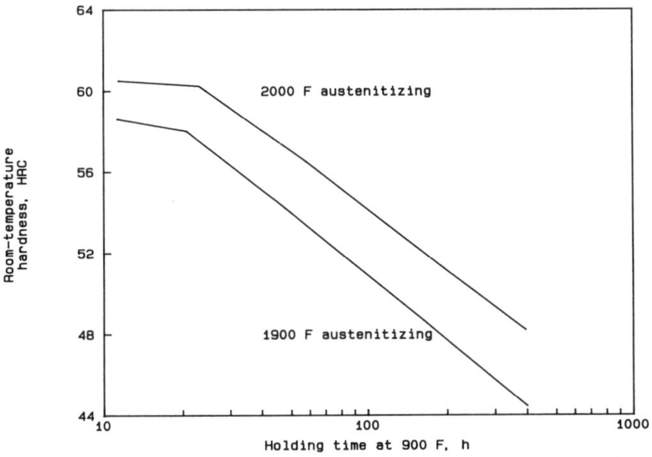

AISI 440B: Softening after extended holding at 900 F. Specimens were double tempered at 1000 F. Source: Ref 9, p 187

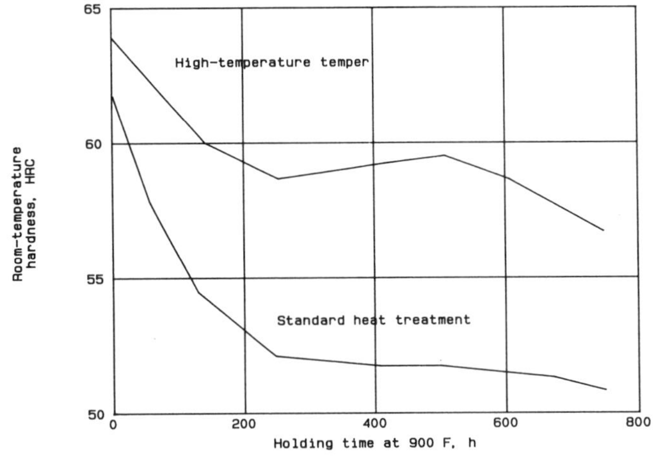

AISI 440B: Softening after extended holding at 900 F. Source: Ref 9, p 187

AISI 440C

UNS number: S44004
Specifications: AMS 5618, AMS 5630, AMS 5880, AMS 7445, ASTM A276, ASTM A314, ASTM A473, ASTM A493, ASTM A580, FED QQ S-763, MIL S-862, SAE J405, NF Z80CSN20-02, JIS G 4303-57SUS57B, DGN B83, BS 970 434S19, GOST 9Kh18, DIN 1.4126 X110CrMo17
Nominal composition: 0.950 to 1.200% C, 16.000 to 18.000% Cr, 1.000% Mn max, 0.750% Mo max, 0.030% S max, 1.000% Si max, 0.040% P max
Product forms: Wrought, bar, wire, sheet, strip, billets, forgings

Description: AISI 440C is a hardenable chromium steel that attains optimum corrosion resistance in the hardened and tempered condition. Type 440C can be heat treated to the highest hardness of any corrosion- or heat-resisting steel. Scaling temperatures in air are 760 C (1400 F) max for continuous service and 815 C (1500 F) max for intermittent service.

Typical applications: Balls, bearings, races, nozzles, balls and seats for oil well pumps, and valve parts.

Physical Properties

AISI 440C: Mean coefficients of linear thermal expansion

TEMP C	TEMP F	THEXP 10^{-6}/C	THEXP 10^{-6}/F
0-100	32-212	10.3	5.7
0-316	32-600	10.8	6.0
0-538	32-1000	11.7	6.5
0-649	32-1200	12.2	6.8

Source: Ref 22, p 3

Mechanical Properties

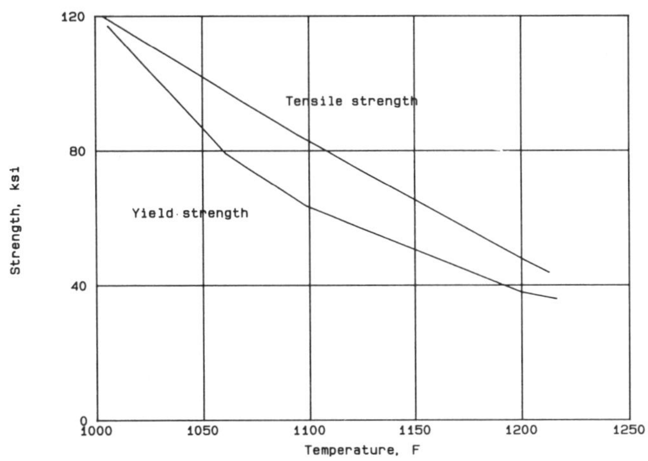

AISI 440C: Tensile and yield strength. Specimens were austenitized at 1900 F, tempered at 50 F above test temperature, and held for 1 h. Source: Ref 22, p 4

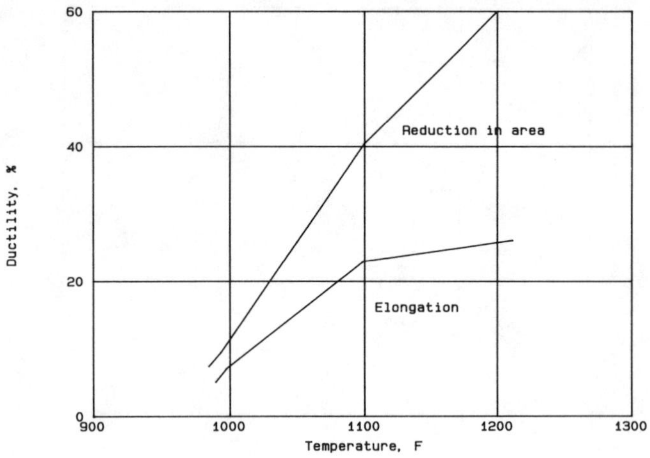

AISI 440C: Tensile elongation and reduction in area. Specimens were austenitized at 1900 F, tempered at 50 F above test temperature, and held for 1 h. Source: Ref 22, p 4

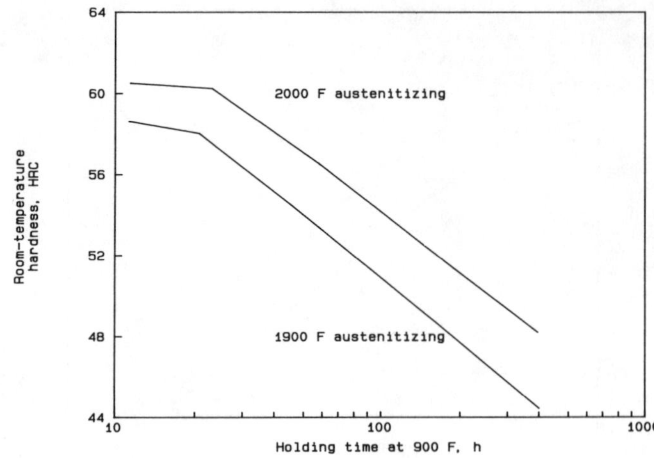

AISI 440C: Softening after extended holding at 900 F. Specimens were double tempered at 1000 F. Source: Ref 9, p 187

AISI 440C: Hot hardness data

TEMP C	TEMP F	HARD HRC
22	72	58
93	200	56.5
149	300	55
204	400	54
260	500	53
316	600	52
371	700	51
427	800	50
482	900	47
538	1000	40

Note: Specimens were austenitized at 1038 C (1900 F), oil quenched, and double tempered at 482 C (900 F). Samples were held at each temperature for 30 min prior to testing.
Source: Ref 28, p 44

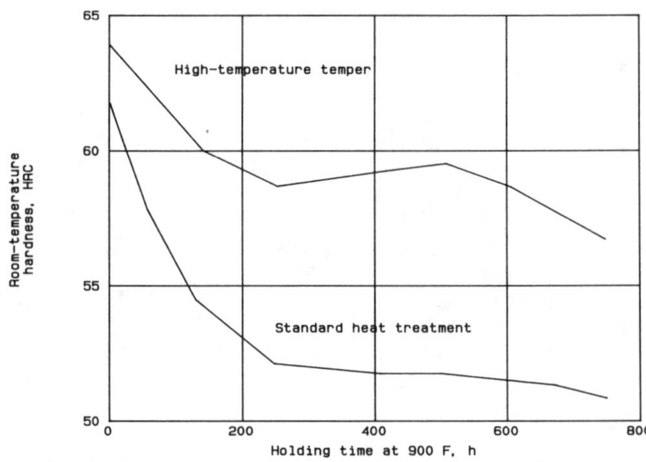

AISI 440C: Softening after extended holding at 900 F. Source: Ref 9, p 187

AISI 440C: Effect of exposure time at 480 C (900 F) on room-temperature hardness

EXP h	HARD HRC
1	58.5
20	58
30	56.5
50	55
100	51
400	45

Specimens were austenitized at 1038 C (1900 F), oil quenched, and double tempered at 482 C (900 F)
Source: Ref 28, p 45

AISI 440C: Tensile property data

TEMP C	TEMP F	UTS MPa	UTS ksi
704	1300	214	31
816	1500	117	17
927	1700	114	16.5

Note: Standard ASTM specimens were taken from 25-mm (1-in.) diam bars, oil quenched, heated to 1038 C (1900 F), and tempered at 316 C (600 F)
Source: Ref 28, p 44

15-5 PH

UNS number: S15500

Specifications: AMS 5658, AMS 5659, AMS 5862, ASME SA705 (XM-12), ASTM A564 (XM-12), ASTM A693, ASTM A705

Nominal composition: 0.070% C, 14.000 to 15.500% Cr, 2.500 to 4.500% Cu, 1.000% Mn max, 3.500 to 5.500% Ni, 0.040% P max, 0.030% S max, 1.000% Si max, 0.150 to 0.450% (Nb + Ta)

Product forms: Wrought, bar, wire, plate, sheet, strip, tube, billets, forgings

Description: 15-5 PH is a martensitic precipitation-hardening (maraging) stainless steel that combines high strength and hardness with excellent corrosion resistance. It is hardened by a single low temperature heat treatment that virtually eliminates scaling and distortion.

Typical applications: Gears, cams, cutlery, shafting, and aircraft parts.

Physical Properties

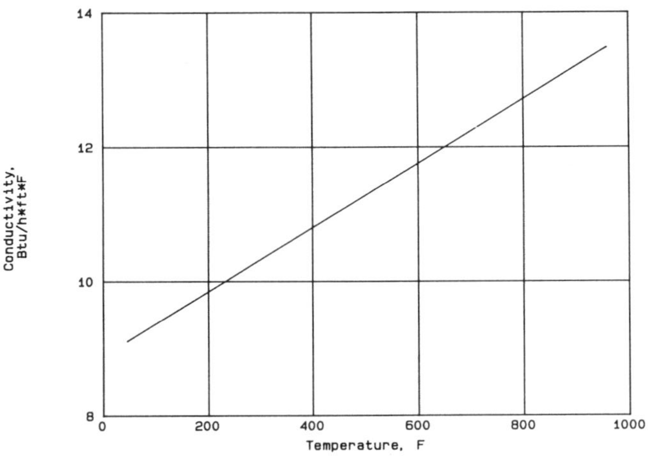

15-5 PH: Thermal conductivity of specimens heat treated at 900 F. Source: Ref 22, p 7

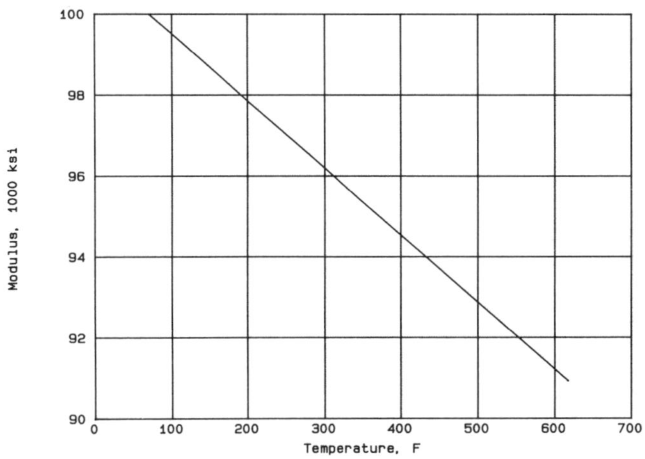

15-5 PH: Dynamic modulus of elasticity of specimens heat treated at 900 F. Source: Ref 22, p 9

15-5 PH: Mean coefficients of linear thermal expansion

TEMP C	TEMP F	THEXP 10^{-6}/C	THEXP 10^{-6}/F
21-93	70-200	10.8	6.0
21-205	70-400	10.8	6.0
21-315	70-600	11.3	6.3
21-425	70-800	11.7	6.5

Source: Ref 15, p 406

Mechanical Properties

15-5 PH: Short-time tensile property data

Condition	UTS MPa	UTS ksi	YS MPa	YS ksi	EL(a) %	RA %
Tested at 24 C (75 F)						
H925	1317	191	1213	176	16	59
H1025	1145	166	1110	161	17	64
H1100	1069	155	1034	150	19	67
H1150M	896	130	717	104	23	75
Tested at 205 C (400 F)						
H925	1158	168	1048	152	15	54
H1025	1014	147	958	139	15	58
H1100	952	138	924	134	16	62
H1150M	765	111	689	100	20	64
Tested at 315 C (600 F)						
H925	1096	159	965	140	14	59
H1025	958	139	903	131	14	57
H1100	910	132	869	126	14	57
H1150M	717	104	662	96	19	70
Tested at 425 C (800 F)						
H925	1027	149	869	126	15	60
H1025	917	133	820	119	15	60
H1100	848	123	786	114	14	60
H1150M	676	98	607	88	17	69
Tested at 540 C (1000 F)						
H925	758	110	634	92	17	70
H1025	724	105	627	91	18	70
H1100	662	96	607	88	18	71
H1150M	552	80	462	67	20	74
Tested at 650 C (1200 F)						
H925	400	58	317	46	26	83
H1025	372	54	283	41	28	83

(a) In 4D
Source: Ref 15, p 406

AM-355

Alternate designation: AISI 634
UNS number: S35500
Specifications: AMS 5547, AMS 5549, AMS 5594, AMS 5780, ASTM A564 (634), ASTM A579, ASTM A693, ASTM A705, MIL S-8840, MIL S-46123, SAE J467
Nominal composition: 0.100 to 0.150% C max, 15.000 to 16.000% Cr, 0.500 to 1.250% Mn, 2.500 to 3.250% Mo, 0.070 to 0.130% N max, 4.000 to 5.000% Ni, 0.040% P max, 0.030% S max, 0.500% Si max
Product forms: Wrought, bar, wire, plate, sheet, strip, tube, billets, forgings
Description: AM-355 is a lower chromium, higher carbon version of AM-350.
Typical applications: Pressure vessels, aircraft structurals, jet engine compressor components, wrapped solid fuel cases for missiles, spot welded sandwich structures, and springs.

Physical Properties

AM-355: Thermal conductivity

TEMP C	TEMP F	THCON W/m*K	THCON Btu/ft*h*F
38	100	15.1	8.7
150	300	16.4	9.5
260	500	17.8	10.3
370	700	19.4	11.2
480	900	20.7	12.0

Source: Ref 15, p 389

AM-355: Mean coefficients of linear thermal expansion

TEMP C	TEMP F	THEXP(a) 10^{-6}/C	THEXP(a) 10^{-6}/F	THEXP(b) 10^{-6}/C	THEXP(b) 10^{-6}/F
25-100	80-212	17.1	9.5	11.5	6.4
25-200	80-390	18.1	10.1	11.9	6.6
25-310	80-590	18.4	10.2	12.2	6.8
25-400	80-750	18.9	10.5	12.6	7.0
25-500	80-930	19.3	10.7	13.0	7.2
25-600	80-1110	19.8	11.0	13.0	7.2
25-1000	80-1830	20.3	11.3	13.3	7.4

(a) Hardened. (b) Subzero cooled and tempered
Source: Ref 15, p 388

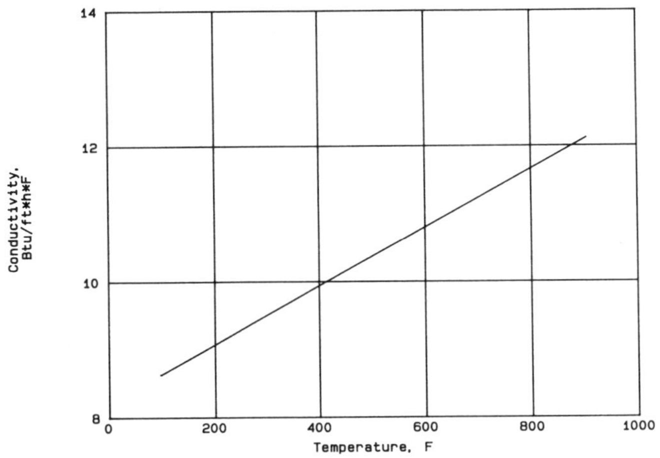

AM-355: Thermal conductivity of specimens subzero cooled and tempered at 850 F. Source: Ref 22, p 5

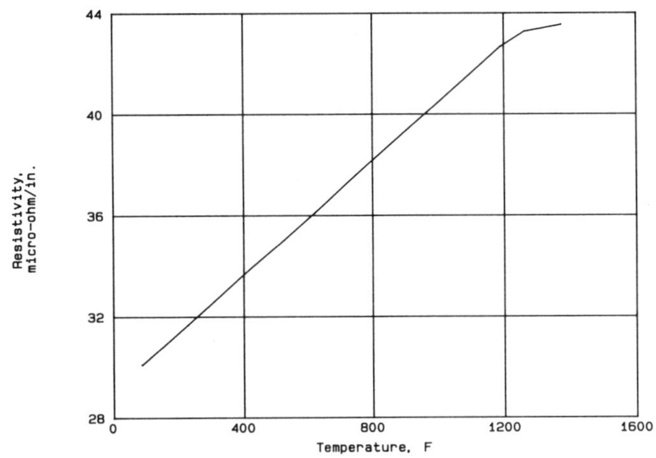

AM-355: Electrical resistivity of specimens subzero cooled and tempered at 850 F. Source: Ref 22, p 5

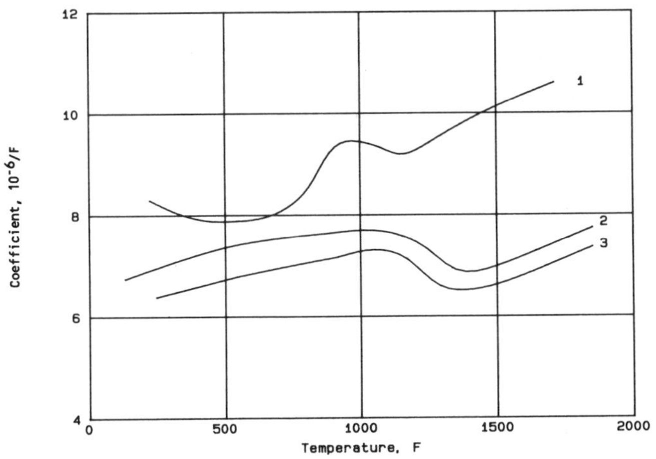

AM-355: Mean coefficient of linear thermal expansion. Curve 1: heat treated. Curve 2: subzero cooled and tempered at 850 F. Curve 3: subzero cooled and tempered at 1000 F. Source: Ref 22, p 5

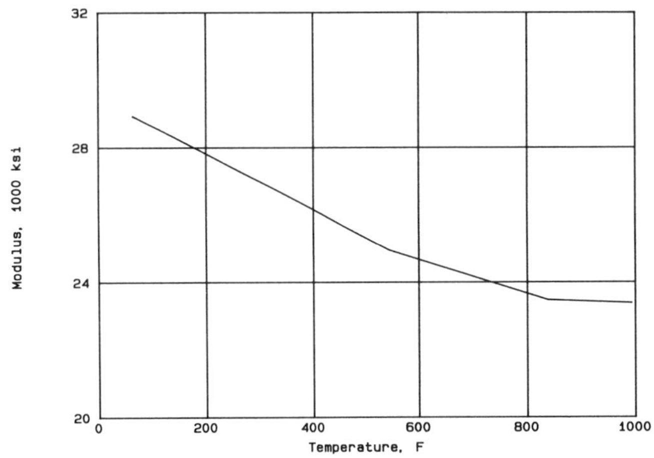

AM-355: Modulus of elasticity of 0.056-in. sheet. Specimens were cold rolled and tempered in both transverse and longitudinal sections and tested in tension and compression. Source: Ref 22, p 15

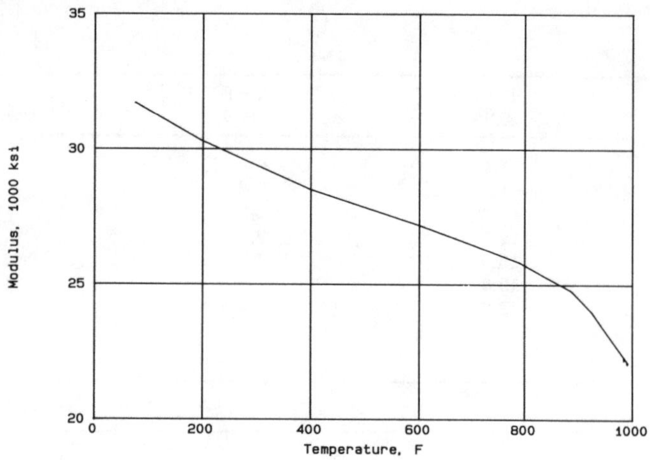

AM-355: Modulus of elasticity of 0.018-in. sheet. Specimens were subzero cooled, cold rolled, and tempered. Source: Ref 22, p 15

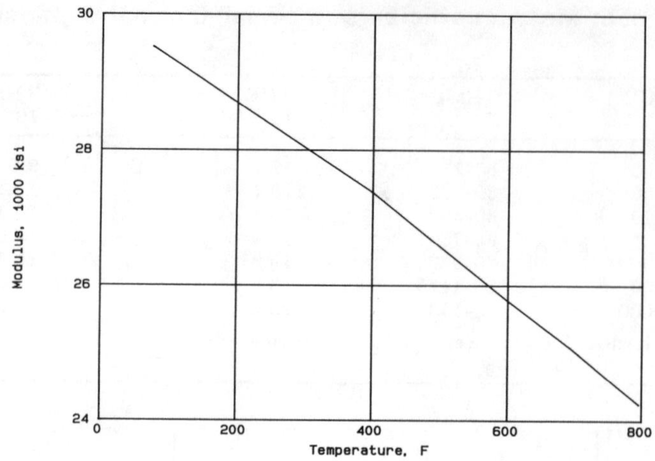

AM-355: Dynamic modulus of elasticity for 1-in. bar subzero cooled and tempered at 850 F. Source: Ref 22, p 15

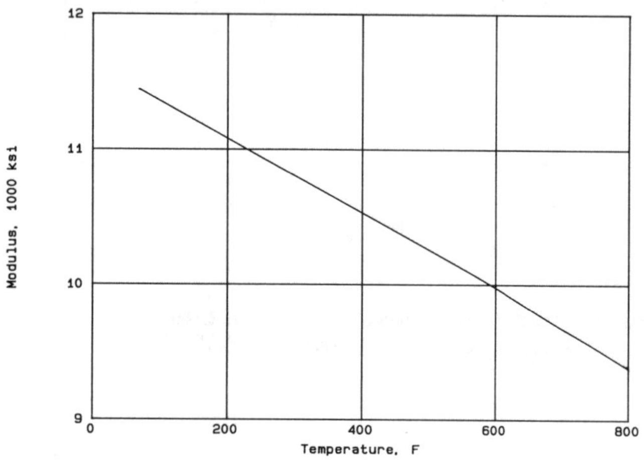

AM-355: Dynamic modulus of rigidity of 1-in. bar. Specimens were subzero cooled and tempered at 850 F. Source: Ref 22, p 16

AM-355: Electrical resistivity

TEMP C	TEMP F	ELRES n{o}*m
25	80	757
100	212	797
245	420	868
390	735	948
565	1050	1036

Note: Specimens were subzero cooled and tempered at 455 C (850 F)
Source: Ref 15, p 388

AM-355: Modulus of elasticity

TEMP C	TEMP F	E(a) GPa	E(a) 10^3 ksi	E(b) GPa	E(b) 10^3 ksi
27	80	...	...	202	29.3
204	400	...	...	188	27.3
316	600	...	...	179	26.0
371	700	172	25	174	25.3
427	800	...	...	170	24.6

Note: Modulus of elasticity was determined by ultrasonic method
(a) Hardened. (b) Subzero cooled and tempered
Source: Ref 15

AM-355: Modulus of rigidity

TEMP C	TEMP F	G(a) GPa	G(a) 10^3 ksi
27	80	79	11.4
204	400	72	10.5
316	600	68	9.93
371	700	66	9.63
427	800	64	9.35

Note: Modulus of rigidity was determined by ultrasonic method
(a) Subzero cooled and tempered
Source: Ref 15

Mechanical Properties

AM-355: Tensile property data for bar

Ttpr C	Ttpr F	TEMP C	TEMP F	YS(a) MPa	YS(a) ksi	YS MPa	YS ksi	UTS MPa	UTS ksi	EL %	RA %
455	850	21	70	980	142	1255	182	1490	216	19.0	38.5
540	1000	21	70	1015	147	1180	171	1280	186	19.0	57.0
455	850	205	400	850	123	1125	163	1425	207	15.5	45.0
540	1000	205	400	885	128	1050	152	1145	166	16.0	59.5
455	850	315	600	760	110	1050	152	1450	210	11.5	35.5
540	1000	315	600	850	123	985	143	1095	159	14.0	49.0
455	850	425	800	675	98	960	139	1365	198	11.0	35.5
540	1000	425	800	740	107	885	128	965	140	15.0	53.5
455	850	540	1000	450	65	670	97	995	144	16.0	57.0
540	1000	540	1000	485	70	660	96	795	115	19.0	65.0

(a) 0.02% offset
Source: Ref 3, p 203

AM-355: Tensile property data for cold rolled strip

TEMP C	TEMP F	YS MPa	YS ksi	UTS MPa	UTS ksi	EL %
Cold rolled and tempered						
21	70	1383	200.6	1616	234.4	16.0
427	800	1028	149.1	1331	193.0	6.5
482	900	936	135.8	1266	183.6	5.0
538	1000	848	123.0	1027	149.0	4.0
Extra hard						
21	70	2310	335.0	2406	349.0	1.0
427	800	1747	253.4	1984	287.7	2.0
482	900	1577	228.7	1813	263.0	1.5
538	1000	1072	155.5	1248	181.0	4.0

Note: Specimens were taken in a transverse direction from rolling
Source: Ref 28, p 75

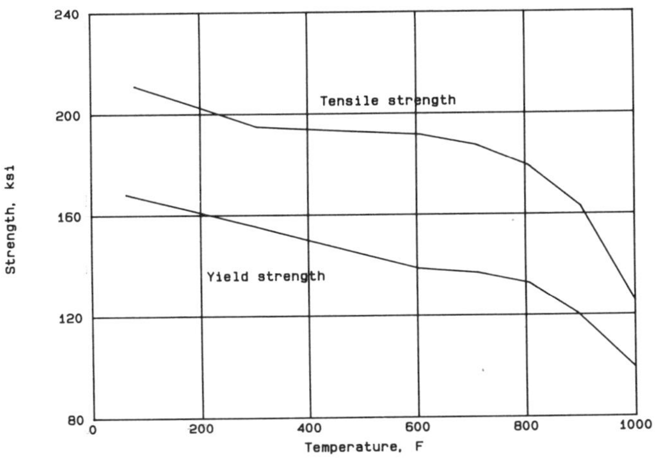

AM-355: Effect of temperature on tensile and yield strength of 0.75-in. bar. Specimens were subzero cooled and tempered at 850 F. Source: Ref 22, p 10

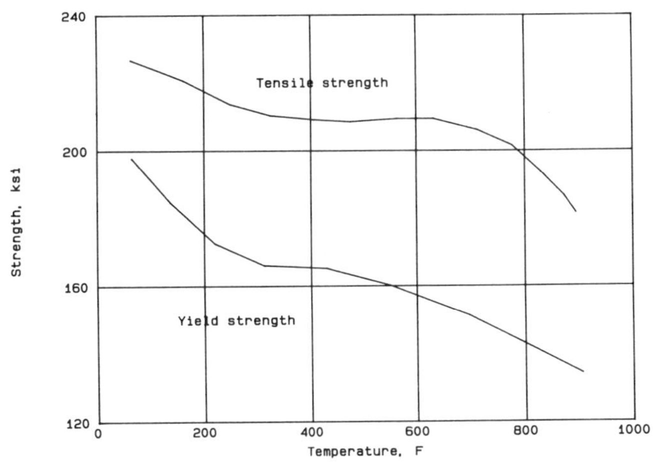

AM-355: Tensile and yield strength of sheet. Specimens were subzero cooled and tempered at 850 F. Source: Ref 22, p 9

AM-355: Room-temperature tensile property data after thermal exposure

EXP TEMP C	EXP TEMP F	EXP TIME h	UTS MPa	UTS ksi	YS MPa	YS ksi	EL %	RA %
Subzero cooled and tempered at 540 C (1000 F)								
Square specimen: 133 mm (5.25 in.)								
...	...	...	1158	168	1103	160	18	52
260	500	1000	1169	169.5	1110	161	18	54
370	700	1000	1344	195	1289	187	15	47
Square specimen: 100 mm (4 in.)								
...	...	...	1134	164.5	1076	156	18	57
260	500	100	1124	163	1072	155.5	19	58
260	500	1000	1127	163.5	1096	159	18	58
370	700	100	1186	172	1141	165.5	18	54
370	700	1000	1227	178	1207	175	17	54
Round specimen: 25 mm (1 in.)								
...	...	...	1148	166.5	1100	159.5	19	57
455	850	24	1217	176.5	1172	170	19	59
455	850	100	1269	184	1241	180	18	55
455	850	1000	1376	199.5	1300	188.5	17	49
Subzero cooled and tempered at 455 C (850 F)								
Round specimen: 25 mm (1 in.)								
...	...	...	1407	204	1207	185	16	50
455	850	24	1396	202.5	1245	180.5	16	52
455	850	100	1413	205	1310	190	15	50
455	850	1000	1441	209	1386	201	16	50

Note: Specimens were 9.53-mm (0.375-in.) diam
(a) In 4D
Source: Ref 15, p 389

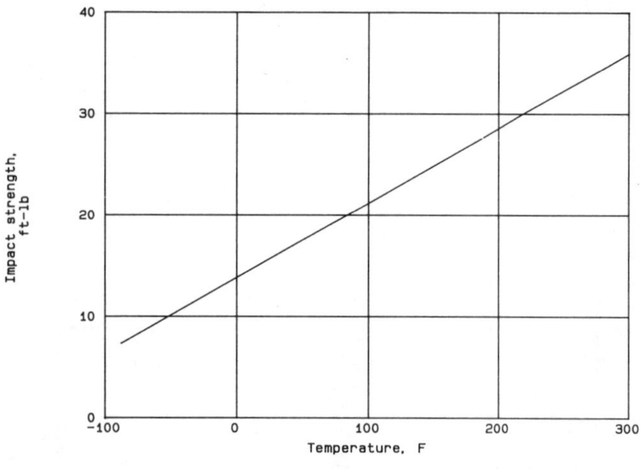

AM-355: Charpy V-notch impact strength of 0.75-in. bar. Specimens were subzero cooled and tempered at 850 and 950 F. Source: Ref 22, p 12

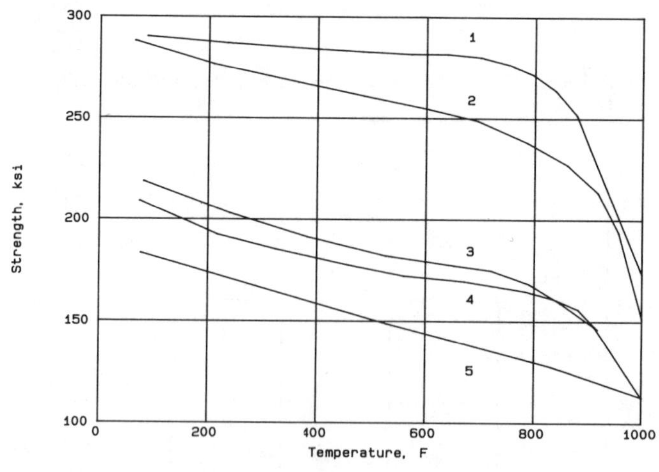

AM-355: Compressive yield strength of sheet. Curve 1: subzero cooled, cold rolled and tempered, transverse. Curve 2: subzero cooled, cold rolled, longitudinal. Curve 3: subzero cooled and tempered at 950 F. Curve 4: subzero cooled and tempered at 850 F, transverse. Curve 5: cold rolled and tempered, longitudinal. Source: Ref 22, p 12

AM-355: Room-temperature tensile property data for sheet and bar after thermal exposure

EXP TEMP C	EXP TEMP F	STREX MPa	STREX ksi	EXP h	UTS MPa	UTS ksi	YS MPa	YS ksi	EL %	RA %
Subzero cooled and tempered at 455 C (850 F)										
Sheet: 0.69 mm (0.027 in.)										
...	...	...	...	...	1520	220.5	1286	186.5	8	...
315	600	689	100	502	1617	234.5	1424	206.5	11	...
315	600	862	125	518	1620	235.0	1469	213.0	7	...
Sheet: 1.27 mm (0.050 in.)										
...	...	...	...	...	1458	211.5	1169	169.5	12	...
315	600	455	66.0	1001	1472	213.5	1186	172.0	12	...
315	600	683	99.0	1001	1475	214.0	1238	179.5	11	...
315	600	1024	148.5	1001	1469	213.0	1424	206.5	8	...
370	700	445	64.5	1003	1496	217.0	1227	178.0	11	...
370	700	669	97.0	1072	1507	218.5	1300	188.5	12	...
370	700	1003	145.5	1072	1503	218.0	1496	217.0	6	...
425	800	427	62.0	1072	1572	228.0	1379	200.0	13	...
425	800	641	93.0	1016	1548	224.5	1410	204.5	13	...
425	800	965	140.0	1007	1565	227.0	1524	221.0	9	...
Round bar: 25 mm (1 in.)										
...	...	...	...	...	1462	212.0	1155	167.5	21	54
315	600	710.1	103	1000	1465	212.5	1265	183.5	9	54
370	700	717.0	104	1000	1500	217.5	1324	192.0	9	43
Subzero cooled and tempered at 540 C (1000 F)										
Round bar: 25 mm (1 in.)										
...	...	...	...	...	1276	185.0	1134	164.5	16	57
370	700	676	98	5400	1489	216.0	1382	200.5	15	61
...	...	...	...	...	1213	176.0	1065	154.5	18	60
370	700	641	93	5400	1403	203.5	1096	159.0	17	51
...	...	...	...	...	1317	191.0	1213	176.0	9	54
(a)	(a)	413	60	(a)	1427	207.0	1365	198.0	11	53
...	...	...	...	...	1372	199.0	1255	182.0	7	53
(a)	(a)	413	60	(a)	1544	224.0	1482	215.0	5	50
(a)	(a)	827	120	(a)	1551	225.0	...	...	5	49

(a) Cycled daily for 15 days between -73 C and 425 C (-100 F and 800 F); held at 425 C (800 F) about 15 h during each cycle
Source: Ref 15, p 390

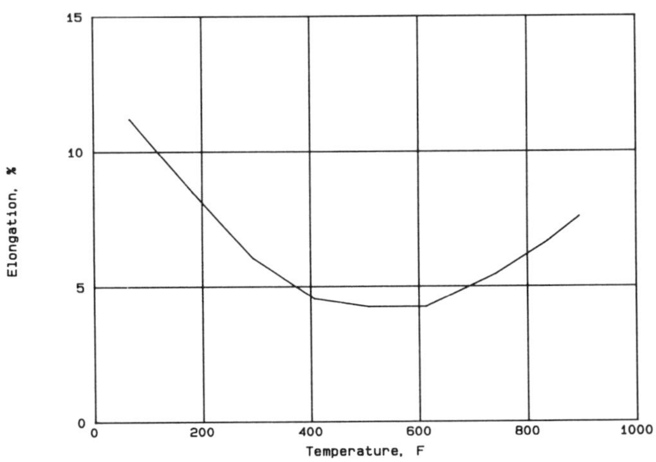

AM-355: Tensile elongation of sheet. Specimens were subzero cooled and tempered at 850 F. Source: Ref 22, p 9

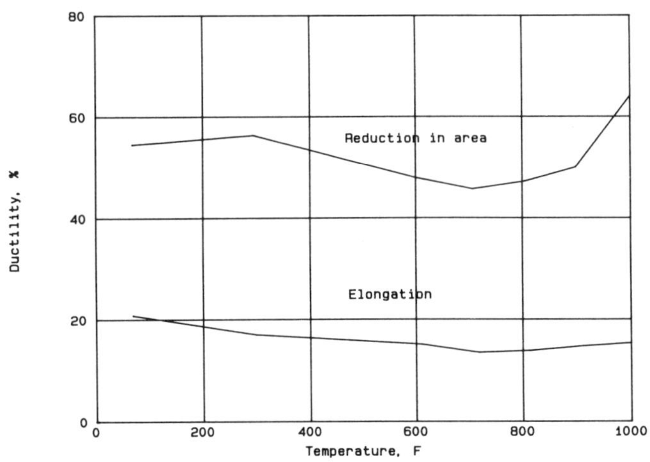

AM-355: Tensile elongation and reduction in area of 0.75-in. bar. Specimens were subzero cooled and tempered at 850 F. Source: Ref 22, p 10

AM-355: Creep strength of sheet

TEMP C	TEMP F	CRSTR 0.1%/1000 h MPa	CRSTR 0.1%/1000 h ksi	CRSTR 0.2%/1000 h MPa	CRSTR 0.2%/1000 h ksi
315	600	848	123	1000	145
370	700	724	105	862	125
425	800	421	61	627	91

Note: Specimens were subzero cooled and tempered at 455 C (850 F)
Source: Ref 15, p 389

AM-355: Stress-rupture property data for heat treated specimens

TEMP C	TEMP F	Ttpr C	Ttpr F	RPSTR 10 h MPa	RPSTR 10 h ksi	RPSTR 100 h MPa	RPSTR 100 h ksi	RPSTR 1000 h MPa	RPSTR 1000 h ksi
425	800	455	850	1305	189	1280	186	1240	180
425	800	540	1000	940	136	925	134	910	132
480	900	455	850	995	144	835	121	675	98
480	900	540	1000	770	112	725	105	675	98
540	1000	455	850	580	84	485	70	400	58
540	1000	540	1000	600	87	505	73	420	61

Source: Ref 3, p 203

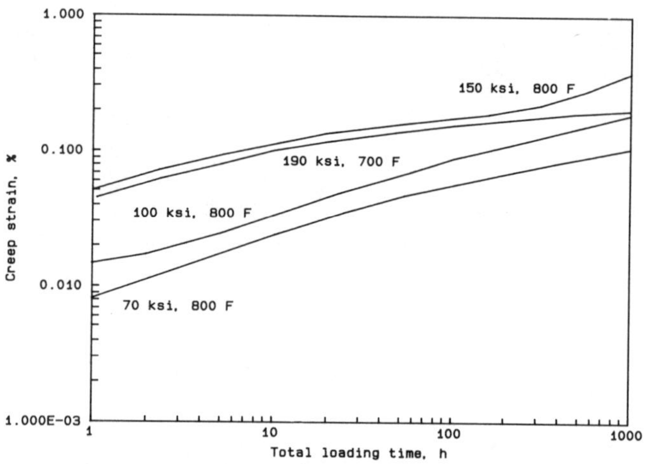

AM-355: Total creep strain at 700 and 800 F for four levels of stress for 0.056-in. sheet. Specimens were cold rolled and tempered. Source: Ref 22, p 14

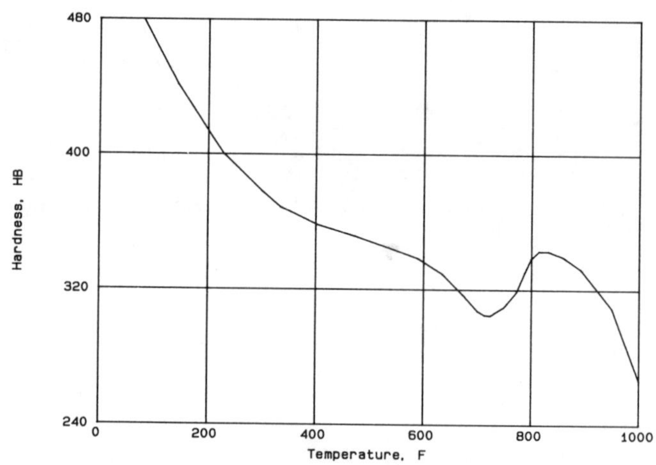

AM-355: Brinell hardness of subzero cooled and tempered specimens. Source: Ref 22, p 4

AM-355: Shear property data for sheet

TEMP C	TEMP F	USS MPa	USS ksi	USS:UTS
21	70	1168	169.4	0.75
204	400	960	139.3	0.69
316	600	878	127.4	0.63
371	700	872	126.4	0.64
427	800	836	121.3	0.64
482	900	834	120.9	...
538	1000	665	96.5	0.8

Note: Specimens were 38 mm (1.5 in.) in diameter and tested at room temperature. Condition: Subzero cooled and tempered at 454 C (850 F)
Source: Ref 28, p 75

AM-355: Stress-rupture property data for flat rolled bar

TEMP C	TEMP F	RPSTR 10 h MPa	RPSTR 10 h ksi	RPSTR 100 h MPa	RPSTR 100 h ksi	RPSTR 1000 h MPa	RPSTR 1000 h ksi
Subzero cooled and tempered at 455 C (850 F)							
425	800	1296	188	1276	185	1255	182
480	900	972	141	827	120	676	98
540	1000	607	88	496	72	400	58
Subzero cooled and tempered at 540 C (1000 F)							
425	800	965	140	951	138	931	135
480	900	758	110	724	105	683	99
540	1000	579	84	490	71	414	60

Source: Ref 15, p 390

Custom 450

UNS number: S45000
Specifications: AMS 5763, ASME SA564 (XM-25), ASME SA705, ASTM A564 (XM-25), ASTM A693, ASTM A705
Nominal composition: 0.050% C max, 14.000 to 16.000% Cr, 1.250 to 1.750% Cu, 1.000% Mn max, 0.500 to 1.000% Mo, 5.000 to 7.000% Ni, 0.030% P max, 0.030% S max, 1.000% Si max
Product forms: Wrought, bar, wire, plate, sheet, strip, tube, billets, forgings
Description: Custom 450 is a martensitic precipitation hardenable steel that exhibits very good corrosion resistance with moderate strength. This alloy has a yield strength somewhat greater than 690 MPa (100 ksi) in the annealed condition, but is easily fabricated. A single-step aging treatment develops higher strength with good ductility and toughness. It will resist atmospheric corrosion, including salt water atmospheres. It exhibits excellent resistance to rusting and pitting in 5 and 20% salt spray at 35 C (95 F) and to oxidation up to 650 C (1200 F). The maximum service temperature for continuous service is 425 C (800 F).
Typical applications: Gears, springs, fasteners, and aircraft and turbine parts.

Physical Properties

Custom 450: Mean coefficients of linear thermal expansion

TEMP C	TEMP F	THEXP(a) 10^{-6}/C	THEXP(a) 10^{-6}/F	THEXP(b) 10^{-6}/C	THEXP(b) 10^{-6}/F
24-95	75-200	10.6	5.9	10.8	6.0
24-150	75-300	10.1	5.6	10.4	5.8
24-205	75-400	10.3	5.7	10.6	5.9
24-260	75-500	10.4	5.8	11.0	6.1
24-315	75-600	10.6	5.9	11.2	6.2
24-370	75-700	10.8	6.0	11.3	6.3
24-425	75-800	11.0	6.1	11.5	6.4
24-480	75-900	11.1	6.2	11.7	6.5
24-540	75-1000	11.0	6.1	11.75	6.53
24-595	75-1100	11.2	6.2	11.75	6.53

(a) Annealed at 1040 C (1900 F) and water quenched. (b) Annealed and aged at 480 C (900 F) for 4 h and air cooled
Source: Ref 15, p 425

Mechanical Properties

Custom 450: Effect of aging temperature on tensile property data

TEMP C	TEMP F	UTS MPa	UTS ksi	YS MPa	YS ksi	EL(a) %	RA %	IS(b) J	IS(b) ft-lb
Aged at 315 C (600 F)									
480	900	1105	160	951	138	12	47.7	54	40
565	1050	1050	152	1005	146	12.4	49.3	68	50
620	1150	917	133	862	125	14.1	53.7	111	82
Aged at 425 C (800 F)									
480	900	1035	150	903	131	12.4	44.5	57	42
565	1050	986	143	896	130	12.2	44.5	73	54
620	1150	834	121	793	115	13.4	49.1	111	82
Aged at 565 C (1050 C)									
480	900	580	84	525	76	24	74.6	89	66
565	1050	585	85	540	78	26.5	73.7	91	67
620	1150	540	78	485	70	30	76.6	113	83

Note: Specimens were taken from 25-mm (1-in.) diam bar
(a) Gage length = 4D. (b) Charpy V-notch
Source: Ref 15, p 425

FERRALIUM Alloy No. 255

UNS number: S32550
Specifications: ASTM A240, ASTM A479
Nominal composition: 24.000 to 27.000% Cr, 2.900 to 3.900% Mo, 4.500 to 6.500% Ni, 1.000% Si max, 1.500% Mn max, 0.040% C max, 0.100 to 0.250% N, 1.500 to 2.500% Cu, 0.040% P max, 0.030% S max
Product forms: Wrought, bar, wire, plate, sheet, strip, tube, billets, forgings, extrusions, structural shapes
Description: FERRALIUM Alloy No. 255 is a duplex ferritic-austenitic stainless steel that combines strength, ductility, and hardness with resistance to corrosion and erosion. Use in the 760 to 870 C (1400 to 1600 F) range should be avoided because of the formation of embrittling sigma phase. This alloy is not recommended for sustained use much above 260 C (500 F) as the alloy's structure can be severely degraded by exposure at higher temperatures.
Typical applications: Pumps, agitators, valves, and piping for handling wet phosphoric acid slurries; marine scrubbers, pumps, and mechanical seals; and equipment for handling hot organic acids, fatty acids, and solutions of nitric and hydrofluoric acids.

Physical Properties

FERRALIUM Alloy No. 255: Mean coefficients of linear thermal expansion

TEMP C	TEMP F	THEXP 10^{-6}/C	THEXP 10^{-6}/F
21-93	70-200	11.0	6.1
21-205	70-400	11.7	6.5
21-315	70-600	12.1	6.7
21-425	70-800	12.4	6.9
21-540	70-1000	13.0	7.2

Source: Ref 15, p 428

Mechanical Properties

FERRALIUM Alloy No. 255: Tensile property data

TEMP C	TEMP F	UTS MPa	UTS ksi	YS MPa	YS ksi	EL %	E GPa	E 10^3 ksi
21	70	867.4	125.8	674	97.8	27	210	30.5
95	200	786.0	114.0	594	86.1	27	206	29.9
205	400	704.0	102.1	515	74.7	27	198	28.7
315	600	706.7	102.5	494	71.7	29	189	27.4
425	800	697.3	100.7	487	70.7	26	179	25.9
540	1000	580.0	84.1	425	61.7	28	165	24.0

Note: Specimens were 10.2 to 19-mm (0.400-to 0.75-in.) thick plate, heat treated at 1120 C (2050 F); cooled rapidly
Source: Ref 15, p 428

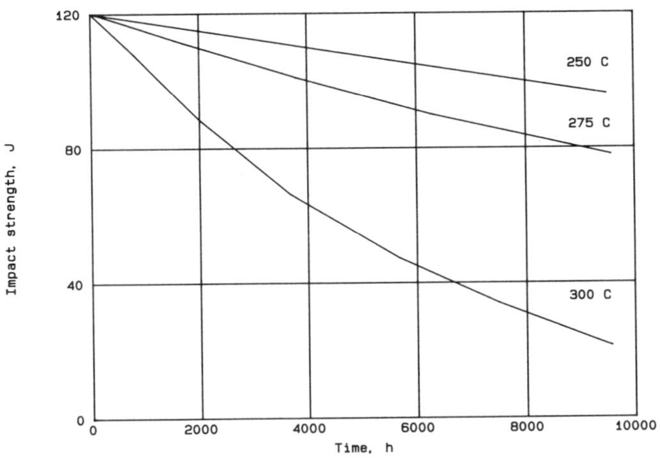

FERRALIUM Alloy No. 255: Impact strength of solution treated cast specimens. Source: Ref 24, p 635

E-Brite 26-1

UNS number: S44627
Specifications: ASME SA240, ASME SA268, ASME SA479, ASME SA731, ASTM A176, ASTM A240, ASTM A268, ASTM A276, ASTM A314, ASTM A479, ASTM A731
Nominal composition: 0.010% C max, 0.050 to 0.200% Cb max, 25.000 to 27.000% Cr, 0.200% Cu max, 0.400% Mn max, 0.750 to 1.500% Mo, 0.015% N max, 0.500% Ni max, 0.020% P max, 0.020% S max, 0.400% Si max
Product forms: Wrought, bar, wire, plate, sheet, strip, tube, billets, forgings, extrusions
Description: E-Brite 26-1 is a chromium-molybdenum ferritic stainless steel with low carbon and low nitrogen. The vacuum melted alloy combines toughness and ductility with good corrosion resistance. It is essentially immune to chloride stress corrosion cracking. The alloy is subject to 475 C (885 F) embrittlement. It can develop notch sensitivity after long exposures in the 400 to 595 C (750 to 1100 F) temperature range.
Typical applications: Heat exchangers, pressure vessels, tanks, and other equipment for use in the chemical processing, food processing, petroleum refining, water treatment, and pollution control industries.

Mechanical Properties

E-Brite 26-1: High strain rate tensile property data

TEMP C	TEMP F	UTS MPa	UTS ksi	RA %
649	1200	277	40.2	82
760	1400	197	28.5	98
871	1600	123	17.9	96
982	1800	77	11.1	98
1093	2000	49	7.1	96
1204	2200	37	5.3	97

Note: Specimens were annealed at 871 C (1600 F), water quenched, heated at 35 C/s (65 F/s) to test temperature, held 5 min, and pulled to failure at 20 in./in.*s
Source: Ref 21, p 15

E-Brite 26-1: Tensile property data for plate and bar

TEMP C	TEMP F	UTS MPa	UTS ksi	YS MPa	YS ksi	EL %	E GPa	E 10^3 ksi
21	70	483	70	345	50	35	215	31.2
66	150	455	66	310	45	30	202	29.3
93	200	448	65	283	41	30	199	28.8
149	300	427	62	262	38	30	188	27.2
204	400	427	62	262	38	30	188	27.2
260	500	496	72	241	35	30	190	27.5
316	600	496	72	228	33	30	196	28.4
371	700	517	75	234	34	30	181	26.2
427	800	517	75	234	34	30	...	...
482	900	517	75	310	45	30	...	...
538	1000	462	67	296	43	23	...	...
649	1200	152	22	103	15	59	...	...
760	1400	83	12	45	6.5	110	...	...
871	1600	34	5	17	2.5	139	...	...

Source: Ref 21, p 16

E-Brite 26-1: Tensile property data for tubing

TEMP C	TEMP F	UTS MPa	UTS ksi	YS MPa	YS ksi	EL %
21	70	490	71	379	55	34
204	400	448	65	359	52	31
260	500	483	70	276	40	...
316	600	531	77	269	39	30
371	700	524	76	276	40	31

Note: Tubing was 15.875-mm (0.625-in.) in diameter with a 1.27-mm (0.050-in.) wall. Specimens were annealed at 816 C (1500 F)
Source: Ref 21, p 16

E-Brite 26-1: Effect of thermal exposure on room-temperature tensile properties

EXP TEMP C	EXP TEMP F	EXP TIME h	UTS MPa	UTS ksi	YS MPa	YS ksi	EL %
None	None	None	479	69.5	358	52.0	35
343	650	5000	485	70.4	348	50.5	30
371	700	5000	556	80.7	438	63.5	24
399	750	5000	576	83.6	491	71.2	13

Note: Sheet was originally annealed at 871 C (1600 F) and water quenched
Source: Ref 29, Jan 1979

E-Brite 26-1: Compression stress-strain curves for specimens prestrained in torsion at 0.006/s to a total strain of 0.04. Source: Ref 21, p 20

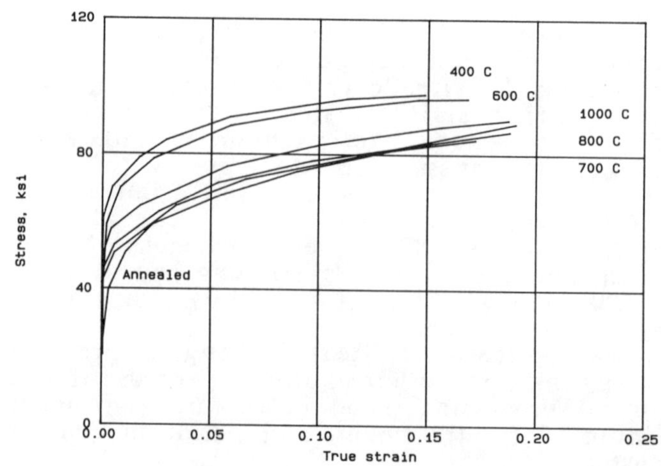

17-4 PH

UNS number: S17400
Specifications: AMS 5604
Nominal composition: 0.070% C max, 15.500 to 17.500% Cr, 3.000 to 5.000% Cu, 1.000% Mn max, 3.000 to 5.000% Ni, 0.040% P max, 0.030% S max, 1.000% Si max, 0.150 to 0.450% (Nb + Ta)
Product forms: Wrought, bar, wire, plate, sheet, strip, tube, billets, forgings
Description: 17-4 PH is similar to 15-5 PH, but has a higher chromium content. It can be hardened using a single low-temperature heat treatment that eliminates scaling and distortion. The martensitic precipitation hardening (maraging) stainless combines high strength and excellent corrosion resistance with good mechanical properties at temperatures up to 315 C (600 F).
Typical applications: Gears, springs, cutlery, fasteners, and aircraft and turbine parts.

Physical Properties

17-4 PH: Mean coefficients of linear thermal expansion

TEMP C	TEMP F	THEXP(a) 10^{-6}/C	THEXP(a) 10^{-6}/F	THEXP(b) 10^{-6}/C	THEXP(b) 10^{-6}/F	THEXP(c) 10^{-6}/C	THEXP(c) 10^{-6}/F
21-93	70-200	10.8	6.0	10.8	6.0	11.9	6.6
21-204	70-400	10.8	6.0	11.0	6.1	12.4	6.9
21-316	70-600	11.2	6.2	11.3	6.3	12.8	7.1
21-427	70-800	11.3	6.3	11.7	6.5	13.0	7.2
21-482	70-900	...	...	...	...	13.1	7.3

(a) Condition A. (b) Condition H900. (c) Condition H1150
Source: Ref 28, p 53

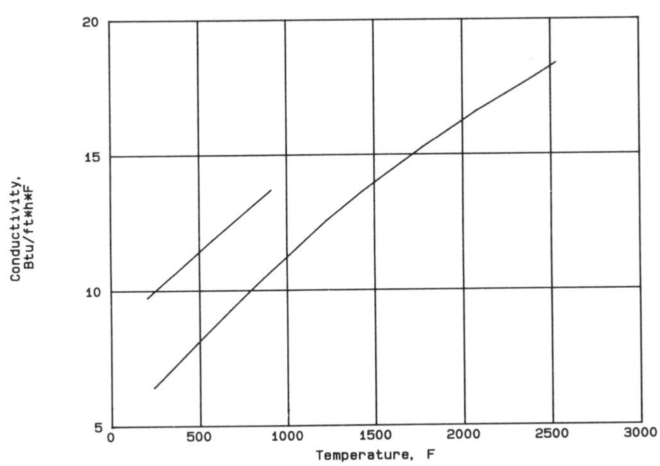

17-4 PH: Thermal conductivity. Specimens were heat treated at 900 F. Source: Ref 22, p 5

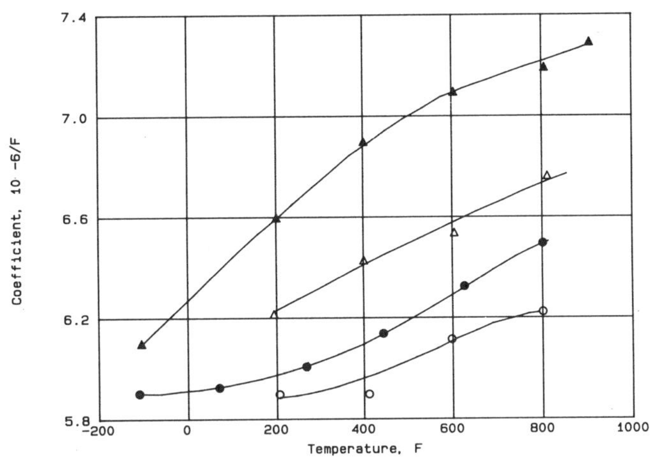

17-4 PH: Coefficient of thermal expansion from room temperature to temperature indicated. Open circle: annealed. Solid circle: heat treated at 900 F. Open triangle: heat treated at 1025 F. Solid triangle: heat treated at 1150 F. Source: Ref 22, p 5

17-4 PH: Modulus of elasticity

TEMP C	TEMP F	E GPa	E 10^3 ksi
22	72	690	100.0
38	100	687	99.6
93	200	674	97.8
149	300	664	96.3
204	400	653	94.7
260	500	641	93.0
316	600	630	91.4

Source: Ref 28, p 54

17-4 PH: Thermal conductivity

TEMP C	TEMP F	THCON W/m*K	THCON Btu/ft*h*F
150	300	17.8	10.3
260	500	19.5	11.3
460	860	22.5	13.0
480	900	22.7	13.1

Source: Ref 15, p 375

Mechanical Properties

17-4 PH: Effect of exposure on tensile properties

EXP s	UTS MPa	UTS ksi	YS MPa	YS ksi	EL %
Tested at 21 C (70 F)					
5	1062	154	896	130	12.0
10	1041	151	876	127	9.0
30	1041	151	841	122	9.0
60	1027	149	793	115	9.0
90	1034	150	772	112	8.5
300	1034	150	807	117	8.0
600	1014	147	724	105	8.0
Tested at 760 C (1400 F)					
5	317	46.0	311	45.2	27.0
10	297	43.4	293	42.6	29.0
30	327	47.5	317	46.0	21.0
60	327	47.5	318	46.2	19.0
90	322	46.8	312	45.3	18.0
300	306	44.5	298	43.3	23.0
600	305	43.3	292	42.3	21.0

Note: Condition H1150
Source: Ref 28, p 10

17-4 PH: Room-temperature tensile property data after thermal exposure

EXP s	UTS MPa	UTS ksi	YS MPa	YS ksi	EL %
Exposure temperature: 538 C (1000 F)					
5	1469	213	1358	197	8
10	1448	210	1303	189	8
30	1448	210	1331	193	7
60	1448	210	1338	194	7
90	1437	207	1290	187	6
300	1414	205	1324	192	5.5
600	1386	201	1297	188	5
Exposure temperature: 593 C (1100 F)					
5	1386	201	1262	183	7.5
10	1407	210	1269	184	8.5
30	1359	197	1228	178	7
60	1317	191	1207	175	6
90	1324	192	1241	180	5.5
300	1214	176	1172	170	5
600	1228	1787	1152	167	6
Exposure temperature: 649 C (1200 F)					
5	1290	187	1193	173	8
10	1290	187	1118	176	8
30	1234	179	1151	167	7
60	1248	181	1159	168	8
90	1138	165	1124	163	6.5
300	1131	164	1027	149	6
600	1124	163	1020	148	6
Exposure temperature: 704 C (1300 F)					
5	1186	172	1007	146	9
10	1186	172	1062	154	8
30	1138	165	945	137	8
60	1115	162	986	143	6
90	1103	160	924	134	6.5
300	1069	155	772	112	7
600	1062	154	903	131	7
Exposure temperature: 760 C (1400 F)					
5	1138	165	910	132	8.5
10	1131	164	848	123	8
30	1110	161	855	124	7
60	1115	162	827	120	6
90	1096	159	876	127	6
300	1103	160	862	125	5
600	1082	157	814	118	6
Exposure temperature: 871 C (1600 F)					
5	1152	167	834	121	6
10	1138	165	834	121	5
30	1138	165	841	122	5
60	1138	165	834	121	5
90	1110	161	869	126	4.5
300	1117	162	882	128	4.5
Exposure temperature: 982 C (1800 F)					
5	1110	161	820	119	6
10	1096	159	820	119	6
30	1089	158	786	114	5.5
60	1082	157	814	118	6
Exposure temperature: 1093 C (2000 F)					
5	1110	161	876	127	6
10	1110	161	800	116	5
30	1089	158	752	109	5

Note: Initial condition of material: H900
Source: Ref 28, p 9

17-4 PH: Tensile property data

EXP TEMP C	EXP TEMP F	EXP h	UTS MPa	UTS ksi	YS MPa	YS ksi	EL(a) %	RA %	HARD HRC
...	...	...	1407	204	1289	187	10	49	43
316	600	10	1427	207	1289	187	10	53	43
316	600	1000	1476	214	1338	194	8	47	45
316	600	5000	1503	218	1379	200	7	47	45

Note: Specimens were taken from 6.35-mm (0.25-in.) diam wire and tested at room temperature after long-time exposure at 316 C (600 F), condition H900
(a) In 25 mm (1 in.)
Source: Ref 28, p 54

17-4 PH: Tensile property data

EXP TEMP C	EXP TEMP F	EXP h	UTS MPa	UTS ksi	YS MPa	YS ksi	EL %
Tested at 21 C (70 F)							
316	600	1000	1041	151	979	142	13
316	600	7000	1082	157	1034	150	11
316	600	7000(a)	972	141	820	119	13
343	650	1000	1069	155	1027	149	13
343	650	7000	1159	168	1117	162	11
343	650	7000(a)	972	141	820	119	12
371	700	100	1048	152	993	144	13
371	700	1000	1117	162	1069	155	11
371	700	7000	1234	179	1200	174	9
371	700	7000(a)	965	140	807	117	13
399	750	100	1062	154	1014	147	13
399	750	1000	1179	171	1138	165	11
399	750	7000	1276	185	1227	178	10
399	750	7000(a)	965	140	807	117	13
...	...	...	1020	148	924	134	13
Tested at 316 C (600 F)							
316	600	1000	848	123	814	118	6
316	600	7000	882	128	848	123	8
316	600	7000(a)	779	113	738	107	7
343	650	1000	882	128	855	124	6
343	650	7000	952	138	931	135	7
343	650	7000(a)	793	115	745	108	7
371	700	100	862	125	827	120	6
371	700	1000	924	134	896	130	7
371	700	7000	1020	148	993	144	6
371	700	7000(a)	786	114	745	108	7
399	750	100	882	128	841	122	6
399	750	1000	965	140	931	135	7
399	750	7000	1027	149	993	144	7
399	750	7000(a)	779	113	738	107	7
...	...	...	834	121	793	115	5
Tested at 339 C (750 F)							
316	600	1000	793	115	758	110	5
316	600	7000	827	120	807	117	5
316	600	7000(a)	738	107	710	103	6
343	650	1000	834	121	800	116	5
343	650	7000	910	132	882	128	5
343	650	7000(a)	745	108	703	102	7
371	700	100	814	118	772	112	4
371	700	1000	896	130	855	124	6
371	700	7000	979	142	945	137	5
371	700	7000(a)	745	108	710	103	6
399	750	100	834	121	793	115	4
399	750	1000	931	135	889	129	6
399	750	7000	993	144	952	138	8
399	750	7000(a)	724	105	696	101	7
...	...	...	800	116	765	111	4

(a) Re-aged at 621 C (1150 F) for 4 h after exposure
Source: Ref 28, p 8

17-4 PH: Tensile property data after thermal exposure

EXP TEMP C	EXP TEMP F	EXP TIME h	UTS MPa	UTS ksi	YS MPa	YS ksi	EL %	RA %	HARD HRC
Condition H900									
...	...	...	1400	203	1269	184	12	52	45
371	700	10	1338	194	1214	176	12	53	43
371	700	1000	1413	205	1310	190	12	52	45
371	700	5000	1496	217	1379	200	11	50	46
400	750	1000	1476	214	1393	202	11	49	47
400	750	2000	1538	223	1441	209	10	45	47
427	800	10	1427	207	1282	186	10	49	45
427	800	1000	1579	229	1476	214	9	40	48
427	800	4000	1607	233	1531	222	9	32	49
427	800	10	1131	164	910	132	6	16	...
427	800	1000	1220	177	1014	147	4	6	...
427	800	4000	1269	184	1103	160	2	2	...
Condition H1000									
...	...	...	1262	183	1186	172	11	56	41
371	700	10	1214	176	1151	167	14	58	40
371	700	1000	1358	197	1317	191	11	53	43
371	700	5000	1400	203	1358	197	10	50	45
400	750	1000	1434	208	1379	200	10	50	45
400	750	2000	1476	214	1420	206	10	47	46
427	800	10	1269	184	1207	175	12	56	42
427	800	1000	1489	216	1427	207	9	45	46
427	800	4000	1551	225	1476	214	8	40	48
427	800	10	1007	146	869	126	8	26	...
427	800	1000	1172	170	1020	148	5	11	...
427	800	4000	1227	178	1083	157	4	10	...
Condition H1100									
...	...	...	1124	163	1096	159	13	58	37
371	700	10	1138	165	1096	159	15	61	37
371	700	1000	1296	188	1269	184	12	55	42
371	700	5000	1365	198	1331	193	12	52	44
400	750	1000	1372	199	1351	196	11	51	43
400	750	2000	1420	206	1393	202	10	47	45
427	800	10	1158	168	1131	164	12	59	38
427	800	1000	1400	203	1372	199	9	47	44
427	800	4000	1455	211	1407	204	11	37	46

Note: Specimens were taken from 19.1-mm (0.75-in.) diam bar and tested at room temperature after long-time exposure at various temperatures
Source: Ref 28, p 55

17-4 PH: Tensile property data

TEMP C	TEMP F	UTS MPa	UTS ksi	YS MPa	YS ksi	EL %
Tested at 21 C (70 F)						
538	1000	1069	155	972	141	10.0
649	1200	1062	154	938	136	11.0
760	1400	1041	151	841	122	9.0
871	1600	1069	155	807	117	7.5
982	1800	1069	155	745	108	6.0
1093	2000	1082	157	779	113	4.5
Tested at exposure temperature						
538	1000	882	128	745	108	6.0
649	1200	528	76.6	493	71.6	13.0
760	1400	327	47.5	317	46.0	21.0
871	1600	262	38.0	258	37.4	26.0
982	1800	183	26.6	179	26.0	28.0
1093	2000	103	15.1	87	12.7	48.0

Note: Specimens were exposed for 30 s, condition H1150
Source: Ref 28, p 10

17-4 PH: Tensile property data

Condition	UTS MPa	UTS ksi	YS MPa	YS ksi	EL %	RA %
Tested at 24 C (75 F)						
H900	1400	203	1280	186	11	50
H925	1315	191	1255	182	18	54
H1025	1200	174	1160	168	15	55
H1075	1130	164	1095	159	16	54
H1150	965	140	900	129	17	61
Tested at 315 C (600 F)						
H900	1190	173	1035	150	10	31
H925	1135	165	1000	145	12	32
H1025	1005	146	931	135	12	42
H1075	950	138	910	132	9	38
H1150	855	124	827	120	12	54
Tested at 370 C (700 F)						
H900	1165	169	1005	146	8	25
H925	1110	161	980	142	12	33
H1025	979	142	903	131	10	38
H1075	924	134	876	127	9	33
H1150	827	120	786	114	12	52
Tested at 425 C (800 F)						
H900	1115	162	972	141	10	21
H925	1070	155	958	139	10	34
H1025	945	137	882	128	11	39
H1075	883	128	834	121	10	30
H1150	800	116	772	112	13	43
Tested at 480 C (900 F)						
H900	1025	149	910	132	10	30
H925	1000	145	883	128	10	35
H1025	869	126	814	118	12	39
H1075	786	114	758	110	11	38
H1150	752	109	717	104	13	51
Tested at 540 C (1000 F)						
H900	820	119	731	106	15	46
H925	800	116	710	103	16	45
H1025	731	106	696	101	15	43
H1075	680	99	650	94	16	55
H1150	660	96	640	93	15	55

Source: Ref 15, p 376

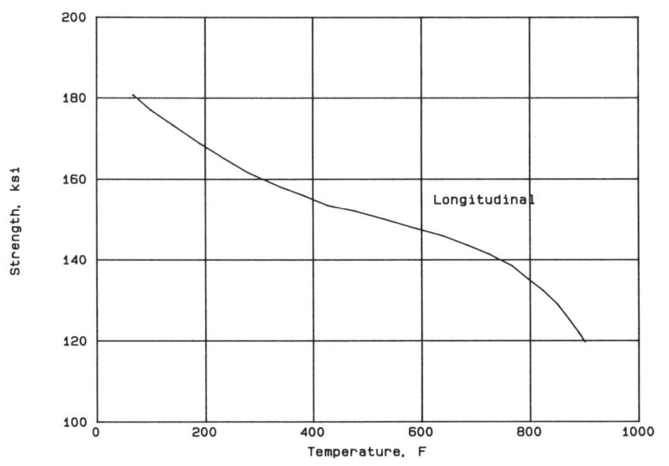

17-4 PH: Compressive yield strength of bar. Specimens were heat treated at 900 F. Source: Ref 22, p 16

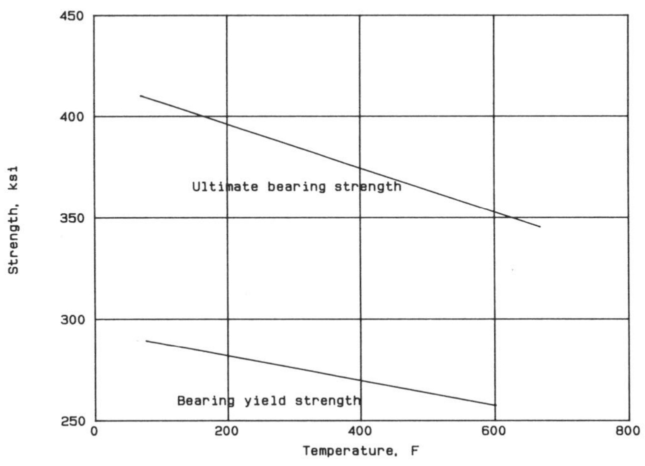

17-4 PH: Bearing ultimate and yield strength of bar. Specimens were heat treated at 900 F. e/D = 2.0. Source: Ref 22, p 17

17-4 PH: Effect of temperature on Izod impact values of bar

EXP TEMP C	EXP TEMP F	EXP h	IS (a) J	IS (a) ft-lb	IS (b) J	IS (b) ft-lb	IS (c) J	IS (c) ft-lb
24	75	...	19	14	61	45	76	56
371	700	1000	9	7	9	7	9	7
371	700	2000	5	4	4	3	5	4
427	800	1000	3	2	3	2	4	3
427	800	2000	3	2	3	2	3	2
482	900	1000	4	3	5	4	8	6
482	900	2000	8	6	11	8	15	11

(a) Condition H900. (b) Condition H1000. (c) Condition H1100
Source: Ref 29, Mar 1979

17-4 PH: Stress-rupture and creep property data

Condition	TEMP C	RPSTR 100 h MPa	RPSTR 100 h ksi	RPSTR 1000 h MPa	RPSTR 1000 h ksi	CRSTR 0.1%/1000 h MPa	CRSTR 0.1%/1000 h ksi	CRSTR 0.01%/1000 h MPa	CRSTR 0.01%/1000 h ksi
H900	315	...	...	...	...	931	135	862	125
H900	330	1117	162	1082	157	...	...	...	...
H1075	330	945	137	924	134	...	...	...	...
H900	370	1076	156	1034	150	724	105	689	100
H1075	370	869	126	848	123	...	...	...	...
H900	425	965	140	883	128	414	60	296	43
H1075	425	745	108	710	103	...	...	...	...
H900	480	...	95	...	60	159	23	...	...

Source: Ref 15, p 376

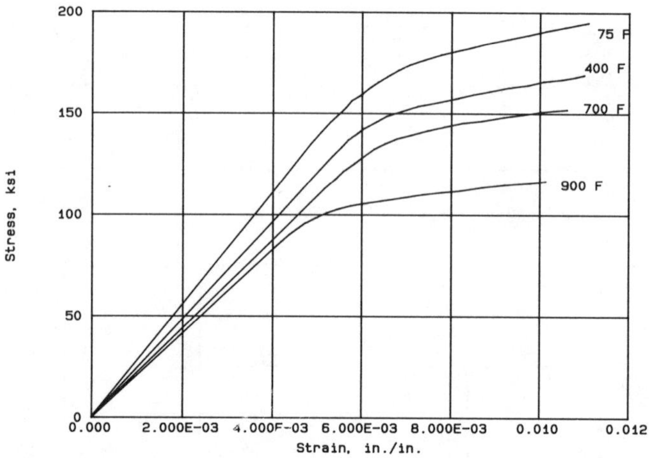

17-4 PH: Tensile stress-strain curves for bar. Specimens were heat treated at 900 F. Source: Ref 22, p 11

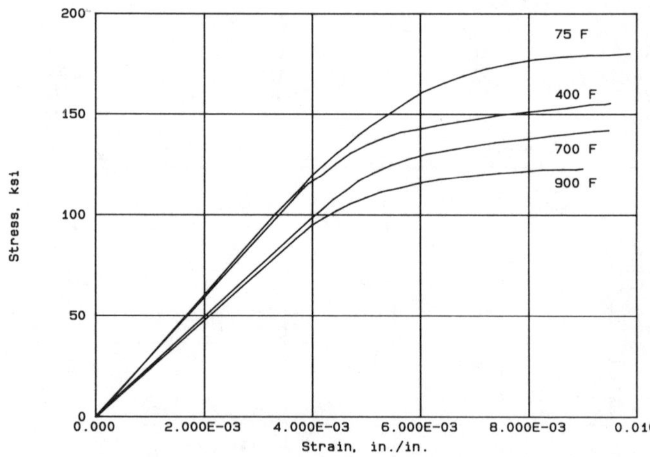

17-4 PH: Compressive stress-strain curves for bar. Specimens were heat treated at 900 F. Source: Ref 22, p 16

17-4 PH: Stress-to-rupture property data

Condition	RPT h	RPSTR MPa	RPSTR ksi	EL %	RA %
Tested at 330 C (625 F)					
H925	100	1125	163	3	13
H1075	100	945	137	3.5	4.5
H1150	100	850	123	5.5	17.5
H925	1000	1105	160	2.5	12
H1075	1000	925	134	3	14
H1150	1000	840	122	4.5	16.5
Tested at 370 C (700 F)					
H900	100	1075	156	3	7
H925	100	1060	154	3	13.5
H1075	100	869	126	4	15.5
H1150	100	786	114	6.5	19
H900	1000	1035	150	2	6
H925	1000	1040	151	2.5	12.5
H1075	1000	848	123	3.5	15
H1150	1000	765	111	5.5	18
Tested at 425 C (800 F)					
H900	100	965	140	4	8
H925	100	883	128	3.5	13.5
H1075	100	745	108	6	16
H1150	100	689	100	6.5	25.5
H900	1000	883	128	4	6
H925	1000	834	121	4.3	13
H1075	1000	710	103	5.5	15
H1150	1000	650	94	6	20
Tested at 480 C (900 F)					
H900	100	655	95	5	9
H1150	100	550	80	9	40
H900	1000	415	60	12	25
H1150	1000	490	71	9	36

Source: Ref 15, p 377

17-4 PH: Fatigue strength

FATST MPa	FATST ksi	FAT.n 10^4 cycles
586	85	65
552	80	770
517	75	2300
503	73	3200
496	72	10000
483	70	22000

Source: Ref 28, p 56

17-7 PH

UNS number: S17700
Specifications: AMS 5528
Nominal composition: 0.090% C max, 16.000 to 18.000% Cr, 1.000% Mn max, 6.500 to 7.750% Ni, 0.040% P max, 0.030% S max, 1.000% Si max, 0.750 to 1.500% Al
Product forms: Wrought, bar, wire, plate, sheet, strip, tube, billets, forgings
Description: 17-7 PH is a semi-austenitic precipitation hardening stainless steel combining strength and hardness with corrosion resistance. It is easily formed in the annealed condition, transformed to martensite by an intermediate heat treatment, and then hardened via a low-temperature heat treatment. In the heat treated condition, 17-7 PH features excellent mechanical properties at temperatures up to 480 C (900 F).
Typical applications: Springs, knives, pressure vessels, and aircraft, missile, and turbine engine parts.

Physical Properties

17-7 PH: Thermal conductivity

TEMP C	TEMP F	THCON W/m*K	THCON Btu/ft*h*F
Transformation hardened at 566 C (1050 F)			
150	300	17.0	9.8
260	500	18.5	10.7
450	840	21.1	12.2
480	900	21.1	12.2
Refrigeration hardened at 516 C (960 F) (estimated values)			
150	300	17.0	9.8
260	500	18.5	10.7
450	840	21.1	12.2
480	900	21.1	12.2
Cold work hardened at 482 C (900 F)			
150	300	16.4	9.5
260	500	18.3	10.6
450	840	21.6	12.5
480	900	21.8	12.6

Source: Ref 15, p 382

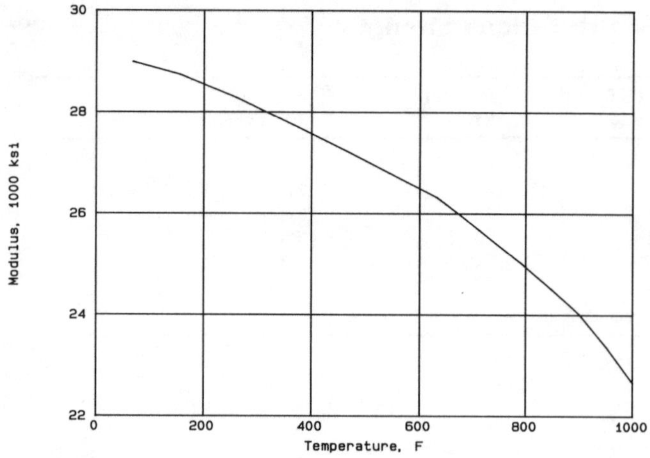

17-7 PH: Dynamic modulus of elasticity of sheet refrigeration hardened at 950 F. Source: Ref 22, p 23

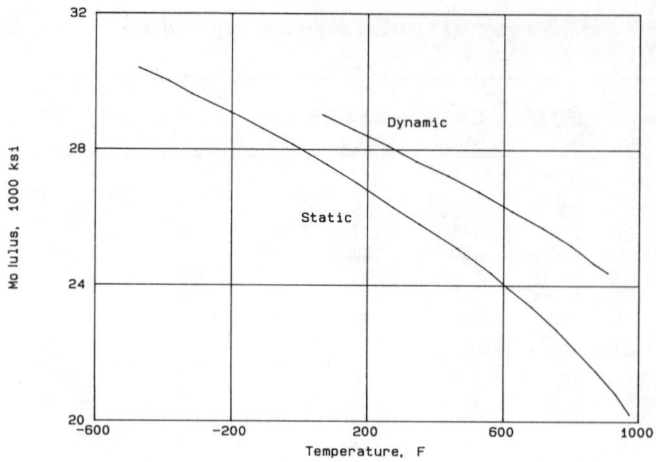

17-7 PH: Modulus of elasticity of sheet transformation hardened at 1050 F. Source: Ref 22, p 23

PH 17-7: Mean coefficients of linear thermal expansion

TEMP C	TEMP F	THEXP 10^{-6}/C	THEXP 10^{-6}/F
Annealed			
21-93	70-200	15.3	8.5
21-205	70-400	16.2	9.0
21-315	70-600	17.1	9.5
21-425	70-800	17.3	9.6
Transformation hardened at 566 C (1050 F)			
21-93	70-200	10.1	5.6
21-205	70-400	11.0	6.1
21-315	70-600	11.3	6.3
21-425	70-800	11.9	6.6
Refrigeration hardened at 510 C (950 F)			
21-93	70-200	10.3	5.7
21-205	70-400	11.9	6.6
21-315	70-600	12.2	6.8
21-425	70-800	12.4	6.9
Cold worked and hardened at 482 C (900 F)			
21-93	70-200	11	6.1
21-205	70-400	11.2	6.2
21-315	70-600	11.5	6.4
21-425	70-800	11.9	6.6

Source: Ref 15, p 382

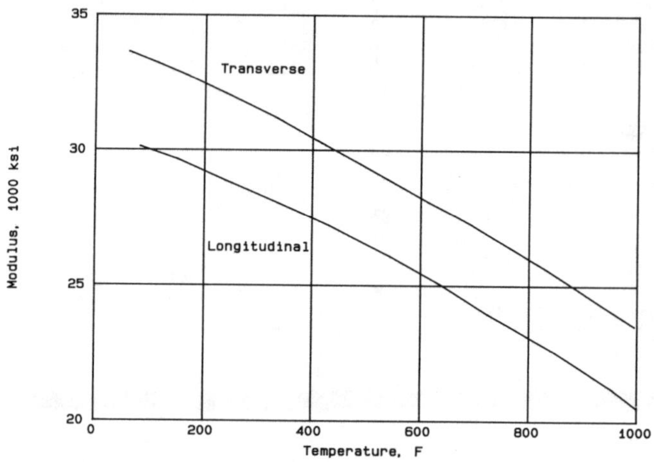

17-7 PH: Static modulus of elasticity of sheet cold worked and hardened at 900 F. Source: Ref 22, p 23

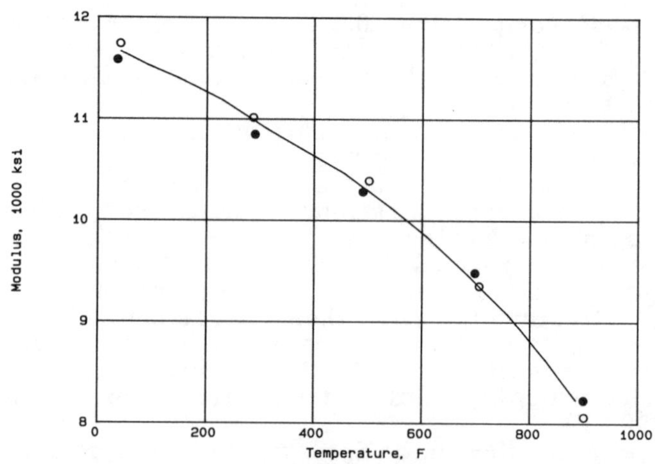

17-7 PH: Dynamic modulus of rigidity of sheet. Open circle: refrigeration hardened at 950 F. Solid circle: transformation hardened at 1050 F. Source: Ref 22, p 23

Mechanical Properties

17-7 PH: Effect of elevated-temperature exposure under load on room-temperature tensile properties of sheet

TEMP C	TEMP F	Load MPa	Load ksi	UTS MPa	UTS ksi	EL %
Heat A						
21	70	...	...	1389.3	201.5	8.0
316	600	550	80	1421.7	206.2	7.5
		827	120	1446.6	209.8	7.0
371	700	345	50	1523.8	221.0	6.0
		621	90	1574.8	228.4	4.5
427	800	345	50	1625.2	235.7	4.5
		550	80	1689.3	245.0	4.0
482	900	103	15	1434.2	208.0	5.5
		241	35	1448.0	210.0	5.0
Heat B						
21	70	..	...	1299.7	188.5	11.0
316	600	550	80	1310.1	190.0	10.0
		827	120	1329.4	192.8	7.5
371	700	345	50	1427.3	207.0	7.5
		621	90	1448.0	210.0	7.0
427	800	345	50	1516.9	220.0	3.5
		550	80	1565.2	227.0	3.5
482	900	103	15	1314.2	190.6	5.5
		241	35	1329.4	192.8	5.0

Note: Specimens were taken from 1.27-mm (0.050-in.) sheet and exposed for 1000 h at various temperatures. Condition: Transformation hardened at 566 C (1050 F)
Source: Ref 29, Mar 1970

17-7 PH: Effect of elevated-temperature exposure under load on room-temperature tensile properties of sheet

TEMP C	TEMP F	Load MPa	Load ksi	UTS MPa	UTS ksi	EL %
21	70	...	...	1603.8	232.6	7.5
316	600	276	40	1627.2	236.0	7.5
		1172	170	1862.3	270.1	2.3
371	700	276	40	1744.4	253.0	5.5
		793	115	1814.8	263.2	3.0
427	800	207	30	1769.9	256.7	2.0
		483	70	1813.4	263.0	1.0
482	900	69	10	1547.9	224.5	5.5
		207	30	1565.2	227.0	4.5

Note: Specimens were taken from 1.27-mm (0.050-in.) sheet after exposure of 1000 h at various temperatures. Condition: Refrigeration hardened at 510 C (950 F)
Source: Ref 29, Mar 1970

17-7 PH: Tensile property data for sheet and strip

Condition	UTS MPa	UTS ksi	YS MPa	YS ksi	EL %
Tested at 24 C (75 F)					
TH566	1330	193	1255	182	10.0
RH510	1585	230	1495	217	6.0
CH482	1805	262	1695	246	5.0
Tested at 150 C (300 F)					
TH566	1235	179	1170	170	8.0
RH510	1435	208	1325	192	4.5
CH482	1710	248	1614	234	4.0
Tested at 260 C (500 F)					
TH566	1150	167	1105	160	4.5
RH510	1345	195	1215	176	4.5
CH482	1570	228	1475	214	3.0
Tested at 315 C (600 F)					
TH566	1115	162	1070	155	4.0
RH510	1305	189	1165	169	5.0
CH482	1530	222	1405	204	3.0
Tested at 370 C (700 F)					
TH566	1075	156	1005	146	4.5
RH510	1250	181	1115	162	7.0
Tested at 425 C (800 F)					
TH566	986	143	896	130	6.2
RH510	1105	160	945	137	12.0
CH482	1425	207	1215	176	5.0
Tested at 482 C (900 F)					
TH566	855	124	689	100	10.0
RH510	915	133	786	114	15.0
CH482	1260	183	993	144	6.0

Note: TH = transformation hardened. RH = refrigeration hardened. CH = cold worked and hardened. All conditions at either 566 C (1050 F), 510 C (950 F), or 482 C (900 F)
Source: Ref 15, p 383

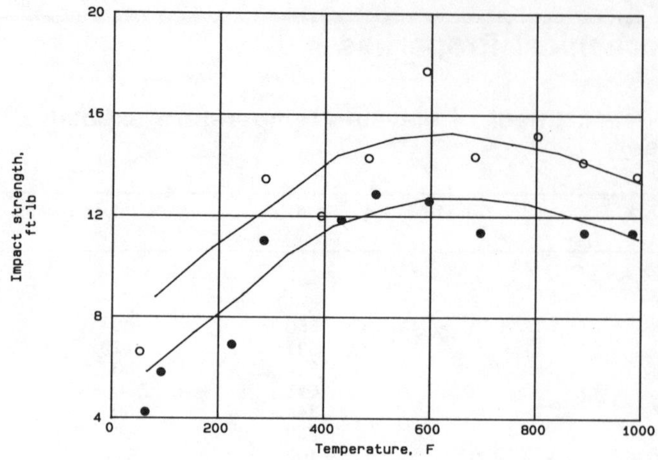

17-7 PH: Charpy V-notch strength of 0.75-in. bar. Open circle: transformation hardened at 1050 F. Solid circle: refrigeration hardened at 950 F. Source: Ref 22, p 18

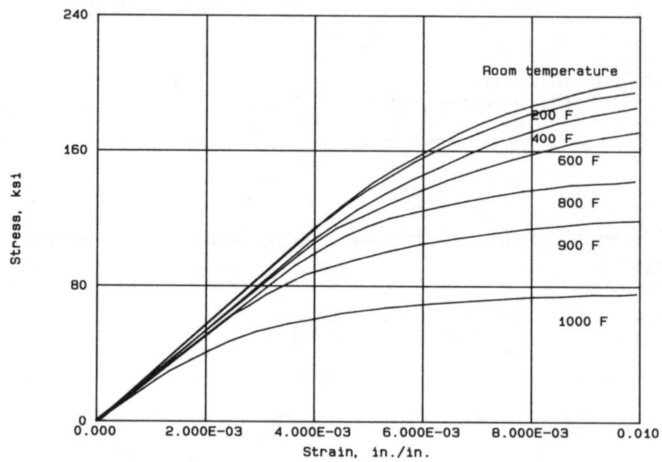

17-7 PH: Stress-strain curves for 0.050-in. sheet refrigeration hardened at 950 F. Source: Ref 22, p 15

17-7 PH: Stress-strain curves for 0.050-in. sheet transformation hardened at 1050 F. Source: Ref 22, p 11

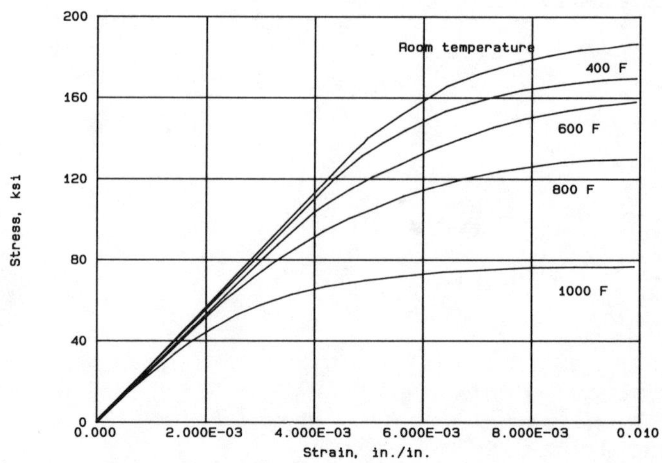

PH 17-7: Stress-rupture and creep-strength property data of sheet and strip

Condition	RPSTR 100h MPa	RPSTR 100h ksi	RPSTR 1000h MPa	RPSTR 1000h ksi	CRSTR 0.1%/1000 h MPa	CRSTR 0.1%/1000 h ksi	CRSTR 0.2%/1000 h MPa	CRSTR 0.2%/1000 h ksi
At 315 C (600 F)								
TH566	1170	170	1090	158	600	87	724	105
RH510	1295	188	1240	180	725	105	869	126
CH482	1515	220	1490	216	...	...	...	...
At 370 C (700 F)								
TH566	895	130	840	122	395	57	483	70
RH510	1165	169	1007	146	415	60	600	87
CH482	1340	194	1240	180	...	...	...	...
At 425 C (800 F)								
TH566	758	110	620	90	275	40	310	45
RH510	779	113	635	92	215	31	250	36
CH482	931	135	505	73	...	...	...	...
At 482 C (900 F)								
TH566	540	78	360	52	105	15	125	18
RH510	420	61	305	44	83	12	97	14
CH482	365	53	250	36	...	...	...	...

Note: TH = transformation hardened. RH = refrigeration hardened. CH = cold worked and hardened. All conditions at either 566 C (1050 F), 510 C (950 F), or 482 C (900 F).
Source; Ref 15, p 384

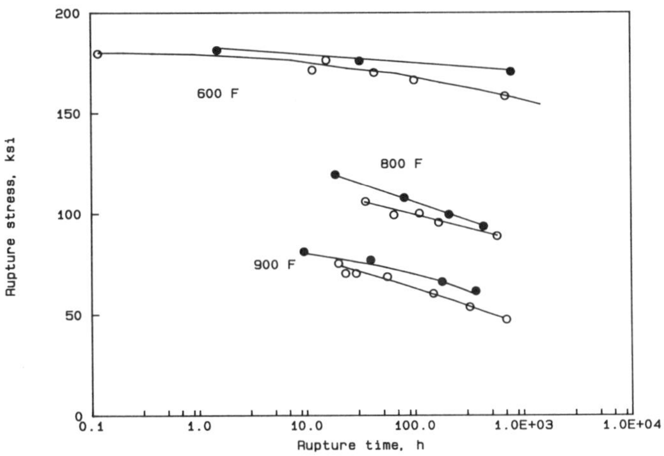

17-7 PH: Stress-rupture properties of specimens transformation hardened at 1050 F. Open circle: unnotched; solid circle: V-notched. Source: Ref 3, p 204

AM-350

Alternate designation: AISI 633
UNS number: S35000
Specifications: AMS 5546, AMS 5548, AMS 5554, AMS 5745, ASTM A579, ASTM A693, MIL S-8840, SAE J467
Nominal composition: 0.070 to 0.110% C, 16.000 to 17.000% Cr, 0.500 to 1.250% Mn, 2.500 to 3.250% Mo, 0.070 to 0.130% N, 4.000 to 5.000% Ni, 0.040% P max, 0.030% S max, 0.500% Si max
Product forms: Wrought, bar, wire, plate, sheet, strip, tube, billets, forgings

Description: AM-350 is a transformation and precipitation-hardening stainless steel. High strength levels can be achieved by heat treating or by cold rolling and tempering. In these hardened conditions, the corrosion resistance of AM-350 approaches that of the Cr-Ni stainless steels.

Typical applications: Pressure vessels, aircraft hydraulic tubing, airframe structurals such as bulkheads and stiffeners, jet engine compressor blades, and high-temperature springs.

Physical Properties

AM-350: Mean coefficients of linear thermal expansion

TEMP C	TEMP F	THEXP (a) 10^{-6}/C	THEXP (a) 10^{-6}/F	THEXP (b) 10^{-6}/C	THEXP (b) 10^{-6}/F
21-100	70-212	17.28	9.6	11.34	6.3
21-314	70-592	17.64	9.8	12.24	6.8
21-400	70-752	17.82	9.9	12.60	7.0
21-500	70-932	18.00	10.0	12.96	7.2
21-621	70-1150	18.72	10.4	12.96	7.2
21-732	70-1358	19.08	10.6	12.06	6.7
21-816	70-1500	19.44	10.8	12.60	7.0
21-927	70-1700	19.8	11.0	13.50	7.5

(a) Hardened. (b) Subzero cooled and tempered
Source: Ref 28, p 69

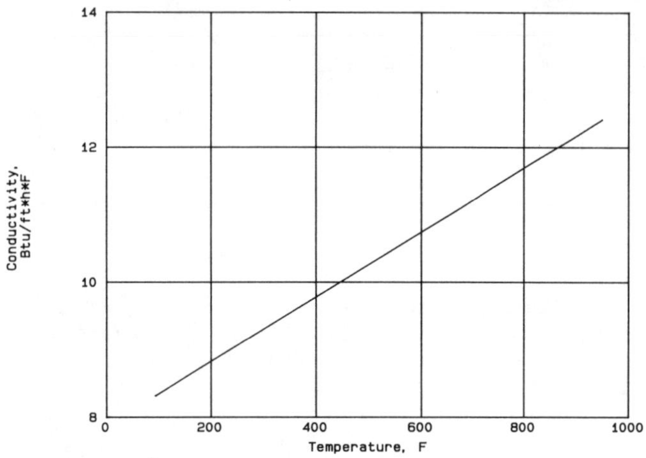

AM-350: Thermal conductivity of subzero cooled and tempered specimens. Source: Ref 22, p 6

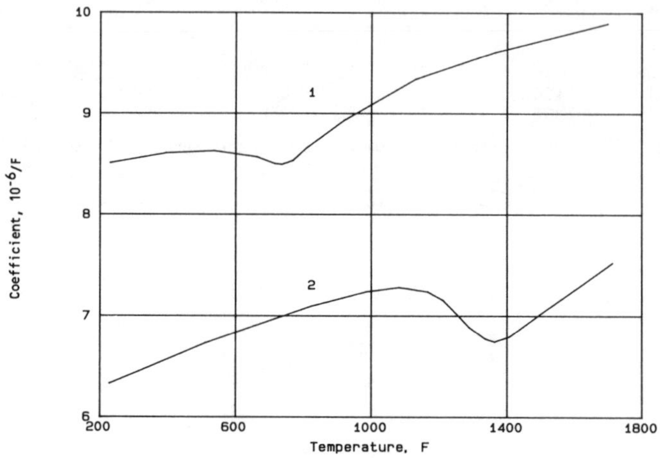

AM-350: Mean coefficient of linear thermal expansion. Curve 1: heat treated. Curve 2: subzero cooled and tempered at 850 F. Source: Ref 22, p 6

AM-350: Thermal conductivity

TEMP C	TEMP F	THCON W/m*K	THCON Btu/ft*h*F
38	100	14.5	8.4
93	200	15.4	8.9
150	300	16.2	9.4
205	400	17.0	9.8
260	500	17.8	10.3
315	600	18.7	10.8
370	700	19.6	11.3
425	800	20.2	11.7
480	900	21.1	12.2

Note: Specimens were subzero cooled and tempered at 455 C (850 F)
Source: Ref 15, p 387

AM-350: Electrical resistivity

TEMP C	TEMP F	ELRES n{o}*m
27	80	788
93	199	826
238	461	912
388	729	998
527	981	1075

Source: Ref 28, p 69

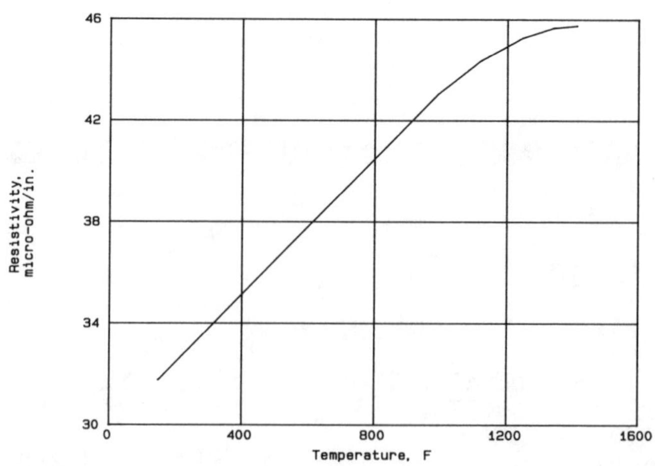

AM-350: Electrical resistivity of 1-in. bar. Specimens were subzero cooled and tempered at 850 F. Source: Ref 22, p 7

AM-350: Modulus of elasticity

TEMP C	TEMP F	E GPa	E 10^3 ksi
27	80	203	29.4
204	400	188	27.3
316	600	179	25.9
371	700	174	25.2
427	800	168	24.3

Note: Subzero cooled and tempered, 932 to 954 C (1710 to 1750 F) annealed for 1 h, -73 C (-100 F) cooled for 3 h, 454 C (850 F) tempered for 3 h. Modulus of elasticity was determined by ultrasonic method
Source: Ref 28, p 69

AM-350: Modulus of rigidity

TEMP C	TEMP F	G GPa	G 10^3 ksi
27	80	78	11.3
204	400	72	10.4
316	600	68	9.8
371	700	66	9.6
427	800	64	9.3

Note: Subzero cooled and tempered, 932 to 954 C (1710 to 1750 F) annealed for 1 h, -73 C (-100 F) cooled for 3 h, 454 C (850 F) tempered for 3 h. Modulus of elasticity determined by ultrasonic method
Source: Ref 28, p 69

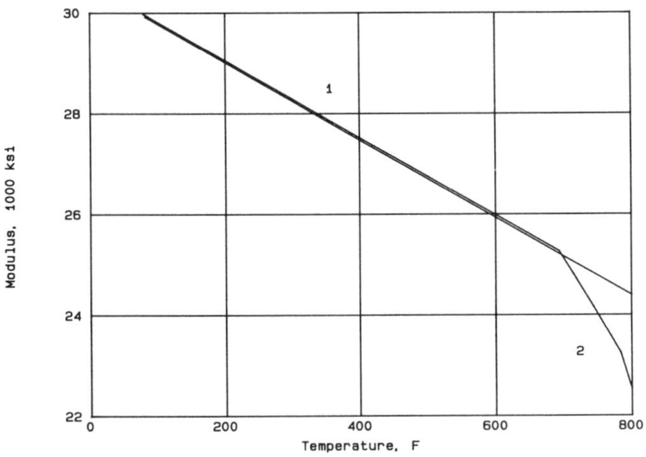

AM-350: Modulus of elasticity of specimens either subzero cooled and tempered at 850 F or double aged. Curve 1: static elasticity in tension and compression of sheet. Curve 2: static elasticity of double aged sheet in compression. Source: Ref 22, p 23

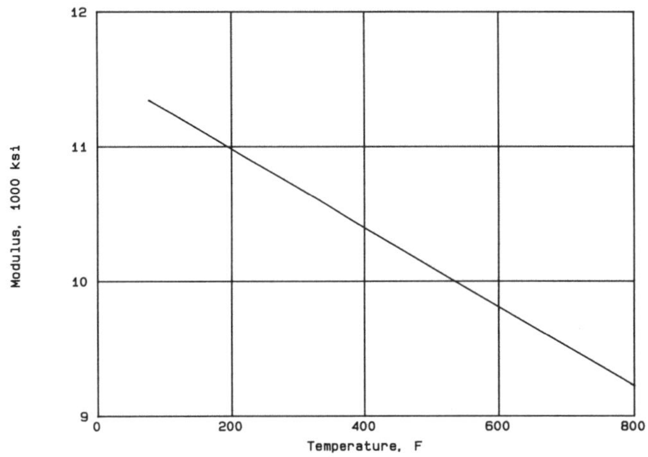

AM-350: Dynamic modulus of rigidity of 1-in. bar. Specimens were subzero cooled and tempered at 850 F. Source: Ref 22, p 23

Mechanical Properties

AM-350: Tensile and compressive property data for sheet

TEMP C	TEMP F	YS MPa	YS ksi	UTS MPa	UTS ksi	EL %	YSC MPa	YSC ksi
27	80	1207	175.0	1420	206.0	12	1317	191
93	200	1099	159.4	1372	199.0	13	...	...
149	300	1051	152.5	1351	195.9	10	...	...
204	400	1020	148.0	1351	195.9	9	1110	161
260	500	991	143.7	1353	196.3	10	...	...
316	600	960	139.2	1356	196.7	7	1089	158
371	700	915	132.7	1358	196.9	10	1020	148
427	800	861	124.9	1308	189.7	11	986	143
482	900	796	115.4	1191	172.7	10	...	...
538	1000	641	93.0	866	125.6	11	...	...

Note: Subzero cooled and tempered at 454 C (850 F). Compression specimens were 68.6 mm x 16.5 mm (2.70 in. x 0.65 in.), with a gage length of 38 mm (1.5 in.). Testing speed was 0.05 in./min
Source: Ref 28, p 70

AM-350: Mechanical property data for welded material

TEMP(a) C	TEMP(a) F	UTS MPa	UTS ksi	YS MPa	YS ksi	EL %
Base metal(b)						
27	80	1470	213	1205	175	12.5
165	300	1345	195	1070	155	10.5
315	600	1340	194	960	139	6.5
370	700	1340	194	910	132	10
425	800	1290	187	855	124	11.5
480	900	1160	168	800	116	10.5
Transverse to weld(c)						
27	80	1315	191	1140	165	4
165	300	1315	191	1050	152	6.5
315	600	1260	183	925	134	5
370	700	1250	181	825	120	4.5
425	800	1185	172	780	113	5
480	900	1105	160	710	103	6

Note: Specimens were taken from 2.0-mm (0.078-in.) thick sheet
(a) Specimens were ground flush and held 20 min at test temperature prior to testing. (b) Heat treated. (c) Heat treated after welding. All specimens broke in the weld.
Source: Ref 3, p 204

AM-350: Tensile property data

TEMP C	TEMP F	UTS MPa	UTS ksi	YS MPa	YS ksi	EL %
27	80	1400	203	1170	170	13.0
205	400	1295	188	970	141	8.5
315	600	1305	189	940	136	7.0
370	700	1310	190	880	128	8.0
425	800	1280	186	860	125	9.5
580	900	1145	166	765	111	9.0
640	1000	730	106	585	85	16.0

Note: Specimens were subzero cooled and tempered at 455 C (850 F)
Source: Ref 15, p 387

AM-350: Effect of long-time exposure under stress at elevated temperatures on tensile properties

STREX MPa	STREX ksi	UTS MPa	UTS ksi	YS MPa	YS ksi	EL %
At room temperature						
...	...	1385	201	1090	158	12
At 315 C (600 F) for 1000 h						
410	60	1365	198	1115	162	14
620	90	1395	202	1220	177	13
970	140	1405	204	1385	201	12
At 370 C (700 F) for 1000 h						
410	60	1405	204	1165	169	11
620	90	1420	206	1240	180	11
1030	150	1570	228	1565	227	5
At 425 C (800 F) for 1000 h						
410	60	1515	220	1310	190	7
620	90	1475	214	1325	192	7.5
900	130	1515	220	1460	212	4.5(a)

Note: Specimens were subzero cooled and tempered at 455 C (850 F)
(a) Broke outside gage marks
Source: Ref 15, p 387

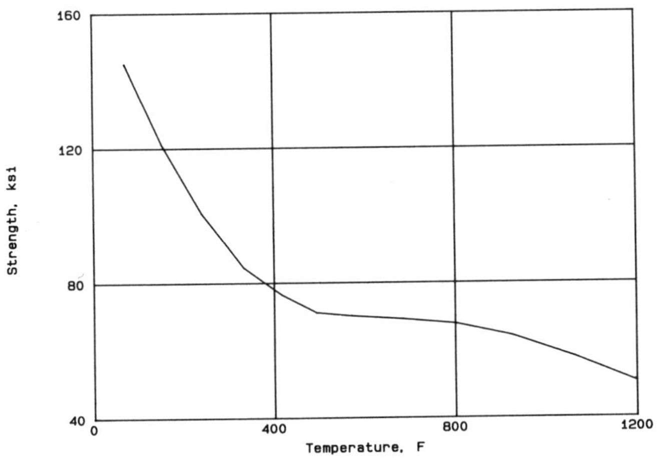

AM-350: Shear strength of 0.064-in. annealed sheet. Source: Ref 22, p 15

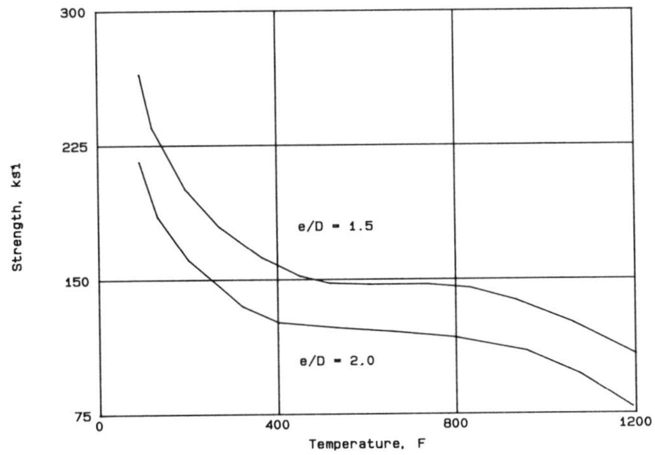

AM-350: Bearing ultimate strength of 0.064-in. annealed sheet. Source: Ref 22, p 16

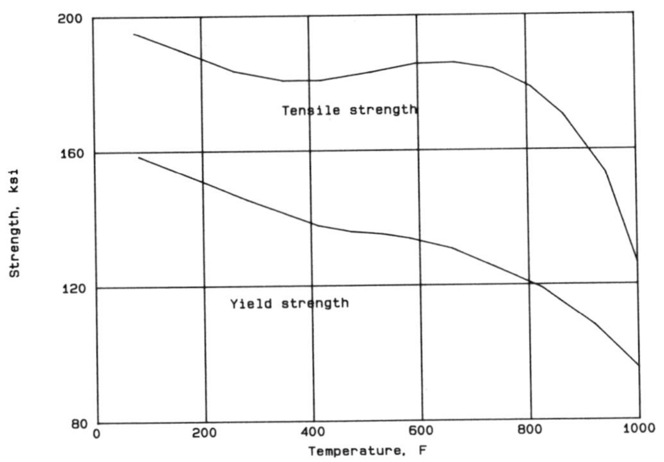

AM-350: Effect of test temperature on tensile and yield strength of double aged 0.064-in. sheet. Source: Ref 22, p 13

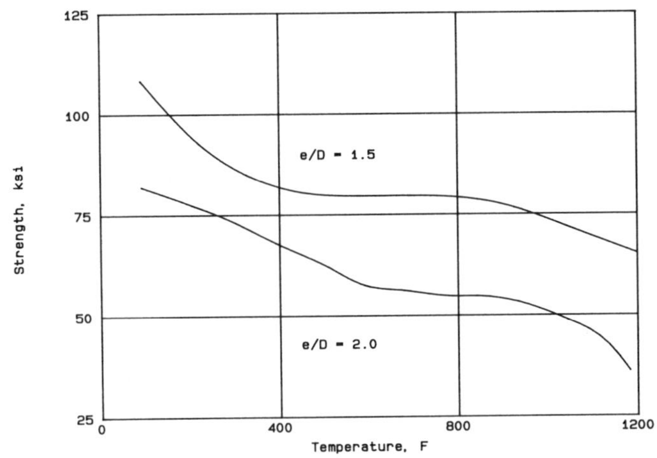

AM-350: Bearing yield strength of 0.064-in. annealed sheet. Source: Ref 22, p 16

AM-350: Effect of temperature on Charpy V-notch impact values of bar

TEMP C	TEMP F	IS J	IS ft-lb
Subzero cooled and tempered at 454 C (850 F)			
24	75	19	14
100	212	33	24
204	400	41	30
316	600	45	33
Subzero cooled and tempered at 538 C (1000 F)			
24	75	34	25

Source: Ref 28, p 70

AM 350: Shear strength of sheet

TEMP C	TEMP F	USS MPa	USS ksi	USS:UTS
21	70	958	139.0	0.67
204	400	831	120.5	0.6
316	600	803	116.5	0.58
427	800	817	118.5	0.62
482	900	734	106.5	0.61
538	1000	607	88.0	0.66

Note: Subzero cooled and tempered at 454 C (850 F). Specimens were 38.1-mm (1.5-in.) diameter disk at room temperature; specially designed specimen at elevated temperatures
Source: Ref 28, p 70

AM-350: Bearing strength

TEMP C	TEMP F	YSB(a) MPa	YSB(a) ksi	UTSB MPa	UTSB ksi	YSB:YS	UTSB:UTS
21	70	2120	307.5	3034	440.0	1.77	2.10
204	400	1910	277.0	2761	400.5	1.84	2.00
316	600	1827	265.0	2620	380.0	1.88	1.90
427	800	1855	269.0	2479	359.5	2.20	1.90
482	900	1803	261.5	2368	343.5	2.19	1.97
538	1000	1413	205.0	2037	295.5	2.08	2.23

Note: Specimens were taken from 6.35-mm (0.250-in.) diameter pin, e/D = 2.0, 12.7 mm (0.500 in.), longitudinal, 1.65-mm (0.065-in.) thick.
(a) Offset value equal to 2% of pin diameter.
Source: Ref 28, p 70

AM-350: Creep strength of sheet

TEMP C	TEMP F	CRSTR 0.1%/1000 h MPa	CRSTR 0.1%/1000 h ksi	CRSTR 0.2%/1000 h MPa	CRSTR 0.2%/1000 h ksi
316	600	765	111	896	130
371	700	696	101	869	126
427	800	379	55	586	85

Note: Subzero cooled and tempered at 454 C (850 F). Specimen dimensions within ASTM recommendations for flat rolled products
Source: Ref 28, p 70

AM-350: Stress-rupture property data for heat treated specimens

Ttpr C	Ttpr F	TEMP C	TEMP F	RPSTR 10 h MPa	RPSTR 10 h ksi	RPSTR 100 h MPa	RPSTR 100 h ksi	RPSTR 1000h MPa	RPSTR 1000h ksi
455	850	425	800	1280	186	1270	184	1255	182
540	1000	425	800	925	134	895	130	885	128
455	850	480	900	1015	147	835	121	655	95
540	1000	480	900	745	108	710	103	625	91

Source: Ref 3, p 203

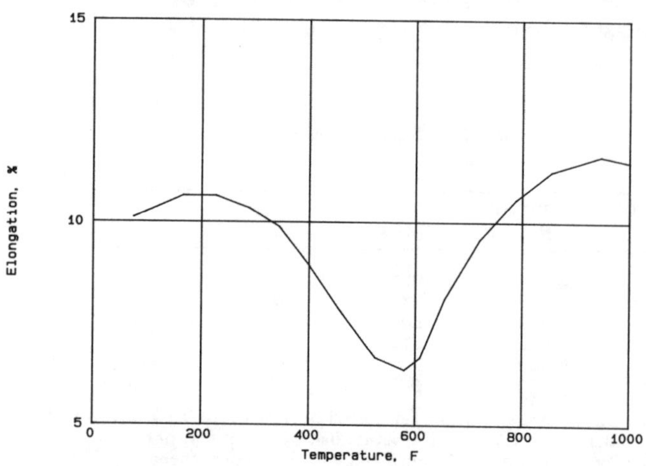

AM-350: Elongation of double aged 0.064-in. sheet. Source: Ref 22, p 13

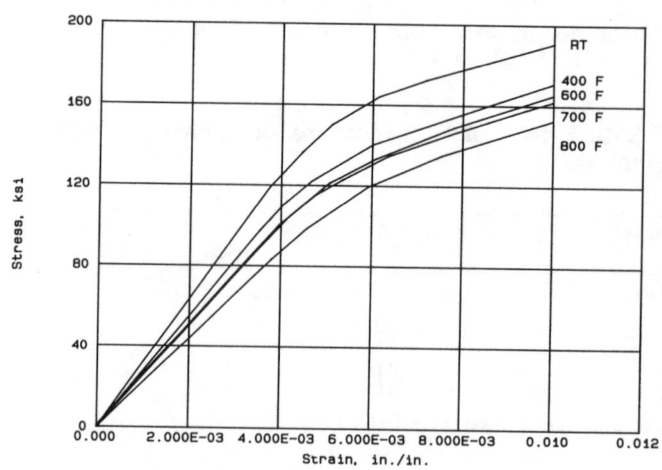

AM-350: Stress-strain curves in compression for double aged 0.064-in. sheet. Source: Ref 22, p 14

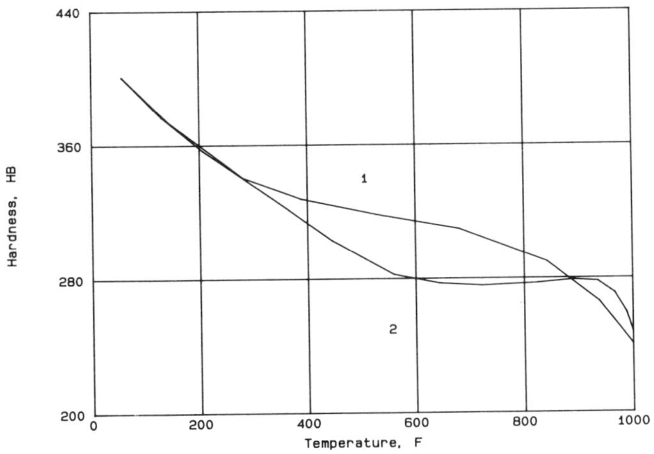

AM-350: Effect of test temperature on Brinell hardness. Curve 1: subzero cooled and tempered. Curve 2: double aged. Source: Ref 22, p 6

Custom 455

UNS number: S45500
Specifications: AMS 5578, AMS 5860, ASTM A313 (XM-16), ASTM A564, ASTM A693, ASTM A705, MIL S-83311
Nominal composition: 0.050% C max, 11.000 to 12.500% Cr, 1.500 to 2.500% Cu, 0.500% Mn max, 0.500% Mo max, 0.100 to 0.500% Nb, 7.500 to 9.500% Ni, 0.040% P max, 0.030% S max, 0.500% Si max, 0.800 to 1.400% Ti

Product forms: Wrought, bar, wire, plate, sheet, strip, tube, billet, forgings

Description: Custom 455 is a martensitic precipitation-hardenable stainless steel with higher strength than Custom 450.

Typical applications: High-temperature bolts, springs, and valves.

Physical Properties

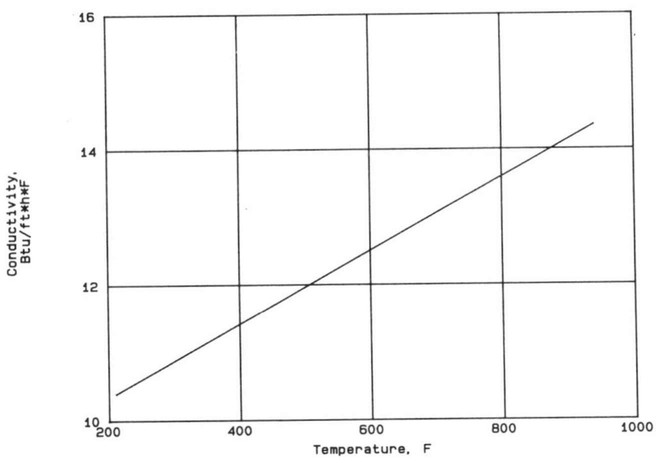

Custom 455: Thermal conductivity. Specimens were annealed and aged at 900 F. Source: Ref 22, p 4

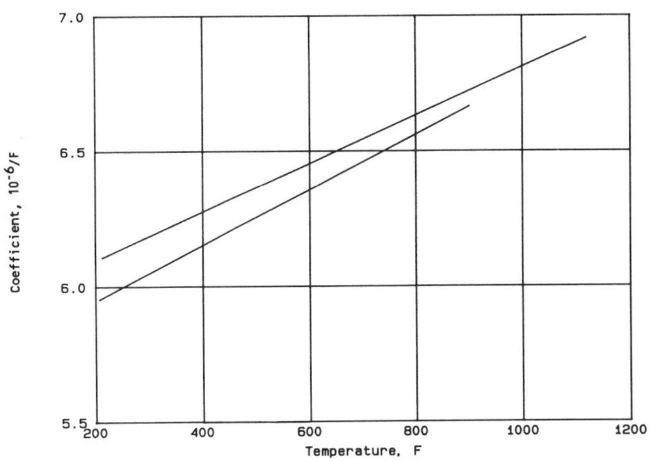

Custom 455: Mean coefficient of linear thermal expansion. Specimens were annealed and aged at 900 F. Source: Ref 22, p 4

Custom 455: Electrical resistivity. Specimens were annealed and aged at 900 F. Source: Ref 22, p 4

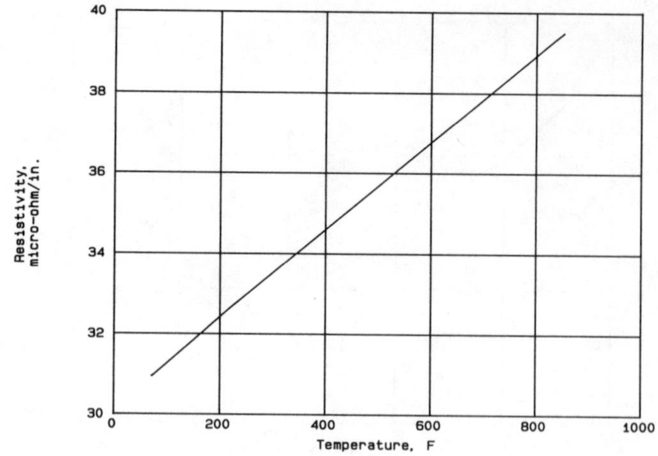

Mechanical Properties

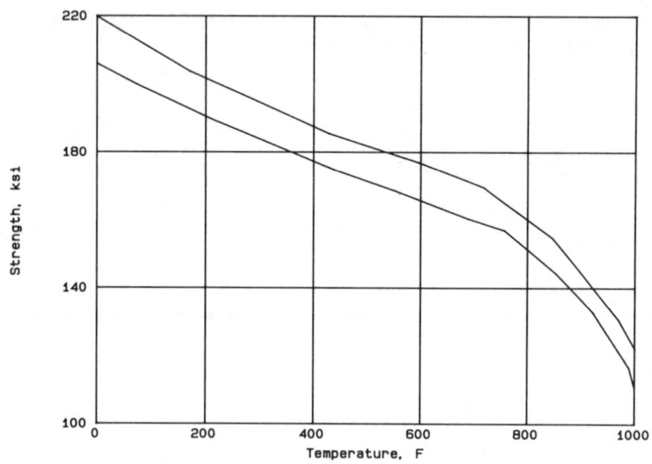

Custom 455: Scatterband for tensile strength of 0.75- to 1-in. diam bar and 4-in. square billet. Specimens were heat treated at 1500 F, water quenched, aged at 1000 F for 4 h, and air cooled. Data from three studies are shown. Source: Ref 22, p 11

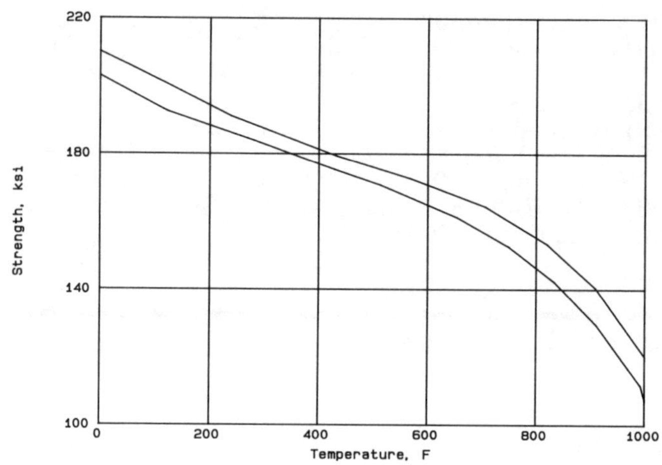

Custom 455: Scatterband for yield strength of 0.75- to 1-in. diam bar and 4-in. square billet. Specimens were heat treated at 1500 F, water quenched, aged at 1000 F for 4 h, and air cooled. Data from three studies are shown. Source: Ref 22, p 11

Custom 455: Tensile property data

TEMP C	TEMP F	UTS MPa	UTS ksi	YS MPa	YS ksi	EL(a) %	RA %	HARD(b) HRC
21	70	1690	245	1635	237	11	48	49
260	500	1475	214	1370	199	10	49	49
315	600	1405	204	1295	188	11	50	49
370	700	1345	195	1240	180	12	52	49
425	800	1240	180	1145	166	14	50	49

Note: Specimens were hardened at 815 C (1500 F) for 30 min; water quenched; aged at 480 C (900 F) for 4 h; and air cooled
(a) In 4D. (b) After testing
Source: Ref 15, p 427

Custom 455: Effect of elevated-temperature exposure on mechanical properties of bar

TEMP(a) C	TEMP(a) F	NTS(b) MPa	NTS(b) ksi	UTS MPa	UTS ksi	YS MPa	YS ksi	EL %	RA %	IS(c) J	IS(c) ft-lb
...	...	1793	260	1724	250	1655	240	11	46	14	10
371	700	1517	220	1875	272	1820	264	10	43	8	6
427	800	1586	230	1862	270	1806	262	10	43	9	7

Note: Specimens were annealed and tempered at 482 C (900 F) for 4 h, air cooled, and exposed
(a) Exposure temperature for 1000 h. (b) Notch tensile strength. (c) Charpy V-notch impact strength
Source: Ref 22, p 9

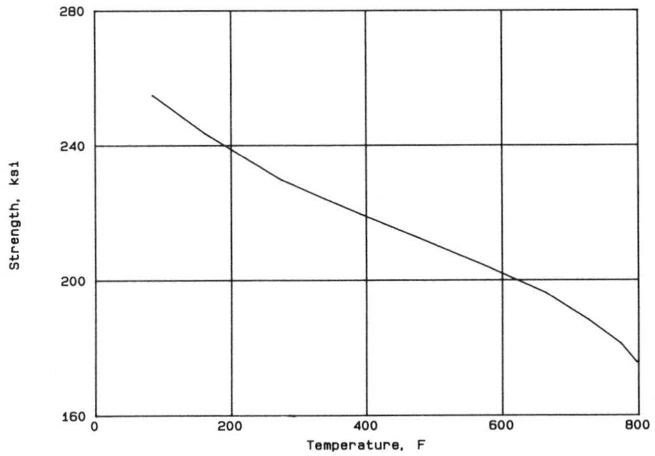

Custom 455: Compressive yield strength of 1-in. diam bar. Specimens were aged and tested longitudinally. Source: Ref 22, p 14

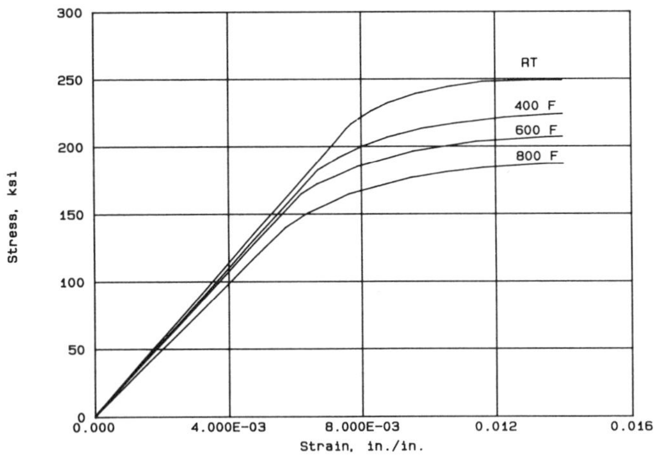

Custom 455: Stress-strain curves for 0.75-in. diam bar. Specimens were annealed and aged at 950 F for 4 h, and air cooled. Specimens were tested longitudinally. Source: Ref 22, p 10

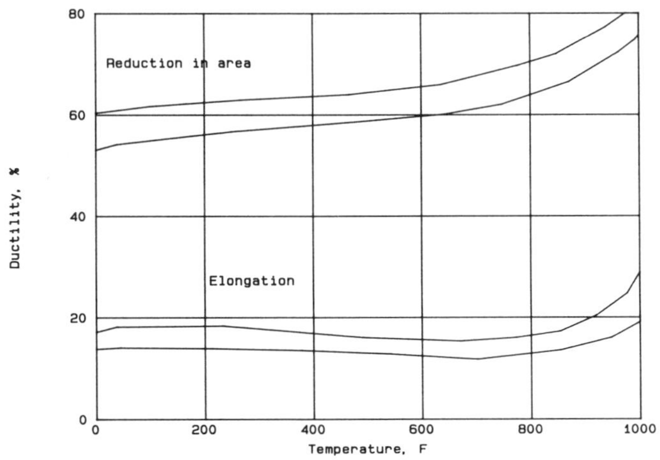

Custom 455: Scatterbands for tensile elongation and reduction in area of 0.75- to 1-in. diam bar and 4-in. square billet. Specimens were heat treated at 1500 F, water quenched, aged at 1000 F for 4 h, and air cooled. Data from three studies are shown. Source: Ref 22, p 11

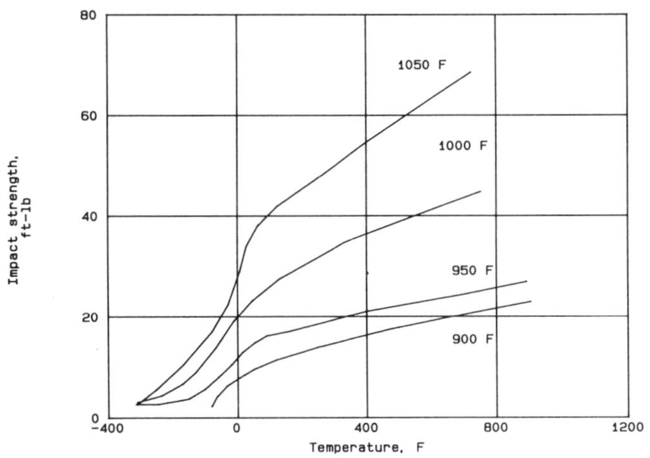

Custom 455: Effect of aging temperature on Charpy V-notch impact strength of 1-in. diam bar. Specimens were heat treated at 1500 F, water quenched, and aged 4 h at indicated temperature. Source: Ref 22, p 14

Custom 455: Stress-rupture property data

TEMP C	TEMP F	RPSTR 10 h MPa	RPSTR 10 h ksi	RPSTR 100 h MPa	RPSTR 100 h ksi	RPSTR 1000 h MPa	RPSTR 1000 h ksi
Heated to 816 C (1500 F) for 30 min, water quenched, heated to 482 C (900 F) for 4 h, and air cooled							
427	800	1034	150	827	120	641	93
Heated to 816 C (1500 F) for 30 min, water quenched, heated to 510 C (950 C) for 4 h, and air cooled							
427	800	979	142	807	117	627	91
482	900	752	109	565	82	372	54

Source: Carpenter Technology Data Sheet

Custom 455: Plastic deformation from 205 to 455 C (400 to 850 F)

TEMP C	TEMP F	RPSTR 100 h MPa	RPSTR 100 h ksi	RPSTR 1000 h MPa	RPSTR 1000 h ksi
204	400	1427	207	1400	203
316	600	1124	163	1069	155
454	850	303	44	207	30

Note: Specimens were taken from 19- to 25-mm (0.75- to 1-in.) diam bar. Specimens were aged at 510 C (950 F) for 4 h and air cooled
Source: Ref 22, p 16

PH 13-8 Mo

UNS number: S13800

Specifications: AMS 5629, AMS 5840, ASME SA705 (XM-13), ASTM A564 (XM-13), ASTM A693, ASTM A705

Nominal composition: 0.050% C max, 12.250 to 13.250% Cr, 0.100% Mn max, 2.000 to 2.500% Mo, 0.010% N max, 7.500 to 8.500% Ni, 0.010% P max, 0.008% S max, 0.100% Si max, 0.900 to 1.350% Al

Product forms: Wrought, bar, wire, plate, sheet, strip, tube, billets, forgings

Description: PH 13-8 Mo is a precipitation hardenable Ni-Cr-Al-Mo stainless steel that combines high strength and hardness with good corrosion resistance. It can be hardened by a single low-temperature heat treatment.

Typical applications: Aircraft parts, and forgings.

Physical Properties

PH 13-8 Mo: Thermal conductivity

TEMP C	TEMP F	THCON W/m*K	THCON Btu/ft*h*F
100	212	14.0	8.1
200	390	15.7	9.1
315	600	17.8	10.3
425	800	20.4	11.8
540	1000	22.3	12.9
595	1110	22.5	13.0

Note: As supplied
Source: Ref 15, p 402

PH 13-8 Mo: Mean coefficients of linear thermal expansion

TEMP C	TEMP F	THEXP 10^{-6}/C	THEXP 10^{-6}/F
21-95	70-200	10.6	5.9
21-205	70-400	10.8	6.0
21-315	70-600	11.2	6.2
21-425	70-800	11.3	6.3
21-540	70-1000	11.9	6.6

Source: Ref 15, p 402

PH 13-8 Mo: Modulus of elasticity

TEMP C	TEMP F	E GPa	E 10^3 ksi
21	70	195.8	28.4
149	300	175.1	25.4
260	500	165.8	23.9
371	700	159.3	23.1
427	800	175.1	25.4
538	1000	140.7	20.4

Source: Carpenter Technology Data Sheet

PH 13-8 Mo: Electrical resistivity

TEMP C	TEMP F	ELRES n{o}*m
25	77	1001
100	212	1019
200	390	1046
315	600	1061
425	800	1081
540	1000	1091
595	1100	1095

Source: Ref 15, p 402

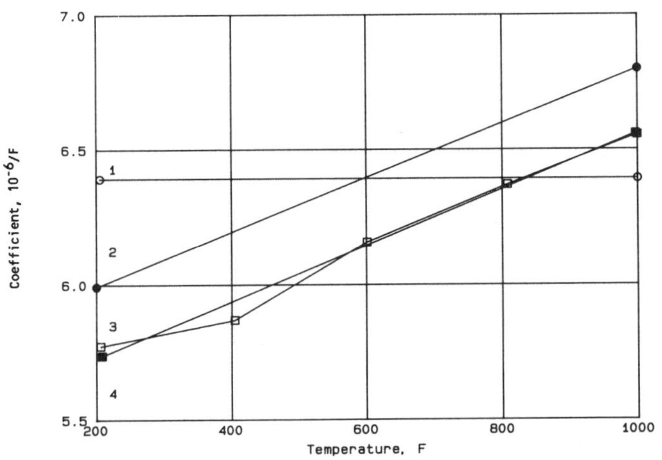

PH 13-8 Mo: Mean coefficient of linear thermal expansion. Curve 1: mill annealed. Curve 2: heat treated at 1000 F. Curve 3: heat treated at 1050 F. Curve 4: heat treated at 950 and 1000 F. Source: Ref 22, p 5

Mechanical Properties

PH 13-8 Mo: Tensile property data

TEMP C	TEMP F	UTS MPa	UTS ksi	YS MPa	YS ksi	EL %	RA %
27	80	1151	167	1000	145	22	69
93	200	1069	155	1007	146	23	69
149	300	1062	154	1007	146	22	70
204	400	1020	148	972	141	20	70
260	500	993	144	951	138	19	69
316	600	979	142	931	135	18	69
371	700	951	138	889	129	18	69

Note: Data represents average of triplicate tests from 3 heats. Specimens were H 1100, 25-mm (1-in.) diam bar
Source: Carpenter Technology Data Sheet

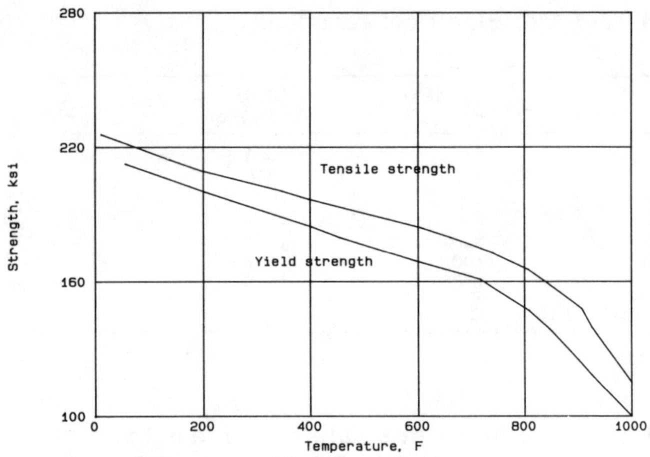

PH 13-8 Mo: Tensile and yield strength of 0.75-in. bar. Specimens were heat treated at 1000 F. Source: Ref 22, p 6

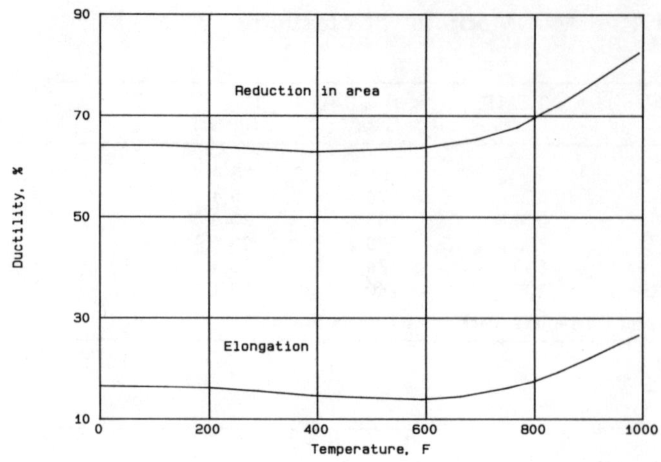

PH 13-8 Mo: Tensile elongation and reduction in area of 0.75-in. bar. Specimens were heat treated at 1000 F. Source: Ref 22, p 6

PH 14-8 Mo

UNS number: S14800

Specifications: AMS 5601, AMS 5603

Nominal composition: 0.050% C max, 13.750 to 15.000% Cr, 1.000% Mn max, 2.000 to 3.000% Mo, 7.750 to 8.750% Ni, 0.015% P max, 0.010% S max, 1.000% Si max, 0.750 to 1.500% Al

Product forms: Wrought, bar, wire, plate, sheet, strip, tube, billets, forgings

Description: PH 14-8 Mo is a precipitation hardenable, semi-austenitic Cr-Ni-Mo-Al stainless steel.

Typical applications: Springs, stampings, watch and instrument parts, and fasteners.

Physical Properties

PH 14-8 Mo: Mean coefficients of linear thermal expansion

TEMP C	TEMP F	THEXP 10^{-6}/C	THEXP 10^{-6}/F
21-93	70-200	9.5	5.3
21-205	70-400	10.6	5.9
21-315	70-600	11.2	6.2
21-425	70-800	11.3	6.3
21-540	70-1000	11.5	6.4

Source: Ref 15, p 404

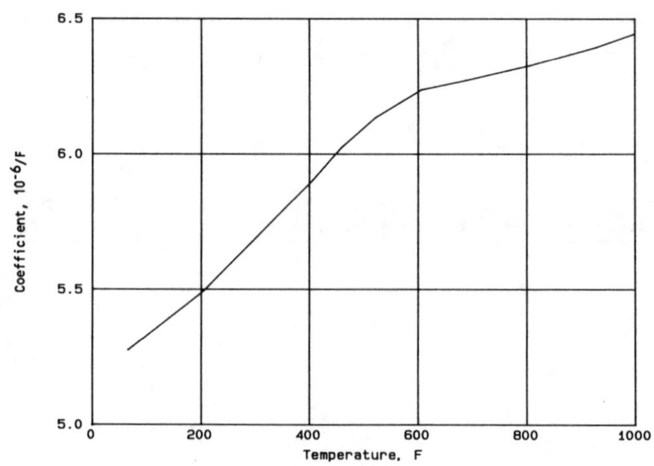

PH 14-8 Mo: Mean coefficient of linear thermal expansion of sheet, stress relieved and hardened at 950 F. Source: Ref 22, p 8

Mechanical Properties

PH 14-8 Mo: Tensile property data for specimens stress relieved and hardened at 566 C (1050 F)

TEMP C	TEMP F	UTS MPa	UTS ksi	YS MPa	YS ksi	EL %	NTS MPa	NTS ksi	NTS:YS	NTS:TS
Not exposed										
24	75	1502	217.9	1447	209.8	6.0	1592	230.9	1.10	1.06
345	650	1252	181.6	1046	151.7	5.0	1065	154.4	1.02	0.85
At 345 C (650 F) for 1000 h; 275-MPa (40-ksi) load										
24	75	1617	234.5	1553	225.3	5.0	1502	217.9	0.97	0.93
345	650	1308	189.7	1229	178.2	4.0	1149	166.7	0.94	0.88

Note: Specimens were vacuum induction melted; tested in transverse direction; 25-mm (1.0-in.) wide, 6.4-mm (0.025-in.) thick, with an edge notch-root radius of 0.018 mm (0.0007 in.)
Source: Ref 15, p 404

PH 14-8 Mo: Tensile property data for specimens stress relieved and hardened at 510 C (950 F)

TEMP C	TEMP C	UTS MPa	UTS ksi	YS MPa	YS ksi	EL %
Longitudinal						
27	80	1615	234	1505	218	5
260	500	1435	208	1255	182	4
290	550	1415	205	1240	180	4
315	600	1370	199	1215	176	4
345	650	1370	199	1205	175	5
425	800	1240	180	1055	153	8
540	1000	910	132	730	106	18
Transverse						
27	80	1670	242	1570	228	5
260	500	1460	212	1315	191	3
290	550	1455	211	1330	193	3
315	600	1425	207	1280	186	3
345	650	1405	204	1255	182	3
425	800	1290	187	1095	159	6
540	1000	930	135	760	110	17

Source: Ref 15, p 404

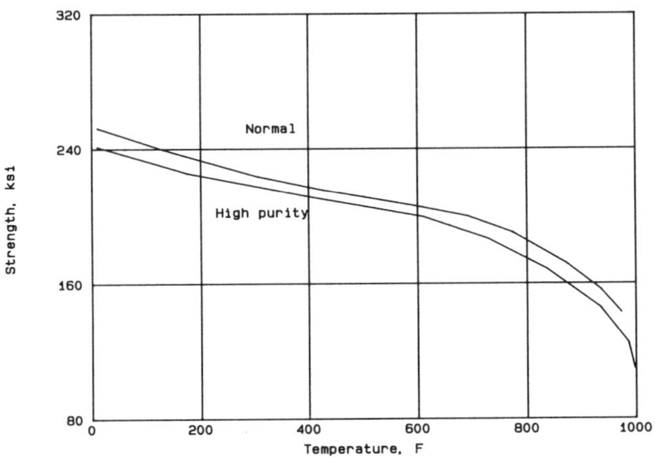

PH 14-8 Mo: Tensile strength of normal and high-purity air melted sheet, stress relieved and hardened at 950 F. Source: Ref 22, p 13

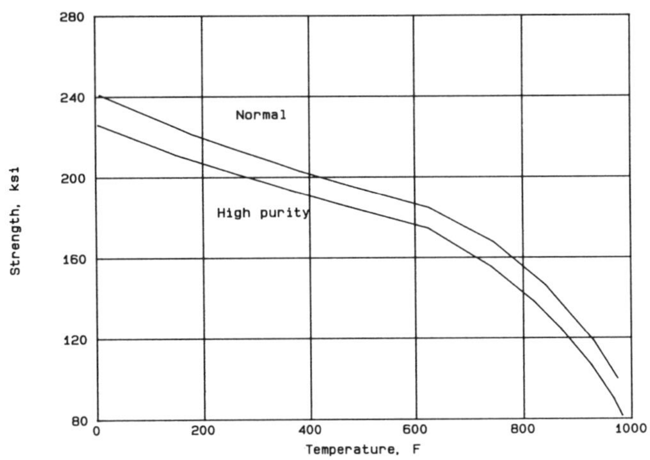

PH 14-8 Mo: Yield strength of normal and high-purity air melted sheet, stress relieved and hardened at 950 F. Source: Ref 22, p 13

PH 14-8 Mo: Effect of test temperature on tensile and notch tensile properties before and after elevated-temperature stress exposure

EXP TEMP C	EXP TEMP F	EXP h	Direction	YS MPa	YS ksi	UTS MPa	UTS ksi	EL %	NTS MPa	NTS ksi	NTS:YS
...	...	...	L	1262	183	1551	225	12	1455	211	1.2
...	...	...	T	1269	184	1565	227	10	1303	189	1.0
288	550	2400	L	1282	186	1565	227	13	1407	204	1.1
288	550	2400	T	1282	186	1620	235	8	...	...	...
288	550	7000	L	1317	191	1593	231	12	1427	207	1.1
288	550	7000	T	1386	201	1607	233	7	1276	185	0.92
288	550	10000	L	1331	193	1579	229	11	1469	213	1.1
288	550	10000	T	1303	189	1558	226	10	1179	171	0.9

Note: Specimens were taken from 1.27-mm (0.050-in.) sheet and were air melted.
Source: Ref 22, p 4

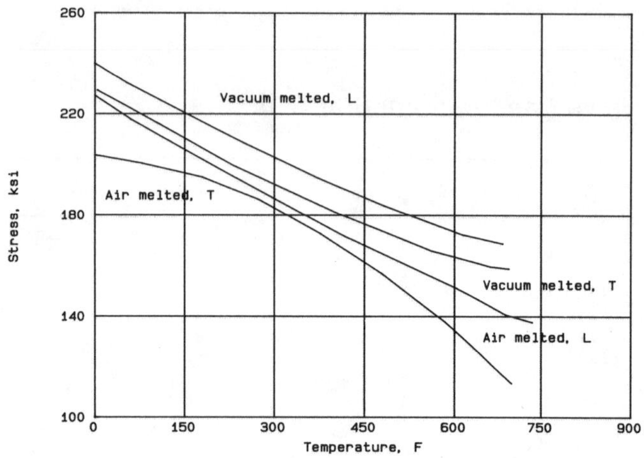

PH 14-8 Mo: Longitudinal and transverse (L and T) sharp notch yield strength of air and vacuum-melted sheet, stress relieved and hardened at 950 F. Source: Ref 22, p 15

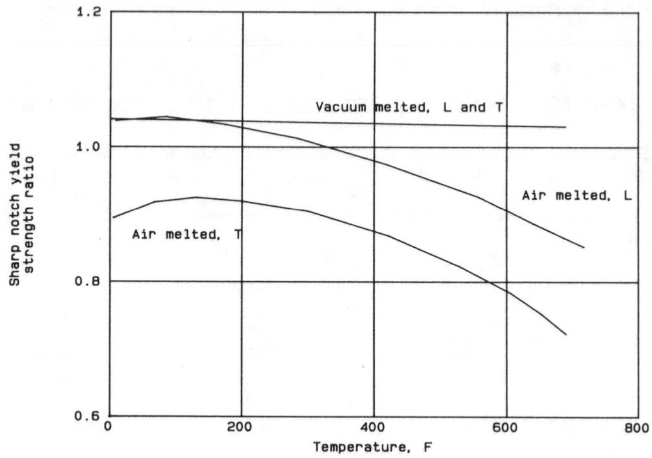

PH 14-8 Mo: Longitudinal and transverse (L and T) notch yield/smooth yield strength ratio of air and vacuum-melted sheet, stress relieved and hardened at 950 F. Source: Ref 22, p 15

PH 14-8 Mo: Tensile yield strength and crack strength

EXP TEMP C	EXP TEMP F	EXP h	Direction	YS MPa	YS ksi	KIC MPa{rt}m	KIC ksi{rt}in.	KIC/YS
Cold worked and hardened at 482 C (900 F)								
...	...	...	L	1841	267	256	233	0.86
...	...	...	T	1869	271	156	142	0.52
260	500	1000	L	1862	270	267	243	0.9
260	500	1000	T	1882	273	148	135	0.5
343	650	1000	L	2000	290	249	227	0.78
343	650	1000	T	2027	294	131	119	0.41
260	500	10000	L	1917	278	211	192	0.69
260	500	10000	T	1972	286	122	111	0.39
Cold worked and hardened at 566 C (1050 F)								
...	...	...	L	1717	249	245	223	0.89
...	...	...	T	1751	254	175	159	0.62
260	500	1000	L	1731	251	252	229	0.91
260	500	1000	T	1779	258	186	169	0.65
343	650	1000	L	1958	284	259	236	0.83
343	650	1000	T	1862	270	168	153	0.57
260	500	10000	L	1703	247	209	190	0.77
260	500	10000	T	1820	264	131	119	0.45

Note: Specimens were taken from 1.27-mm (0.050-in.) sheet and were air melted
Source: Ref 22, p 4

PH 14-8 Mo: Tensile yield strength and fracture toughness

EXP TEMP C	EXP TEMP F	EXP h	Direction	YS MPa	YS ksi	KIC MPa{rt}m	KIC ksi{rt}in.	KIC:YS
Stress relieved and hardened at 510 C (950 F)								
...	...	...	L	1482	215	222	202	0.94
...	...	...	T	1489	216	179	163	0.75
260	500	1000	L	1538	223	225	205	0.92
260	500	1000	T	1524	221	186	169	0.76
343	650	1000	L	1655	240	217	197	0.82
343	650	1000	T	1600	232	162	147	0.63
Stress relieved and hardened at 566 C (1050 F)								
...	...	...	L	1427	207	222	202	0.98
...	...	...	T	1407	204	192	175	0.86
260	500	1000	L	1448	210	226	206	0.98
260	500	1000	T	1455	211	200	182	0.86
343	650	1000	L	1524	221	230	209	0.94
343	650	1000	T	1517	220	192	175	0.79
Cold worked and hardened and tempered 566 C (1050 F)								
...	...	...	L	1420	206	208	189	0.92
...	...	...	T	1420	206	184	167	0.81
260	500	1000	L	1455	211	219	199	0.94
260	500	1000	T	1427	207	188	171	0.82
343	650	1000	L	1586	230	220	200	0.87
343	650	1000	T	1551	225	170	155	0.69

Note: Specimens were taken from 1.27-mm (0.05-in.) sheet and were air melted. Simulated braze cycle: heated to 913 C (1674 F), held 15 min, reheated to 538 C (1000 F) in 30 min, air cooled, held at -73 C (-100 F) for 8 h, reheated to 566 C (1050 F), held 8 h, and air cooled
Source: Ref 22, p 3

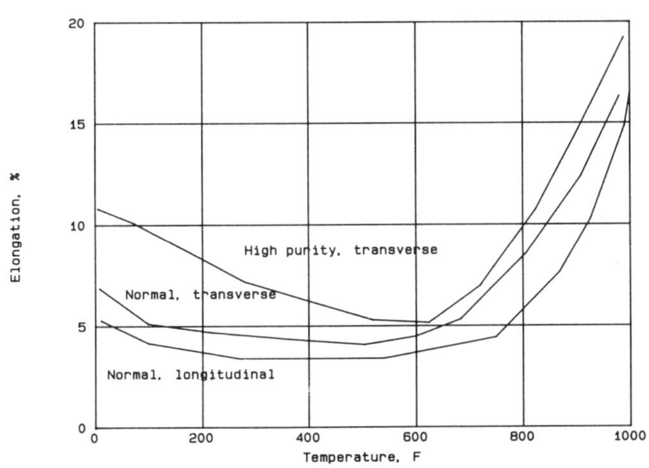

PH 14-8 Mo: Tensile elongation of normal and high-purity air melted sheet, stress relieved and hardened at 950 F. Source: Ref 22, p 13

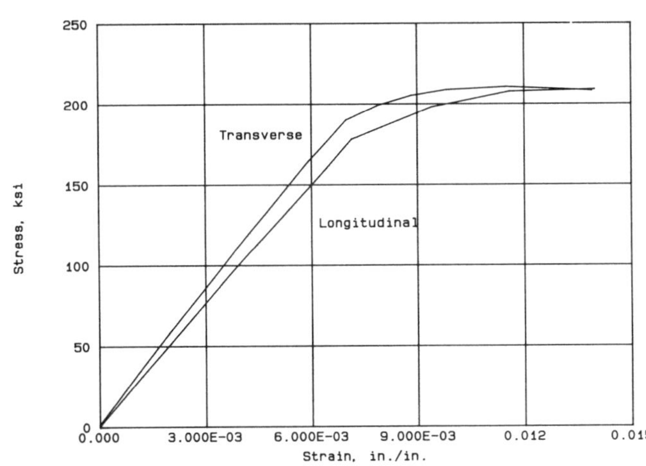

PH 14-8 Mo: Stress-strain curves for 0.05-in. sheet, stress relieved and hardened at 1050 F. Source: Ref 22, p 8

PH 15-7 Mo

Alternate designation: AISI 632
UNS number: S15700
Specifications: AMS 5520, ASTM A564(632), ASTM A579, ASTM A693, ASTM A705, MIL S-8955, SAE J467
Nominal composition: 0.090% C max, 14.000 to 16.000% Cr, 1.000% Mn max, 2.000 to 3.000% Mo, 6.500 to 7.750% Ni, 0.040% P max, 0.030% S max, 1.000% Si max, 0.750 to 1.500% Al

Product forms: Wrought, bar, wire, plate, sheet, strip, tube, billets, forgings, structural shapes

Description: 15-7 Mo is a precipitation-hardenable, semi-austenitic Cr-Ni-Mo-Al stainless steel combining high strength and corrosion resistance.

Typical applications: Springs, fasteners, diaphragms, bellows, and watch and instrument parts such as gears, cams, and levers.

Physical Properties

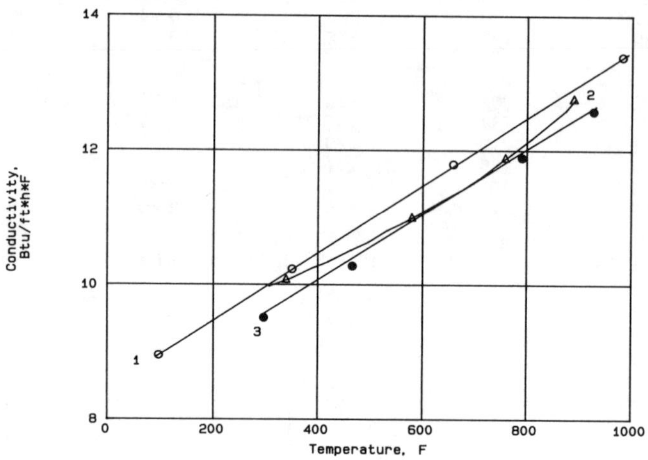

PH 15-7 Mo: Thermal conductivity. Curve 1: transformation hardened at 1050 F. Curve 2: cold worked and hardened at 900 F. Curve 3: refrigeration hardened at 950 F. Source: Ref 22, p 7

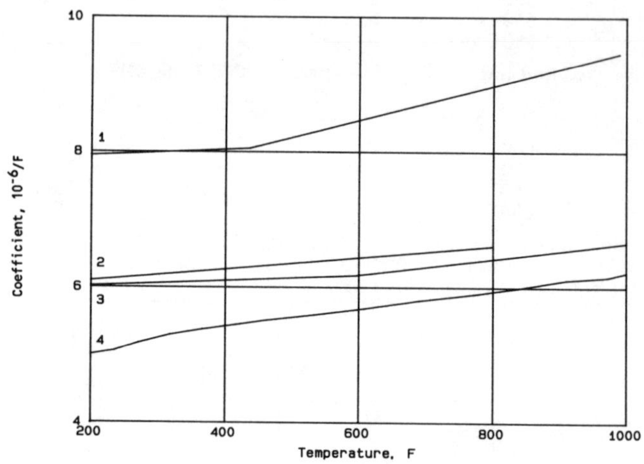

PH 15-7 Mo: Mean coefficient of linear thermal expansion. Curve 1: annealed. Curve 2: cold worked and hardened at 900 F. Curve 3: transformation hardened at 1050 F. Curve 4: refrigeration hardened at 950 F. Source: Ref 22, p 7

PH 15-7 Mo: Thermal conductivity

TEMP C	TEMP F	THCON W/m*K	THCON Btu/ft*h*F
21	70	15.1	8.7
95	200	16.1	9.3
205	400	17.6	10.2
315	600	19.2	11.1
425	800	20.8	12.0
480	900	21.6	12.5

Source: Ref 15, p 385

PH 15-7 Mo: Mean coefficients of linear thermal expansion

TEMP C	TEMP F	THEXP 10^{-6}/C	THEXP 10^{-6}/F
21-93	70-200	9.0	5.0
21-205	70-400	9.7	5.4
21-316	70-600	10.1	5.6
21-425	70-800	10.6	5.9
21-480	70-900	10.8	6.0
21-540	70-1000	11.1	6.1

Source: Ref 15, p 385

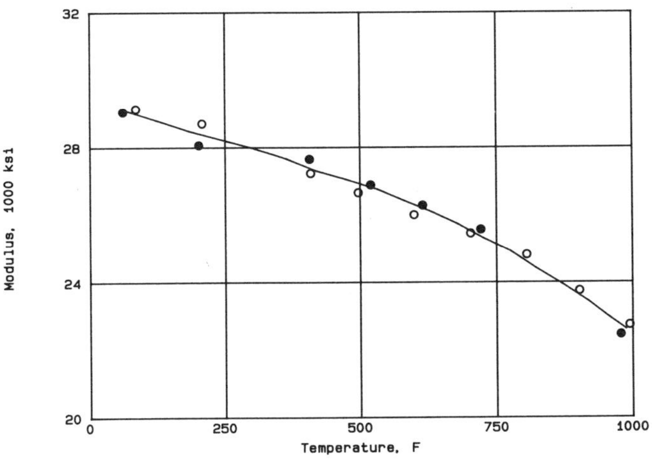

PH 15-7 Mo: Static modulus of elasticity of sheet, strip, and plate. Open circle: transformation hardened at 1050 F. Solid circle: refrigeration hardened at 950 F. Source: Ref 22, p 22

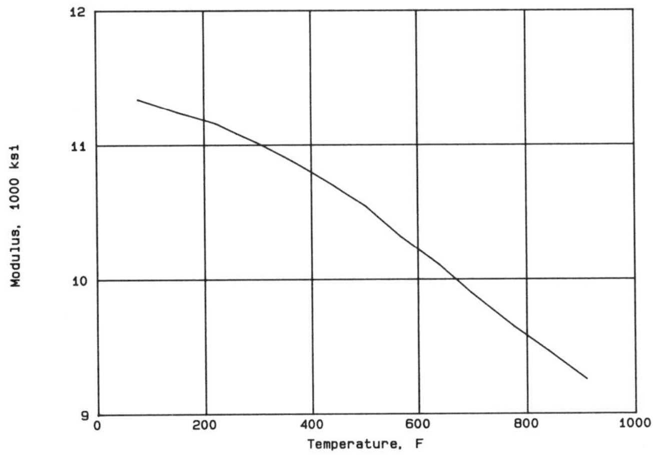

PH 15-7 Mo: Dynamic modulus of rigidity of sheet, strip, and plate refrigeration hardened at 950 F. Source: Ref 22, p 22

Mechanical Properties

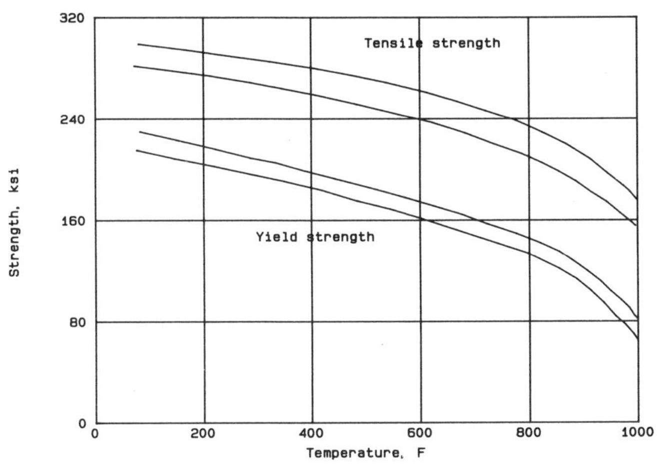

PH 15-7 Mo: Tensile and yield strength of 0.041-in. sheet cold worked and hardened at 900 F. Top curves of each set were transverse and bottom curves were longitudinal specimens. Source: Ref 22, p 14

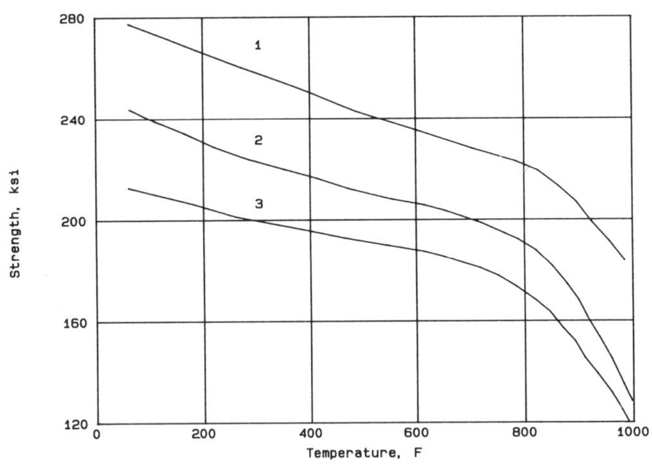

PH 15-7 Mo: Compressive yield strength of sheet. Curve 1: 0.034-in., cold worked and hardened at 900 F. Curve 2: 0.052-in., refrigeration hardened at 950 F. Curve 3: 0.052-in., transformation hardened at 1050 F. Source: Ref 22, p 15

PH 15-7 Mo: Short-time tensile property data

Condition	UTS MPa	UTS ksi	YS MPa	YS ksi	EL %	USS MPa	USS ksi
At 24 C (75 F)							
TH1050	1462	212	1413	205	7.0	986	143
RH950	1634	237	1517	220	5.0	1117	162
CH900(a)	1744	253	1675	243	3.0	...	...
CH900(b)	1800	261	1758	255	3.0	...	...
At 150 C (300 F)							
TH1050	1379	200	1344	195	4.5	896	130
RH950	1517	220	1379	200	4.0	1000	145
CH900(a)	1655	240	1551	225	2.0	...	...
CH900(b)	1779	258	1606	233	2.0	...	...
At 315 C (600 F)							
TH1050	1255	182	1186	172	4.5	800	116
RH950	1379	200	1200	174	5.0	883	128
CH900(a)	1517	220	1407	204	1.5	...	...
CH900(b)	1641	238	1455	211	1.5	...	...
At 370 C (700 F)							
TH1050	1207	175	1131	164	6.0	758	110
RH950	1344	195	1138	165	6.0	855	124
CH900(a)	1434	208	1331	193	1.5	...	...
CH900(b)	1572	228	1379	200	1.5	...	...
At 425 C (800 F)							
TH1050	1124	163	1034	150	9.0	717	104
RH950	1255	182	1034	150	8.0	800	116
CH900(a)	1372	199	1248	181	1.5	...	...
CH900(b)	1510	219	1310	190	1.5	...	...
At 480 C (900 F)							
TH1050	979	142	876	127	14.0	662	96
RH950	1103	160	896	130	10.0	710	103
CH900(a)	1269	184	1138	165	2.5	...	...
CH900(b)	1393	202	1207	175	2.5	...	...
At 540 C (1000 F)							
TH1050	793	115	724	105	19.0	552	80
TH950	896	130	724	105	14.0	607	88
CH900(a)	1089	158	903	131	4.0	...	...
CH900(b)	1193	173	986	143	4.0	...	...

Note: TH = transformation hardened. RH = refrigeration hardened. CH = cold worked and hardened. All conditions at either 566 C (1050 F), 510 C (950 F), or 482 C (900 F)
(a) Longitudinal. (b) Transverse
Source: Ref 15, p 385

PH 15-7 Mo: Creep-rupture curves at 1000 F for sheet, refrigeration hardened at 950 F. Creep: curve 1: 0.1%, curve 2: 0.3%, curve 3: 0.5%, curve 4: 1.0%, and curve 5: rupture. Source: Ref 22, p 19

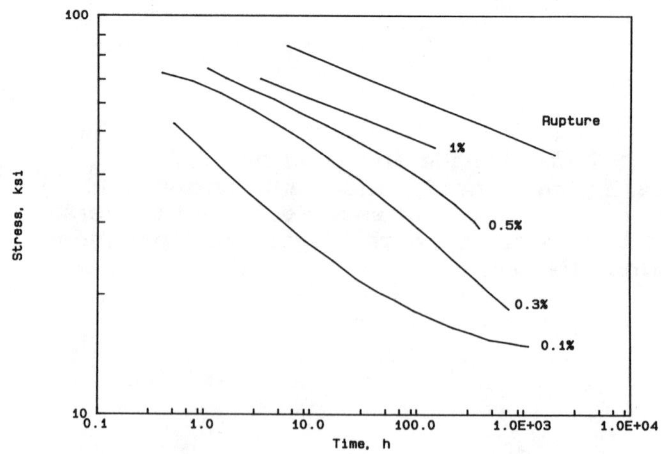

PH 15-7 Mo: Stress-rupture and creep property data

Condition	TEMP C	TEMP F	RPSTR 100 h MPa	RPSTR 100 h ksi	RPSTR 1000 h MPa	RPSTR 1000 h ksi	CRSTR 1%/1000 h MPa	CRSTR 1%/1000 h ksi	CRSTR 2%/1000 h MPa	CRSTR 2%/1000 h ksi
RH950	315	600	1395	202	1380	200	906.6	131.5	1035	150.1
TH1050	315	600	1235	179	1225	178	...	...	...	...
RH950	370	700	1330	193	1315	191	830.8	120.5	979.0	142.0
TH1050	370	700	1110	161	1040	151	...	...	...	...
RH950	425	800	1200	174	1180	171	655	95.0	752.9	109.2
TH1050	425	800	960	139	945	137	...	...	...	...
RH950	480	900	860	125	745	108	248	36.0	279.0	40.5
TH1050	480	900	745	108	675	98	...	...	...	...

Note: RH = refrigeration hardened. TH = transformation hardened. All conditions at either 510 C (950 F) or 566 C (1050 F)
Source: Ref 15, p 385

Stainless W

Alternate designation: AISI 635
UNS number: S17600
Specifications: ASTM A564 (635), ASTM A693, ASTM A705
Nominal composition: 0.080% C max, 15.000 to 17.500% Cr, 1.000% Mn max, 6.000 to 7.500% Ni, 0.040% P max, 0.030% S max, 1.000% Si max, 0.400 to 1.200% Ti, 0.400% Al max
Product forms: Wrought, bar, wire, plate, sheet, strip, billets, forgings

Description: Stainless W is a precipitation-hardenable Cr-Ni-Al-Ti stainless steel. Stainless W combines high corrosion resistance with high strength and hardness. After fabrication, it can be converted into a wide range of high strength conditions by use of a single heat treatment.
Typical applications: Knives in the paper and pulp industry, pump shafts, valve parts, springs, bearing plates on bridges, and rollers and tracks on dams.

Physical Properties

Stainless W: Thermal conductivity

TEMP C	TEMP F	THCON(a) W/m*K	THCON(a) Btu/ft*h*F	THCON(b) W/m*K	THCON(b) Btu/ft*h*F
100	212	19	10.8	21	12.1
200	392	20	11.6	23	13.0
400	752	23	13.5	25	14.2

(a) Annealed. (b) Aged
Source: Ref 28, p 84

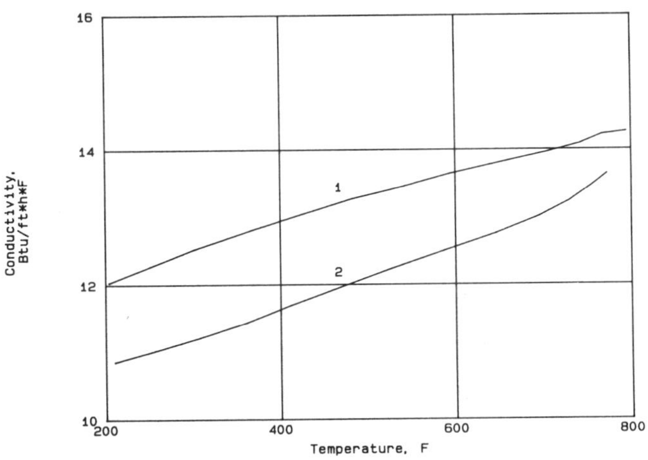

Stainless W: Thermal conductivity. Curve 1: estimated data for similar steels aged, held at 950 F for 1 h, and air cooled. Curve 2: experimental data for annealed steels. Source: Ref 22, p 3

Stainless W: Moduli of elasticity, rigidity, and Poisson's ratio

Condition	TEMP C	TEMP F	E GPa	E 10^3 ksi	G GPa	G 10^3 ksi	PR
(a)	24	75	185	26.9	77	11.2	0.20
(a)	149	300	182	26.4	73	10.6	0.24
(a)	260	500	177	25.6	71	10.3	0.24
(a)	371	700	169	24.5	67	9.7	0.26
(b)	24	75	190	27.6	79	11.4	0.21
(b)	149	300	185	26.8	75	10.9	0.23
(b)	260	500	179	26.0	73	10.6	0.23
(b)	371	700	172	24.9	69	10.0	0.25
(c)	24	75	192	27.8	80	11.6	0.20
(c)	149	300	182	26.4	77	11.1	0.20
(c)	260	500	177	25.6	74	10.7	0.20
(c)	371	700	168	24.4	68	9.8	0.24

(a) Solution annealed at 1038 C (1900 F) and air cooled. (b) Solution annealed at 1038 (1900 F), air cooled, reheated to 649 C (1200 F) and air cooled. (c) Solution annealed at 1066 C (1950 F), air cooled and then precipitation hardened by heating to 538 C (1000 F), held for 30 min, and air cooled
Source: Ref 28, p 87

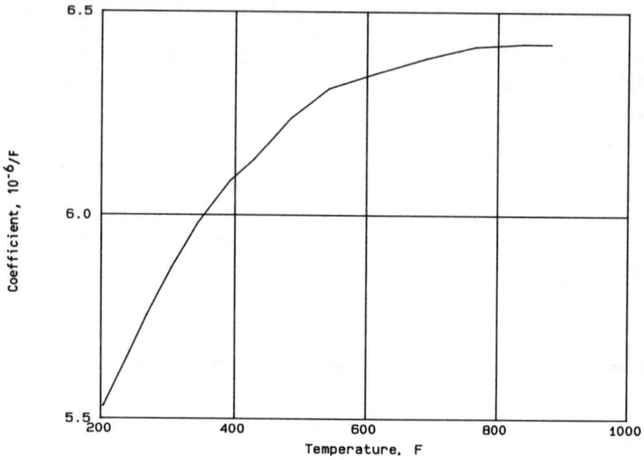

Stainless W: Mean coefficient of linear thermal expansion. Specimens were aged at 950 F for 1 h and air cooled. Source: Ref 22, p 4

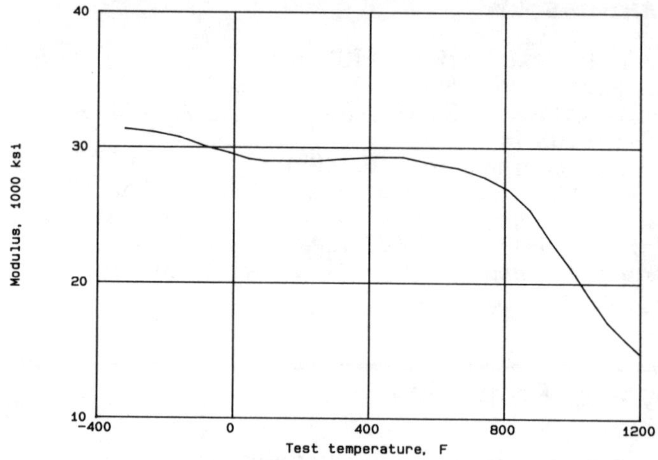

Stainless W: Static modulus of elasticity. Specimens were solution annealed and aged. Source: Ref 22, p 7

Stainless W: Mean coefficients of linear thermal expansion of annealed specimens

TEMP C	TEMP F	THEXP 10^{-6}/C	THEXP 10^{-6}/F
0-100	32-212	9.9	5.5
0-400	32-572	11.2	6.2
0-500	32-932	11.3	6.3

Source: Ref 28, p 84

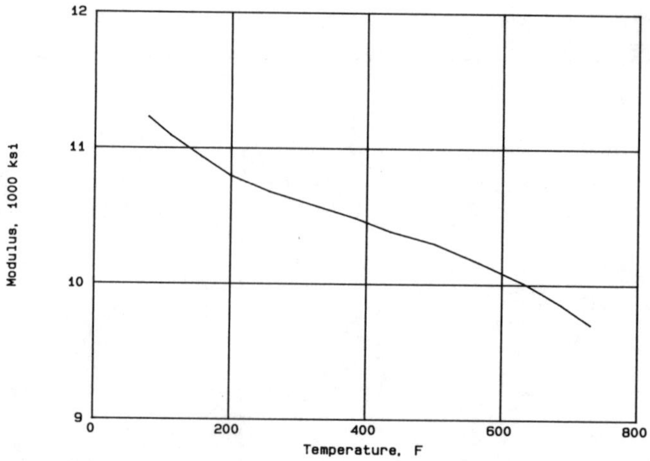

Stainless W: Modulus of rigidity. Specimens were solution annealed or solution annealed and aged. Source: Ref 22, p 8

Mechanical Properties

Stainless W: Tensile property data

Condition (a)	UTS MPa	UTS ksi	YS 0.2% MPa	YS 0.2% ksi	EL %	E GPa	E 10^3 ksi
At 27 C (80 F)							
116W	1544	224	1489	216	1.1	203	29.5
118W	1525	221.2	1482	215	5.3	201	29.1
112X	1553	225.2	1491	216.2	2.3	220	31.9
113X	1553	225.2	1510	219	1.4	209	30.3
At 316 C (600 F)							
113W	1255	182	1190	172.6	...	184	26.7
115W	1282	186	1211	175.6	3	181	26.3
115X	1289	187	1217	176.5	2.5	198	28.7
117X	1295	187.8	1196	173.4	3.1	188	27.2
At 538 C (1000 F)							
111W	523	75.8	346	50.2	47	115	16.7
117W	556	80.6	254	36.8	58	117	17
113X	536	77.7	340	49.3	50	119	17.2

(a) W indicates with rolling direction; X indicates transverse to rolling direction
Source: Ref 28, p 88

Stainless W: Compressive yield strength and modulus of elasticity

Condition(a)	YSC 0.2% MPa	YSC 0.2% ksi	E GPa	E 10^3 ksi
At 27 C (80 F)				
1113W	1510	219.0	210	30.4
1115W	1520	220.5	211	30.6
1116W	1520	220.5	208	30.2
114X	1527	221.5	143	20.8
115X	1527	221.5	213	30.9
At 204 C (400 F)				
113W	1345	195.1	199	28.8
114W	1346	195.2	197	28.6
117X	1353	196.2	205	29.8
118X	1365	198.0	205	29.8
At 316 C (600 F)				
119W	1242	180.1	192	27.8
1111W	1371	198.8	192	27.8
119X	1234	179.0	194	28.2
1110X	1246	180.7	196	28.4
At 427 C (800 F)				
1118W	912	132.2	171	24.8
1119W	914	132.6	177	25.6
1111X	930	134.9	172	25.0
At 538 C (1000 F)				
115W	388	56.3	143	20.8
116W	390	56.5	131	19.0
113X	360	52.2	143	20.8
116X	261	37.8	142	20.6
At 649 C (1200 F)				
117W	165	24.0	89	12.9
118W	175	25.4	91	13.2
112X	183	26.5	84	12.2

(a) W indicates with rolling direction; X indicates transverse to rolling direction
Source: Ref 28, p 88

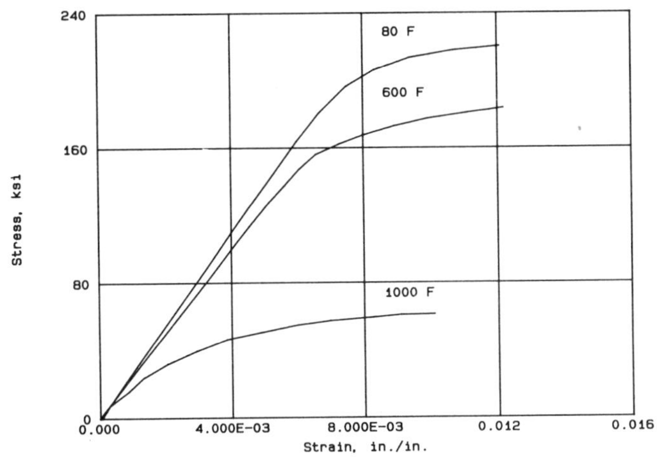

Stainless W: Stress-strain curves in tension for 0.064-in. sheet, solution annealed and held at 1000 F for 0.5 h. Longitudinal and transverse specimens were tested. Source: Ref 22, p 5

Stainless W: Tensile property data

TEMP C	UTS MPa	UTS ksi	YS MPa	YS ksi	EL %	RA %	E GPa	E 10E3 ksi	RPSTR 100 h MPa	RPSTR 100 h ksi	CRSTR 0.2%/500 h MPa	CRSTR 0.2%/500 h ksi
21	1324	192	1289	187	13.5	53.5	200	29	...	...	...	...
204	...	...	...	...	...	29	...	...	...	...	...	...
316	1117	162	1076	156	12	51.5	...	...	...	...	862	125
427	1007	146	1000	145	13.5	53	186	27	...	...	276	40
538	648	94	372	54	22.5	75	138	20	217	31.5	...	...
649	283	41	165	24	40	88	97	14	86	12.5	...	...

Note: Bar stock was solution annealed at 899 C (1650 F), aged at 538 C (1000 F) for 30 min, and air cooled
Source: Ref 28, p 83

Greek Ascoloy

UNS number: S41800
Specifications: AMS 5508, AMS 5616, AMS 5817, ASTM A565 (615), SAE J467
Nominal composition: 0.150 to 0.200% C, 12.000 to 14.000% Cr, 0.500% Mn max, 1.800 to 2.200% Ni, 0.040% P max, 0.030% S max, 0.500% Si max, 2.500 to 3.500% W
Product forms: Wrought, bar, wire, plate, sheet, strip, billets, forgings
Description: Greek Ascoloy is a hardenable Cr-Ni-W martensitic stainless steel designed for service at temperatures up to 650 C (1200 F) in highly stressed parts. It has good resistance to scaling and oxidation at temperatures up to 760 C (1400 F). Greek Ascoloy is chosen over other 12% Cr steels when better creep strength and resistance to tempering and stress corrosion cracking are needed.

Typical applications: Steam turbine buckets and blades, gas turbine compressor parts, and high temperature bolts.

Physical Properties

Greek Ascoloy: Mean coefficients of linear thermal expansion

TEMP C	TEMP F	THEXP 10^{-6}/C	THEXP 10^{-6}/F
25-95	80-200	10.4	5.8
25-205	80-400	11.0	6.1
25-315	80-600	11.2	6.2
25-425	80-800	11.5	6.4
25-540	80-1000	11.7	6.5
25-595	80-1100	11.9	6.6

Source: Ref 15, p 373

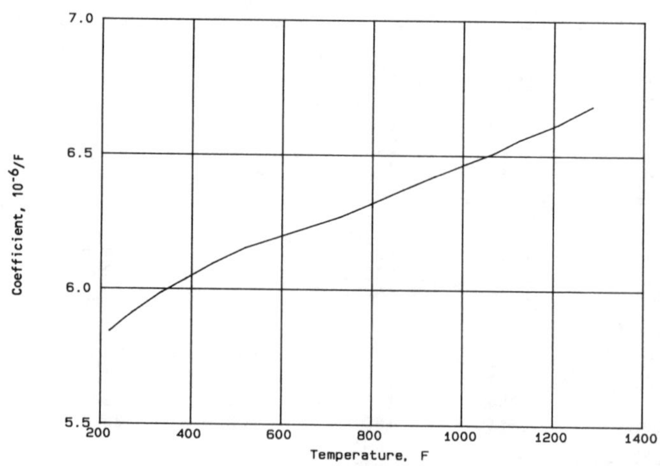

Greek Ascoloy: Mean coefficient of linear thermal expansion. Source: Ref 22, p 4

Greek Ascoloy: Electrical resistivity and conductivity

TEMP C	TEMP F	ELCON %IACS	ELRES n{o}*m
25	77	2.12	787
371	700	2.8	617
427	800	2.84	607
482	900	2.86	602

Source: Ref 29, Mar 1965

Greek Ascoloy: Modulus of elasticity

TEMP C	TEMP F	E GPa	E 10^3 ksi
21	70	200	29(a)
538	1000	148	21.5
593	1100	117	17

(a) Approximate
Source: Ref 28, p 35

Mechanical Properties

Greek Ascoloy: Tensile property data

TEMP C	TEMP F	UTS MPa	UTS ksi	YS MPa	YS ksi	EL %	RA %
21	70	971	140.8	793	115	21.0	58.0
482	900	689	100	565	82	16.0	63.0
538	1000	579	84	483	70	22.0	74.0
593	1100	462	67	379	55	30.0	83.0

Note: Specimens were heat treated by hardening at 954 C (1750 F), oil quenched followed by tempering at 649 C (1200 F), and air cooled
Source: Carpenter Technology

Greek Ascoloy: Tensile property data

TEMP C	TEMP F	UTS MPa	UTS ksi	YS MPa	YS ksi	EL %	RA %
Tempered at 565 C (1050 F)							
425	800	929	135.0	839	122.0	13.0	67.0
538	1000	711	103.0	673	97.6	17.0	71.0
650	1200	319	46.2	303	44.0	29.4	89.6
760	1400	177	25.6	147	21.3	84.6	89.9
Tempered at 650 C (1200 F)							
425	800	817	118.5	776	112.5	16.0	55.0
540	1000	621	90.0	614	89.0	24.0	72.5
650	1200	355	51.5	348	50.5	30.0	86.0

Note: Specimens were hardened at 980 C (1800 F), held for 30 min, oil quenched, tempered at temperature indicated, held for 2 h, and air cooled
Source: Ref 15, p 374

Greek Ascoloy: Tensile property data

TEMP C	TEMP F	UTS MPa	UTS ksi	YS MPa	YS ksi	EL %	RA %
25	77	971	140.8	793	115	21	58
482	900	690	100	565	82	16	63
538	1000	579	84	483	70	22	74
593	1100	462	67	379	55	30	83

Note: Specimens were taken from forged coupons and bars. They were oil quenched at 954 C (1750 F), reheated to 649 C (1200 F), and air cooled to 28 to 29 HRC
Source: Ref 28, p 35

Greek Ascoloy: Tensile property data

TEMP C	TEMP F	UTS MPa	UTS ksi	YS MPa	YS ksi	EL %	RA %
21	70	1103	160	931	135	16	45
260	500	1034	150	889	129	12	43
316	600	993	144	855	124	11	43
427	800	876	127	772	112	10	43
482	900	779	113	696	101	13	52
538	1000	655	95	579	84	17	64

Note: Specimens were taken from bars and forgings. They were oil quenched at 954 C (1750 F), reheated to 566 C (1050 F) and air cooled to 34 to 36 HRC
Source: Ref 28, p 35

Greek Ascoloy: Creep property data

TEMP C	TEMP F	CRSTR 0.1%/100 h MPa	CRSTR 0.1%/100 h ksi	CRSTR 0.1%/150 h MPa	CRSTR 0.1%/150 h ksi	CRSTR 1%/100 h MPa	CRSTR 1%/100 h ksi	CRSTR 1%/150 h MPa	CRSTR 1%/150 h ksi
370	700	503	73.0	483	70.0	...	...	689	100.0
425	800	324	47.0	303	44.0	...	...	...	...
480	900	228	33.0	214	31.0	552	80.0	538	78.0
540	1000	131	19.0	117	17.0	331	48.0	324	47.0
590	1100	45	6.5	36	5.2	200	29.0	193	28.0

Note: Specimens were hardened at 955 C (1750 F), oil quenched, tempered at 565 C (1050 F) to 33.5 to 35.5 HRC
Source: Ref 15, p 373

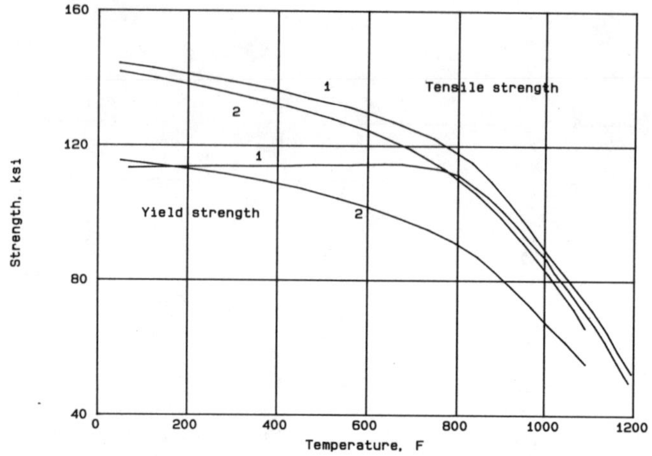

Greek Ascoloy: Tensile and yield strength of bar. Curve 1: treated at 1800 F, oil quenched, tempered at 1200 F for 2 h, and air cooled. Curve 2: treated at 1750 F, oil quenched, tempered at 1200 F for 4 h, and air cooled. Source: Ref 22, p 6

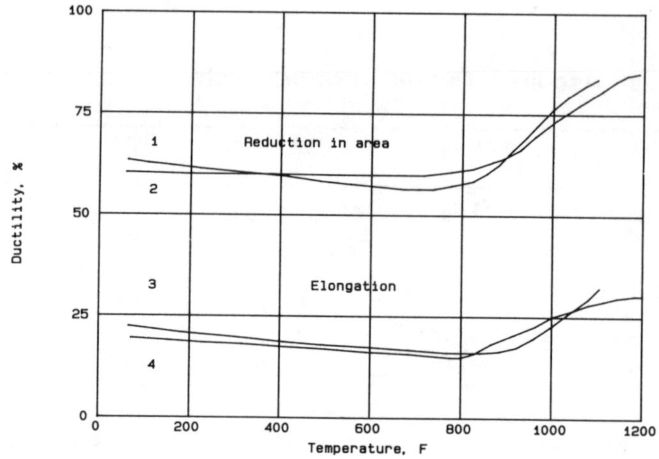

Greek Ascoloy: Tensile elongation and reduction in area of bar. Curves 1 and 4: treated at 1800 F, oil quenched, tempered at 1200 F for 4 h, and air cooled. Curves 2 and 3: treated at 1750 F, oil quenched, tempered at 1200 F for 4 h, and air cooled. Source: Ref 22, p 6

Greek Ascoloy: Stress-rupture property data

TEMP C	TEMP F	RPT h	RPSTR MPa	RPSTR ksi
538	1000	10	483	70
538	1000	100	400	58
538	1000	1000	290	42
593	1100	10	345	50
593	1100	100	262	38
593	1100	1000	159	23
649	1200	10	193	28
649	1200	100	138	20
649	1200	1000	69	10

Note: Specimens were oil quenched at 954 C (1750 F), tempered at 566 C (1050 F), and hardened to 34 to 36 HRC. Specimens were taken from bars and forgings
Source: Ref 28, p 36

Greek Ascoloy: Effect of tempering temperature on Izod impact values

Ttpr C	Ttpr F	IS(a) J	IS(a) ft-lb
260	500	26	19
316	600	22	16
371	700	14	10
427	800	12	9
482	900	14	10
538	1000	23	17
593	1100	54	40
649	1200	104	77
704	1300	88	65

Note: Specimens were heat treated by hardening at 954 C (1750 F) and oil quenched
(a) At room temperature
Source: Carpenter Technology

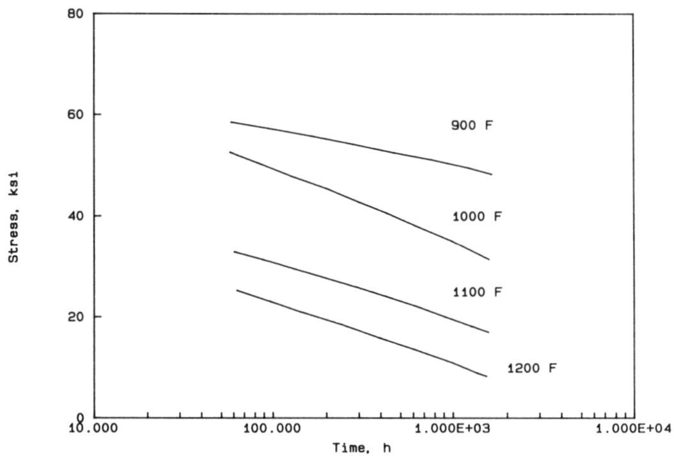

Greek Ascoloy: Creep-rupture curves for bar. Specimens were treated at 1750 F, oil quenched, tempered at 1200 F for 1.5 h, and air cooled. Source: Ref 22, p 7

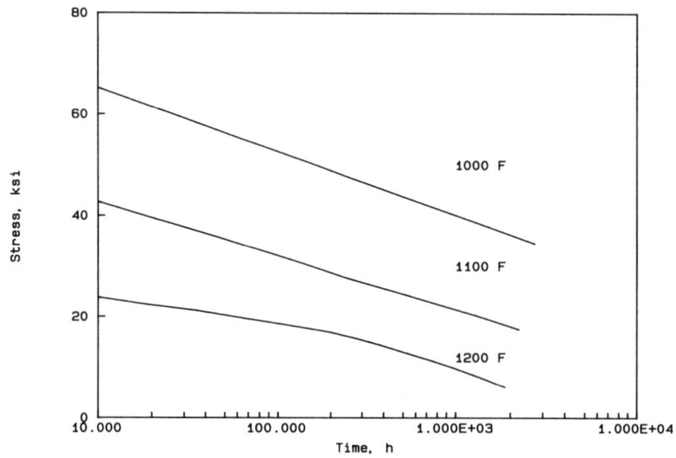

Greek Ascoloy: Creep-rupture curves for bar. Specimens were treated at 1800 F, air cooled, tempered at 1050 F for 2 h, and air cooled. Source: Ref 22, p 7

Greek Ascoloy: Stress-rupture property data

TEMP C	TEMP F	RPSTR 10 h MPa	RPSTR 10 h ksi	RPSTR 100 h MPa	RPSTR 100 h ksi	RPSTR 1000 h MPa	RPSTR 1000 h ksi	RPSTR(a) 10000 h MPa	RPSTR(a) 10000 h ksi
Hardened at 980 C (1800 F); held 1/2 h; oil quenched; tempered at 680 C (1260 F); held 1-1/2 h; air cooled									
480	900	448	65	379	55.0	331	48.0	290	42.0
538	1000	331	48	290	42.0	248	36.0	210	30.5
593	1100	262	38	216	31.3	172	25.0	128	18.5
649	1200	186	27	141	20.5	79	11.5	41	6.0
Hardened at 955 C (1750 F); oil quenched; tempered at 650 C (1200 F) to 27 to 29 HRC									
480	900	...	...	386	56.0	345	50.0	...	...
538	1000	...	...	338	49.0	241	35.0	...	...
593	1100	...	...	207	30.0	138	20.0	...	...
649	1200	...	...	131	19.0	69	10.0	...	...

(a) Extrapolated
Source: Ref 15, p 374

Greek Ascoloy: Stress-rupture property data

TEMP C	TEMP F	RPSTR 10 h MPa	RPSTR 10 h ksi	RPSTR 100 h MPa	RPSTR 100 h ksi	RPSTR 1000 h MPa	RPSTR 1000 h ksi
538	1000	448	65	359	52	283	41
593	1100	290	42	228	33	152	21
649	1200	159	23	131	19	76	11

Note: Specimens were heat treated by hardening at 982 C (1800 F), air cooled, tempered at 566 C (1050 F), heated for 2 h, and air cooled
Source: Carpenter Technology

Greek Ascoloy: Stress relaxation

TEMP C	TEMP F	CRIS MPa	CRIS ksi	CRSRD 1000 h MPa	CRSRD 1000 h ksi	CRSRD 10000 h MPa	CRSRD 10000 h ksi
538	1000	296	43	81	11.8	43	6.2
593	1100	234	34	45	6.5	...	...

Note: Specimens were heated to 1093 C (2000 F) for 4 h, air cooled, tempered to 649 C (1200 F) for 4 h, reheated to 593 C (1100 F) for 2 h, and air cooled
Source: Ref 28, p 36

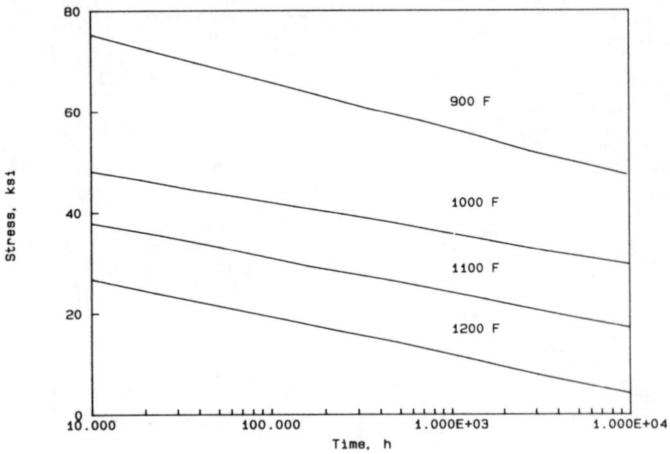

Greek Ascoloy: Creep-rupture curves for bar. Specimens were treated at 1800 F, air cooled, tempered at 1260 F for 1.5 h, and air cooled. Source: Ref 22, p 7

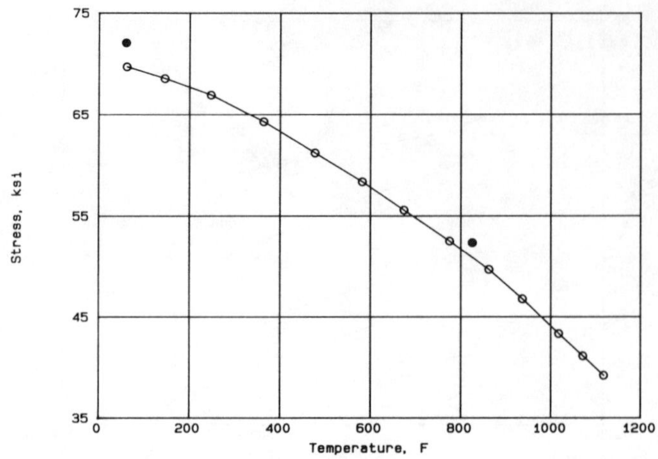

Greek Ascoloy: Fatigue endurance limit at 10^7 cycles of bar. Specimens were hardened and tempered to 36 HRC (open circle), and 34 HRC (solid circle). R = -1. Source: Ref 22, p 9

Greek Ascoloy: Hot hardness data

TEMP C	TEMP F	HARD HRC
21	70	37.5
316	600	34.0
427	800	31.0
538	1000	27.0
649	1200	84(a)

Note: Specimens were taken from wrought bar stock and heat treated at 954 C (1750 F) for 30 min and air cooled followed by tempering at 538 C (1000 F) for 2 h and air cooled. The Rockwell C values are actual Rockwell C readings made with a special apparatus permitting the specimen and indenter to be heated and held at temperature on the hardness testing machine. The specimen was held at temperature 30 min prior to the hardness test. All values represent an average of three individual readings. The Rockwell B value was converted from Rockwell A readings.
(a) HRB
Source: Ref 28, p 36

Greek Ascoloy: Fatigue strength of smooth specimens

TEMP C	TEMP F	FATST MPa	FATST ksi
21	70	517	75
93	200	503	73
204	400	469	68
316	600	421	61
427	800	379	55
538	1000	310	45
593	1100	276	40

Note: Specimens were taken from bars and forgings and were hardened to 36 HRC and tested at 120 cycles/s.
Source: Ref 28, p 35

Greek Ascoloy: Brinell hardness of bar. Specimens were treated at 1750 F for 0.5 h, air cooled, tempered at 1000 F for 2 h, and air cooled. Source: Ref 22, p 3

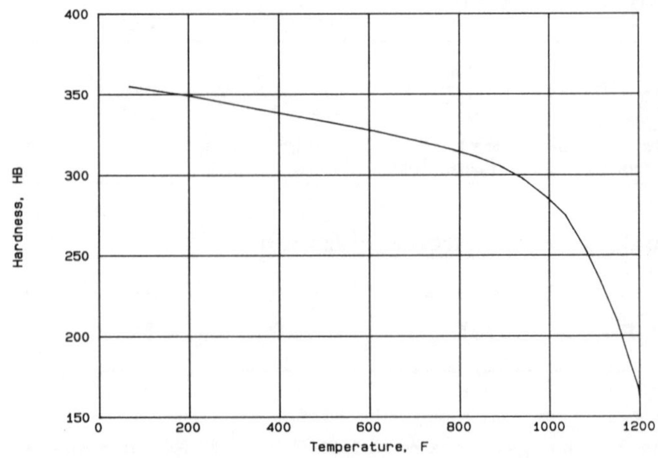

Nitronic 40

Alternate designation: 21-6-9

UNS number: S21900

Specifications: AMS 5561, ASTM A276 (XM-10), ASTM A314, ASTM A412, ASTM A429, ASTM A473, ASTM A580

Nominal composition: 0.080% C max, 19.000 to 21.500% Cr, 8.000 to 10.000% Mn, 0.150 to 0.400% N, 5.500 to 7.500% Ni, 0.060% P max, 0.030% S max, 1.000% Si max

Description: Nitronic 40 combines good elevated temperature properties with corrosion resistance and a yield strength about twice that of AISI Types 304, 321, and 347. The austenitic Cr-Mn-Ni stainless steel also boasts good resistance to high-temperature oxidation in air. An extralow carbon version, 0.04% C max, is available.

Typical applications: Aircraft parts such as ducting and bellows systems, tailpipes and exhaust systems, clamps, flanges, and hydraulic tubing; chemical processing and pollution control equipment; and steam service and autoclave applications.

Physical Properties

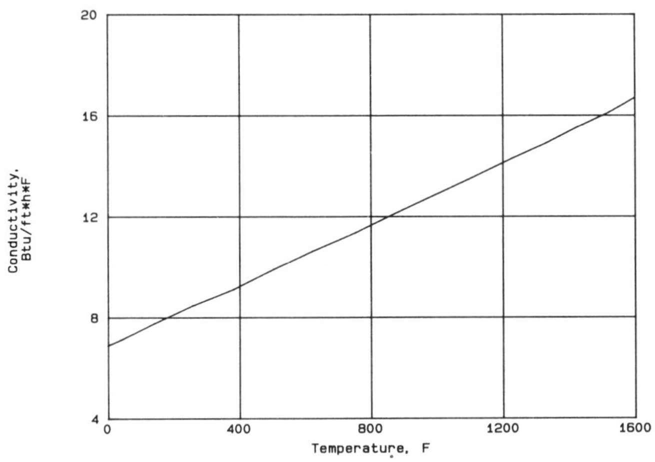

Nitronic 40: Thermal conductivity of annealed specimens. Source: Ref 22, p 3

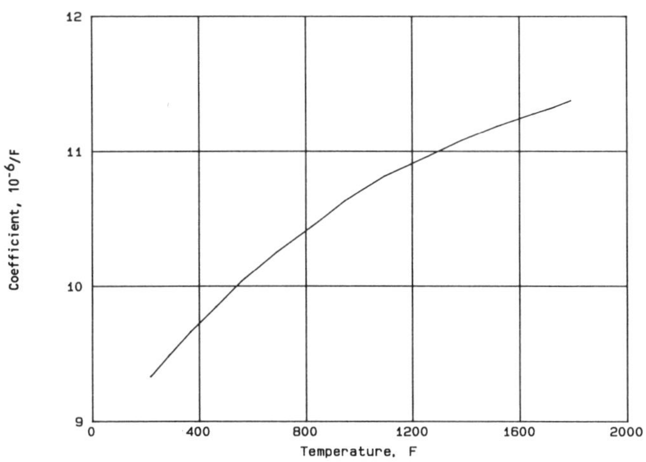

Nitronic 40: Mean coefficient of linear thermal expansion of annealed specimens. Source: Ref 22, p 3

Nitronic 40: Thermal conductivity

TEMP C	TEMP F	THCON W/m*K	THCON Btu/h*ft*F
93	200	13.8	8.0
204	400	16.1	9.3
316	600	18.2	10.5
427	800	20.2	11.7
538	1000	22.5	13.0
649	1200	24.8	14.3
760	1400	26.8	15.5
871	1600	28.8	16.7

Source: Armco Data Sheet

Nitronic 40: Mean coefficients of linear thermal expansion of annealed specimens

TEMP C	TEMP F	THEXP 10^{-6}/C	THEXP 10^{-6}/F
27-93	80-200	16.7	9.3
27-204	80-400	17.3	9.6
27-316	80-600	18.2	10.1
27-538	80-1000	18.6	10.6
27-760	80-1400	20.0	11.1
27-899	80-1600	20.2	11.2
27-982	80-1800	20.5	11.4

Source: Armco Data Sheet

Mechanical Properties

Nitronic 40: Tensile property data for annealed sheet and strip

TEMP C	TEMP F	UTS MPa	UTS ksi	YS MPa	YS ksi	EL %
24	75	765	111.0	444	64.4	42.5
95	200	693	100.5	353	51.2	41.5
205	400	616	89.4	289	41.9	40.0
315	600	594	86.1	259	37.5	33.0(a)
425	800	549	79.6	221	32.0	33.0(a)
480	900	518	75.1	202	29.3	41.0
540	1000	490	71.1	203	29.4	35.0
595	1100	467	67.8	193	28.0	30.0
650	1200	416	60.3	184	26.7	26.0
705	1300	354	51.3	185	26.8	24.0

(a) Broke on gage mark
Source: Ref 15, p 416

Nitronic 40: Tensile property data for cold rolled sheet

TEMP C	TEMP F	UTS MPa	UTS ksi	YS MPa	YS ksi	EL %	HARD HRC
10% cold reduction							
24	75	870	126.2	630	91.4	27.0	26
95	200	779	113.0	611	88.6	28.0	...
205	400	696	100.9	500	72.5	25.0	...
315	600	681	98.7	470	68.1	25.5	...
425	800	624	90.5	430	62.4	26.5	...
540	1000	552	80.0	377	54.7	24.0	...
650	1200	469	68.0	351	50.9	17.0	...
25% cold reduction							
24	75	1097	159.1	876	127.0	14.5	35
95	200	998	144.8	881	127.8	12.0	...
205	400	931	135.1	836	121.3	5.5	...
315	600	877	127.2	759	110.1	6.0	...
425	800	798	115.8	667	96.8	10.5	...
540	1000	715	103.7	601	87.1	9.0	...
650	1200	592	85.8	500	72.5	8.0	...
60% cold reduction							
24	75	1407	204.0	1162	168.6	6.0	43
95	200	1321	191.6	1189	172.5	4.5	...
205	400	1256	182.2	1122	162.8	3.5	...
315	600	1214	176.1	1045	151.6	4.0	...

Source: Ref 15, p 417

Nitronic 40: Tensile and yield strength of transverse specimens of sheet annealed to 93 HRB. Source: Ref 22, p 7

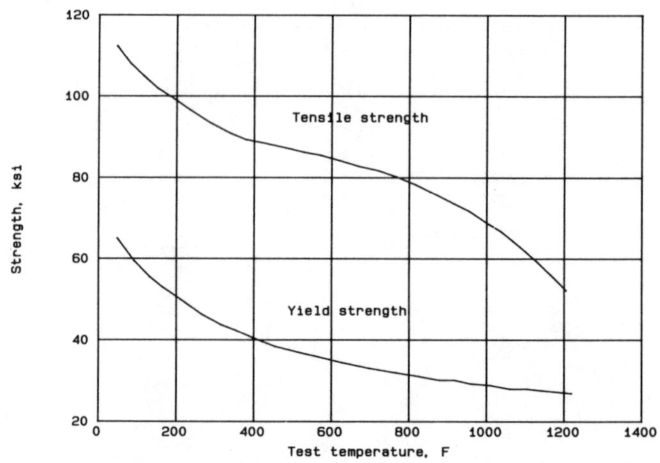

Nitronic 40: Tensile property data

TIME	YS MPa	YS ksi	UTS MPa	UTS ksi	EL(a) %	EL(b) %
Annealed at 677 C (1250 F)						
5 s	484	70.2	786	114.0	64.0	43.5
30 s	470	68.3	772	112.0	70.0	45.5
30 s(c)	490	71.2	811	117.7	67.0	44.0
20 min	469	68.0	782	113.3	67.5	44.0
Annealed at 760 C (1400 F)						
5 s	471	69.8	785	113.9	66.0	43.5
30 s	462	67.0	763	110.7	68.5	45.5
30 s(c)	482	69.9	787	114.1	66.0	43.5
Cold rolled 15% at 677 C (1250 F)						
5 s	885	128.3	1008	146.2	37.0	18.5
30 s	858	124.5	690	143.6	38.5	18.0
30 s(c)	860	124.8	1002	145.3	38.5	18.0
20 min	832	120.7	976	141.6	39.5	17.0
Cold rolled 15% at 760 C (1400 F)						
5 s	857	124.3	996	144.5	38.0	16.5
30 s	852	123.5	993	144.0	39.0	16.0
30 s(c)	848	123.0	990	143.7	37.5	14.5
Cold rolled 30% at 677 C (1250 F)						
5 s	1095	158.8	1203	171.7	26.0	8.0
30 s	1110	161.0	1223	177.3	24.5	7.0
30 s(c)	1111	161.4	1218	176.7	24.0	7.0
20 min	1110	161.0	1225	177.6	27.0	8.0
Cold rolled 30% at 760 C (1400 F)						
5 s	1098	159.3	1179	171.3	26.0	6.5
30 s	1066	154.5	1166	169.3	25.0	7.0
30 s(c)	1104	160.2	1185	173.3	25.0	7.0

(a) In 13 mm (0.5 in.). (b) In 50 mm (2 in.). (c) 5 cycles
Source: Armco Data Sheet

Nitronic 40: Effect of cold reduction and stress relieving on sheet and strip

.%CW	UTS MPa	UTS ksi	YS MPa	YS ksi	EL %	HARD
Tested at room temperature						
As rolled						
0	794	115.2	482	69.9	44.5	94 HRB
15	978	141.9	905	131.2	22.5	34 HRC
30	1222	177.2	1099	159.4	12.0	40 HRC
Stress relieved at 480 C (900 F) for 1 h						
0	803	116.5	494	71.6	44.0	94 HRB
15	1007	146.0	912	132.3	23.0	34 HRC
30	1278	185.6	122	178.3	11.0	42 HRC
Stress relieved at 675 C (1250 F) for 1 h						
0	798	115.8	492	71.4	44.0	94 HRB
15	977	141.7	814	118.0	26.0	32 HRC
30	1151	167.0	106	154.2	18.5	39 HRC
Tested at 480 C (900 F)						
As rolled						
0	537	77.9	234	33.9	41.5	...
15	654	94.9	529	76.7	18.0	...
30	836	121.2	717	104.0	7.0	...
Stress relieved at 675 C (1250 F)						
0	538	78.1	239	34.7	42.0	...
15	631	91.5	490	71.0	21.5	...
30	774	112.2	669	97.0	13.0	...

Source: Ref 15, p 417

Nitronic 40: Properties of welded joints in sheet and strip

TEMP C	TEMP F	UTS MPa	UTS ksi	YS MPa	YS ksi	EL(a) %	EL(b)
24	75	794	115.2	459	66.6	37.0	36.0
425	800	553	80.2	231	33.5	37.5	...
480	900	536	77.8	227	32.9	33.0	...
540	1000	507	73.5	216	31.4	28.5	...
595	1100	461	66.8	199	28.9	24.5	25.0
650	1200	365	52.9	192	27.9	20.0	20.0
705	1300	332	48.2	183	26.5	11.0	12.5

(a) In 50 mm (2 in.). (b) In 13 mm (0.5 in.)
Source: Ref 15, p 417

Nitronic 40: Stress-rupture property data for annealed bar and sheet

TEMP C	TEMP F	RPSTR 100 h MPa	RPSTR 100 h ksi	RPSTR 1000 h MPa	RPSTR 1000 h ksi
Bar					
650	1200	231	33.5	190	27.5
730	1350	145	21.0	93	13.5
815	1500	70	10.2	42	6.1
Sheet					
650	1200	221	32.0	148	21.5
815	1500	44	6.4	19	2.8

Source: Ref 15, p 416

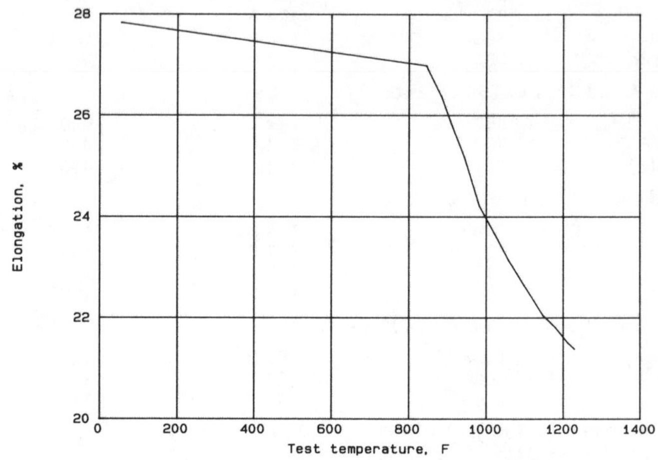

Nitronic 40: Tensile elongation of transverse specimens of sheet annealed to 93 HRB. Source: Ref 22, p 7

19-9 DL

Alternate designation: AISI 651
UNS number: S63198
Specifications: AMS 5526, AMS 5579, ASTM A453 (651), ASTM A457 (651), ASTM A458 (651), ASTM A477 (651), SAE J467 (19-9DL)
Nominal composition: 19.500% Cr, 9.500% Ni, 1.400% Mo, 1.400% W, 0.420% (Nb + Ta), 0.220% Ti, 0.300% C, 1.000% Mn, 0.600% Si
Product forms: Wrought, bar, rod, wire, plate, sheet, strip, tube, pipe, billets, forgings, structural shapes
Description: Austenitic alloy for applications requiring high strength and resistance to corrosion and oxidation at temperatures up to 675 C (1250 F). It is frequently used in the annealed condition. Higher properties can be developed by hot/cold working: forge or roll at 1095 to 1150 C (2000 to 2100 F), followed by a finish work at 650 to 870 C (1200 to 1600 F). This alloy has good oxidation resistance to 955 C (1750 F) in continuous service and satisfactory oxidation resistance to 790 C (1450 F) in intermittent service.
Typical applications: Turbine wheels, supercharger wheels, buckets, frames, casings, and afterburner parts.

Physical Properties

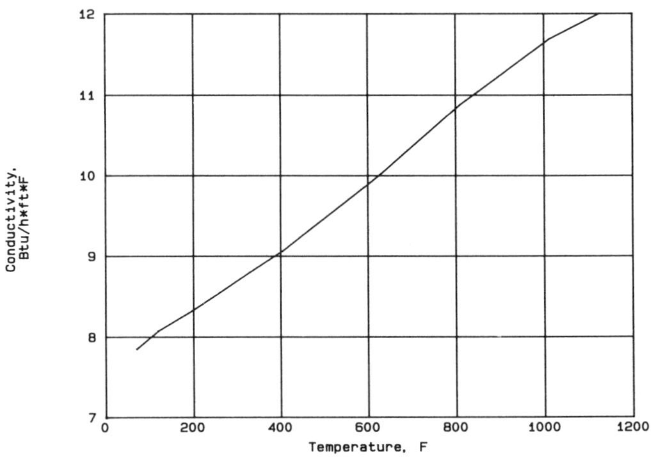

19-9 DL: Thermal conductivity. Source: Ref 22, p 3

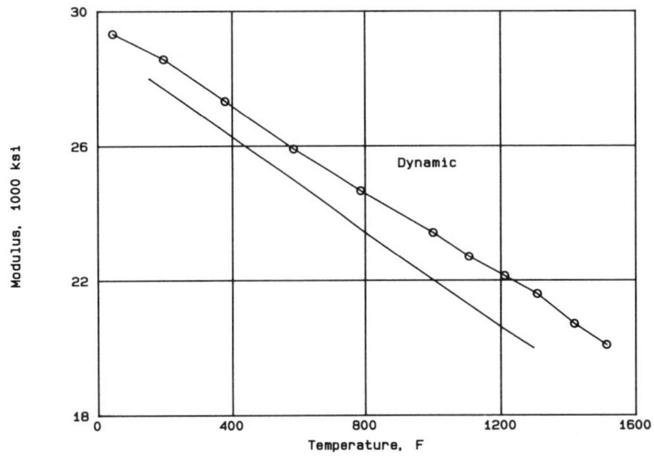

19-9 DL: Modulus of elasticity. Source: Ref 22, p 5

19-9 DL: Thermal conductivity

TEMP C	TEMP F	THCON W/m*K	THCON Btu/ft*h*F
21	70	13.5	7.80
425	800	18.5	10.7
650	1200	21.3	12.3

Source: Ref 15, p 391

19-9 DL: Mean coefficients of linear thermal expansion

TEMP C	TEMP F	THEXP 10^{-6}/C	THEXP 10^{-6}/F
21-93	70-200	15.3	8.50
21-205	70-400	16.3	9.10
21-315	70-600	16.7	9.30
21-425	70-800	17.3	9.60
21-540	70-1000	17.6	9.80
21-815	70-1500	18.0	10.0

Source: Ref 15, p 391

19-9 DL: Modulus of elasticity in tension

TEMP C	TEMP F	E GPa	E 10^3 ksi
21	70	203	29.5
538	1000	161	23.3
816	1500	138	20.0

Source: Ref 28, p 97

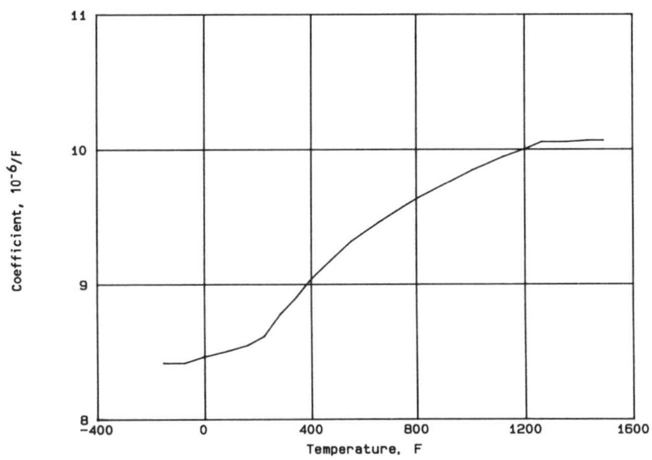

19-9 DL: Mean coefficient of thermal expansion from room temperature to indicated temperature. Source: Ref 22, p 3

19-9 DL: Poisson's ratio

TEMP C	TEMP F	PR
21	70	0.286
538	1000	0.318
816	1500	0.336

Source: Ref 28, p 97

19-9 DL: Modulus of rigidity. Source: Ref 22, p 5

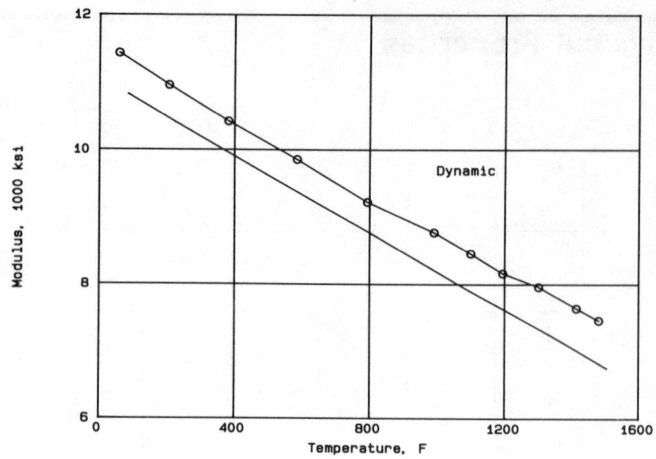

Mechanical Properties

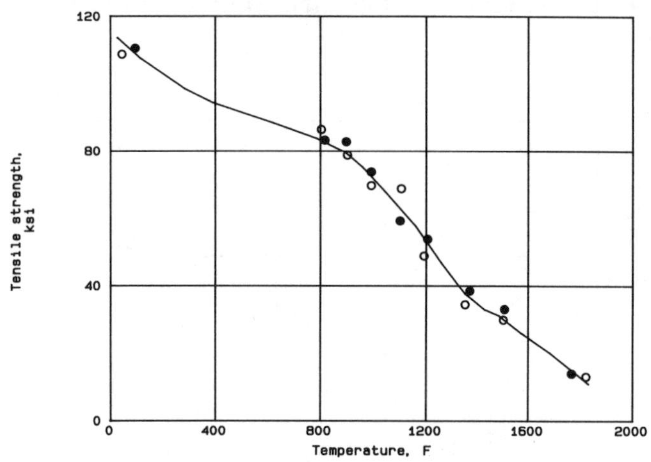

19-9 DL: Tensile strength of sheet and plate annealed at 1800 F. Open circle: 0.10-in. sheet. Solid circle: 0.25-in. plate. Source: Ref 22, p 4

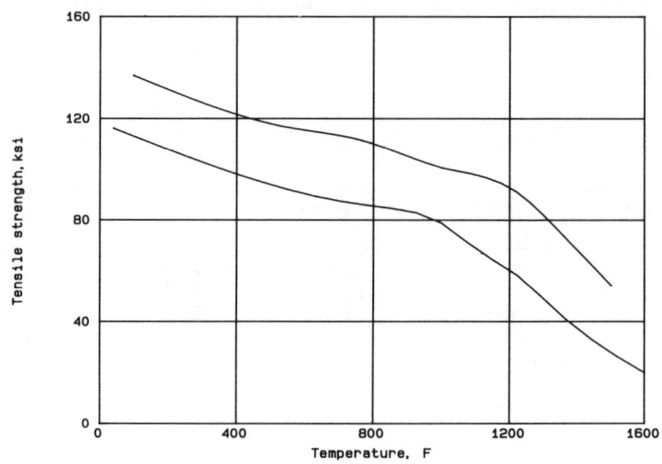

19-9 DL: Scatterband for tensile strength of sheet, plate, and bar, hot worked and stress relieved at 1200 F. Source: Ref 22, p 3

19-9 DL: Tensile property data

TEMP C	TEMP F	UTS MPa	UTS ksi	YS(a) MPa	YS(a) ksi	EL(b) %	RA %
24	75	838	121.5	579	84.0	43.0	55.5
150	300	703	102.0	517	75.0	30.0	52.5
260	500	707	102.5	496	72.0	27.5	49.5
370	700	672	97.5	455	66.0	31.5	49.5
480	900	634	92.0	427	62.0	29.5	48.5
540	1000	614	89.0	396	57.5	29.5	46.0
650	1200	534	77.5	362	52.5	30.5	47.5
705	1300	459	66.5	352	51.0	29.0	49.0
760	1400	365	53.0	317	46.0	30.5	56.5
815	1500	331	48.0	279	40.5	29.0	62.0

Note: Specimens were 60-mm (2.4-in.) diameter bar; warm rolled and stress relieved at 730 C (1350 F) for 8 h, with 229-241 HB
(a) 0.2% offset (b) In 50 mm (2 in.)
Source: Ref 15, p 392

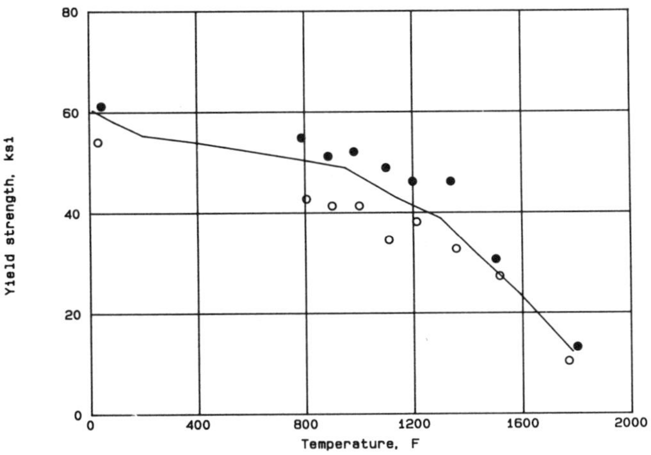

19-9 DL: Tensile yield strength of sheet and plate annealed at 1800 F. Open circle: 0.10-in. sheet. Solid circle: 0.25-in. plate. Source: Ref 22, p 4

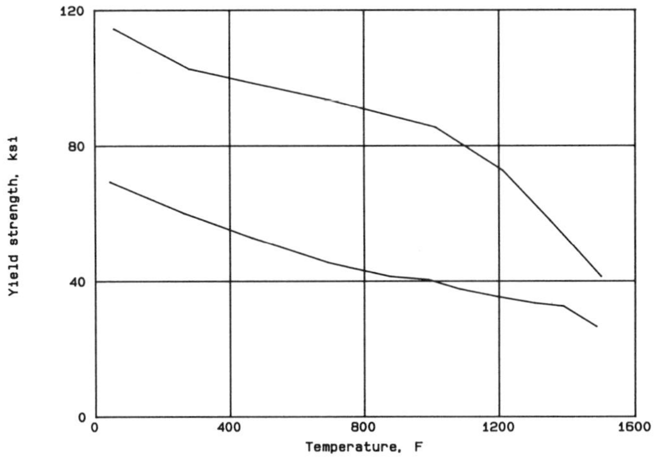

19-9 DL: Scatterband for tensile yield strength of sheet, plate, and bar, hot worked and stress relieved at 1200 F. Source: Ref 22, p 3

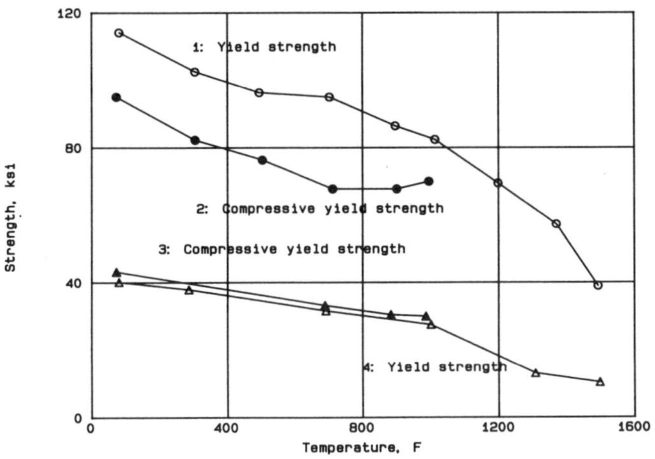

19-9 DL: Tensile and compressive yield strength of annealed and hot worked sheet. Curves 1 and 2: hot worked and held at 1200 F with a tensile strength of 134 ksi. Curves 3 and 4: annealed at 1800 F with a tensile strength of 93 ksi. Source: Ref 22, p 4

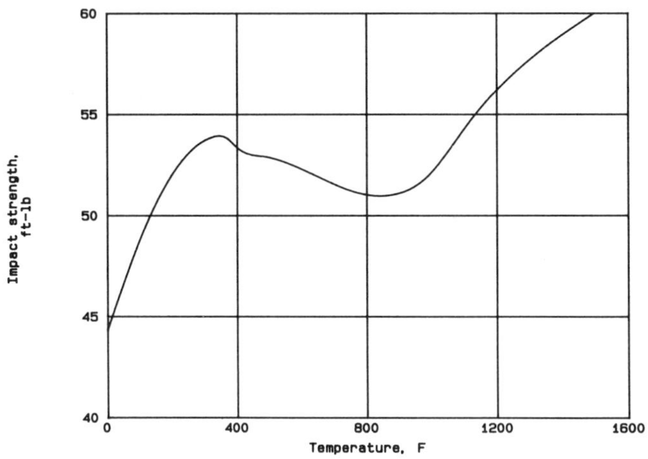

19-9 DL: Charpy V-notch impact strength of bar, hot rolled and aged at 1200 F. Source: Ref 22, p 4

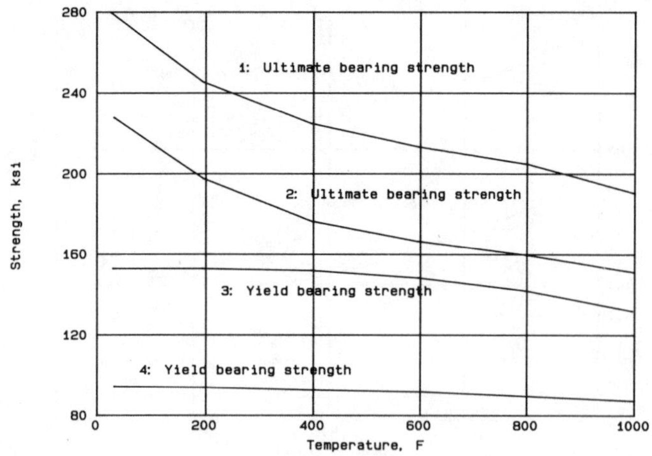

19-9 DL: Bearing strength of sheet. Curves 1 and 3: hot worked at 1200 F. Curves 2 and 4: annealed at 1800 F. e/D = 2.0. Source: Ref 22, p 4

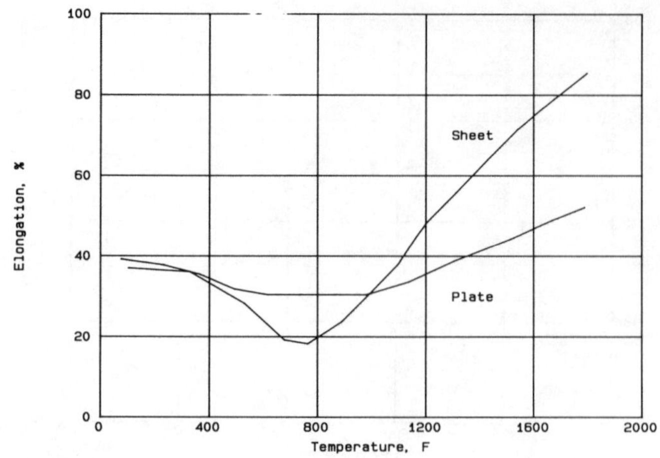

19-9 DL: Tensile elongation of 0.10-in. sheet and 0.25-in. plate annealed at 1800 F. Source: Ref 22, p 4

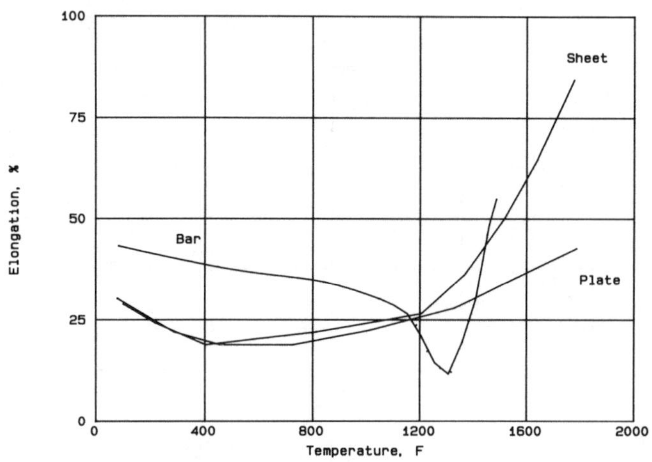

19-9 DL: Tensile elongation of sheet, plate, and bar. Specimens were hot worked and stress relieved at 1200 F. Source: Ref 22, p 3

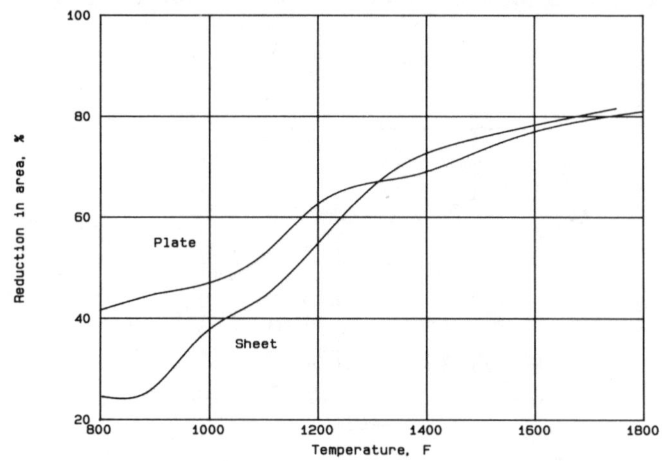

19-9 DL: Tensile reduction in area of 0.10-in. sheet and 0.25-in. plate annealed at 1800 F. Source: Ref 22, p 4

19-9 DL: Minimum creep rates

TEMP C	TEMP F	CRSTR 0.0001%/h MPa	CRSTR 0.0001%/h ksi	CRSTR 0.00001%/h MPa	CRSTR 0.00001%/h ksi
425	800	524	76	228	33
538	1000	276	40	131	19
649	1200	134	19.5	68	9.8
732	1350	69	10	36	5.2
816	1500	38	5.5	...	...

Note: For tests below 704 C (1300 F), specimens were warm worked and stress relieved. For tests above 704 C (1300 F), specimens were solution treated and aged. These data represent bar sizes from 20- to 75-mm (0.75- to 3.0-in.) diameter with test specimens taken at one-half radius location
Source: Ref 28, p 98

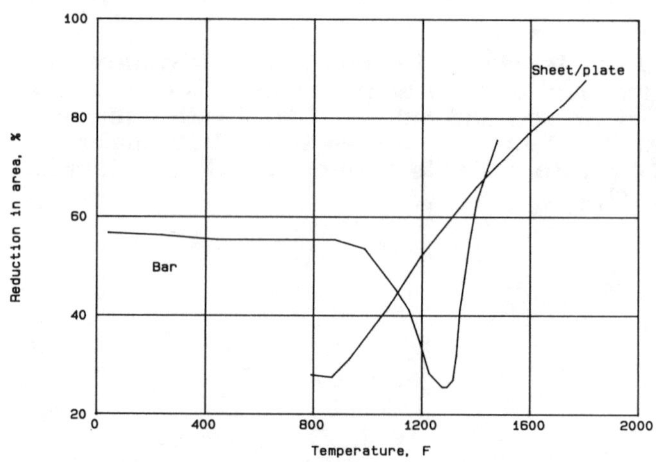

19-9 DL: Tensile reduction in area of sheet, plate, and bar, hot worked and stress relieved at 1200 F. Source: Ref 22, p 3

19-9 DL: Stress relaxation

TEMP C	TEMP F	CRIS(a) MPa	CRIS(a) ksi	CRSRD 1000 h MPa	CRSRD 1000 h ksi	CRSRD 10000 h MPa	CRSRD 10000 h ksi
566	1050	196	28.43	155	22.5	139	20.1
649	1200	149	21.68	105	15.2	90	13
733	1350	103	14.99	48	6.9	28	4.1
816	1500	57	8.27	11	1.6	6	0.8

Note: Specimens were hot rolled and stress relieved. These data represent bar sizes from 20- to 76-mm (0.75- to 3.0-in.) diameter with test specimens taken at one-half radius location.
(a) Initial stress required to produce a limiting deformation (strain) of 0.173% at test temperature
Source: Ref 28, p 98

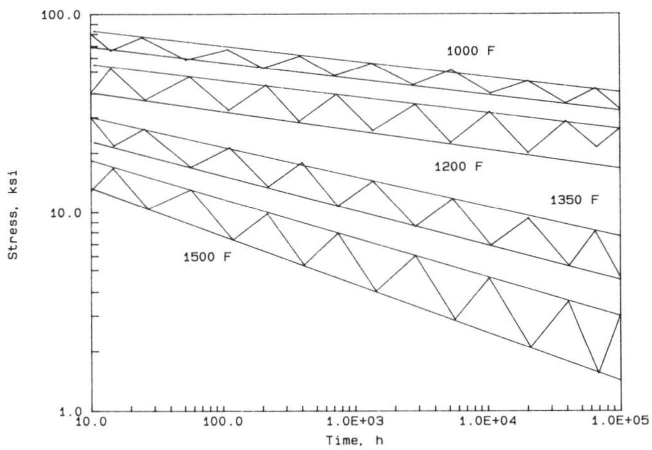

19-9 DL: Creep-rupture scatterbands for bar at 1000 to 1500 F. Rupture below 1300 F: maximum value for hot worked and stress relieved; above 1300 F, maximum value for solution treated and aged. Source: Ref 22, p 5

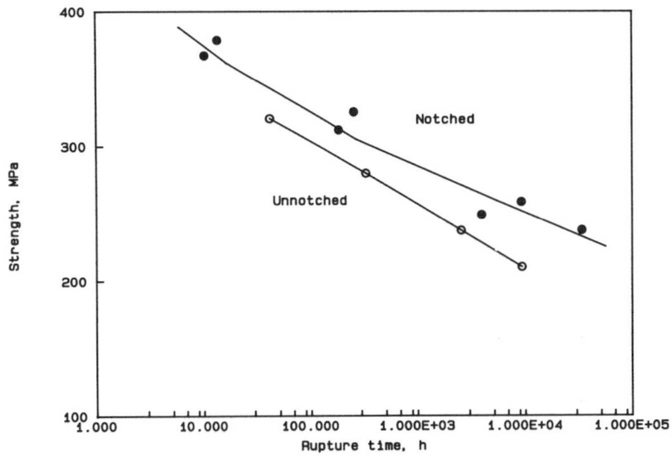

19-9 DL: Notch strengthening of material heat treated for 50 h at 650 C and air cooled. Notched depth = (D-d)/2 = 1.5 mm. Root radius = 0.13 mm. Source: Ref 8, p 315

19-9 DL: Fatigue strength

TEMP C	TEMP F	FATST MPa	FATST ksi
21	70	558	81
538	1000	427	62
649	1200	359	52

Note: Specimens were hot rolled and stress relieved at 649 C (1200 F) to 241 to 288 HB. These data represent bar sizes from 20- to 75-mm (0.75-in. to 3-in.) diameter with test specimens taken at one-half radius location
Source: Ref 28, p 98

19-9 DL: Damping capacity

TEMP C	TEMP F	DMPCP(a) at 5 ksi	DMPCP(a) at 15 ksi	DMPCP(a) at 25 ksi
21	70	0.0012	0.0031	0.0073
316	600	0.0003	0.0018	0.0042
482	900	0.0003	0.0006	0.001
649	1200	0.0009	0.0012	0.0018
816	1500	0.0076	0.008	0.012

Note: Specimens were stabilized for 50 h at 816 C (1500 F)
(a) Logarithmic decrement was determined by tuning fork method
Source: Ref 28, p 98

17-14 CuMo

Alternate designation: AISI 653
Nominal composition: 16.000% Cr, 14.000% Ni, 2.500% Mo, 0.400% Nb, 0.300% Ti, 62.400% Fe, 0.120% C, 0.750% Mn, 0.500% Si, 3.000% Cu
Product forms: Wrought, bar, rod, wire, plate, sheet, strip, tube, pipe, billets, forgings, structural shapes
Description: Iron-base alloy that combines fabricability with useful strength after long-time exposures to temperatures as high as 815 C (1500 F). When used at 730 C (1350 F) or above for long times, creep and stress rupture properties are optimized by annealing for 0.5 h at 1230 C (2250 F), water quenching, and then aging for 5 h at 730 C (1350 F). For service at lower temperatures and for shorter times at 730 C (1350 F) or above, an 1120 C (2050 F) anneal and 730 C (1350 F) age will give properties nearly equal to those obtained using the higher temperature solution anneal. The latter treatment may be a more economical choice, particularly if scaling poses a problem. For service temperatures up to 650 C (1200 F), strength can be enhanced by cold or hot work prior to aging.
Typical applications: Superheater tubing in power plants, blowers, or fans for handling high-temperature gases, high-temperature bolting, tubing for chemical and petrochemical plant use, salt bath hooks, high-temperature conveyors, and sintering plant conveyors.

Physical Properties

17-14 CuMo: Modulus of elasticity

TEMP C	TEMP F	%E at 21 C (70 F)
260	500	91
371	700	87
482	900	82
593	1100	77
704	1300	73
816	1500	69

Note: Modulus of elasticity of specimens solution annealed at 1232 C (2250 F) and aged at 727 C (1340 F) was 28 x 10^3 ksi
Source: Ref 28, p 99

17-14 CuMo: Thermal conductivity

TEMP C	TEMP F	THCON W/m*K	THCON Btu/h*ft*F
38	100	14	8.2
93	200	15	8.7
149	300	16	9.2
204	400	17	9.7
260	500	18	10.2
316	600	19	10.7

Source: Ref 28, p 99

17-14 CuMo: Mean coefficients of linear thermal expansion

TEMP C	TEMP F	THEXP 10^{-6}/C	THEXP 10^{-6}/F
21-149	70-300	14.81	8.23
21-260	70-500	15.75	8.75
21-371	70-700	16.51	9.17
21-482	70-900	16.99	9.44
21-538	70-1000	17.44	9.69
21-704	70-1300	17.84	9.91

Source: Ref 28, p 99

Mechanical Properties

17-14 CuMo: Tensile property data for sheet

TEMP C	TEMP F	UTS MPa	UTS ksi	YS 0.2% MPa	YS 0.2% ksi	EL %
24	75	605	87.8	292	42.3	40
565	1050	495	71.8	232	33.6	31
649	1200	422	61.2	215	31.2	35
732	1350	320	46.4	201	29.1	46
815	1500	223	32.3	152	22.1	32

Note: Specimens treated at 1121 C (2050 F) for 5 min, at 732 C (1350 F) for 5 h, and air cooled. All tensile specimens were held at test temperature 20 min before loading
Source: Ref 28, p 100

17-14 CuMo: Tensile property data for bar

TEMP C	TEMP F	UTS MPa	UTS ksi	YS 0.2% MPa	YS 0.2% ksi	EL %	RA %
1232 C (2250 F) for 30 min, water quenched, 1010 C (1850 F) for 5 h, water quenched							
24	75	593	86	290	42	45	63
149	300	531	77	214	31	40	63
316	600	517	75	207	30	35	55
482	900	503	73	186	27	35	41
538	1000	496	72	193	28	32	45
649	1200	441	64	186	27	30	40
732	1350	365	53	179	26	29	37
815	1500	234	34	176	25.5	29	34
871	1600	172	25	...	...	21	20
927	1700	138	20	107	15.5	19	20
1121 C (2050 F) for 30 min, water quenched, 732 C (1350 F) for 5 h, air cooled							
24	75	614	89	283	41	45	59
149	300	545	79	234	34	42	68
316	600	517	75	200	29	40	65
482	900	510	74	...	...	39	60
538	1000	490	71	179	26	38	60
649	1200	427	62	179	26	37	65
732	1350	324	47	165	24	52	71
815	1500	228	33	145	21	64	72
871	1600	165	24	90	13	67	70
927	1700	131	19	83	12	66	81

Note: Specimens were held at test temperature 20 min before loading
Source: Ref 28, p 100

17-14 CuMo: Effect of thermal exposure on room-temperature tensile properties of bar

EXP h	UTS MPa	UTS ksi	YS 0.2% MPa	YS 0.2% ksi	EL %	RA %
Exposed at 593 C (1100 F)						
0	601	87.2	360	52.2	46	61
500	599	86.9	331	48	49	62
1000	626	90.8	359	52	45	59
3000	670	97.1	421	61	33	56
5000	659	95.6	396	57.5	39	54
Exposed at 649 C (1200 F)						
0	601	87.2	360	52.2	46	61
500	636	92.3	360	52.2	38	55
1000	635	92.1	362	52.5	38	57
3000	660	95.7	384	55.7	34	49
5000	702	101.8	379	55	31	42
Exposed at 704 C (1300 F)						
0	601	87.2	360	52.2	46	61
500	633	91.8	379	55	38	55
1000	650	94.3	386	56	36	54
3000	676	98.1	414	60	31	45
5000	686	99.5	365	53	32	39
Exposed at 760 C (1400 F)						
0	601	87.2	360	52.2	46	61
500	610	88.4	339	49.2	36	53
1000	610	88.5	294	42.6	35	49
3000	627	91	308	44.7	35	48
5000	621	90	321	46.5	32	45
Exposed at 816 C (1500 F)						
0	601	87.2	360	52.2	46	61
500	603	87.5	339	49.2	37	55
1000	611	88.6	344	49.9	36	53
3000	586	85	319	46.2	36	46
5000	586	85	303	44	34	44

Note: Initial condition: annealed at 1232 C (2250 F) and aged at 732 C (1350 F)
Source: Ref 28, p 101

17-14 CuMo: Effect of long-term thermal exposure on tensile properties of bar

EXP h	UTS MPa	UTS ksi	YS 0.2% MPa	YS 0.2% ksi	EL %	RA %
Exposed at 593 C (1100 F)						
500	490	71	262	38	28	32
1000	493	71.5	321	46.5	28	32
3000	483	70	319	46.3	24	32
5000	453	65.7	349	50.6	19	35
Exposed at 649 C (1200 F)						
500	402	58.3	305	44.2	24	35
1000	401	58.2	328	47.6	26	36
3000	391	56.7	308	44.7	22	34
5000	368	53.3	290	42	21	36
Exposed at 704 C (1300 F)						
500	346	50.2	284	41.2	27	35
1000	336	48.8	276	40.1	27	32
3000	336	48.7	308	44.6	25	32
5000	338	49	248	36	24	33
Exposed at 760 C (1400 F)						
500	250	36.2	214	31	31	37
1000	252	36.6	231	33.5	20	34
3000	241	35	198	28.7	25	36
5000	248	36	231	33.5	25	34
Exposed at 816 C (1500 F)						
500	200	29	190	27.5	27	35
1000	201	29.1	190	27.5	25	33
3000	169	24.5	169	24.5	12	14
5000	163	23.7	163	23.7	9	15

Note: Initial condition: annealed at 1232 C (2250 F) and aged at 732 C (1350 F)
Source: Ref 28, p 101

17-14 CuMo: Izod impact strength after long-term thermal exposure at 649 C (1200 F)

EXP h	IS J	IS ft-lb	HARD HB
5	142-155	105-114	80
25	136-140	100-103	81
100	129-140	95-103	85
500	106-108	78-80	89
1000	95-104	70-77	91

Note: Specimens were annealed at 1232 C (2250 F) and tested at 21 C (70 F)
Source: Ref 28, p 102

17-14 CuMo: Fatigue strength of bar products

TEMP C	TEMP F	FATST(a) MPa	FATST(a) ksi
R.R. Moore rotating beam test			
649	1200	241	35
727	1350	179	26
816	1500	131	19
Westinghouse 120-cycle electromagnetic bending test			
649	1200	269	39
816	1500	131	19

Note: Specimens were solution annealed at 1232 C (2250 F) and aged at 732 C (1350 F)
(a) At 10^8 cycles
Source: Ref 28, p 101

17-14 CuMo: Charpy V-notch impact strength after long-term thermal exposure

EXP TEMP C	EXP TEMP F	EXP h	IS(a) J	IS(a) ft-lb	IS(b) J	IS(b) ft-lb	HARD HB
(c)	(c)	...	...	...	35-35	26-26	77
593	1100	100	22-23	16-17	26-33	19-24	84
593	1100	500	20-20	15-15	27-27	20-20	86
593	1100	1000	19-20	14-15	27-37	20-20	84
593	1100	3000	19-19	14-14	23-24	17-18	93
593	1100	5000	16-18	12-13	20-22	15-16	89
649	1200	100	18-18	13-13	27-27	20-20	89
649	1200	500	15-16	11-12	20-20	15-15	95
649	1200	1000	16-16	12-12	19-19	14-14	92
649	1200	3000	15-15	11-11	19-19	14-14	93
649	1200	5000	14-15	10-11	14-15	10-11	94
704	1300	100	19-22	14-16	23-23	17-17	89
704	1300	500	20-20	15-15	22-22	16-16	87
704	1300	1000	19-19	14-14	19-19	14-14	88
704	1300	3000	16-18	12-13	14-14	10-10	90
704	1300	5000	12-14	9-10	14-15	10-11	92
760	1400	100	15-15	11-11	22-26	16-19	88
760	1400	500	14-18	10-13	19-22	14-16	87
760	1400	1000	15-16	11-12	18-22	13-14	89
760	1400	3000	12-14	9-10	14-14	10-10	87
760	1400	5000	11-14	8-9	12-12	9-9	91
816	1500	100	...	...	23-23	17-17	84
816	1500	500	...	...	22-22	16-16	84
816	1500	1000	...	...	19-19	14-14	86
816	1500	3000	...	...	14-14	10-10	86
816	1500	5000	11-11	8-8	11-11	8-8	84

Note: Specimens were 7.62 mm^2 (0.300 in.^{2}) annealed at 1232 C (2250 F) and aged at 732 C (1350 F).
(a) Tested at exposure temperature. (b) Tested at 21 C (70 F).
(c) As received
Source: Ref 28, p 102

17-14 CuMo: Creep strength of bar and sheet

TEMP C	TEMP F	CRSTR(a) 0.1% 1000 h MPa	CRSTR(a) 0.1% 1000 h ksi	CRSTR(a) 0.01% 1000 h MPa	CRSTR(a) 0.01% 1000 h ksi	CRSTR(b) 0.1% 1000 h MPa	CRSTR(b) 0.1% 1000 h ksi	CRSTR(b) 0.01% 1000 h MPa	CRSTR(b) 0.01% 1000 h ksi
Bar									
482	900	...	...	Over 290	Over 42	...	...	...	...
566	1050	...	...	Over 310	Over 45	269	39	214	31
649	1200	165	24	131	19	165	24	117	17
732	1350	110	16	69	10	76	11	41	6
816	1500	76	11	47	6.8	...	...	...	...
Sheet									
649	1200	...	...	...	...	165(c)	24(c)	117(c)	17(c)
732	1350	...	...	...	...	76(c)	11(c)	41(c)	6(c)

Note: All creep specimens held at temperature for 24 h before loading.
(a) Held at 1232 C (2250 F) for 30 min, water quenched, at 732 C (1350 F) for 5 h, water quenched. (b) Held at 1121 C (2050 F) for 30 min, water quenched, at 732 C (1350 F) for 5 h, water quenched. (c) Sheet specimens treated at 1121 C (2050 F) for 5 min, air cooled, treated at 732 C (1350 F) for 5 h, air cooled
Source: Ref 28, p 100

17-14 CuMo: Stress-to-rupture property data for sheet

TEMP C	TEMP F	RPSTR 100 h MPa	RPSTR 100 h ksi	RPSTR 1000 h MPa	RPSTR 1000 h ksi	RPSTR 10000 h MPa	RPSTR 10000 h ksi	EL 100 h %	EL 1000 h %	RA 100 h %	RA 1000 h %
Condition 1											
649	1200	276	40	234	34	200	29	22	16	...	...
727	1350	165	24	116	16.8	77	11.2	19	12	...	...
816	1500	83	12	50	7.2	30	4.4	17	8	...	...
Condition 2											
649	1200	345	50	283	41	228	33	11	7.5	20.2	15

Note: Condition 1: Held at 1121 C (2050 F) for 5 min, air cooled, 732 C (1350 F) for 5 h, air cooled. Condition 2: Hot rolled, at 1121 C (2050 F) for 30 min, water quenched, hot/cold rolled 20% reduction from 732 C (1350 F). Test specimens held at temperature 24 h before loading
Source: Ref 28, p 100

17-14 CuMo: Stress relaxation

TEMP C	TEMP F	CRSTN %	CRIS MPa	CRIS ksi	CRSRD 100 h MPa	CRSRD 100 h ksi	CRSRD 1000 h MPa	CRSRD 1000 h ksi	CRSRD 10000 h MPa	CRSRD 10000 h ksi
At 1232 C (2250 F) for 30 min, water quenched, 732 C (1350 F) for 5 h, water quenched										
649	1200	0.356	296	43.0	...	...	145	21.0	123	17.9
727	1350	0.175	248	36.0	...	...	79	11.5	50	7.2
816	1500	0.100	176	25.5	...	...	48	6.9	29	4.2
At 1232 C (2250 F) for 30 min, water quenched, cold drawn 16%, 649 C (1200 F) for 5 h, water quenched										
566	1050	0.278	476	69.0	...	...	310	45.0	268	38.8
649	1200	0.278	476	69.0	...	...	189	27.4	175	25.4
At 1121 C (2050 F) for 30 min, water quenched, hot/cold rolled 20% from 732 C (1350 F)										
649	1200	0.247	477	69.16	160	23.22	154	22.35	136	19.7

Source: Ref 28, p 101

10. ACI Casting Alloys

ACI Casting Alloys

Castings are classified as heat resistant if they can withstand continuous or intermittent exposure to operating temperatures in excess of 650 C (1200 F). Cast ferrous metals that qualify for this designation include the Fe-Cr, Fe-Cr-Ni, and Fe-Ni-Cr grades in the Alloy Casting Institute "H" series.

Fe-Cr alloys contain 10 to 30% Cr and little or no nickel. These alloys are useful chiefly for resistance to oxidation; they have low strength at elevated temperatures. Use of these alloys is restricted to conditions, either oxidizing or reducing, that involve low static loads and uniform heating. Chromium content depends on anticipated service temperature.

Fe-Cr-Ni alloys contain more than 13% Cr and more than 7% Ni (always more chromium than nickel). These austenitic alloys ordinarily are used under oxidizing or reducing conditions similar to those withstood by the ferritic iron-chromium alloys, but in service they have greater strength and ductility than the straight chromium alloys. They are used, therefore, to withstand greater loads and moderate changes of temperature. These alloys also are used in the presence of oxidizing and reducing gases that are high in sulfur content.

Fe-Ni-Cr alloys contain more than 25% Ni and more than 10% Cr (always more nickel than chromium). These austenitic alloys are used for withstanding reducing as well as oxidizing atmospheres, except where sulfur content is appreciable. (In atmospheres containing 0.05% or more hydrogen sulfide, for example, Fe-Cr-Ni alloys are recommended.) In contrast with Fe-Cr-Ni alloys, Fe-Ni-Cr alloys do not carburize rapidly or become brittle and do not take up nitrogen in nitriding atmospheres. These characteristics become enhanced as nickel content is increased, and in carburizing and nitriding atmospheres casting life increases with nickel content. Austenitic Fe-Ni-Cr alloys are used extensively under conditions of severe temperature fluctuations such as those encountered by fixtures used in quenching and by parts that are not heated uniformly or that are heated and cooled intermittently. In addition, these alloys have characteristics that make them suitable for electrical-resistance heating elements.

Physical Properties

Compositions of ACI castings

ACI designation	UNS No.	ASTM specification	C, %	Cr, %	Ni, %	Si, % max
HA	J82090	A217	0.20 max	8-10	...	1.00
HC	J92605	A297, A608	0.50 max	26-30	4 max	2.00
HD	J93005	A297, A608	0.50 max	26-30	4-7	2.00
HE	J93403	A297, A608	0.20-0.50	26-30	8-11	2.00
HF	J92603	A297, A608	0.20-0.40	19-23	9-12	2.00
HH	J93503	A297, A608	0.20-0.50	24-28	11-14	2.00
HI	J94003	A297, A567, A608	0.20-0.50	26-30	14-18	2.00
HK	J94224	A297, A351, A567, A608	0.20-0.60	24-28	18-22	2.00
HL	J94604	A297, A608	0.20-0.60	28-32	18-22	2.00
HN	J94213	A297, A608	0.20-0.50	19-23	23-27	2.00
HP	J95705	A297	0.35-0.75	24-28	33-37	2.00
HP-50WZ	...	...	0.45-0.55	24-28	33-37	2.50
HT	J94605	A297	0.35-0.75	15-19	33-37	2.50
HU	J95405	A297, A608	0.35-0.75	17-21	37-41	2.50
HW	...	A297, A608	0.35-0.75	10-14	58-62	2.50
HX	...	A297, A608	0.35-0.75	15-19	64-68	2.50

Note: ASTM designations are the same as ACI designations. Rem Fe in all compositions. Manganese content: 0.35 to 0.65% for HA; 1% for HC; 1.5% of HD and 2% for the other alloys. Phosphorus and sulfur contents: 0.04% max for all but HP-50WZ. Molybdenum is intentionally added only to HA, which has 0.90 to 1.20% Mo; maximum for other alloys is set at 0.5% Mo. HH also contains 0.2% max N. HP-50WZ also contains 4 to 6% W, 0.1 to 1.0% Zr, and 0.035% max S and P
Source: Ref 30, p 16-18

Mechanical Properties

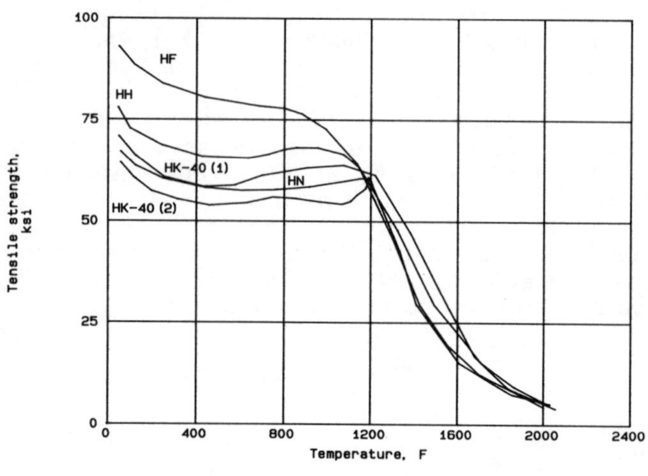

ACI casting alloys: Tensile strength of heat-resistant casting alloys. HK-40 (1): 0.35% C. HK-40 (2): 0.44% C. Source: Ref 21, p 16

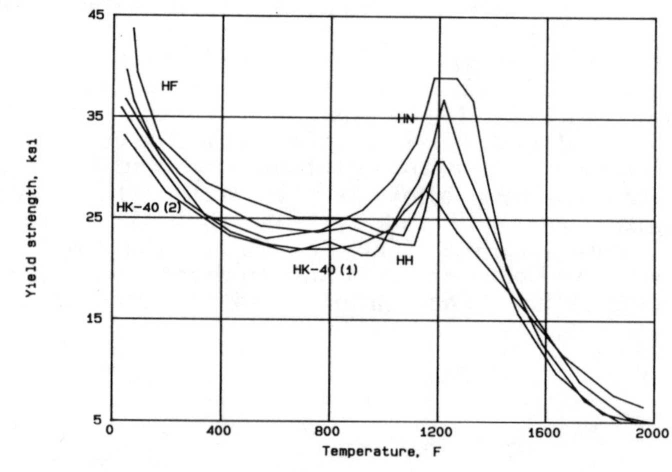

ACI casting alloys: Yield strength of heat-resistant casting alloys. HK-40 (1): 0.35% C. HK-40 (2): 0.44%. Source: Ref 21, p 16

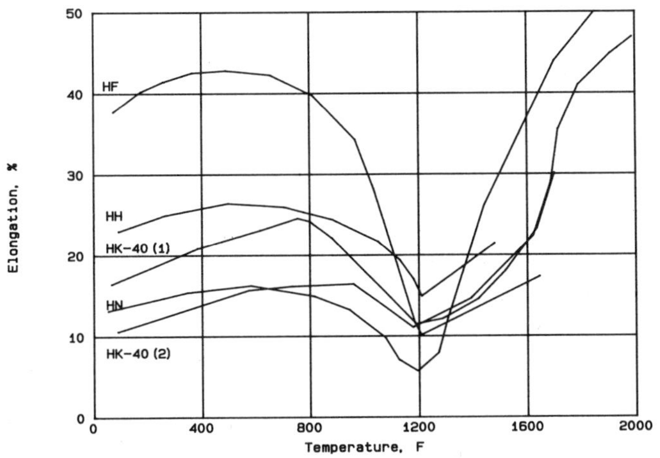

ACI casting alloys: Tensile elongation of heat-resistant casting alloys. HK-40 (1): 0.35% C. HK-40 (2): 0.44% C. Source: Ref 21, p 16

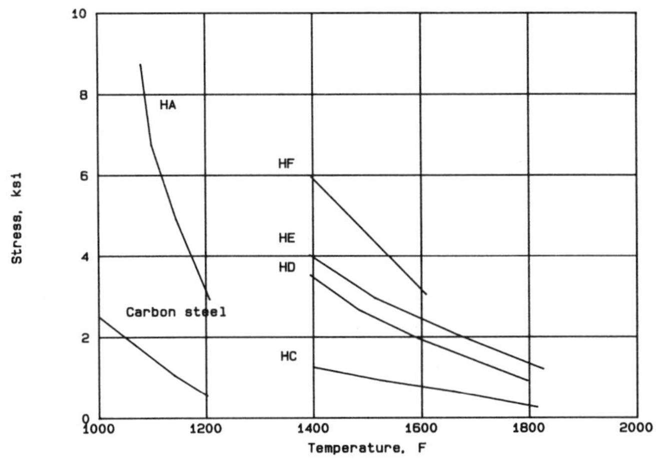

ACI casting alloys: Stress to produce 0.0001%/h minimum creep rate in various heat-resistant casting alloys. Source: Ref 21, p 25

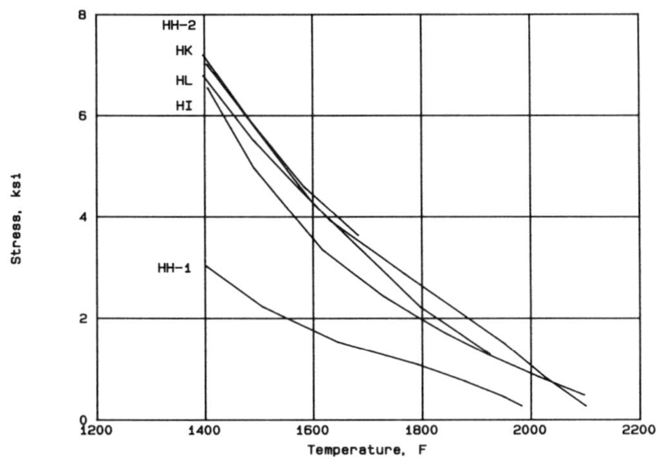

ACI casting alloys: Stress to produce 0.0001%/h minimum creep rate in various heat-resistant casting alloys. HH-1: partially ferritic. HH-2: wholly austenitic. Source: Ref 21, p 25

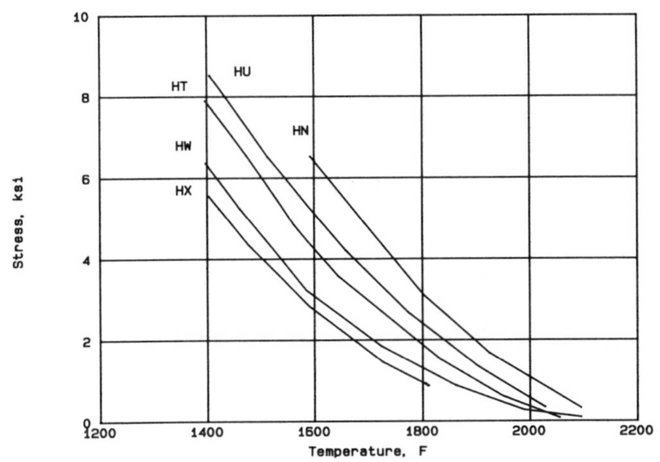

ACI casting alloys: Stress to produce 0.0001%/h minimum creep rate in various heat-resistant casting alloys. Source: Ref 21, p 25

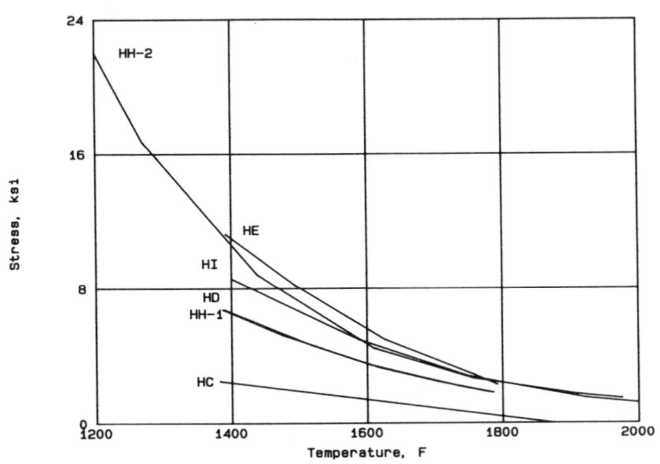

ACI casting alloys: Stress required to produce rupture in 1000 h. HH-1: partially ferritic. HH-2: wholly austenitic. Source: Ref 21, p 26

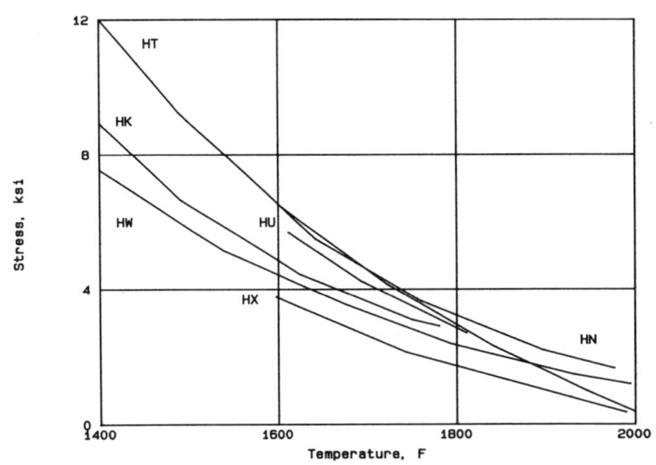

ACI casting alloys: Stress required to produce rupture in 1000 h. Source: Ref 21, p 26

HA

Alternate designation: 9Cr-1Mo
UNS number: J82090
Specification: ASTM A217 (C12), ASTM A426 (CP9)
Nominal composition: 0.200% C max, 8.000 to 10.000% Cr, 0.900 to 1.200% Mo, 1.000% Si max
Product forms: Cast, sand, investment
Description: 9Cr-1Mo is a heat treatable "straight chromium" alloy with good oxidation resistance at temperatures up to 650 C (1200 F). In the normalized and tempered condition, HA exhibits satisfactory toughness throughout its useful temperature range. A 12 to 14% Cr modification is widely used in the glass industry.

Typical applications: Many used in the oil refining industry; also fan blades, furnace roller, Lehr rolls, and fittings and trunnions.

Physical Properties

HA: Thermal conductivity

TEMP C	TEMP F	THCON W/m*K	THCON Btu/ft*h*F
100	212	25.9	15.0
316	600	26.7	15.4
538	1000	27.1	15.7
650	1200	27.3	15.8

Source: Ref 3, p 286

HA: Mean coefficients of linear thermal expansion

TEMP C	TEMP F	THEXP 10^{-6}/C	THEXP 10^{-6}/F
21-100	70-212	11	6.1
21-316	70-600	12	6.5
21-538	70-1000	13	7.1
21-650	70-1200	14	7.5

Source: Ref 3, p 285

HC

Alternate designation: 28Cr
UNS number: J92605
Specification: ASTM A-297-76
Nominal composition: 0.500% C max, 1.000% Mn max, 26.000 to 30.000% Cr, 4.000% Ni max, 2.000% Si max, 0.500% Mo max, 0.040% P max, 0.040% S max
Product forms: Cast, sand, investment
Description: This "straight chromium" alloy is ferritic at all temperatures and resists oxidation and the effects of high sulfur flue gases up to 1100 C (2000 F). It is used for applications where strength is not a consideration, or where only moderate loads are involved, at temperatures as high as 650 C (1200 F). This alloy will embrittle, however, during long-term exposure at 400 to 550 C (750 to 1025 F). A modification of HC with more than 2% Ni and at least 0.15% N features improved room-temperature ductility and impact resistance and enhanced elevated-temperature creep strength.
Typical applications: Fan blades, furnace rollers and blowers, refinery fittings, kiln parts and grate bars, boiler baffles, electrodes, gas outlet dampers, lute rings, rabble blades and holders, recuperators, salt pots, soot blower tubes, support skids, tuyeres. HC-30 alloy is used for tubing in pressure applications at high temperature.

Physical Properties

HC: Mean coefficients of linear thermal expansion

TEMP C	TEMP F	THEXP 10^{-6}/C	THEXP 10^{-6}/F
21-538	70-1000	11.3	6.3
21-650	70-1200	11.5	6.4
21-760	70-1400	11.9	6.6
21-870	70-1600	12.6	7.0
21-980	70-1800	13.3	7.4
21-1093	70-2000	13.9	7.7
650-871	1200-1600	15.7	8.7
650-980	1200-1800	16.7	9.3

Source: Ref 3, p 287

HC: Thermal conductivity

TEMP C	TEMP F	THCON W/m*K	THCON Btu/ft*h*F
100	212	21.8	12.6
538	1000	30.9	17.9
816	1500	35.1	20.3
1093	2000	41.9	24.2

Source: Ref 3, p 287

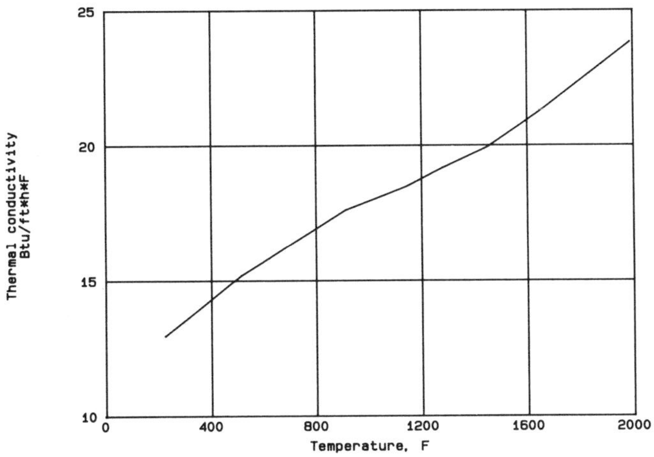

HC: Thermal conductivity. Source: Ref 3, p 286

Mechanical Properties

HC: Creep-rupture property data

TEMP C	TEMP F	CRSTR 0.0001%/h MPa	CRSTR 0.0001%/h ksi	RPSTR 10 h MPa	RPSTR 10 h ksi	RPSTR 100 h MPa	RPSTR 100 h ksi	RPSTR 1000 h MPa	RPSTR 1000 h ksi
760	1400	8.96	1.30	31.7	4.6	22.8	3.3	15.5	2.3
870	1600	5.17	0.75	13.8	2.0	11.7	1.7	8.96	1.3
980	1800	2.48	0.36	7.6	1.1	5.86	0.85	4.27	0.62

Note: For constant temperature; for cyclic temperature, lower values would apply. Values represent test samples containing over 2.0% Ni with 0.15% N
Source: Ref 3, p 287

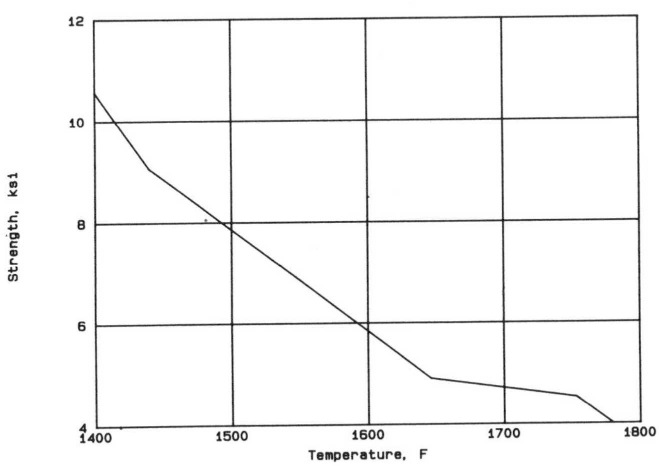

HC: Tensile strength. Source: Ref 3, p 286

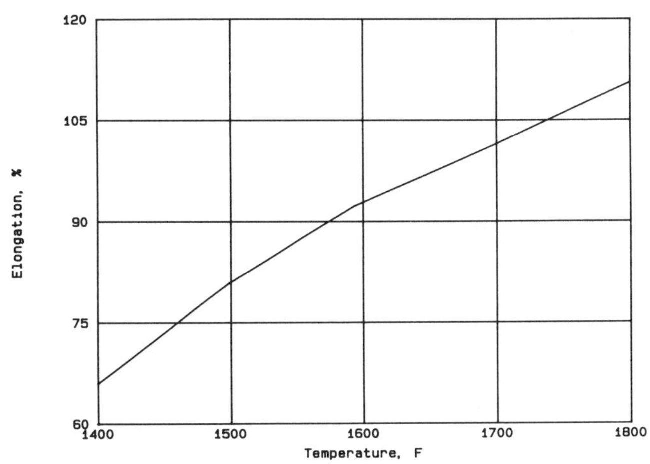

HC: Short-time tensile elongation. Source: Ref 3, p 286

HC: Stress-rupture strength. Source: Ref 3, p 286

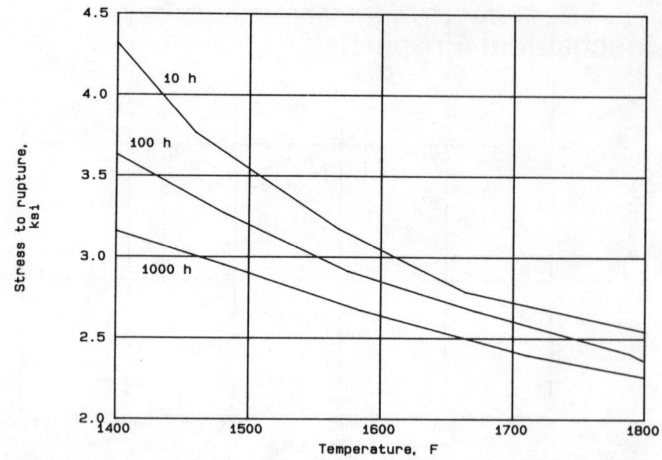

HD

Alternate designation: 28Cr-5Ni
UNS number: J93005
Specification: ASTM A-297-76
Nominal composition: 0.500% C max, 1.500% Mn max, 26.000 to 30.000% Cr, 4.000 to 7.000% Ni, 2.000% Si max, 0.500% Mo max, 0.040% P max, 0.040% S max
Product forms: Cast, sand, investment

Description: HD is an iron-chromium alloy with properties similar to those of HC, but with somewhat higher elevated-temperature strength.
Typical applications: Brazing furnace parts, cracking equipment, furnace blowers, gas burner parts, holding pots, kiln parts, pouring spouts, rabble shoes and arms, and recuperator sections.

Physical Properties

HD: Mean coefficients of linear thermal expansion

TEMP C	TEMP F	THEXP 10^{-6}/C	THEXP 10^{-6}/F
21-538	70-1000	13.9	7.7
21-650	70-1200	14.4	8.0
21-760	70-1400	14.9	8.3
21-870	70-1600	15.5	8.6
21-980	70-1800	16.0	8.9
21-1090	70-2000	16.6	9.2
650-870	1200-1600	18.5	10.3
650-980	1200-1800	19.1	10.6

Source: Ref 3, p 288

HD: Thermal conductivity

TEMP C	TEMP F	THCON W/m*K	THCON Btu/ft*h*F
100	212	21.8	12.6
538	1000	30.9	17.9
816	1500	35.1	20.3
1093	2000	41.9	24.2

Source: Ref 3, p 288

HD: Thermal conductivity. Source: Ref 3, p 286

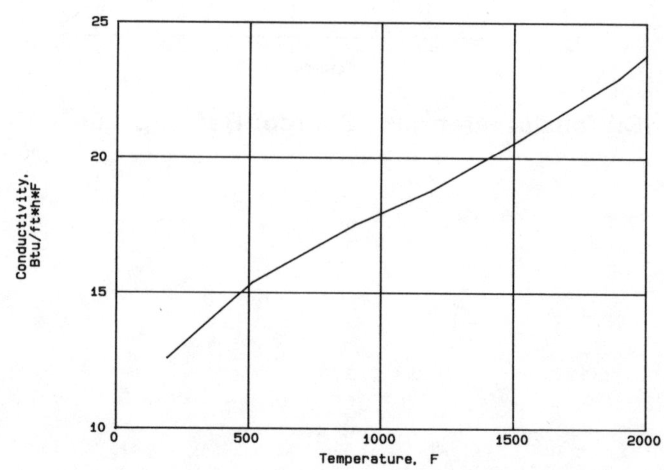

Mechanical Properties

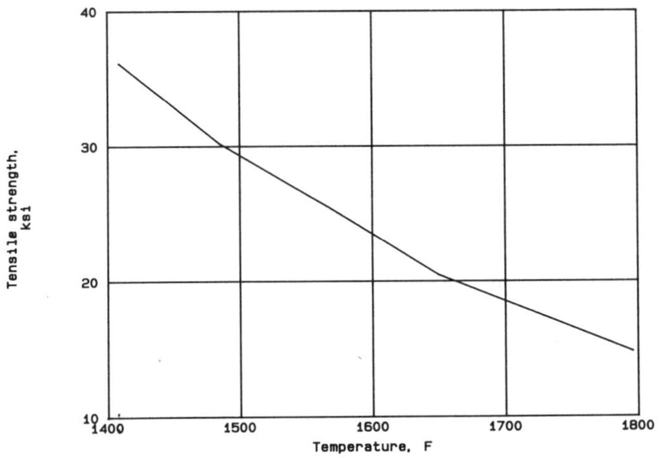

HD: Tensile strength. Source: Ref 3, p 288

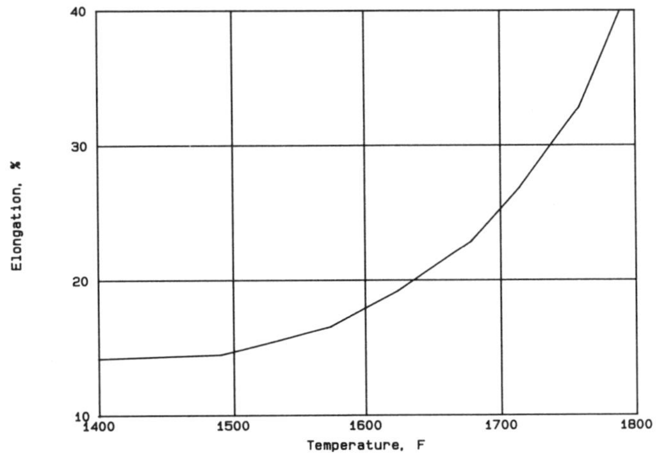

HD: Tensile elongation. Source: Ref 3, p 288

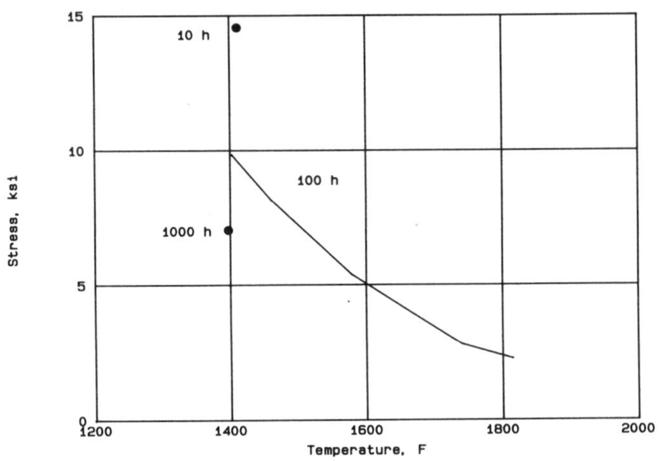

HD: Stress-rupture strength. Source: Ref 3, p 288

HE

Alternate designation: 28Cr-10Ni
UNS number: J93403
Specification: ASTM A297
Nominal composition: 0.200 to 0.500% C, 2.000% Mn max, 26.000 to 30.000% Cr, 8.000 to 11.000% Ni, 2.000% Si max, 0.500% Mo max, 0.040% P max, 0.040% S
Product forms: Cast, sand, investment
Description: This Fe-Cr-Ni alloy has excellent resistance to corrosion at elevated temperatures. It is suitable for applications up to 1100 C (2000 F) and is stronger and more ductile at room temperature than the "straight chromium" grades. Long exposure to temperatures near 815 C (1500 F) promotes formation of sigma phase and consequent embrittlement.

Typical applications: Used extensively in ore roasting equipment. Other applications include billet skids, burner nozzles, dampers, furnace chains and conveyors, furnace door frames, oil burner parts, rabble arms and blades, recuperators, rotating shafts, soot blower elements, steam generator parts, and tube supports.

Physical Properties

HE: Thermal conductivity

TEMP C	TEMP F	THCON W/m*K	THCON Btu/ft*h*F
100	212	14.7	8.5
316	600	18.2	10.5
538	1000	21.5	12.4
650	1200	23.4	13.5
760	1400	25.3	14.6
870	1600	27.5	15.9
980	1800	29.2	16.9
1090	2000	31.5	18.2

Source: Ref 3, p 290

HE: Mean coefficients of linear thermal expansion

TEMP C	TEMP F	THEXP 10^{-6}/C	THEXP 10^{-6}/F
21-538	70-1000	17.3	9.6
21-650	70-1200	17.8	9.9
21-760	70-1400	18.4	10.2
21-870	70-1600	18.9	10.5
21-980	70-1800	19.4	10.8
21-1090	70-2000	19.9	11.1
650-870	1200-1600	21.9	12.2
650-980	1200-1800	22.5	12.5

Source: Ref 3, p 290

Mechanical Properties

HE: Long-time creep-rupture property data

TEMP C	TEMP F	CRSTR 0.0001%/h MPa	CRSTR 0.0001%/h ksi	RPSTR 100 h MPa	RPSTR 100 h ksi
760	1400	27.6	4.0	75.8	11.0
870	1600	16.5	2.4	36.5	5.3
980	1800	9.65	1.4	17.2	2.5
1090	2000	2.76	0.4	...	...

Note: For constant temperature; lower values would apply for cyclic temperature
Source: Ref 3, p 290

HF

Alternate designation: 20Cr-10Ni
UNS number: J92603
Specification: ASTM A297
Nominal composition: 0.200 to 0.400% C, 2.000% Mn max, 18.000 to 23.000% Cr, 8.000 to 12.000% Ni, 2.000% Si max, 0.500% Mo max, 0.040% P max, 0.040% S max
Product forms: Cast, sand, investment
Description: HF is a cast modification of 18-8 stainless steel for applications requiring strength and corrosion resistance at 650 to 870 C (1200 to 1600 F). When used to resist oxidation at elevated temperatures, carbon content should be no more than about 0.20%. Molybdenum, tungsten, niobium, and titanium are sometimes added to improve elevated-temperature strength. HF is susceptible to sigma phase embrittlement after prolonged exposure at 760 to 815 C (1400 to 1500 F).
Typical applications: Arc furnace electrode arms, annealing boxes and trays, baskets, brazing channels, burner tips, burnishing rolls, conveyor belts and chains, fan housings, furnace rails, gas burner rings, hardening retorts, hearth plates, Lehr rolls, pier caps, soaking pit dampers, tempering baskets, and wear plates.

Physical Properties

HF: Mean coefficients of linear thermal expansion

TEMP C	TEMP F	THEXP 10^{-6}/C	THEXP 10^{-6}/F
21-100	70-212	13	7.2
21-538	70-1000	17.8	9.9
21-650	70-1200	18.2	10.1
21-760	70-1400	18.5	10.3
21-871	70-1600	18.7	10.4
21-982	70-1800	18.9	10.5
21-1093	70-2000	19.6	10.9
21-871	1200-1600	19.9	11.1

Source: Ref 3, p 292

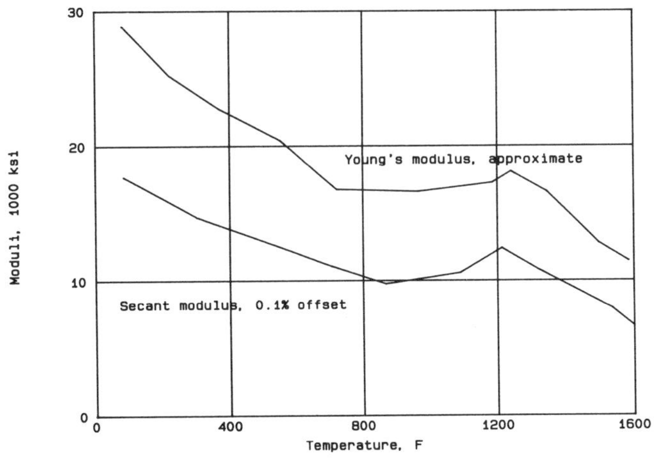

HF: Elastic moduli. Source: Ref 3, p 291

Mechanical Properties

HF: Creep-rupture property data

TEMP C	CRSTR 0.0001% MPa	CRSTR 0.0001% ksi	CRSTR (a) MPa	CRSTR (a) ksi	RPSTR 100 h MPa	RPSTR 100 h ksi	RPSTR 1000 h MPa	RPSTR 1000 h ksi	RPSTR 10000 h MPa	RPSTR 10000 h ksi	RPSTR 1000000 h MPa	RPSTR 1000000 h ksi
649	124	18	77.9	11.3	228	33.0	172	25	114	16.5	75.8	11
760	46.9	6.8	30.3	4.4	93.1	13.5	62.7	9.1	42.1	6.1	27.6	4
871	26.9	3.9	9.65	1.4	49.6	7.2	30.3	4.4	18.6	2.7	11.7	1.7

Note: Representative values for constant temperature; for cyclic temperature, lower values would apply
(a) Limiting stress for 1% total creep in 100,000 h
Source: Ref 3, p 291

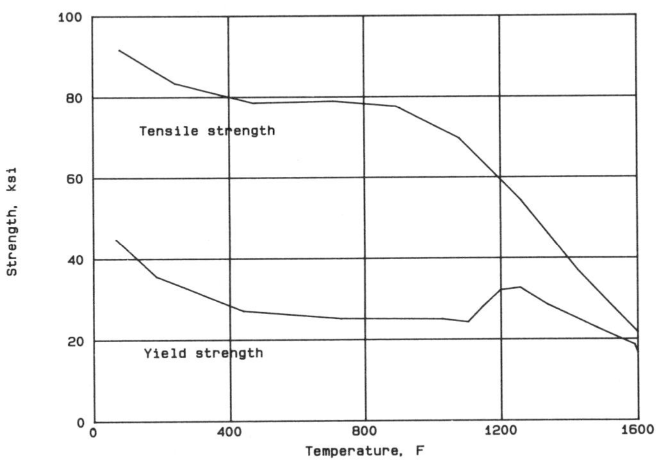

HF: Tensile and yield strength. Source: Ref 3, p 291

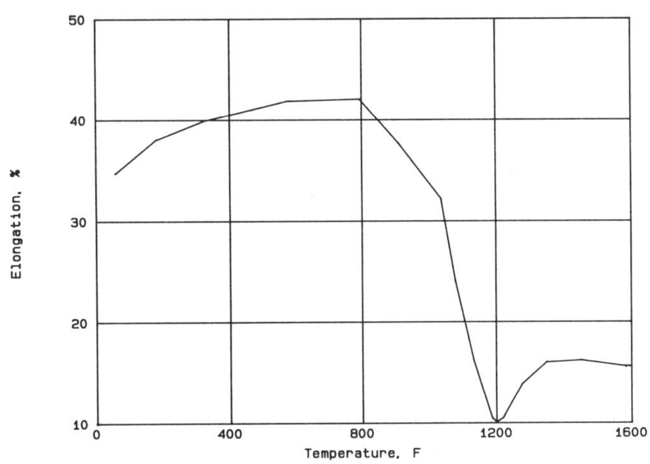

HF: Tensile elongation. Source: Ref 3, p 291

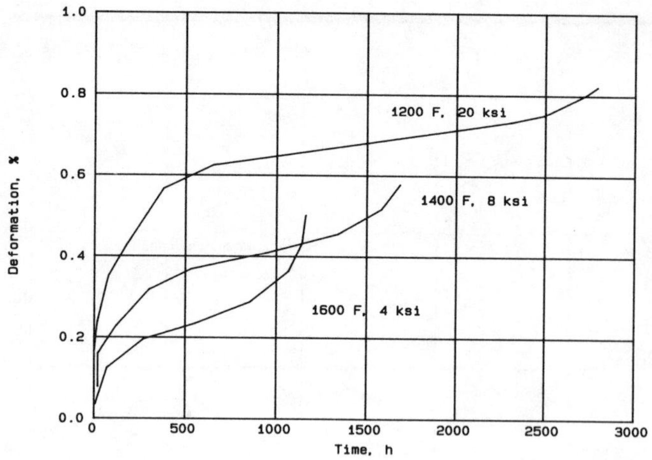

HF: Creep curves for cast specimens. Source: Ref 21, p 23

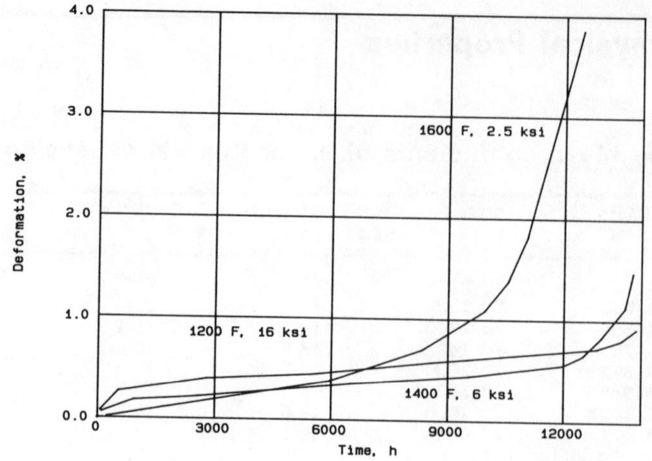

HF: Creep curves for cast specimens. Source: Ref 21, p 23

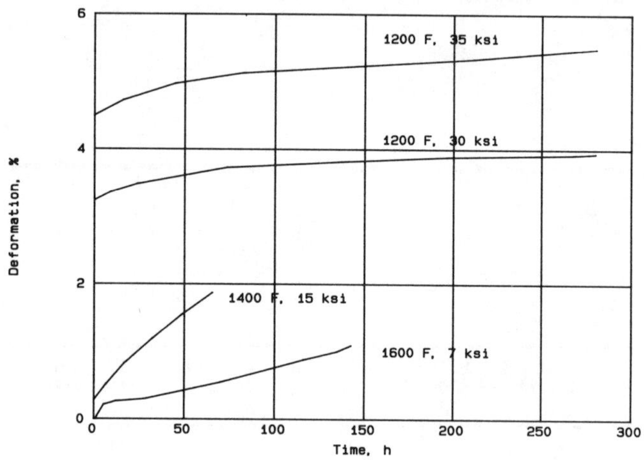

HF: Creep curves for cast specimens. Source: Ref 21, p 23

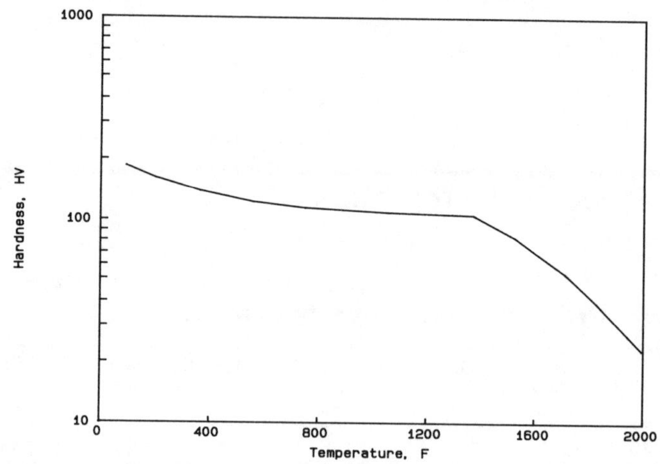

HF: Vickers hardness. Source: Ref 21, p 23

HH

Alternate designation: 26Cr-12Ni
UNS number: J93503
Specifications: ASTM A297
Nominal composition: 0.200 to 0.500% C, 2.000% Mn max, 24.000 to 28.000% Cr, 11.000 to 14.000% Ni, 2.000% Si max, 0.500% Mo max, 0.040% P max, 0.040% S max
Product forms: Cast, sand, investment
Description: This Fe-Cr-Ni alloy is used extensively in temperature applications because of its combination of relatively high strength and good resistance to oxidation at temperatures up to 1100 C (2000 F). To achieve maximum elevated-temperature strength and creep resistance, HH must be fully austenitic. A properly balanced composition is needed to obtain this desirable microstructure. For example, castings for continuous service well above 870 C (1600 F) require a composition of 0.50% C max (0.35 to 0.40% C preferred), 10 to 12% Ni, and 24 to 27% Cr. If service temperatures are from 650 to 870 C (1200 to 1600 F), the optimum composition is 0.40% C max, 11 to 14% Ni, and 23 to 27% Cr.
Typical applications: Annealing trays, billet skids, burner nozzles, carburizing boxes, convection tube supports, dampers, exhaust manifolds, flue gas stacks, grate supports, hardening trays, kiln nose ring segments, muffles, normalizing discs, pier caps, quenching trays, rabble arms and blades, radiant tubes and supports, refractory supports, roller hearths and rails, stoker parts, and tube hangers.

Physical Properties

HH: Thermal conductivity

TEMP C	TEMP F	THCON W/m*K	THCON Btu/ft*h*F
100	212	14.2	8.2
316	600	17.5	10.1
538	1000	20.8	12.0
650	1200	22.5	13.0
760	1400	24.4	14.1
870	1600	26.5	15.3
980	1800	28.2(a)	16.3(a)
1090	2000	30.3(a)	17.5(a)

(a) Estimated
Source: Ref 3, p 295

HH: Mean coefficients of linear thermal expansion

TEMP C	TEMP F	THEXP 10^{-6}/C	THEXP 10^{-6}/F
21-538	70-1000	17.1	9.5
21-650	70-1200	17.5	9.7
21-760	70-1400	17.8	9.9
21-870	70-1600	18.4	10.2
21-980	70-1800	18.9	10.5
21-1090	70-2000	19.3	10.7
650-870	1200-1600	20.5	11.4
650-980	1200-1800	21.1	11.7

Source: Ref 3, p 295

HH: Elastic moduli. Source: Ref 3, p 294

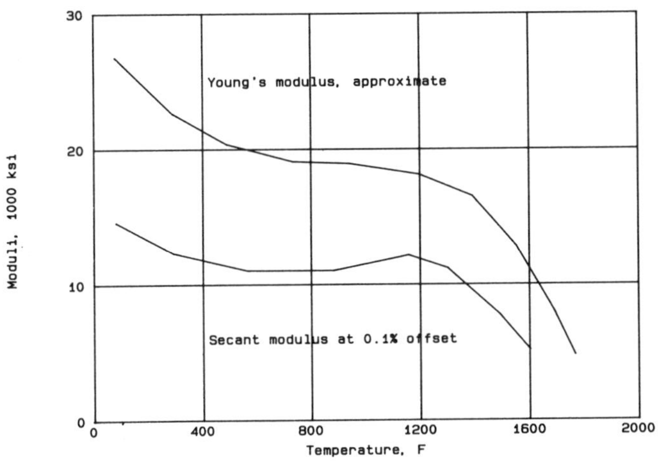

Mechanical Properties

HH: Creep-rupture properties of type I partially ferritic grades

TEMP C	TEMP F	CRSTR 0.0001%/h MPa	CRSTR 0.0001%/h ksi	RPSTR 10 h MPa	RPSTR 10 h ksi	RPSTR 100 h MPa	RPSTR 100 h ksi	RPSTR 1000 h MPa	RPSTR 1000 h ksi
760	1389	20.7	3.0	...	...	96.5	14.0	44.8	6.5
871	1587	11.7	1.7	...	...	44.1	6.4	26.2	3.8
982	1786	7.6	1.1	32	4.7	21.4	3.1	14.5	2.1
1093	1984	2.1	0.3	...	...	10.3	1.5	...	...

Note: Representative long-time values for constant temperature; for cyclic temperature, lower values would apply
Source: Ref 3, p 294

HH: Creep-rupture properties of type II austenitic grades

TEMP C	CRSTR 0.0001%/ h MPa	CRSTR 0.0001%/ h ksi	CRSTR (a) MPa	CRSTR (a) ksi	RPSTR 100 h MPa	RPSTR 100 h ksi	RPSTR 1000 h MPa	RPSTR 1000 h ksi	RPSTR 10000 h MPa	RPSTR 10000 h ksi	RPSTR 100000 h MPa	RPSTR 100000 h ksi
649	124	18	65.5	9.5	241	35	152	22	96.5	14	62.1	9.0(b)
760	43	6.3	13.8	2	96.5	14	55.2	8	33.1	4.8	19.3	2.8(b)
871	26.9	3.9	7.6	1.1	46.9	6.8	26.2	3.8	14.8	2.15	8.3	1.2(b)
982	14.5	2.1	...	...	22.1	3.2	11.4	1.65	5.9(b)	0.86(b)	3.0(b)	0.44(b)
1093	5.5	0.8	...	...	9.7	1.4	4.7	0.68	2.34(b)	0.34(b)	1.0(b)	0.15(b)

Note: Values are representative of long-time tests at constant temperature; for cyclic temperatures, lower values would apply
(a) Limiting stress for 1% total creep in 100,000 h. (b) Extrapolated data
Source: Ref 3, p 294

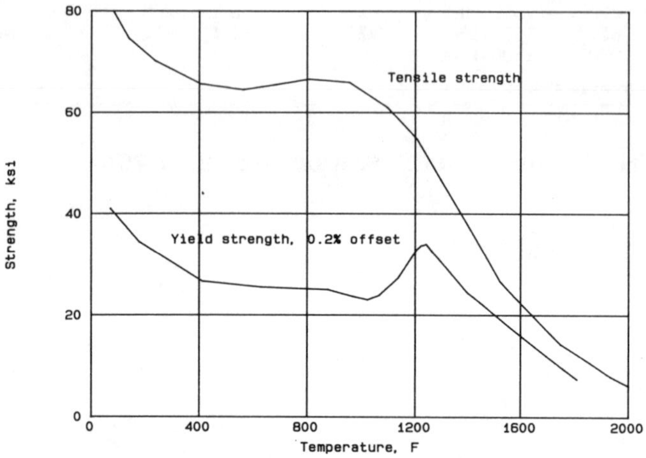

HH: Short-time tensile and yield strength. Source: Ref 3, p 293

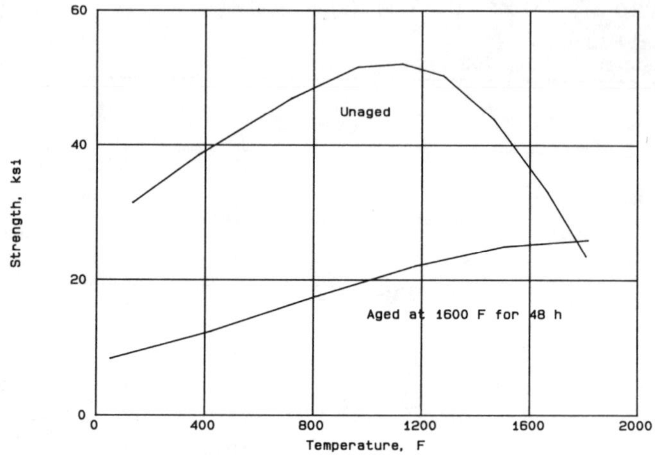

HH: Charpy impact strength of wholly austenitic specimens. Source: Ref 3, p 293

HI

Alternate designation: 28Cr-15Ni
UNS number: J94003
Specification: ASTM A297
Nominal composition: 0.200 to 0.500% C, 2.000% Mn max, 26.000 to 30.000% Cr, 14.000 to 18.000% Ni, 2.000% Si max, 0.500% Mo max, 0.040% P max, 0.040% S max
Product forms: Cast, sand, investment
Description: Similar to HH, but with improved elevated-temperature strength and oxidation resistance due to higher chromium and nickel contents. Maximum service temperature exceeds 1100 C (2000 F). For example, HH has enjoyed extensive use for retorts operating with an internal vacuum at a continuous temperature of 1175 C (2150 F). Room-temperature ductility suffers after service in the 760 to 870 C (1400 to 1600 F) range due to carbide precipitation.
Typical applications: Retorts, billet skids, brazing fixtures, conveyor rollers, furnace rails, hearth plates, lead pots, pier caps, and tube spacers.

Mechanical Properties

HI: Thermal conductivity

TEMP C	TEMP F	THCON W/m*K	THCON Btu/ft*h*F
100	212	14.2	8.2
316	600	17.5	10.1
538	1000	20.8	12.0
650	1200	22.5	13.0
760	1400	24.4	14.1
870	1600	26.5	15.3
980	1800	28.2	16.3
1090	2000	29.8	17.5

Source: Ref 3, p 297

HI: Mean coefficients of linear thermal expansion

TEMP C	TEMP F	THEXP 10^{-6}/C	THEXP 10^{-6}/F
21-538	70-1000	17.8	9.9
21-650	70-1200	18.0	10.0
21-760	70-1400	18.2	10.1
21-870	70-1600	18.5	10.3
21-980	70-1800	18.9	10.5
21-1090	70-2000	19.4	10.8
650-870	1200-1600	19.8	11.0
650-980	1200-1800	21.6	12.0

Source: Ref 3, p 297

HI: Creep-rupture property data

TEMP C	TEMP F	CRSTR 0.0001%/h MPa	CRSTR 0.0001%/h ksi	RPSTR 100 h MPa	RPSTR 100 h ksi	RPSTR 1000 h MPa	RPSTR 1000 h ksi
760	1400	45.5	6.6	89.6	13.0	58.6	8.50
871	1600	24.8	3.6	51.7	7.5	33.0	4.80
982	1800	13.1	1.9	28.3	4.1	17.9	2.60
1093	2000	5.5	0.8	13.1	1.9	8.62	1.25
1177	2150	1.03	0.15	...	...	...	...

Note: Representative long-time values for constant temperature; for cyclic temperature, lower values would apply
Source: Ref 15, p 296

HK

Alternate designation: 26Cr-20Ni
UNS number: J94224
Specifications: ANSI A297, ASTM A297
Nominal composition: 0.200 to 0.600% C, 2.000% Mn max, 24.000 to 28.000% Cr, 18.000 to 22.000% Ni, 2.000% Si max, 0.500% Mo max, 0.040% P max, 0.040% S max
Product forms: Cast, sand, investment
Description: Similar to wholly austenitic HH, HK contains enough chromium to ensure good resistance to corrosion by hot gases, including sulfur-bearing gases, under both oxidizing and reducing conditions. HK alloy castings are widely used for stressed parts in structural applications at temperatures up to 1150 C (2100 F). It is one of the strongest heat-resistant cast alloys at temperatures above 1040 C (1900 F). Embrittling sigma phase can form in the 760 to 870 C (1400 to 1600 F) range. The various available carbon contents help users optimize properties. Lower creep rates and longer rupture lives are favored by higher carbon. On the other hand, high carbon also results in a decrease in room-temperature ductility after elevated-temperature aging.
Typical applications: Billet skids, brazing fixtures, calcining tubes, cement kiln nose segments, conveyor rolls, furnace door arches and lintels, heat treating trays and fixtures, pier caps, rabble arms and blades, radiant tubes, reformer tubes, retorts, rotating shafts, skid rails, sprockets, and stack dampers.

Physical Properties

HK: Thermal conductivity

TEMP C	TEMP F	THCON W/m*K	THCON Btu/ft*h*F
100	212	13.7	7.9
316	600	16.9	9.8
538	1000	20.4	11.8
650	1200	22.3	12.9
760	1400	24.6	14.2
870	1600	27.2	15.7
980	1800	29.6(a)	17.1(a)
1090	2000	32.2(a)	18.6(a)

(a) Estimated
Source: Ref 3, p 299

HK: Mean coefficients of linear thermal expansion

TEMP C	TEMP F	THEXP 10^{-6}/C	THEXP 10^{-6}/F
21-538	70-1000	16.9	9.4
21-650	70-1200	17.3	9.6
21-760	70-1400	17.6	9.8
21-870	70-1600	18.0	10.0
21-980	70-1800	18.4	10.2
650-870	1200-1800	20.5	11.4
650-980	1200-2000	20.7	11.5

Source: Ref 3, p 299

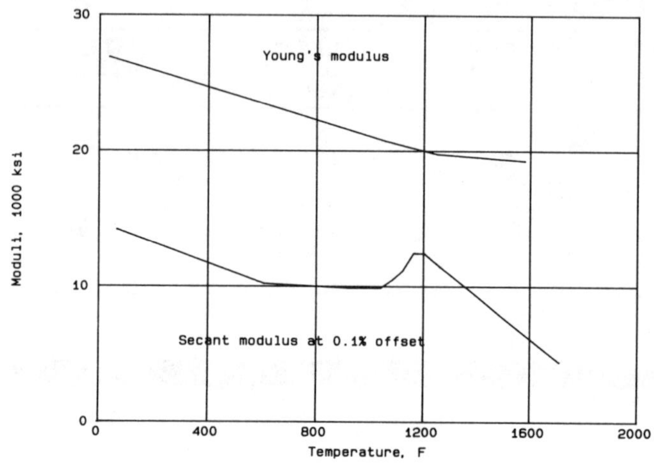

HK-40: Elastic moduli. Source: Ref 3, p 299

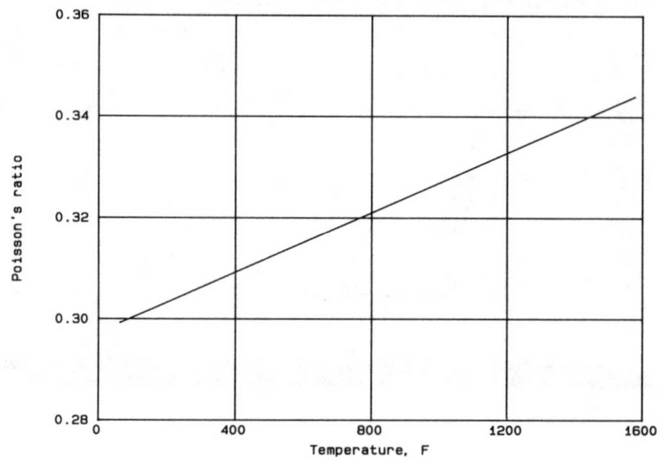

HK-40: Poisson's ratio. Source: Ref 3, p 299

Mechanical Properties

HK: Creep-rupture property data for grade HK-40

TEMP C	TEMP F	CRSTR 0.0001% MPa	CRSTR 0.0001% ksi	CRSTR (a) MPa	CRSTR (a) ksi	RPSTR 100 h MPa	RPSTR 100 h ksi	RPSTR 1000 h MPa	RPSTR 1000 h ksi	RPSTR 10000 h MPa	RPSTR 10000 h ksi	RPSTR 100000 h MPa	RPSTR 100000 h ksi
760	1400	70.3	10.2	43.4	6.3	108.0	15.6	82.7	12.0	60.7	8.8	42.7	6.2
871	1600	41.2	6.0	17.2	2.5	63.4	9.2	41.4	6.0	26.2	3.8	17.2	2.5
982	1800	17.2	2.5	6.20	0.90	32.8	4.75	19.3	2.8	11.7	1.7	6.89	1.0
1038	1900	9.66	1.4	2.69	0.39	22.1	3.2	13.1	1.9	7.58	1.1	4.55	0.66
1093	2000	4.48	0.65	1.57	0.23	15.2	2.2	8.62	1.25	4.96	0.72	2.90	0.42

Note: Representative long-time values for constant temperature; for cyclic temperatures, lower values would apply
(a) Limiting stress for 1% total creep in 100,000 h
Source: Ref 3, p 298

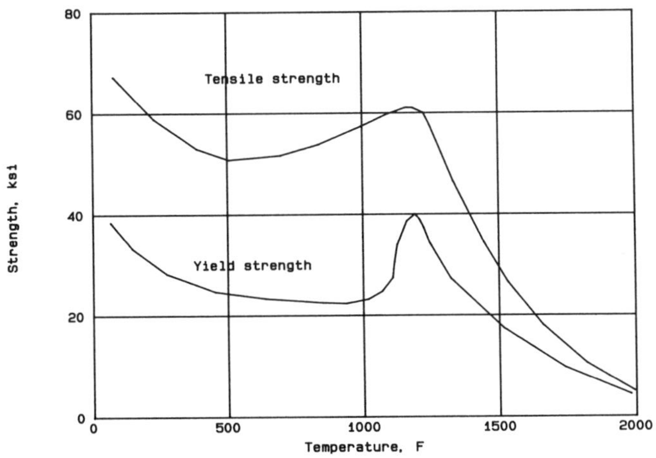

HK-40: Short-time tensile and yield strength. Source: Ref 3, p 298

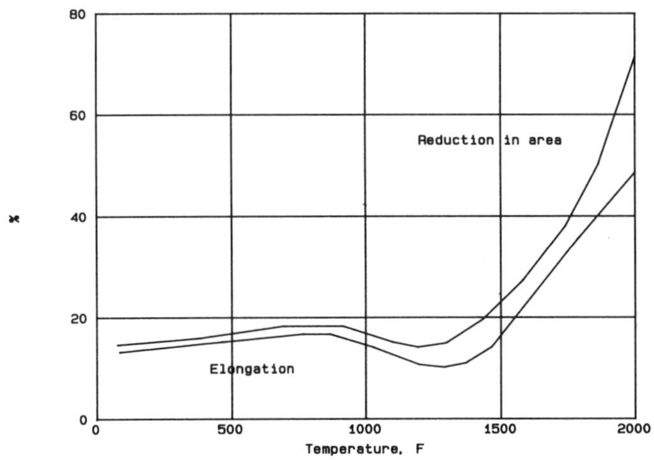

HK-40: Tensile elongation and reduction in area. Source: Ref 3, p 298

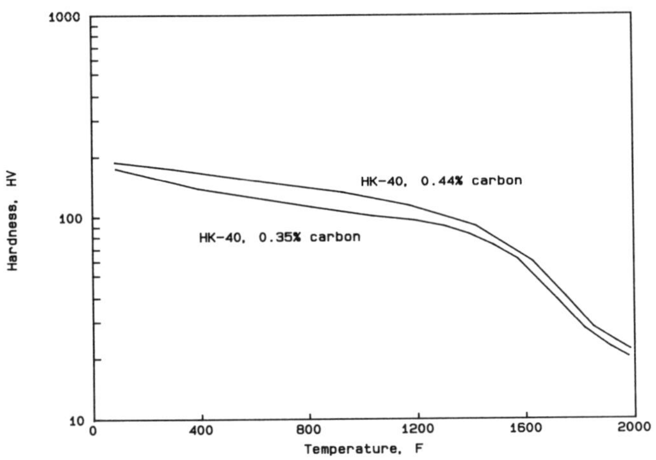

HK-40: Vickers hardness. Source: Ref 3, p 298

HL

Alternate designation: 30Cr-20Ni
UNS number: J94604
Specifications: ANSI A297, ASTM A297
Nominal composition: 0.200 to 0.600% C, 2.000% Mn max, 28.000 to 32.000% Cr, 18.000 to 22.000% Ni, 2.000% Si max, 0.500% Mo max, 0.040% P max, 0.040% S max
Product forms: Cast, sand, investment
Description: Similar to HK, but more resistant to corrosion by hot gases, particularly those containing appreciable amounts of sulfur. HL is often the choice for high-strength, elevated-temperature applications that cannot tolerate excessive scaling.

Typical applications: Carrier fingers, enameling furnace fixtures, furnace skids for slabs and bars, radiant tubes, and stack dampers. HL-30 and HL-40 are used for pressure pipe applications at high temperatures.

Physical Properties

HL: Thermal conductivity

TEMP C	TEMP F	THCON W/m*K	THCON Btu/ft*h*F
100	212	14.2	8.2
316	600	17.7	10.2
538	1000	21.1	12.2
650	1200	23.2	13.4
760	1400	25.4	14.7
870	1600	28.2	16.3
980	1800	30.6	17.7
1090	2000	33.4	19.3

Source: Ref 3, p 301

HL: Mean coefficients of linear thermal expansion

TEMP C	TEMP F	THEXP 10^{-6}/C	THEXP 10^{-6}/F
21-538	70-1000	16.6	9.2
21-650	70-1200	16.9	9.4
21-760	70-1400	17.3	9.6
21-870	70-1600	17.5	9.7
21-980	70-1800	17.8	9.9
21-1090	70-2000	18.2	10.1
650-870	1200-1600	18.9	10.5
650-980	1200-1800	19.3	10.7

Source: Ref 3, p 300

Mechanical Properties

HL: Creep-stress rupture property data

TEMP C	TEMP F	CRSTR 0.0001%/h MPa	CRSTR 0.0001%/h ksi	RPSTR 100 h MPa	RPSTR 100 h ksi
760	1400	48	7.0	103	15.0
871	1600	30	4.3	63	9.2
982	1800	15	2.2	36	5.2

Note: Representative values, long time for constant temperatures, for cyclic temperature lower values would apply.
Source: Ref 3, p 300

HN

Alternate designation: 25Ni-20Cr
UNS number: J94213
Specifications: ANSI A297, ASTM A297
Nominal composition: 0.200 to 0.500% C, 2.000% Mn max, 19.000 to 23.000% Cr, 23.000 to 27.000% Ni, 2.000% Si max, 0.500% Mo max, 0.040% P max, 0.040% S max
Product forms: Cast, sand, investment
Description: HN is a Fe-Ni-Cr alloy used for highly stressed components in the 980 to 1100 C (1800 to 2000 F) range. Properties are similar to those of the more widely used HT, but the ductility of HN is better. In several specialized applications (notably, brazing fixtures), it has given satisfactory service at temperatures from 1100 to 1150 C (2000 to 2100 F). Its composition limits lie well within the stable austenite field.

Typical applications: Brazing fixtures, chain, furnace beams and parts, pier caps, radiant tubes and tube supports, sill plate brackets, torch nozzles, trays, and tubes.

Physical Properties

HN: Thermal conductivity

TEMP C	TEMP F	THCON W/m*K	THCON Btu/ft*h*F
100	212	13	7.5
316	600	16	9.2
538	1000	19	11.0
650	1200	21	12.1
760	1400	23	13.2
870	1600	25	14.5
980	1800	27(a)	15.7(a)
1090	2000	29(a)	17.0(a)

(a) Estimated
Source: Ref 3, p 302

HN: Mean coefficients of linear thermal expansion

TEMP C	TEMP F	THEXP 10^{-6}/C	THEXP 10^{-6}/F
21-538	70-1000	16.7	9.3
21-650	70-1200	17.1	9.5
21-760	70-1400	17.2	9.7
21-870	70-1600	17.8	9.9
21-980	70-1800	18.1	10.1
21-1090	70-2000	18.3	10.2
650-980	1200-1800	19.8	11.0

Source: Ref 3, p 302

HN: Elastic moduli. Source: Ref 3, p 303

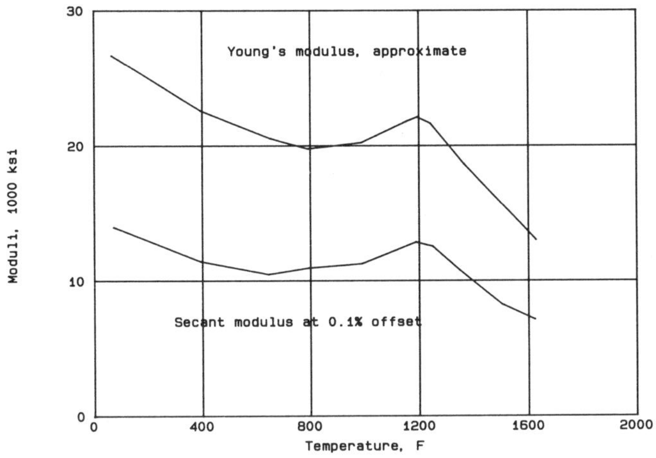

Mechanical Properties

HN: Short-time tensile property data

TEMP C	TEMP F	UTS MPa	UTS ksi	YS MPa	YS ksi	EL %
21	70	468.84	68.00	262.00	38.00	13
871	1600	139.62	20.25	99.97	14.50	37
982	1800	82.39	11.95	66.71	9.68	51
1038	1900	58.61	8.50	37.92	5.50	53
1093	2000	42.51	6.17	33.96	4.93	55

Source: Ref 29, Jun 1982

HN: Long-time creep-rupture property data

TEMP C	CRSTR 0.0001%/h MPa	CRSTR 0.0001%/h ksi	CRSTR (a) MPa	CRSTR (a) ksi	RPSTR 100 h MPa	RPSTR 100 h ksi	RPSTR 1000 h MPa	RPSTR 1000 h ksi	RPSTR 10000 h MPa	RPSTR 10000 h ksi	RPSTR 100000 h MPa	RPSTR 100000 h ksi
871	43	6.3	21	3.0	76	11	51	7.4	33	4.8	22.0	3.2
982	17	2.4	8	1.1	39	5.6	23	3.4	14	2.1	9	1.3
1038	11	1.6	3	0.45	32	4.6	14	2.1	6.6	0.96	3	0.44
1093	7	1.04	1.2	0.17	20	2.9	9	1.25	3.6	0.52	1.5	0.22

Note: For constant temperature; for cyclic temperature lower values would apply
(a) Limiting stress for 1% total creep in 100,000 h
Source: Ref 3, p 302

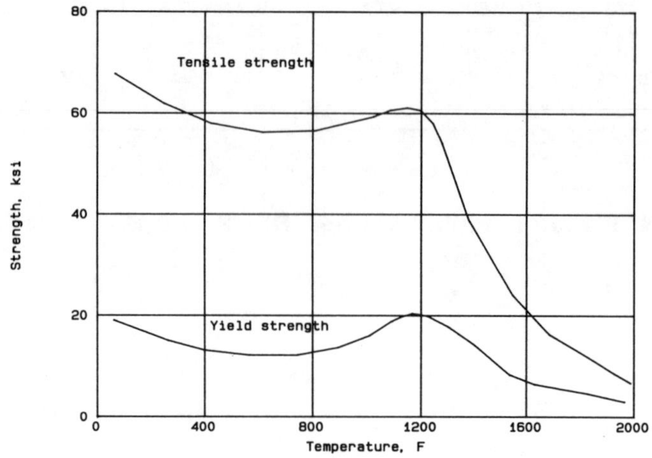

HN: Short-time tensile and yield strength. Source: Ref 3, p 302

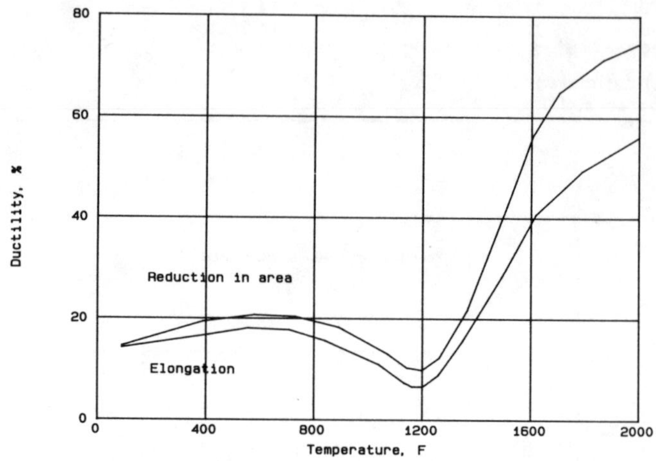

HN: Short-time tensile elongation and reduction in area. Source: Ref 3, p 302

HN: Vickers hardness. Source: Ref 3, p 302

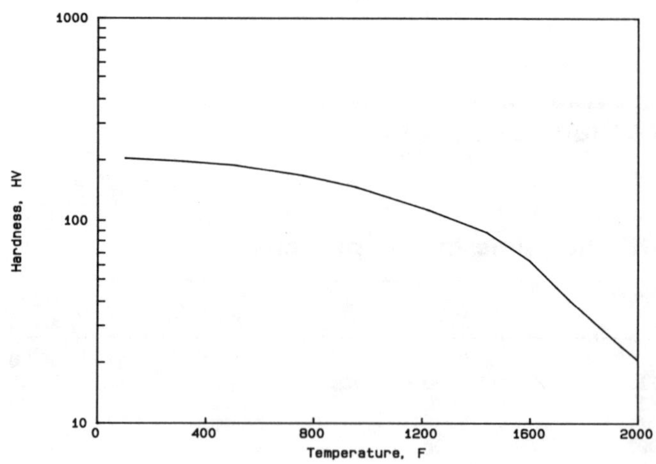

HP

Alternate designation: 35Ni-26Cr
UNS number: J95705
Specifications: ANSI A297, ASTM A297
Nominal composition: 0.350 to 0.750% C, 2.000% Mn max, 24.000 to 28.000% Cr, 33.000 to 37.000% Ni, 2.500% Si max, 0.500% Mo max, 0.040% P max, 0.040% S max
Product forms: Cast, sand, investment
Description: HP is a Fe-Ni-Cr alloy that is resistant to both oxidizing and carburizing atmospheres at high temperatures. Its creep-rupture properties at 980 to 1100 C (1800 to 2000 F) are at least equal to those of HK-40 and HN. HP alloy is austenitic at all temperatures and is not susceptible to sigma phase formation.
Typical applications: Castings for ethylene pyrolysis heaters, heat treating fixtures, radiant tubes, and refinery tubes.

Physical Properties

HP: Thermal conductivity

TEMP C	TEMP F	THCON W/m*K	THCON Btu/ft*h*F
100	212	12.9	7.5
316	600	15.9(a)	9.2(a)
538	1000	19.0(a)	11.0(a)
650	1200	20.9(a)	12.1(a)
760	1400	22.8(a)	13.2(a)
870	1600	25.1(a)	14.5(a)
980	1800	27.2(a)	15.7(a)
1090	2000	29.4(a)	17.0(a)

(a) Estimated
Source: Ref 3, p 304

HP: Mean coefficients of linear thermal expansion

TEMP C	TEMP F	THEXP 10^{-6}/C	THEXP 10^{-6}/F
21-358	70-1000	16.6(a)	9.2(a)
21-650	70-1200	17.1	9.5
21-760	70-1400	17.6(a)	9.8(a)
21-870	70-1600	18.0(a)	10.0(a)
21-980	70-1800	18.5(a)	10.3(a)
21-1090	70-2000	19.1(a)	10.6(a)
650-870	1200-1600	20.5(a)	11.4(a)
650-980	1200-1800	21.4(a)	11.9(a)
870-1090	1600-2000	23.6(a)	13.1(a)

(a) Estimated
Source: Ref 3, p 304

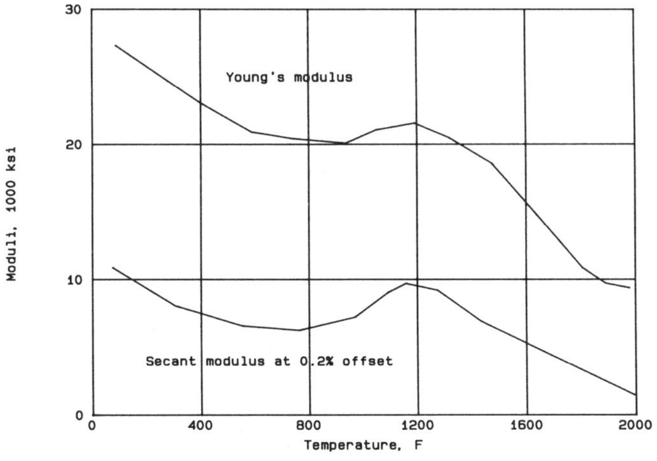

HP: Elastic moduli. Source: Ref 3, p 305

Mechanical Properties

HP: Long-time creep-rupture property data

TEMP C	CRSTR 0.0001%/h MPa	CRSTR 0.0001%/h ksi	CRSTR (a) MPa	CRSTR (a) ksi	RPSTR 100 h MPa	RPSTR 100 h ksi	RPSTR 1000 h MPa	RPSTR 1000 h ksi	RPSTR 10000 h MPa	RPSTR 10000 h ksi	RPSTR 100000 h MPa	RPSTR 100000 h ksi
871	40.0	5.8	34	4.9	69	10	52	7.5	35	5.1	23(b)	3.3(b)
982	19.0	1.8	14	2.1	41	5.9	25	3.6	15	2.2	7.6(b)	1.1(b)
1093	6.9	1.0	2.8	0.4	19	2.8	10	1.5	4.1(b)	0.6(b)	...	...

Note: For constant temperature; for cyclic temperature, lower values would apply
(a) Limting stress for 1% total creep in 100,000 h. (b) Extrapolated
Source: Ref 15, p 304

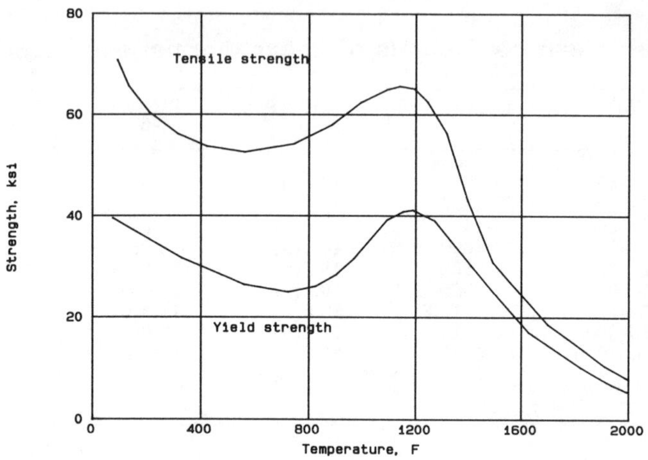

HP: Short-time tensile and yield strength. Source: Ref 3, p 305

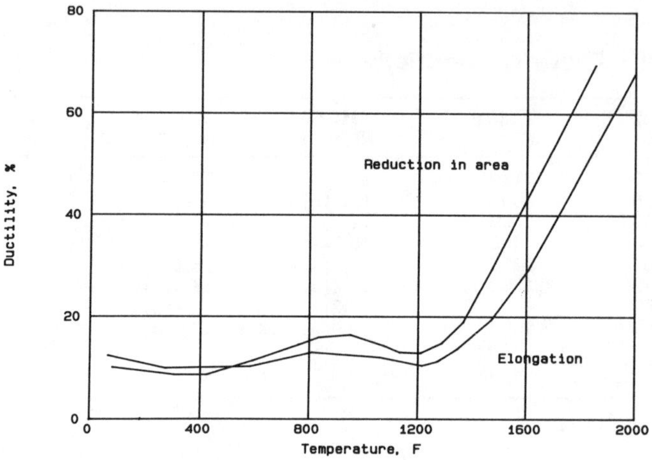

HP: Short-time tensile elongation and reduction in area. Source: Ref 3, p 305

HT

Alternate designation: 35Ni-17Cr
UNS number: J94605
Specifications: ANSI A297, ASTM A297
Nominal composition: 0.350 to 0.750% C, 2.000% Mn max, 15.000 to 19.000% Cr, 33.000 to 37.000% Ni, 2.500% Si max, 0.500% Mo max, 0.040% P max, 0.040% S
Product forms: Cast, sand, investment
Description: In general, this Fe-Ni-Cr alloy performs satisfactorily in oxidizing atmospheres at temperatures up to 1150 C (2100 F) and in reducing atmospheres up to 1100 C (2000 F). HT is also resistant to thermal shock and carburization and has good strength at the temperatures used to treat steel. This alloy is widely specified for highly stressed parts in general heat-resistant applications. Additions of up to 2% Nb can further enhance its already excellent thermal shock resistance.

Typical applications: Air ducts, brazing trays, carburizing containers, chain, cyanide pots, dampers, dippers, door frames, enameling bars and supports, fan blades, feed screws, gear spacers, glass molds, glass rolls, hearth plates, heat treating fixtures and trays, idler drums, kiln nose rings, lead pots, malleablizing baskets, muffles, oil burner nozzles, point bars, radiant tubes, resistor guides, retorts, roller rails, rolling mill guides, salt pots, and tube supports.

Physical Properties

HT: Thermal conductivity

TEMP C	TEMP F	THCON W/m*K	THCON Btu/ft*h*F
100	212	12.1	7.0
316	600	15.4	8.9
538	1000	18.7	10.8
650	1200	20.6	11.9
760	1400	22.3	12.9
870	1600	24.2	14.0
980	1800	26.5(a)	15.3(a)
1090	2000	28.2(a)	16.3(a)

(a) Estimated
Source: Ref 3, p 306

HT: Mean coefficients of linear thermal expansion

TEMP C	TEMP F	THEXP 10^{-6}/C	THEXP 10^{-6}/F
21-93	70-200	14.2	7.9
21-204	70-400	14.6	8.1
21-316	70-600	15.1	8.4
21-427	70-800	15.2	8.6
21-358	70-1000	16.0	8.9
21-650	70-1200	16.2	9.1
21-760	70-1400	16.7	9.3
21-870	70-1600	17.3	9.6
21-980	70-1800	17.6	9.8
21-1090	70-2000	18.0	10.0
650-870	1200-1600	19.4	10.8
650-980	1200-1800	19.8	11.0

Source: Ref 3, p 306

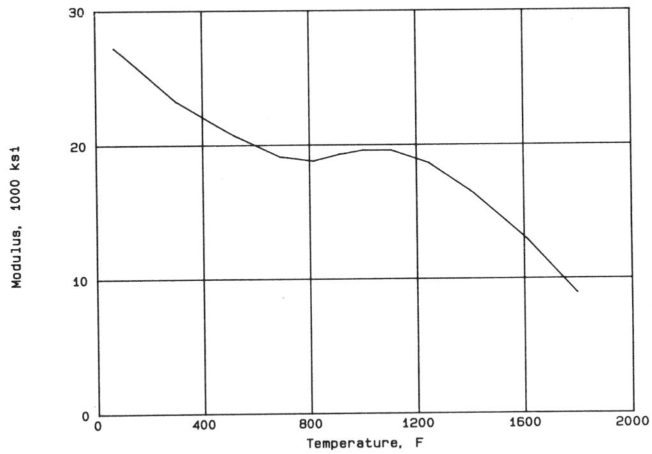

HT: Elastic moduli. Source: Ref 3, p 308

Mechanical Properties

HT: Short-time tensile and Charpy keyhole notch impact property data

TEMP C	TEMP F	UTS MPa	UTS ksi	YS MPa	YS ksi	EL %	IS J	IS ft-lb
21	70	482.6	70.0	275.8	40.0	10	5.4	4
649	1200	292.3	42.4	193.1	28.0	5	9.5	7
760	1400	241.3	35.0	179.3	26.0	10	10.8	8
871	1600	129.6	18.8	103.4	15.0	26	12.2	9
982	1800	75.8	11.0	55.2	8.0	28	13.6	10
1093	2000	41.4	6.0	...	...	...	...	...

Source: Ref 29, Feb 1979

10.22 High-Temperature Property Data: Ferrous Alloys

HT: Long-time creep-rupture property data

TEMP C	TEMP F	CRSTR 0.0001%/h MPa	CRSTR 0.0001%/h ksi	RPSTR 100 h MPa	RPSTR 100 h ksi	RPSTR 1000 h MPa	RPSTR 1000 h ksi	RPSTR 10000 h MPa	RPSTR 10000 h ksi	RPSTR 100000 h MPa	RPSTR 100000 h ksi
760	1400	55	8.0	110	16.0	83	12.0	58	8.4	39	5.6
871	1600	31	4.5	61	8.9	40	5.8	26	3.7	17	2.4
982	1800	14	2.0	30	4.4	19	2.7	12	1.7	7.2	1.05
1093	2000	3	0.5	14	2.1	8.9	1.3	...	...	...	...
1177	2150	1	0.15	...	...	...	...	...	...	...	...

Note: For constant temperature; for cyclic temperature, lower values would apply
Source: Ref 3, p 306

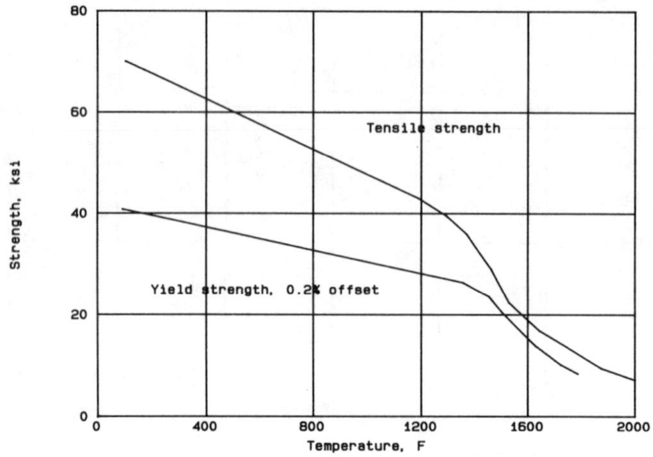

HT: Short-time tensile and yield strength. Source: Ref 3, p 308

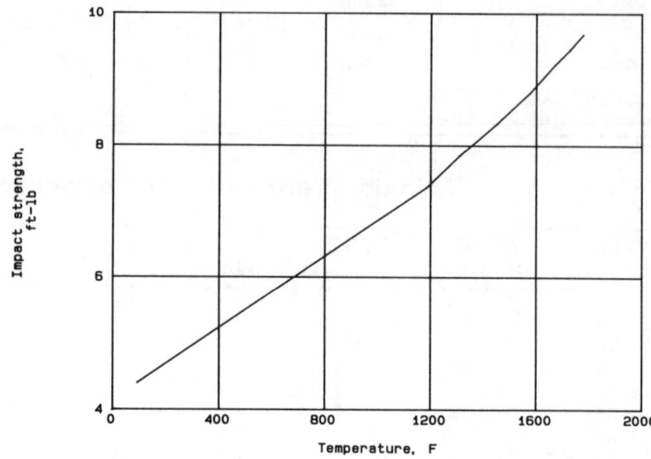

HT: Short-time impact strength. Source: Ref 3, p 308

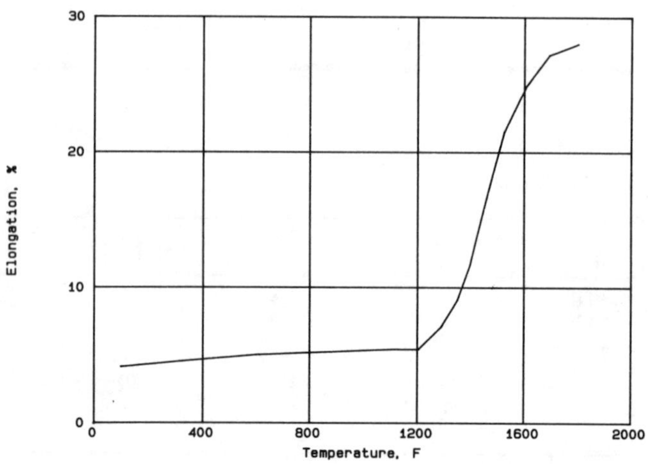

HT: Short-time tensile elongation. Source: Ref 3, p 308

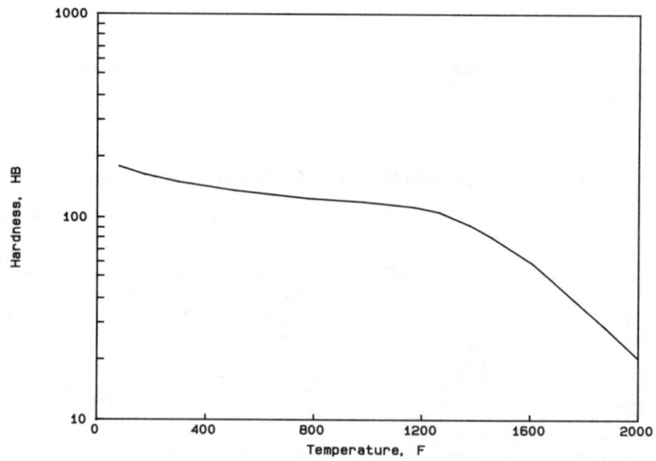

HT: Brinell hardness. Source: Ref 3, p 308

HU

Alternate designation: 39Ni-19Cr
UNS number: J95405
Specifications: ANSI A297, ASTM A297
Nominal composition: 0.350 to 0.750% C, 2.000% Mn max, 17.000 to 21.000% Cr, 37.000 to 41.000% Ni, 2.500% Si max, 0.500% Mo max, 0.040% P max, 0.040% S max,
Product form: Cast, sand, investment
Description: Similar to HT in terms of high-temperature strength and resistance to carburization, but superior in resistance to corrosion by either oxidizing or reducing hot gases, including those that contain sulfur. These properties make HU an excellent candidate for severe service involving high stresses and/or rapid thermal cycling in combination with an aggressive environment.

Typical applications: Articulated trays, burner tubes, carburizing retorts, conveyor screws and chains, cyanide pots, dipping baskets, furnace rolls, lead pots, muffles, pouring spouts, radiant tubes, and resistor guides.

Physical Properties

HU: Thermal conductivity

TEMP C	TEMP F	THCON W/m*K	THCON Btu/ft*h*F
100	212	12.1	7.0
316	600	15.4	8.9
538	1000	18.7	10.8
650	1200	20.6	11.9
760	1400	22.3	12.9
870	1600	24.2	14.0
980	1800	26.5	16.3

Source: Ref 3, p 3 10

HU: Mean coefficients of linear thermal expansion

TEMP C	TEMP F	THEXP 10^{-6}/C	THEXP 10^{-6}/F
21-358	70-1000	15.8	8.8
21-650	70-1200	16.2	9.0
21-760	70-1400	16.6	9.2
21-870	70-1600	16.9	9.4
21-980	70-1800	17.3	9.6
21-1090	70-2000	17.5	9.7
650-870	1200-1600	18.9	10.5
650-980	1200-1800	19.1	10.6

Source: Ref 3, p 310

Mechanical Properties

HU: Short-time tensile property data

TEMP C	TEMP F	UTS MPa	UTS ksi	YS MPa	YS ksi	EL %
760	1400	275.8	40.0	...	...	...
871	1600	135.1	19.6	...	...	20
982	1800	68.9	10.0	42.7	6.2	28

Source: Ref 3, p 310

HU: Creep-rupture property data

TEMP C	TEMP F	CRSTR 0.0001%/h MPa	CRSTR 0.0001%/h ksi	RPSTR 100 h MPa	RPSTR 100 h ksi	RPSTR 1000 h MPa	RPSTR 1000 h ksi	RPSTR 10000 h MPa	RPSTR 10000 h ksi
760	1400	58.6	8.5	103	15.0	...	...	...	...
870	1600	34.5	5.0	55	8.0	35.9	5.2	22.8	3.3
980	1800	15.2	2.2	31	4.5	19.9	2.9	12.4	1.8
1090	2000	4.1	0.6	...	...	...	...	...	...

Note: For constant temperature; for cyclic temperature, lower values would apply
Source: Ref 3, p 308

HW

Alternate designation: 60Ni-12Cr
Specifications: ASTM A297
Nominal composition: 0.350 to 0.750% C, 2.000% Mn max, 10.000 to 14.000% Cr, 58.000 to 62.000% Ni, 2.500% Si max, 0.500% Mo max, 0.040% P max, 0.040% S max
Product forms: Cast, sand, investment
Description: 60Ni-12Cr is well suited to applications involving thermal shock, steep temperature gradients, and high stresses. HW also resists carburization and has good strength at steel treating temperatures. This alloy performs satisfactorily at temperatures up to 1120 C (2050 F) in strongly oxidizing atmospheres and up to 1040 C (1900 F) in oxidizing or reducing products of combustion (if sulfur free). Its tightly adherent oxide scale makes it suitable for enameling furnace service.
Typical applications: Cyanide pots, electric heating elements, enameling tools, gas retorts, hardening fixtures, hearth plates, lead pots, and muffles.

Physical Properties

HW: Thermal conductivity

TEMP C	TEMP F	THCON W/m*K	THCON Btu/ft*h*F
100	212	12.5	7.2
316	600	15.6	9.0
538	1000	19.2	11.1
650	1200	21.1	12.2
760	1400	23.0	13.3
870	1600	25.1	14.5
980	1800	27.2	15.7
1090	2000	29.4	17.0

Source: Ref 3, p 312

HW: Mean coefficients of linear thermal expansion

TEMP C	TEMP F	THEXP 10^{-6}/C	THEXP 10^{-6}/F
21-93	70-200	12.6	7.00
21-204	70-400	12.9	7.22
21-316	70-600	13.4	7.45
21-427	70-800	13.9	7.70
21-358	70-1000	14.3	7.95
21-650	70-1200	14.8	8.20
21-760	70-1400	15.2	8.47
21-870	70-1600	15.7	8.74
21-980	70-1800	16.2	9.07
21-1090	70-2000	16.7	9.28
650-870	1200-1600	18.0	10.00
650-980	1200-1800	18.6	10.33

Source: Ref 3, p 311

Mechanical Properties

HW: Creep-rupture property data

TEMP C	TEMP F	CRSTR 0.0001%/h MPa	CRSTR 0.0001%/h ksi	RPSTR 10 h MPa	RPSTR 10 h ksi	RPSTR 100 h MPa	RPSTR 100 h ksi	RPSTR 1000 h MPa	RPSTR 1000 h ksi
760	1400	41.3	6.0	110	16	69.0	10	53.8	7.8
870	1600	20.7	3.0	56.5	8.2	41.3	6	31	4.5
900	1800	9.7	1.4	29.6	4.3	24.8	3.6	17.9	2.6

Note: For constant temperature; for cyclic temperature, lower values would apply
Source: Ref 3, p 311

HX

Alternate designation: 66Ni-17Cr
Specifications: ANSI A297, ASTM A297
Nominal composition: 0.350 to 0.750% C, 2.000% Mn max, 15.000 to 19.000% Cr, 64.000 to 68.000% Ni, 2.500% Si max, 0.500% Mo max, 0.040% P max, 0.040% S max
Product forms: Cast, sand, investment
Description: 66Ni-17Cr is an iron-nickel-chromium Fe-Ni-Cr alloy similar to HW, but its higher chromium content increases resistance to corrosion by hot gases (including those containing sulfur). HX is intended for severe service applications at temperatures up to 1150 C (2100 F). However, rapid decarburization may be a problem in the 1100 to 1150 C (2000 to 2100 F) range.

Typical applications: Autoclaves, brazing furnace rails and doors, calciner tubes, carburizing boxes, cyanide pots, enameling tools, heating elements, hearth plates, heat treating trays and fixtures, lead pots, muffles, retorts, roller hearths, salt bath electrodes, salt pots, and shaker hearths.

Physical Properties

HX: Thermal conductivity

TEMP C	TEMP F	THCON W/m*K	THCON Btu/ft*h*F
100	212	12.5	7.2
316	600	15.6	9.0
538	1000	19.2	11.1
650	1200	21.1	12.2
760	1400	23.0	13.3
870	1600	25.1	14.5
980	1800	27.2	15.7
1090	2000	29.4	17.0

Source: Ref 3, p 312

HX: Mean coefficients of linear thermal expansion

TEMP C	TEMP F	THEXP 10^{-6}/C	THEXP 10^{-6}/F
21-358	70-1000	14.0	7.8
21-650	70-1200	14.6	8.1
21-760	70-1400	15.3	8.5
21-870	70-1600	15.8	8.8
21-980	70-1800	16.6	9.2
21-1090	70-2000	17.1	9.5
650-870	1200-1600	19.3	10.7
650-980	1200-1800	20.3	11.3

Source: Ref 3, p 312

Mechanical Properties

HX: Creep-rupture property data

TEMP C	TEMP F	CRSTR 0.0001%/h MPa	CRSTR 0.0001%/h ksi	RPSTR 10 h MPa	RPSTR 10 h ksi	RPSTR 100 h MPa	RPSTR 100 h ksi	RPSTR 1000 h MPa	RPSTR 1000 h ksi
760	1400	44	6.4	125	18	90	13	...	...
870	1600	22	3.2	70	10	46	6.7	28	4
980	1800	11	1.6	37	5.4	24	3.5	14	2
1090	2000	4	0.6	17	2.5	12	1.7	6	0.9

Note: For constant temperature; for cyclic temperatures, lower values would apply
Source: Ref 3, p 313

11. Wrought Iron-Nickel Alloys/ Iron-Nickel Superalloys

Wrought Iron-Nickel Alloys/Iron-Nickel Superalloys

Iron and iron-nickel superalloys (as well as other types of superalloys) contain chromium for increased resistance to oxidation and hot corrosion, as well as other elements for elevated-temperature strength. High-temperature applications of these materials are numerous, including components for aerospace, chemical, and petrochemical industries.

The iron-nickel alloys and iron-nickel wrought superalloys covered in this section generally contain about 30% or more iron, 20% or more nickel, and are strengthened by precipitation or solid solution hardening. Typical elements added to provide precipitation-strengthening for these alloys include aluminum, titanium, and niobium. Hardening via use of carbon to form general carbide precipitates or grain boundary carbide phases is also common. Nitrogen is sometimes added to enhance this effect. Refractory metals, such as molybdenum and tungsten, are also common additions made to enhance solid solution strength.

Oxidation resistance is provided by judicious use of elements such as chromium, silicon, and manganese. Nickel stabilizes theaustenitic matrix and retards formation of the generally undesirable topologically close packed phases such as sigma, mu, Laves, and chi. Boron and lanthanum in fairly small amounts greatly improve elevated-temperature properties.

The usual maximum operating temperature for precipitation hardened iron-nickel superalloys is about 750 C (1380 F). Above this temperature, the strengthening precipitates coarsen and overage. For lower stress/higher temperature uses, iron-nickel alloys and solid solution strengthened iron-nickel alloys and solid solution strengthened iron-nickel superalloys are used at temperatures as high as about 1150 C (2100 F).

16-25-6

Alternate designation: AISI 650
Specification: AMS 5727B
Nominal composition: 16.000% Cr max, 25.000% Ni max, 6.000% Mo max, 50.700% Fe max, 0.060% C max, 1.350% Mn max, 0.700% Si max, 0.150% N max
Product form: Wrought
Description: 16-25-6 is an austenitic nonmagnetic steel, strengthened by hot/cold work, for stressed applications at temperatures up to 760 C (1400 F). Work hardening by hot/cold work requires plastic deformation below the recrystallization temperature (815 C, or 1500 F). The amount of plastic deformation may correspond to as much as 40% elongation or 30% reduction in area. Corrosion resistance of 16-25-6 is superior to that of 18-8 stainless steel; its oxidation resistance is comparable.

Typical applications: Turbo-superchargers, turbine wheels and other jet engine components, shafts, and bolting.

Physical Properties

16-25-6: Moduli of elasticity, rigidity, and Poisson's ratio

TEMP C	TEMP F	E GPa	E 10^3 ksi	G GPa	G 10^3 ksi	PR
21	70	197	28.5	76	11.0	0.295
93	200	192	27.8	74	10.7	0.3
149	300	188	27.2	72	10.4	0.303
204	400	183	26.6	70	10.2	0.306
260	500	179	26	68	9.9	0.31
316	600	175	25.4	66	9.6	0.314
371	700	172	24.9	65	9.4	0.317
427	800	168	24.4	63	9.1	0.32
482	900	165	23.9	61	8.8	0.323
538	1000	161	23.4	59	8.5	0.326
593	1100	157	22.8	57	8.2	0.33
649	1200	154	22.4	54	7.9	0.334
704	1300	152	22	53	7.7	0.336
760	1400	149	21.6	52	7.5	0.34
816	1500	147	21.3	50	7.3	0.344

Note: Those elastic values were determined by means of ultrasonic pulse techniques. At the more elevated temperature, they apply only under instantaneous loading because metals in the plastic condition cannot possess true elastic properties when the time factor is considered.
Source: Ref 28, p 94

16-25-6: Mean coefficients of linear thermal expansion

TEMP C	TEMP F	THEXP 10^{-6}/C	THEXP 10^{-6}/F
21-316	70-600	16.2	9.0
21-427	70-800	16.9	9.4
21-538	70-1000	17.3	9.6
21-649	70-1200	18.4	10.2
21-760	70-1400	19.4	10.8
21-871	70-1600	21.1	11.7

Source: Ref 28, p 93

Mechanical Properties

16-25-6: Tensile property data

TEMP C	TEMP F	YS MPa	YS ksi	UTS MPa	UTS ksi	EL %	RA %
25-mm (1-in.) diameter bar solution quenched from 1177 C (2150 F)							
21	70	283	41.0	752	109	46.0	65.0
538	1000	224	32.5	621	90	58.0	60.0
593	1100	224	32.5	596	86.5	51.0	51.0
649	1200	224	32.5	438	63.5	15.0	55.0
704	1300	224	32.5	429	62.25	17.0	18.0
760	1400	224	32.5	345	50	25.0	30.0
816	1500	203	29.5	269	39	30.0	42.0
871	1600	...	...	221	32	51.0	59.0
927	1700	...	...	172	25	61.0	67.0
Cold worked (20% elongation) and tempered at 677 C (1250 F)							
21	70	772	112	965	140	20.0	35.0
649	1200	496	72	621	90	20.0	35.0
704	1300	434	63	538	78	21.0	33.0
760	1400	345	50	407	59	23.0	40.0
816	1500	248	36	310	45	30.0	41.0
871	1600	...	...	221	32	38.0	40.0
927	1700	...	...	172	25	49.0	45.0

Source: Ref 28, p 94

16-25-6: Room-temperature mechanical properties before and after creep tests

TEMP C	TEMP F	STRESS MPa	STRESS ksi	TIME h	YS 0.2% MPa	YS 0.2% ksi	UTS MPa	UTS ksi	EL %	RA %	IS Z J	IS Z ft-lb	HARD HV
(a)	(a)	(a)	(a)	(a)	358	51.9	750	108.75	46.0	65.0	144	106	181
649	1200	86	12.5	12358	510	74.0	946	137.20	14.0	17.0	...	...	...
649	1200	121	17.5	11365	...	...	...	...	...	...	16	12	275
(b)	(b)	(b)	(b)	(b)	839	121.75	988	143.25	17.0	34.0	22	16	309
649	1200	138	20.0	11260	679	98.5	948	137.50	8.0	12.0	...	...	...
649	1200	172	25.0	11419	...	...	...	...	...	...	7	5	290
704	1300	86	12.5	11873	690	100	906	131.375	10.0	17.0	...	...	...

(a) Original solution quenched from 1177 C (2150 F). (b) Original hot-cold worked
Source: Ref 28, p 94

16-25-6: Creep and rupture strength of 25-mm (1-in.) diameter bar

TEMP C	TEMP F	CRSTR 0.01%/ 1000 h MPa	CRSTR 0.01%/ 1000 h ksi	CRSTR 0.10%/ 1000 h MPa	CRSTR 0.10%/ 1000 h ksi	RPSTR 100 h MPa	RPSTR 100 h ksi	RPSTR 1000 h MPa	RPSTR 1000 h ksi	RPSTR 10000 h MPa	RPSTR 10000 h ksi	RPSTR 100000 h MPa	RPSTR 100000 h ksi
Solution quenched from 1177 C (2150 F)													
538	1000	179	26.0	310	45.0	537	78	524	76.0	510	74.0	...	...
593	1100	...	...	...	...	358	52	303	44.0	248	36.0	214	31.0
649	1200	124	18.0	152	22.0	303	44	220	32.0	158	23.0	114	16.5
704	1300	79.3	11.5	114	16.5	207	30	145	21.0	90	13.0	54	7.8
760	1400	...	...	...	...	138	20	86	12.5	52	7.5	32	4.7
816	1500	21	3.0	32	4.7	96	14	62	9.0	40	5.8	23	3.3
Hot-cold worked 15% at 816 C (1500 F), tempered at 690 C (1275 F) for 6 h													
649	1200	128	18.5	172	25.0	345	50.0	248	36.0	193	28.0	134	19.5
704	1300	79.3	11.5	103	15.0	234	34.0	152	22.0	96.5	14.0	66.2	9.6
760	1400	...	...	...	...	124	18.0	75.8	11.0	48	7.0	...	...

Source: Ref 28, p 94

A-286

Alternate designation: AISI 660

UNS number: S66286

Specifications: AMS 5525, AMS 5728, AMS 5731, AMS 5732, AMS 5734, AMS 5735, AMS 5736, AMS 5737, AMS 5804, AMS 5805, AMS 5858, ASME SA638 (660), ASTM A453 (660), ASTM A638 (660), SAE J467 (A286)

Nominal composition: 15.000% Cr, 26.000% Ni, 1.250% Mo, 2.000% Ti, 0.200% Al max, 55.200% Fe, 0.040% C, 0.005% B, 0.300% V

Product forms: Wrought, bar, rod, wire, plate, sheet, strip, tube, pipe, billets, forgings, structural shapes

Description: Precipitation hardening, austenitic Fe-Ni-Cr steel recommended for applications requiring high strength and corrosion resistance at temperatures up to 705 C (1300 F). A-286 offers high ductility in notched sections. Its notched rupture strength, for example, is superior to that of any other commercial alloy with comparable high-temperature properties. The oxidation resistance of A-286 is equivalent to that of AISI type 310 stainless steel at temperatures up to 980 C (1800 F).

Typical applications: Jet engine and supercharger parts, turbine wheels and blades, frames, casings, shafts, housings, afterburner and tail cone parts, and bolts.

Physical Properties

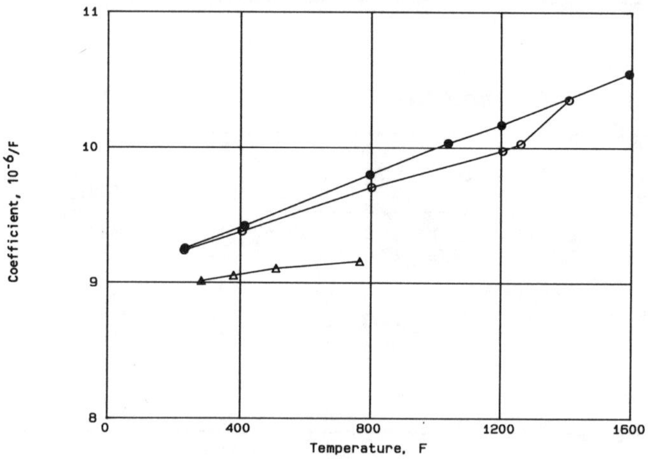

A-286: Mean coefficient of linear thermal expansion from room temperature to indicated temperature. Data from several tests. Source: Ref 22, p 6

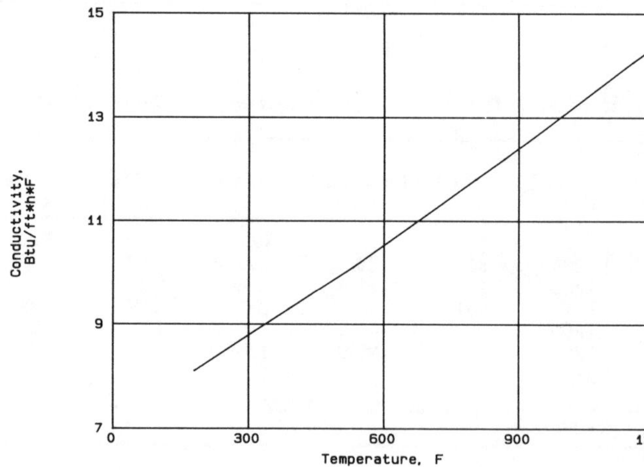

A-286: Thermal conductivity. Source: Ref 22, p 6

A-286: Thermal conductivity

TEMP C	TEMP F	THCON W/m*K	THCON Btu/ft*h*F
21	70	12.6	7.3
150	300	15.1	8.7
425	800	20.2	11.7
600	1110	23.79	13.75
650	1200	24.7	14.3

Source: Ref 15, p 396

A-286: Mean coefficients of linear thermal expansion

TEMP C	TEMP F	THEXP(a) 10^{-6}/C	THEXP(a) 10^{-6}/F	THEXP(b) 10^{-6}/C	THEXP(b) 10^{-6}/F
21-100	70-212	16.9	9.4	16.7	9.3
21-500	70-930	17.4	9.67	17.6	9.8
21-650	70-1200	17.6	9.79	...	...
21-750	70-1380	17.7	9.81	18.5	10.3
21-900	70-1650	19.4	10.8	...	...

(a) Solution treated. (b) Solution treated and aged
Source: Ref 15, p 396

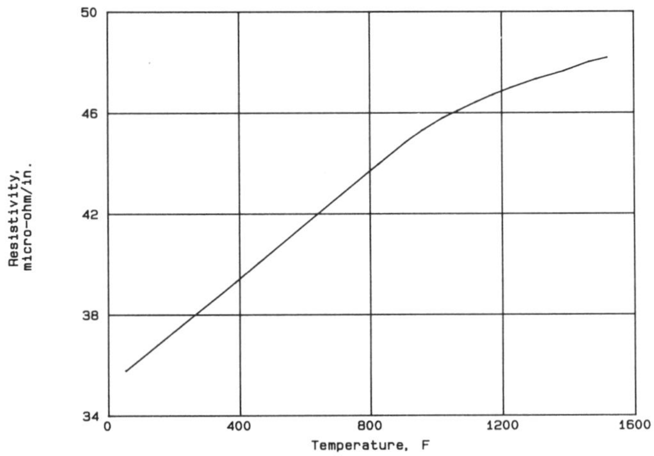

A-286: Electrical resistivity. Source: Ref 22, p 6

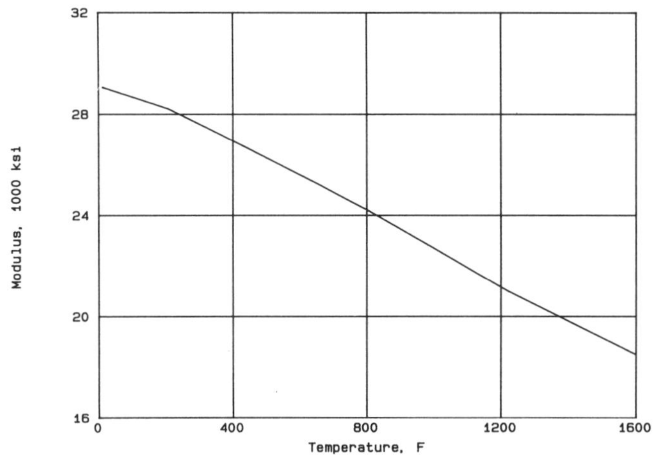

A-286: Dynamic modulus of elasticity. Source: Ref 22, p 15

A-286: Electrical resistivity

TEMP C	TEMP F	ELRES n{o}*m
25	77	910
538	1000	1156
649	1200	1188
732	1350	1201
816	1500	1224

Source: Ref 28, p 105

A-286: Modulus of elasticity

TEMP C	TEMP F	E GPa	E 10^3 ksi
24	75	201	29.1
538	1000	162	23.5
649	1200	153	22.2
760	1400	142	20.6
816	1500	137	19.8
871	1600	130	18.9

Source: Ref 29, Feb 1981

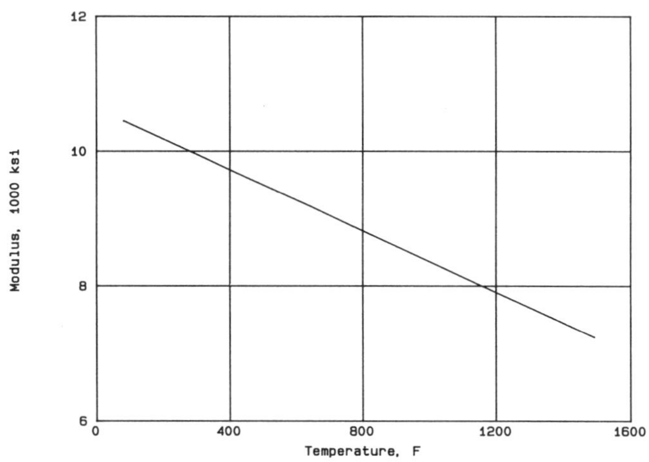

A-286: Dynamic modulus of rigidity. Source: Ref 22, p 15

A-286: Modulus of rigidity

TEMP C	TEMP F	G GPa	G 10^3 ksi
24	75	71.7	10.4
538	1000	57.9	8.4
649	1200	54.5	7.9
760	1400	51.7	7.5
816	1500	50.3	7.3

Source: Ref 29, Feb 1981

A-286: Poisson's ratio

TEMP C	TEMP F	PR
24	75	0.306
538	1000	0.328
649	1200	0.336
760	1400	0.34
816	1500	0.344

Source: Ref 29, Feb 1981

Mechanical Properties

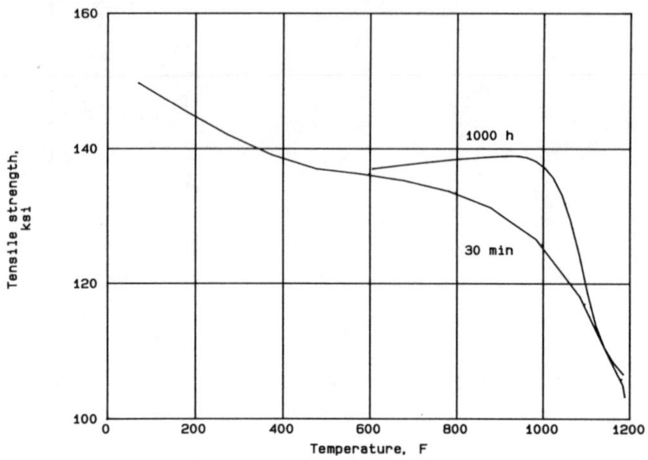

A-286: Tensile strength of sheet. Specimens were heat treated to 1800 F for 1 h in argon, oil quenched, held at 1325 F for 16 h and exposed for 30 min and 1000 h at temperature. Source: Ref 22, p 9

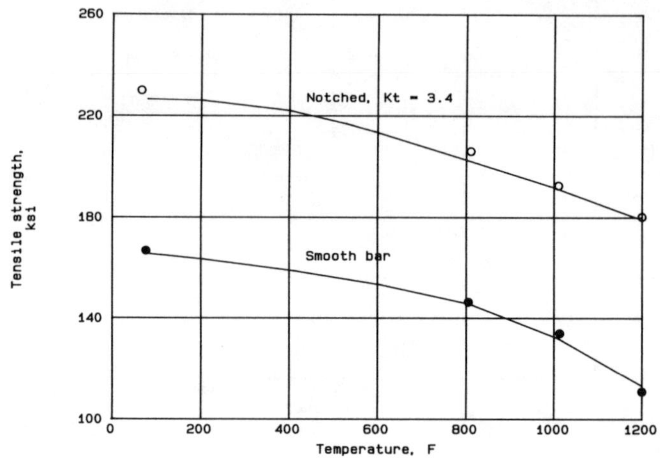

A-286: Tensile strength of smooth and notched 0.125-in. diam bar subjected to double aging as follows: heat treated at 1650 F, held for 1 h, air cooled, aged at 1325 F, held 16 h, air cooled, aged at 1200 F, held 16 h and air cooled. Source: Ref 22, p 11

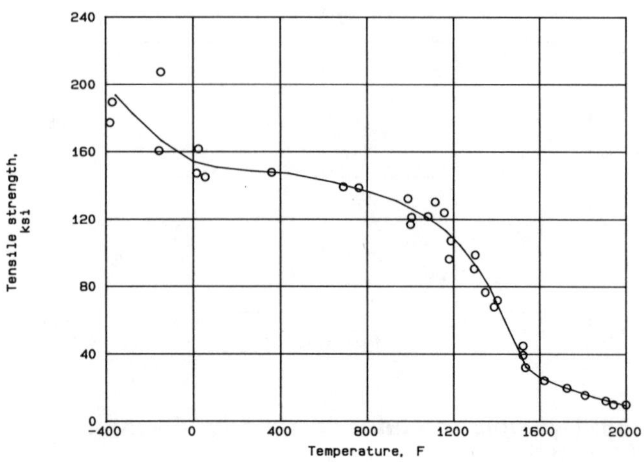

A-286: Tensile strength of 0.875-in. bar. Specimens were heated to 1800 F for 1 h, oil quenched, held at 1325 F for 16 h, and air cooled. Source: Ref 22, p 8

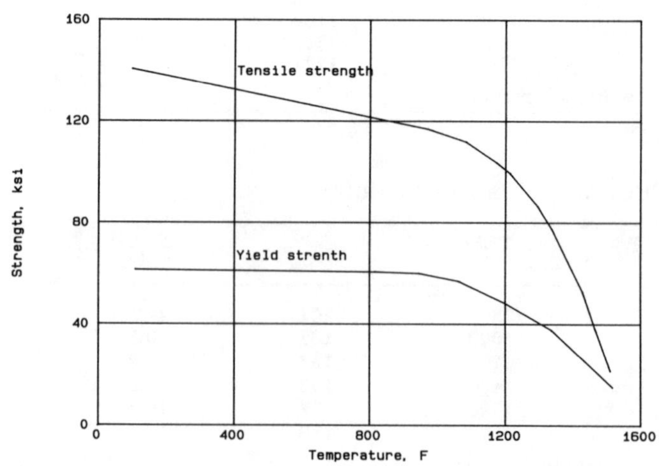

A-286: Tensile and yield strength fo 0.040-in. sheet. Specimens were solution treated at 1800 F for 1 h, oil quenched, held at 1325 F for 16 h, and air cooled. Source: Ref 22, p 11

A-286: Tensile property data

TEMP C	TEMP F	UTS MPa	UTS ksi	YS 0.02% MPa	YS 0.02% ksi	YS 0.2% MPa	YS 0.2% ksi	EL %	RA %
20	68	1007	146.0	569	82.5	703.2	102.0	25.0	36.8
205	400	1000	145.0	524	76.0	645	93.5	21.5	52.8
370	700	948.0	137.5	490	71.0	645	93.5	22.0	45.0
425	800	951.4	138.0	496	72.0	645	93.5	18.5	35.0
540	1000	903.2	131.0	427	62.0	603	87.5	18.5	31.2
595	1100	841.2	122.0	445	64.5	621	90.0	21.0	23.0
650	1200	713.6	103.5	431	62.5	607	88.0	13.0	14.5
705	1300	596	86.5	472	68.5	...	...	11.0	9.6
760	1400	441	64.0	307	44.5	...	...	18.5	23.4
815	1500	252	36.5	213	31.0	...	...	68.5	37.5

Note: A 22-mm (0.875-in.) round hot rolled bar was heated to 982 C (1800 F) for 1 h, oil quenched, heated to 718 C (1325 F) for 16 h, and air cooled
Source: Ref 15, p 397

A-286: Effect of cooling rate on tensile properties

RATE C/min	RATE F/min	YS MPa	YS ksi	UTS MPa	UTS ksi	EL %	RA %
Tested at 21 C (70 F)							
(a)	(a)	680	99	1030	149	24	39
15	27	690	100	1030	150	22	42
7	15	680	99	1020	148	24	39
5	9	700	102	990	143	22	36
1	2	660	96	1030	149	23	33
Tested at 650 C (1200 F)							
(a)	(a)	650	94	780	113	8	10
5	27	650	94	770	112	6	7
7	15	630	92	770	112	9	16
5	9	590	97	740	108	7	17
1	2	220	90	740	108	11	18

Note: After cooling from 980 to 535 C (1800 to 1000 F), specimens were aged at 720 C (1325 F) for 16 h and air cooled
(a) Oil quenched
Source: Ref 4, p 661

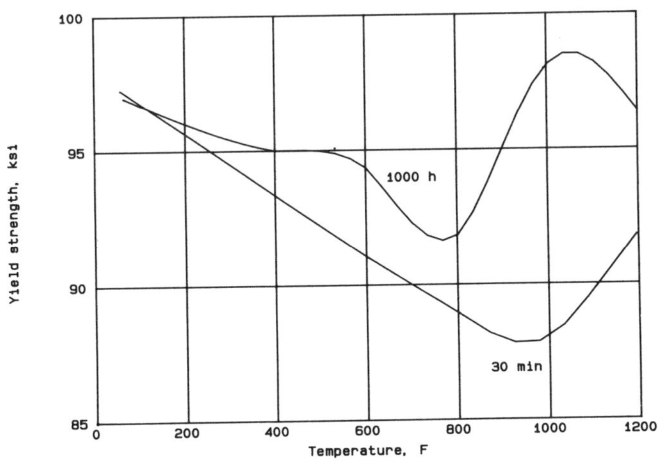

A-286: Yield strength of 0.062-in. sheet. Specimens were heated to 1800 F for 1 h in argon, oil quenched, held at 1325 F for 16 h, and exposed for 30 min and 1000 h at temperature. Source: Ref 22, p 9

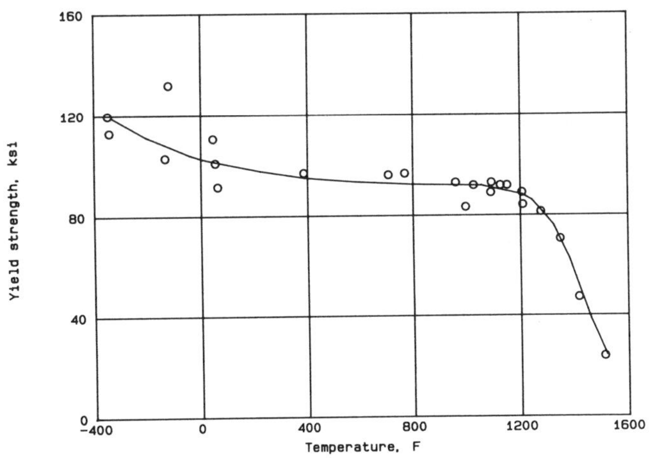

A-286: Yield strength of 0.875-in. diam bar. Specimens were heated to 1800 F for 1 h, oil quenched, held at 1325 F for 16 h, and air cooled. Source: Ref 22, p 8

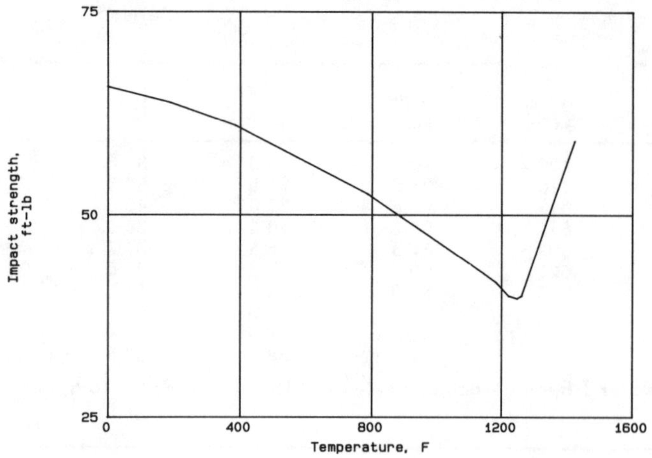

A-286: Charpy V-notch impact strength of 0.875-in. diam bar. Specimens were heated to 1800 F for 1 h, oil quenched, and held at 1325 F for 16 h. Source: Ref 22, p 10

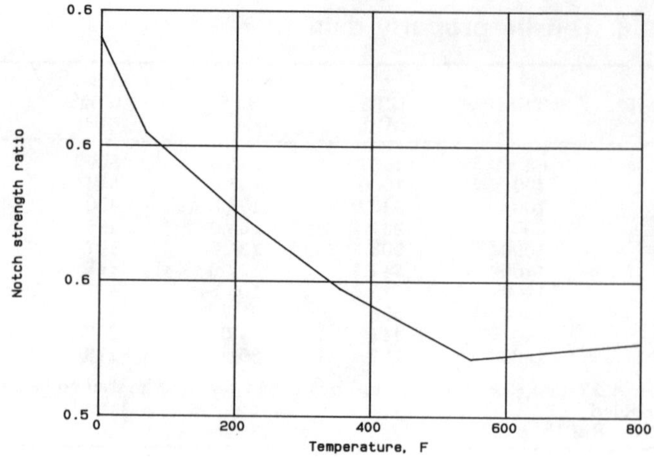

A-286: Notch strength ratio of longitudinal specimens of 0.025-in. sheet, cold rolled 80%, and aged at 1100 F for 16 h. Source: Ref 22, p 11

A-286: Effect of temperature on Charpy V-notch impact values

TEMP C	TEMP F	IS J	IS ft-lb
27	80	86.8	64.0
210	410	80.7	59.5
430	810	69.8	51.5
545	1010	61.7	45.5
600	1115	59.7	44.0
655	1215	47.5	35.0
675	1250	55.6	41.0
690	1275	50.8	37.5
712	1315	59.7	44.0
740	1365	81.3	60.0
790	1450	70.5	52.0

Note: Specimens were 22-mm (0.875-in.) diam bar; solution treated at 980 C (1800 F); oil quenched; heated to 720 C (1325 F); held for 16 h; air cooled
Source: Ref 15, p 398

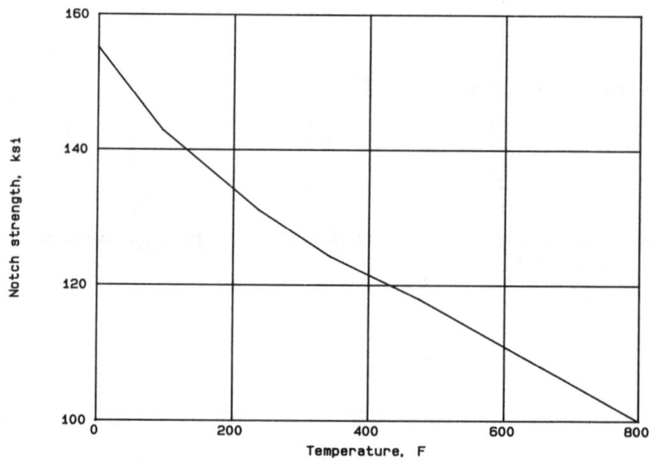

A-286: Notch strength of longitudinal specimens of 0.025-in. sheet, cold rolled 80%, and aged at 1100 F for 16 h. Source: Ref 22, p 11

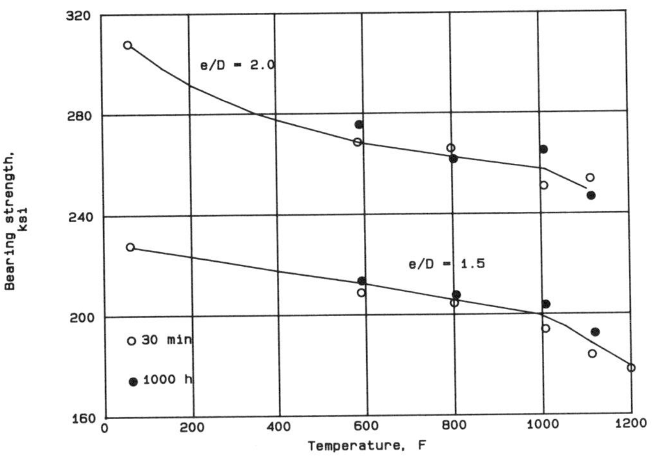

A-286: Effect of exposure time and test temperature on ultimate bearing strength of 0.062-in. sheet. Specimens were heated to 1800 F for 1 h, oil quenched, held at 1325 F for 16 h. Source: Ref 22, p 10

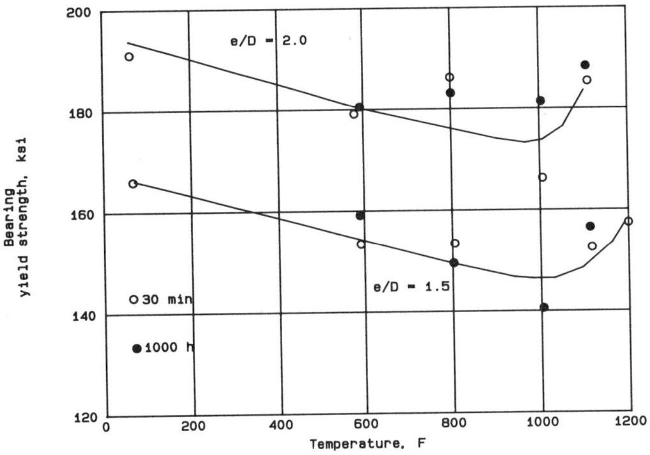

A-286: Effect of exposure time and test temperature on bearing yield strength of 0.062-in. sheet. Specimens were heated to 1800 F for 1 h, oil quenched, held at 1325 F for 16 h. Source: Ref 22, p 10

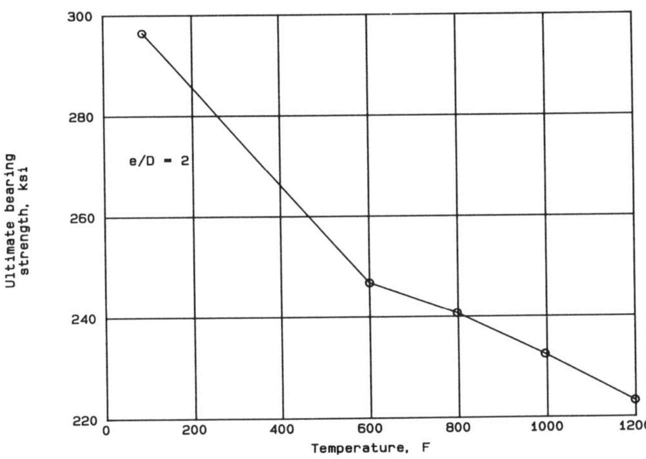

A-286: Ultimate bearing strength of 0.125-in. sheet. Specimens were heat treated at 1800 F for 1 h, oil quenched and held at 1325 F for 16 h. Source: Ref 22, p 10

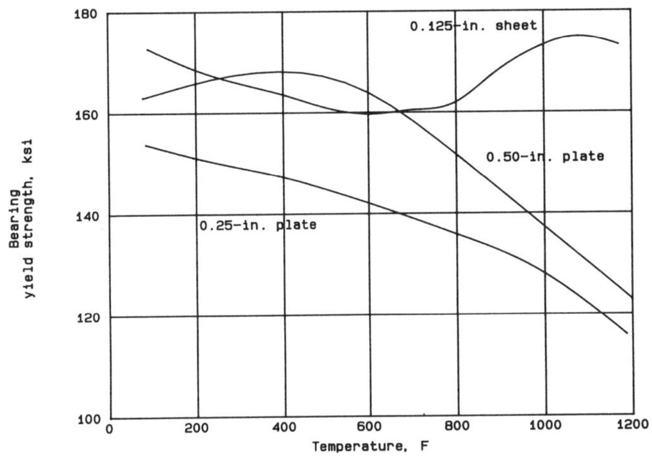

A-286: Bearing yield strength of plate. Specimens were heated to 1800 F for 1 h, oil quenched, and held at 1325 F for 16 h. Source: Ref 22, p 10

A-286: Minimum creep rates for bar

TEMP C	TEMP F	CRSTR 0.01%/h MPa	CRSTR 0.01%/h ksi	CRSTR 0.001%/h MPa	CRSTR 0.001%/h ksi	CRSTR 0.0001%/h MPa	CRSTR 0.0001%/h ksi
538	1000	669	97	614	89	558	81
593	1100	621	90	524	76	441	64
649	1200	448	65	314	45.5	224	32.5
704	1300	286	41.5	200	29	141	20.5
732	1350	231	33.5	159	23	110	16

Note: Solution treated at 982 C (1800 F) for 1 h, rapidly cooled. Large sections are oil quenched, but thin sections may be air cooled.
Aging treatment at 718 C (1325 F) for 16 h, air cooled
Source: Ref 29, Feb 1981

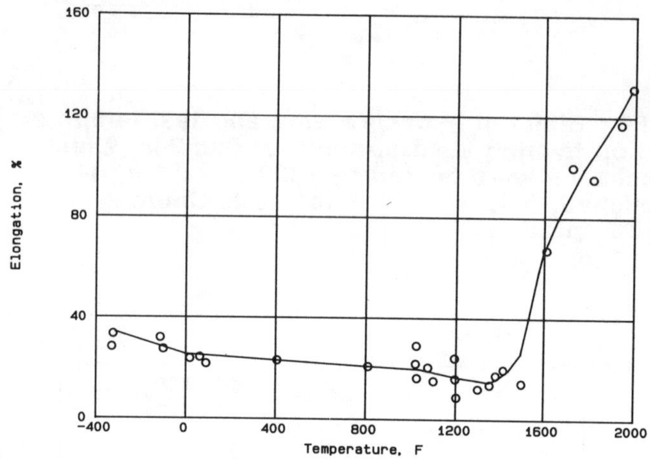

A-286: Tensile elongation of 0.875-in. bar. Specimens were heated to 1800 F for 1 h, oil quenched, held at 1325 F for 16 h, and air cooled. Source: Ref 22, p 8

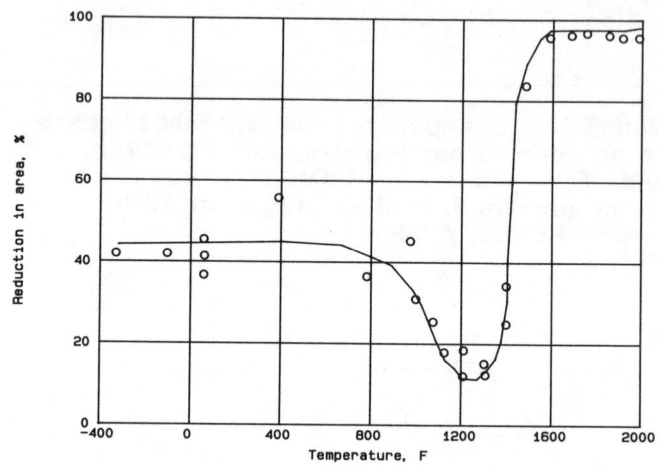

A-286: Tensile reduction in area of 0.875-in. diam bar. Specimens were heated to 1800 F for 1 h, oil quenched, held at 1325 F for 16 h, and air cooled. Source: Ref 22, p 8

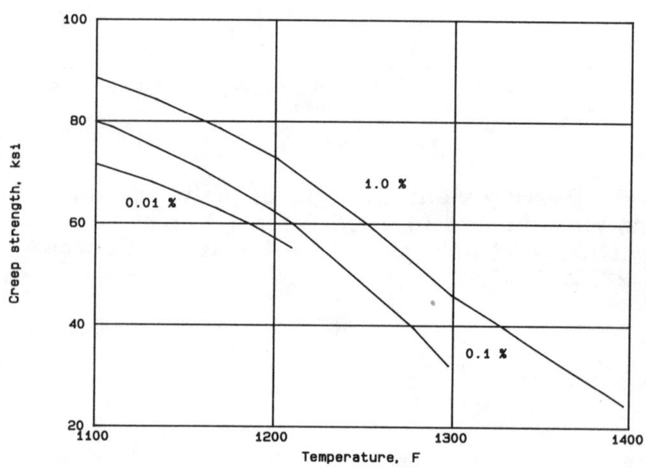

A-286: Typical 1000-h creep strength of bar and forgings. Source: Ref 22, p 13

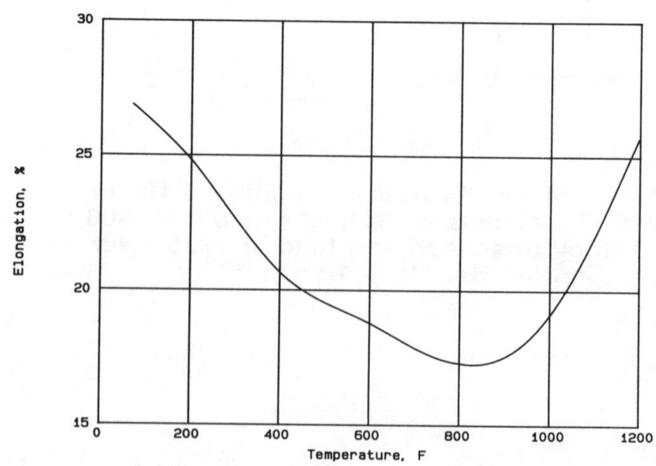

A-286: Tensile elongation of 0.062-in. sheet. Specimens were heated to 1800 F for 1 h in argon, oil quenched, and held at 1325 F for 16 h. Specimens were exposed for 1000 h. Source: Ref 22, p 9

A-286: Creep strength property data

TEMP C	TEMP F	CRSTR 0.5%/100 h MPa	CRSTR 0.5%/100 h ksi	CRSTR 1%/100 h MPa	CRSTR 1%/100 h ksi	CRSTR 0.5%/1000 h MPa	CRSTR 0.5%/1000 h ksi	CRSTR 1%/1000 h MPa	CRSTR 1%/1000 h ksi
540	1000	560	81	634	92	540	78	586	85
595	1100	525	76	552	80	470	68	483	70
650	1200	365	53	414	60	240	35	283	41
705	1300	205	30	245	35.5			155	22.5

Source: Ref 15, p 398

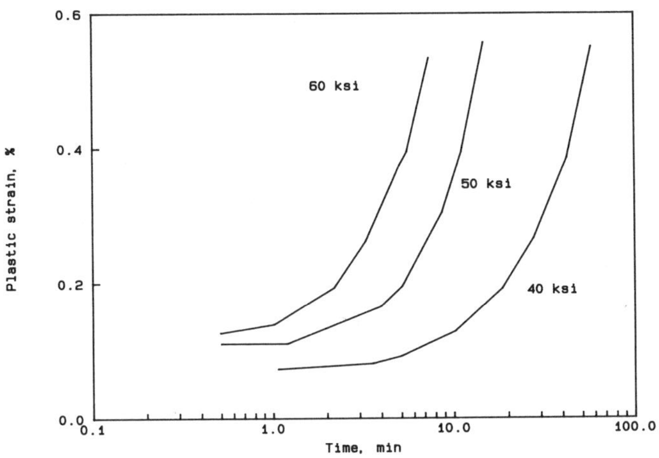

A-286: Creep curves at 1350 F for bar. Specimens were heated to 1650 F for 2 h, oil quenched, and held at 1300 F for 16 h, and air cooled. Source: Ref 22, p 13

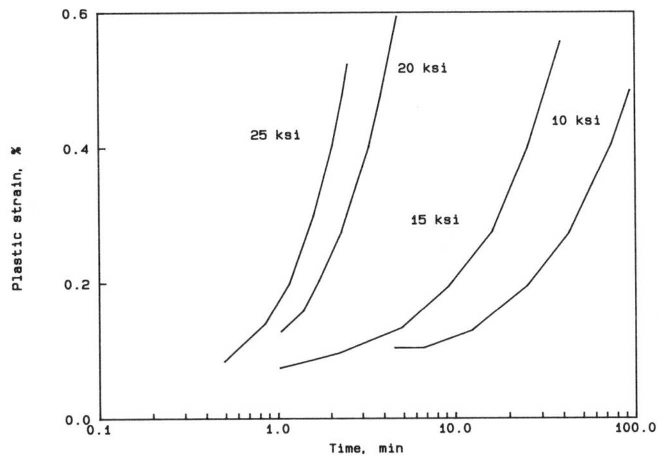

A-286: Creep curves at 1500 F for bar. Specimens were heated to 1650 F for 2 h, oil quenched, held at 1300 F for 16 h, and air cooled. Source: Ref 22, p 13

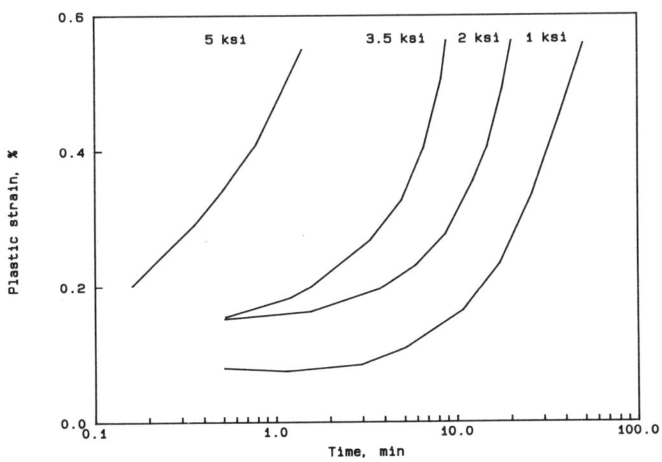

A-286: Creep curves at 1700 F for bar. Specimens were heated to 1650 F for 2 h, oil quenched, held at 1300 F for 16 h, and air cooled. Source: Ref 22, p 13

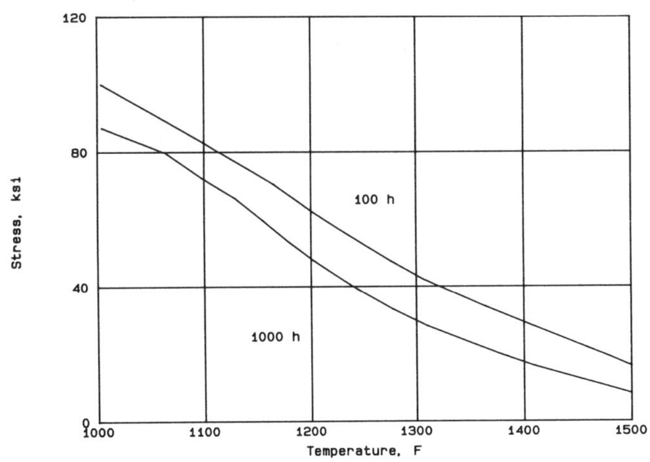

A-286: Typical 100- and 1000-h stress-rupture strength of bar and forgings. Source: Ref 22, p 12

A-286: Stress-to-rupture property data

TEMP C	TEMP F	RPSTR 10 h MPa	RPSTR 10 h ksi	RPSTR 100 h MPa	RPSTR 100 h ksi	RPSTR 1000 h MPa	RPSTR 1000 h ksi	RPSTR(a) 10000 h MPa	RPSTR(a) 10000 h ksi	EL(b) 100 h %	EL(b) 1000 h %
540	1000	827	120	689	100	600	87	525	76	3	3
595	1100	634	92	565	82	490	71	370	54	3	3
650	1200	552	80	425	62	315	46	235	34	5	9
705	1300	...	...	305	44	205	30	105	15	12	24
730	1350	...	...	240	35	150	22	...	...	28	35
815	1500	...	...	90	13	...	...	...	...	55	...

(a) Extrapolated data. (b) Gage length = 4D
Source: Ref 15, p 397

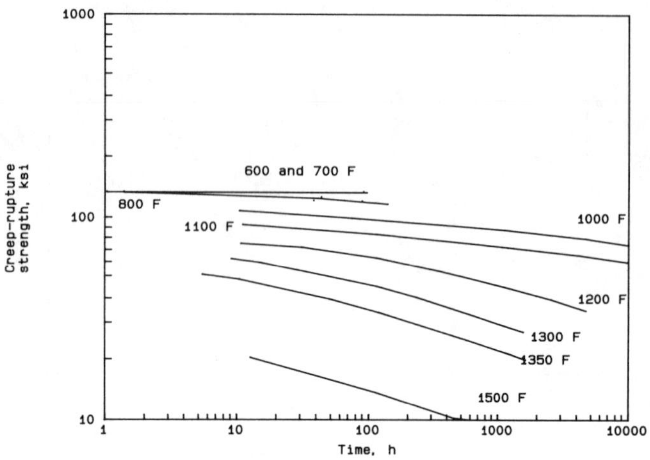

A-286: Creep-rupture strength of bar. Specimens were heated to 1800 F for 1 h, oil quenched, and held at 1325 F for 16 h. Source: Ref 22, p 12

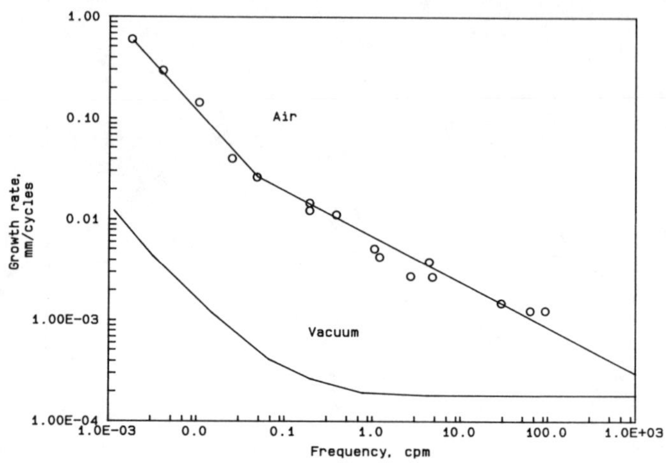

A-286: Comparison of fatigue crack growth rates in air and vacuum. Specimens were tested at 595 C. Plastic strain range: 0.2%. Source: Ref 8, p 413

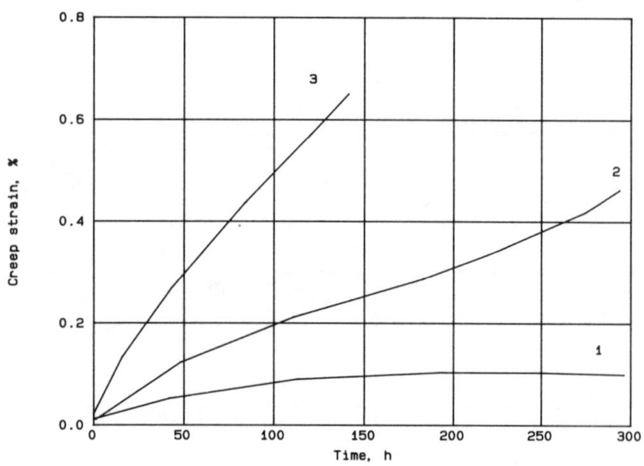

A-286: Creep curves at 480 C and 705 N/mm^2. Curve 1: solution treated at 980 C and aged at 730 C. Curve 2: solution treated at 980 C and aged at 730, 760, and 650 F. Curve 3: solution treated at 980 C, aged at 730, 775, and 650 F. Source: Ref 12, p 139

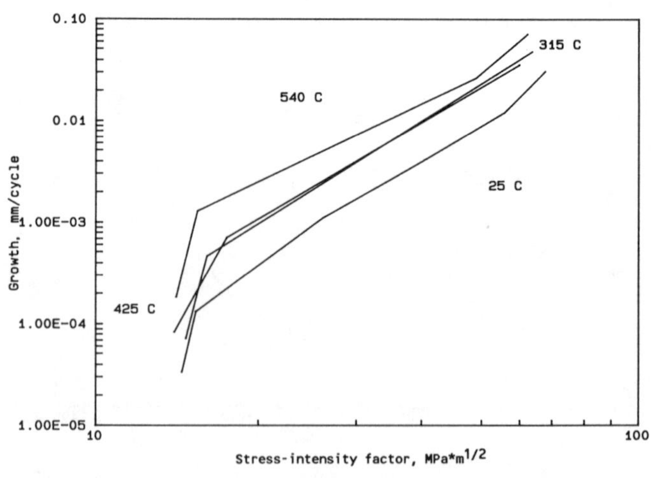

A-286: Fatigue crack propagation rate for tests in air at 3 Hz (room temperature) and 0.67 Hz (elevated temperature). R ratio of 0.05 and L-T, T-L, R-L, and R-C orientations. Source: Ref 23, p 274

A-286: Fatigue strength of bar

TEMP C	TEMP F	FATST MPa 10^6 cycles	FATST ksi 10^6 cycles	FATST MPa 10^7 cycles	FATST ksi 10^7 cycles	FATST MPa 10^8 cycles	FATST ksi 10^8 cycles
538	1000	273	39.6	247	35.8	234	34
649	1200	291	42.2	268	38.8	259	37.5
732	1350	330	47.8	308	44.6	285	41.3

Note: Solution treated at 982 C (1800 F), held for 1 h, and cooled rapidly. Large sections are oil quenched, but thin sections may be air cooled. Aged at 718 C (1325 F) for 16 h and air cooled
Source: Ref 29, Feb 1981

A-286: Hot hardness data

TEMP C	TEMP F	HARD(a) HB
538	1000	265
593	1100	252
649	1200	240
704	1300	196
760	1400	116
816	1500	91
871	1600	65

Note: Specimens solution treated at 1177 C (2150 F) for 1 h, water quenched, then aged at 760 C (1400 F) for 12 h and air cooled. Specimens were held for 20 min at test temperature before test, then tested with a cold ball.
(a) Using a 1500-kg load and a 5-mm chromium carbide ball
Source: Ref 28, p 106

A-286: Hot hardness data

TEMP C	TEMP F	HARD HRC
21	70	33
93	200	31
204	400	29
316	600	27
427	800	24
538	1000	22
649	1200	20

Note: Oil quenched at 982 C (1800 F) and aged at 718 C (1325 F) for 16 h
Source: Ref 29, Apr 1964

MULTIMET Alloy

Alternate designation: N-155, AISI 661

UNS number: R30155

Specifications: AMS 5376, AMS 5531, AMS 5532, AMS 5585, AMS 5769, AMS 5794, AMS 5795, ASTM A639 (661), ASTM A567 (661)

Nominal composition: 21.000% Cr, 20.000% Ni, 20.000% Co, 3.000% Mo, 2.500% W, 1.000% Nb, 32.200% Fe, 0.150% C, 0.150% N, 1.500% Mn, 1.000% Si max

Product forms: Wrought, bar, rod, wire, plate, sheet, strip, tube, pipe, billets, forgings, structural shapes

Description: Austenitic steel recommended for high stress applications at temperatures up to 815 C (1500 F) and for moderate stress applications up to 1095 C (2000 F). Oxidation resistance is good: up to 870 C (1600 F) for intermittent service, up to 1040 C (1900 F) for continuous service. Although the high-temperature properties are due primarily to solid solution and multiphase strengthening, it also has a limited response to precipitation treatments.

Typical applications: Turbine rotors, shafts, blades, buckets, bolts, tailpipes and tail cones, afterburner parts, exhaust manifolds, and combustion chambers.

Physical Properties

MULTIMET Alloy: Thermal conductivity

TEMP C	TEMP F	THCON W/m*K	THCON Btu/ft*h*F
21	70	12.3	7.1
95	200	13.1	7.6
205	400	14.5	8.4
315	600	16.1	9.3
425	800	17.6	10.2
540	1000	19.2	11.1
650	1200	20.8	12.0

Source: Ref 15, p 399

MULTIMET Alloy: Mean coefficients of linear thermal expansion

TEMP C	TEMP F	THEXP 10^{-6}/C	THEXP 10^{-6}/F
21-95	70-200	14.0	7.8
21-205	70-400	15.3	8.5
21-315	70-600	15.3	8.5
21-540	70-1000	17.4	9.1
21-650	70-1200	17.1	9.5
21-760	70-1400	17.5	9.7
21-870	70-1600	17.8	9.9
21-980	70-1800	18.2	10.1

Source: Ref 15, p 399

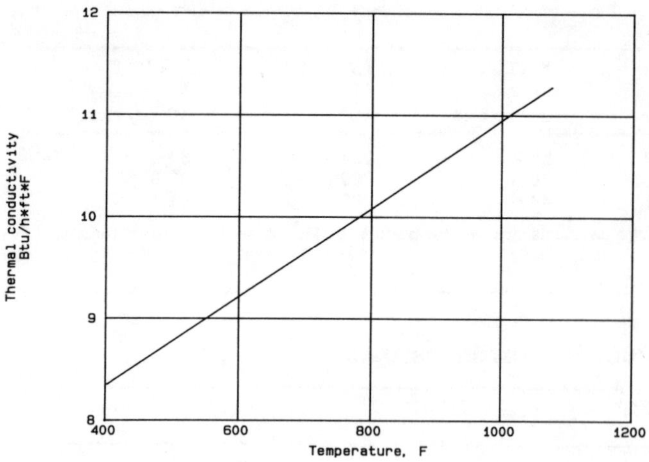

MULTIMET Alloy: Thermal conductivity. Source: Ref 22, p 3

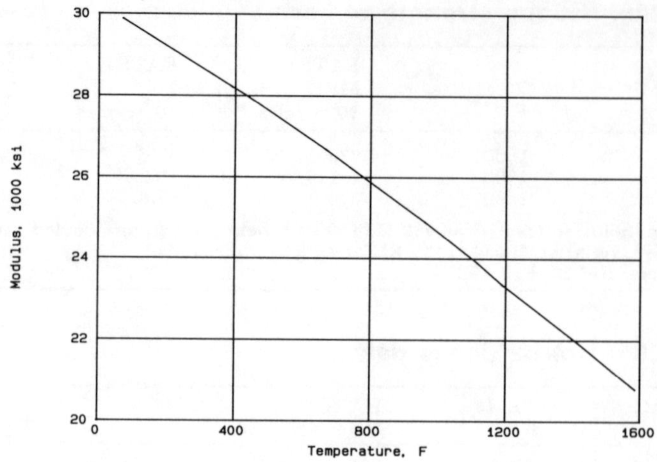

MULTIMET Alloy: Modulus of elasticity. Source: Ref 22, p 5

MULTIMET Alloy: Electrical resistivity

TEMP C	TEMP F	ELRES n{o}*m
25	77	942
200	390	1017
400	750	1092
600	1110	1150
800	1470	1180
1000	1830	1227

Source: Ref 15, p 399

MULTIMET Alloy: Poisson's ratio

TEMP C	TEMP F	PR
21	70	0.298
427	800	0.315
649	1200	0.325
816	1500	0.339

Note: Specimens were held at test temperature a minimum of 30 min prior to start of test
Source: Ref 28, p 109

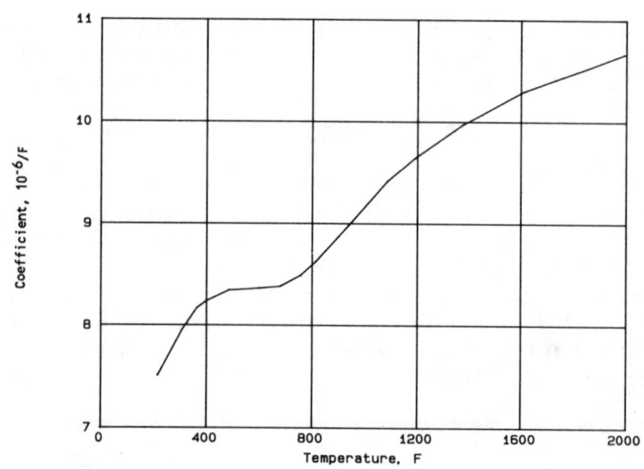

MULTIMET Alloy: Mean coefficient of linear thermal expansion from room temperature to indicated temperature. Source: Ref 22, p 3

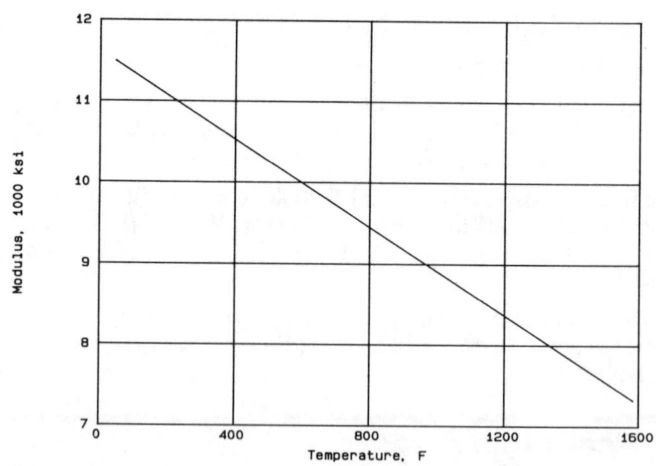

MULTIMET Alloy: Modulus of rigidity. Source: Ref 22, p 5

MULTIMET Alloy: Modulus of elasticity

TEMP C	TEMP F	E GPa	E 10^3 ksi
21	70	202	29.3
538	1000	167	24.2
816	1500	143	20.8

Note: Specimens were held at test temperature a minimum of 30 min prior to start of test
Source: Ref 28, p 109

Mechanical Properties

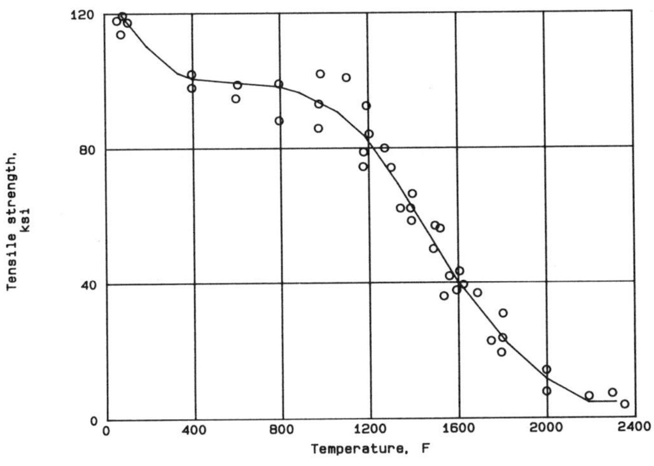

MULTIMET Alloy: Tensile strength of 0.062- to 0.123-in. sheet, heated to 2030 to 2150 F and rapid air cooled. Source: Ref 22, p 3

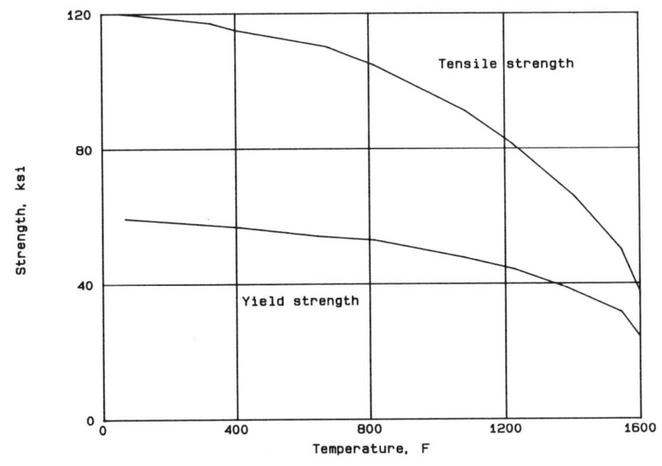

MULTIMET Alloy: Tensile and yield strength of solution treated and aged specimens. Source: Ref 22, p 3

MULTIMET Alloy: Tensile property data

TEMP C	TEMP F	UTS MPa	UTS ksi	YS 0.2% MPa	YS 0.2% ksi	EL %	RA %
21	70	314	118	407	59	40	50
538	1000	641	93	303	44	37	48
649	1200	552	80	290	42	35	47
760	1400	414	60	262	38	35	42
871	1600	262	38	179	26	37	42
982	1800	131	19	124	18	38	50
1093	2000	69	10	...	...	45	60

Note: Solution treated at 1177 C (2150 F) and air cooled
Source: Carpenter Technology

MULTIMET Alloy: Tensile property data for sheet

TEMP C	TEMP F	UTS MPa	UTS ksi	YS MPa	YS ksi	EL %
21	70	820.5	119.0	407	59.0	42.5
425	800	689.5	100.0	272	39.5	42.5
540	1000	676	98.0	248	36.0	42.5
595	1100	655	95.0	269	39.0	40.0
650	1200	600	87.0	255	37.0	37.5
705	1300	514	74.5	259	37.5	31.5
760	1400	452	65.5	234	34.0	25.5
815	1500	348	50.5	210	30.5	38.0
870	1600	293	42.5	179	26.0	29.5
925	1700	217	31.5	131	19.0	50.0
980	1800	162	23.5	93.1	13.5	39.5

Note: Specimens were 0.075- to 3.4-mm (0.030- to 0.135-in.) thick sheet; heated to 1175 C (2150 F) for 10 min; water quenched
Source: Ref 15, p 400

MULTIMET Alloy: Tensile property data for bar

TEMP C	TEMP F	UTS MPa	UTS ksi	YS 0.02% MPa	YS 0.02% ksi	YS 0.2% MPa	YS 0.2% ksi	EL %	RA %
540	1000	621	90.0	217	31.5	265	38.5	47	48.5
650	1200	562	81.5	217	31.5	248	36.0	44	46.0
760	1400	403	58.5	186	27.0	238	34.5	36	36.0
870	1600	248	36.0	193	28.0	224	32.5	44	42.0

Note: Specimens were 33-mm (1.3-in.) square bar; heated to 1175 C (2150 F) for 1 h; water quenched; heated to 816 C (1500 F) for 4 h; air cooled
Source: Ref 15, p 399

MULTIMET Alloy: Tensile property data

TEMP C	TEMP F	UTS MPa	UTS ksi	YS 0.2% MPa	YS 0.2% ksi	EL %
21	70	800	116	393	57.0	43
649	1200	505	73.2	...	...	35
927	1700	137	19.8	...	...	39
1038	1900	69	10.0	...	...	34

Note: 1.32-mm (0.052-in.) thick sheet was heat treated at 1177 C (2150 F) and rapidly air cooled
Source: Ref 28, p 109

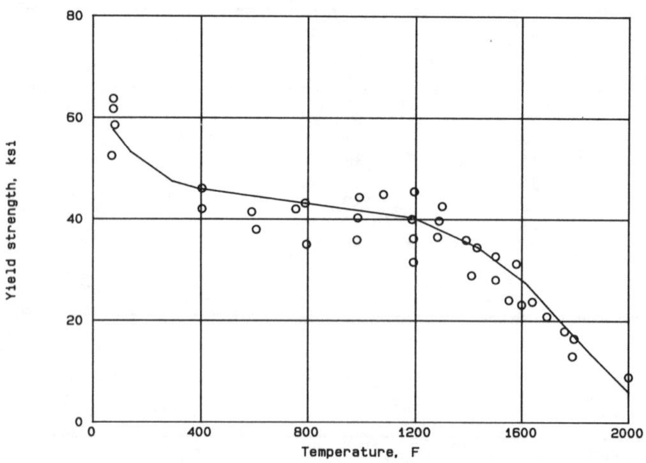

MULTIMET Alloy: Yield strength of 0.062- to 0.123-in. sheet, heated to 2030 to 2150 F and rapid air cooled. Source: Ref 22, p 3

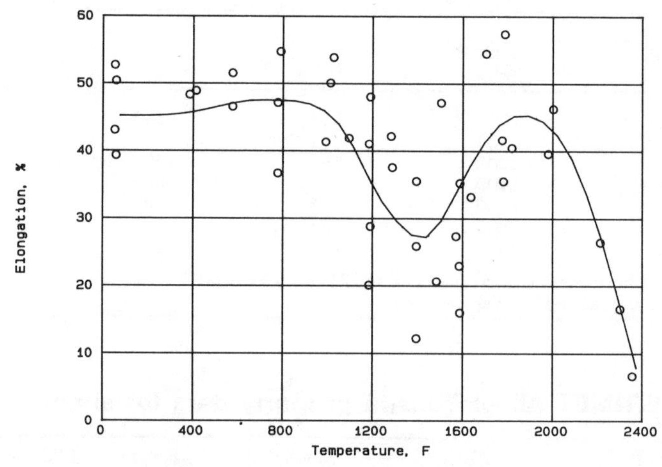

MULTIMET Alloy: Tensile elongation of 0.062- to 0.123-in. sheet, heat treated at 2030 to 2150 F and rapid air cooled. Source: Ref 22, p 3

MULTIMET Alloy: Creep strength of bar and sheet

TEMP C	TEMP F	CRSTR 1%/100 h MPa	CRSTR 1%/100 h ksi	CRSTR 1%/1000 h MPa	CRSTR 1%/1000 h ksi	CRSTR 1%/10000 h MPa	CRSTR 1%/10000 h ksi	CRSTR(a) 1%/100000 h MPa	CRSTR(a) 1%/100000 h ksi
650	1200	276	40.0	234	34.0	200	29.0	152	22.0
705	1300	...	...	...	...	121	17.5	97	14.0
730	1350	155	22.5	128	18.5	96.5	14.0	79.2	11.5
815	1500	100	14.5	79.3	11.5	62	9.0	45	6.5
900	1650	48	7.0	34	5.0	24	3.5	17	2.5

Note: Specimens were solution treated at 1175 C (2150 F) and water quenched
(a) Extrapolated
Source: Ref 15, p 399

MULTIMET Alloy: Effect of temperature on Charpy keyhole notch impact strength of bar

AGING TEMP C	AGING TEMP F	AGING TIME h	TEST TEMP C	TEST TEMP F	IS J	IS ft-lb
Heat treated 2 h at 1093 C (2000 F) and air cooled						
...	...	...	21	70	83	61
649	1200	2	21	70	77	57
649	1200	2	649	1200	85	63
649	1200	168	21	70	42	31
649	1200	168	649	1200	56	41
732	1350	2	21	70	58	43
732	1350	2	732	1350	66	49
732	1350	168	21	70	26	19
732	1350	168	732	1350	47	35
816	1500	2	21	70	49	36
816	1500	2	816	1500	64	47
816	1500	168	21	70	22	16
816	1500	168	816	1500	43	32
Heat treated 1 h at 1204 C (2200 F) and air cooled						
...	...	...	21	70	88	65
649	1200	2	21	70	79	58
649	1200	2	649	1200	83	61
649	1200	168	21	70	35	26
649	1200	168	649	1200	61	45
732	1350	2	21	70	60	44
732	1350	2	732	1350	69	51
732	1350	168	21	70	15	11
732	1350	168	732	1350	46	34
816	1500	2	21	70	42	31
816	1500	2	816	1500	66	49
816	1500	168	21	70	12	9
816	1500	168	816	1500	38	28

Source: Ref 28, p 110

MULTIMET Alloy: Effect of temperature on Charpy V-notch impact values

TEMP C	TEMP F	IS J	IS ft-lb
24	75	81	60
205	400	104	77
425	800	108	80
540	1000	108	80
650	1200	89	66
760	1400	87	64
870	1600	87	64

Note: Specimens were 16-mm (0.63-in.) diam bar; heat treated at 1190 C (2175 F) for 1 h; air cooled, plus 730 C (1350 F) for 16 h; air cooled
Source: Ref 15, p 400

MULTIMET Alloy: Creep data

TEMP C	CRSTR MPa	CRSTR ksi	EXP h	EL %	CRT 500 h %/h	CRT 1000 h %/h	CRT 5000 h %/h	CRT 9000 h %/h	CRT 14000 h %/h
Sheet: 1.6-mm (0.063-in.) thick; heat treated at 1177 C (2150 F); rapid air cooled									
927	31	4.5	2070	0.098	0.001356	0.001013	...	...	...
927	35	5.0	990	0.042	0.002178	...	...	...	...
982	21	3.0	1020	0.007	0.002114	0.0066874	...	...	...
982	24	3.5	390	0.035	...	...	...	...	...
982	31	4.5	250	0.091	...	...	...	...	...
Forging: 533-mm x 82.6-mm (21-in. x 3.25-in.) diam and thickness; stress relieved for 2 h at 649 C (1200 F); air cooled									
649	103	15	14000	0.070	0.00018	0.00006	0.000037	0.000009	0.000003
649	138	20	14000	0.090	0.00016	0.000087	0.000048	0.000014	0.000003
649	172	25	14000	0.113	0.00018	0.00018	0.000083	0.000039	0.000025
649	207	30	1530	0.135	0.00033	0.00015	...	...	...
649	241	35	1060	0.191	0.00056	0.00044	...	...	...
Bar: 25-mm (1-in.) diam; heat treated for 1 h at 1204 C (2200 F); water quenched; aged 50 h at 732 C (1350 F)									
732	69	10	16000	...	0.00016	0.000017	0.000017	0.000003	0.000009
732	86	12.5	29000	...	0.00031	0.000067	0.000037	0.000037	0.000037(a)
732	103	15	12000	...	0.00035	0.00020	0.00012	0.00034	...
732	103	15	2062	0.061	0.00030	0.00011	...	...	...
732	138	20	1320	0.116	0.00123	0.00130	...	...	...
Bar: 25-mm (1-in.) diam: heat treated for 1 h at 1204 C (2200 F); water quenched; aged for 50 h at 816 C (1500 F)									
816	28	4	6000	...	0.00013	0.000008	0.000008	...	...
816	41	6	6000	...	0.00007	0.000024	0.000024	...	...
816	55	8	5500	...	0.000075	0.00002	0.00045	...	...
816	69	10	1640	...	0.0022	0.0045	...	...	...

(a) Test continued to 25000 h with a creep rate of 0.00007%/h at 25000 h
Source: Ref 28, p 110

MULTIMET Alloy: Stress-to-rupture data

TEMP C	TEMP F	RPSTR 10 h MPa	RPSTR 10 h ksi	RPSTR 100 h MPa	RPSTR 100 h ksi	RPSTR 1000 h MPa	RPSTR 1000 h ksi
649	1200	448	65	359	52	262	38
732	1350	262	38	207	30	159	23
816	1500	179	26	124	18	97	14
899	1650	124	18	69	10	48	7

Note: Solution treated at 1204 C (2200 F) for 1 h; water quenched; aged for 16 h at 760 C (1400 F); air cooled
Source: Carpenter Technology

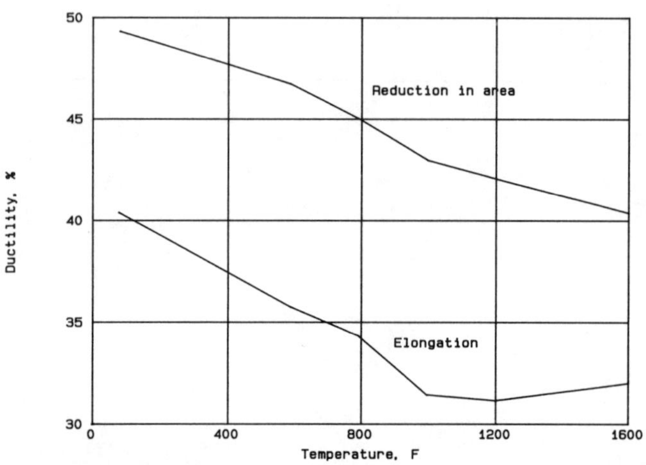

MULTIMET Alloy: Tensile reduction in area and elongation of solution treated and aged specimens. Source: Ref 22, p 3

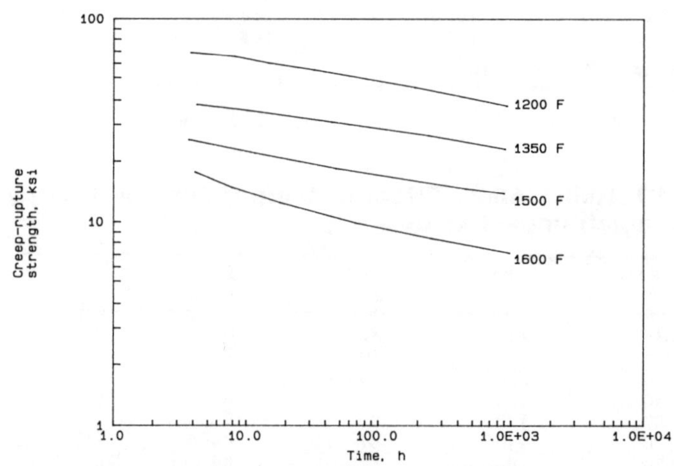

MULTIMET Alloy: Creep-rupture strength. Specimens were heated to 2200 F for 1 h, water quenched, held at 1400 F for 16 h, and air cooled. Source: Ref 22, p 5

MULTIMET Alloy: Stress-to-rupture data for sheet

TEMP C	TEMP F	RPSTR 10 h MPa	RPSTR 10 h ksi	RPSTR 100 h MPa	RPSTR 100 h ksi	RPSTR 1000 h MPa	RPSTR 1000 h ksi	RPSTR(a) 10000 h MPa	RPSTR(a) 10000 h ksi
650	1200	365	53.0	317	46.0	265	38.5	207	30.0
705	1300	307	44.5	245	35.5	186	27.0	138	20.0
760	1400	231	33.5	172	25.0	128	18.5	93.1	13.5
815	1500	169	24.5	121	17.5	86.2	12.5	73	9.1
870	1600	117	17.0	84.8	12.3	59	8.5	40	5.8
925	1700	82.7	12.0	59	8.5	38	5.5	...	...

Note: Specimens were 1.9-mm (0.075-in.) thick; heat treated to 1175 C (2150 F) for 10 min; water quenched
(a) Extrapolated data
Source: Ref 14, p 400

MULTIMET Alloy: Stress-to-rupture data for bar

TEMP C	TEMP F	RPSTR 10 h MPa	RPSTR 10 h ksi	RPSTR 100 h MPa	RPSTR 100 h ksi	RPSTR 1000 h MPa	RPSTR 1000 h ksi	RPSTR(a) 10000 h MPa	RPSTR(a) 10000 h ksi
650	1200	427	62.0	345	50.0	276	40.0	217	31.5
705	1300	324	47.0	259	37.5	200	29.0	148	21.5
760	1400	245	35.5	186	27.0	138	20.0	100	14.5
815	1500	176	25.5	128	18.5	90	13.0	64	9.3
870	1600	128	18.5	88.3	12.8	60	8.7	40	5.8
925	1700	86.2	12.5	60	8.7	38	5.5	24	3.5

Note: Specimens were 33-mm (1.3-in.) square; hot rolled bar; heat treated at 1175 C (2150 F) for 1 h; water quenched; plus 815 C (1500 F) for 4 h; air cooled
(a) Extrapolated data
Source: Ref 15, p 400

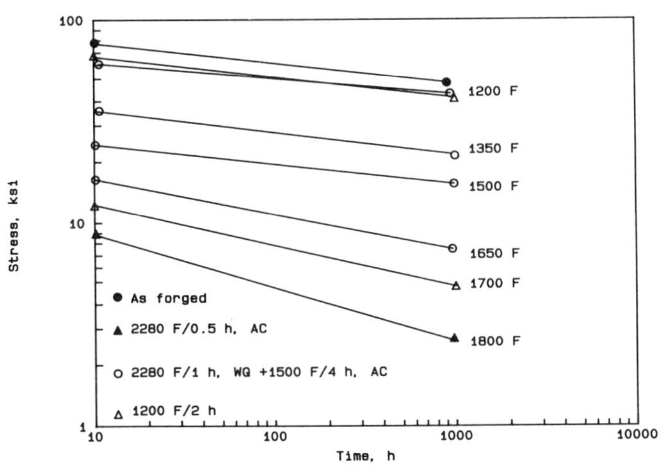

MULTIMET Alloy: Creep-rupture strength of bar and forgings. Source: Ref 22, p 4

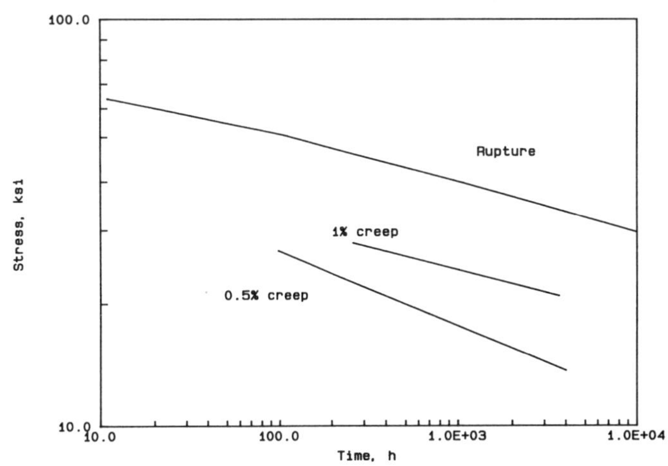

MULTIMET Alloy: Creep and rupture strength of bar tested at 1200 F, heat treated at 2150 F and rapid air cooled. Source: Ref 22, p 4

MULTIMET Alloy: Fatigue strength

TEMP C	TEMP F	FATST MPa 10^8 cycles	FATST ksi 10^8 cycles
649	1200	455	66
816	1500	228	33

Note: 25- to 38.1-mm (1- to 1.5-in.) diam bar was aged 50 h at 649 C (1200 F)
Source: Ref 28, p 110

MULTIMET Alloy: Stress-to-rupture data

TEMP C	TEMP F	RPSTR 10 h MPa	RPSTR 10 h ksi	RPSTR 50 h MPa	RPSTR 50 h ksi	RPSTR 100 h MPa	RPSTR 100 h ksi	RPSTR 500 h MPa	RPSTR 500 h ksi	RPSTR 1000 h MPa	RPSTR 1000 h ksi
Sheet: 1.6-mm (0.063-in.) thick; heat treated at 1177 C (2150 F); rapid air cooled											
927	1700	83	12	66	9.6	54	7.8	45	6.5	33	4.8
982	1800	72	10.5	48	6.9	39	5.6	24	3.5	20	2.9
Forging: 533-mm x 82.6-mm (21-in. diam x 3.25-in. thick); stress relieved for 2 h at 649 C (1200 F)											
649	1200	448	65	...	...	379	55	...	...	290	42
Bar: 25-mm (1-in.) diam; as forged											
649	1200	538	78	469	68	414	60	359	52	338	49
Bar: 25-mm (1-in.) diam; heat treated for 1 h at 1204 C (2200 F); water quenched; aged 50 h at 732 C (1350 F)											
649	1200	427	62	379	55	359	52	317	46	296	43
732	1350	248	36	207	30	193	28	165	24	152	22
816	1500	172	25	152	22	138	20	117	17	110	16
899	1650	117	17	90	13	83	12	59	8.6	50	7.2
Bar: 25-mm (1-in.) diam; heat treated for 1 h at 1204 C (2200 F); water quenched; aged 50 h at 816 C (1500 F)											
927	1700	86	12.5	62	9	52	7.6	39	5.6	33	4.8
982	1800	61	8.8	40	5.8	34	4.9	21	3.1	18	2.6

Source: Ref 28, p 110

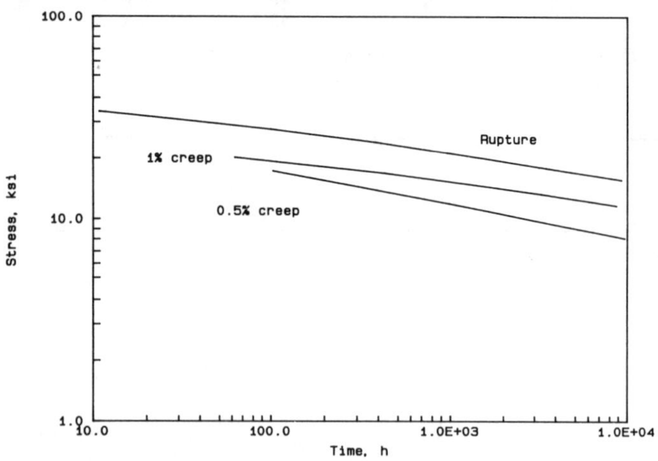

MULTIMET Alloy: Creep and rupture strength of bar tested at 1350 F, treated at 2150 F and rapid air cooled. Source: Ref 22, p 4

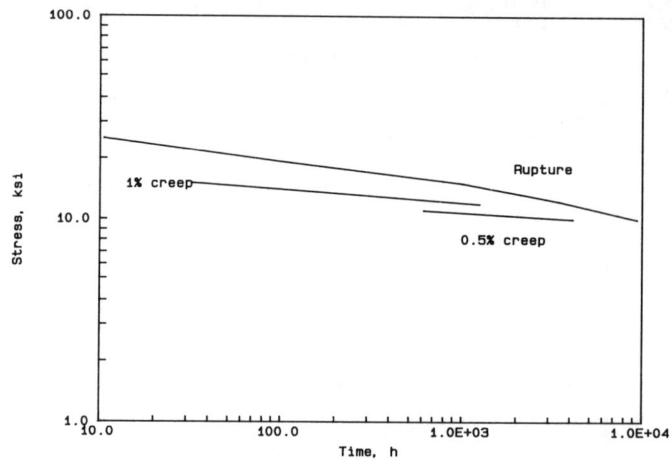

MULTIMET Alloy: Creep and rupture strength of bar tested at 1500 F, treated at 2150 F, and rapid air cooled. Source: Ref 22, p 4

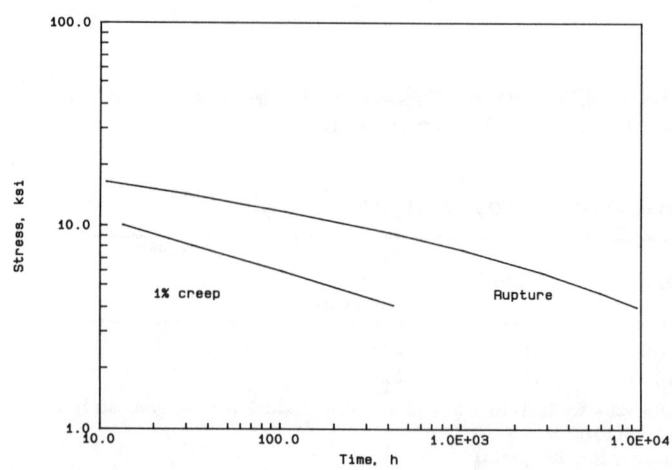

MULTIMET Alloy: Creep and rupture strength of bar tested at 1650 F, treated at 2150 F, and rapid air cooled. Source: Ref 22, p 4

MULTIMET Alloy: Hot hardness data for hot rolled bar

TEMP C	TEMP F	HARD HB(a)	HARD HB (b)
400	752	138	...
500	932	130	...
600	1112	130	...
700	1292	120	...
800	1472	105	...
900	1652	60	...
400	752	...	190
500	932	...	192
600	1112	...	187
700	1292	...	192
800	1472	...	192
900	1652	...	218

Note: Heat treated at 1184 C (2165 F) and water quenched; aged 20 min at temperature
(a) Special Brinell hot hardness tests were made using 2000-kg load and a 10-mm tungsten carbide ball. Both the ball and the specimen were held at temperature for 20 min before testing.
(b) Special Brinell hardness measured at room temperature using 2000-kg load and a 10-mm tungsten carbide ball
Source: Ref 28, p 111

Discaloy

Alternate designation: AISI 662

UNS number: S66220

Specifications: AMS 5733, ASTM A453 (662), ASTM A638 (662), SAE J467 (Discaloy)

Nominal composition: 14.000-14.000% Cr max, 26.000-26.000% Ni max, 3.000-3.000% Mo max, 1.700-1.700% Ti max, 0.250-0.250% Al max, 55.000-55.000% Fe max, 0.060-0.060% C max

Product forms: Wrought, bar, rod, wire, plate, sheet, strip, tube, pipe, billets, forgings, structural shapes

Description: Precipitation hardening iron-base alloy developed for aircraft gas turbine discs. It combines high creep strength with good ductility, low notch sensitivity, and good oxidation resistance in the 540 to 705 C (1000 to 1300 F) temperature range. Maximum operating temperature is not limited by oxidation resistance, but by a decrease in strength due to overaging at temperatures above 730 C (1350 F).

Typical applications: Aircraft gas turbine discs, rotors, bolts, and stationary gas and steam turbine components.

Physical Properties

Discaloy: Thermal conductivity

TEMP C	TEMP F	THCON W/m*K	THCON Btu/h*ft*F
100	212	14.2	8.20
300	572	17.1	9.88
500	932	20.4	11.8
700	1292	23.3	13.5

Source: Ref 28, p 113

Discaloy: Modulus of elasticity

TEMP C	TEMP F	E GPa	E 10^3 ksi
24	75	196	28.4
410	770	163	23.6
538	1000	154	22.3
649	1200	145	21.0

Source: Ref 28, p 113

Discaloy: Mean coefficients of linear thermal expansion

TEMP C	TEMP F	THEXP 10^{-6}/C	THEXP 10^{-6}/F
21-93	70-200	15.3	8.5
21-204	70-400	15.7	8.7
21-316	70-600	16.4	9.1
21-427	70-800	17.0	9.4
21-538	70-1000	17.1	9.5
21-649	70-1200	17.3	9.6

Source: Ref 28, p 113

Mechanical Properties

Discaloy: Tensile property data

TEMP C	TEMP F	YS 0.2% MPa	YS 0.2% ksi	UTS MPa	UTS ksi	EL %	RA %
21	70	678	98.3	1000	145.0	24	23
204	400	609	88.3	900	130.5	19	31
316	600	487	70.6	912	132.2	16	20
427	800	601	87.2	818	118.7	14	15
538	1000	603	87.4	840	121.8	14	31
649	1200	574	83.2	754	109.3	15	22
704	1300	466	67.6	601	87.2	10	17
760	1400	446	64.7	453	65.7	11	12
816	1500	245	35.5	246	35.7	19	28

Note: 2.67-mm (0.1052-in.) thick sheet was solution treated at 995 C (1825 F) for 1 h, oil quenched, and aged 20 h at 732 C (1350 F), plus 20 h at 649 C (1200 F)
Source: Ref 28, p 113

Discaloy: Tensile property data

TEMP C	TEMP F	YS 0.2% MPa	YS 0.2% ksi	UTS MPa	UTS ksi	EL %	RA %
21	70	731	106	1000	145	19	23
538	1000	648	94.0	861	125	16	34
649	1200	628	91.0	717	104	19	24
732	1350	510	74.0	562	81.5	14	26

Note: Specimens were cut from a disc forging 343-mm x 75-mm (13.5-in. diam x 3-in. thick), with a uniform hardness of 293 HB on both faces
Source: Ref 28, p 113

Discaloy: Effect of temperature on Charpy V-notch impact values

TEMP C	TEMP F	IS J	IS ft-lb
27	80	49	36
93	200	58	43
204	400	58	43
316	600	53	39
427	800	46	34
538	1000	46	34
649	1200	47	35

Note: Specimens were machined from 14.3-mm (0.563-in.) diam bars, heat treated 1 h at 995 C (1825 F), water quenched, aged 20 h at 732 C (1350 F) and 20 h at 649 C (1200 F). The material used was from one heat with an average hardness of 299 HB. Grain size was a uniform ASTM No. 7
Source: Ref 28, p 114

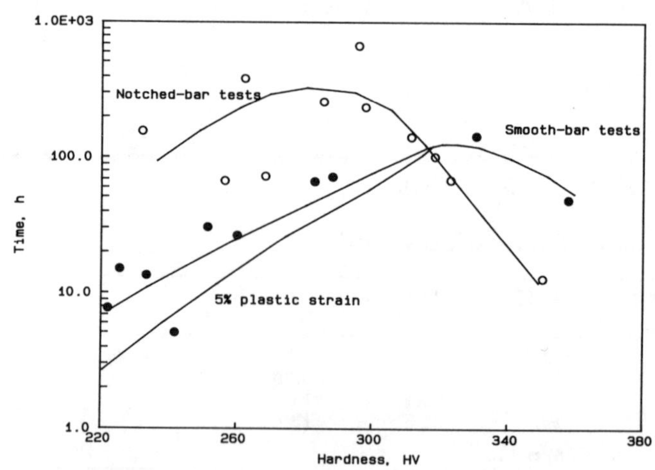

Discaloy: Variation of rupture time at 650 C with initial hardness. Kt = 3.9 for notched bar. Source: Ref 8, p 317

Discaloy: Rupture strength property data

TEMP C	TEMP F	RPSTR(a) 100 h MPa	RPSTR(a) 100 h ksi	RPSTR(b) 100 h MPa	RPSTR(b) 100 h ksi
538	1000	621	90	745	108
649	1200	359	52	496	72

Note: Specimens were obtained from a disc forging 343-mm (13.5-in.) diam x 75-mm (3-in.) thick, with a uniform hardness of 293 HB on both faces. Gage sections of all test specimens were 12.8-mm (0.505-in.) diam x 50-mm (2-in.) long
(a) Unnotched specimens. (b) Notched specimens
Source: Ref 28, p 114

Discaloy: Fatigue strength of pancake disc

TEMP C	TEMP F	FATST(a) 10^6 cycles ksi	FATST(b) 10^6 cycles ksi	FATST(a) 10^7 cycles ksi	FATST(b) 10^7 cycles ksi	FATST(a) 10^8 cycles ksi	FATST(b) 10^8 cycles ksi
21	70	67	43	56	41	55	41
649	1200	58	37	55	36	51	36

Note: Specimens were taken from a 1.75% titanium disc forging, solution treated 2 h at 995 C (1825 F), oil quenched, and aged 20 h at 732 C (1350 F) and 20 h at 649 C (1200 F). The disc had a uniform hardness of 269 HB on both faces and a grain size was ASTM No. 7
(a) Unnotched specimens. (b) Notched specimens
Source: Ref 28, p 113

Discaloy: Fatigue strength of rolled bar

ASTM grain size	FATST(a) 10^8 cycles MPa	FATST(a) 10^8 cycles ksi	FATST(b) 10^8 cycles MPa	FATST(b) 10^8 cycles ksi
Tested at 21 C (70 F)				
0.3	186	27.0	186	27.0
1.5	248	36.0	224	32.5
6.7	310	45.0	193	28.0
7.9	331	48.0	193	28.0
10.5	448	65.0	224	32.5
8-duplex	396	57.5	...	...
Tested at 649 C (1200 F)				
3	279	40.5	...	...
5.5	321	46.5	...	...
8	345	50.0	...	...
Tested at 704 C (1300 F)				
3	259	37.5	...	...
5.5	286	41.5	...	...
8	286	41.5	...	...

Note: Specimens were taken from a 1.84% titanium 15.9-mm (0.625-in.) diam rolled bar with various solution treatments to obtain grain sizes of ASTM No. 0 to 10.5. Material was aged 20 h at 732 C (1350 F) and 20 h at 649 C (1200 F).
(a) Unnotched specimens. (b) Notched specimens
Source: Ref 28, p 114

Discaloy: Stress-rupture life test results for disc forgings at 649 C (1200 F) and 414 MPa (60 ksi)

Test	RPT h	EL %
1	37.3	8
2	93.0	16
3	39.0	6
4	55.0	9
5	76.5	6
6	62.5	13

Note: Test specimens were obtained from a disc forging 343-mm (13.5-in.) diam by 75-mm (3-in.) thick, with a uniform hardness of 293 HB on both faces. Gage sections of all test specimens were 12.8-mm (0.505-in.) by 50-mm (2-in.) long
Source: Ref 28, p 114

V-57

Alternate designation: AISI 663
Nominal composition: 14.800% Cr, 27.000% Ni, 1.250% Mo, 3.000% Ti, 0.250% Al, 48.600% Fe, 0.080% C, 0.010% B, 0.500% V
Product forms: Wrought, bar, rod, wire, plate, sheet, strip, tube, pipe, billets, forgings, structural shapes

Description: Modified A-286 with higher titanium and boron contents. This austenitic, precipitation-hardening alloy isdesigned for applications requiring strength and corrosion resistance at temperatures to 760 C (1400 F).
Typical applications: Turbine wheels, spacers, buckets, and high-temperature bolting.

Physical Properties

V-57: Mean coefficients of linear thermal expansion

TEMP C	TEMP F	THEXP(a) 10^{-6}/C	THEXP(a) 10^{-6}/F	THEXP(b) 10^{-6}/C	THEXP(b) 10^{-6}/F
21-50	70-122	13.9	7.7	15.1	8.4
21-100	70-212	16.2	9.0	16.4	9.1
21-200	70-392	16.6	9.2	16.6	9.2
21-300	70-572	16.7	9.3	16.9	9.4
21-400	70-752	17.1	9.5	17.1	9.5
21-450	70-842	17.3	9.6	17.3	9.6
21-500	70-932	17.3	9.6	17.3	9.6
21-550	70-1022	17.1	9.5	17.5	9.7
21-600	70-1112	17.1	9.5	17.5	9.7
21-650	70-1202	16.7	9.3	17.5	9.7
21-700	70-1292	16.2	9.0	17.8	9.9
21-750	70-1382	16.7	9.3	18.5	10.3
21-800	70-1472	17.5	9.7	19.4	10.8
21-850	70-1562	18.4	10.2	20.3	11.3
21-900	70-1652	18.5	10.3	...	...

(a) Heat treated at 982 C (1800 F) for 2 h and oil quenched. (b) Heat treated at 982 C (1800 F) for 2 h, oil quenched, reheated to 732 C (1350 F) for 16 h, and air cooled
Source: Ref 28, p 115

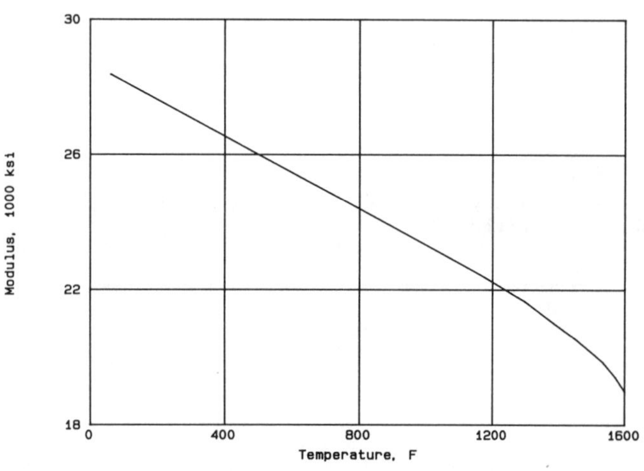

V-57: **Modulus of elasticity. Source: Ref 22, p 3**

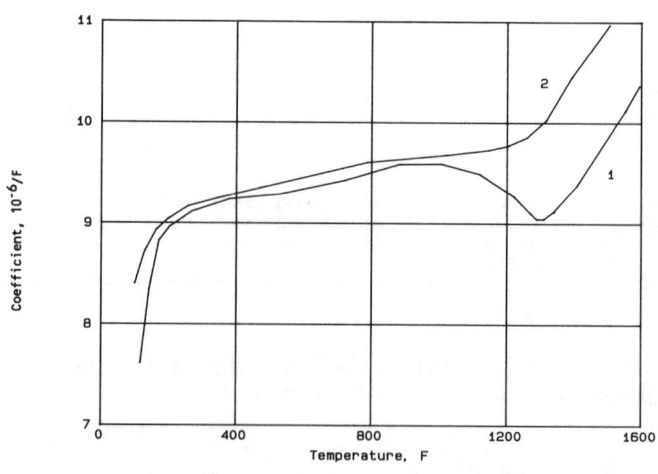

V-57: **Mean coefficient of linear thermal expansion. Curve 1: specimens were heat treated at 1800 F for 2 h and oil quenched. Curve 2: specimens were also aged at 1350 F for 16 h and air cooled. Source: Ref 22, p 1**

V-57: Moduli of elasticity, rigidity, and Poisson's ratio

TEMP C	TEMP F	E GPa	E 10^3 ksi	G GPa	G 10^3 ksi	PR
27	80	196.5	28.5	75.8	11.0	0.29
93	200	191.7	27.8	...	...	...
149	300	187.5	27.2	72.4	10.5	0.30
204	400	184.1	26.7	...	...	...
260	500	180.0	26.1	69	10.0	0.31
316	600	176.5	25.6	...	...	...
371	700	172.4	25.0	65.5	9.5	0.31
427	800	169.0	24.5	...	...	...
482	900	165.5	24.0	62.7	9.1	0.32
538	1000	161.3	23.4	...	...	...
593	1100	157.2	22.8	59.3	8.6	0.33
649	1200	153.1	22.2	57.9	8.4	...
677	1250	...	...	...	...	0.33
704	1300	148.2	21.5	55.8	8.1	...
760	1400	144.1	20.9	53.8	7.8	0.34
816	1500	138.6	20.1	51.7	7.5	...
871	1600	129.6	18.8	47.6	6.9	0.35

Source: Ref 28, p 115

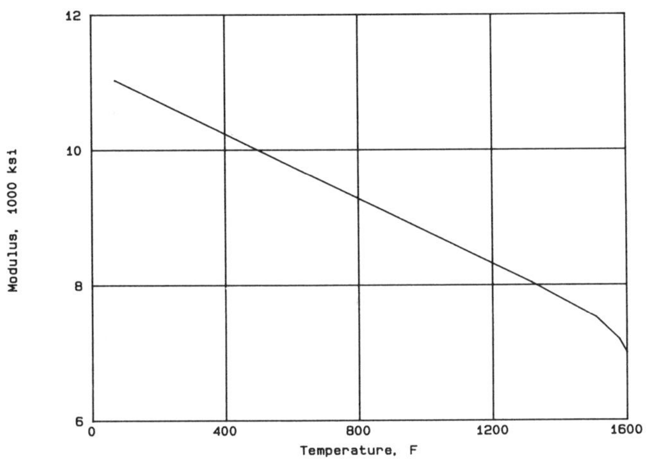

V-57: Modulus of rigidity. Source: Ref 22, p 3

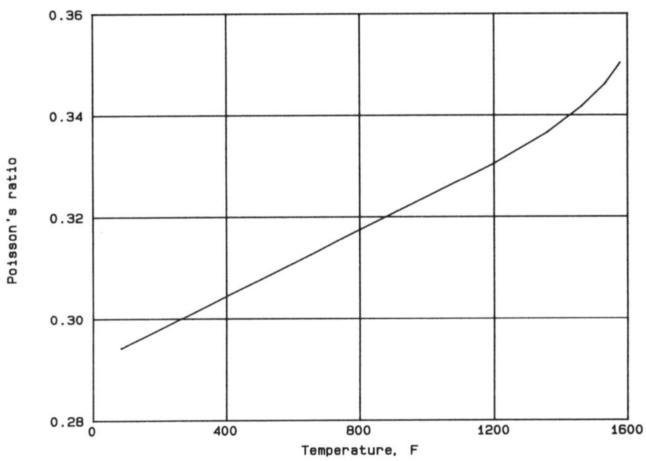

V-57: Poisson's ratio. Source: Ref 22, p 3

Mechanical Properties

V-57: Tensile property data for rolled bar

TEMP C	TEMP F	UTS MPa	UTS ksi	YS 0.02% MPa	YS 0.02% ksi	YS 0.2% MPa	YS 0.2% ksi	EL %	RA %
21	70	1227	178	696	101.0	848	123.0	24	42
93	200	1200	174	690	100.0	834	121.0	22	42
204	400	1172	170	655	95.0	814	118.0	22	42
316	600	1138	165	634	92.0	800	116.0	20	42
427	800	1089	158	621	90.0	779	113.0	19	42
538	1000	1055	153	607	88.0	765	111.0	16	43
649	1200	910	132	593	86.0	758	110.0	20	30
732	1350	696	101	427	62.0	614	89.0	18	27
816	1500	414	60.0	214	31.0	338	49.0	40	71

Note: Heat treated 2 h at 982 C (1800 F), oil quenched, aged 16 h at 732 C (1350 F), and air cooled
Source: Ref 28, p 116

V-57: Stress-to-rupture property data

TEMP C	TEMP F	RPSTR 10 h MPa	RPSTR 10 h ksi	RPSTR 100 h MPa	RPSTR 100 h ksi	RPSTR 1000 h MPa	RPSTR 1000 h ksi
538	1000	...	...	772	112	703	102
593	1100	745	108	655	95	572	83
649	1200	627	91	531	77	441	64
704	1300	517	75	414	60	290	42
732	1350	...	...	276	40	172	25
816	1500	...	...	97	14	59	8.5

Note: Specimens were heat treated at 982 C (1800 F), held for 2 h, and oil quenched. Then they were aged at 732 C (1350 F) for 16 h and air cooled
Source: Carpenter Technology

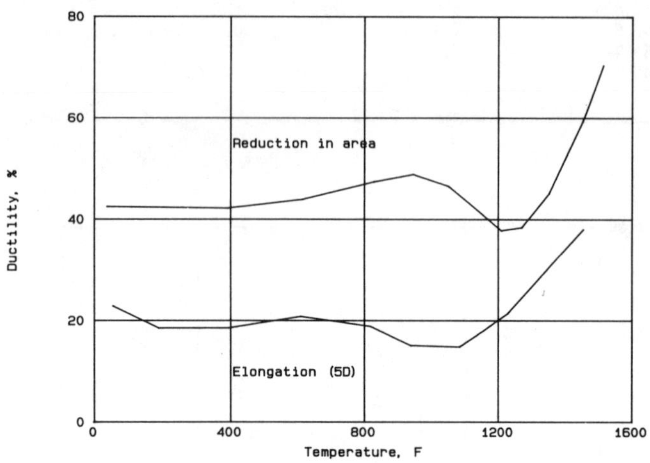

V-57: Tensile elongation and reduction in area of rolled bar. Specimens were heat treated at 1800 F for 2 h, oil quenched, aged at 1350 F for 16 h, and air cooled. Source: Ref 22, p 2

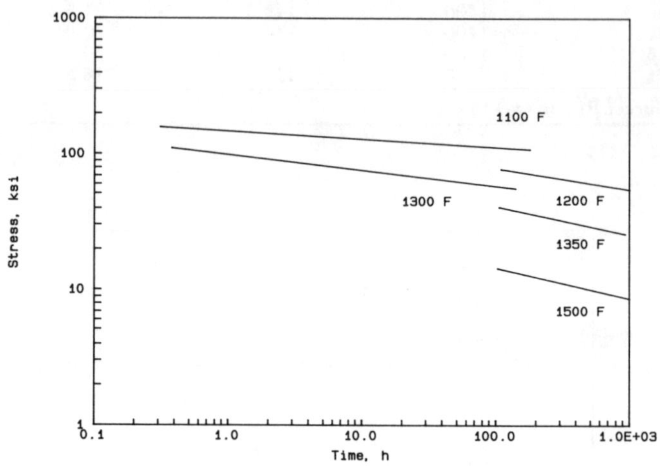

V-57: Creep and stress-rupture strength of 0.75-in. diam bar. Specimens were heat treated at 1800 F, oil quenched, aged at 1350 F for 16 h, and air cooled. Source: Ref 22, p 2

V-57: Creep strength

TEMP C	TEMP F	CRSTR 0.2%/ 10 h MPa	CRSTR 0.2%/ 10 h ksi	CRSTR 0.2%/ 100 h MPa	CRSTR 0.2%/ 100 h ksi
538	1000	...	...	689	100
593	1100	655	95	586	85
649	1200	558	81	455	66
677	1250	483	70	379	55
704	1300	427	62	...	...

Note: Specimens were heat treated at 982 C (1800 F) for 2 h, oil quenched, aged at 732 C (1350 F) for 16 h and air cooled
Source: Carpenter Technology

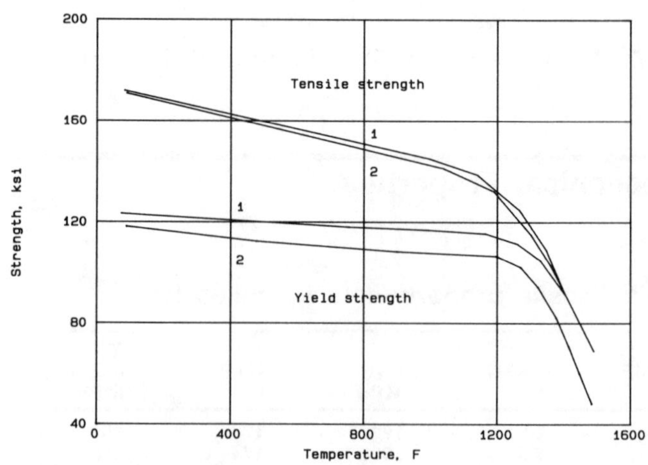

V-57: Tensile and yield strength of forgings (1) and rolled bar (2). Specimens were heat treated at 1800 F for 2 h, oil quenched, aged at 1350 F for 16 h, and air cooled. Rate: 0.025 in./min to yield strength; 0.05 in./min to tensile strength. Source: Ref 22, p 2

V-57: Stress-to-rupture property data

TEMP C	TEMP F	RPSTR 70 h MPa	RPSTR 70 h ksi	RPSTR 60 h MPa	RPSTR 60 h ksi	EL %
621	1150	586	85	...	...	10
704	1300	...	...	427	62	14

Source: Carpenter Technology

V-57: Stress-to-rupture of wheel forgings

TEMP C	TEMP C	RPSTR 10 h MPa	RPSTR 10 h ksi	RPSTR 100 h MPa	RPSTR 100 h ksi	RPSTR 1000 h MPa	RPSTR 1000 h ksi
538	1000	...	...	772	112	703	102
593	1100	745	108	655	95	572	83
649	1200	627	91	531	77	441	64
704	1300	503	73	393	57	276	40

Note: Specimens were heat treated at 1010 C (1850 F) for 2 h, oil quenched, then aged 16 h at 732 C (1350 F), and air cooled. The approximate sizes of the wheel forgings were 609.6-mm (24-in.) diam and 25- to 75-mm (1- to 3-in.) thick. Typical results were accomplished after eight heats
Source: Ref 28, p 116

D979

Alternate designation: AISI 664
UNS number: N09979
Specifications: AMS 5509, AMS 5746, SAE J467 (D979)
Nominal composition: 15.000-15.000% Cr max, 45.000-45.000% Ni max, 4.000-4.000% Mo max, 4.000-4.000% W max, 3.000-3.000% Ti max, 1.000-1.000% Al max, 27.000-27.000% Fe max, 0.050-0.050% C max, 0.010-0.010% B max
Product forms: Wrought, bar, rod, wire, plate, sheet, strip, tube, pipe, billets, forgings, structural shapes

Description: Precipitation-hardening iron-nickel-base alloy that combines good creep strength and rupture properties at temperatures up to 815 C (1500 F). D979 also has good oxidation and corrosion resistance up to 980 C (1800 F). Two heat treatments are used. One optimizes tensile and rupture properties and is generally recommended. The second is specified when creep strength is of prime importance.

Typical applications: Turbine and compressor wheels, bolts, and screws.

Physical Properties

D979: Mean coefficients of linear thermal expansion

TEMP C	TEMP F	THEXP 10^{-6}/C	THEXP 10^{-6}/F
21-100	70-212	13.7	7.6
21-200	70-392	14.0	7.8
21-300	70-572	14.2	7.9
21-400	70-752	14.8	8.2
21-500	70-932	14.8	8.2
21-600	70-1112	15.1	8.4
21-700	70-1292	15.8	8.8
21-800	70-1472	17.1	9.5
21-900	70-1652	18.0	10.0

Source: Ref 28, p 117

D979: Thermal conductivity

TEMP C	TEMP F	THCON W/m*K	THCON Btu/h*ft*F
500	932	218	126
600	1112	235	134
800	1472	261	151

Source: Ref 28, p 117

D979: Moduli of elasticity, rigidity, and Poisson's ratio

TEMP C	TEMP F	E GPa	E 10^3 ksi	G GPa	G 10^3 ksi	PR
24	75	206.9	30.0	80.0	11.6	0.29
204	400	271.0	39.3	77.9	11.3	0.30
427	800	189.6	27.5	72.4	10.5	0.30
649	1200	165.5	24.0	62.7	9.1	0.33
732	1350	160.0	23.2	60.0	8.7	0.33
816	1500	151.0	21.9	56.5	8.2	0.34
871	1600	146.2	21.2	54.5	7.9	0.34

Note: Specimens were heat treated at 1038 C (1900 F) for 1 h, oil quenched, reheated at 843 C (1550 F) for 6 h, air cooled, reheated at 704 C (1300 F) for 16 h, and air cooled. Moduli were determined from ultrasonic measurements
Source: Ref 28, p 118

Mechanical Properties

D979: Tensile property data

TEMP C	TEMP F	YS 0.02% MPa	YS 0.02% ksi	YS 0.2% MPa	YS 0.2% ksi	UTS MPa	UTS ksi	EL %	RA %
21	70	839.8	121.8	1005.3	145.8	1405.9	203.9	15	23
427	800	896.4	130.0	1006.7	146.0	1351.4	196.0	14	24
538	1000	790.2	114.6	925.3	134.2	1301.8	188.8	15	28
593	1100	813.6	118.0	958.4	139.0	1248.0	181.0	18	28
649	1200	728.1	105.6	888.1	128.8	1107.3	160.6	21	36
704	1300	583.3	84.6	756.4	109.7	856.4	124.2	22	41
760	1400	518.5	75.2	652.3	94.6	720.5	104.5	17	43
816	1500	322.7	46.8	437.1	63.4	512.3	74.3	28	57
871	1600	216.5	31.4	301.3	43.7	342.0	49.6	18	54

Note: All tensile tests were pulled at a crosshead speed of 0.635 mm/min (0.025 in./min) to the 0.2% offset yield strength and 1.27 mm/min (0.05 in./min) to the tensile strength
Source: Ref 28, p 118

D979: Creep strength

TEMP C	TEMP F	CRSTR 0.5% 10 h MPa	CRSTR 0.5% 10 h ksi	CRSTR 0.5% 100 h MPa	CRSTR 0.5% 100 h ksi	CRSTR 0.5% 1000 h MPa	CRSTR 0.5% 1000 h ksi	CRSTR 1.0% 10 h MPa	CRSTR 1.0% 10 h ksi	CRSTR 1.0% 100 h MPa	CRSTR 1.0% 100 h ksi	CRSTR 1.0% 1000 h MPa	CRSTR 1.0% 1000 h ksi
649	1200	483	70	379	55	296	43	621	90	510	74	407	59
704	1300	345	50	262	38	200	29	490	71	359	52	262	38
760	1400	248	36	186	27	131	19	345	50	241	35	152	22
816	1500	179	26	124	18	83	12	221	32	152	22	...	...

Source: Ref 28, p 118

D979: Fatigue strength

TEMP C	TEMP F	FATST 10^8 cycles MPa	FATST 10^8 cycles ksi
649	1200	413.7	60.0

Source: Ref 28, p 118

D979: Effect of temperature on Charpy V-notch impact values

TEMP C	TEMP F	IS J	IS ft-lb
21	70	10.8	8
204	400	16.3	12
427	800	16.3	12
538	1000	16.3	12
649	1200	17.6	13
816	1500	24.4	18

Source: Ref 28, p 118

D979: Stress rupture strength

TEMP C	TEMP F	RPSTR 100 h MPa	RPSTR 100 h ksi	RPSTR 1000 h MPa	RPSTR 1000 h ksi
649	1200	606.8	88	482.7	70
704	1300	434.4	63	331.0	48
760	1400	296.5	43	206.9	30
816	1500	193.1	28	131.0	19

Source: Ref 28, p 118

W545

Alternate designation: AISI 665

UNS number: S66545

Specifications: AMS 5543, AMS 5741, ASTM A453 (665), SAE J467 (W545)

Nominal composition: 13.500% Cr max, 26.000% Ni max, 1.500% Mo max, 2.850% Ti max, 0.200% Al max, 55.800% Fe max, 0.080% C max, 0.050% B max

Product forms: Wrought, bar, rod, wire, plate, sheet, strip, tube, pipe, billets, forgings, structural shapes

Description: Austenitic, precipitation-hardening alloy combining high creep strength with ductility, low notch sensitivity, and oxidation resistance in the 540 to 760 C (1000 to 1400 F) temperature range. Oxidation resistance equals that of AISI type 310 stainless steel at temperatures up to 980 C (1800 F).

Typical applications: Aircraft gas turbine discs, rotors, bolts, and other hardware; lower stress applications in stationary gas and steam turbines operaturing between 565 and 705 C (1050 and 1300 F).

Physical Properties

W545: Thermal conductivity

TEMP C	TEMP F	THCON W/m*K	THCON Btu/h*ft*F
93	200	10.7	6.2
204	400	12.1	7.0
316	600	13.5	7.8
427	800	15.1	8.8
538	1000	16.7	9.6
649	1200	18.4	10.6
760	1400	20.3	11.7
871	1600	22.5	13.0

Source: Ref 28, p 120

W545: Mean coefficients of linear thermal expansion

TEMP C	TEMP F	THEXP 10^{-6}/C	THEXP 10^{-6}/F
27-149	80-300	16.78	9.32
27-316	80-600	17.41	9.67
27-482	80-900	17.78	9.88
27-649	80-1200	18.18	10.1
27-816	80-1500	19.26	10.7
27-871	80-1600	19.80	11.0

Source: Ref 28, p 120

W545: Modulus of elasticity

TEMP C	TEMP F	E GPa	E 10^3 ksi
27	80	195.8	28.4
260	500	180.6	26.2
538	1000	162.0	23.5
649	1200	146.9	21.3
760	1400	120.7	17.5

Note: Specimens were cut from 25-mm (1-in.) diam bar. The specimen gage section was 12.8-mm (0.505-in.) diam by 75-mm (3-in.) long, with a hardness of 36 HRC
Source: Ref 28, p 120

W545: Modulus of rigidity

TEMP C	TEMP F	G GPa	G 10^3 ksi
27	80	79.3	11.50
260	500	72.4	10.50
538	1000	63.1	9.15
649	1200	58.6	8.50
760	1400	45.9	6.65

Note: Specimens were cut from 25-mm (1-in.) diam bar. The specimen gage section was 12.8-mm (0.505-in.) diam by 75-mm (3-in.) long, with a hardness of 36 HRC
Source: Ref 28, p 120

Mechanical Properties

W545: Tensile property data for bar

TEMP C	TEMP F	YS 0.02% MPa	YS 0.02% ksi	YS 0.2% MPa	YS 0.2% ksi	UTS MPa	UTS ksi	EL %	RA %	Ti %
21	70	683	99	848	123	1214	176	19	32	2.7
21	70	848	123	979	142	1289	187	19	28	3.3
260	500	634	92	814	118	1117	162	15	32	2.7
260	500	821	119	931	135	1172	170	15	33	3.3
538	1000	607	88	772	112	1007	146	15	37	2.7
538	1000	786	114	896	130	1117	162	12	31	3.3
649	1200	565	82	724	105	876	127	21	46	2.7
649	1200	724	105	855	124	965	140	15	33	3.3
704	1300	448	65	648	94	758	110	27	43	2.7
704	1300	676	98	793	115	834	121	19	39	3.3
760	1400	352	51	483	70	572	83	34	63	2.7
760	1400	538	78	655	95	690	100	23	50	3.3
816	1500	193	28	324	47	379	55	35	79	2.7
816	1500	338	49	400	58	414	60	24	66	3.3
871	1600	103	15	117	17	155	22.5	83	98	2.7
982	1800	62	9	72	10.5	86	12.5	120	99	2.7
1093	2000	31	4.5	34	5	41	6	136	99	2.7
1093	2000	31	4.5	34	5	41	6	114	99	3.3

Note: Specimens were obtained from solution heat treated and aged bars with the maximum and minimum titanium hardener contents. The hardness was 330 HB for 2.70% Ti content alloy and 400 HB for 3.30% Ti content alloy
Source: Ref 28, p 120

W545: Tensile property data for sheet

TEMP C	TEMP F	UTS MPa	UTS ksi	YS 0.2% MPa	YS 0.2% ksi	EL %	RA %
Longitudinal							
21	70	1207	175	855	124	17	27
649	1200	855	124	745	108	15	36
704	1300	690	100	641	93	15	42
Transverse							
21	70	1200	174	910	132	14	21
649	1200	876	127	772	112	12	23
704	1300	765	111	690	100	13	28

Note: Specimens were taken from a 1.27-mm (0.050-in.) thick sheet
Source: Ref 28, p 121

W545: Creep strength for various forms

TEMP C	TEMP F	CRSTR 0.1%/100 h MPa	CRSTR 0.1%/100 h ksi
593	1100	627	91
649	1200	524	76
704	1300	400	58

Note: Specimens were taken from disc forgings, rolled bar, and sheet with hardness between 34 to 39 HRC. Gage sections of round test specimens were 9.08-mm (0.357-in.) diam x 4D long minimum. All samples were heat treated by solution treating 3 h at 1093 C (2000 F), water quenched, aged at 732 to 745 C (1350 to 1375 F) for 20 h and at 649 C (1200 F) for 20 h, and air cooled
Source: Ref 28, p 121

W545: Effect of temperature on Charpy V-notch impact values

TEMP C	TEMP F	IS J	IS ft-lb
27	80	40.4	29.8
149	300	38.4	28.3
316	600	36.3	26.8
538	1000	31.6	23.3
649	1200	31.6	23.3
704	1300	32.5	24.0
760	1400	36.6	27.0

Note: Specimens were machined from 14.3-mm (0.563-in.) diam bars. All bars were heat treated 3 h at 1093 C (2000 F), water quenched, aged 20 h at 745 C (1375 F), 20 h at 649 C (1200 F). The alloy used was from one heat with an average hardness of 350 HB. Grain size was ASTM No. 3 to 5
Source: Ref 28, p 121

W545: Stress-to-rupture property data for sheet

TEMP C	TEMP F	RPSTR MPa	RPSTR ksi	RPT h	EL(a) %
649	1200	517	75	397	5.5
649	1200	552	80	152	7
649	1200	586	85	55	10
704	1300	379	55	269	39
704	1300	414	60	212	28
704	1300	448	65	60	13

Note: 1.27-mm (0.050-in.) thick sheet was used
(a) In 4D
Source: Ref 28, p 122

W545: Stress-to-rupture property data for bar

TEMP C	TEMP F	RPSTR 10 h MPa	RPSTR 10 h ksi	RPSTR 100 h MPa	RPSTR 100 h ksi	RPSTR 1000 h MPa	RPSTR 1000 h ksi
Unnotched specimens							
538	1000	896	130	827	120	724	105
593	1100	724	105	648	94	538	78
649	1200	655	95	565	82	455	66
704	1300	538	78	414	60	317	46
760	1400	400	58	276	40	159	23
Notched specimens							
538	1000	...	...	1034	150	827	120
593	1100	...	...	758	110	634	92
649	1200	...	...	621	90	517	75
704	1300	...	...	469	68	359	52
760	1400	...	...	...	...	...	...

Note: Specimens were taken from rolled bar with hardness between 34 and 39 HRC. Gage sections of round test spedimens were 9.08-mm (0.357-in.) diam x 4D long minimum. All samples were heat treated by solution treating 3 h at 1093 C (2000 F), water quenched, and aged at 732 to 745 C (1350 to 1375 F) for 20 h and at 649 C (1200 F) for 20 h, and air cooled
Source: Ref 28, p 122

W545: Fatigue strength

TEMP C	TEMP F	Ti %	FAT.n cycles	FATST MPa	FATST ksi
21	70	2.7	10^8	469	68
21	70	2.7	10^7	476	69
21	70	2.7	10^6	531	77
21	70	3.3	10^8	476	69
21	70	3.3	10^7	483	70
21	70	3.3	10^6	538	78
649	1200	2.7	10^8	359	52
649	1200	2.7	10^7	386	56
649	1200	2.7	10^6	427	62
649	1200	3.3	10^8	400	58
649	1200	3.3	10^7	427	62
649	1200	3.3	10^6	476	69
704	1300	2.7	10^8	331	48
704	1300	2.7	10^7	372	54
704	1300	2.7	10^6	421	61
704	1300	3.3	10^8	359	52
704	1300	3.3	10^7	407	59
704	1300	3.3	10^6	455	66

Note: Unnotched specimens were taken from 22.2-mm (0.875-in.) diam bars. The specimens were solution treated 3 h at 1093 C (2000 F), water quenched, and aged 20 h at 745 C (1375 F). Then they were aged another 20 h at 649 C (1200 C) and air cooled. The hardness on the 2.70% Ti samples was 36 to 37 HRC and 39 to 40 HRC on the 3.30% Ti samples. Grain size of the bars was ASTM No. 3 to 5. Tests were made in a completely reversing stress cycle with a zero mean stress in an electromagnetic bending fatigue machine
Source: Ref 28, p 121

W545: Stress-to-rupture property data

RPT(a) h	EL(a) %	RPT(b) h	EL(b) %
108	6.0	129	7.7
260	6.4	104	7.4
615	6.9	73	8.5
217	7.4	91	18.3
72	15.4	49	18.5
271	6.2	84	13.7
61	12.4	115	15.2
84	9.3	54	5.3

Note: All stress rupture tests were made on combination smooth and notch forging specimens (3.9 K_t). Specimens ruptured in the smooth section
(a) At 552 MPa (80 ksi), 649 C (1200 F). (b) At 448 MPa (65 ksi), 982 C (1800 F). Elongation in 4D
Source: Ref 28, p 122

W545: Stress relaxation data

TEMP C	TEMP F	CRIS MPa	CRIS ksi	CRSRD 10 h MPa	CRSRD 10 h ksi	CRSRD 100 h MPa	CRSRD 100 h ksi	CRSRD 1000 h MPa	CRSRD 1000 h ksi
593	1100	270.3	39.2	262.0	38.0	262.0	38.0	241.3	35.0
621	1150	223.4	32.4	220.0	31.9	216.5	31.4	204.1	29.6
649	1200	222.0	32.2	222.0	32.2	221.3	32.1	206.9	30.0

Source: Ref 28, p 122

Haynes 556

UNS number: R30556

Specifications: MIL S-163, ASTM B-435, ASTM B-572, ASTM B-619, ASTM B-622, ASTM B-626, AMS 5874

Nominal composition: 22.000% Cr, 20.000% Ni, 18.000% Co, 32.000% Fe, 3.000% Mo, 2.500% W, 0.600% Ta, 0.200% N, 0.200% Al, 0.100% C, 0.020% La, 0.020% Zr, 0.400% Si, 1.000% Mn, 0.300% Cb max

Product forms: Wrought, bar, rod, wire, plate, sheet, strip, tube, pipe, billets, forgings, structural shapes

Description: Fe-Cr-Ni-Co alloy with basic composition similar to that of Multimet alloy (N-155, AISI 661). Closer control of alloying elements gives Haynes Alloy No. 556 improved oxidation resistance, hot corrosion resistance, weldability, high-temperature ductility, and thermal shock and fatigue resistance.

Typical applications: Structural applications up to 1100 C (2000 F), such as in gas turbines, waste heat boilers, incinerators, industrial fans, and furnace hardware and structures. Ideal for stressed components where high-temperature corrosion resistance, creep resistance, and high-temperature ductility are all important.

Physical Properties

Haynes 556: Specific heat

TEMP C	TEMP F	SHTCP J/Kg*K	SHTCP Btu/lb*F
21	70	464	0.111
100	200	475	0.113
200	400	493	0.118
300	600	508	0.122
400	800	523	0.126
500	1000	538	0.130
600	1200	552	0.133
700	1400	561	0.135
800	1600	570	0.140
900	1800	595	0.147
1000	2000	618	0.152

Source: Haynes Alloy Data Sheet, Haynes International, Inc.

Haynes 556: Thermal conductivity

TEMP C	TEMP F	THCON W/m*K	THCON Btu/h*ft*F
21	70	11.1	6.41
100	200	13.1	7.57
200	400	15.4	8.90
300	600	17.3	9.99
400	800	19.0	10.98
500	1000	20.8	12.02
600	1200	22.4	12.94
700	1400	24.0	13.86
800	1600	25.5	14.73
900	1800	27.2	15.71
1000	2000	28.9	16.70

Source: Haynes Alloy Data Sheet, Haynes International, Inc.

Haynes 556: Thermal diffusivity

TEMP C	TEMP F	THDIF cm^2/s	THDIF $in.^2/s$
21	70	28.7	4.5
100	200	31.2	4.8
200	400	34.2	5.3
300	600	37.0	5.8
400	800	39.7	6.3
500	1000	42.3	6.7
600	1200	44.8	7.1
700	1400	47.0	7.5
800	1600	48.8	7.7
900	1800	50.3	8.0
1000	2000	51.6	8.2

Note: Thermal diffusivity values are multiplied by 10^{-3}
Source: Haynes Alloy Data Sheet, Haynes International, Inc.

Haynes 556: Electrical resistivity

TEMP C	TEMP F	ELRES n{o}*m
21	70	952
100	200	986
200	400	1026
300	600	1065
400	800	1095
500	1000	1125
600	1200	1151
700	1400	1172
800	1600	1190
900	1800	1207
1000	2000	1223

Source: Haynes Alloy Data Sheet, Haynes International, Inc.

Haynes 556: Mean coefficients of linear thermal expansion

TEMP C	TEMP F	THEXP $10^{-6}/C$	THEXP $10^{-6}/F$
21-100	70-200	14.7	8.1
21-200	70-400	14.9	8.2
21-300	70-600	15.1	8.4
21-400	70-800	15.4	8.6
21-500	70-1000	15.7	8.8
21-600	70-1200	16.1	9.0
21-700	70-1400	16.4	9.2
21-800	70-1600	16.7	9.4
21-900	70-1800	17.0	9.5
21-1000	70-2000	17.1	9.6

Source: Haynes Alloy Data Sheet, Haynes International, Inc.

Haynes 556: Modulus of elasticity

TEMP C	TEMP F	E GPa	E 10^3 ksi
21	70	205	29.7
100	200	200	29.1
200	400	195	28.2
300	600	187	26.9
400	800	179	25.6
500	1000	172	24.4
600	1200	164	23.1
700	1400	155	21.8
800	1600	148	20.9
900	1800	143	20.1

Source: Haynes Alloy Data Sheet, Haynes International, Inc.

Mechanical Properties

Haynes 556: Tensile property data for sheet and plate

TEMP C	TEMP F	UTS MPa	UTS ksi	YS MPa	YS ksi	EL %
Cold-rolled and solution annealed sheet, 0.8 to 2.8 mm (0.033 to 0.109 in.) thick						
21	70	815	118.1	410	59.5	47.7
540	1000	645	93.4	240	34.9	54.4
650	1200	590	85.4	225	32.8	52.4
760	1400	470	68.5	220	32.0	49.1
870	1600	330	47.6	195	28.6	52.6
980	1800	195	28.0	105	15.5	63.3
1095	2000	100	14.8	55	8.0	55.4
Hot-rolled and solution annealed plate, 6.4 to 12.7 mm (0.250 to 0.500 in.) thick						
21	70	805	116.4	375	54.6	51.4
540	1000	625	90.3	210	30.6	60.3
650	1200	575	83.1	210	30.6	57.4
760	1400	470	68.5	200	29.3	52.6
870	1600	340	49.3	190	27.9	69.1
980	1800	210	30.7	130	18.5	83.9
1095	2000	110	16.1	60	8.7	95.2

Source: Haynes Alloy Data Sheet, Haynes International, Inc.

Haynes 556: Tensile property data for bar after 16,000 h thermal exposure

TEMP C	TEMP F	UTS MPa	UTS ksi	YS MPa	YS ksi	EL %
540	1000	660	95.7	260	37.4	48.0
650	1200	610	88.8	260	37.8	23.4
760	1400	500	72.3	240	35.1	25.3
870	1600	290	42.1	150	21.9	29.5

Note: Results from an average of two or more tests
Source: Haynes Alloy Data Sheet, Haynes International, Inc.

Haynes 556: Creep and rupture property data for sheet

CRSTN %	CRIS 10 h MPa	CRIS 10 h ksi	CRIS 100 h MPa	CRIS 100 h ksi	CRIS 1000 h MPa	CRIS 1000 h ksi
Tested at 650 C (1200 F)						
0.5	324	47.0	223	32.4	152	22.0
1.0	328	47.5	231	33.5	168	24.3
Rupture	...	...	379	55.0	276	40.0
Tested at 705 C (1300 F)						
0.5	200	29.0	136	19.7	102	14.8
1.0	210	30.4	153	22.2	113	16.4
Rupture	348	50.5	254	36.8	184	26.7
Tested at 760 C (1400 F)						
0.5	125	18.2	88	12.8	65	9.4
1.0	144	20.9	105	15.3	77	11.2
Rupture	234	34.0	172	25.0	124	18.0
Tested at 815 C (1500 F)						
0.5	84	12.2	62	9.0	46	6.6
1.0	100	14.5	74	10.7	54	7.8
Rupture	165	24.0	118	17.1	81	11.8
Tested at 870 C (1600 F)						
0.5	60	8.7	44	6.4	32	4.6
1.0	71	10.3	52	7.5	37	5.3
Rupture	113	16.4	78	11.3	52	7.6
Tested at 925 C (1700 F)						
0.5	43	6.3	31	4.5	21	3.1
1.0	51	7.4	36	5.2	24	3.5
Rupture	77	11.2	51	7.4	33	4.8
Tested at 980 C (1800 F)						
0.5	31	4.5	21	3.1	13	1.9
1.0	36	5.2	24	3.5	14	2.1
Rupture	51	7.4	33	4.8	19	2.8

Note: Results from more than 200 tests. Sheet was 0.8 to 2.8 mm (0.029 to 0.109 in.) thick
Source: Haynes Alloy Data Sheet, Haynes International, Inc.

Haynes 556: Hot hardness data

TEMP C	TEMP F	HARD DPH
21	70	214
540	1000	131
650	1200	124
760	1400	116
870	1600	73
980	1800	44

Source: Haynes Alloy Data Sheet, Haynes International, Inc.

Alloy 800

UNS number: N08800
Specifications: AMS 5871, ASME SB163, ASME SB407, ASME SB408, ASME SB409, ASME SB564, ASTM B163, ASTM B366, ASTM B407, ASTM B408, ASTM B409, ASTM B514, ASTM B515, ASTM B564
Nominal composition: 21.000% Cr, 32.500% Ni, 46.000% Fe, 0.400% Ti, 0.400% Al, 0.100% C max
Product forms: Wrought, bar, rod, wire, plate, sheet, strip, tube, pipe, billets, forgings, structural shapes

Description: High-strength Fe-Ni-Cr alloy resistant to carburization and oxidation at elevated temperatures. Incoloy 800 resists sulfur attack, internal oxidation, scaling, and corrosion in a wide variety of atmospheres.

Typical applications: Heat exchangers, process piping, and carburizing fixtures and retorts, heating element sheathing, and nuclear steam generator tubing.

Physical Properties

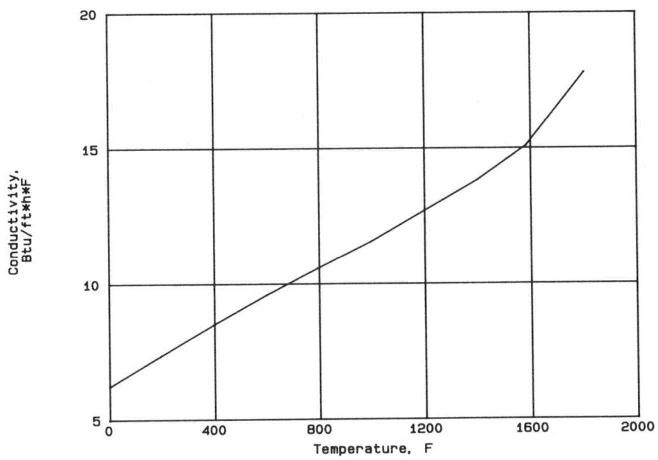

Alloy 800: Thermal conductivity of specimens annealed at 1800 F for 0.25 h. Source: Ref 22, p 5

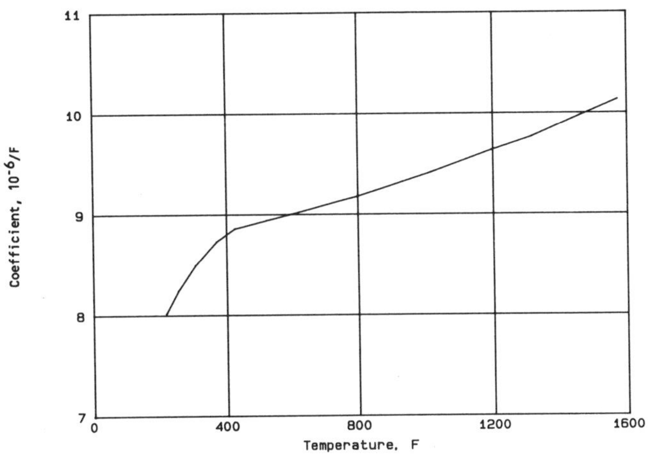

Alloy 800: Mean coefficient of linear thermal expansion of specimens annealed at 1800 F for 0.25 h. Source: Ref 22, p 5

Alloy 800: Thermal conductivity

TEMP C	TEMP F	THCON W/m*K	THCON Btu/h*ft*F
21	70	11.5	6.6
38	100	12.0	6.9
93	200	12.8	7.4
204	400	14.8	8.5
316	600	16.6	9.6
427	800	18.3	10.6
538	1000	20.0	11.6
649	1200	21.9	12.7
760	1400	23.9	13.3
871	1600	26.1	15.1
982	1800	30.8	17.8

Source: Ref 14, p 329

Alloy 800: Mean coefficients of linear thermal expansion

TEMP C	TEMP F	THEXP 10^{-6}/C	THEXP 10^{-6}/F
93	200	14.2	7.9
204	400	15.8	8.8
316	600	16.2	9.0
427	800	16.6	9.2
538	1000	16.9	9.4
649	1200	17.3	9.6
760	1400	17.8	9.9
871	1600	18.4	10.2

Source: Ref 14, p 329

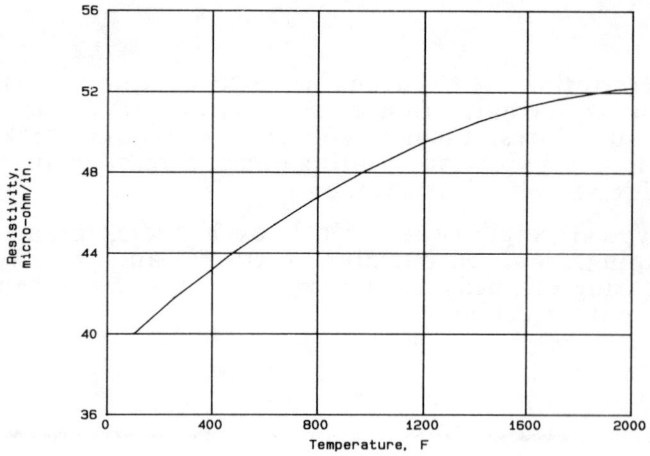

Alloy 800: Electrical resistivity of specimens annealed at 1800 F for 0.25 h. Source: Ref 22, p 6

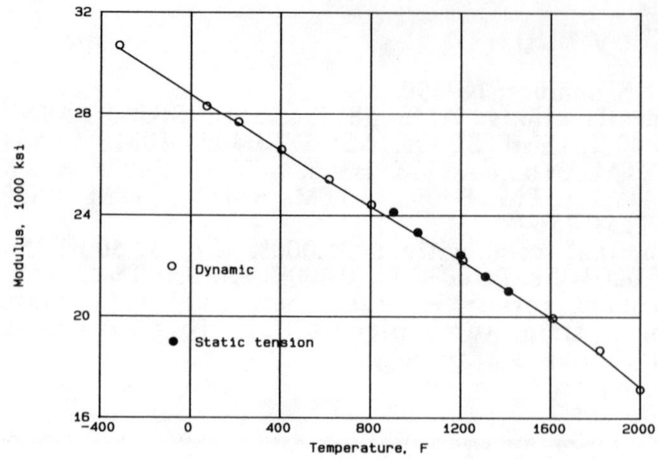

Alloy 800: Modulus of elasticity of specimens annealed at 1800 F for 0.25 h. Source: Ref 22, p 23

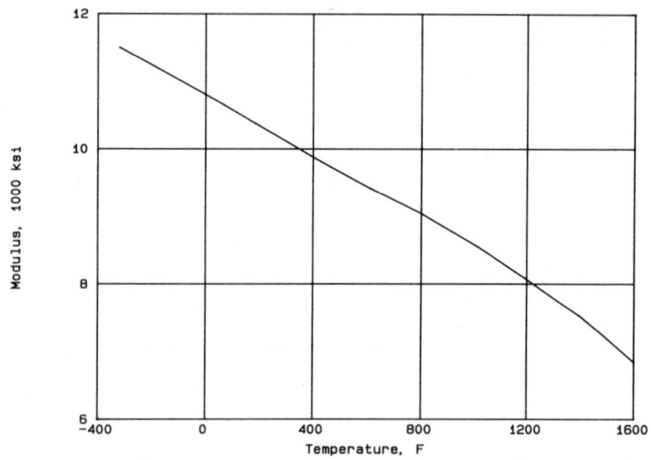

Alloy 800: Modulus of rigidity of specimens annealed at 1800 F for 0.25 h. Source: Ref 22, p 23

Alloy 800: Modulus of rigidity

TEMP C	TEMP F	G GPa	G 10^3 ksi
24	75	73.4	10.64
93	200	71.5	10.37
204	400	68.3	9.91
316	600	65.3	9.47
427	800	62.3	9.04
538	1000	59.3	8.60
649	1200	56.0	8.12
760	1400	52.3	7.58
871	1600	47.0	6.82

Note: Modulus of rigidity was determined by the dynamic method for annealed material
Source: Ref 14, p 329

Alloy 800: Electrical resistivity

TEMP C	TEMP F	ELRES n{o}*m
21	70	989
38	100	997
93	200	1030
204	400	1091
316	600	1110
427	800	1170
538	1000	1199
649	1200	1239
760	1400	1259
871	1600	1279
982	1800	1289
1093	2000	1309

Source: Ref 14, p 329

Alloy 800: Modulus of elasticity

TEMP C	TEMP F	E GPa	E 10^3 ksi
24	75	195.47	28.35
38	100	194.30	28.18
93	200	190.65	27.65
149	300	186.79	27.09
204	400	183.20	26.57
260	500	179.41	26.02
316	600	175.41	25.44
371	700	171.96	24.94
427	800	168.51	24.44
482	900	165.14	23.95
538	1000	161.55	23.43
593	1100	157.76	22.88
649	1200	153.83	22.31
704	1300	149.69	21.71
760	1400	145.62	21.12
816	1500	142.04	20.60
871	1600	137.90	20.00
927	1700	133.83	19.41
982	1800	129.14	18.73
1038	1900	124.45	18.05
1093	2000	118.80	17.23

Note: Modulus of elasticity was determined by the dynamic method for annealed material
Source: Ref 14, p 329

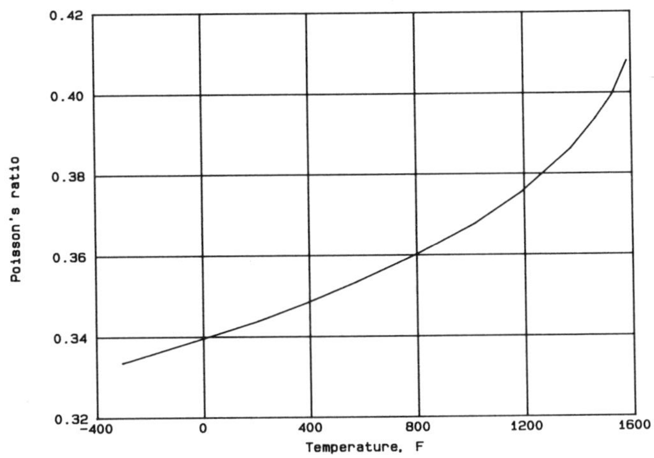

Alloy 800: Poisson's ratio of specimens annealed at 1800 F for 0.25 h. Data calculated from dynamic elasticity and rigidity. Source: Ref 22, p 23

Alloy 800: Poisson's ratio

TEMP C	TEMP F	PR
24	75	0.339
93	200	0.341
204	400	0.353
316	600	0.357
427	800	0.363
538	1000	0.367
649	1200	0.377
760	1400	0.389
871	1600	0.408

Note: Poisson's ratio was calculated from modulus of elasticity.
Source: Ref 14, p 329

Mechanical Properties

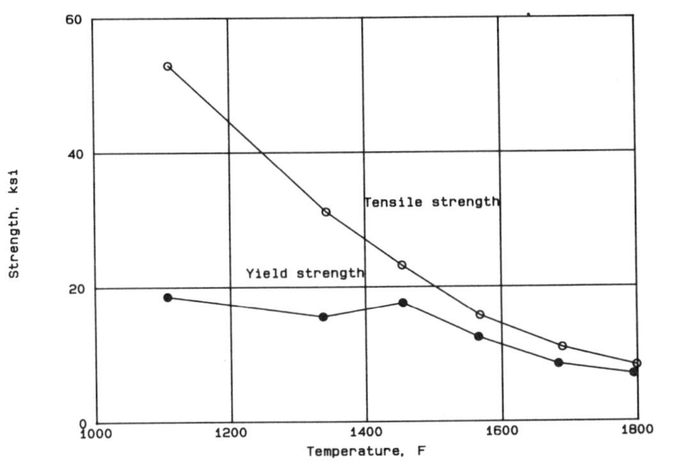

Alloy 800: Tensile and yield strength of extruded tubing. Tubing had 5-in. outside diameter, 0.5-in. wall. Test specimen was 0.5 in. diameter. Room temperature tensile strength: 70 ksi. Yield strength: 26.6 ksi. Source: Ref 22, p 12

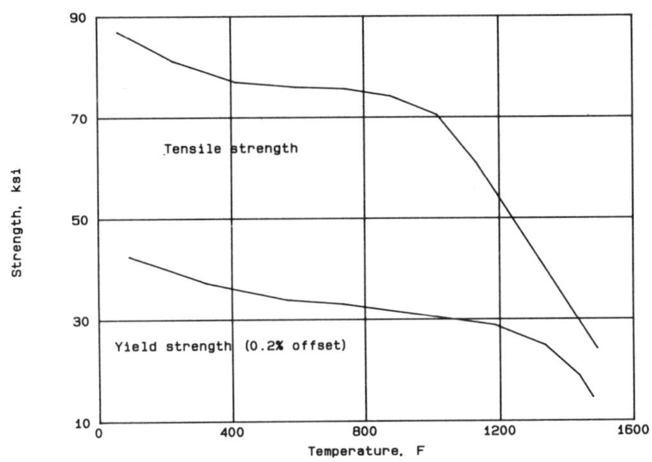

Alloy 800: Tensile and yield strength of hot rolled rod, annealed at 1800 F for 0.25 h. Source: Ref 14, p 330

Alloy 800: Tensile and hardness data

TEMP C	TEMP F	HARD HB	UTS MPa	UTS ksi	YS 0.2% MPa	YS 0.2% ksi
Solution treated at 1149 C (2100 F)						
27	80	126	536	77.8	150	27.7
425	800	...	465	67.5	130	18.8
540	1000	90	432	62.7	90	13.0
650	1200	84	378	54.8	93	13.5
705	1300	82	329	47.7	109	15.8
760	1400	74	236	34.2	90	13.1
Annealed at 982 C (1800 F)						
27	80	138	590	85.5	250	36.2
425	800	120	511	74.1	172	24.9
540	1000	119	508	73.7	178	25.8
640	1200	110	405	58.7	176	25.5
705	1300	97	319	46.3	174	25.3
760	1400	66	238	34.5	149	21.6
As hot rolled						
27	80	198	665	96.4	445	64.6
425	800	170	583	84.5	359	52.0
540	1000	161	579	84.0	361	52.4
640	1200	145	450	65.3	333	48.3
705	1300	120	369	53.5	323	46.8
760	1400	91	307	44.5	284	41.2

Source: Ref 14, p 330

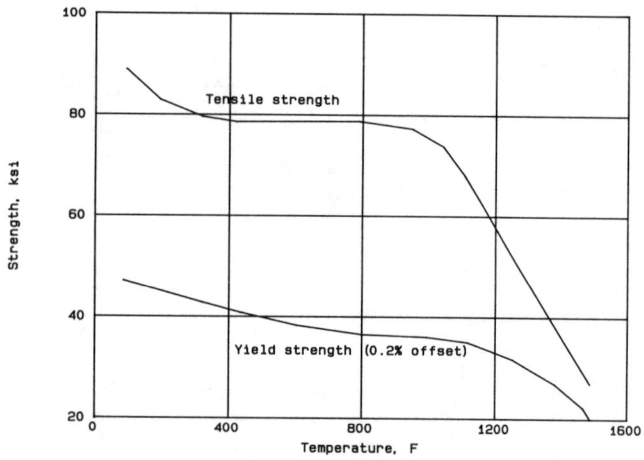

Alloy 800: Tensile and yield strength of hot rolled rod, as rolled. Source: Ref 14, p 330

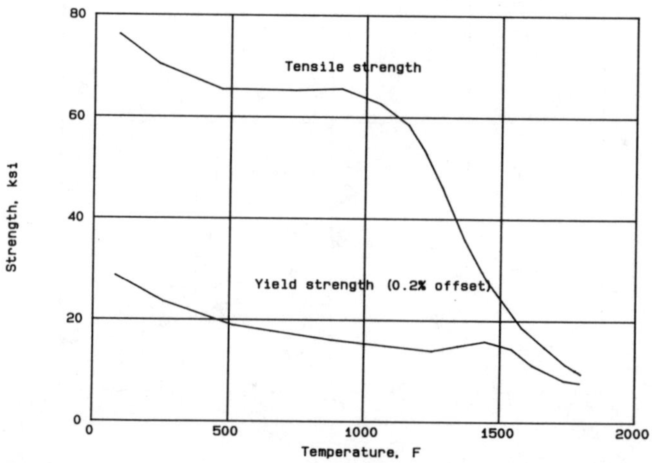

Alloy 800: Tensile and yield strength of hot rolled rod, solution treated at 2100 F for 1 h. Source: Ref 14, p 331

Alloy 800: Tensile property data

TEMP C	TEMP F	UTS MPa	UTS ksi	YS 0.2% MPa	YS 0.2% ksi	EL %	RA %
Extruded tubing: 127 mm (5 in.) O.D. x 12.7 mm (0.500 in.) wall							
29	85	524	76	183.4	26.6	60	...
649	1200	362	52.5	124.1	18	47	59
760	1400	208.9	30.3	108.3	15.7	85	73
816	1500	161.1	23.36	119.3	17.3	98	79.5
871	1600	110.3	16	93.1	13.5	109.5	92.5
927	1700	81.4	11.8	63.4	9.2	111.5	93
982	1800	61.4	8.9	49.6	7.2	131.5	94
Cold drawn rod: 25 mm (1 in.) diameter							
29	85	770.9	111.8	689.5	100	17	64
93	200	741.2	107.5	655	95	16	63.3
204	400	706.7	102.5	649.5	94.2	13	58.8
316	600	686.1	99.5	641.2	93	12	56.6
371	700	664	96.3	630.9	91.5	15	53.2
482	900	641.2	93	620.6	90	15	52.5
538	1000	601.9	87.3	595	86.3	16.5	54
593	1100	541.3	78.5	556.4	80.7	15	50
649	1200	444	64.4	460.6	66.8	19.5	44.5
704	1300	...	...	423.4	61.4	28	42.5

Source: Ref 14, p 330

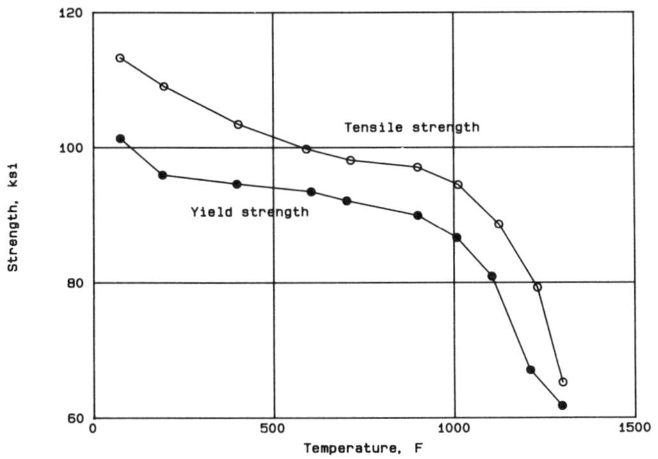

Alloy 800: Tensile and yield strength of 1-in. diam cold drawn rod. Source: Ref 22, p 12

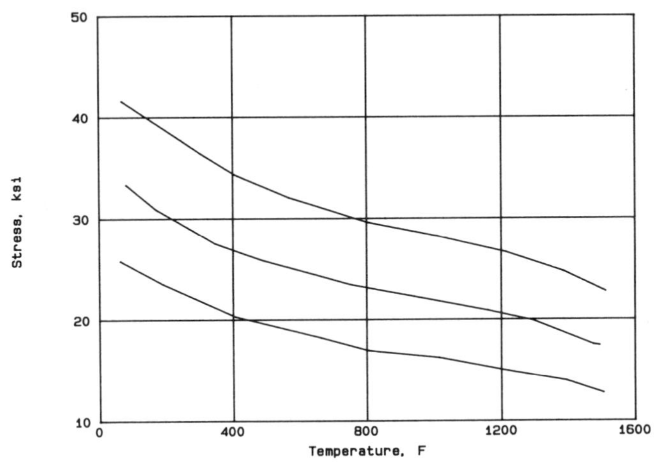

Alloy 800: Scatterband of yield strength of bar. Specimens were solution annealed at 2100 F for 1 h. Source: Ref 22, p 12

Alloy 800: Effect of temperature on Charpy V-notch impact values

TEMP C	TEMP F	IS(a) J	IS(a) ft-lb
21	70	281	207
649	1200	245	181
704	1300	239	176
816	1500	239	176

Note: Specimens were solution treated at 1121 C (2050 F)
(a) Average of three tests
Source: Ref 14, p 331

Alloy 800: Room-temperature Charpy V-notch impact strength after exposure at 760 C (1400 F)

TIME h	IS J	IS ft-lb
0	145	107
500	133	98
1000	135	99
1500	133	98

Note: Cold rolled, annealed bar
Source: Ref 14, p 331

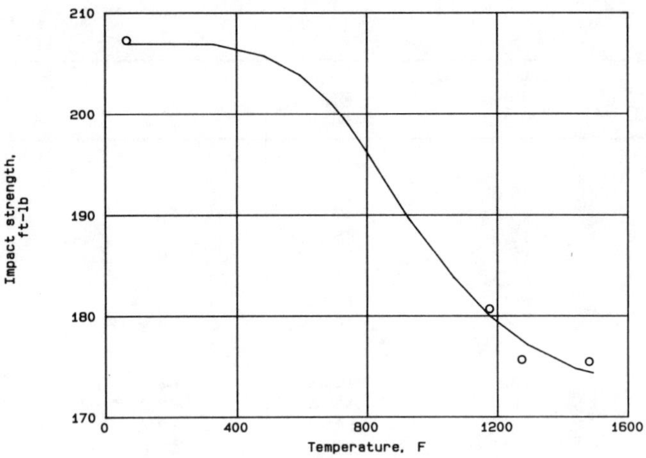

Alloy 800: Charpy V-notch impact strength of a standard specimen, solution treated at 2050 F for 1 h. Each value shown is an average of 3 tests. Source: Ref 22, p 14

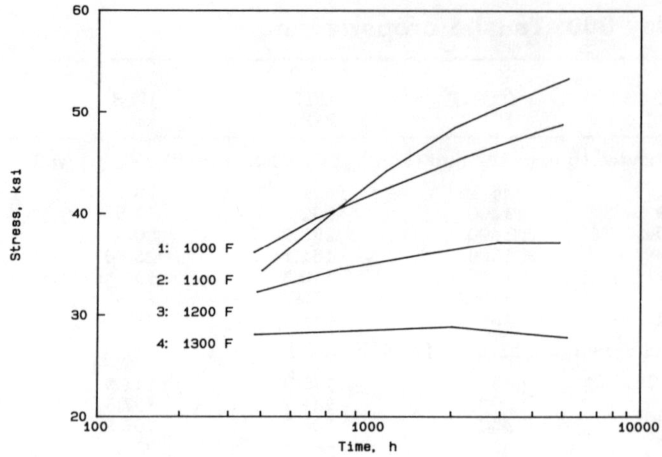

Alloy 800: Effect of prolonged exposure at temperatures from 1000 to 1300 F on yield strength of 0.5-in. diam rod. Pre-exposure tensile strengths: 33.45 ksi (curve 1), 32.40 ksi (curve 2), 30.90 ksi (curve 3), and 29.45 ksi (curve 4). Source: Ref 22, p 14

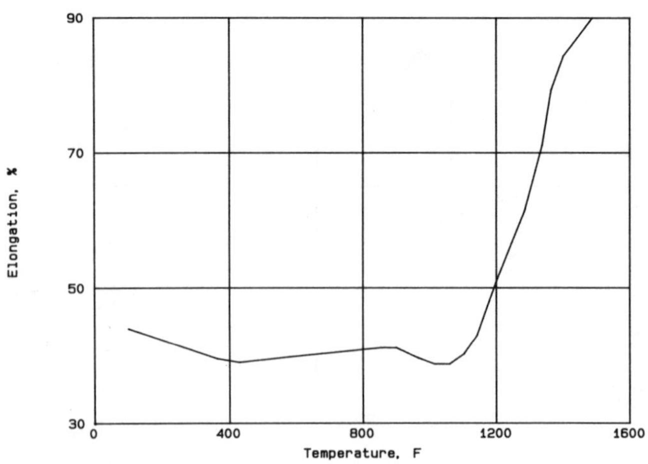

Alloy 800: Tensile elongation of hot rolled rod, annealed at 1800 F for 0.25 h. Source: Ref 14, p 330

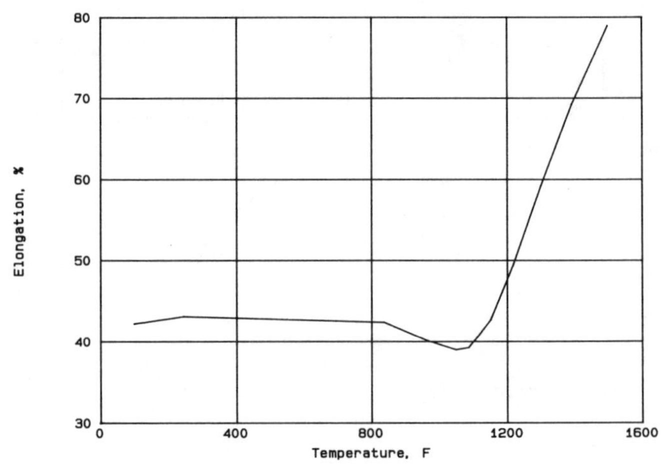

Alloy 800: Tensile elongation of hot rolled rod, as rolled. Source: Ref 14, p 330

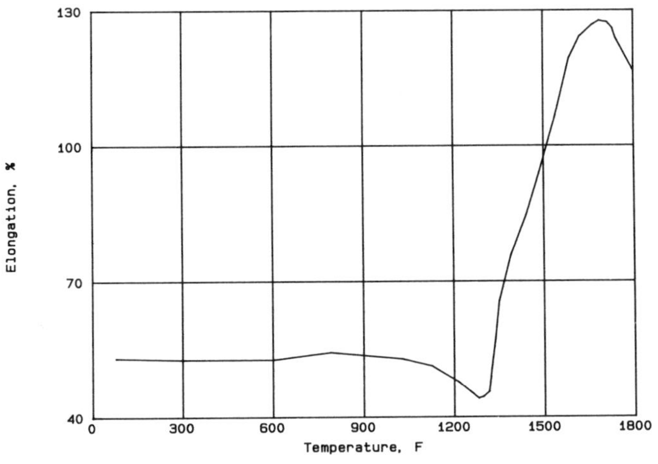

Alloy 800: Tensile elongation of hot rolled rod, solution treated at 2100 F for 1 h. Source: Ref 14, p 331

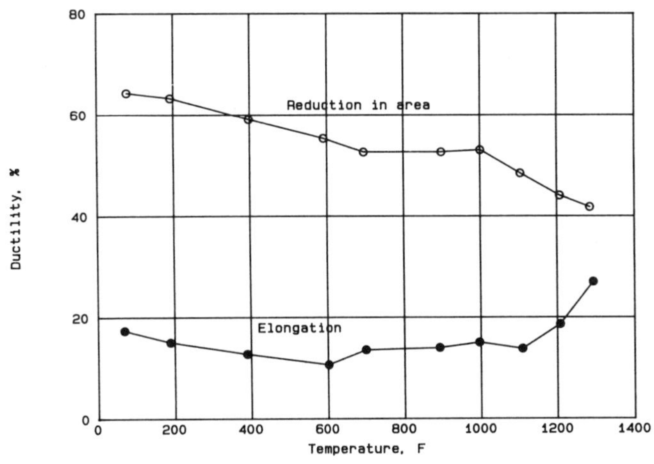

Alloy 800: Tensile elongation and reduction in area of 1-in. diam cold drawn rod. Source: Ref 22, p 12

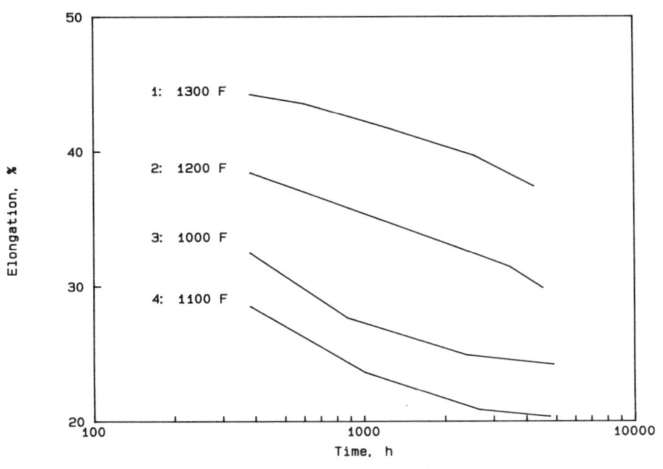

Alloy 800: Effect of prolonged exposure at temperatures from 1000 to 1300 F on tensile elongation of 0.5-in. diam rod. Pre-exposure tensile elongation: 59.0% (curve 1), 42.5% (curve 2), 37.5% (curve 3), 31.0% (curve 4). Source: Ref 22, p 14

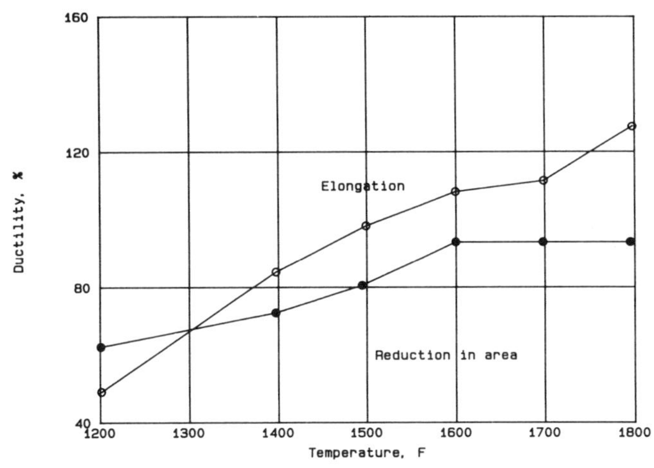

Alloy 800: Tensile elongation and reduction in area of extruded tubing with a 5-in. outside diameter and a 0.5-in. wall. Test specimen was 0.5 in. diameter. Room-temperature tensile elongation was 60%. Source: Ref 22, p 12

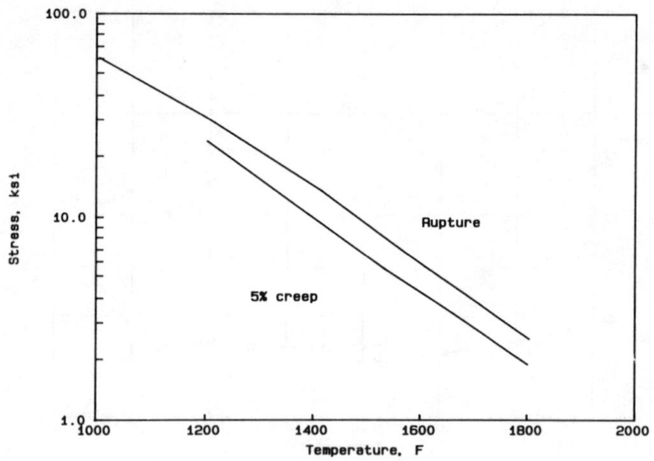

Alloy 800: Typical 100-h creep-rupture strength of solution treated specimens. Source: Ref 14, p 334

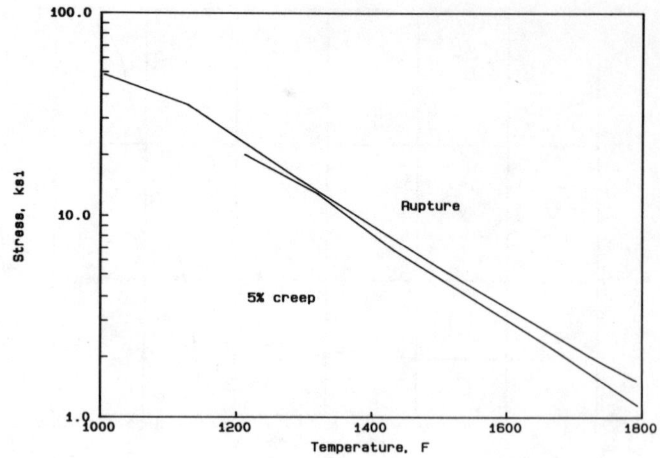

Alloy 800: Typical 1000-h creep-rupture strength of solution treated specimens. Source: Ref 14, p 334

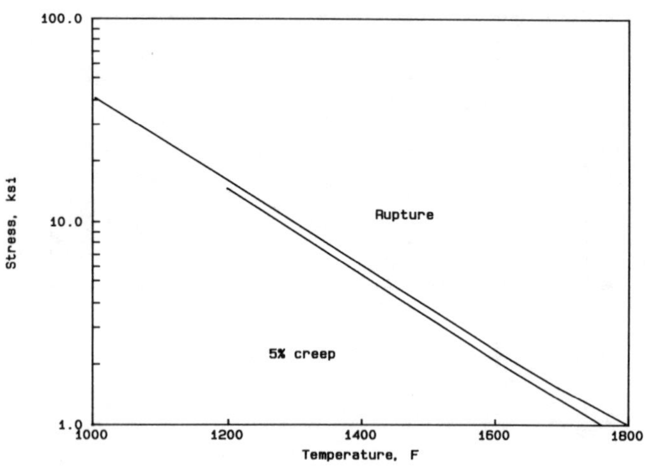

Alloy 800: Typical 10,000-h creep-rupture strength of solution treated specimens. Source: Ref 14, p 334

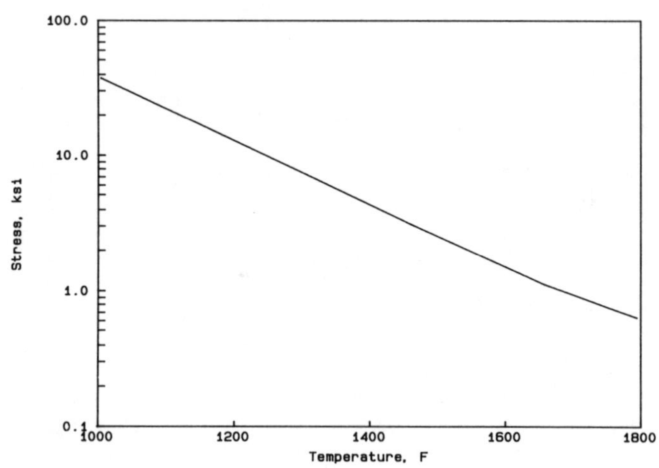

Alloy 800: Typical 100,000-h creep-rupture strength of solution treated specimens. Source: Ref 14, p 334

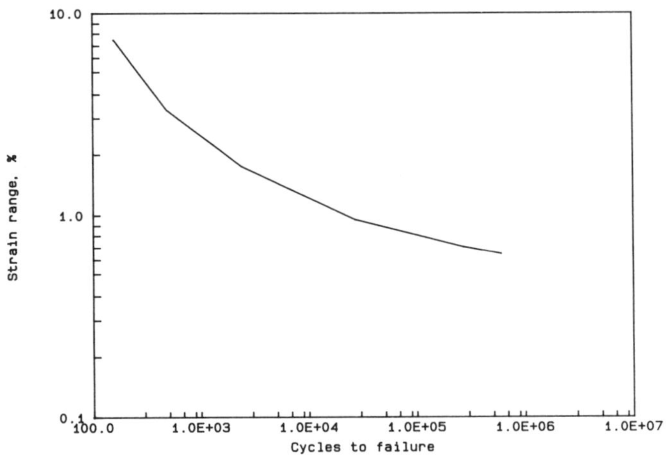

Alloy 800: Low-cycle fatigue tests at 800 F for bar. Specimens were mill annealed at 1800 F for 0.25 h and tested in air. Specimens were 0.25-in. diam hourglass shape. Strain rates were 4 X 10^{-3}/s and 4 X 10^{-4}/s. Axial loading, R = -1. Source: Ref 22, p 20

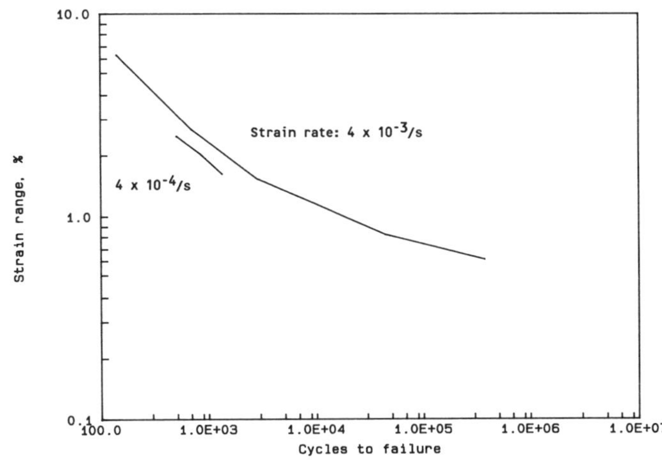

Alloy 800: Low-cycle fatigue tests at 1000 F for bar. Specimens were mill annealed at 1800 F for 0.25 h and tested in air. Specimens were 0.25-in. diam hourglass shape. Strain rates were 4 X 10^{-3}/s and 4 X 10^{-4}/s. Axial loading, R = -1. Source: Ref 22, p 20

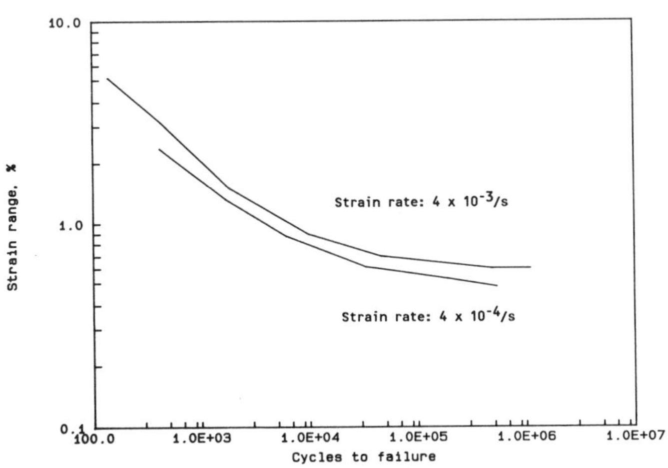

Alloy 800: Low-cycle fatigue tests at 1100 F for bar. Specimens were mill annealed at 1800 F for 0.25 h and tested in air. Specimens were 0.25-in. diam hourglass shape. Strain rates were 4 X 10^{-3}/s and 4 X 10^{-4}/s. Axial loading, R = -1. Source: Ref 22, p 20

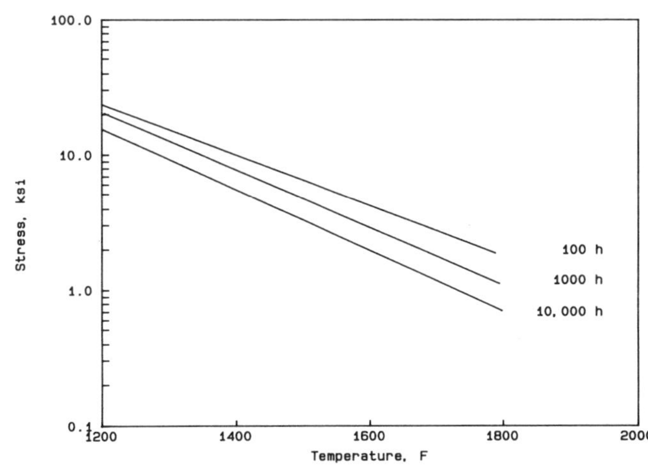

Alloy 800: Stress required to produce 5% creep in 100 to 10,000 h at temperatures from 1200 to 1800 F for specimens solution treated at 2100 F for 1 h. Source: Ref 22, p 15

Alloy 800: Fatigue strength

TEMP C	TEMP F	FATST 10^5 cyc MPa	FATST 10^5 cyc ksi	FATST 10^6 cyc MPa	FATST 10^6 cyc ksi	FATST 10^7 cyc MPa	FATST 10^7 cyc ksi	FATST 10^8 cyc MPa	FATST 10^8 cyc ksi
29	85	324.1	47.0	296.5	43.0	293.0	42.5	289.6	42.0
427	800	324.1	47.0	296.5	43.0	293.0	42.5	289.6	42.0
538	1000	275.8	40.0	268.9	39.0	265.5	38.5	262.0	38.0
760	1400	...	...	203.4	29.5	175.8	25.5	151.7	22.0
871	1600	...	...	137.9	20.0	113.8	16.5	93.1	13.5

Note: Specimens were cold drawn, annealed at 871 C (1600 F) for 30 min, and air cooled
Source: Ref 14, p 331

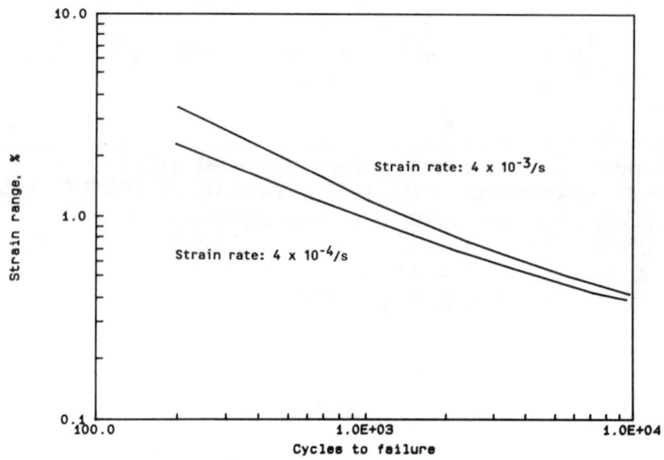

Alloy 800: Low-cycle fatigue tests at 1200 F for plate. Specimens were mill annealed at 2100 F for 0.25 h and tested in air. Specimens were 0.25-in. diam hourglass shape. Strain rates were 4 X 10^{-3}/s and 4 X 10^{-4}/s. Axial loading, R = -1. Source: Ref 22, p 20

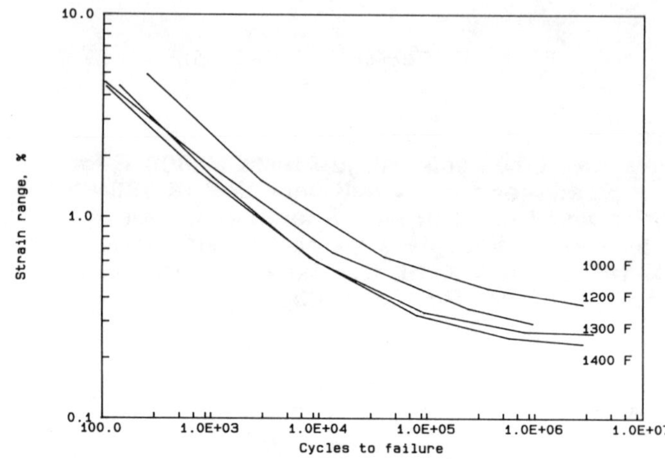

Alloy 800: Low-cycle fatigue tests for bar solution annealed at 2100 F for 1 h. Specimens were 0.25-in. diam hourglass shape. Strain rates were 4 X 10^{-3}/s and 4 X 10^{-4}/s. Hold time at maximum tensile strain varied from 0 to 60 min. Source: Ref 22, p 20

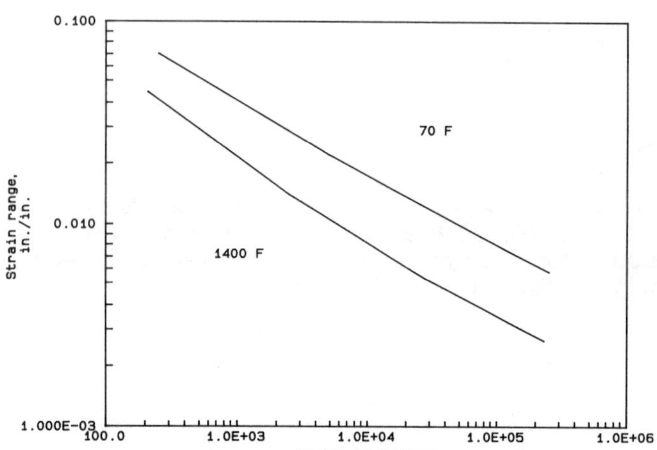

Alloy 800: Low-cycle fatigue strength of solution treated specimens. Source: Ref 14, p 331

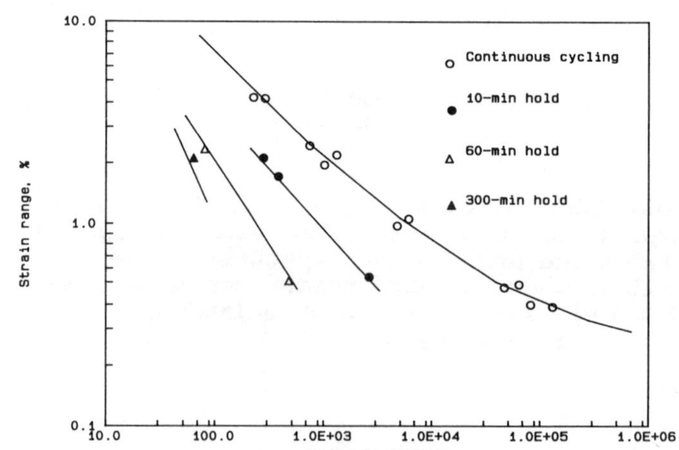

Alloy 800: Effect of hold time at maximum tensile strain on low-cycle fatigue life at 1200 F for bar, solution annealed at 2100 F for 1 h. Specimens had a 0.25-in. diam hourglass shape, tested in air. Strain rate was 4 X 10^{-3}/s. Source: Ref 22, p 21

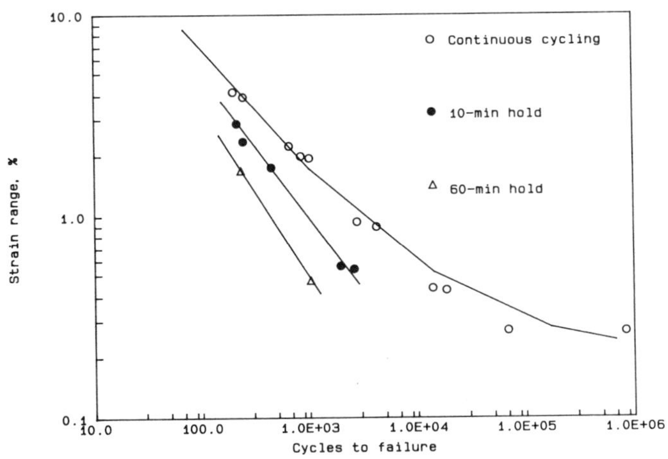

Alloy 800: Effect of hold time at maximum tensile strain on low-cycle fatigue life at 1400 F for bar, solution annealed at 2100 F for 1 h. Specimens had a 0.25-in. diam hourglass shape, tested in air. Strain rate: 4 X 10^{-3}/s. Source: Ref 22, p 21

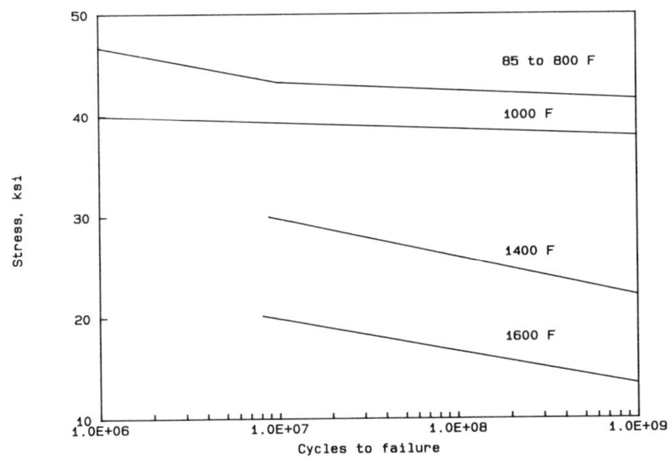

Alloy 800: Rotating beam fatigue strength. Specimens were cold drawn, annealed at 1600 F, held 30 min, and air cooled. R = -1. Source: Ref 22, p 19

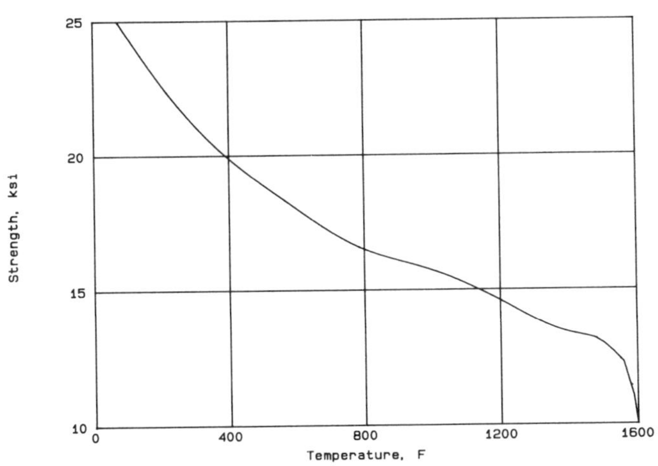

Alloy 800: Minimum yield strength of sheet and bar at temperatures as specified by the ASME Boiler and Pressure Vessel Code. Specimens were solution treated at 2100 F for 1 h. Source: Ref 22, p 10

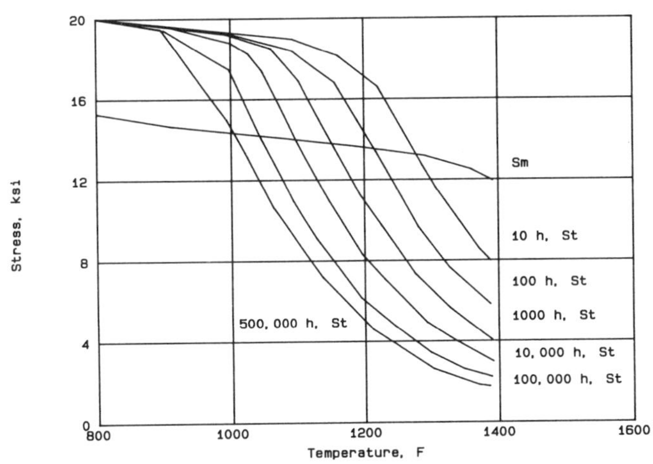

Alloy 800: Allowable stress values for data of actual service life and for normal-plus-upset conditions, according to the ASME Boiler and Pressure Vessel Code. Sheet was solution treated at 2100 F for 1 h. Stress components were time-dependent (St) and time-independent (Sm). Source: Ref 22, p 10

Alloy 800: Brinell hardness. Source: Ref 22, p 5

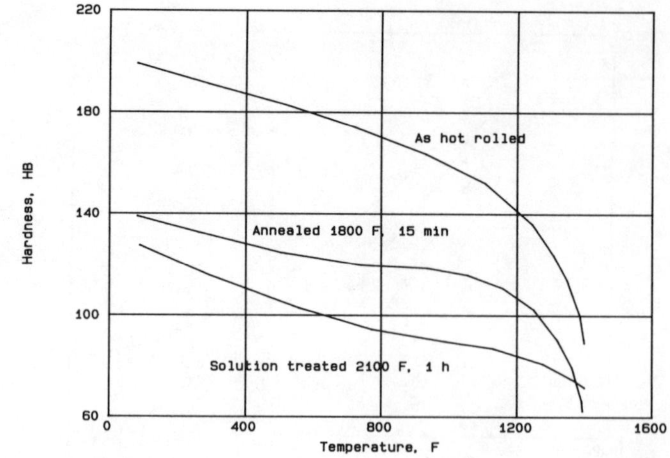

Alloy 800H

UNS number: N08810

Specifications: ASME SB163, ASME SB407, ASME SB408, ASME SB409, ASME SB564, ASTM B163, ASTM B407, ASTM B408, ASTM B409, ASTM B564

Nominal composition: 21.000% Cr, 32.500% Ni, 46.000% Fe, 0.400% Ti, 0.400% Al, 0.080% C

Product forms: Wrought, bar, rod, wire, plate, sheet, strip, tube, pipe, billets, forgings, structural shapes

Description: Similar to Incoloy 800, but with superior high-temperature strength due to carbon content and grain size control. This alloy features higher design strength values for use above 620 C (1150 F) and improved creep and stress-rupture properties in the 595 to 980 C (1100 to 1800 F) temperature range.

Typical applications: Chemical and power plant superheater and reheater tubing, steam methane reformer pigtails, headers, furnace tubing, and process piping.

Physical Properties

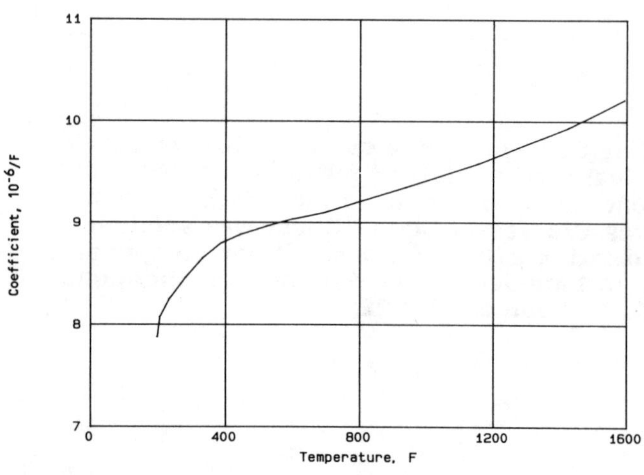

Alloy 800H: Mean coefficient of linear thermal expansion of solution annealed specimens. Source: Ref 22, p 7

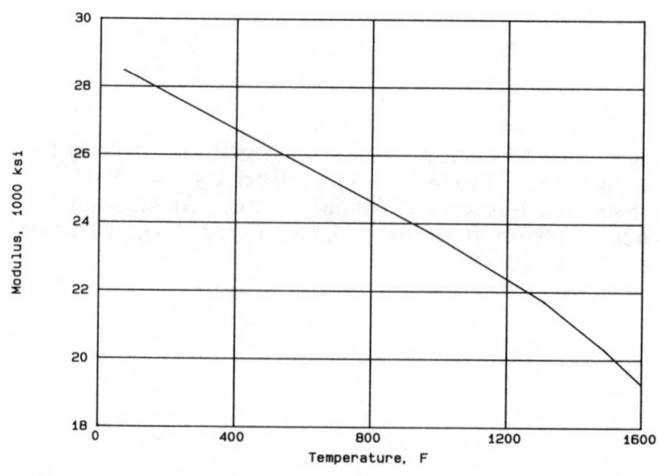

Alloy 800H: Modulus of elasticity of solution annealed tubing. Source: Ref 22, p 33

Alloy 800H: Effect of working and aging treatments on elevated-temperature yield strength of bar

Condition	YS 538 C (1000 F) MPa	YS 538 C (1000 F) ksi	YS 593 C (1100 F) MPa	YS 593 C (1100 F) ksi	YS 649 C (1200 F) MPa	YS 649 C (1200 F) ksi	YS 704 C (1300 F) MPa	YS 704 C (1300 F) ksi	YS 760 C (1400 F) MPa	YS 760 C (1400 F) ksi
1	131.0	19.0	124.1	18.0	124.1	18.0	124.1	18.0	108.9	15.8
2	276.5	40.1	248.2	36.0	231.0	33.5	217.2	31.5	224.1	32.5
3	331.0	48.0	303.4	44.0	306.1	44.4	293.0	42.5	272.4	39.5
4	214.4	31.1	282.2	40.93	217.2(a)	31.5(a)	143.4(a)	20.8(a)	124.1	18.0
5	226.8	32.9	...	...	272.4	39.5	...	...	124.1	18.0
6	317.2	46.0	393.0(a)	57.0(a)	258.6(b)	37.5(b)	204.8	29.7	193.1	28.0
7	414.4	60.1	456.4	66.2	320.0(b)	46.4(b)	272.4	39.5	234.4	34.0

Note: Condition 1: solution annealed. Condition 2: cold worked 5%. Condition 3: cold worked 10%. Condition 4: aged 4000 h at test temperature. Condition 5: aged 8000 h at test temperature. Condition 6: cold worked 5%, aged 4000 h at test temperature. Condition 7: cold worked 10%, aged 4000 h at test temperature
(a) Gamma prime in grain boundaries. (b) Gamma prime in matrix
Source: Ref 29, Dec 1979

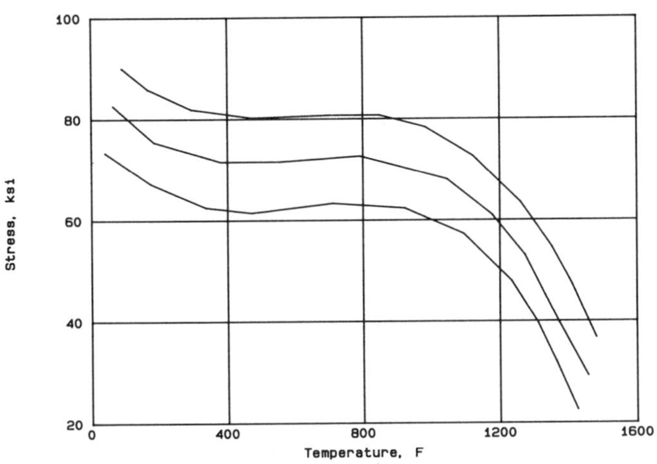

Alloy 800H: Scatterband for tensile strength from room temperature to 1400 F. Area between curves includes +/- two standard deviations from the mean. Source: Ref 22, p 11

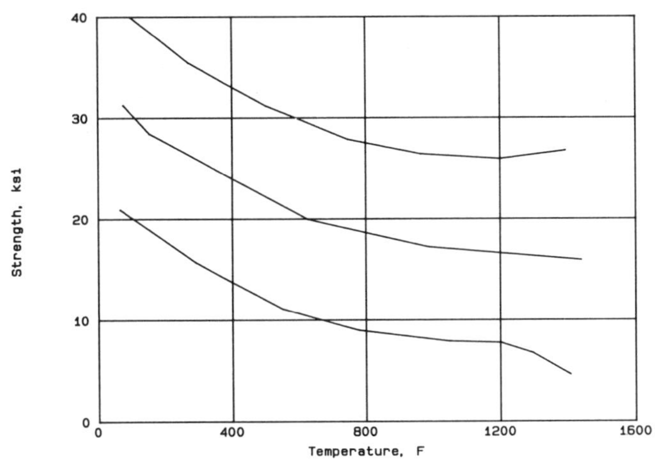

Alloy 800H: Scatterband for yield strength of various product forms and heat treatments. Area between curves includes +/- two standard deviations from the mean. Source: Ref 22, p 12

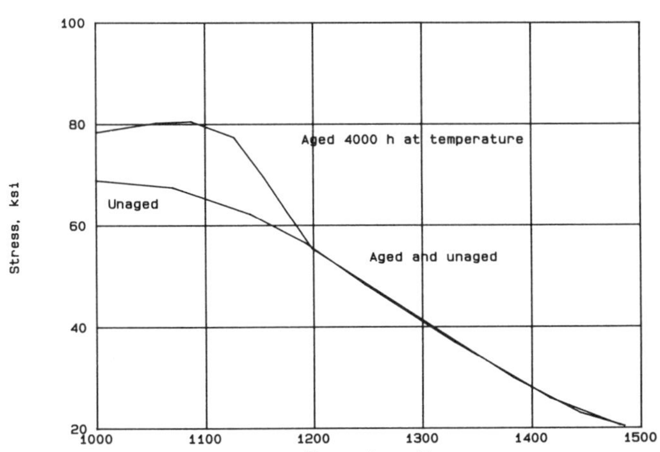

Alloy 800H: Effect of aging for 4000 h at various temperatures on tensile strength of solution annealed bar. Specimens were solution treated only or solution treated and aged 4000 h at test temperature, as indicated. Source: Ref 22, p 12

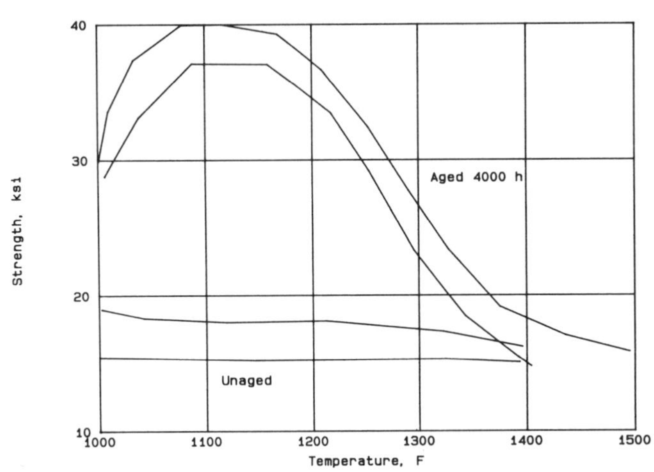

Alloy 800H: Effect of aging for 4000 h at various temperatures on yield strength of solution annealed bar. Source: Ref 22, p 13

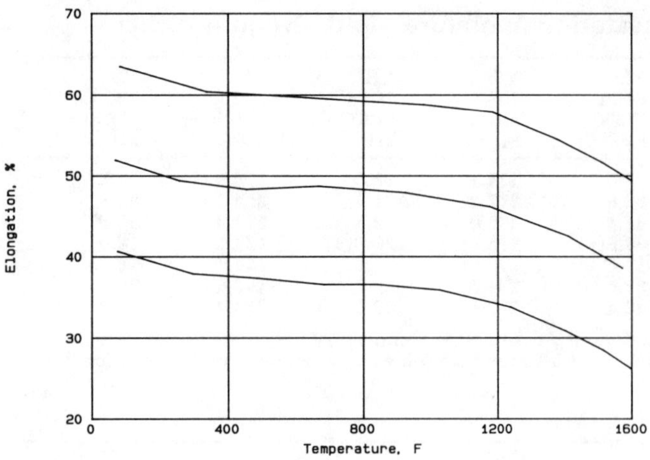

Alloy 800H: Scatterband for tensile elongation of various product forms and heat treatments. Area between curves includes +/- two standard deviations from the mean. Source: Ref 22, p 12

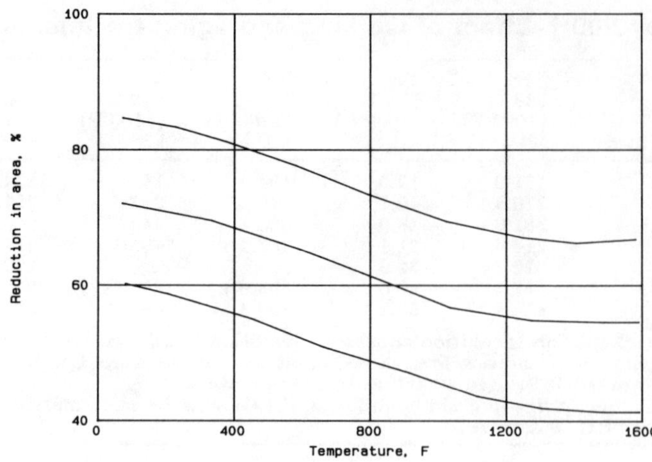

Alloy 800H: Scatterband for tensile reduction in area. Area between curves includes +/- two standard deviations from the mean. Source: Ref 22, p 12

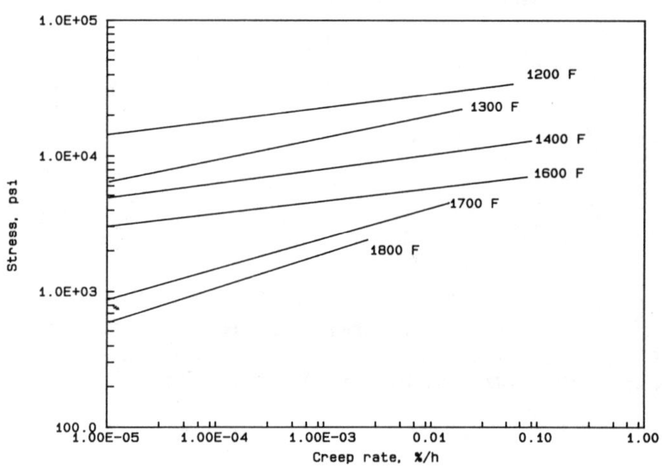

Alloy 800H: Minimum creep strength of solution treated specimens. Source: Ref 14, p 335

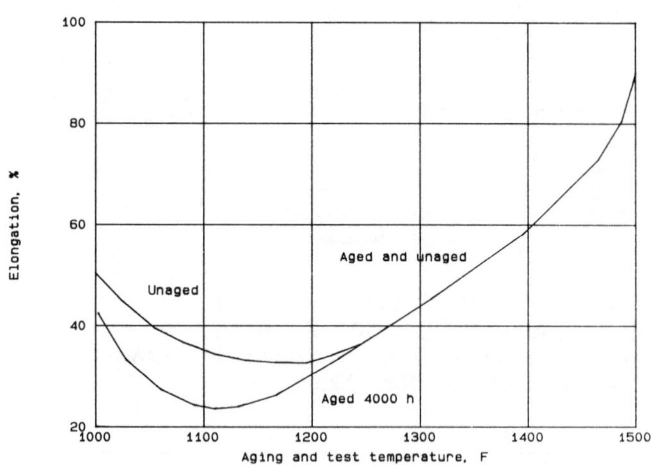

Alloy 800H: Effect of aging for 4000 h at various temperatures on tensile elongation. Source: Ref 22, p 13

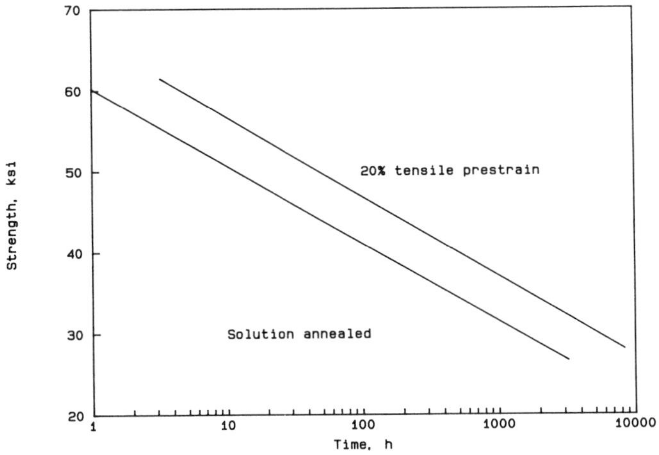

Alloy 800H: Creep-rupture strength at 1100 F for bar. Specimens were solution annealed or solution annealed and 20% cold prestrained. Source: Ref 22, p 16

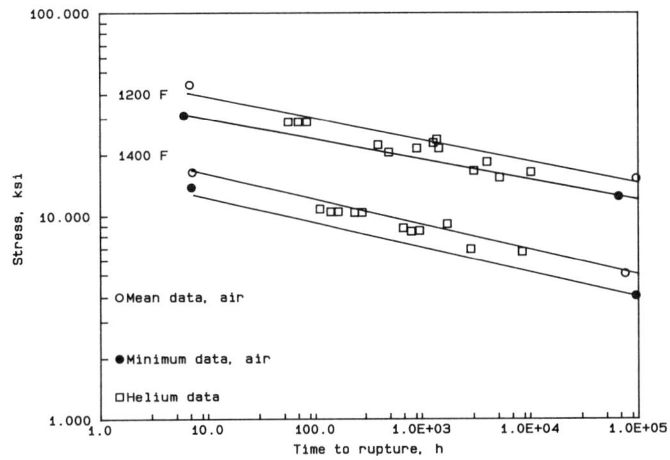

Alloy 800H: Creep-rupture for bar solution annealed at 1200 and 1400 F in helium, compared to ASME Code Case 1592 specifications for air. Source: Ref 22, p 17

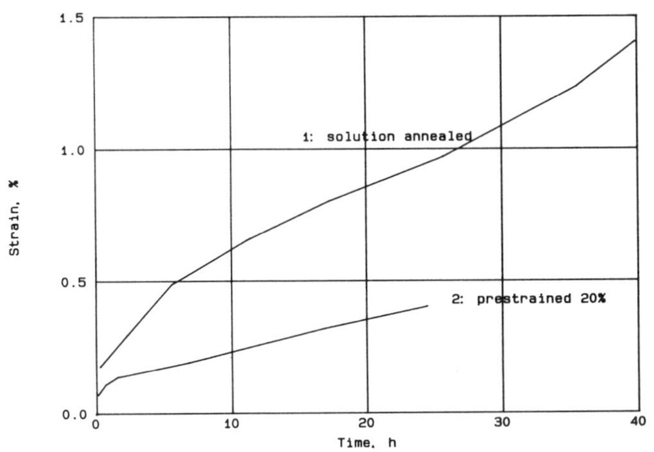

Alloy 800H: Creep curves at 1340 F and 9 ksi for solution annealed and 20% prestrained bar. Curves show no evidence of recrystallization for times up to 4000 h. Steady-state creep rate for curve 1 was 2.2 X 10^{-4} %/h and 1.3 X 10^{-4} %/h for curve 2. Source: Ref 22, p 16

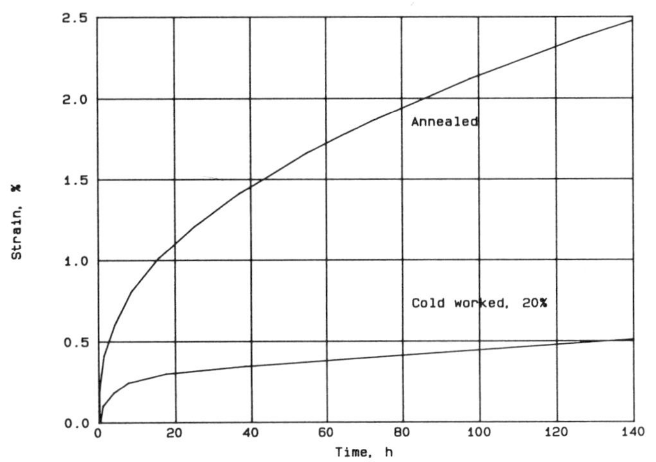

Alloy 800H: Creep curves at 1100 F and 40 ksi for annealed and 20% cold worked bar. Source: Ref 22, p 15

11.50 High-Temperature Property Data: Ferrous Alloys

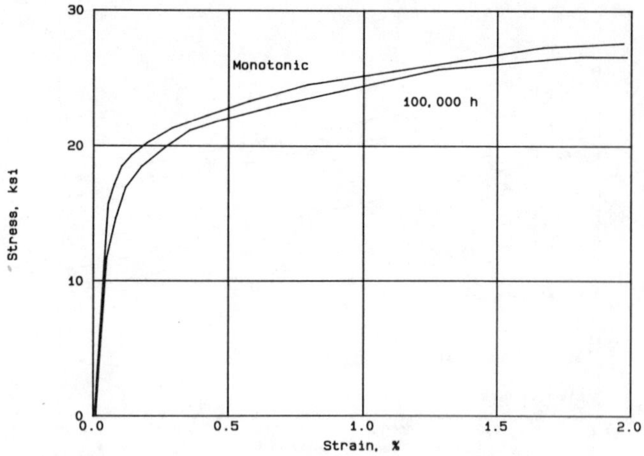

Alloy 800H: Isochronous stress-strain curves tested at 900 F for solution annealed bar. Source: Ref 22, p 10

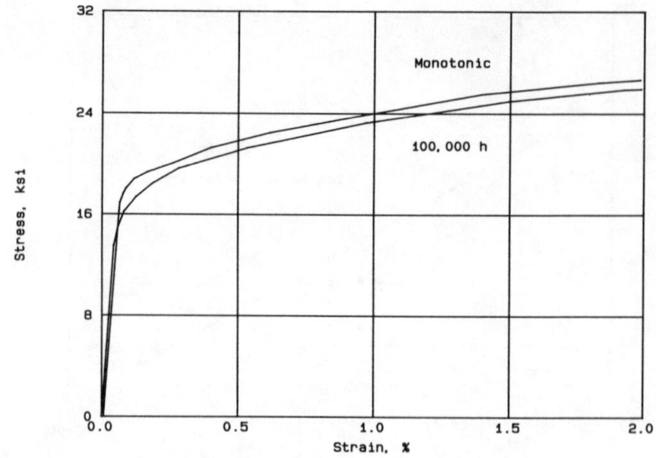

Alloy 800H: Isochronous stress-strain curves tested at 1000 F for solution annealed bar. Source: Ref 22, p 10

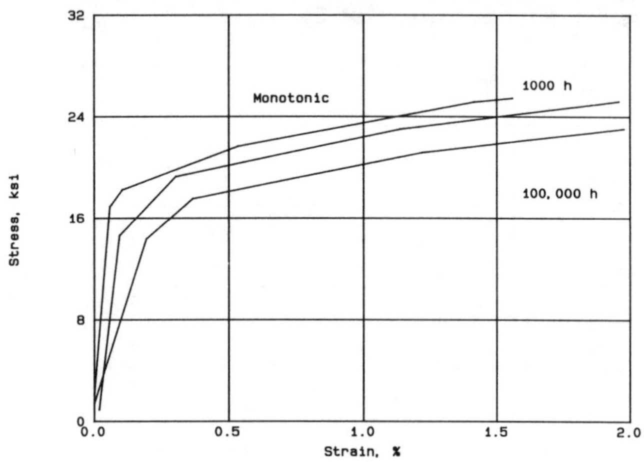

Alloy 800H: Isochronous stress-strain curves tested at 1100 F for solution annealed bar. Source: Ref 22, p 10

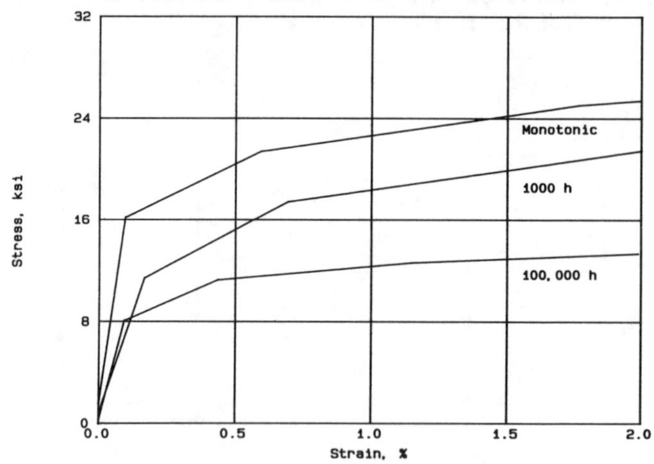

Alloy 800H: Isochronous stress-strain curves tested at 1200 F for solution annealed bar. Source: Ref 22, p 10

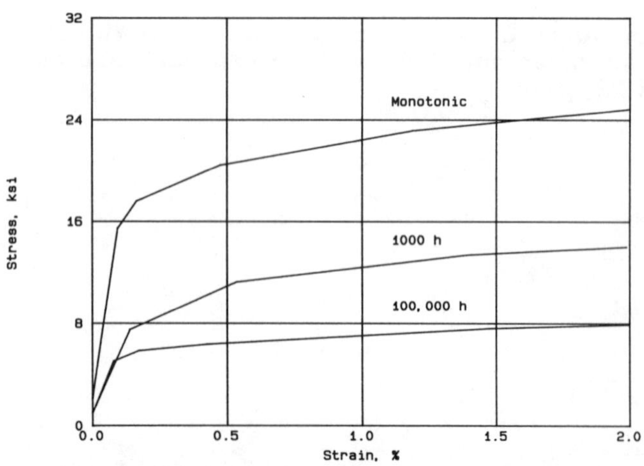

Alloy 800H: Isochronous stress-strain curves tested at 1300 F for solution annealed bar. Source: Ref 22, p 11

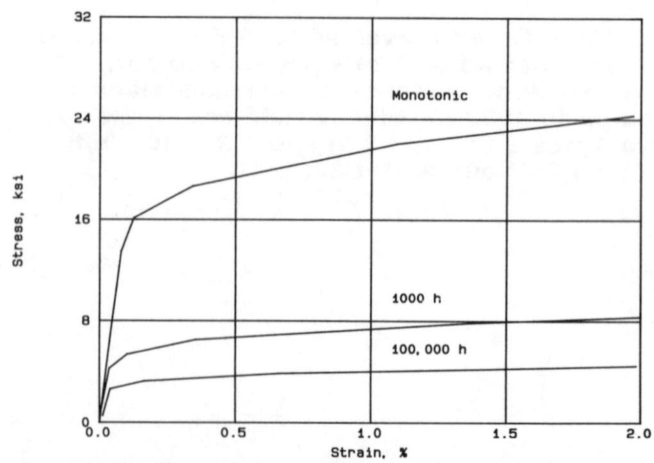

Alloy 800H: Isochronous stress-strain curves tested at 1400 F for solution annealed bar. Source: Ref 22, p 11

Alloy 800H: Stress-to-rupture property data for bar

TEMP C	TEMP F	RPSTR 1 h MPa (ksi)	RPSTR 3 h MPa (ksi)	RPSTR 10 h MPa (ksi)	RPSTR 30 h MPa (ksi)	RPSTR 100 h MPa (ksi)	RPSTR 300 h MPa (ksi)	RPSTR 1000 h MPa (ksi)	RPSTR 3000 h MPa (ksi)	RPSTR 10000 h MPa (ksi)	RPSTR 30000 h MPa (ksi)	RPSTR 100000 h MPa (ksi)	RPSTR 300000 h MPa (ksi)
427	800	418.5	418.5	418.5	418.5	418.5	418.5	418.5	418.5	418.5	418.5	413.7	379.2
427	800	(60.7)	(60.7)	(60.7)	(60.7)	(60.7)	(60.7)	(60.7)	(60.7)	(60.7)	(60.7)	(60.0)	(55.0)
455	850	418.5	418.5	418.5	418.5	418.5	418.5	418.5	418.5	396.5	368.9	399.9	314.4
455	850	(60.7)	(60.7)	(60.7)	(60.7)	(60.7)	(60.7)	(60.7)	(60.7)	(57.5)	(53.5)	(49.3)	(45.6)
482	900	418.5	418.5	418.5	418.5	418.5	418.5	399.2	371.0	340.6	313.7	286.1	262.0
482	900	(60.7)	(60.7)	(60.7)	(60.7)	(60.7)	(60.7)	(57.9)	(53.8)	(49.4)	(45.5)	(41.5)	(38.0)
510	950	418.5	418.5	418.5	418.5	406.8	377.2	345.4	317.9	288.9	263.4	237.2	215.1
510	950	(60.7)	(60.7)	(60.7)	(60.7)	(59.0)	(54.7)	(50.1)	(46.1)	(41.9)	(38.2)	(34.4)	(31.2)
538	1000	418.5	418.5	418.5	388.2	355.1	325.4	295.1	268.9	241.3	217.9	194.4	175.1
538	1000	(60.7)	(60.7)	(60.7)	(56.3)	(51.5)	(47.2)	(42.8)	(39.0)	(35.0)	(31.6)	(28.2)	(25.4)
566	1050	418.5	403.4	368.9	337.9	305.4	277.9	248.9	224.8	200.0	178.6	157.9	140.7
566	1050	(60.7)	(58.5)	(53.5)	(49.0)	(44.3)	(40.3)	(36.1)	(32.6)	(29.0)	(25.9)	(22.9)	(20.4)
593	1100	386.8	354.4	320.6	291.0	260.6	234.4	207.5	185.5	163.4	145.5	126.9	112.4
593	1100	(56.1)	(51.4)	(46.5)	(42.2)	(37.8)	(34.0)	(30.1)	(26.9)	(23.7)	(21.1)	(18.4)	(16.3)
621	1150	339.9	308.2	275.8	248.2	219.3	195.8	171.7	152.4	133.1	117.2	101.4	88.9
621	1150	(49.3)	(44.7)	(40.0)	(36.0)	(31.8)	(28.4)	(24.9)	(22.1)	(19.3)	(17.0)	(14.7)	(12.9)
649	1200	295.8	265.5	235.1	208.9	183.4	162.0	141.3	124.0	106.9	93.8	80.7	70.3
649	1200	(42.9)	(38.5)	(34.1)	(30.3)	(26.6)	(23.5)	(20.5)	(18.0)	(15.5)	(13.6)	(11.7)	(10.2)
677	1250	255.1	226.8	198.6	175.1	152.4	133.1	115.1	100.0	86.2	75.2	64.1	55.8
677	1250	(37.0)	(32.9)	(28.8)	(25.4)	(22.1)	(19.3)	(16.7)	(14.5)	(12.5)	(10.9)	(9.3)	(8.1)
704	1300	217.9	191.7	166.2	144.8	125.5	108.9	93.1	80.7	69.0	59.3	51.0	44.1
704	1300	(31.6)	(27.8)	(24.1)	(21.0)	(18.2)	(15.8)	(13.5)	(11.7)	(10.0)	(8.6)	(7.4)	(6.4)
732	1350	184.8	161.3	138.6	120.7	102.7	88.9	75.2	64.8	55.2	47.6	40.0	34.5
732	1350	(26.8)	(23.4)	(20.1)	(17.5)	(14.9)	(12.9)	(10.9)	(9.4)	(8.0)	(6.9)	(5.8)	(5.0)
760	1400	155.8	135.1	115.1	99.3	84.1	71.7	60.7	51.7	44.1	37.9	31.7	27.6
760	1400	(22.6)	19.6	(16.7)	(14.4)	(12.2)	(10.4)	(8.8)	(7.5)	6.4	(5.5)	(4.6)	(4.0)

Note: Specimens were taken from solution annealed bar
Source: Ref 29, Dec 1979

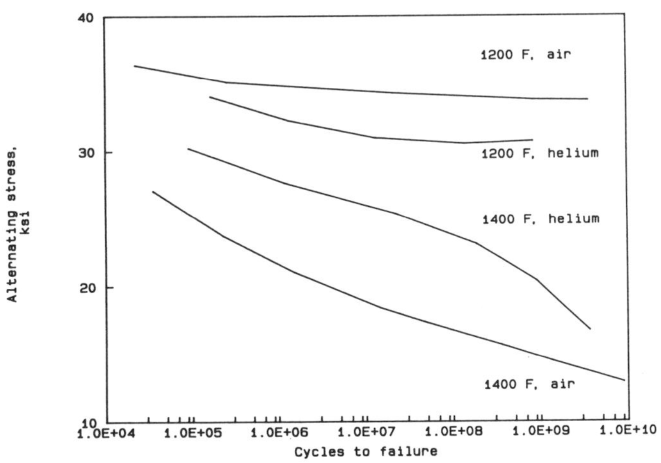

Alloy 800H: High-cycle fatigue at 1200 and 1400 F in air and in helium for solution annealed tubing. Source: Ref 22, p 33

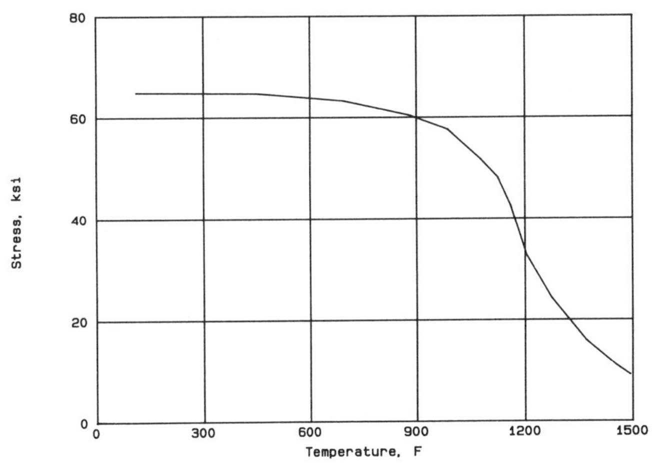

Alloy 800H: Allowable primary tensile stress from room temperature to 1500 F for solution annealed specimens, according to ASME Boiler and Pressure Vessel Code. Source: Ref 22, p 9

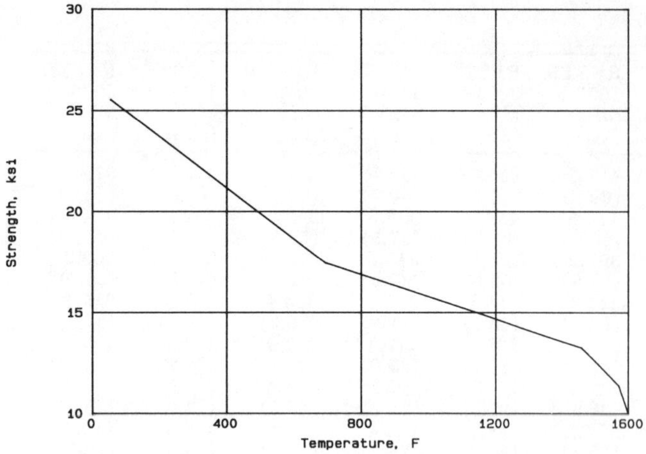

Alloy 800H: Minimum yield strength of sheet and bar. Specimens were solution treated at 2100 F for 1 h, according to ASME Boiler and Pressure Vessel Code. Source: Ref 22, p 8

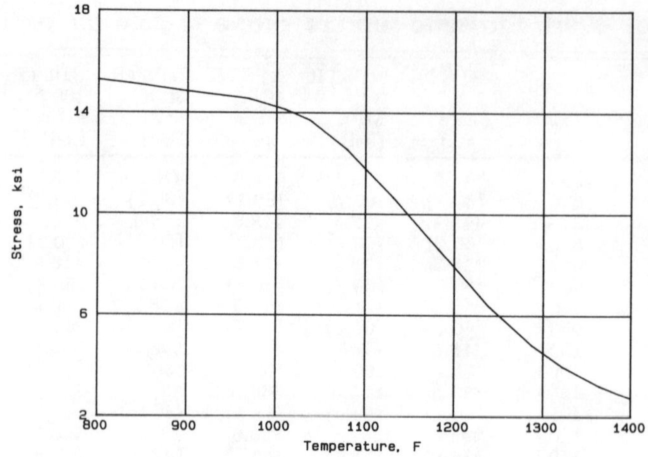

Alloy 800H: Allowable stress values for primary general membrane calculations for design conditions, as defined by ASME Boiler and Pressure Vessel Code for sheet. Specimens were solution treated at 2100 F for 1 h. Source: Ref 22, p 8

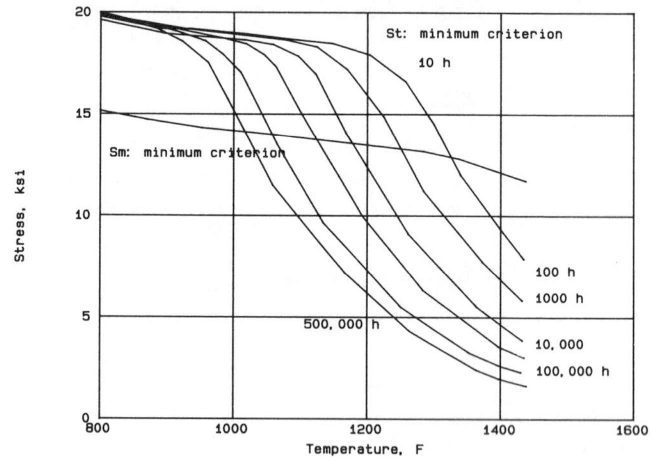

Alloy 800H: Allowable stress per ASME Boiler and Pressure Vessel Code for sheet solution treated at 2100 F for 1 h. Data represent actual service life/normal + upset coonditions.
St: time-dependent stress components.
Sm: time-independent stress components.
Source: Ref 22, p 8

Incoloy 800HT

Product forms: Wrought, tube
Description: An Incoloy 800 variation with superior creep and rupture strengths during extended exposures to elevated temperatures. A restricted composition (within the Incoloy 800 specification) and use of a high-temperature anneal account for increased properties.
Typical applications: Chemical and power plant superheater and reheater tubing, steam methane reformer pigtails, headers, furnace tubing, and process piping.

Physical Properties

Incoloy 800HT: Thermal conductivity

TEMP C	TEMP F	THCON W/m*K	THCON Btu/h*ft*F
21	70	11.5	6.7
38	100	12.0	6.9
93	200	12.8	7.4
204	400	14.8	8.6
316	600	16.6	9.6
427	800	18.3	10.6
538	1000	20.0	11.6
649	1200	21.9	12.6
760	1400	23.9	13.8
871	1600	26.1	15.1
982	1800	30.8	17.8

Source: Incoloy alloy 800HT Data Sheet, INCO Alloys Inc.

Incoloy 800HT: Electrical resistivity

TEMP C	TEMP F	ELRES n{o}*m
21	70	989
38	100	997
93	200	1030
204	400	1092
316	600	1133
427	800	1170
538	1000	1200
649	1200	1240
760	1400	1260
871	1600	1280
1038	1800	1290
1093	2000	1310

Source: Incoloy alloy 800HT Data Sheet, INCO Alloys Inc.

Incoloy 800HT: Modulus of elasticity

TEMP C	TEMP F	E GPa	E 10^3 ksi
24	75	196.51	28.50
93	200	191.82	27.82
204	400	184.85	26.81
316	600	177.27	25.71
427	800	169.89	24.64
538	1000	162.17	23.52
649	1200	154.24	22.37
760	1400	145.21	21.06
871	1600	132.38	19.20

Note: Modulus of elasticity was determined by the dynamic method.
Source: Incoloy alloy 800HT Data Sheet, INCO Alloys Inc.

Incoloy 800HT: Mean coefficients of linear thermal expansion

TEMP C	TEMP F	THEXP 10^{-6}/C	THEXP 10^{-6}/F
93	200	14.2	7.9
204	400	15.8	8.8
316	600	16.2	9.0
427	800	16.6	9.2
538	1000	16.9	9.4
649	1200	17.3	9.6
760	1400	17.8	9.9
871	1600	18.4	10.2

Source: Incoloy alloy 800HT Data Sheet, INCO Alloys Inc.

Incoloy 800HT: Modulus of rigidity

TEMP C	TEMP F	G GPa	G 10^3 ksi
24	75	73.36	10.64
93	200	71.50	10.37
204	400	68.33	9.91
316	600	65.30	9.47
427	800	62.33	9.04
538	1000	59.30	8.60
649	1200	55.99	8.12
760	1400	52.26	7.58
871	1600	47.02	6.82

Note: Modulus of rigidity was determined by the dynamic method
Source: Incoloy alloy 800HT Data Sheet, INCO Alloys Inc.

Incoloy 800HT: Poisson's ratio

TEMP C	TEMP F	PR
24	75	0.339
93	200	0.341
204	400	0.353
316	600	0.357
427	800	0.363
538	1000	0.367
649	1200	0.377
760	1400	0.389
871	1600	0.408

Note: Poisson's ratio was calculated from modulus of elasticity
Source: Incoloy alloy 800HT Data Sheet, INCO Alloys Inc.

Mechanical Properties

Incoloy 800HT: Hardness and tensile property data

TEMP C	TEMP F	HARD HB	UTS MPa	UTS ksi	YS 0.2% MPa	YS 0.2% ksi
27	80	126	536	77.8	150	27.7
425	800	...	465	67.5	130	18.8
540	1000	90	432	62.7	90	13.0
650	1200	84	378	54.8	93	13.5
705	1300	82	329	47.7	109	15.8
760	1400	74	236	34.2	90	13.1

Source: Incoloy alloy 800HT Data Sheet, INCO Alloys Inc.

Incoloy 800HT: Room-temperature tensile and Charpy V-notch properties of cold rolled (20%) specimens after high-temperature exposure

EXP TEMP C	EXP TEMP F	EXP TIME h	IS J	IS ft-lb	YS 0.2% MPa	YS 0.2% ksi	UTS MPa	UTS ksi	EL %	RA %
...	...	...	152	112	779	113.0	786	114.0	15.5	58.0
540	1000	1000	85	63	789	114.5	879	127.5	18.5	50.5
540	1000	4000	106	78	776	112.5	865	125.5	20.0	52.5
540	1000	8000	83	61	783	113.5	886	128.5	20.0	47.0
540	1000	12000	83	61	783	113.5	886	128.5	20.0	52.0
650	1200	1000	118	87	624	90.5	752	109.0	23.0	46.5
650	1200	4000	88	65	547	79.4	738	107.0	21.5	43.0
650	1200	8000	84	62	561	81.4	734	106.5	25.5	52.5
650	1200	12000	85	63	544	78.9	724	105.0	24.0	50.0

Source: Incoloy alloy 800HT Data Sheet, INCO Alloys Inc.

Incoloy 800HT: Tensile property data

TEMP C	TEMP F	Stress (a) MPa	Stress (a) ksi	Stress (b) MPa	Stress (b) ksi
38	100	111.6	16.2	111.6	16.2
93	200	106.1	15.4	111.6	16.2
149	300	100.0	14.5	111.6	16.2
204	400	93.1	13.5	111.6	16.2
260	500	88.9	12.9	110.3	16.0
316	600	84.1	12.2	110.3	16.0
343	650	82.1	11.9	110.3	16.0
371	700	80.7	11.7	108.2	15.7
399	750	78.6	11.4	106.1	15.4
427	800	76.5	11.1	105.4	15.3
454	850	75.2	10.9	104.1	15.1
482	900	73.8	10.7	102.0	14.8
510	950	72.4	10.5	100.6	14.6
538	1000	71.0	10.3	99.2	14.4
566	1050	69.6	10.1	94.4	13.7
593	1100	69.0	10	88.9	12.9
621	1150	67.6	9.8	71.7	10.4
649	1200	57.2	8.3	57.2	8.3
677	1250	46.2	6.7	46.2	6.7
704	1300	37.2	5.4	37.2	5.4
732	1350	29.6	4.3	29.6	4.3
760	1400	23.4	3.4	23.4	3.4
788	1450	18.6	2.7	18.6	2.7
816	1500	15.2	2.2	15.2	2.2
843	1550	11.7	1.7	11.7	1.7
871	1600	9.6	1.4	9.6	1.4
899	1650	7.6	1.1	7.6	1.1

Note: Design stress taken from ASME Boiler and Pressure Vessel Code, Code Case 1987 Section 1 and Section VIII, Division 1

(a) Maximum allowable stress in tension. (b) Maximum allowable stress in tension. Conditional: These higher stress values of up to 90% of yield strength at temperature may be used where slightly greater deformation is acceptable. The stresses may result in dimensional changes due to permanent strain and are not recommended for applications such as flanges of gasketed joints

Source: Incoloy alloy 800HT Data Sheet, INCO Alloys Inc.

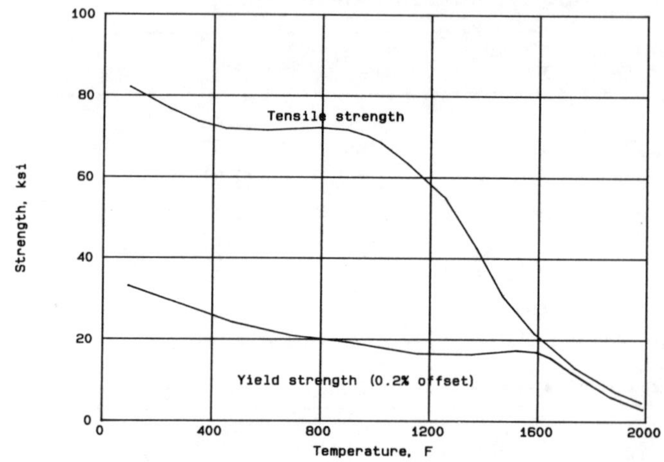

Incoloy 800HT: Tensile and yield strength of annealed extruded tubing with a 5-in. outside diameter and 0.5-in. thick wall. Source: Incoloy Alloy 800HT Data Sheet, INCO Alloys, Inc.

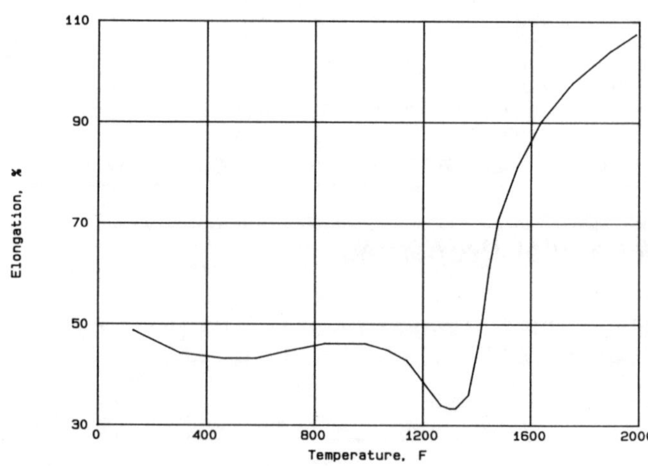

Incoloy 800HT: Tensile elongation of annealed extruded tubing with a 5-in. outside diameter and a 0.5-in. thick wall. Source: Incoloy Alloy 800HT Data Sheet, INCO Alloys, Inc.

Incoloy 800HT: Stress-to-rupture property data

TEMP C	TEMP F	RPSTR 10000 h MPa	RPSTR 10000 h ksi	RPSTR 30000 h MPa	RPSTR 30000 h ksi	RPSTR 50000 h MPa	RPSTR 50000 h ksi	RPSTR 100000 h MPa	RPSTR 100000 h ksi
650	1200	121	17.5	103	15	97	14	90	13
705	1300	76	11.0	66	9.5	61	8.8	55	8.0
760	1400	50	7.3	43	6.3	40	5.8	37	5.3
815	1500	36	5.2	30	4.4	28	4.1	26	3.7
870	1600	24	3.5	21	3.0	19	2.8	17	2.5
925	1700	13	1.9	11	1.6	10	1.4	8.3	1.2
980	1800	8.3	1.2	6.9	1.0	6.2	9.0	5.5	8.0

Source: Incoloy alloy 800HT Data Sheet, INCO Alloys Inc.

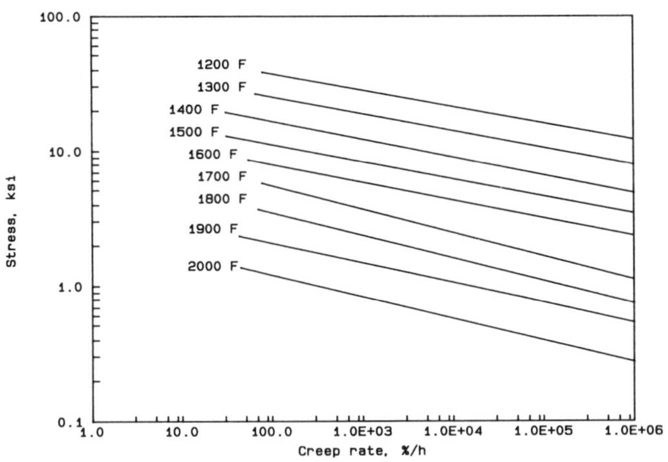

Incoloy 800HT: Stress-rupture strength at various temperatures. Source: Incoloy Alloy 800HT Data Sheet, INCO Alloys, Inc.

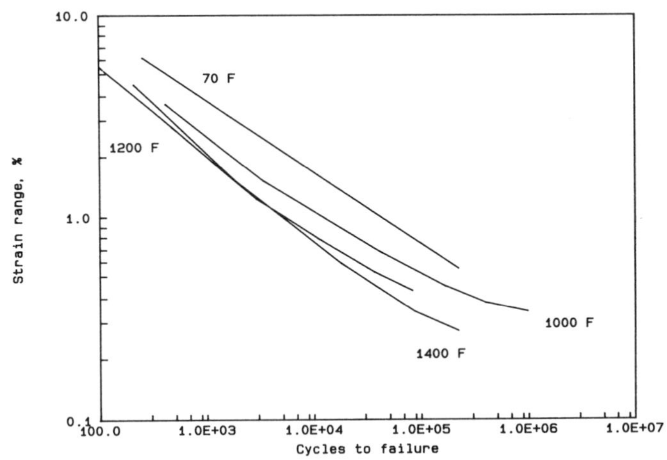

Incoloy 800HT: Low-cycle fatigue strength, tested using axial strain. Source: Incoloy Alloy 800HT Data Sheet, INCO Alloys, Inc.

Incoloy 802

Nominal composition: 0.358% C, 21.000% Cr, 32.500% Ni, 46.000% Fe, 0.750% Ti max, 0.580% Al max
Product forms: Wrought, bar, rod, wire, plate, sheet, strip, tube, pipe, billets, forgings, structural shapes
Description: Incoloy 800 modification with a relatively high carbon content. Hardening obtained via carbide precipitation.
Typical applications: Dies for creep forming of titanium, ethylene furnace tubes.

Physical Properties

Incoloy 802: Thermal conductivity of annealed specimens

TEMP C	TEMP F	THCON W/m*K	THCON Btu/h*ft*F
24	76	11.8	6.82
93	200	13.1	7.57
204	400	14.8	8.57
316	600	16.4	9.48
427	800	18.1	10.48
538	1000	19.7	11.40
649	1200	21.3	12.30
760	1400	22.8	13.14
871	1600	24.2	13.98
982	1800	25.5	14.70

Note: Thermal conductivity was calculated from electrical resistivity
Source: Ref 14, p 336

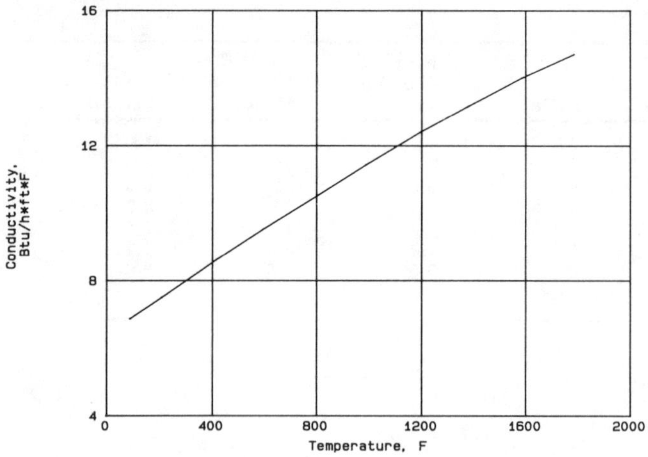

Incoloy 802: Thermal conductivity. Specimens were annealed at 1800 F for 0.25 h and air cooled. Source: Ref 22, p 3

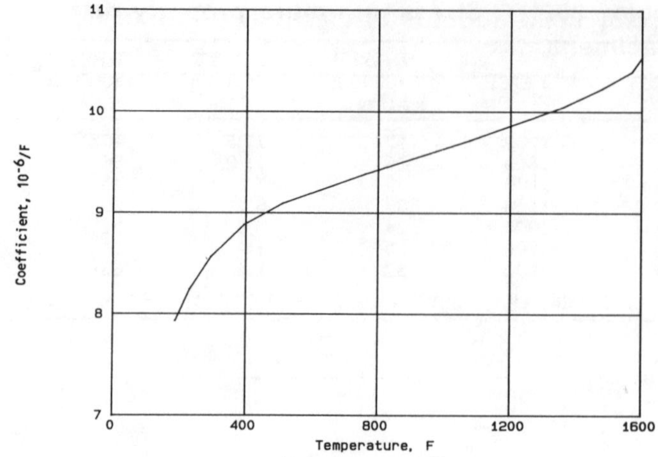

Incoloy 802: Mean coefficient of linear thermal expansion. Specimens were annealed at 1800 F for 0.25 h and air cooled. Source: Ref 22, p 3

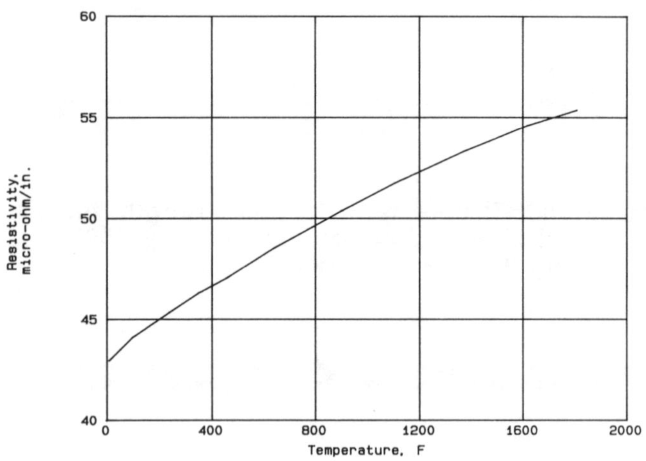

Incoloy 802: Electrical resistivity. Specimens were annealed at 1800 F for 0.25 h and air cooled. Source: Ref 22, p 4

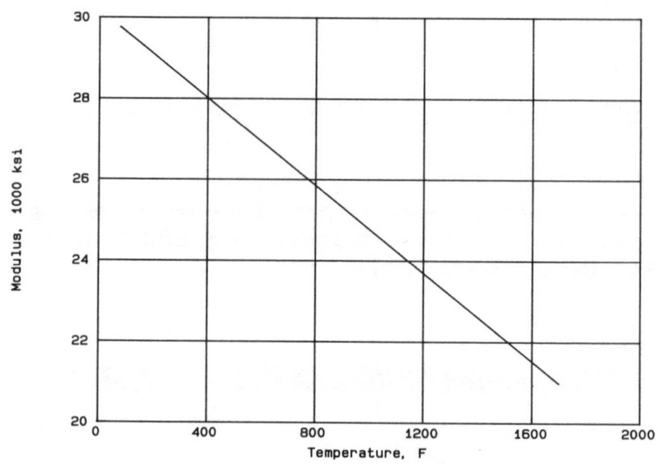

Incoloy 802: Modulus of elasticity. Specimens were annealed at 1800 F for 0.25 h and air cooled. Source: Ref 22, p 8

Incoloy 802: Electrical resistivity

TEMP C	TEMP F	ELRES n{o}*m
24	76	1111
93	200	1140
204	400	1185
316	600	1228
427	800	1266
538	1000	1301
649	1200	1332
760	1400	1362
871	1600	1387
982	1800	1407

Source: Ref 14, p 336

Incoloy 802: Mean coefficients of linear thermal expansion

TEMP C	TEMP F	THEXP 10^{-6}/C	THEXP 10^{-6}/F
93	200	14.0	7.8
204	400	15.7	8.7
316	600	16.0	8.9
427	800	16.4	9.1
538	1000	16.7	9.3
649	1200	17.1	9.5
760	1400	17.6	9.8
871	1600	18.2	10.1

Source: Ref 14, p 336

Incoloy 802: Modulus of elasticity of annealed specimens

TEMP C	TEMP F	E GPa	E 10^3 ksi
28	82	205	29.7
38	100	204	29.6
93	200	200	29.0
149	300	196	28.4
204	400	192	27.9
260	500	189	27.4
316	600	185	26.8
371	700	181	26.2
427	800	177	25.7
482	900	173	25.1
538	1000	169	24.5
593	1100	165	23.9
649	1200	161	23.4
704	1300	158	22.9
760	1400	156	22.6
816	1500	154	22.4
871	1600	152	22.0
927	1700	145	21.1

Note: Modulus of elasticity was determined by the dynamic method
Source: Ref 14, p 336

Mechanical Properties

Incoloy 802: Room-temperature tensile and Charpy impact properties after thermal exposure for 2000 h

EXP TEMP C	EXP TEMP F	UTS MPa	UTS ksi	YS 0.2% MPa	YS 0.2% ksi	EL %	HARD HRB	IS J	IS ft-lb
...	...	681.9	98.9	231.0	33.5	43.5	81	62.0	46.0
649	1200	865.3	125.5	433.0	62.8	20.0	97	31.0	23.0
760	1400	775.7	112.5	457.1	66.3	17.0	94	26.0	19.0
871	1600	699.8	101.5	358.5	52.0	24.0	90	17.0	12.5

Source: Ref 14, p 336

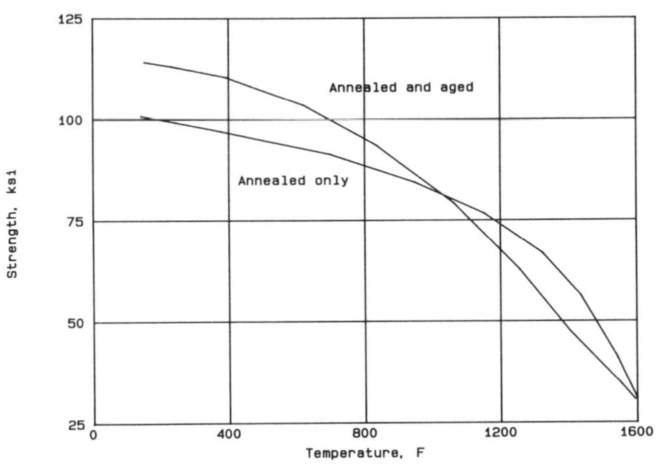

Incoloy 802: Effect of temperature and aging on tensile strength of specimens annealed at 1800 F for 0.25 h and air cooled. Some specimens were aged for 1000 h at 1400 F, as indicated. Source: Ref 22, p 6

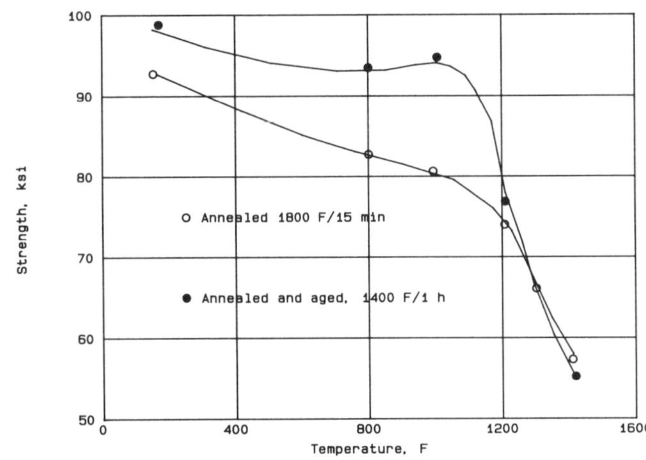

Incoloy 802: Tensile strength of 2.5-in. hot rolled plate in two conditions. Curve 1: annealed at 1800 F for 0.25 h. Curve 2: annealed at 1800 F for 0.25 h, air cooled, aged at 1400 F for 1 h, and air cooled. Source: Ref 22, p 6

Incoloy 802: Hardness and tensile property data

TEMP C	TEMP F	HARD HB	UTS MPa	UTS ksi	YS MPa	YS ksi
Annealed						
27	80	158	638.5	92.6	290.3	42.1
427	800	138	572.3	83.0	238.6	34.6
538	1000	128	557.1	80.8	191.7	27.8
649	1200	125	512.3	74.3	195.1	28.3
704	1300	122	455.1	66.0	183.4	26.6
760	1400	119	396.5	57.5	262.0	38.0
Annealed and heat treated at 760 C (1400 F) for 1 h						
27	80	215	677.8	98.3	386.1	56.0
800	800	190	643.3	93.3	320.6	46.5
538	1000	190	653.0	94.7	358.5	52.0
649	1200	174	530.9	77.0	313.7	45.5
704	1300	152	...	...	...	...
760	1400	128	382.0	55.4	292.3	42.4

Note: A 2000-kg load and 10-mm diam titanium carbide ball was used to measure hardness
Source: Ref 14, p 337

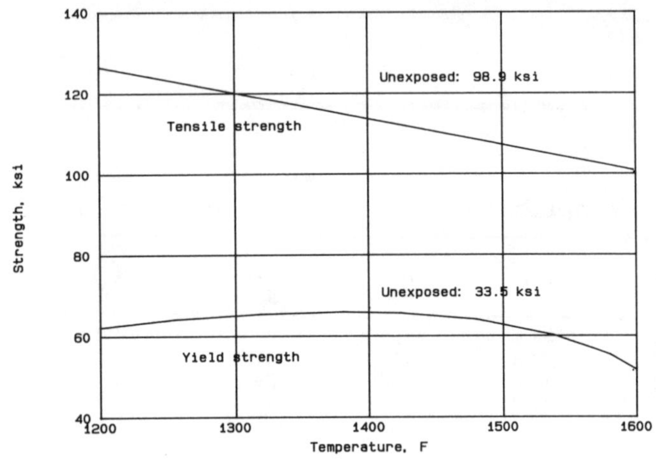

Incoloy 802: Effect of 2000-h exposure on room-temperature tensile and yield strength of plate annealed at 1800 F for 0.25 h, air cooled, and exposed at temperature indicated. Source: Ref 22, p 5

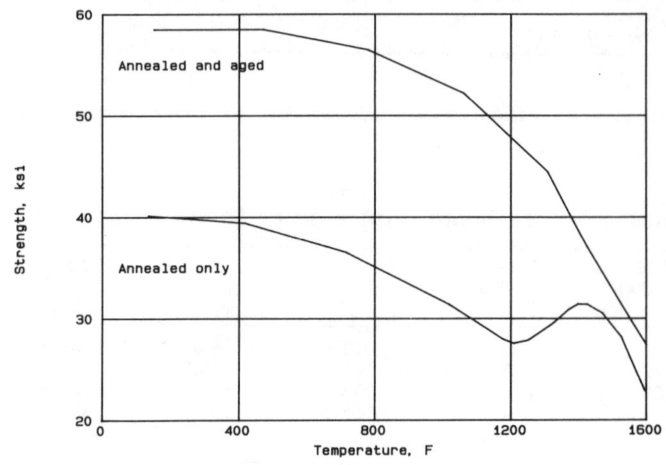

Incoloy 802: Effect of temperature and aging on yield strength of specimens annealed at 1800 F for 0.25 h and air cooled. Specimens were aged for 1000 h at 1400 F, as indicated. Source: Ref 22, p 6

Incoloy 802: Tensile and yield strength of annealed specimens. Source: Ref 14, p 337

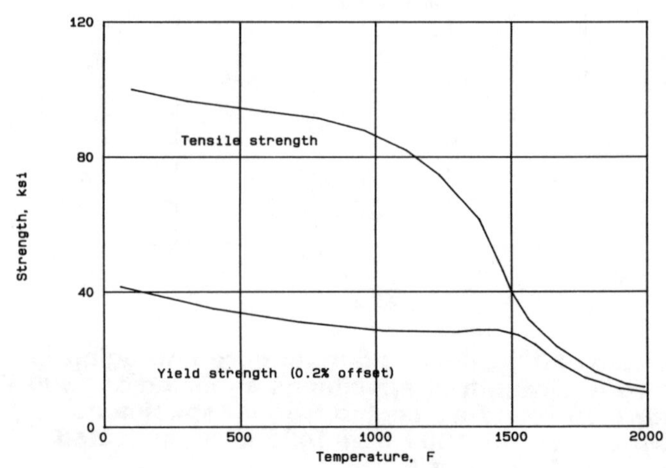

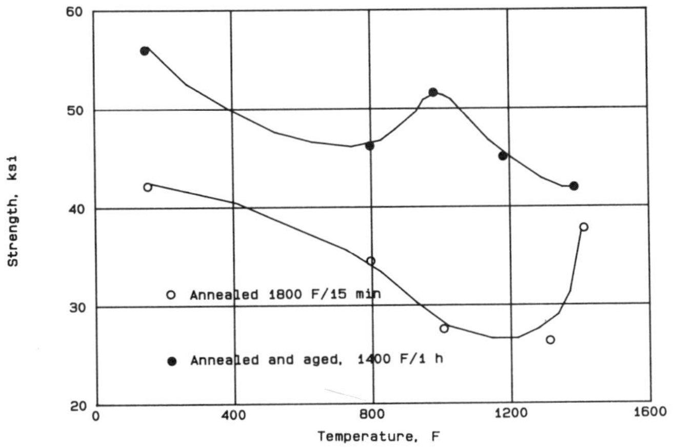

Incoloy 802: Yield strength of 2.5-in. hot rolled plate in two conditions. Curve 1: annealed at 1800 F for 0.25 h. Curve 2: annealed at 1800 F for 0.25 h, air cooled, aged at 1400 F for 1 h, and air cooled. Source: Ref 22, p 6

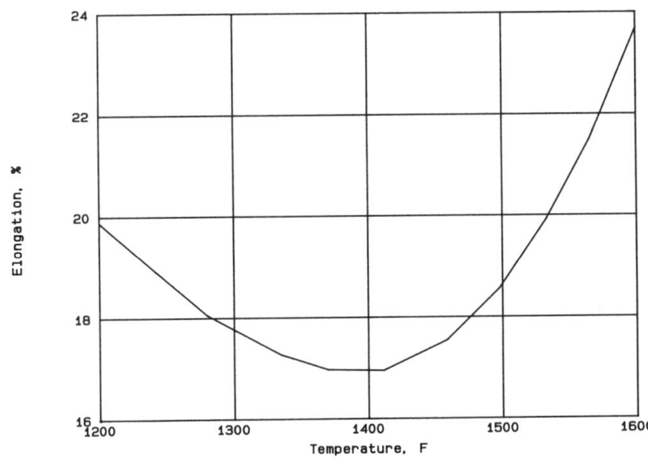

Incoloy 802: Effect of 2000-h exposure on room-temperature tensile elongation of plate, annealed at 1800 F for 0.25 h, air cooled, and exposed at the temperature indicated. Source: Ref 22, p 5

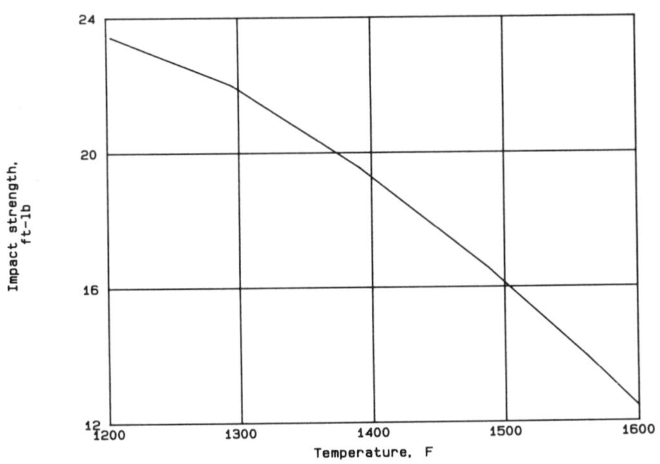

Incoloy 802: Effect of 2000-h exposure on room-temperature Charpy V-notch impact strength of plate, annealed at 1800 F for 0.25 h, air cooled, and exposed at the temperature indicated. Initial strength: 46 ft-lb. Source: Ref 22, p 5

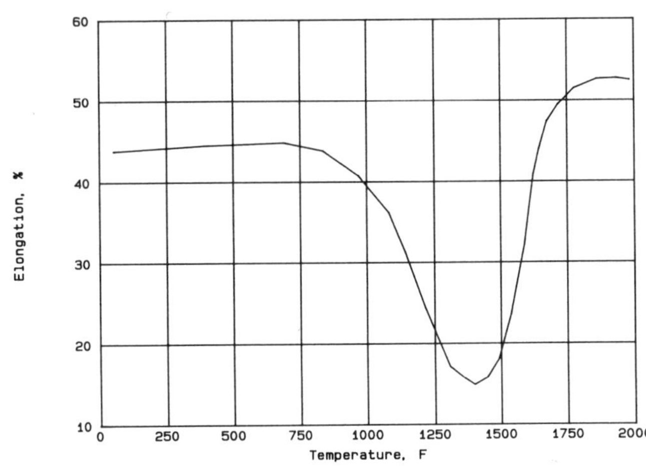

Incoloy 802: Tensile elongation of annealed specimens. Source: Ref 14, p 337

Incoloy 802: Stress-to-rupture property data

TEMP C	TEMP F	RPSTR 10000 h MPa	RPSTR 10000 h ksi	RPSTR 30000 h MPa	RPSTR 30000 h ksi	RPSTR 50000 h MPa	RPSTR 50000 h ksi	RPSTR 100000 h MPa	RPSTR 100000 h ksi
649	1200	131	19.0	110.3	16.0	96.5	14.0	89.6	13.0
760	1400	81.4	11.8	72.4	10.5	67.6	9.8	59.3	8.6
871	1600	34.5	5.0	24.1	3.5	21.4	3.1	17.2	2.5
927	1700	20.7	3.0	15.2	2.2	12.8	1.85	10.3	1.5
982	1800	11.7	1.7	8.3	1.2	6.9	1.0	5.5	0.8
1093	2000	3.7	0.54	2.6	0.37	2.1	0.3	1.7	0.24

Source: Ref 14, p 339

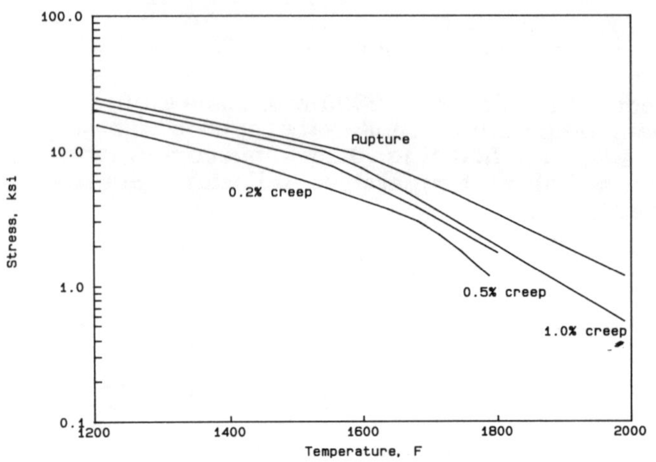

Incoloy 802: Creep strength for 1000 h life of annealed specimens at the temperature indicated. Source: Ref 14, p 339

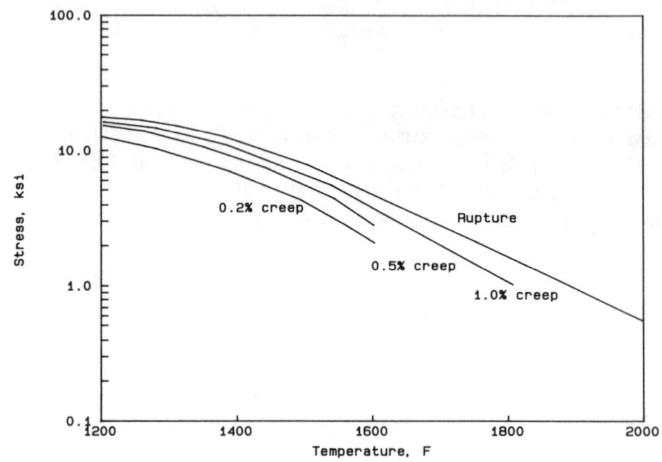

Incoloy 802: Creep strength for 10,000 h life of annealed specimens at the temperature indicated. Source: Ref 14, p 339

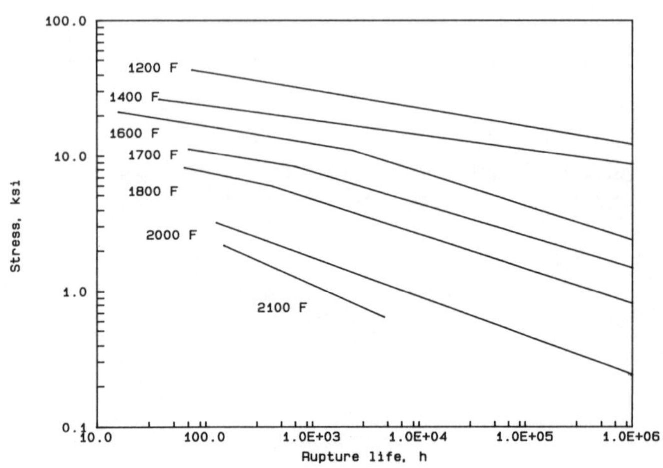

Incoloy 802: Rupture strength at various temperatures. Source: Ref 14, p 338

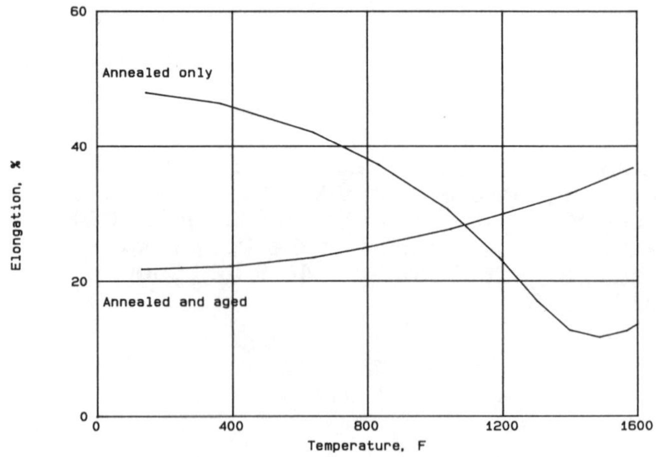

Incoloy 802: Effect of temperature and aging on tensile elongation of specimens annealed at 1800 F for 0.25 h and air cooled. Some specimens were aged for 1000 h at 1400 F, as indicated. Source: Ref 22, p 6

Incoloy 802: Stress-rupture property data

TEMP C	TEMP F	RPSTR MPa	RPSTR ksi	TIME h	EL %	RA %
649	1200	137.9	20.0	6200.0	10.0	7.5
760	1400	82.7	12.0	6923.9	11.5	13.5
871	1600	31.0	4.5	15492.2	14.5	13.0
982	1800	13.8	2.0	5216.8	12.0	9.5
1093	2000	6.9	1.0	2123.6	39.0	51.5

Source: Ref 14, p 339

Incoloy 802: Fatigue strength

TEMP C	TEMP F	FATST 10^5 cyc MPa	FATST 10^5 cyc ksi	FATST 10^6 cyc MPa	FATST 10^6 cyc ksi	FATST 10^7 cyc MPa	FATST 10^7 cyc ksi	FATST 10^8 cyc MPa	FATST 10^8 cyc ksi
27	80	324	47	262	38	241	35	228	33
538	1000	365	53	352	51	338	49	331	48
649	1200	338	49	324	47	317	46	310	45
760	1400	290	42	276	40	262	38	255	37
871	1600	172	25	152	22	131	19	117	17
982	1800	145	21	117	17	97	14	69	10

Note: Specimens were hot rolled and annealed. Rotating-beam tests were used to determine fatigue strength
Source: Ref 14, p 337

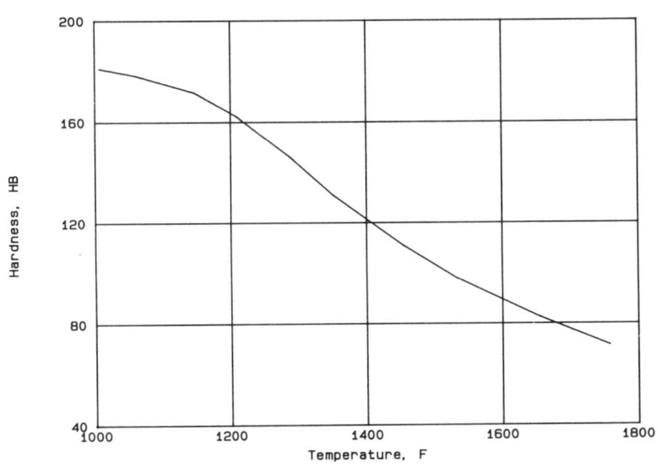

Incoloy 802: Brinell hardness tested by the mutual indentation method for specimens annealed at 1800 F for 0.25 h, air cooled, and aged at 1400 F for 1 h. At 70 F, hardness was 229 HB. Mutual indentation hardness was 220 HB and was obtained by pressing two cylinders together. Source: Ref 22, p 3

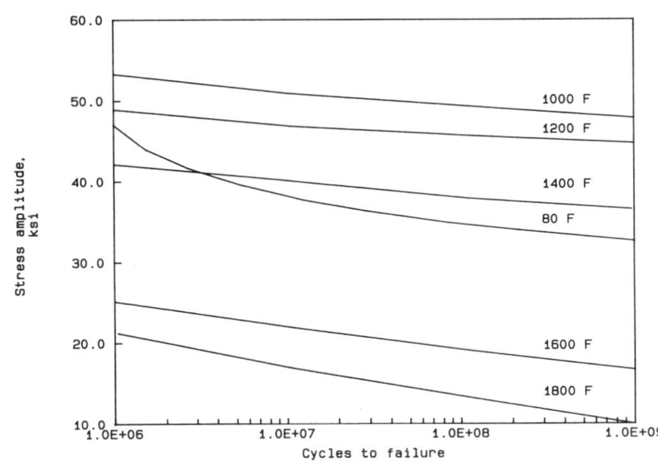

Incoloy 802: S-N curves in rotating bending from room temperature to 1800 F for specimens hot rolled and annealed at 1800 F for 0.25 h and air cooled. Rotating beam, R = -1. Source: Ref 22, p 8

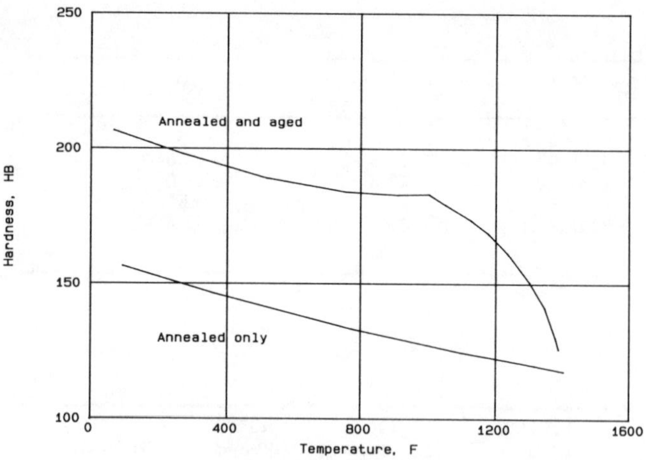

Incoloy 802: Effect of temperature and aging on Brinell hardness of specimens annealed at 1800 F for 0.25 h, air cooled, aged at 1400 F for 1 h, and air cooled, as indicated. Hardness was tested with 2000-kg load on a 10-mm titanium carbide ball. Source: Ref 22, p 2

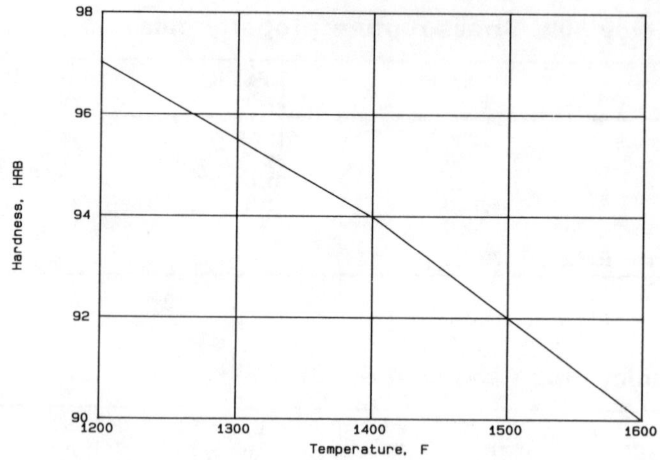

Incoloy 802: Effect of 2000-h exposure on room-temperature Rockwell B hardness of plate annealed at 1800 F for 0.25 h, air cooled, and exposed at the temperature indicated. Initial hardness prior to exposure was 81 HRB. Source: Ref 22, p 2

Incoloy 901

Alternate designation: AISI 681/682
UNS number: N09901
Specification: AMS 5661
Nominal composition: 12.500% Cr, 42.500% Ni, 35.000% Fe, 5.800% Mo, 2.900% Ti, 0.015% B, 0.100% C max, 0.350% Al max
Product forms: Wrought, bar, rod, wire, plate, sheet, strip, tube, pipe, billets, forgings, structural shapes

Description: Nickel-iron-base superalloy designed for high strength and corrosion resistance in the 540 to 760 C (1000 to 1400 F) temperature range. This alloy is hardened by the precipitation of the Ni3(Al,Ti)-type compounds. The scaling resistance of Incoloy 901 is slightly less than that of AISI type 310 stainless steel.
Typical applications: Aircraft turbine rotors, compressor discs, hubs, and shafts.

Physical Properties

Incoloy 901: Thermal conductivity

TEMP C	TEMP F	THCON W/m*K	THCON Btu/h*ft*F
43	109	13.4	7.74
167	333	14.3	8.24
246	476	15.4	8.90
360	680	16.3	9.40
457	855	17.6	10.15
538	1000	18.3	10.56
649	1200	19.3	11.15
760	1400	20.5	11.80

Source: Carpenter Technology

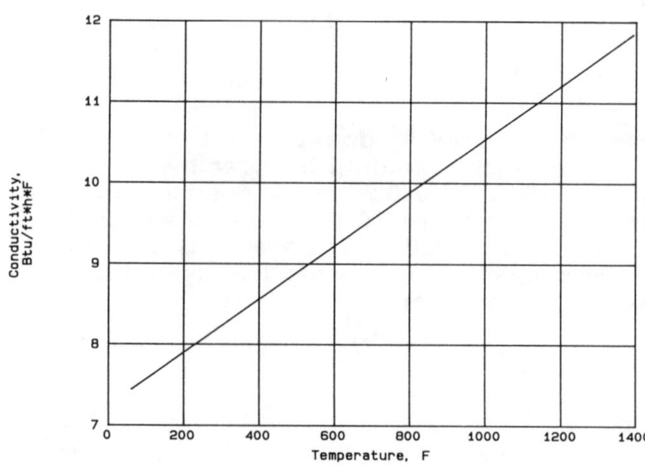

Incoloy 901: Thermal conductivity. Source: Ref 22, p 3

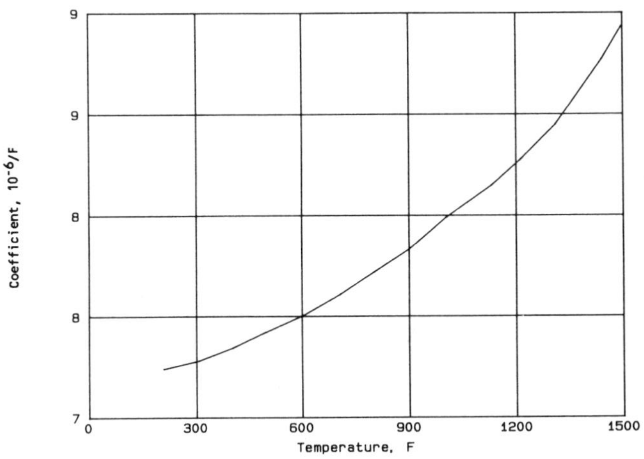

Incoloy 901: Mean coefficient of linear thermal expansion. Source: Ref 22, p 4

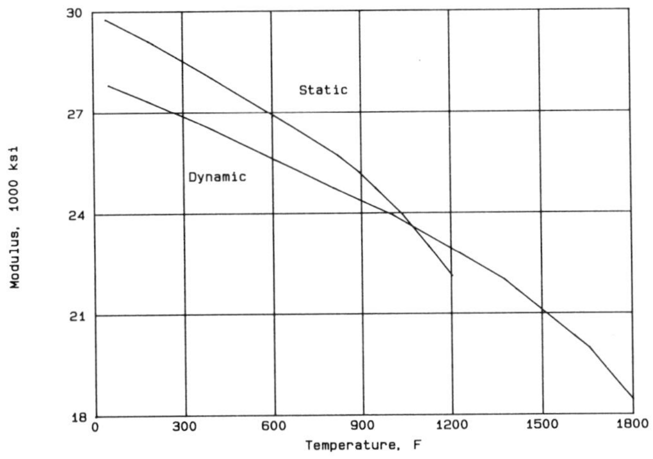

Incoloy 901: Modulus of elasticity. Source: Ref 22, p 4

Incoloy 901: Mean coefficients of linear thermal expansion

TEMP C	TEMP F	THEXP $10^{-6}/C$	THEXP $10^{-6}/F$
27-93	80-200	13.9	7.75
27-204	80-400	14.1	7.85
27-316	80-600	14.4	8.02
27-427	80-800	14.9	8.27
27-538	80-1000	15.3	8.50
27-649	80-1200	15.8	8.79
27-760	80-1400	16.5	9.15
27-816	80-1500	17.0	9.43

Source: Carpenter Technology

Incoloy 901: Modulus of elasticity

TEMP C	TEMP F	E GPa	E 10^3 ksi
24	75	206	29.9
204	400	197	28.5
316	600	183	26.5
427	800	179	25.9
538	1000	167	24.2
649	1200	152	22.1

Source: Carpenter Technology

Mechanical Properties

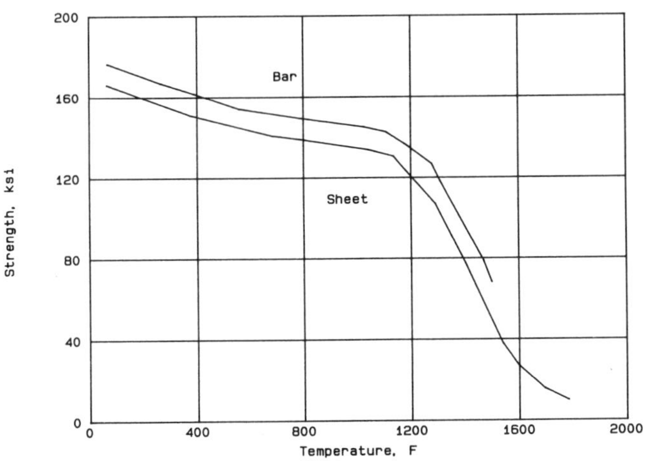

Incoloy 901: Tensile strength of sheet and bar. Sheet was solution treated and aged at 1400 F for 2 h. Bar was heat treated at 2050 F for 2 h, oil quenched, aged at 1300 F for 16 h, and air cooled. Source: Ref 22, p 7

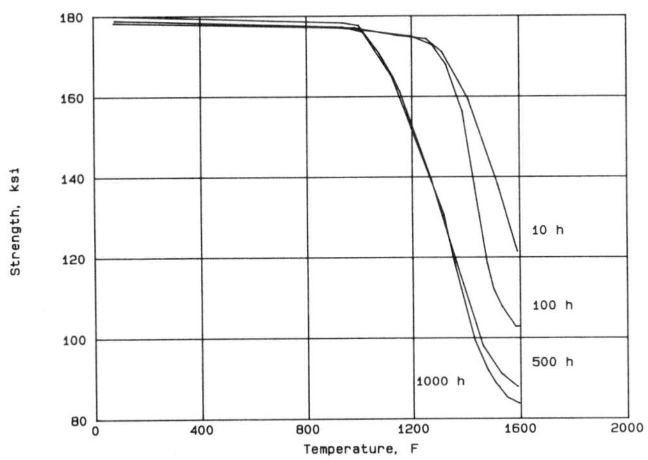

Incoloy 901: Effect of exposure on tensile strength of bar. Specimens were heat treated at 2000 F for 2 h, water quenched, aged at 1450 F for 2 h, aged at 1325 F for 24 h, and air cooled. Source: Ref 22, p 4

Incoloy 901: Tensile property data

TEMP C	TEMP F	YS 0.2% MPa	YS 0.2% ksi	UTS MPa	UTS ksi	EL %	RA %
21	70	862	125.0	1207	175	15.0	19.0
427	800	803	116.5	1096	159	16.0	25.5
538	1000	779	113	1076	156	17.0	29.0
649	1200	807	117	1034	150	14.0	21.0
704	1300	786	114	896	130	11.5	15.5
760	1400	707	102.5	741	107.5	9.0	14.5
816	1500	545	79	558	81	13.0	20.0

Note: Specimens were solution treated at 1093 C (2000 F), held 2 h, and water quenched. Then they were stabilized at 774 C (1425 F), held for 4 h, and air cooled. Specimens were aged at 732 C (1350 F), held 24 h, and air cooled
Source: Carpenter Technology

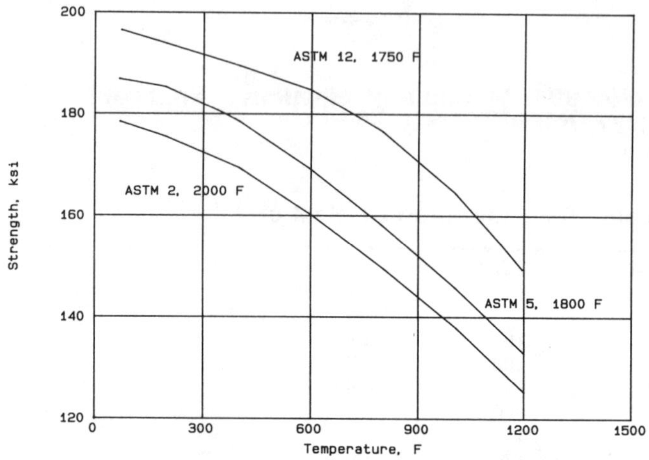

Incoloy 901: Effect of temperature and grain size on tensile strength of forgings. Specimens were forged by the mini-grain thermal-mechanical treatment, solution treated for 1 h at the indicated temperature, stabilized at 1325 F for 4 h, air cooled, aged at 1200 F for 12 h, and air cooled. Source: Ref 22, p 6

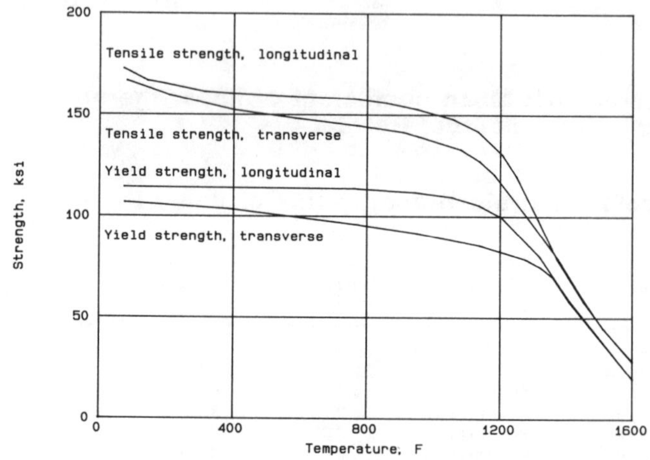

Incoloy 901: Effect of temperature and orientation on tensile and yield strength of bar and forgings. Specimens were solution treated at 2000 F for 2 h, water quenched, aged at 1450 for 2 h, air cooled, aged at 1325 F for 24 h, and air cooled. Source: Ref 22, p 5

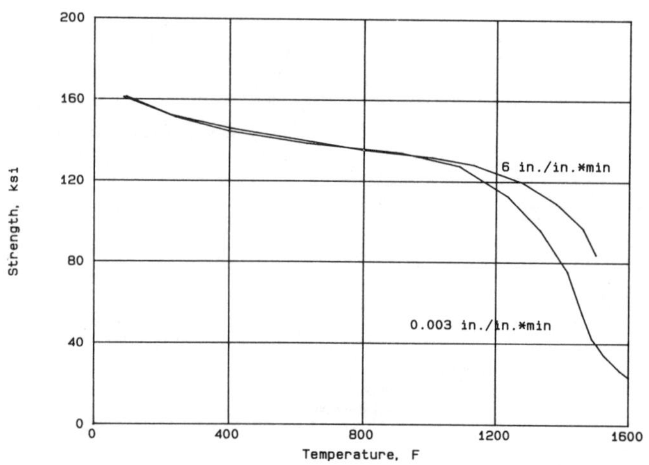

Incoloy 901: Effect of temperature and strain rate on tensile strength of 0.063-in. sheet. Specimens were solution treated, aged at 1400 F for 2 h, and air cooled. Source: Ref 22, p 8

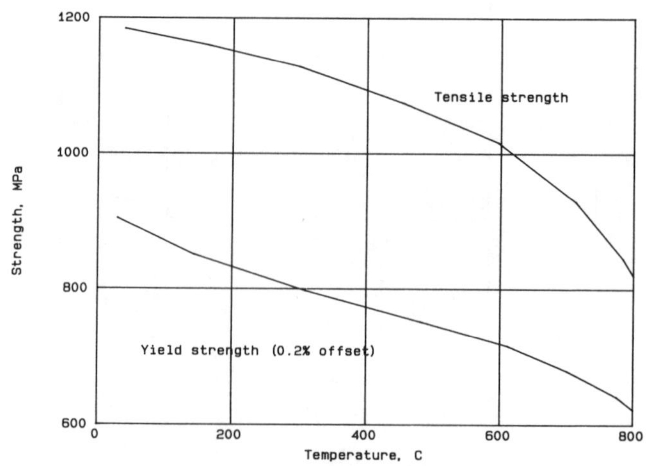

Incoloy 901: Tensile and yield strength of specimens solution treated at 1093 C, stabilized at 771 C, and aged at 704 C. Source: Ref 12, p 243

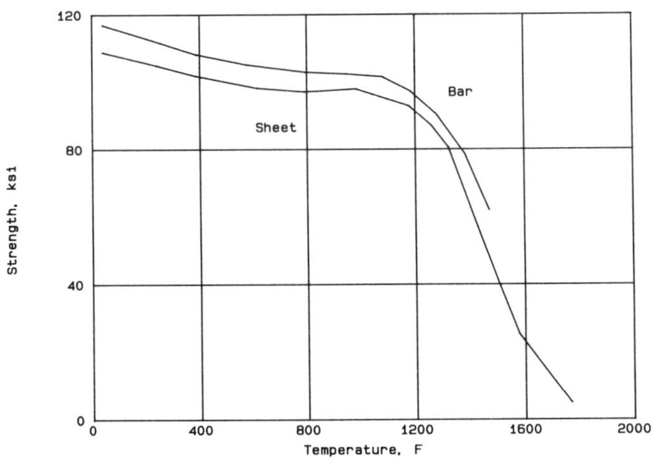

Incoloy 901: Yield strength of sheet and bar. Sheet was solution treated and aged at 1400 F for 2 h. Bar was heat treated at 2050 F for 2 h, oil quenched, aged at 1300 F for 16 h, and air cooled. Source: Ref 22, p 7

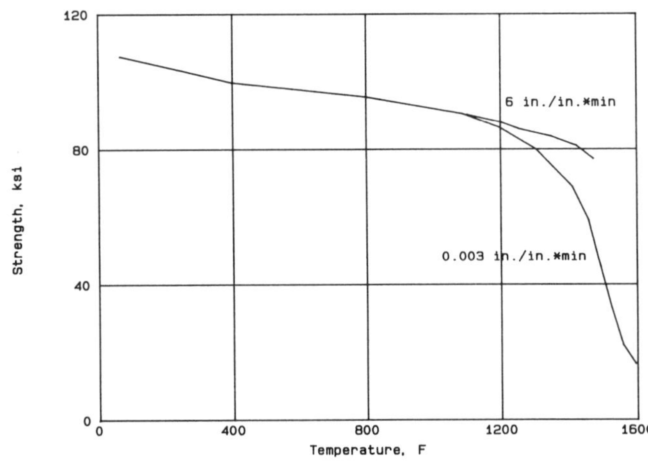

Incoloy 901: Effect of temperature and strain rate on yield strength of 0.063-in. sheet. Specimens were solution treated, aged at 1400 F for 2 h, and air cooled. Hold time at test temperature was either 10 s or 30 min. Source: Ref 22, p 7

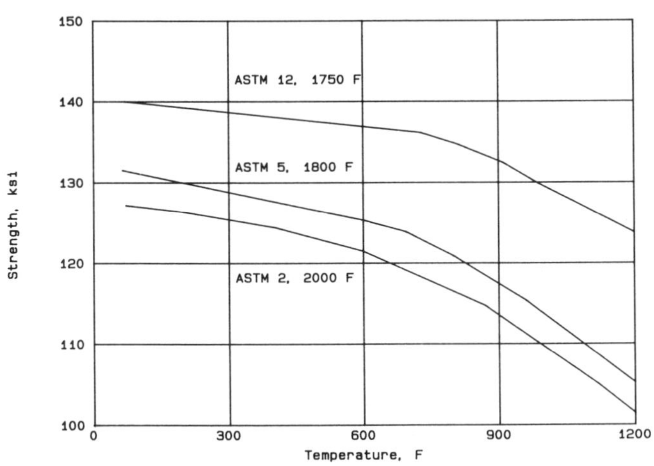

Incoloy 901: Effect of temperature and grain size on yield strength of forgings. Specimens were forged by the mini-grain thermal-mechanical treatment, solution treated for 1 h at the indicated temperature, stabilized at 1325 F for 4 h, air cooled, aged at 1200 F for 12 h, and air cooled. Source: Ref 22, p 6

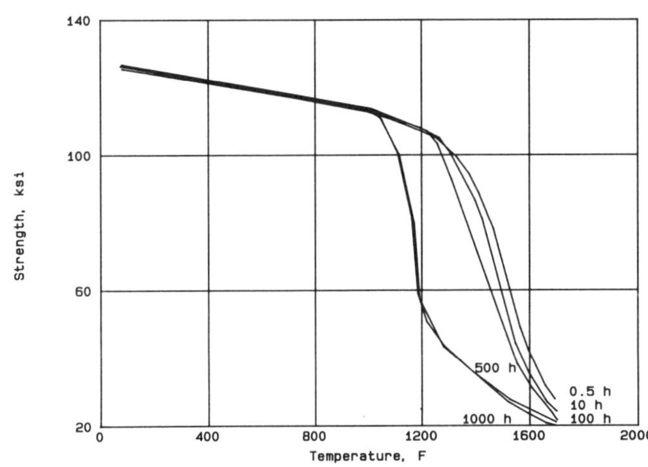

Incoloy 901: Effect of exposure on yield strength of bar. Specimens were heat treated at 2000 F, water quenched, aged at 1325 F for 24 h, and air cooled. Source: Ref 22, p 4

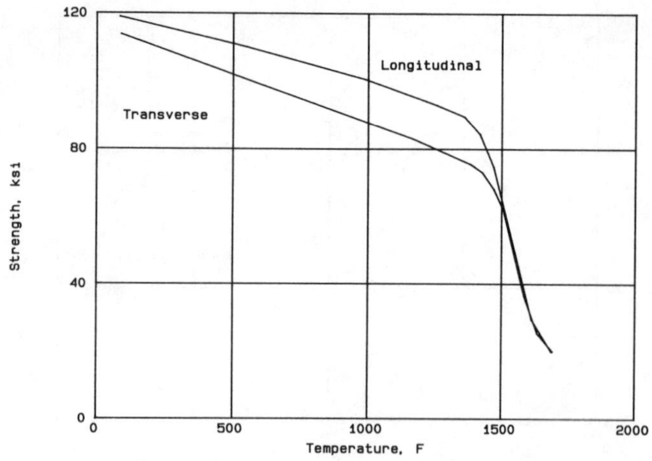

Incoloy 901: Effect of temperature and orientation on compressive yield strength of bar and forgings. Specimens were heat treated at 2000 F for 2 h, water quenched, aged at 1450 for 2 h, air cooled, aged at 1325 F for 24 h, and air cooled. Source: Ref 22, p 8

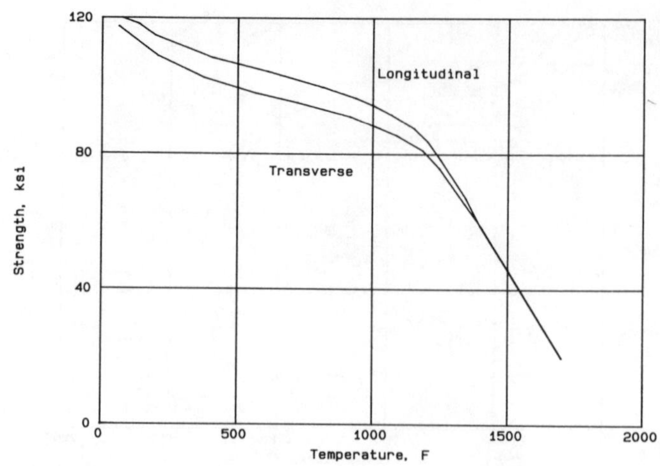

Incoloy 901: Effect of temperature and orientation on shear strength of bar and forgings. Specimens were heat treated at 2000 F for 3 h, water quenched, aged at 1450 for 2 h, air cooled, aged at 1325 F for 24 h, and air cooled. Source: Ref 22, p 9

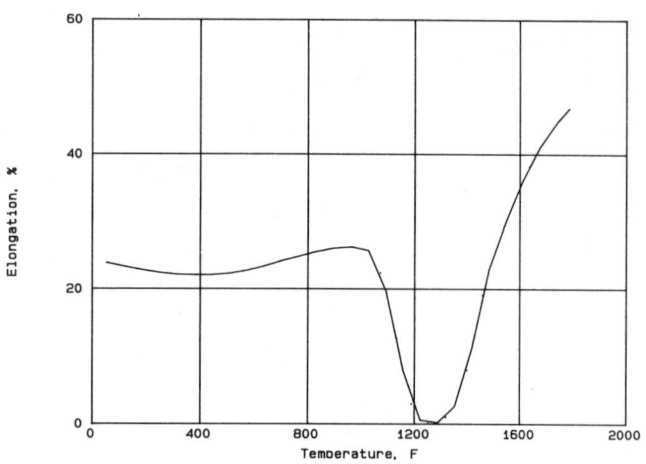

Incoloy 901: Tensile elongation of sheet solution treated and aged at 1400 F for 2 h. Source: Ref 22, p 7

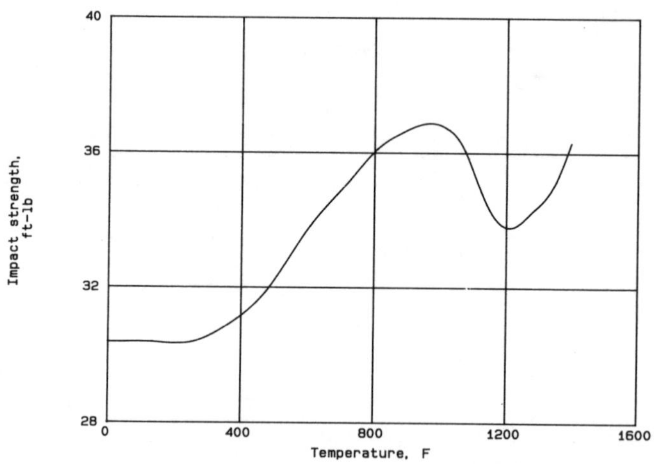

Incoloy 901: Charpy V-notch impact strength of 0.625-in. diam bar. Specimens were heat treated at 1965 F for 2 h, water quenched, aged at 1500 F for 2 h, air cooled, aged at 1325 F for 24 h, and air cooled. Source: Ref 22, p 9

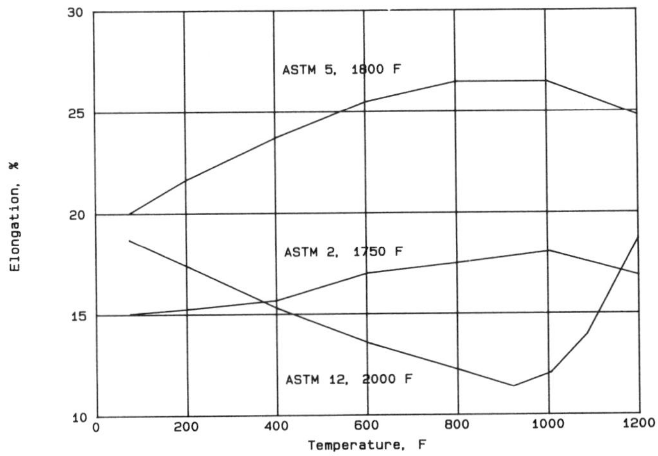

Incoloy 901: Effect of temperature and grain size on tensile elongation of forgings. Specimens were forged by the mini-grain thermal-mechanical treatment, solution treated for 1 h at the indicated temperature, stabilized at 1325 F for 4 h, air cooled, aged at 1200 F for 12 h, and air cooled. Source: Ref 22, p 6

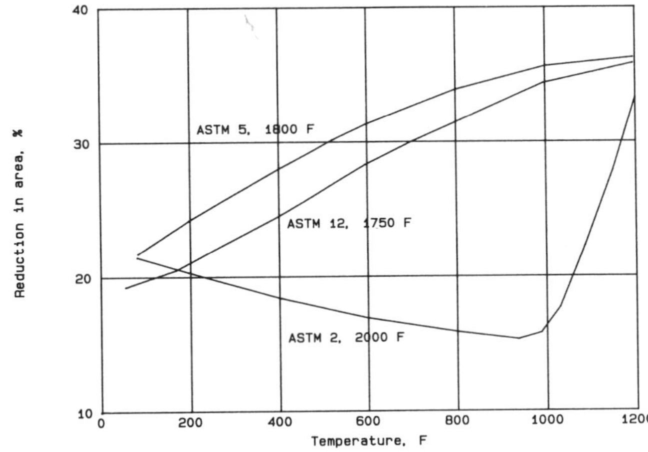

Incoloy 901: Effect of temperature and grain size on tensile reduction in area of forgings. Specimens were forged by the mini-grain thermal-mechanical treatment, solution treated for 1 h at the indicated temperature, stabilized at 1325 F for 4 h, air cooled, aged at 1200 F for 12 h, and air cooled. Source: Ref 22, p 6

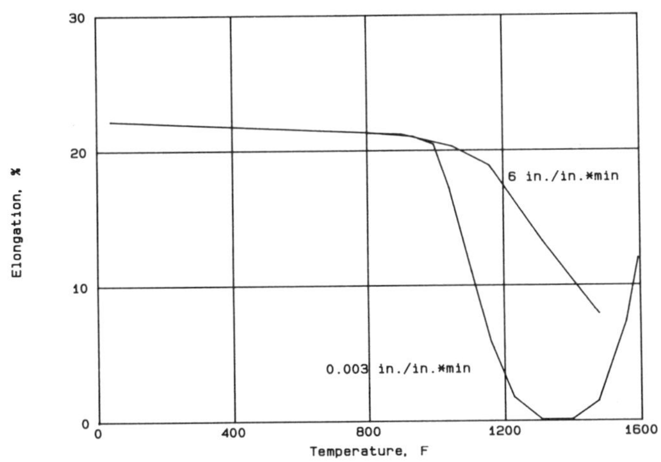

Incoloy 901: Effect of temperature and strain rate on tensile elongation of 0.063-in. sheet. Specimens were solution treated, aged at 1400 F for 2 h, and air cooled. Source: Ref 22, p 7

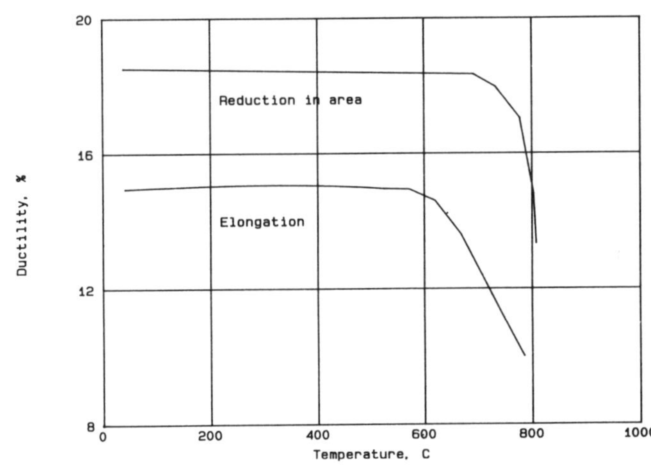

Incoloy 901: Tensile elongation and reduction in area of specimens solution treated at 1093 C, stabilized at 771 C, and aged at 704 C. Source: Ref 12, p 243

Incoloy 901: Creep strength

TEMP C	TEMP F	CRSTR 0.1%/100 h MPa	CRSTR 0.1%/100 h ksi
538	1000	724	105
649	1200	503	73
760	1400	186	27

Note: Specimens were solution treated at 1093 C (2000 F), held for 2 h, and water quenched. Then they were stabilized at 774 C (1425 F), held for 4 h, and air cooled. Specimens were aged at 732 C (1350 F), held for 24 h, and air cooled
Source: Carpenter Technology

Incoloy 901: Stress-to-rupture property data

TEMP C	TEMP F	RPSTR 10 h MPa	RPSTR 10 h ksi	RPSTR 100 h MPa	RPSTR 100 h ksi	RPSTR 1000 h MPa	RPSTR 1000 h ksi
538	1000	...	...	827	120	689	100
649	1200	665	96.5	621	90	517	75
732	1350	414	60	352	51	228	33
816	1500	241	30	155	22.5	76	11

Note: Specimens were solution treated at 1093 C (2000 F), held for 2 h, and water quenched. Then they were stabilized at 774 C (1425 F), held for 4 h, and air cooled. Specimens were aged at 732 C (1350 F), held for 24 h, and air cooled
Source: Carpenter Technology

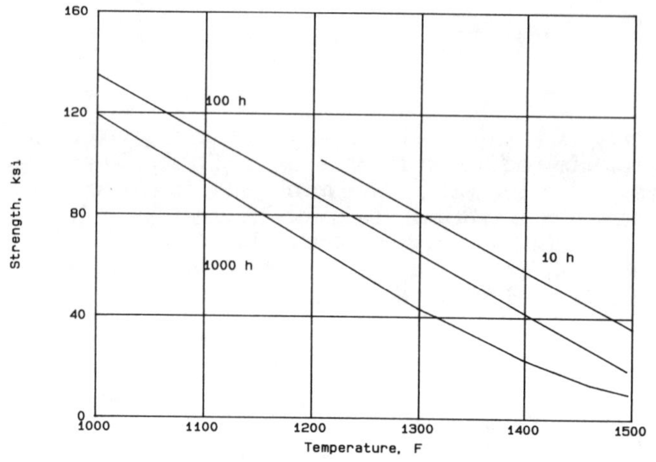

Incoloy 901: Creep-rupture strength of bar. Specimens were heat treated at 2000 F for 2 h, water quenched, aged at 1450 F for 2 h, air cooled, aged at 1325 F for 24 h, and air cooled. Source: Ref 22, p 11

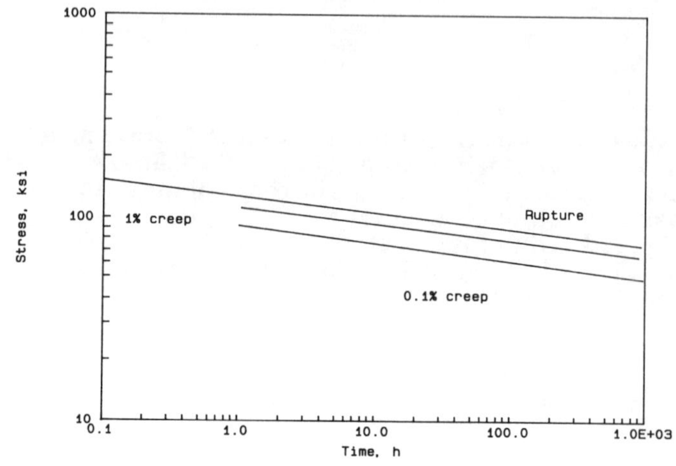

Incoloy 901: Creep and rupture lives at 1200 F for 0.5-in. bar. Specimens were heat treated at 2000 F for 2 h, air cooled, aged at 1400 F for 2 h, air cooled, aged at 1325 F for 24 h, and air cooled. Source: Ref 22, p 10

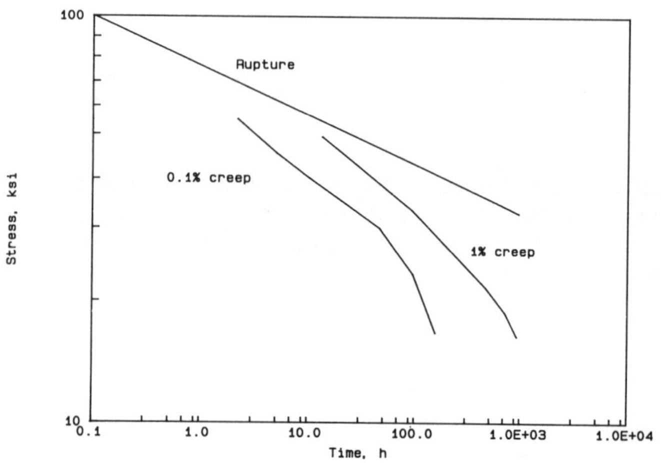

Incoloy 901: Creep and rupture lives at 1400 F for 0.5-in. bar. Specimens were heat treated at 2000 F for 2 h, air cooled, aged at 1400 F for 2 h, air cooled, aged at 1325 F for 24 h, and air cooled. Source: Ref 22, p 10

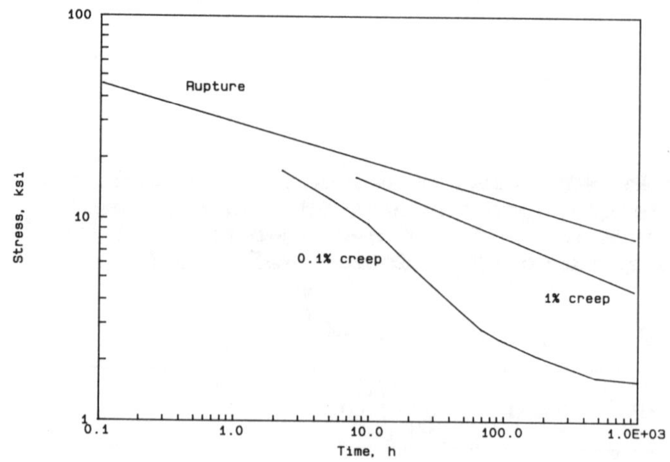

Incoloy 901: Creep and rupture lives at 1600 F for 0.5-in. bar. Specimens were heat treated at 2000 F for 2 h, air cooled, aged at 1400 F for 2 h, air cooled, aged at 1325 F for 24 h, and air cooled. Source: Ref 22, p 10

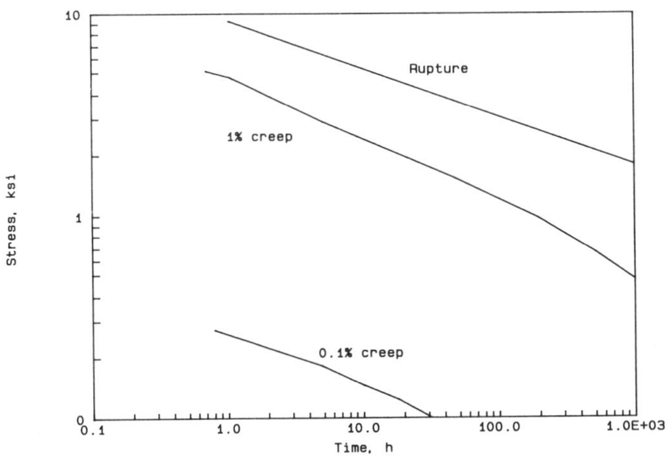

Incoloy 901: Creep and rupture lives at 1800 F for 0.5-in. bar. Specimens were heat treated at 2000 F for 2 h, air cooled, aged at 1400 F for 2 h, air cooled, aged at 1325 F for 24 h, and air cooled. Source: Ref 22, p 10

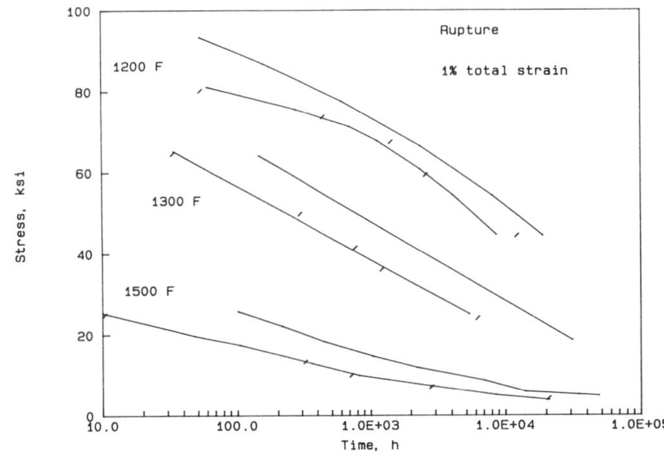

Incoloy 901: Creep and rupture lives for 0.875-in. bar. Specimens were heat treated at 2000 F for 2 h, water quenched, aged at 1450 F for 2 h, air cooled, aged at 1325 F for 24 h, and air cooled. Source: Ref 22, p 10

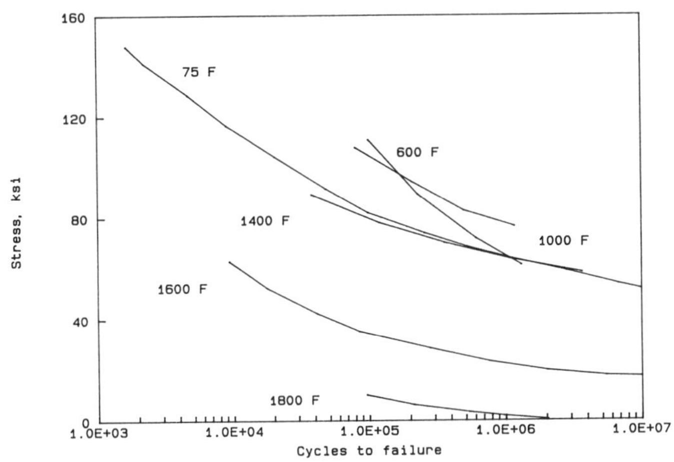

Incoloy 901: Fatigue life for 1.0-in. bar. Specimens were treated at 2000 F for 2 h, water quenched, aged at 1450 F for 2 h, air cooled, aged at 1325 F for 24 h, and air cooled. Bar was axially loaded with a stress-amplitude ratio of 0.67. Source: Ref 22, p 11

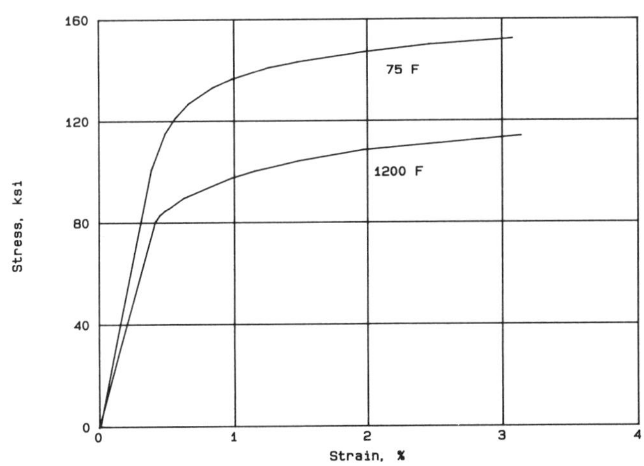

Incoloy 901: Stress-strain curves for bar. Specimens were solution treated and aged. Source: Ref 22, p 5

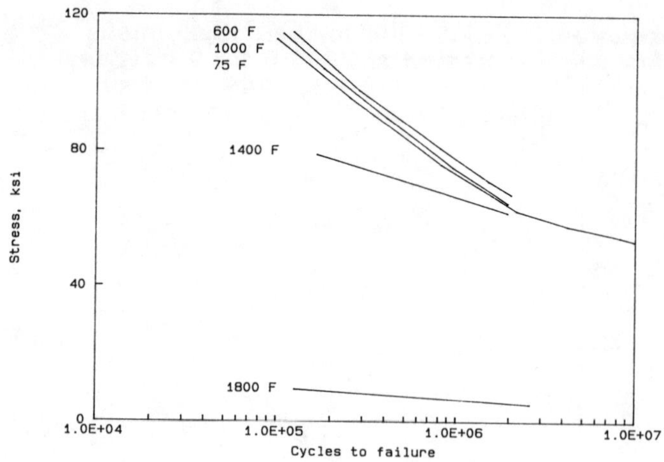

Incoloy 901: Fatigue life for 1.0-in. bar. Specimens were treated at 2000 F for 2 h, water quenched, aged at 1450 F for 2 h, air cooled, aged at 1325 F for 24 h, and air cooled. Bar was axially loaded with a stress-amplitude ratio of 0.98. Source: Ref 22, p 12

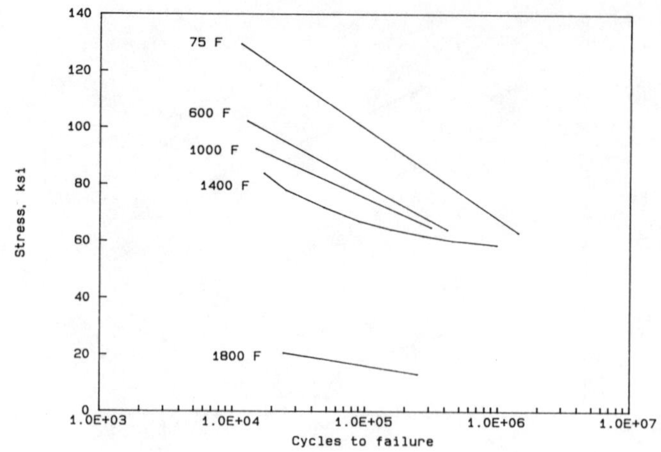

Incoloy 901: Fatigue life for 1.0-in. bar. Specimens were treated at 2000 F for 2 h, water quenched, aged at 1450 F for 2 h, air cooled, aged at 1325 F for 24 h, and air cooled. Bar was axially loaded with a stress-amplitude ratio of 2.0. Source: Ref 22, p 12

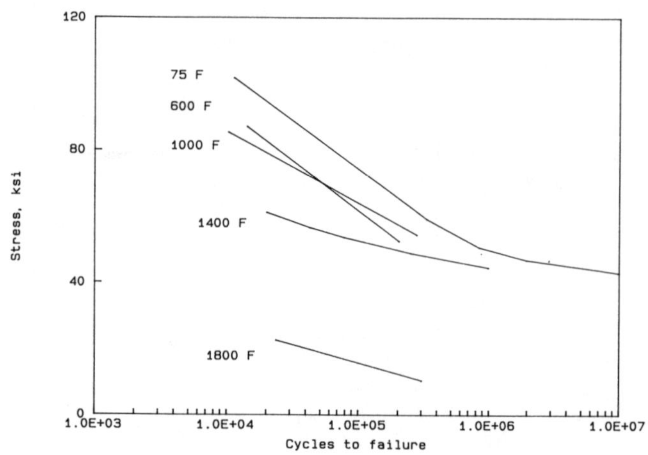

Incoloy 901: Fatigue life for 1.0-in. bar. Specimens were treated at 2000 F for 2 h, water quenched, aged at 1450 F for 2 h, air cooled, aged at 1325 F for 24 h, and air cooled. Bar was axially loaded with a stress-amplitude ratio of infinity. Source: Ref 22, p 12

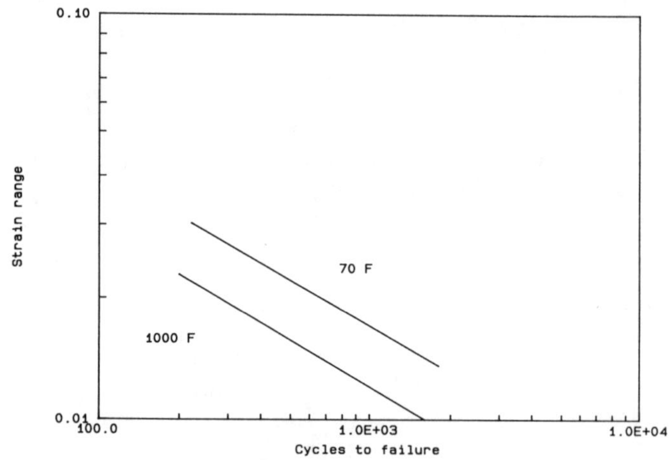

Incoloy 901: Low-cycle fatigue life as a function of strain range at room temperature and 1000 F for 0.375-in. square bar. Specimens were heat treated at 2000 F for 2 h, water quenched, aged at 1450 F for 2 h, air cooled, aged at 1325 F for 24 h, and air cooled. Source: Ref 22, p 14

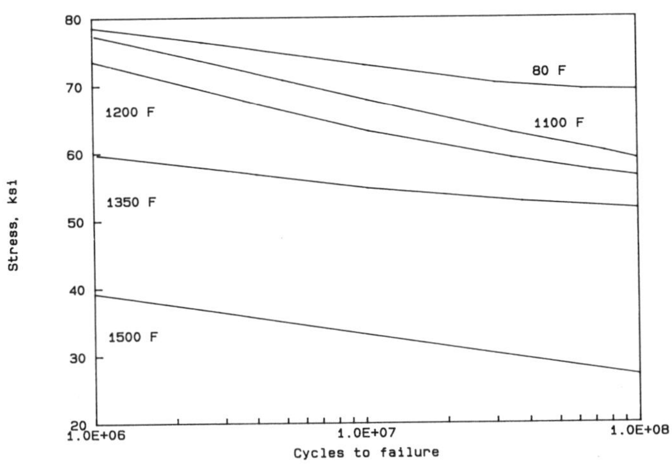

Incoloy 901: Fatigue life for bar. Specimens were solution treated at 2000 F for 2 h, water quenched, aged at 1375 F for 24 h, and air cooled. Rotating beam, R = -1. Source: Ref 22, p 12

RA-330

UNS number: N08330
Specifications: AMS 5592, AMS 5716, ASTM B511, ASTM B512, ASTM B535, ASTM B536, ASTM B546, SAE J405 (30330), SAE J412 (30330)

Nominal composition: 19.000% Cr, 35.000% Ni, 43.000% Fe, 0.050% C, 1.25% Si

Product forms: Wrought, bar, rod, wire, plate, sheet, strip, tube, pipe, billets, forgings, structural shapes
Description: Ni-Fe-Cr alloy, solid solution strengthened.
Typical applications: Heat exchangers and radiant tubes.

Physical Properties

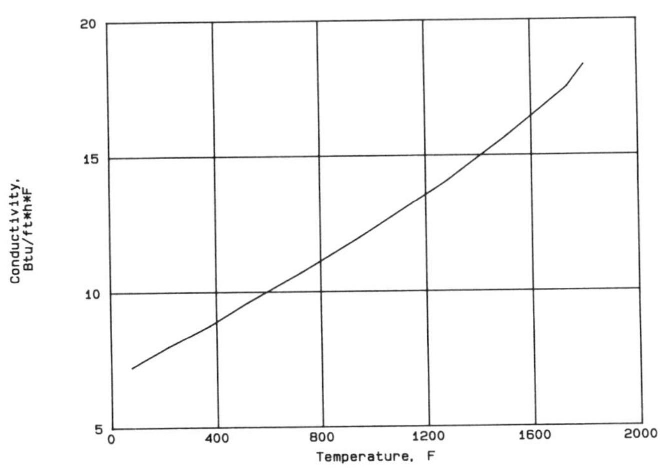

RA-330: Thermal conductivity of bar. Source: Ref 22, p 3

RA-330: Thermal conductivity

TEMP C	TEMP F	THCON W/m*K	THCON Btu/ft*h*F
24	75	12.5	7.2
100	212	16.3	9.4
425	800	19.4	11.2
500	930	21.6	12.5
870	1600	28.5	16.5

Source: Ref 15, p 309

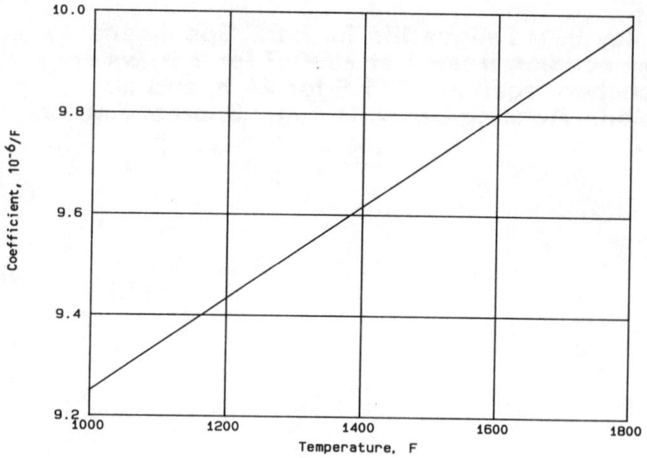

RA-330: Mean coefficient of linear thermal expansion. Source: Ref 22, p 3

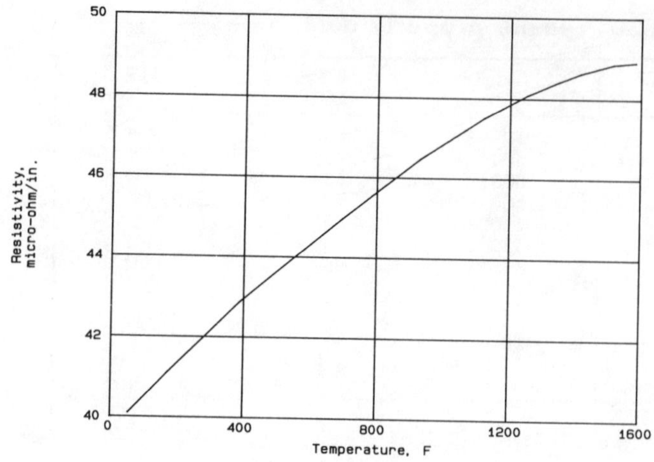

RA-330: Electrical resistivity of bar. Source: Ref 22, p 3

RA-330: Mean coefficients of linear thermal expansion

TEMP C	TEMP F	THEXP 10^{-6}/C	THEXP 10^{-6}/F
0-100	32-212	14.4	8.0
0-315	32-600	16.0	8.9
0-540	32-1000	16.7	9.3
0-650	32-1200	17.3	9.6
0-870	32-1600	18.0	10.0

Source: Ref 15, p 309

RA-330: Modulus of elasticity

TEMP C	TEMP F	E GPa	E 10^3 ksi
21	70	197	28.5
315	600	179	26.0
540	1000	164	23.8
760	1400	145	21.0
870	1600	134	19.5
980	1800	124	18.0

Source: Ref 15, p 309

RA-330: Modulus of elasticity. Source: Ref 22, p 6

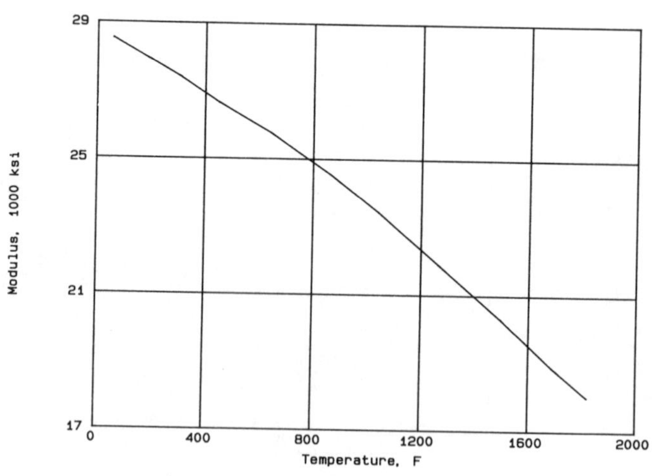

RA-330: Tensile property data

TEMP C	TEMP F	UTS MPa	UTS ksi	YS MPa	YS ksi	EL %	RA %
21	70	586	85.0	290	42.0	43	68
205	400	552	80.0	241	35.0	40	65
315	600	524	76.0	217	31.5	...	...
425	800	490	71.0	193	28.0	...	...
540	1000	448	65.0	172	25.0	...	...
650	1200	393	57.0	155	22.5	28	32
760	1400	276	40.0	138	20.0	22	27
870	1600	152	22.0	90	13.0	30	34
980	1800	76	11.0	50	7.2	53	48
1095	2000	35	5.1	27	3.9	72	50
1150	2100	30	4.4	23	3.4	93	75

Source: Ref 15, p 310

RA-330: Creep-rupture property data for bar

TEMP C	TEMP F	CRSTR MPa	CRSTR ksi	RPT(a) h	RPT(b) h	RPT(c) h
Solution treated at 1204 C (2200 F), 30 min water quenched						
760	1400	82.7	12	79.9	92.2	...
871	1600	41.4	6	137.4	107.9	...
982	1800	20.7	3	78.8	76.8	...
Mill annealed at 1121 C (2050 F), 30 min water quenched						
760	1400	69	10	167.7	175.5	141
871	1600	41.4	6	50.9	65.7	45.5
982	1800	20.7	3	37.9	44	39.1

(a) Heat A. (b) Heat B. (c) Heat C
Source: Ref 29, Jun 1972

RA-330: Stress-rupture and creep property data

TEMP C	TEMP F	RPSTR 1000 h MPa	RPSTR 1000 h ksi	RPSTR 10000 h MPa	RPSTR 10000 h ksi	CRSTR 1%/10000 h MPa	CRSTR 1%/10000 h ksi
815	1500	39	5.70	24	3.50	19	2.80
870	1600	26	3.80	17	2.40	14	2.00
925	1700	19	2.70	11	1.60	9	1.30
980	1800	12	1.70	7	1.00	6	0.80
1040	1900	8	1.20	5	0.68	3	0.42
1095	2000	5	0.78	3	0.45	2	0.27

Source: Ref 15, p 310

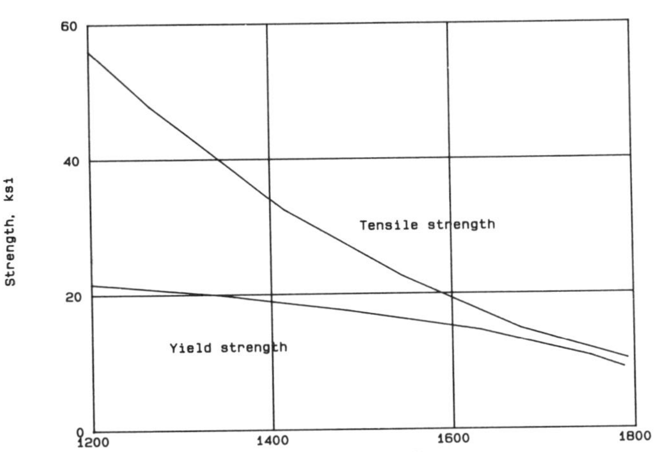

RA-330: Tensile and yield strength of 1-in. mill annealed bar. Source: Ref 22, p 4

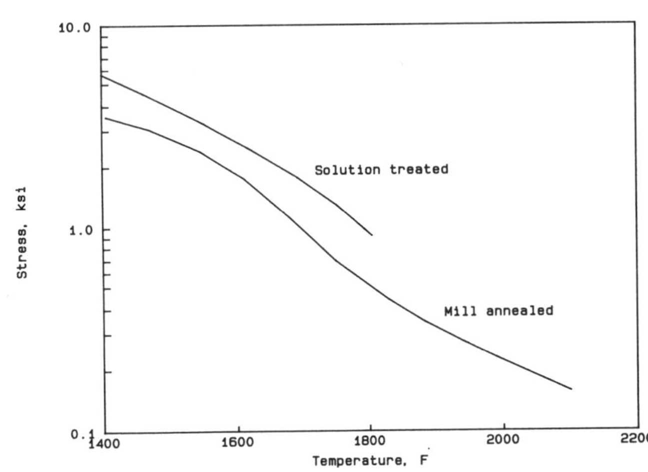

RA-330: Stress to produce 0.0001%/h minimum creep rate for bar. Specimens were solution treated at 2200 F and mill annealed at 2050 F, as indicated. Source: Ref 22, p 5

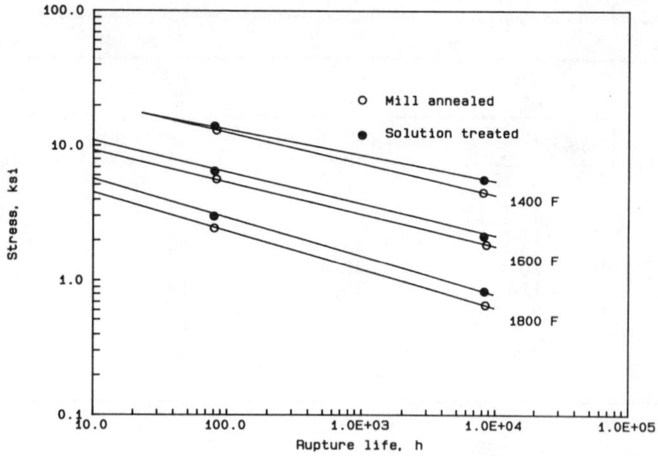

RA-330: Creep-rupture lives for bar in two conditions. Curve 1: solution treated at 2200 F and water quenched. Curve 2: mill annealed at 2050 F and water quenched. Source: Ref 22, p 4

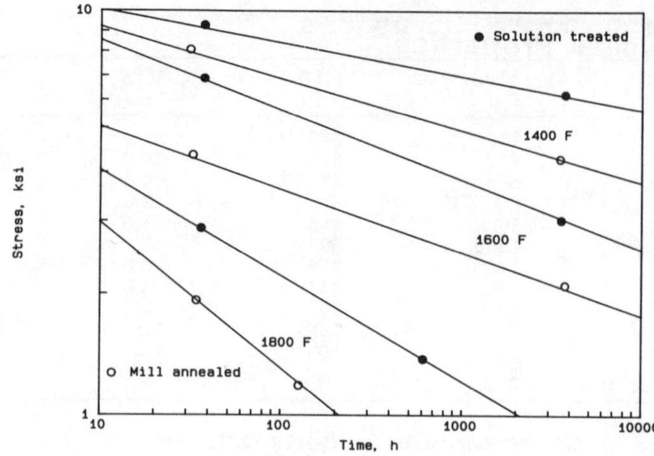

RA-330: Stress to produce 1% creep for bar. Curve 1: mill annealed at 2050 F and water quenched. Curve 2: solution treated at 2200 F and water quenched. Source: Ref 22, p 4

RA-330: Stress to produce 2% creep for bar. Curve 1: mill annealed at 2050 F and water quenched. Curve 2: solution treated at 2200 F and water quenched. Source: Ref 22, p 4

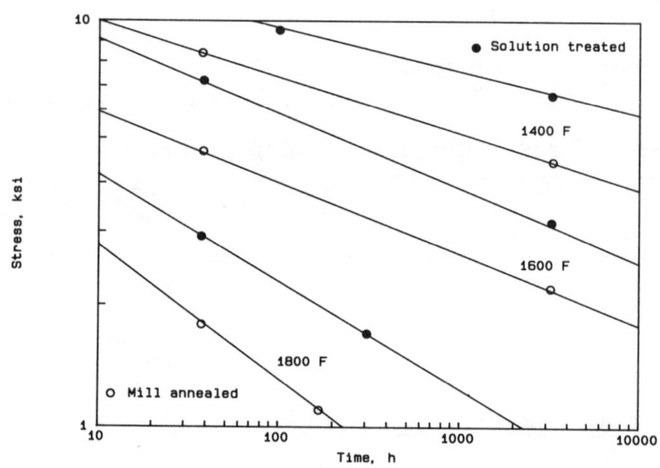

RA-330 HC

Nominal composition: 19.000% Cr, 35.000% Ni, 43.000% Fe, 0.400% C, 1.25% Si
Product forms: Wrought, bar, plate, sheet, strip, billets, forgings, structural shapes
Description: Higher carbon version of RA-330, combining the same elevated-temperature corrosion resistance with greater strength in the 760 to 980 C (1400 to 1800 F) range. The strength increase derives from dispersion hardening by carbon at service temperatures.
Typical applications: Belt pins and bolting, and platens and dies used in hot forming of titanium.

Physical Properties

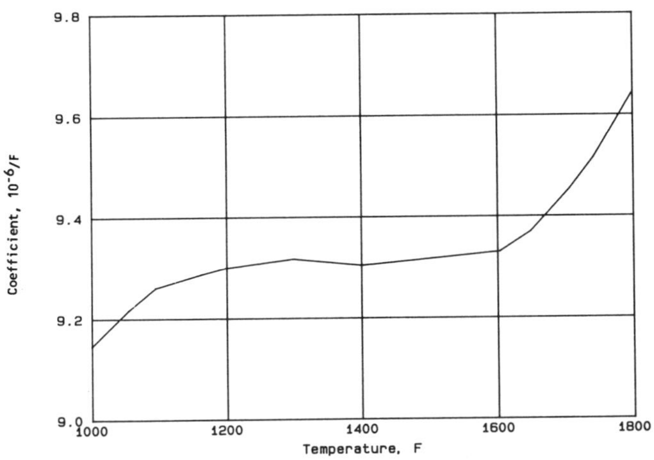

RA-330HC: Mean coefficient of linear thermal expansion. Source: Ref 22, p 2

Mechanical Properties

RA-330HC: Compressive yield strength for bar

TEMP C	TEMP F	YSC(a) MPa	YSC(a) ksi
Annealed at 1204 C (2200 F), held 30 min, water quenched			
760	1400	210.3	30.5
871	1600	135.8	19.7
982	1800	88.9	12.9
Annealed at 1204 C (2200 F), held 30 min, water quenched, held at 704 C (1300 F) for 4 h, and air cooled			
760	1400	215.1	31.2
871	1600	143.4	20.8
982	1800	94.5	13.7

(a) Speed of testing, strain rate 0.005 in./in.*min to yield point. Values are averages of three tests
Source: Ref 22, p 3

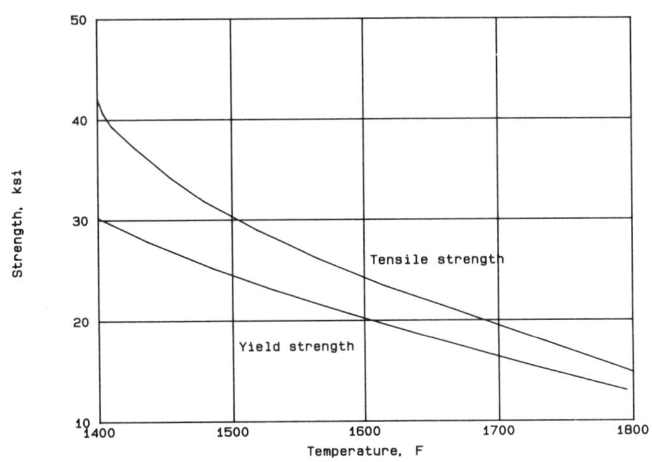

RA-330HC: Tensile and yield strength of 1-in. diam mill annealed bar. Source: Ref 22, p 3

RA-330HC: Creep-rupture property data for bar

TEMP C	TEMP F	CRSTR MPa	CRSTR ksi	RPT h
Mill annealed at 1177 C (2150 F), 30 min, water quenched				
760	1400	82.7	12.0	960.7
871	1600	42.7	6.2	380.2
982	1800	24.1	3.5	104.7
Solution treated at 1260 C (2300 F), 2 h, water quenched				
760	1400	96.5	14.0	353.4
871	1600	41.4	6.0	623.1
982	1800	27.6	4.0	134.0

Source: Ref 22, p 5

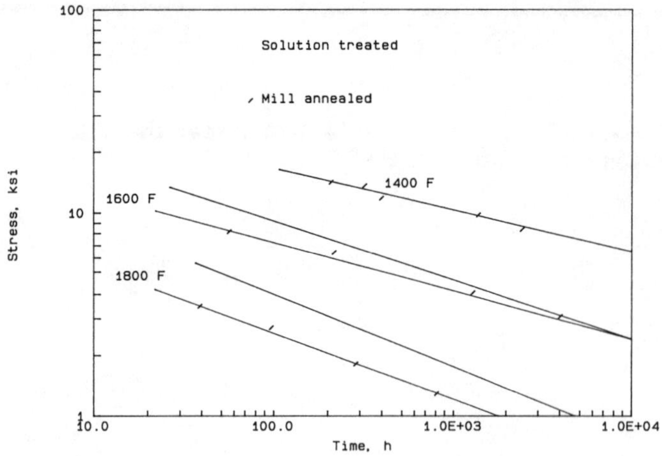

RA-330HC: Stress to produce 1% creep for bar. Curve 1: solution treated at 2300 F and water quenched. Curve 2: mill annealed at 2150 F and water quenched. Source: Ref 22, p 3

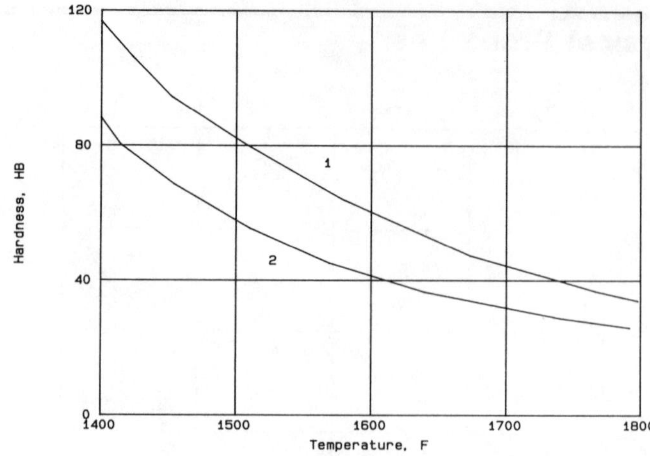

RA-330HC: Brinell hardness of bar. Curve 1: solution treated at 2300 F and water quenched. Curve 2: mill annealed at 2150 F and water quenched. Source: Ref 22, p 2

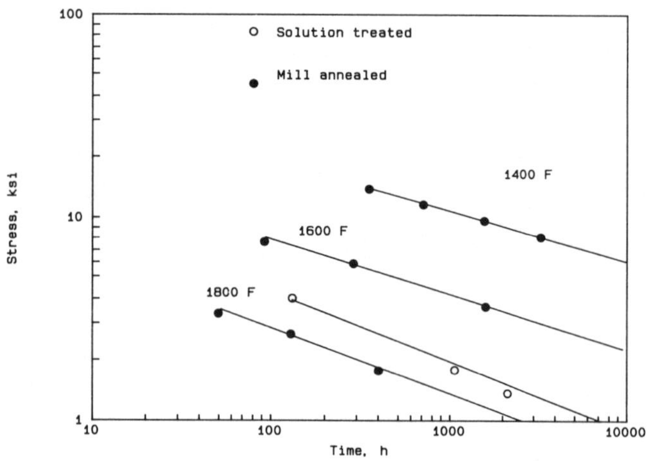

RA-330HC: Stress to produce 2% creep for bar. Curve 1: solution treated at 2300 F and water quenched. Curve 2: mill annealed at 2150 F and water quenched. Source: Ref 22, p 4

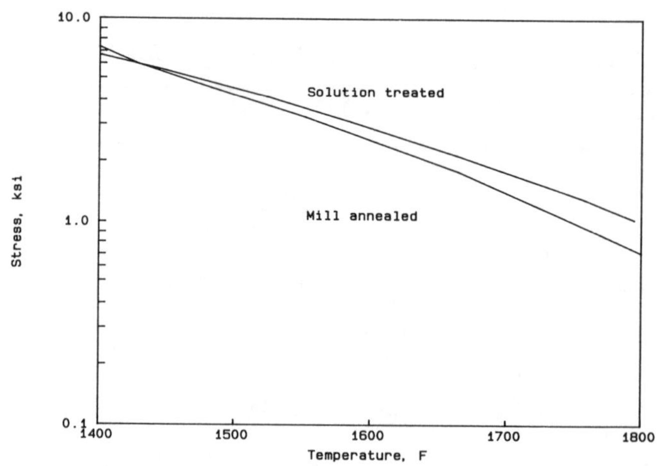

RA-330HC: Stress to produce a minimum creep of 0.0001%/h for bar. Curve 1: solution treated at 2300 F and water quenched. Curve 2: mill annealed at 2150 F and water quenched. Source: Ref 22, p 4

Abbreviations

.%CW	Percent Cold Work
BLKMD	Bulk Modulus
CORR	Corrosion Rate
CRIS	Initial Stress before Creep
CRSRD	Residual Stress after Creep
CRSTN	Creep Strain
CRSTR	Creep Strength
CRT	Creep Rate
CRTA	Average Creep Rate
CRTMN	Minimum Creep Rate
CSHX	Cyclic Strain Hardening Exponent
CSTRC	Cyclic Strength Coefficient
DENS	Density
DMPCP	Specific Damping Capacity
da/dN	Fatigue Crack Propagation Rate
E	Modulus of Elasticity
E_C	Modulus of Elasticity (Compression)
EL	Tensile Elongation
ELCON	Electrical Conductivity
ELRES	Electrical Resistivity
EMITT	Emittance
FAT.n	Fatigue Cycles

FATDC	Fatigue Ductility Coefficient
FATDX	Fatigue Ductility Exponent
FATL	Fatigue Resistance
FATSC	Fatigue Strength Coefficient
FATSR	Fatigue Stress Ratio
FATST	Fatigue Strength at N Cycles
FATSX	Fatigue Strength Exponent
FRSTN	Elastic Strain at Fracture
FRSTR	Fracture Strength
G	Modulus of Rigidity
GRT	Growth Rate
HARD	Hardness
IS	Impact Strength
KIC	Fracture Toughness
NTS	Notch Tensile Strength
PR	Poisson's Ratio
RA	Tensile Reduction in Area
RMIBR	Residual Magnetic Inductance
RPSTR	Stress-Rupture Strength
RPT	Rupture Life
SHTCP	Specific Heat
SHX	Strain Hardening Exponent
SRX	Strain Rate Exponent
STREX	Exposure Stress
STRN	Natural Stress at Maximum Load
STRTR	True Stress at Maximum Load
THCON	Thermal Conductivity
THDIF	Thermal Diffusivity
THEXP	Thermal Coefficient of Expansion
TSR	True Total Strain Rate
TSRA	Axial Strain Rate
TSRGX	Maximum Total Strain Range
USB	Ultimate Bearing Strength

USS	Ultimate Strength in Shear
UTS	Tensile Ultimate Strength
WTGN	Weight Gain
YS	Tensile Yield Strength
YSB	Bearing Yield Strength
YSC	Compressive Yield Strength
YSS	Yield Strength in Shear

References

1. *Metals Handbook,* Vol 1, *Properties and Selection: Irons and Steels,* 9th ed., American Society for Metals, 1978.

2. *Metals Handbook,* Vol 2, *Properties and Selection: Nonferrous Alloys and Pure Metals,* 9th ed., American Society for Metals, 1979.

3. *Metals Handbook,* Vol 3, *Properties and Selection: Stainless Steels, Tool Materials and Special-Purpose Metals,* 9th ed., American Society for Metals, 1980.

4. *Metals Handbook,* Vol 4, *Heat Treating,* 9th ed., American Society for Metals, 1981.

5. *Metals Handbook,* Vol 5, *Surface Cleaning, Finishing, and Coating,* 9th ed., American Society for Metals, 1982.

6. *Metals Handbook,* Vol 6, *Welding, Brazing, and Soldering,* 9th ed., American Society for Metals, 1983.

7. *Metals Handbook,* Vol 7, *Powder Metallurgy,* 9th ed., American Society for Metals, 1984.

8. *Metals Handbook,* Vol 8, *Mechanical Testing,* 9th ed., American Society for Metals, 1985.

9. *Source Book on Materials for Elevated Temperature Applications,* E.F. Bradley, Ed., American Society for Metals, 1979.

10. *Source Book on Stainless Steels,* American Society for Metals, 1976.

11. *Source Book on the Ferritic Stainless Steels,* R.A. Lula, Ed., American Society for Metals, 1982.

12. *Superalloys: Source Book,* M.J. Donachie, Ed., American Society for Metals, 1983.

13. *Titanium and Titanium Alloys: Source Book,* M.J. Donachie, Ed., American Society for Metals, 1982.

14. *Source Book on Industrial Alloy and Engineering Data,* American Society for Metals, 1978.

15. *Engineering Properties of Steel,* P.D. Harvey, Ed., American Society for Metals, 1982.

16. *Flow and Fracture at Elevated Temperatures,* R. Raj, Ed., American Society for Metals, 1985.

17. *Principles of Physical and Chemical Metallurgy,* G.F. Carter, American Society for Metals, 1979.

18. *Heat Treatment, Structure and Properties of Nonferrous Alloys,* C.R. Brooks, American Society for Metals, June 1982.

19. *Metal Progress 1985 Materials and Processing Data Book,* Vol 128, No. 1, American Society for Metals, June 1985.

20. *Forging Handbook,* T.G. Byrer, Ed., Forging Industry Association and American Society for Metals, 1985.

21. *Structural Alloys Handbook,* Vol 1 and 2, D.J. Maykuth, Ed., Mechanical Properties Data Center, Battelle Columbus Laboratories, 1980.

22. *Aerospace Structural Materials Handbook,* Vol 1 to 5, W.F. Brown, Ed., Mechanical Properties Data Center, Battelle Columbus Laboratories, 1981.

23. *Atlas of Fatigue Curves,* H. Boyer, Ed., American Society for Metals, 1986.

24. *Duplex Stainless Steels,* R.A. Lula, Ed., American Society for Metals, 1983.

25. *Standards Handbook,* Part 2 -- Alloy Data, "Wrought Copper and Copper Alloy Mill Products," 8th ed., Copper Development Association, Inc., Greenwich, CT, 1985.

26. *Steel Products Manual,* "Tool Steels," American Iron and Steel Institute, New York, January 1970.

27. *Steel Products Manual,* "Stainless and Heat Resisting Steels," American Iron and Steel Institute, Washington DC, December 1974.

28. *Steel Products Manual,* "High Temperature High Strength Alloys," American Iron and Steel Institute, New York, February 1963.

29. *Alloy Digest,* Engineering Alloys Digest, Inc., Upper Montclair, NJ.

30. *Metals Handbook Desk Edition,* H.E. Boyer and T.L. Gall, Ed., American Society for Metals, 1985.